# Geological Risk and Uncertainty in Oil Exploration

# Geological Risk and Uncertainty in Oil Exploration

Ian Lerche

*Department of Geological Sciences*
*University of South Carolina*
*Columbia, South Carolina*

ACADEMIC PRESS

San Diego London Boston New York Sydney Tokyo Toronto

This book is printed on acid-free paper. ∞



Academic Press
*a division of Harcourt Brace & Company*
525 B Street, Suite 1900, San Diego, California 92101-4495, USA
http://www.apnet.com

Academic Press Limited
24-28 Oval Road, London NW1 7DX, UK
http://www.hbuk.co.uk/ap/

Library of Congress Cataloging-in-Publication Data

Lerche, I. (Ian)
Geological risk and uncertainty in oil exploration / by Ian Lerche.
p. cm.
Includes bibliographical references and index.
ISBN 0-12-444174-2
1. Petroleum--Prospecting--Statistical methods. 2. Petroleum reserves--Forecasting--Statistical methods. 3. Risk assessment.
I. Title.
TN271.P4L45 1997
622'.1828--DC21 97-9787
CIP

PRINTED IN THE UNITED STATES OF AMERICA
97 98 99 00 01 02 BB 9 8 7 6 5 4 3 2 1

# CONTENTS

# PART I

## MODELS AND THEIR UNCERTAINTIES

## 2

### FLEXURAL PLATE MOTION: ELASTIC, THERMO-ELASTIC, AND PLASTIC-ELASTIC MODELS



## 3

### DYNAMIC MODELS OF SEDIMENT EVOLUTION AND THEIR UNCERTAINTIES

# 4

## THERMAL MODELS OF SEDIMENT MATURATION AND THEIR UNCERTAINTIES



# 5

## HYDROCARBON KINETIC MODELS AND THEIR UNCERTAINTIES

# 6

## VARIABLE TOTAL ORGANIC CARBON AND VARIABLE SOLUBILITY CONSIDERATIONS

# 7

## FAULTS: OPEN OR SHUT?



# 8

## MODELS OF BASINAL SEDIMENT FILL, TURBIDITE FLOWS, AND THEIR UNCERTAINTIES

# PART II

## SCIENTIFIC PROBABILITY, RISK, AND STRATEGY

## 9

## THE UNCERTAINTY OF AGE



## 10

## THE UNCERTAINTY OF INFORMATION



## 11

## ASSIGNING RISK TO SCIENTIFIC RESULTS

# 12

## STRATEGIES FOR MODEL USE

# Preface

Quantitative methods for handling basin evolution problems have burgeoned in the past few years, as shown by the volume of literature. The advent of fast computers has helped this growth. Problems that used to take hours or days on PC 486-level machines now take minutes to hours on SunSparc 10 machines, thereby enabling rapid turnaround and improving the potential for investigation of many possible scenarios regarding basin evolution. And yet we find ourselves in a more uncertain state than, say, five years ago. There are three reasons for this uncertainty: (1) No matter what we do, observations of sedimentary basins are made only at the present time. Somehow inferences concerning paleo-development must be coaxed out of present-day imprecise information. (2) No matter what we do, it is a fact that some model (or models) of particular facets of basin development must be invoked. These models not only must honor geological, geophysical, and geochemical data at the present day, but also must be capable of providing the required paleo-evolution. (3) No matter what we do, the paleo-evolution determined in some manner from a model will carry with it the assumptions of the model, the parameters of the model, and the consequences of limited data, nonuniformly sampled data, and inaccurate data. As a result the modeled paleo-evolution is not unique, is not precise, is not perfectly resolved, and is sensitive (to varying degrees) to the assumptions made, parameters used, and data incorporated. Such factors must be quantified in order to evaluate elements of risk and uncertainty in reconstructed paleo-evolution scenarios. Two previous volumes (*Basin Analysis: Quantitative Methods,* Volumes 1 and 2) have dealt with quantitative procedures for constructing models for different components of basinal evolution, from which combined dynamic, thermal, and hydrocarbon histories can then be constructed.

The concern in this volume is on uncertainty. The majority of the book's audience will already be familiar with the subject of basin analysis, and likely be involved in the quantification of geologic processes. Graduate students, research scientists, oil expolorationists, and strategic economists should enjoy this systematic approach to quantifying geological risk and uncertainty.

Two extreme viewpoints can be taken: We can consider paleo-uncertainty as leading to the inability to determine factors of interest or we can use the knowledge of the degree of uncertainty of paleo-evolution to provide a probabilistic assessment of a range of answers. Clearly, the philosophy in a quantitative basin analysis mode of operation must be to ascertain what can be obtained without equivocation, what can be determined with an acceptable degree of equivocation (and some definition of "acceptable" must also be provided), and what cannot be determined without further information—and some idea of the amount and types of information needed must also be provided.

From a hydrocarbon exploration perspective, the ability to bracket the time and amount of oil and gas generation in relation to the then-existing structural and stratigraphic state of the basin, to bracket the time of hydrocarbon migration (and fraction of hydrocarbons migrated) in the basin in relation to the basin's state at that time, to bracket the accumulation locations and amounts of hydrocarbons, to bracket the capability of retaining hydrocarbons until the present day, and to allow for uncertainties in the dynamic evolution of the basin at each of these stages are all aspects that require quantification, but which must include the uncertainties as necessary components. Thus, scientific risk and scientific uncertainty factors must be of paramount concern in assessing the likelihood of a basin containing hydrocarbons.

Precisely similar problems have arisen over the years in relation to economic worth predictions, in which future inflation rates, future hydrocarbon prices, future development costs, and so on, are all unknown. In that arena the ability to provide expected values and expected ranges of uncertainties has been well developed based on different economic projection and prediction models, each of which contains its own assumptions and its own parameters and is beholden to past and present-day data. Crudely put, the tools that economic models use to make predictions with present-day economic data can be carried over, nearly wholesale, to scientific retrodiction with present-day scientific data. It is this close parallel between the two arenas of hydrocarbon exploration that is exploited here.

As in the economic theater, where models are tested for accuracy, reproducibility, and ability to account for known data prior to being used in a forecast mode (so the limitations, strengths, and weaknesses of each model can then be determined) so, too, in the realm of quantitative basin analysis must the models be subjected to the same level of scrutiny and rigor prior to being used to evaluate the likelihood of a basin containing hydrocarbons. The interleaved development of scientific uncertainty and scientific risk in quantitative basin analysis models is the main thread examined throughout this book.

The emphasis is on determining the extent to which one could be uncertain in assessment of a basin, rather than on declaring categorically how correct a given assessment of a basin must be. To paraphrase the legendary Joe Warren: "Tell me how wrong you could be, not how right you want to be."

The structure of the book is set in two parts. The first part illustrates how to handle a variety of geological, geophysical, and geochemical problems quantitatively with methods for determining uncertainties of results based on information at hand. The purpose of this section is to illustrate the degree to which various aspects of basin analysis models are sensitive to the assumptions made, the parameters used, and the data available. Problems investigated range from flexural motion of a basement, to thermal and hydrocarbon generation determinations, through faults as well as migration of fluids and turbidite slumps. Case history examples are given so that some appreciation of the vagaries of real basins is available.

The second part of the book is concerned with using the knowledge of model behavior uncertainty—gleaned from the first part—to provide probabilistic ranges of uncertainties of behaviors. The thrust here is to bring the scientific uncertainty to a point at which the integration of both scientific and economic risk can be handled on an equal footing, with each beholden to the uncertainty of the other—an integrated hydrocarbon exploration strategy, if you will.

A large number of oil companies and individuals have made this volume possible. I am indebted, as always, to the Industrial Associates of the Basin Modeling Group at the University of South Carolina. Unocal, Saga Petroleum, Phillips Petroleum, and Texaco are singled out for their major support, both financial and scientific. I am also truly indebted to my post-docs and graduate students over the years; without their work I really would have little to write about. Those who made contributions are Song Cao, Alan Yu, Y. Y. Li, Z. Kai, A. Abbott, Kenneth Petersen, Frédéric Maubeuge, Olivier Bour, J. Liu, J. Tang, Z. He, and Kay Carter. Professionals who have provided ideas, advice, and data are far too numerous to mention individually, but Brett Mudford, Marty Perlmutter, Dan Tetzlaff, Chuck Shearer, René Thomsen, David Worsley, Tony Doré, Jim MacKay, and Alan Burnham have often given me many a sleepless night in attempts to sort out particularly puzzling problems. To you all I extend my thanks.

The work reported here would not have been possible without the assistance of my secretary, Donna Black, who, once again, converted my illegible hieroglyphics into rational English. Perhaps family and friends suffer the most as one labors, seemingly endlessly, at putting a book manuscript together, and I thank them for their tolerance and support.

*Ian Lerche*

# 1

# INTRODUCTION

## I. GENERAL

Two of the major aims of quantitative economic models for hydrocarbon exploration are the assessment of the likely total value to a corporation if production is enabled and the assessment of the range of uncertainty in the likely total value resulting from vagaries in the future, such as unknown selling price of the product, inflation, escalation, production costs, and so on. Thus, exploration economic models are predictive devices, controlled to some extent by modeling of prior completed situations to ascertain forecast accuracy against known templates. Incorporated into such economic models are the estimated reserves in a reservoir being evaluated and the uncertainty ranges of such reserves.

One of the major aims of quantitative basin analysis models is to assess the likely generation, migration, and present-day accumulation amounts of hydrocarbons at reservoir sites (and to estimate the likely locations of reservoir sites in a basin), so that economic arguments applied to each such potential reservoir site can then be invoked to assess some form of ranking and priority.

Basin analysis models must also assess the range of uncertainty in accumulation amounts at the potential reservoir sites, as a result of uncertainties in (1) intrinsic assumptions of the models, (2) parameters used in the models, (3) numerical discretizations, and (4) data quality and quantity and the sampling frequency used to control model outputs. Thus, basin analysis models are retrodictive devices, controlled to some extent by availability of present-day data and by the ability of the models to account for observed hydrocarbon occurrences in well-explored basins analogous to the one being investigated. Incorporated into such models should be models of the physical, chemical, geological, and transport processes being evaluated in a basin and the uncertainty ranges of such processes.

The point about these two types of models (economic and basin analysis) is that both try to predict or retrodict patterns of behavior based on incomplete information, with uncertain parameters, and with different model assumptions and process behaviors.

Of concern is the assessment of the relative strengths and weaknesses of models in relation to their predictive and retrodictive outputs. In this way one has some idea of the resolution, accuracy, uniqueness, and sensitivity of results obtained for each and every model with respect to variations in the intrinsic assumptions, parameter uncertainty, and control data used. The availability of such quantitative measures of uncertainty on model outputs can then be used in a probabilistic sense to evaluate the risk, and the relevant strategy to be used, in decisions for advancing exploration and production and the associated likely economic returns of one potential reservoir in relation to others; that is, an objective, reproducible risking and ranking strategy can be formulated and applied correctly.

The purpose of this volume is to provide quantitative operational procedures to show how to perform the tasks mentioned, and to show how to use the results in strategic and economic senses, with allowance being made for the uncertainties intrinsic to basin model calculations. An integrated exploration assessment can then be provided, underpinned by the degree to which one has used all available information to obtain robust measures of both scientific and economic worth, with quantitative prescriptions of the uncertainties on robustness.

One can, of course, question whether such an effort is necessary in relation to supply and demand of hydrocarbons. Perhaps one of the more obvious ways to answer that is to look at (1) examples of reserve assessments, and their uncertainties, in order to illustrate the current ranges of variation and (2) examples of models of basin evolution in order to illustrate their ranges of variation and uncertainty. The need to address quantitatively the concerns of uncertainty is then made manifest.

## II. EXAMPLES OF RESERVE ASSESSMENTS AND UNCERTAINTIES

### A. U.S. HYDROCARBON SHORTFALL BY THE YEAR 2000

In April 1991, an Energy Modeling Forum was held at Stanford University to provide assessments of the shortfall expected in the U.S. supply of total hydrocarbon liquids (crude oil plus natural gas liquids) through the year 2000 (Porter, 1992). More than an order of magnitude separated the different model output results, carried out under identical price scenarios, as shown in Fig. 1.1. Porter (1992) has commented that "the range of estimates from prominent forecasters, and the troubling inconsistency of those estimates with currently observed behavior, leads to skepticism that the state of knowledge about U.S. supply has improved much over the last decade." It is not the purpose here to dispute the 10 estimates of U.S. supply shortfall obtained at the Stanford meeting; rather we take the position that each model output was produced objectively, rationally, and in good faith, to the best of each model's intrinsic assumptions and intrinsic parameter values used to assess the next decade's needs.

Indeed, the basic problem of prediction is *precisely* that some assumptions must be made on unknown factors of supply and demand and, in addition, the assumptions made must be incorporated appropriately within each model—and each model has its own rules and functional behaviors for treating processes considered relevant to the predictions to be made. Such differences have occasionally given rise to interesting discussions in the economic and scientific arenas.

In exploration economic models, many parameters occur, either alone or in combination with other parameters, and each of the parameters has its own uncertainty. Presumably this aspect of models is part of the reason for the range of results from the Stanford experiment. We need to have available practical procedures for estimating the combined effects of uncertainty of parameters on an exploration project.

However, the point is that 10 estimates of future shortfall made with

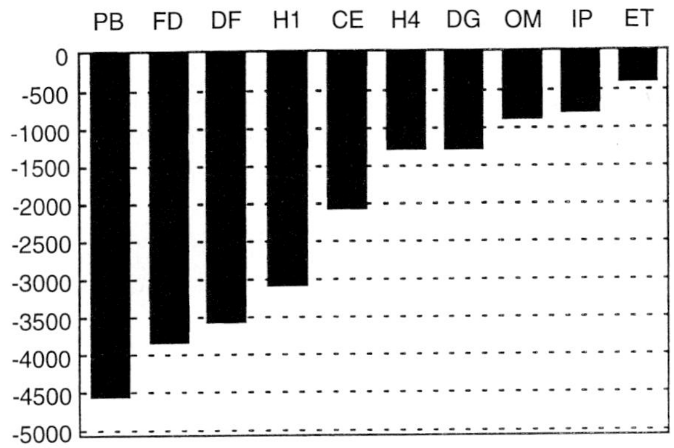


FIGURE 1.1 U.S. shortfall supply estimates for the period 1990–2000 in MBD. The 10 models used in the Stanford exercise, identified by abbreviation, were PB, University of Pennsylvania/Boston University; FD, Federal Reserve Bank–Dallas; DF, Decision Focus, Inc.; H1,4, Harvard Oil Market Simulation Model(s); CE, Canadian Energy Research Institute; DG, Dermot Gately, New York University; OM, U.S. DOE, EIA, Oil Market Simulation Model; IP, International Petroleum Exchange, MIT; ET, Alan Manne, Stanford University. Supply is defined here as total liquids (i.e., crude oil and natural gas liquids) (after Porter, 1992).

different models are available. We would like to be able to use all of the model results (so that systematic bias, conscious or unconscious, does not sway an outcome) in order to provide some quantitative measure of likely shortfall, together with an appreciation of the uncertainty of the measure at some confidence level. The purpose of this section is to demonstrate that such a desire is an eminently achievable goal in a practical sense.

This type of problem has arisen in numerous aspects of the oil industry over the years, ranging from determining probable physical rock properties from downhole logging devices, where the formulas connecting log response to physical characteristics sought contain parameters whose precise values and ranges are not too well known (Hearst and Nelson, 1985; Wahlstrom *et al.*, 1967; Carpentier *et al.*, 1989; Lerche, 1992a,b); to estimates of probable oil field production and lifetime, where secondary and tertiary recovery efficiencies are unknown ahead of implementation (Harbaugh, 1977; Nind, 1981; Megill, 1971); to predictions of bid strategies, where the number of corporate competitors and their bids are unknown (Lerche, 1992c); and to assessments of remaining potential, recoverable reserves in particular sedimentary basins (Warren, 1978; Megill, 1971; Lerche, 1992c; Murtha, 1995; Caldwell and Heather, 1995; Moore and Tucker, 1995).

The objective here is to apply similar probabilistic techniques to assess the U.S. supply shortfall based on the 10 model results of the Stanford Energy Modeling Forum experiment. Although a Monte Carlo procedure is an effective method for assessing error, it suffers from the drawback that a series of Monte Carlo runs must be made to assess error range on a physically required parameter. Such a numerical investigation can be extremely time consuming and may be part of the reason that such error determination is not yet routinely applied. In its favor, we note that the Monte Carlo procedure is completely general and can be applied to any response equation once the range and underlying probability distributions for each parameter and for the measurement uncertainties have been provided.

The question we address here is whether it is possible to produce a probabilistic assessment of error and uncertainty on quantities of physical interest without having to perform the tedious, time-consuming Monte Carlo calculations for each measurement value with assigned ranges for intrinsic parameters and an assigned range of measurement uncertainty. Clearly, if such a procedure can be developed, then it becomes a relatively simple matter to assign error ranges to quantities of relevance.

To make full use of the powerful techniques of probability measures, knowledge of the intrinsic probability of obtaining an event is required. This knowledge is, by and large, based on data dealing with scientific conditions. But the data available to work with are often very limited at the assessment stage, usually imprecise or derived by analogy, or dependent on conditions about which little knowledge (as opposed to surmise) is available. For these reasons, approximations and assumptions are introduced in attempts to obtain relatively robust estimates of data-related quantities from which some form of assessment can be made.

Generally, parameters or variables that are not too well known (and we will be more precise later) are usually treated as randomly varying in some manner around a mean value. The random component is customarily represented by a frequency distribution histogram, which provides the relative number of times a parameter has been observed in a given interval range compared to all interval ranges. Customarily, the area under the histogram is normalized to unity so that the frequency distribution then provides an approximate empirical assessment of the probability of occurrence of a parameter in each and every interval range, based on the data at hand.

Interest often centers not so much on the frequency distribution [e.g., what is the probability that the supply shortfall is between 500 and 1000 thousand barrels of liquid hydrocarbons per day (MBD)], but rather on the cumulative frequency distribution [e.g., what is the probability that the shortfall is greater (less) than 1000 MBD?].

In one case (greater than), the cumulative frequency distribution histogram is usually the residual fractional area still lying beyond a parameter value. In the other case (less than), the measure is the fractional area contained up to the parameter value. This set is mutually exclusive (i.e., the probability of a shortfall of >1000 MBD plus the probability of a shortfall of <1000 MBD must sum to unity). Thus if $p(x)\,dx$ measures the frequency distribution (normalized) of occurrence of $x$ in the range $x$ to $x + dx$, then

$$P(y > x) = \int_y^\infty p(x)\,dx \tag{1.1}$$

measures the cumulative frequency distribution, that is, the chance of exceeding a particular value $y$, whereas

$$P(y < x) = \int_0^y p(x)\,dx \equiv \int_0^\infty p(x)\,dx - \int_y^\infty p(x)\,dx \equiv 1 - P(y > x) \tag{1.2}$$

measures the chance of not exceeding a particular value $y$.

In a large number of circumstances in exploration assessments, it is often difficult to provide the frequency distribution even in a rough form. For that reason, moments of the underlying distribution are often used as approximations. The mean value, $E_1(x)$, of $x$ for a frequency distribution $p(x)\,dx$ is

$$E_1(x) = \int_{-\infty}^\infty xp(x)\,dx, \tag{1.3}$$

and the mean square value $E_2(x)$ is

$$E_2(x) = \int_{-\infty}^{\infty} x^2 p(x)\, dx. \tag{1.4}$$

The variance, $\sigma^2$, around the mean is given by

$$\sigma^2 = E_2(x) - E_1(x)^2 \geq 0, \tag{1.5}$$

where $\sigma$ is the standard deviation.

In most of the situations we discuss in this book, only multiple powers of distributions will be needed, as defined by

$$E_j(x) = \int_{-\infty}^{\infty} x^j p(x)\, dx. \tag{1.6}$$

We will also deal with the median value, $x_{1/2}$, of a frequency distribution defined as that value of $x$ such that

$$P(y < x) = P(y > x) = 1/2 \tag{1.7}$$

and with the mode $x_m$ (for a unimodal distribution) defined as that value of $x$ at which $p(x)$ has its maximum value.

The log-normal distribution occurs physically in many situations ranging from the areal size distribution of sunspots to lease sale bid distributions. The normal distribution cannot be appropriate when there is a constraint on a variable; for example, area cannot be negative, bid values must be positive. Under such conditions, empirical evidence suggests that an approximate measure of cumulative frequency distribution is provided by a log-normal behavior with

$$P(x|x_{1/2}, \mu) = \tfrac{1}{2}[1 + \mathrm{erf}(\ln(x/x_{1/2})/2^{1/2}\mu)], \tag{1.8}$$

with the mean value of $x$, $E_1(x)$, given through

$$E_1(x) = x_{1/2} \exp(\mu^2/2), \tag{1.9a}$$

the mode value by

$$x_m = x_{1/2} \exp(-\mu), \tag{1.9b}$$

and the variance in $x$, $E_2(x) - E_1(x)^2 \equiv \sigma^2$, given by

$$\sigma^2 = E_1(x)^2[\exp(\mu^2) - 1], \tag{1.9c}$$

where $x_{1/2}$ is the median value. At $x = x_\sigma \equiv x_{1/2} \exp(\mu)$ we have $P(x_\sigma|x_{1/2}, \mu) = 0.84$, while on $x = x_m$ [$\equiv x_{1/2} \exp(-\mu)$] we have $P(x_m|x_{1/2}, \mu) = 0.16$. In this case note that $P = 1/2$ on $x_{1/2}$, but $P \cong 0.68$ on $x = E_1(x) > x_{1/2}$. (See Appendix A at the end of the chapter.)

Empirically, it is often difficult, if not impossible at the assessment stage, to obtain enough information to determine the precise shape of the frequency distribution of a particular parameter or variable. Indeed, quite often it is considered a fairly good achievement to be able to estimate a likely minimum, $x_{\min}$, a likely maximum, $x_{\max}$, and a likely most probable value, $x_p$, for a parameter. If the variable of interest is approximately log-normally distributed, it is possible to work Eqs. (1.9) in reverse to obtain estimates of $\mu$, $x$, $x_{1/2}$, and $x_m$. Thus, from Eq. (1.9c) we obtain

$$\mu = [\ln\{1 + \sigma^2/E_1(x)^2\}]^{1/2}, \tag{1.10a}$$

and then, from Eqs. (1.9a) and (1.9b), we can estimate

$$x_{1/2} = E_1(x)\exp(-\mu^2/2) \equiv E_1(x)[1 + \sigma^2/E_1(x)^2]^{-1/2} < E_1(x), \tag{1.10b}$$

and from Eq. (1.9b) we have

$$x_m \equiv x_{1/2}\exp(-\mu), \tag{1.10c}$$

and

$$x_\sigma = x_{1/2}\exp[(\ln(1 + \sigma^2/E_1(x)^2))^{1/2}] > x_{1/2}. \tag{1.10d}$$

It is this cumulative probability procedure that we now apply to the 10 results of the U.S. supply shortfall experiment. The 10 data of Fig. 1.1 presumably reflect empirical determination of the intrinsic frequency of occurrence of the supply shortfall. The assumption is that if the experiments were repeated many times, again under identical price scenarios, then the outcomes of the experiments (with internal parameters for each model allowed to vary over what the model developer considers an acceptable range around the "best" values) would reinforce the distribution of results given in Fig. 1.1. Thus, the assumption is that the 10 observed outcomes fairly represent the true frequency of occurrence. On this assumption, it then follows that an unbiased estimate of the expected value, $E_1(x)$, of the supply shortfall can be obtained from

$$E_1(x) = (1/10)\sum_{i=1}^{10} x_i, \tag{1.11a}$$

where $x_i$ is the shortfall for the $i$th model outcome. Reading from Fig. 1.1, we obtain the value

$$E_1(x) \cong 2.6 \text{ MMBD}, \tag{1.11b}$$

where MMBD stands for million barrels of liquid hydrocarbons per day. Likewise, an estimate of the second moment, again unbiased, is

$$E_2(x) \cong 6.8(\text{MMBD})^2. \tag{1.11c}$$

Then, from Eqs. (1.10) we have the estimate

$$\mu^2 = \ln(E_2/E_1^2) = \ln[6.8/(2.6)^2]. \tag{1.12}$$

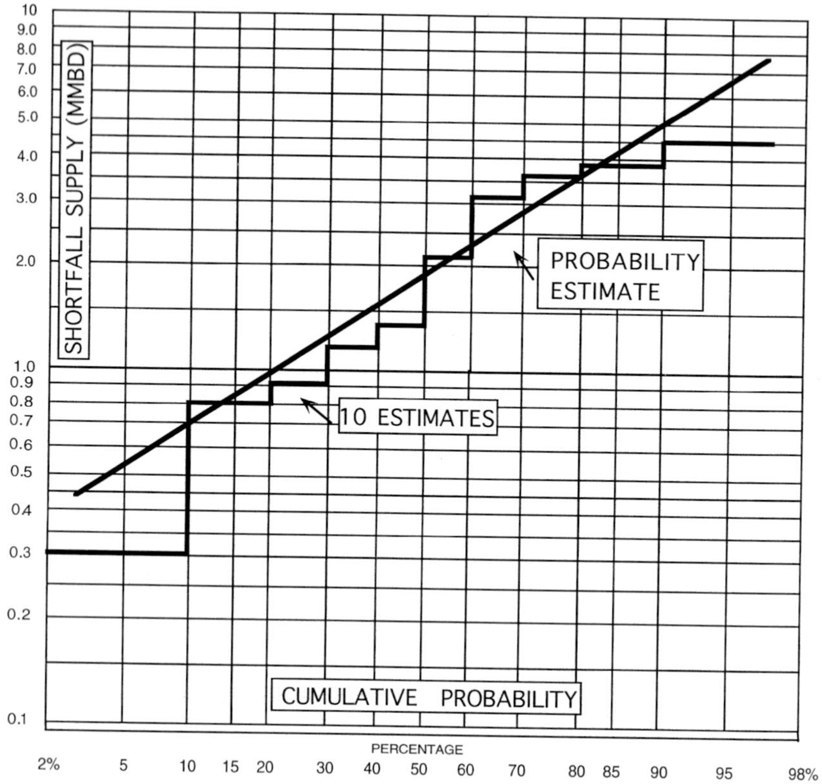


FIGURE 1.2 Cumulative probability estimate (straight line) of shortfall supply based on the 10 results of the Stanford experiment ("staircase" line).

Thus, for a cumulative probability estimate one would assign

$$P(16\%) \text{ occurs on } x = 0.8 \text{ MMBD}, \tag{1.13a}$$

$$P(68\%) \text{ occurs on } x = 2.6 \text{ MMBD}, \tag{1.13b}$$

$$P(84\%) \text{ occurs on } x = 4.0 \text{ MMBD}. \tag{1.13c}$$

A cumulative probability plot can then be drawn as shown in Fig. 1.2, from which one infers that the likely shortfall range of estimates (based on 90% confidence) is

$$\text{Shortfall} = 2.6^{+2.2}_{-1.9} \text{ MMBD}. \tag{1.14}$$

Thus, there is a 68% chance of the shortfall being less than 2.6 MMBD, a 90% chance of the shortfall being less than 4.8 MMBD, but only a 10% chance of the shortfall being less than 0.7 MMBD. It would seem less than prudent to plan on a best-case eventuality.

But the problem now moves from one of quantitative assessment of supply shortfall to one of economic and political health of a nation. Precisely how the U.S. oil industry, the U.S. government, and other interested parties will react to the perceived shifting strengths and weaknesses of the nation in relation to the requirement of a continued uninterrupted supply of liquid hydrocarbons, and precisely how such actions would impact and be impacted by global economic and political considerations, will be a fascinating pattern to watch develop during the next decade.

The ability to place the estimates of U.S. supply shortfall in probabilistic terms, as given here, enables the application of corporate and government policies and actions to be placed on a firmer footing than might otherwise have been appreciated from a cursory overview of the 10 results from the Stanford experiment—and that is, indeed, the point of this example. The results presented here are dependent on two major conditions: (1) that the 10 individual models yield estimates of the U.S. shortfall position, each of which is as precise as the individual model assumptions, parameters, and control information allow it to be (i.e., no resolution or uncertainty was given of each shortfall estimate); and (2) that the model estimates are not especially biased by the use of different control information in different cases so that the 10 estimates, when taken together, fairly represent the correct distribution, even though each individual estimate contains some uncertainty. However, given the broad range of values reported in the 10 cases, it is likely that at least one, and possibly both, of these major conditions is in error. It is also extremely likely that different intrinsic model assumptions and different values of parameters used in the models are both affecting the range of values reported.

Clearly, we need some assessment of the manner in which each estimate is made; in the sensitivity of output results to changes in assumptions and parameter values; and in the quality, quantity, and sampling frequency of control information. Only with such an analysis can we have confidence in any individual estimate.

## B. RANGES OF PROBABLE UNCERTAINTY IN U.S. GAS RESOURCE ESTIMATES

The estimations of likely future reserves of oil and gas in a nation, relative to current and projected usage, are measures by which political, social, and economic assessments are made of the need to address these topics: (1) securing supplies from foreign sources or (2) exploring domestically for further reserves in a timely manner. To maintain or enhance the future energy health of a nation, in relation to that nation's perceived personal, residential, and industrial requirements, decisions for capital investment in different possible energy futures must be made, with the attendant consequences of construction and development timescales for each such decision undertaken (Ethetton and Brumbaugh, 1995).

There is little point in planning a future energy development program to replace an existing energy source that may take, say, 20 years to come into use, when the current energy source/usage lifetime is considerably less than 20 years. One wants to plan well ahead of depletion lifetime, so that a smooth transition from one energy source to another takes place. Thus, both the private corporate sector and the government sector are concerned with the energy needs of a nation. Accordingly, estimates of proven, potential, and possible reserves of a given energy base are used to evaluate projections of national need.

Our concern in this section is to address estimates made of future gas reserves for the United States. Current proven reserves of gas are about 200 trillion cubic feet (Tcf) and the nation uses about 20 Tcf/year, so that a lifetime estimate to depletion is only about 10 years. Clearly, some assessments of potential gas reserves, their likelihood of being found, and the conditions under which gas can be found must be made, so that decisions for investment in and development of such likely reserves can be made. In addition, the assessments must be evaluated for their probability of being correct.

**1. Basic Data and Probabilistic Interpretation**

During the past 6 years, several groups have made reserve estimates organized in different categories, using different methods, and with different qualities and quantities of data. The Potential Gas Agency (1993) has summarized the estimates made and that summary is given here as Table 1.1, taken directly from the PGA report. Note from Table 1.1 that all reported potential gas resource estimates have been adjusted to the end of 1992. The degree of reporting recorded in Table 1.1 by the different assessment groups is variable. For instance, the NRA (USGS/MMS) assessment of potential gas resources in the lower 48 states for existing gas fields is recorded as zero—despite ongoing production from such fields—and the ENRON Corporation does not report a single category, merely a total, so that it is not possible to evaluate the quality of their assessment. In several categories no assessment was made (marked by superscript *b*) by individual groups.

A perusal of the other numbers in Table 1.1 is quite interesting. Each individual group gives *precise* statements, of three and four digits, on their categories covering conventional and unconventional assessments of potential reserves; yet it is doubtful that such precision is significant because the range of values reported by all the groups spans a factor of over 3—thus the *accuracy* of each assessment can be called into question. The same difficulty arises when individual categories are examined in Table 1.1; for example, the entries under Conventional resources (lower 48): Undis-

TABLE 1.1 Reported Potential Gas Resource Estimates in Trillion Cubic Feet (Tcf), Adjusted to 12/31/92

| Resource category (area) | DOE 1988 | NRA (USGS/MMS) 1989 | NES reference 1990 | NES access + adv. 1990 | NPC current 1992 | NPC advanced 1992 | ENRON 1993 | GRI current 1993 | GRI advanced 1993 | Potential gas committee 1993 |
|---|---|---|---|---|---|---|---|---|---|---|
| 1. Conventional resources (lower 48 states) | | | | | | | | | | |
| a. Existing fields | 186 | 0 | 108 | 196 | 149 | 168 | [c] | 209 | 209 | 167 |
| b. Undiscovered fields | 353 | 324 | 256 | 327 | 375 | 413 | [c] | 608 | 667 | 493 |
| 2. Conventional resources (Alaska) | | | | | | | | | | |
| a. Existing fields[a] | 28 | 28 | 33 | 33 | 30 | 30 | [b] | [b] | [b] | 37 |
| b. Undiscovered fields | 93 | 75 | [b] | 75 | 76 | 84 | [b] | [b] | [b] | 157 |
| 3. Unconventional resources (lower 48 states) | | | | | | | | | | |
| a. Tight sands | 180 | [b] | 230 | 383 | 232 | 349 | [c] | 156 | 241 | [c] |
| b. Coalbed methane | 48 | [b] | 50 | 90 | 62 | 98 | [c] | 81 | 103 | 90 |
| c. Shale gas | 31 | [b] | 15 | 30 | 37 | 57 | [c] | 48 | 127 | [c] |
| d. Other | [b] | [b] | [b] | [b] | 15 | 15 | [c] | 0 | 43 | [b] |
| 4. Unconventional resources (Alaska) | | | | | | | | | | |
| a. Coalbed methane | [b] | [b] | [b] | [b] | 37 | 57 | [b] | [b] | [b] | 57 |
| 5. a. Total conventional resources | 660 | 427 | 397 | 631 | 630 | 695 | [c] | 817 | 876 | 854 |
| b. Total unconventional resources | 259 | [b] | 295 | 503 | 383 | 576 | [c] | 285 | 514 | 147 |
| c. Total lower 48 states | 798 | 324 | 659 | 1026 | 870 | 1100 | 1145 | 1102 | 1390 | 750 |
| d. **Total United States** | 919 | 427 | 692 | 1134 | 1013 | 1271 | 1145 | 1102 | 1390 | 1001 |

[a] Includes 25 Tcf of gas reported as proved reserves prior to 1988.
[b] Not assessed.
[c] Not reported as separate category.

covered fields vary from 256 to 667 Tcf, a range of over a factor of 2.5. At a fixed current usage of 20 Tcf/year, this category alone would yield a lifetime estimate to depletion of between 13 and 33 years, a fairly broad range.

Such a broad range of uncertainty suggests that different assumptions and/or data and/or interpretations and/or individual bias may be coloring the assessments made to differing, but unknown, degrees. The difficulty, then, is to figure out how to plan for the future when faced with such broad ranges of uncertainty. Clearly, some measures of quality and confidence in any and all assessments are required if one is to estimate the likely amounts of potential gas resources by category, and also cumulatively, in order to have a rational, risk-weighted basis for planning.

To provide some measure of the accuracy of potential gas resources, here we use cumulative probability methods, which have proven to be extremely powerful devices in the exploration and economic theaters (see, e.g., the many references in Lerche, 1992c). The basic framework and detailed mathematical arguments for evaluating measures of a quantity within a probabilistic setting are well developed within the oil and gas exploration sector (Warren, 1981, 1983a,b, 1988) and have been used extensively in this field and others. The reader is referred to the preceding references for a detailed appreciation of the technical aspects of the methods (see also Appendix A at the end of this chapter). The cumulative probability provides an estimate of the chance of *not* exceeding a particular magnitude of a quantity under consideration.

For instance, for the values recorded in Table 1.1 under the category Conventional resources (lower 48): Existing fields, the positive values (excluding the value of zero given by NRA (USGS/MMS) can be arranged in increasing sequence as 108, 149, 167, 168, 186, 196, 209, and 209 Tcf, for a total of eight values. A cumulative probability histogram of the values versus cumulative probability can then be drawn on log-probability paper as is done in Fig. 1.3, based *solely* on the data at hand, from Table 1.1.

Superimposed on Fig. 1.3 is a best-fit straight line through the histogram representing a log-normal probability distribution. Note that except at the extreme ends of range of the cumulative probability plot, the log-normal distribution is a remarkably good fit to the data. This empirical point has been noted many times previously in other arenas of exploration and economic assessment (Warren, 1983a), which is why the log-normal probability method (on a cumulative probability basis) is used most often to categorize uncertainty—and will be so used here, too.

The important points about a log-normal cumulative distribution are (1) that the expected value, $E(x)$, of a quantity $x$ occurs close to 68% [i.e., there is a two-thirds chance of the actual value being less than or equal to $E(x)$] and (2) that the 10 and 90% cumulative probability values are stable measures of the range within which an exactly log-normal distribution satisfies as best as it can a set of histogram values (Feller, 1957). Accordingly the 10 and 90% cumulative probability points are often used to provide a measure of the "volatility" of an estimate of a quantity, $x$. For instance, in Fig. 1.3, the 68% [$P(68)$] cumulative probability value is 200 Tcf, the 10% [$P(10)$] cumulative probability value is 137 Tcf, and the 90% [$P(90)$] cumulative probability value is 215 Tcf. Thus, there is a two-thirds chance of a value of *less* than 200 Tcf, a 90% chance of a value of *less* than 215 Tcf, and a 90% chance of a value *greater* than 137 Tcf. The estimate is then often written as $200^{+15}_{-63}$ Tcf to show specifically the range. A volatility measure, $v$, is often written (Warren, 1978) as

$$v = [P(90) - P(10)]/P(68),$$

which provides an estimate of how uncertain the $P(68)$ value is. In the case of Fig. 1.3 we have $v = (215 - 137)/200 = 78/200 \equiv 39\%$, a 39% volatility to the two-thirds chance of having less than 200 Tcf.

This volatility factor is useful with more than one category of potential gas reserve for two reasons: If the volatility is small compared to unity, then the estimate is fairly accurate because the $P(90)$ and $P(10)$ values [which bracket the $P(68)$ value] are then close, so the fractional error in $P(68)$ is small. If the volatility is large compared to unity, then a large fractional error exists in the $P(68)$ estimate, and that estimate is then not as reliable as one with a low volatility. To include the volatility, $v$, as a measure of reliance for an expected value, $E$, different weighting schemes are often employed; the simplest is inverse weighting, $E/v$, while some simple weightings that allow for the fact that the volatility can be small compared to unity include $E\exp(-v)$, $E(1 + v^2)^{-1/2}$, etc. Here, for ease of illustration we use the weighting $E/v$ with the data of Table 1.2, although one should, perhaps, also use other weighting schemes to obtain information on the stability of each weighted estimated with respect to the choice of each weighting procedure.

Two simple calculations can be performed on a set of categories. Suppose $N$ categories are being investigated, each with its expected value $E(x_i)$, corresponding to $P(68)$, and each with its volatility, $v_i$. Then a risk-weighted measure of potential reserves can be made from each category as

$$R_i = (E(x_i)/v_i), \tag{1.15}$$

with the total as

$$R = \sum_{i=1}^{N} R_i, \tag{1.16}$$

which weights the contribution of each category to the total according to its relative volatility.

The second simple calculation that can be made is the risk-weighted relative importance, RI($i$) (in a percentage) of the $i$th category in relation to the other categories, with

$$\mathrm{RI}(i) = 100(E(x_i)/v_i) \bigg/ \left( \sum_{i=1}^{N} E(x_i)/v_i \right). \tag{1.17}$$

This relative importance factor measures the likelihood of a particular category contributing potential resources compared to all categories. As such RI is a useful measure of where most exploration effort should be put to enhance the likelihood of reserve appreciation being realized, with allowance made for the speculative nature of individual categories. Such a device is very useful in the decision-making process of where to commit capital investment and resource exploration.

**2. Risk-Weighted Estimates of Likely Potential Reserves**

For each individual category (labeled 1a, 1b, 2a, 2b, 3a, 3b, and 3c on Table 1.1) for which sufficient individual estimates were made, a detailed

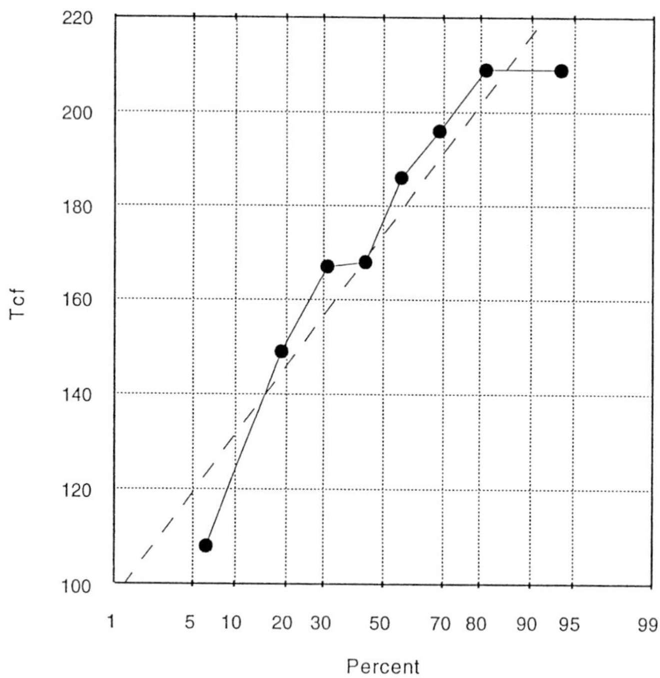


FIGURE 1.3 Cumulative probability histogram and superposed log-normal probability curve for category 1a of Table 1.1.

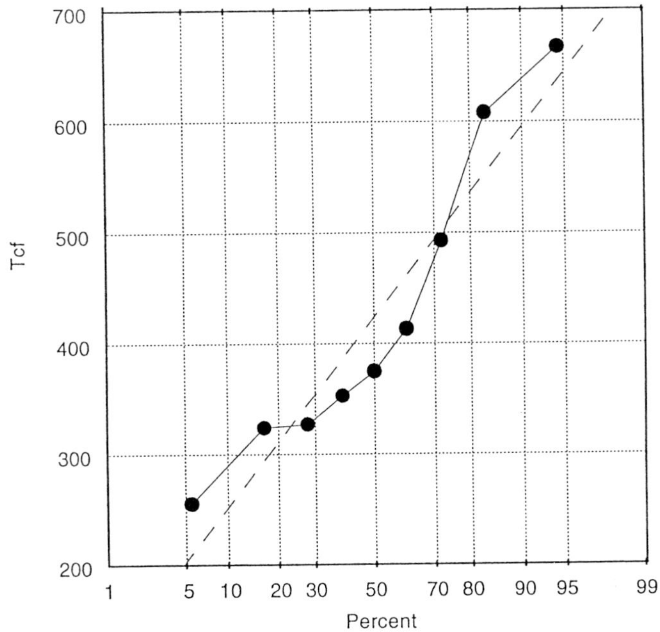


FIGURE 1.4 Cumulative probability histogram and superposed log-normal probability curve for category 1b of Table 1.1.

probability histogram can be drawn and best-fit cumulative log-normal probability curves superposed as shown in Figs. 1.3 through 1.9, respectively. Reading from those figures enables one to construct Table 1.2 showing probability values at $P(10)$, $P(68)$, and $P(90)$ for each category as well as the volatility of each category.

Inspection of Table 1.2 shows that category 1a (lower 48, existing fields) for which most data exist, has the lowest volatility, which is not surprising. It is surprising, however, that the volatility is as high as nearly 40%, and it is even more surprising that the remaining categories have between about 80 and 160% volatilities, suggesting that they are extremely poorly determined relative to category 1a—a factor of between 2 to 4 more uncertainty based on relative volatilities. It would seem that assessments of the categories are less than accurate. Similar arguments can be put forward for the four total resource estimates given in categories 5a (Total conventional resources), 5b (Total unconventional resources), 5c (Total lower 48 states), and 5d (Total United States), where between 8 and 10 of the total of 10 assessment groups provide precise, but inaccurate, numerical values. Figures 1.10, 1.11, 1.12, and 1.13 record the cumulative probability histograms and superposed best-fit log-probability curves to categories 5a, 5b, 5c, and

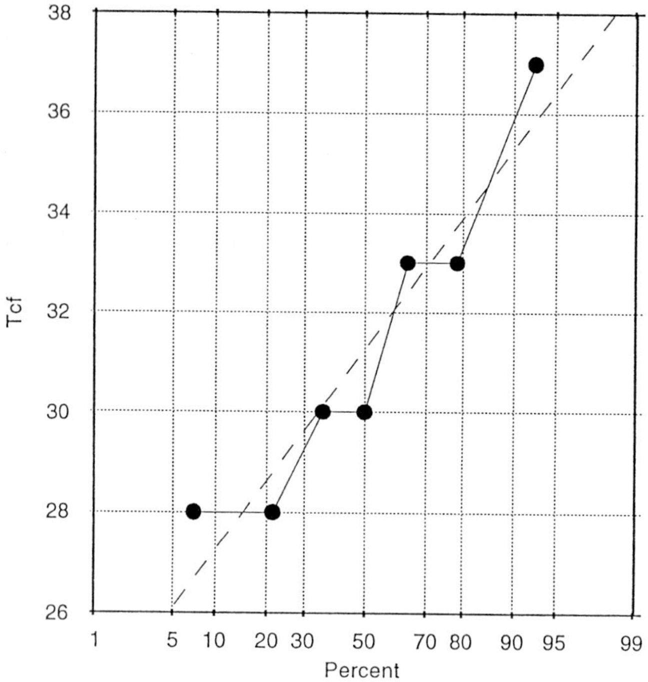


FIGURE 1.5 Cumulative probability histogram and superposed log-normal probability curve for category 2a of Table 1.1.

5d, respectively, and Table 1.3 presents the $P(10)$, $P(68)$, $P(90)$, and volatility measures for the same categories.

Note from Table 1.3 that, even without including the volatilities of each subcategory (1 through 3) that make up the Total resource estimates, there is a high volatility of 70–90% on an unweighted estimate of total reserves. A risk-weighted estimate of potential gas reserves in the sum of categories 1–3 can also be made and gives

$$R = 740^{+270}_{-310}\ \text{Tcf}. \tag{1.18}$$

This risk-weighted estimate argues for a two-thirds chance of *less* than about 740 Tcf likely to be found, with a 90% chance of *less* than 1010 Tcf likely to be found, and a 90% chance of *greater* than 430 Tcf likely to be found.

Thus, instead of just adding all the categories as though each category estimate had the same level of uncertainty as all others, if one allows for the relative worth of each estimated category, then a more representative measure can be given of what fraction is *likely* to be found of the total estimated potential reserves. It would seem that a total range of 430–1010

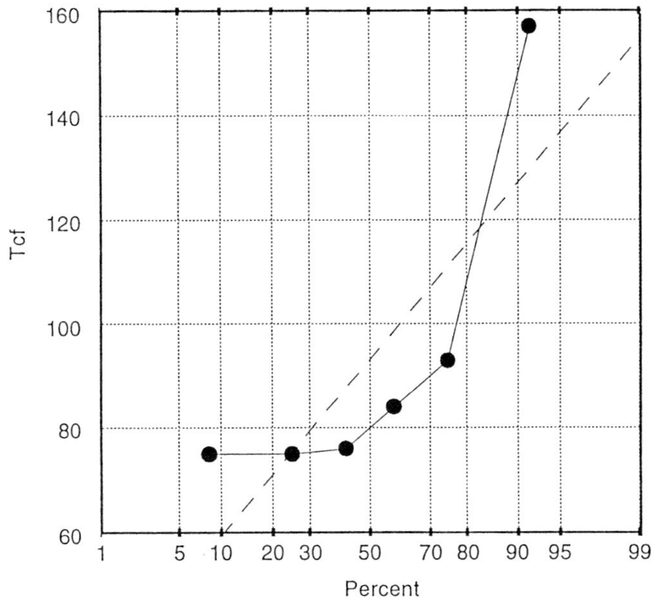


FIGURE 1.6 Cumulative probability histogram and superposed log-normal probability curve for category 2b of Table 1.1.

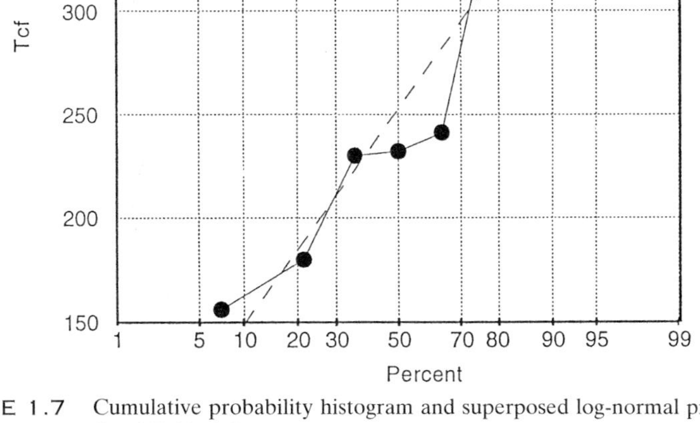


FIGURE 1.7 Cumulative probability histogram and superposed log-normal probability curve for category 3a of Table 1.1.

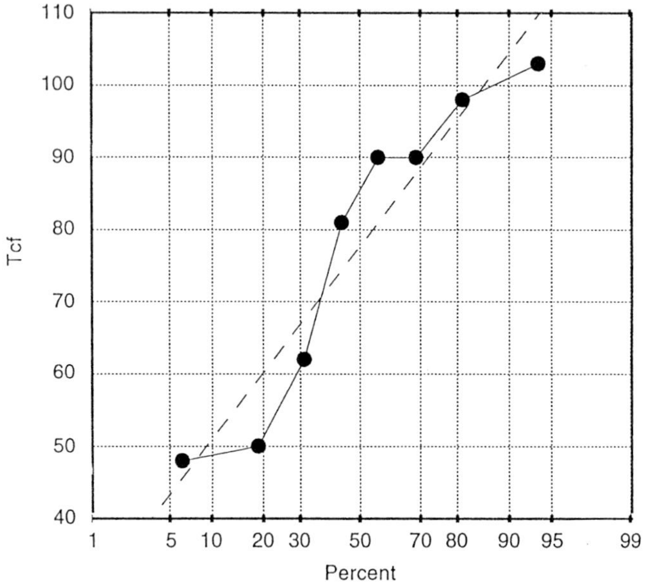


FIGURE 1.8 Cumulative probability histogram and superposed log-normal probability curve for category 3b of Table 1.1.

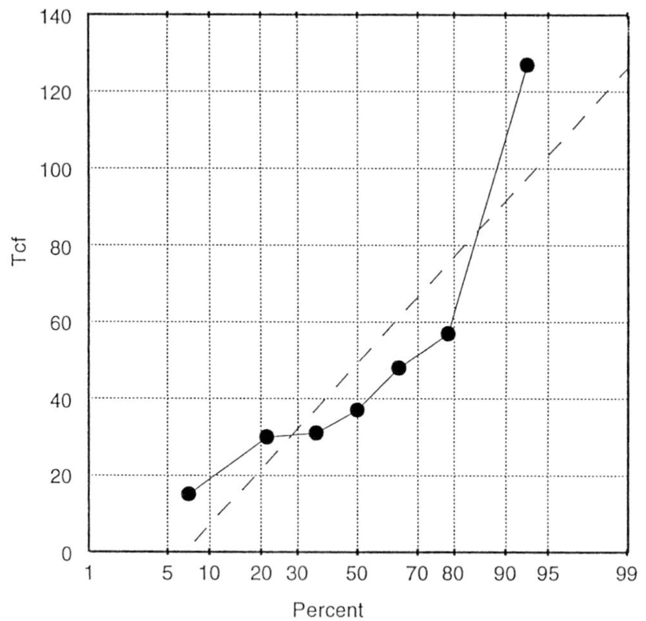


FIGURE 1.9 Cumulative probability histogram and superposed log-normal probability curve for category 3c of Table 1.1.

TABLE 1.2 Probability Values and Volatility Measures for Categories 1a, 1b, 2a, 2b, 3a, 3b, and 3c of Table 1.1

| Category | Value (Tcf) at $P(10)$ | Value (Tcf) at $P(68)$ | Value (Tcf) at $P(90)$ | Volatility (%) |
|---|---|---|---|---|
| 1a | 137 | 200 | 215 | 39 |
| 1b | 270 | 500 | 650 | 76 |
| 2a | 26 | 32 | 55 | 90 |
| 2b | 70 | 80 | 125 | 70 |
| 3a | 150 | 280 | 400 | 89 |
| 3b | 47 | 80 | 118 | 87 |
| 3c | 13 | 57 | 110 | 160 |

Tcf, with a likelihood of less than about 740 Tcf, more than generously covers the statistics of Table 1.1. The suggestion, then, is that not all categories are equally attractive based on the individual sets of assessments made.

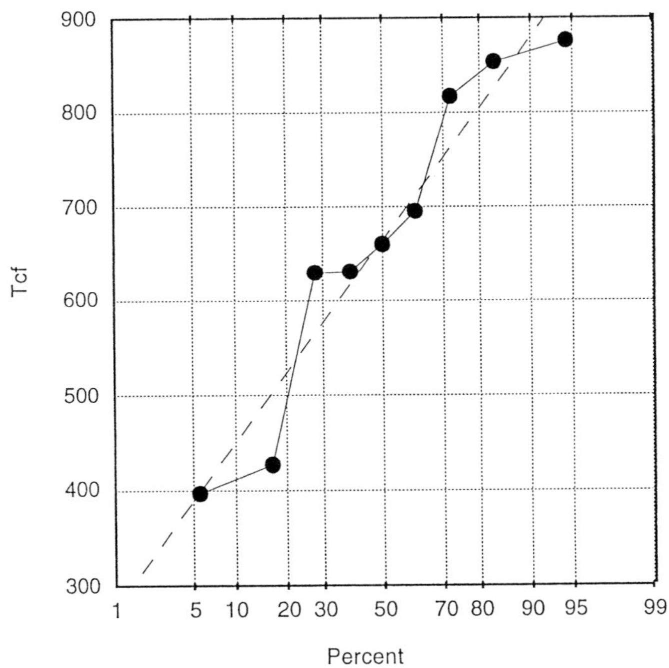


FIGURE 1.10 Cumulative probability histogram and superposed log-normal probability curve for category 5a of Table 1.1.

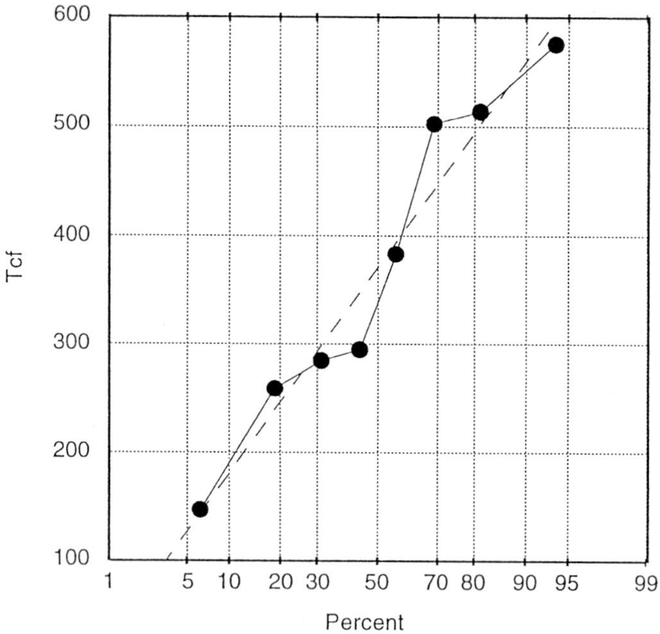


FIGURE 1.11 Cumulative probability histogram and superposed log-normal probability curve for category 5b of Table 1.1.

### 3. Relative Importance of Each Category

In addition to estimates of the total potential gas reserves likely to be found, it is of equal importance to know which categories carry the most likely chance of containing discoverable reserves. Such assessments then provide information on where to concentrate resources and capital to maximize the probability of actually finding the reserves.

Figure 1.14 provides the risk-weighted relative importance (RI) for each of categories 1a, 1b, 2a, 2b, 3a, 3b, and 3c of Table 1.1, using the volatility measures given in Table 1.2. Each RI is evaluated using values at $P(68)$, $P(10)$, and $P(90)$, so that one has an appreciation of the stability and rank of each category's relative importance. Inspection of Fig. 1.14 strongly suggests that, at any level, the relative importance of finding likely gas reserves is roughly equally split between the Lower 48 existing and undiscovered fields categories, with the Tight sands category a distant third.

The categories of Unconventional resources (lower 48): Coalbed methane, and Conventional resources (Alaska): Undiscovered fields are roughly equally fifth in importance, with the categories of Conventional resources (Alaska): Existing fields and Unconventional resources (lower 48): Shale gas being tied at seventh in importance.

A ranking directly proportional to risk-weighted relative importance then yields:

| *Category* | *Ranking order* | *Rank value* |
|---|---|---|
| 1a | Equal 1 and 2 | 6 units |
| 1b | | 6 units |
| 3a | 3 | 3 units |
| 2b | Equal 4 and 5 | 1 unit |
| 3b | | 1 unit |
| 2a | Equal 6 and 7 | 0.8 unit |
| 3c | | 0.8 unit |

This rank and the relative importance ranking values strongly suggest that the most likely categories in which we will actually find significant gas reserves are Existing and Undiscovered fields in the lower 48 states and in Tight sands. The combined rank values of these three categories are 15 units compared to 2.8 units for all other categories. Thus the likelihood of finding gas reserves in the three top categories is five times as great as it is in all other categories. It is likely that between 430 and 1010 Tcf will actually be discovered, with a two-thirds chance of less than 740 Tcf being discovered.

The purpose of this section has been to show how the very large ranges

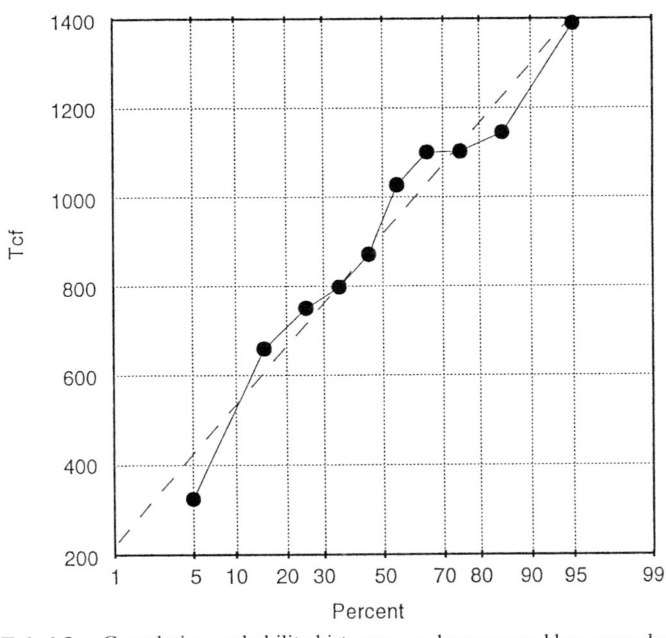


FIGURE 1.12 Cumulative probability histogram and superposed log-normal probability curve for category 5c of Table 1.1.

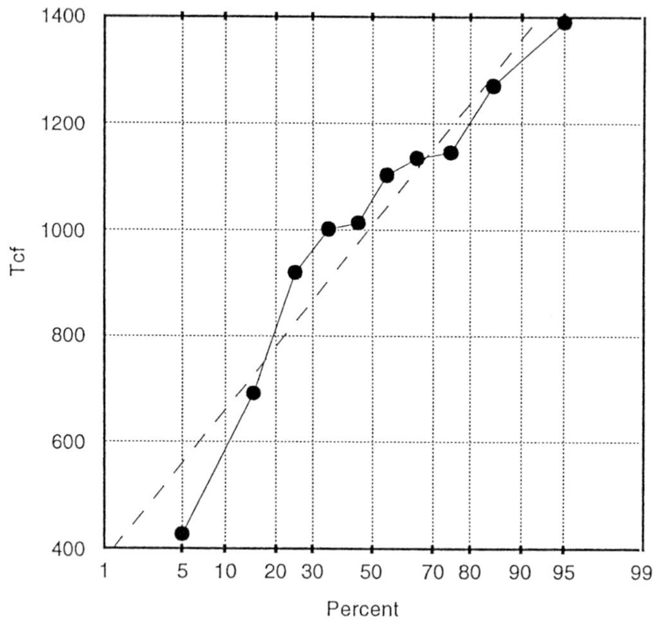


FIGURE 1.13 Cumulative probability histogram and superposed log-normal probability curve for category 5d of Table 1.1.

of uncertainty in individual assessments of categories of potential gas resources can be brought into some form of order, allowing both for the vagaries of individual assessments and for the speculative nature of some of the categories. Probabilistic methods provide an eminently satisfactory vehicle for enabling both the absolute uncertainty within a category and the relative ranking between categories to be assessed: (1) for the relative importance of each category of potential gas reserves; (2) for providing a risk-weighted likelihood of actually finding the potential reserves; and (3) for determining which categories are likely to yield the potential reserves.

TABLE 1.3 Probability Values and Volatility Measures for Categories 5a, 5b, 5c, and 5d of Table 1.1

| Category | Value (Tcf) at *P*(10) | Value (Tcf) at *P*(68) | Value (Tcf) at *P*(90) | Volatility (%) |
|---|---|---|---|---|
| 5a | 390 | 790 | 1050 | 88 |
| 5b | 180 | 370 | 520 | 83 |
| 5c | 550 | 950 | 1200 | 71 |
| 5d | 600 | 1000 | 1300 | 70 |

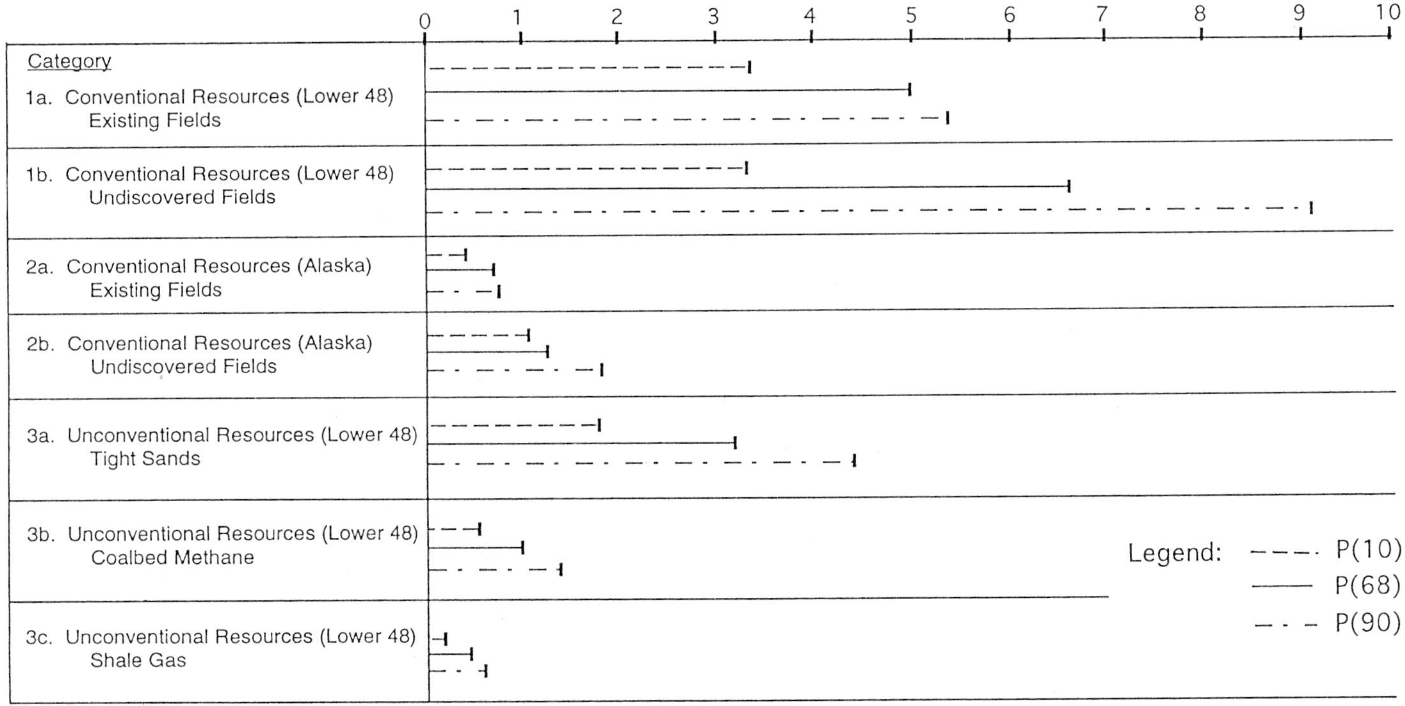


**FIGURE 1.14** Risk-weighted relative importance of each category in terms of likely discoverable reserves. Relative importance lines are drawn at $P(10)$, $P(68)$, and $P(90)$ as shown, using the volatility measures of Table 1.2.

In this way a more accurate appreciation of how, where, and when to commit capital investment and exploration effort can be obtained.

For the 10 assessments reported during the past 6 years in several categories by different groups, it would seem that there is about an 80% likelihood that the majority of the discoverable gas reserves will be found in the lower 48 states in the categories of Existing and Discoverable fields and, to a lesser extent, in Tight sands. The likely, risk-weighted amount that is discoverable would appear to lie in the 90% probability range of 430 to 1010 Tcf, with a most likely discoverable amount of less than about 740 Tcf.

## III. EXAMPLES OF BASIN ANALYSIS AND UNCERTAINTIES

Basin analysis is the study of the interactions of sediment deposition and compaction, thermal changes and fluid flow, tectonic forces and sea-level variations, and hydrocarbon production and accumulation in evolving sedimentary basins. A quantitative understanding of the dynamic interrelationships of these processes, and the controls exerted by present-day data, is sought to improve basic geologic knowledge and to foster oil exploration for future generations.

Well-advanced computer programs simulate fluid flow; sediment compaction; thermal history; and hydrocarbon generation, migration, and accumulation—in one, two, and even three spatial dimensions (Yukler, 1979; Welte and Yukler, 1981; Bethke, 1985; Lerche, 1990a,b; Statoil, 1994). The variously available programs have technical differences in their methods of calculation (grid sizes, time steps, smoothing, and interpolation techniques, etc.) that, of themselves, would make for a fascinating review. However, the emphasis of this section is not so much on the differences of individual computer manipulations, but rather on fundamental difficulties and problems in any endeavor to codify procedures for quantitative basin analysis assessment. The guiding motivation is to provide an appreciation of the strengths, weaknesses, accuracies, and uncertainties of basin analysis models as they relate, directly or indirectly, to potential hydrocarbon accumulation sites in a basin. In addition, major fundamental problems yet to be addressed will be considered.

Three input factors control the veracity and usefulness of a quantitative model: intrinsic assumptions of the model; implicit or explicit parameter values or functional dependences in required equations of state; and quality, quantity, and sampling frequency of data used as controls on the procedure. These factors may be of a forgivable nature in particular applications or they may be of paramount importance. Part of the task of basin analysis procedures is to determine under what conditions each of the input factors plays a dominant role and to delimit the ranges of uncertainty of parameter values that are allowed, consistent with observations and inferences from observations.

Five subsections illustrate some of the difficulties, problems, and consequences of each input factor being invoked in different manners. Subsection A provides an overview of the interrelationship of burial, thermal, and hydrocarbon histories in order to set a basic framework for the later examination of individual facets in this volume. Then Subsection B examines the consequences of dynamic factors in impacting the sedimentary evolution component of the system. This discussion is followed, in Subsection C, by some considerations related to determinations of both present-day and paleoheat flux and temperature. Subsection D discusses the impact of hydrocarbon kinetic models and of primary and secondary migration processes in influencing the assessments of potential hydrocarbon accumulation amounts and sites. In Subsection E the data controls available on models, both pre- and post-borehole drilling, are considered.

Finally, some fundamental outstanding concerns for later investigations are reviewed, as well as suggestions for linking the model behaviors to arenas impacted by the ability to construct models, such as risk analysis and exploration economics, to be discussed later in this volume.

## A. OVERVIEW OF BASIN ANALYSIS

Hydrocarbon accumulations in a basin are massively dependent on (1) the sediment fill and structural evolution over time of a basin, including fluid escape; (2) the thermal history of the sediments; and (3) the organic geochemistry controlling the generation of hydrocarbons from kerogen and the subsequent migration and accumulation in sealed traps of the generated hydrocarbons. We consider each in turn.

### 1. Burial History

#### *a. Overview*

We concentrate first on dynamic evolution. At deposition, sediments typically have porosities in the range 50–75%. When drilled at depth, say, 5000 rm, porosities are typically of order 10% or less. Thus an enormous amount of water has escaped from the sediments during their deposition and evolution. The grain-to-grain contact pressure must then increase in order to support the overlying sediment weight. A problem of fundamental concern, underlying all other aspects of basin analysis, is to understand the dynamic evolution of fluid escape and compaction in a sedimentary basin. In addition, the sediments must be supported on an underlying basement rock, which must respond to the applied load. If the basement material behaves as an incompressible fluid, then isostatic compensation occurs, creating a surficial volume accommodation for further sediments. However, if the basement behaves as a rigid unmoving plate, then no response occurs. Reality must lie between these two extremes, leading to basinal evolution models that encompass tectonic effects due to flexural plate behavior, tectonic rifting, wrenching, compression, and so on.

Dynamical fluid escape depends intricately on the permeability behavior of the deposited, and evolving, lithologic units. As fluid loss proceeds, porosity decreases and permeabilities become smaller, leading to an ever-increasing delay in extracting the residual fluids. The addition of more sedimentary overburden is then compensated for by an increase in the retained fluid pressure over the hydrostatic value, leading to an overpressured situation. If the sedimentary deposition rate is about 30 m per million years or more, then there is a very strong possibility of significant overpressuring. Because, on average, about 75% of all deposition is low permeability shale, overpressure is dominant in burial history calculations. Several consequences are immediate. First, compaction of overpressured formations is less than that without overpressure, so formations retain thicknesses closer to their depositional values for a longer period of time. In turn, this effect changes the structural and stratigraphic shaping of sedimentary units as well as providing a wider (laterally) thoroughfare for fluid migration. Second, the thermal conductivity of an overpressured region is lowered, thereby increasing the vertical temperature gradient (often by as much as about 1°C/100 m from the "normal" value of around 2–2.5°C/100 m) in order to preserve the heat flux. Thus overpressured regions tend to be hotter than nonoverpressured regions at the same subsurface depth. Third, as a consequence of the increased temperatures, the solubilities of carbonates and sulfates are lowered, leading to calcite, gypsum, marl, and anhydrite precipitation, which tend to strengthen formations by diagenetic cementation. Fourth, isolated sand bodies in a surrounding shale matrix will also be overpressured so that a parametric geometry dependence on overpressuring exists. Fifth, the thermal expansion of hot water in the overpressured formations possibly also provides an excess pressure acting to divert fluid flow and to pressure the encompassing rocks. If the total fluid pressure exceeds a limiting value, a rock must fracture, thereby releasing fluid. Empirically the total fluid pressure limit appears to be roughly 0.8–0.85 psi/ft (excess fluid pressure of 0.3–0.35 psi/ft) (1 psi/ft = 0.2 atm/m). Thus the dynamic evolution of sediments is firmly tied to the dynamic fluid flow. The ability to account for this behavior is crucial to any model. In addition, an increased temperature will also speed hydrocarbon generation from kerogen. These mobile components provide a generated pressure and can flow either in solution with the prevailing waters or as separate phases, thus affecting the dynamic burial history of the sediments.

Apart from fluid effects, major components in burial evolution are provided through depositional variation laterally in a basin, by erosional events, overthrusting, rotation of fault blocks, by intrinsic faulting of the sediments as syndepositional growth faults, tectonically induced faulting, or shear failure caused by differential stress on consolidated, lithified formations as the basin evolves. The presence of faults can lead to fluid escape up the faults, if the faults stay open for any length of time, thereby influencing the sediment evolution.

Two other lithological fabrics, salt and carbonates, can also be dominant controls on basinal evolution. Salt normally occurs in an evaporitic basin that is later open to clastic input. The evaporation can leave behind great thicknesses (several kilometers) of salt overcoated with other evaporites. Later sediment deposition on the underlying mobile, constant-density salt causes salt movement laterally and vertically, resulting in sediment distortion, faulting, and changes in the pattern of later sediment deposition through topographic expression caused by the salt and from dissolution.

Carbonates play a different role. Deposited as the skeletal remains of marine organisms, carbonates basically "grow" by secretion of calcite. Hence, carbonate reefs, platforms, pinnacle reefs, and so on represent shallow marine warm water where nutrient supply is abundant and the optical depth to sunlight penetration is high. Large, laterally extensive, carbonate deposition leads to the weight of the deposited carbonates contributing to basinal sag, leaving a volume accommodation for further carbonate deposition. Apart from the continuing impact of such a weight on the evolution of basinal structure, the stress on neighboring formations can often be sufficient to produce major faulting. In addition, the porous carbonates can act as good reservoir sites, can alter further depositional patterns of sedimentation, and can act as aquifers for groundwater transport into the deep subsurface and for deep basinal waters to reach the surface.

In any attempt to determine the dynamic burial of sediments in a basin we have four major controls at each well, or pseudo-well, location. These controls are (1) age-dated formation lithologies and thicknesses, (2) porosity with depth, (3) fluid pressure with depth, and (4) formation permeability with depth. A model must honor these four controls at each location.

### *b. Models and Consequences*

*An Inverse Procedure: Isostatic Reconstruction.* Geohistory (Fig. 1.15a) and burial history plots (Fig. 1.15b) can be used to predict the evolving thickness of formations (van Hinte, 1978), estimate the amount of material removed at unconformities by erosion (Fig. 1.15c), and estimate the rate of sediment deposition as a function of time (Fig. 1.15d).

The data needed to construct burial curves are the geologic ages of the layers, their depths, and the water depth at the time of deposition of each layer. The burial history plot can be obtained by linearly interpolating between the depth today for the first layer and the time at which this layer was at the surface. Ignoring compaction effects, all lower layers move up at the same rate as the first layer. When the first layer

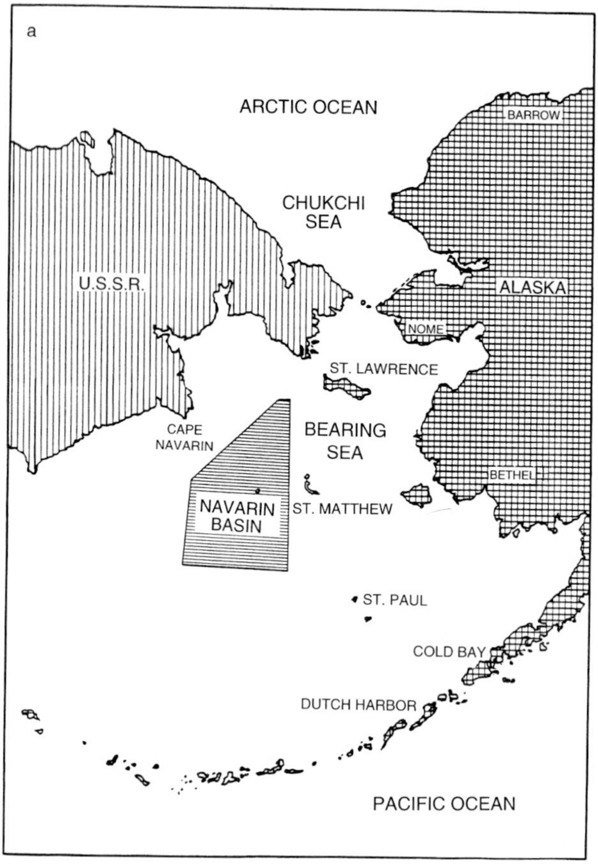
a
ARCTIC OCEAN
BARROW
CHUKCHI
SEA
U.S.S.R.
ALASKA
NOME
ST. LAWRENCE
CAPE
NAVARIN
BEARING
SEA
BETHEL
NAVARIN
BASIN
ST. MATTHEW
ST. PAUL
COLD BAY
DUTCH HARBOR
PACIFIC OCEAN

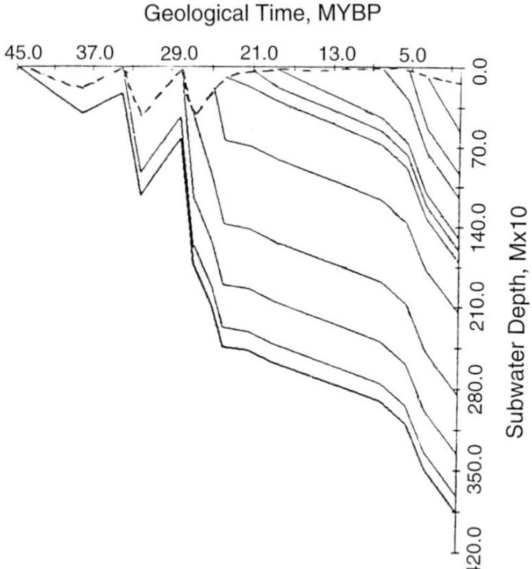
Geological Time, MYBP
45.0
37.0
29.0
21.0
13.0
5.0
0.0
70.0
140.0
210.0
280.0
350.0
420.0
Subwater Depth, Mx10

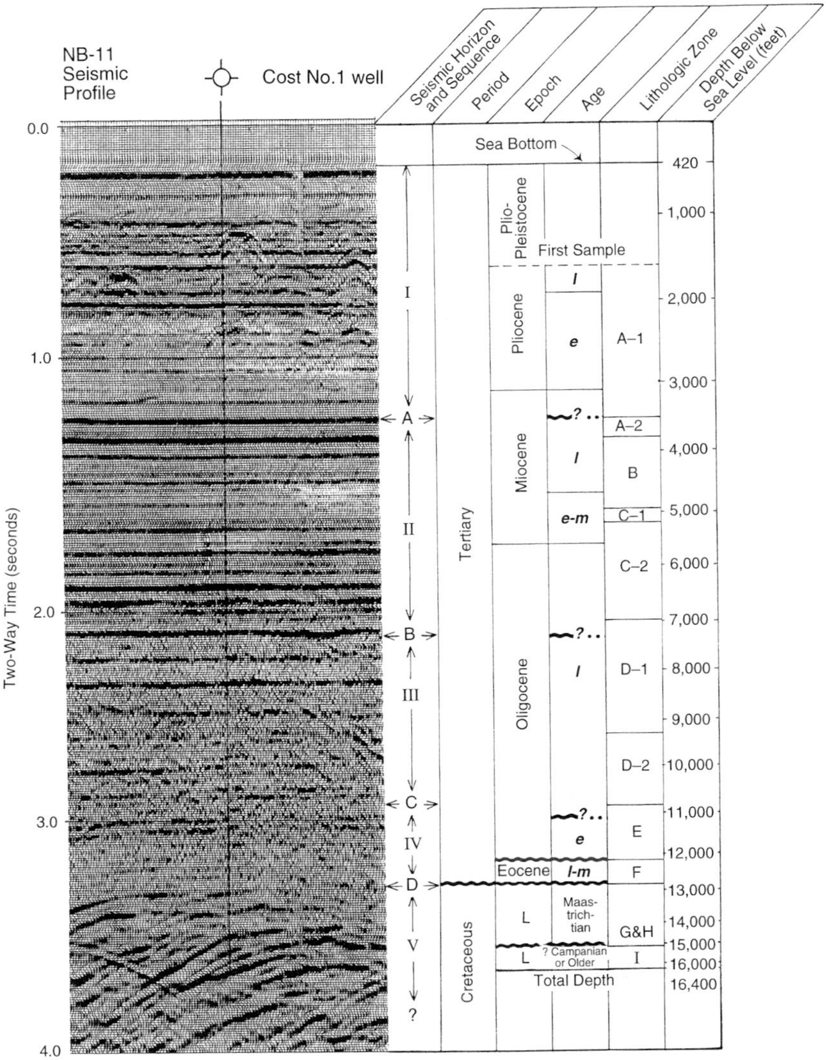


FIGURE 1.15 (a) Reconstructed geohistory of COST No. 1 well (offshore western Alaska) allowing for overpressuring with time. Note the rapid subsidence event at about 28 MYBP, with relatively slow subsidence thereafter, and two major unconformities prior to the rapid subsidence. (b) Superposition of burial history of the COST No. 1 well and isoreflectance lines, obtained by using the best paleoheat-flux variation from thermal inversion. Note that the $R_0 = 0.6$ isoreflectance line (occasionally considered indicative of formations entering the "oil window"), varies from 2800 m at about 20 MYBP to about 2500 m at the present day, suggesting that hydrocarbon maturity has occurred only for formations deeper than about 2500 m and deposited earlier than about 27 MYBP. (c) Geohistory diagram with unconformities handled in different ways. (d) The three curves show (1) the rate of subsidence including sediment load (light dashed line), (2) the rate of basement subsidence with isostatic compensation due to sediment load removed (heavy dashed line), and (3) the sedimentary deposition rate (light dotted line).

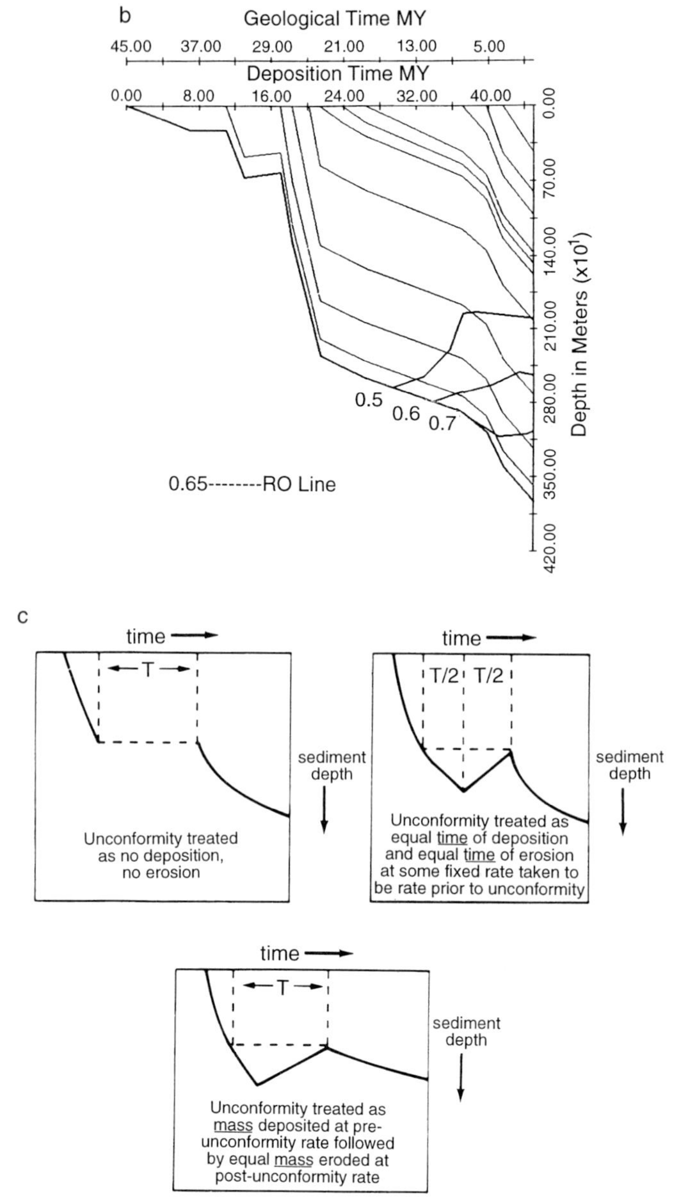


FIGURE 1.15 *(Continued)*

is on the surface, depths to all lower layers are calculated. The process is repeated until all layers have been moved to the surface. The geohistory plot is then constructed by adding the depth of water at each time of deposition.

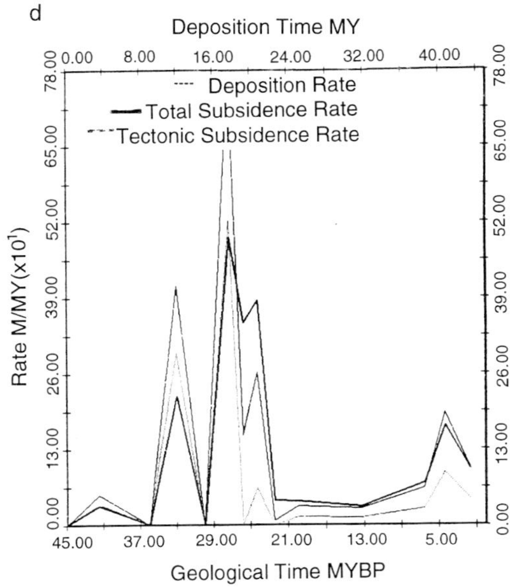


FIGURE 1.15 *(Continued)*

To model compaction history, it is necessary to model the changes in rock porosity with depth and time. One approach is to consider the rate of sediment accumulation ($R$) with respect to time, as given by van Hinte (1978),

$$R = T_\rho(1 - \phi)/A(1 - \phi_0), \tag{1.19}$$

where $T_\rho$ is the thickness at present porosity ($\phi$), $A$ is the time duration, and $\phi_0$ is porosity at time of deposition. To model layer thickness as a function of time, van Hinte (1978) illustrated a procedure using observed present porosities and back-calculating to arbitrarily chosen initial porosities. In the absence of porosity data from wells, phenomenological models (see later) can be used to relate porosity to depth or load for various lithologies. Actual values of porosity are variable and depend on compositional variations of lithology, overpressure or undercompaction, and diagenesis, but they generally decrease with depth. For normal pressures, porosity, $\phi$, can be approximated by

$$\phi = \phi_0 \exp(-cz), \tag{1.20}$$

where $\phi_0$ is the surface ($z = 0$) porosity. Values of $\phi_0$ and $c$ from borehole porosity measurements can be obtained by fitting via least-squares techniques (Sclater and Christie, 1980). An alternate phenomenological model of porosity has the porosity decrease as a function of the load on a layer, which leads to an empirical relation of the form

$$1/\phi = 1/\phi_0 + kz, \tag{1.21}$$

where $\phi_0$ is the initial uncompacted porosity, $z$ is the depth, and $k$ is a lithology-dependent constant.

Well data may be corrected for the effects of compaction as follows. First, the total conserved amount ($h_s$) of solid material in an interval ($z_1$ to $z_2$) is given in the Sclater–Christie model by

$$h_s = (z_2 - z_1) - (\phi_0/c)(\exp(-cz_1) - \exp(-cz_2)). \tag{1.22}$$

When the top of a stratal unit is on the surface ($z_1 = 0$), its base ($z_2$) can be calculated by solving Eq. (1.22). Similarly depths and thicknesses of all lower layers can be obtained sequentially and self-consistently. A similar equation is obtained in the porosity-load model of Eq. (1.21):

$$z_2 - z_1 = h_s + (1/k)\ln([1/\phi_0 + kz_2]/[1/\phi_0 + kz_1]). \tag{1.23}$$

Given the compaction parameters and drill-hole values of depth, age, and paleowater depth, a burial history curve (Fig. 1.15b) can be obtained.

Unconformities disrupt any regular porosity versus time or depth model, regardless of the model assumptions or procedures. There are at least three possible ways to model an unconformity: (1) as a depositional hiatus; (2) as a period of deposition followed by erosion in which the thickness of eroded section and the absolute age marking the beginning of erosion are determined from a knowledge of local geology (see van Hinte, 1978)—parameters can be varied to see which model(s) best fit observed data (e.g., present-day vitrinite reflectance and/or behavior of velocity or density logs); or (3) as a period of deposition and erosion in which the thickness of eroded section and the age at the beginning of erosion are calculated from sedimentation rates in the immediately overlying and underlying sediment layers.

From the depth and thickness of each layer as a function of time, and from the lithologies and densities of the layers, basement subsidence history is inferred by applying isostasy calculations to replace successive backstripped sediment layers by water. The basement subsidence ($Y$) in terms of the water depth ($W_d$) and sediment thickness ($S$) is given by

$$Y = S[(\rho_m - \rho_s)/(\rho_m - \rho_w)] + W_d - \Delta_{SL}[\rho_m/(\rho_m - \rho_w)], \tag{1.24}$$

where $\rho_m(\rho_w)$ is the average mantle (water) density, $\rho_s$ is the average sediment density, $\Delta_{SL}$ is the change in elevation of mean sea level, and $Y$ is the depth to basement without sediment and water loads and represents the subsidence caused by tectonic effects. The terms $W_d$ and $\Delta_{SL}$ must be determined by paleobathymetry. The isostatic reconstruction assumes that the observed present-day porosity variation with depth has been constant through time; effects of fluid overpressuring are not incorporated.

*A Pseudo-Inverse Procedure: Fluid Flow/Compaction.* To incorporate overpressuring, compaction of sediments is taken to be the result of fluid expulsion caused by increasing overburden. Vertical fluid movement through sediment is described by Darcy's equation:

$$\partial(\rho_f e)/\partial t = \partial/\partial\xi(K_z \rho_f \mu^{-1}(1 + e)^{-1}\, \partial P/\partial\xi), \tag{1.25}$$

where $e$ is the void ratio [related to porosity $\phi$ by $e(1 - \phi) = \phi$], $\rho_f(\mu)$ the density (viscosity) of the fluid, $K_z$ the vertical permeability, and $P$ the fluid pressure in excess of hydrostatic. The fully compacted depth coordinate, $\xi$, represents the net thickness of the rock. Darcy's law can then be applied correctly. The physical depth, $z$, is related to $\xi$ by $d\xi/dz = (1 - \phi)$. A multidimensional equation for nonvertical flow is also available (Lerche, 1990a).

Excess fluid pressure and frame pressure, $P_f$, are related by

$$P = \int_{\xi}^{\xi} g(\rho_s - \rho_f)\, d\xi - (P_f - P_f) \tag{1.26}$$

where $g$ is gravitational acceleration, $\rho_s$ is the solid density, $P_f$ is the frame pressure (a prescribed function of void ratio for each lithology), and denotes the appropriate value on the sediment surface $\xi$ ($P = 0$ on $\xi$). All variables are time dependent and related to the sediment accumulation rate. Solving Eq. (1.25) determines the changes of sediment thickness, porosity, permeability, pressure, and fluid flow. Current values constrain the model predictions (Fig. 1.16 and Table 1.4, given in a later section). In the quasi-empirical equations of state for shales,

$$K_z = K\,(e/e\,)^{A}; \qquad P_f = P_f(e/e\,)^{-B}, \tag{1.27}$$

the two numbers $A$ and $B$ are varied until the observational constraints are met. Empirical evidence from many wells suggests $A = 3 \pm 2$ and $B = 3 \pm 3$ (Lerche, 1990a). The procedure is pseudo-inverse because only parameters (not functions) are adjusted until predicted values are in accord with observations as shown in Fig. 1.16 for the Navarin Basin COST No. 1 well (offshore eastern Alaska); Fig. 1.15b shows the burial history.

## 2. Thermal History

### *a. Overview*

We concentrate next on the determination of sedimentary thermal history, which rests on two fundamental factors: variation of sources of heat flux through, and within, the sediments; and heat transfer (conduction, convection) through the sediments. Significant internal heat sources can be highly radioactive layers (e.g., Kimmeridge clay, North Sea basin), igneous intrusives, frictional heating by overthrusting, and so on. Such heat sources, dominant locally, appear secondary in magnitude and regional character to basement heat flux (exceptions occur at greater depths in some sedimentary basins, depending on lithology types).

Present-day estimates of sedimentary rock thermal conductivities vary in the rough range 7 ± 3 mcal/cm · $s^{-1}$ · °$C^{-1}$ with low values for high porosity, unconsolidated rocks and high values for low porosity, nearly fully compacted sediments (Hunt, 1979). Ignoring lithology effects, an empirical estimate connecting fractional porosity, $\phi$, and bulk thermal conductivity, $\kappa$, can be obtained from (Hunt, 1979)

$$\kappa = K_{\mathrm{R}}^{1-\phi} K_{\mathrm{W}}^{\phi}, \tag{1.28}$$

where $K_{\mathrm{w}}$ is thermal conductivity of water (~1 mcal/$cm^{-1}$ · $s^{-1}$ · °$C^{-1}$) and $K_{\mathrm{R}}$ is the thermal conductivity of fully consolidated sediments (~10 $K_{\mathrm{w}}$). Present-day temperature gradients are around 2°C/100 m so that, if conduction dominates, estimates of basement heat flux today are around $Q = 1$ heat flux unit (HFU) (= 42 mW · $m^{-2}$). A dynamical range of about 0.5–3 HFU is not inappropriate for most sedimentary basins today. The spatial variation of present-day heat flux is estimated from measurements, but paleoheat flux is also required. Two end-member approaches are possible: (1) From the dynamic evolution of a basin, posit a *model* of basement motion and a *model* relation to heat flux variations, determine any parameters in the proposed basement motion, and apply these directly to the heat flux. Such schemes give rise to a variety of popular models of the stretching-rift sort (McKenzie, 1978), to phase change models, and the like. Most such models end up predicting some sort of high heating in the past (Q ≈ 4–10 HFU) with a cooldown to the present day (see later). The weakness is that both a dynamic model/thermal model connection and a dynamic model from which to extract salient parameters must be made. (2) From various thermal indicators measured in wells and if the temperature-dependent physics/chemistry of the indicators along the burial paths is known, the temporal evolution of temperature can be determined. Lopatin (1971) and later Waples (1980) and Brooks and Welte (1984) popularized the idea that vitrinite reflectance, $R(z)$, measured at depth, $z$, can be related to a time, $t$, temperature, $T(t)$, integral (TTI) of the form

$$\mathrm{TTI} = \int_0^t \exp[(T(t') - T_c)/T_{\mathrm{D}}]\, dt', \tag{1.29}$$

where the integral is taken along the burial path of the sediments from deposition to the present day, and $T_c$ and $T_{\mathrm{D}}$ are two scaling constants. To make TTI a hydrocarbon measure, Waples (1980) suggested that $T_c$ is best set at 105°C and $T_{\mathrm{D}}$ at $10/\ln_e 2$°C. But a connection still needs to be found between $R(z)$ and TTI. If the geothermal gradient is constant for all time, then TTI can be *calculated* for each sedimentary layer and plotted versus the *measured* $R(z)$ values. A quasi-empirical relation is then generated connecting TTI and $R$. Unfortunately, there is no way of knowing whether the geothermal gradient is constant with time, nor is there any way of assessing the choice of parameter values for $T_c$ and $T_D$. Indeed, a *constant* geothermal gradient with depth today is not possible because of thermal conductivity variations, and paleogeothermal gradients relevant to a specific layer cannot be identical to present gradients. Waples (1989) has strongly pleaded for kinetic descriptions to replace the Lopatin (1971) TTI. Every thermal indicator (vitrinite, steranes, pollen translucency, hopanes, $Ar^{39}/Ar^{30}$, apatite fission tracks, etc.) has its own particular TTI, peculiarly related to the nature of the individual indicator (Lerche, 1988a,b). Inverse schemes, based on multiple thermal indicators, are rapidly replacing the thermal maturity procedure of "calibrating" the Lopatin–Waples TTI to vitrinite reflectance [values of TTI = 15 and TTI = 75 are often taken as marking the onset of entry to the "oil window" and "gas window," respectively (or $R = 0.6$ and $R = 1.1$)] (Waples, 1980).

Even with paleoheat flux known throughout a basin, paleotemperature determination is still confounded by two major problems. First, thermal conductivity variations due to lithology and porosity differences act to focus and defocus conductive heat transport. Second, high-permeability domains provide pathways for convective transport of heat, thereby heating and/or cooling regions in amounts considerably different than conduction alone provides. Hydrocarbon generation is affected by such convective heat transfer, as is hydrocarbon transport and further generation of oil to gas after location in a reservoir site. In addition, mineral solubility is changed by high-temperature fluid flow leading to cementation and/or dissolution, which, in turn, influences the migration pathways by diagenetic evolution. And, in turn, the transport of heat impacts fluid expansion and rock failure, as well as porosity. But the porosity variations relate directly to the thermal conductivity, thereby altering the temperature gradient, which then changes the convective heat transfer, which reflects directly on the burial history. Thus, thermal and burial histories are intertwined and must be addressed simultaneously. The only direct constraints on the thermal history models are measurements of multiple thermal indicators and present-day temperature gradient (Figs. 1.16a and 1.16c).

### *b. Thermal History Models*

Two end-member models are often used: forward models with undetermined parameters and pseudo-inverse models based on present-day observations.

*Forward Models of Paleoheat Flux.* McKenzie (1978) argued that rapid stretching of continental lithosphere produces thinning and passive upwelling of hot asthenosphere, with associated block faulting and subsidence.

The lithosphere then thickens by cooling, and the surface slowly subsides. The decreasing rate of subsidence and heat flow depend on the amount of stretching. Sclater and Christie (1980) modified the McKenzie model by adding a heat flow component from the decay of radioactive elements within the crust. An additional thermal blanketing by low thermal conductivity sediments slows the cooling of the lithosphere, which is perhaps important to studies of older rifted margins, such as the North Sea, where sediment thicknesses occur of the order of 15 km (49,000 ft). Modified stretching models encompass two layers to explain uplift and erosion on the flanks of a rift basin. More recent work has been concerned with nonuniform extension, depth-dependent extension, simple shear models, and so on (see, e.g., references in Lerche, 1990b, pp. 358–360). Royden *et al.* (1980a,b) proposed a model of continental lithosphere fracturing and intrusion by dikes from the mantle. Replacement of light crustal rocks by denser ultramafic materials results in initial subsidence and avoids the general problem of uplift.

The subsidence is *inferred* from the burial history. Discordance between predictions and inferred observations is minimized by parameter adjustment, and such parameters are then used directly in the posited heat flux models. The subsidence components of these models are pseudo-inverse, and the heat flux components are forward models.

These theories of basin formation rely on thermal contraction of the crust to produce subsidence. Elevation of the geotherm during continental margin rifting may produce margin subsidence as a result of deep crustal metamorphism, which occurs predominantly at the greenschist–amphibolite facies boundary. The model requires a period of heating followed by a period of cooling. Subsidence due to deep crustal metamorphism occurs during the later part of the heating period and is followed immediately by subsidence due to thermal contraction during the cooling period. The distinguishing features of this model are that the heat flow *increases* up to the point of breakup and then *decreases,* as opposed to other models in which heat flow always *decreases* after the time of stretching. The heat flow from a metamorphic model is of the order of half that of a crustal extension model, caused by the different model *postulates* that connect heat flux to subsidence (elevation).

*Inverse and Pseudo-Inverse Models.* Dissatisfaction with forward heat flux models has focused attention on inverse methods applied to measured thermal maturity indicators (Lerche, 1990a). Thermal indicator inversion schemes for 1-D systems are constructed as follows. In a one-dimensional burial history, variations of temperature, $T(z, t)$ occur only with depth, $z$, and with time, $t$. A thermal indicator, $\mathrm{TI}(z)$, is then related through a kinetic evolution model to the temperature history of the stratum in which it is measured. The problem of determining the entire temperature field is replaced with the problem of determining the heat flux, $Q(t)$, at one depth in the basin. The inverse problem takes the thermal indicators back-stripped through the layers and determines the $Q(t)$ consistent with observations of TI($z$). Modern basin analysis codes include convective and conductive effects in dealing with real basins to handle the inverse problem appropriately (Lerche, 1988a,b).

The inversion procedures use present-day downhole measurements to determine paleoheat flux variations subject to several constraints:

1. The underlying thermal evolution process and temperature dependence controlling maturation of each thermal indicator must be known.
2. The chemical and physical parameters associated with the evolution process must be provided. (However, inversion procedures normalize out as many such parameters as possible so that paleoheat flux determinations are minimally affected by uncertainties in the parameters.)
3. The burial history of each sedimentary layer is assumed to be known with precision in order that inversion procedures can be used to extract paleoheat flux variations with maximum sharpness.

With geological data and their attendant "scruffiness," not too much improvement can be made to the underlying thermal evolution processes. Such functional determinations are determined more properly by controlled laboratory experiments. But chemical and/or physical parameters of such processes determined in the laboratory, over human lifetimes and under pristine conditions, do not have to be the appropriate values for maturation of thermal indicators over geological timescales and in the uncontrolled environment of an evolving sedimentary basin. So such parameters should be regarded as unknowns and should be determined by inversion procedures. In a similar vein, geological parameters related to unknown variations in burial history information should also be determinable from inversion procedures.

Thus three basic types of parameters enter the methods used for thermal indicators:

1. Parameters such as activation energies, critical temperatures, and rate constants directly associated with the underlying physics and/or chemistry peculiar to each thermal indicator
2. Parameters such as paleoheat flux variations with time, radioactive heating with time, directly associated with changing the thermal maturity of the sedimentary layers
3. Parameters such as erosional unconformities and paleo-overpressur-

ing, which directly change the burial path of the sedimentary layers with time, and so the temperature, or which directly change the thermal conductivity and so the temperature even if the paleoheat flux does not vary with time. Such parameters therefore have a role to play in changing the time–temperature integrals and so the thermal maturity of the individual thermal indicators with depth at the present day.

Out of this array of parameters, one or two in a given geological setting are easy to determine by trial-and-error. But if more than a few parameters are involved (such as would happen with, say, multiple unconformities where the erosional thicknesses, and/or onset time of erosion and paleoheat flux variations with time are to be determined), exhaustive trial-and-error searches are not only inefficient but extremely difficult.

The difficulties are clear: how to determine simultaneously a large number (say, four or more to fix ideas) of a mixture of chemical, thermal, and geological parameters by inversion of downhole thermal indicator data? And how to know when resolution capability of the data is reached so that it is pointless to search for further parameters?

Fortunately, precisely this type of parameter determination problem has arisen in a variety of other scientific disciplines over the years and a substantial body of information has been culled on appropriate methods. In general, the most powerful and accurate techniques are those based on some form of tomographic analysis.

Thermal indicator tomography is a nonlinear inverse problem. Nonlinearity arises because thermal maturity along the paths taken by sediments, and the paths themselves, is affected by parameters to be determined (e.g., paleoheat flux, insertion times, unconformity erosion thicknesses). Generally, enough gross information about the paleostructure of a region is available to approximate roughly the thermal indicators; to this extent, thermal indicator tomography is "almost linear." As usual in nonlinear problems, an initial estimate is made for the parameters, the linear inversion is solved for an improved set of parameters, and the process is reiterated. Experience from many disciplines suggests that if the (linear) inversion has been properly regularized (stabilized), and if the intermediate solutions are constrained to be reasonably smooth, this procedure converges satisfactorily in a few iterations (typically 5 to 10—although extreme cases are known when many tens of nonlinear iterations are necessary).

The essence of thermal indicator tomography, then, is to construct a model that approximates past thermal behavior of the subsurface, to compute theoretical thermal indicators to each sedimentary layer by forward tracing in time through the model, to measure the actual thermal indicators from the appropriate depths, and to modify the thermal model through which the theoretical thermal indicators traveled in time to minimize the difference between measured and computed maturations. The model is modified by inverting a version of a least-squares equation. Finally, the entire process is iterated by tracing through the modified model and proceeding as before (Lerche, 1988a,b).

### 3. Hydrocarbon Generation, Migration, and Accumulation

#### *a. Overview*

We now concentrate on the hydrocarbons. General considerations (Hunt, 1979; Brooks and Welte, 1984) suggest that the dominant mode of hydrocarbon production is by thermal degradation of kerogen after sedimentary burial. A "rich" source rock might contain around 1–5% TOC (total organic carbon) capable of producing hydrocarbons. Three major problems need to be addressed: (1) How are hydrocarbons produced from kerogen? (2) How do hydrocarbons migrate out of the (dominantly shale) source beds, and then how do they move thereafter (primary and secondary migration)? (3) What sort of accumulation sites and sealing conditions are available for the hydrocarbons, and do those sites persist through to the present?

Hydrocarbon production is a peculiar phase change in which solid material (kerogen) is turned into mobile liquids (oils) and gases (mainly methane but with some ethane, $H_2S$, $CO_2$, $N_2$). Oils are virtually incompressible and normally have a specific gravity around 0.8–1.01, usually less than subsurface brines (~1.0–1.3). Gases are highly compressible and normally have a specific gravity of around 0.1–0.3 at production. Thus oils and gas are buoyant relative to water (buoyancy pressure ~0.1 psi/ft for oils, ~0.3–0.5 psi/ft for gas). Some oils are completely miscible with water; others are not (Hunt, 1979); and gas tends to be completely miscible until the "bubble point" pressure and temperature are reached when gas exsolves and transports as a separate phase.

The ability to provide a detailed kinetic scheme relating the thermocatalytic breakdown of kerogen to hydrocarbons is difficult, perhaps because such a scheme must handle organic macromolecule chemistry (kerogen) breaking down to methane (molecular weight 16) in the subsurface when ionic, mineral, and surficial chemistry influence and modify the chemical kinetics. From modified van Krevelen diagrams of H/C and O/C (Hunt, 1979) there are at least three types of kerogen—presumably each having its own kinetic breakdown scheme and yielding different production efficiencies for oil and gas generation. As the subsurface temperature increases, oil converts to gas (Hunt, 1979).

To bypass detailed kinetics, a "lumped" kinetic scheme was introduced (Tissot, 1969; Tissot and Espitalie, 1975; Tissot and Welte, 1978). Since then, many other "lumped" kinetic frameworks have been suggested. A common theme among them is multichannel reactions, in series or parallel, with specification of many reaction rate constants and activation energies that relate kerogen to oil and gas, and oil to gas. All are temperature dependent and most involve first-order chemical kinetics. Typical *total* production estimates range from 1 to 500 mg of hydrocarbons per gram of kerogen, whereas typical estimates of the *onset* of significant oil (gas) production *rates* are round 100°C (150°C).

The question arises as to how the mobile hydrocarbons escape the fine-grained clastic source beds of generally low permeability (primary migration). Several mechanisms have been proposed. Thermal expansion coefficients of fluid hydrocarbons are larger than rock values. Hence the conversion of kerogen to fluid puts an additional pressure on the encompassing source rock leading to microfracturing and fluid loss through the high-permeability fractures. An alternative suggestion is that the volume occupied by produced hydrocarbons is larger than that of the original kerogen, leading directly to an increase in pressure, again driving fluid motion and microfracturing. A third suggestion is that capillary pressure at sand/shale interfaces has a dominant role to play. Typical pore radii in fine-grained materials are around $10^{-5}$ to $10^{-6}$ cm, whereas in sands the size is around $10^{-4}$ cm. The surface tension difference, $T$, between oil and water is around 40 dyn/cm so that the capillary pressure, $p$, at a sand/shale interface is $p \approx 2T\,(1/r_{\text{fine}} - 1/r_{\text{coarse}}) \approx$ 10–100 atmospheres ($\equiv$ 150–1500 psi). Thus capillary pressure is the largest force at, but decreases rapidly away from, the boundary. Hence, oils displace water in coarse-grained beds abutting a fine-grained source, and water displaces oil in the source.

Oil and gas then find themselves in a relatively high permeability bed and move (secondary migration) under several forces. Any excess fluid pressure provides a hydrodynamic drive. Those fractions of oil and gas that go into solution move with the prevailing fluid. The separate phase gas/oil fractions have a vertical buoyancy pressure drive acting on them of around 0.1–0.5 psi/ft (1 psi/ft = 0.2 atm/m), causing motion along the path of least resistance, which is not necessarily vertical. Evolution of the structural aspects of bedding geometry now come into play; up-dip migration to structural culminations being favored by oil and gas.

On the other hand, Darcy's law notes that the speed of motion is proportional to the permeability as well as to the driving pressure gradient. As the hydrocarbons move any significant decrease in permeability will effectively bar their further migration. Such decreases can be caused by compaction as the basin develops, by diagenetic cementation, by spatial variations in lithology, and, of course, by tectonic forces altering the basin. Thus there are also evolving stratigraphic sites of potential accumulation.

Continued evolution of the basin can lead to aquifer flow attempting to flush hydrocarbons from earlier accumulation sites. Dynamic hydrodynamic transport plays a major role in some hydrocarbon regions. In addition, the sealing conditions around a filled reservoir can change significantly as later deposition and basinal development impact the stress conditions of the seal, leading to fracturing and faulting and consequent loss of hydrocarbons.

### *b. Hydrocarbon Models: Forward and Pseudo-Inverse*

To determine the amounts of hydrocarbons that have been generated and accumulated, and their locations, the following information is needed: (1) the amount of oil and gas generated in each source rock and in every part of the basin; (2) the timing of hydrocarbon formation, timing of development of impermeable seals, and timing of folding and faulting, so traps can be identified; (3) the amounts of oil and gas that have been expelled from the source rock(s) and how much is still retained; and (4) an evaluation of the ultimate oil and gas reserves of the sedimentary basin. A quantitative approach provides this information. However, because many of the intermediate steps in petroleum and gas generation from kerogen degradation are unknown, such a scheme has to be phenomenological; parameters have to be determined by least-squares fitting to basin and laboratory data.

The validity of a phenomenological model depends on two factors: the validity of the physics and chemistry involved in the calculation and the values of the parameters chosen. The Tissot scheme considers three types of initial kerogens that differ in their hydrogen-to-carbon ratio. Type 1 (algal) has a high liquid-generating potential, type 2 (amorphous) has an intermediate liquid-generating potential, and type 3 (woody) has a low petroleum potential. Each type is assumed to have six different bonds that have activation energies in the range 10–80 kcal/mol. The amount of organic material ($x_i$) reacting in the $i$th reaction (breaking of the $i$th bond type) for $i = 1$ to 6 is specified for each kerogen type. Calibration for the formation of gas has been made from laboratory experiments on hydrocarbon cracking. A single reaction with an activation energy of 50 kcal/mol was convenient to account for gas generation in the deep parts of sedimentary basins. Tables of the activation energies ($E_i$), the reaction constants ($A_i$), and the initial values ($x_i$) for each kerogen type are given by Tissot and Welte (1978, Table V.4.1).

Two problems are apparent. Is the structure of the kinetic model adequate? How are the best kinetic parameters determined? Other kinetic schemes have been proposed (see references in Lerche, 1990b) that also are consistent with limited observational data. At the time of writing, "lumped" kinetic models can be handled in a forward-modeling sense or a pseudo-inverse sense. The generation parameters are often determined from measurements made *after* the produced hydrocarbons have moved from their source "kitchen" to reservoirs, confounding the problem even more.

## 4. Basement Motion

Basement response to loading (and, correspondingly, loading response to basement movements!) is one of the major factors that determines the shape and sedimentation fill in basins. Early models of lithospheric compensation were based on the principle of isostasy—the idea that crustal loads are in hydrostatic equilibrium with the upper mantle. In effect, the crust is assumed to be supported entirely by buoyant forces in the mantle. Fundamental to this model is the assumption that the lithosphere has zero rigidity. However, the presence of regional free-air gravity anomalies over large-scale features indicates only partial compensation by isostatic forces; some portion of the load is supported by the rigidity of the lithosphere (Turcotte and Schubert, 1982). Two end-members are possible: zero rigidity, in which loads are entirely compensated by isostatic forces, and infinite rigidity, in which loads are completely supported by lithospheric rigidity.

Proposed flexural models fall into one of three general categories: elastic, viscoelastic, and thermally controlled flexure (Keen et al., 1981). Linear elastic models are based on Hooke's law (Turcotte and Schubert, 1982). Deformation occurs instantaneously under an applied stress with complete recovery when the stress is removed (no permanent deformation). Viscoelastic models express strain as a function of both stress and time. Following Maxwell's law, stress is taken to decrease exponentially with time (Turcotte and Schubert, 1982). There is a relaxation of stress with time, resulting in permanent deformation. Thermal flexure models are based on elastic or viscoelastic theory with the additional assumption that the strength of the material decreases with increasing temperature.

All three models have been applied to geologic systems with varying degrees of success. Flexural models have been applied to a wide variety of situations: loading due to ice sheets, seamounts, intracratonic basin fill, and thrust sheets in a foreland setting. Foreland basins represent the most clear-cut example of flexure; they are also among the most productive hydrocarbon basins (see references in Lerche, 1990b). The basement is assumed to act as an elastic plate of finite length and constant thickness, anchored at one end in cantilever fashion. Thrust and sediment loading cause the plate to bend. Immediately in front of the thrust front, a depression, or foredeep, develops in which sediments are deposited. Cratonward of the foredeep, the basement is flexed upward, forming a peripheral bulge that can act as a locus for erosion or nondeposition. This configuration is not static because, as the thrust front moves, the peripheral bulge migrates ahead of the front in the direction of thrusting. The shape and position of the bulge depend on the loading history and the properties of the elastic plate.

The development of the peripheral bulge can have a significant effect on the basinal sediment fill. Periods of erosion and/or nondeposition will occur in times and places where the forebulge is exposed above sea level. In cases where the bulge does not rise above sea level there may be appreciable shallowing, an effect that can contribute to the formation of carbonate build-ups. In addition the forebulge can act as a sediment source for the foredeep depression. If the mechanical properties of the basement are known, the locations of forebulge unconformities and carbonate build-ups can be estimated, even though such locations may be below the resolution of available geological and geophysical data. The properties of an elastic plate in a foreland setting are calculated directly from seismic and well data using quantitative forward and inverse methods (Lerche, 1990b).

## 5. Sea Level

A sedimentary basin's gross morphology depends on (1) basement movement, (2) sediment accumulation and compaction, and (3) eustatic sea-level variations. An accurate history of eustasy would aid in basin stratigraphy and help in the prediction of sediment geometry.

Some techniques that attempt to determine the magnitude of eustatic sea-level excursions include measurement of (1) the amount of sedimentary onlap onto the continental margins with and without the use of hypsometric curves, (2) the thickness of marine sedimentary cycles and the elevation and distance between indicators of old strandlines, (3) the perturbations on individual thermo-tectonic subsidence curves and stacked crustal subsidence curves, (4) the variations in deep-ocean oxygen isotopes found in sediments, and (5) the size of variables used in graphical and numerical simulations, which "invert" the problem of basin fill in terms of tectonic behavior, rates of sediment accumulation, and eustasy. A unique solution for *absolute* eustatic variations is not yet available (Wilgus, 1988). Quantitative modeling shows that, at best, only the *sum* of tectonic basement subsidence and sea-level variations can be obtained. Every proposed scheme to measure sea-level excursions assumes some behavior for two of the three underlying processes: tectonic movement of the basement, sedimentary accumulation, and eustatic sea level. Each scheme determines the third process relative to the assumed model behavior of the other two.

An accurate eustatic sea-level variation chart would be useful in frontier basin exploration, particularly where seismic cross-sections and occasional stratigraphic and/or wildcat wells are the only data sources. However, such a sea-level chart cannot yet be made. When *relative* sea-level charts (combining tectonic and eustatic effects) are tied to wells, sedimentary sequences on seismic cross-sections can be projected (Vail *et al.,* 1988, in Wilgus, 1988). "Relative" (tectono-eustatic) sea-level curves generate a "family" of solutions that are the product of a variety of absolute sea-level curves and tectonic models. Each solution has to be assessed in terms of geologic setting and the hypotheses it generates. Recent work (Cloetingh and Wortel, 1988) on the impact of intraplate stress in relation to plate boundary formation makes the difficulty of unraveling the sea-level problem all the more severe.

## 6. Summary

The simple recognition that hydrocarbons are produced by thermal degradation of organic material elicits an examination of the interlocked sediment and fluid evolutions with the evolving thermal regime in the subsurface, and with hydrocarbon generation, migration, and accumulation.

Quantitative methods in basin analysis examine the interrelations among the various components, quantify the dynamic behavior so that basinal processes can be modeled, use present-day data to constrain generic processes and their applications to particular basins, and provide assessments of the most likely sites of hydrocarbon accumulations ahead of drilling. Such assessments must also provide a measure of their probability of being right based on the best sensitivity studies possible.

## B. PROBLEMS ASSOCIATED WITH BURIAL HISTORY

The ability to fill a basin with sediments rests on four major factors: (1) the rate of supply and compaction of sediments; (2) the motion of the underlying basement, which determines the structural evolution of the basin as a whole; (3) the influence of faults and fractures, which partially determines the structural evolution of sedimentary sections and which partially determines the flow of fluids through a sedimentary sequence depending on whether, and for how long, a fault is both active and open to flow; and (4) the impact of extraneous factors, such as salt diapirism and salt sheet tectonics, which can be dominant influences throughout the history of particular basins (e.g., the northern Gulf of Mexico). Here we examine facets of the four factors in order to illuminate concerns that are often crucial to a consistent description of basin evolution.

### 1. Rate of Sediment Supply and Compaction

Neither the rate of sediment supply nor the evolution of compaction in a sedimentary basin is directly measurable. Inferences must be made from present-day observations; the best information is on present-day thicknesses, stratigraphic ages, and porosity variations with depth. This information can be used, as shown later, to determine the *net* amount of residual solid material deposited during a given time. But complications and uncertainties arise if the information is interpreted further in attempts to deduce the physical thicknesses of sedimentary data at prior times.

First is the problem that there is, most often, limited resolution on geologic ages. Tops and bases of stratal intervals are often the only age information at hand. Thus, if erosion occurred *within* an interval then, because only the net present-day solid thickness is available, no knowledge of such an event would be recorded. On this basis the observed (inferred) rate of solid sediment supply is, most likely, an underestimate. Equally, if postdepositional diagenetic cementation has occurred, then the measured amount of solid material is likely greater than at deposition. (But redistribution/dissolution and/or reprecipitation confound even this interpretation.) On this basis an overestimate of sediment supply would be inferred if diagenetic cementation is not considered. A problem that is becoming of considerable concern is due to the increasing recognition that the estimates of sedimentary ages are themselves uncertain. For instance, for an Ordovician section in the Cooper Basin the age at total depth (TD) is variously quoted as 400–500 Ma (Kantsler et al., 1983), 350–700 Ma (Battersby, 1976), and 450 ± 50 Ma (Pantano and Lerche, 1990). The youngest to oldest spread in age of TD is, therefore, from 700 to 350 Ma, which, taken at face value, would imply a considerable variation (by a factor of 2 or so) in the rate of deposition of solid material. Clearly, precise and accurate ages are of fundamental importance in basin analysis—and as many such ages as possible.

But even if this major problem is discounted, a second major problem that limits the ability to reconstruct the stratigraphic and structural variations of sediment evolution with time is the variation of paleocompaction. Here the problem, as shown in the next section, is the ability to describe correctly the loss of fluids from the deposited sediments. Pseudo-inverse models, which connect parameters in functional equations of state with present-day observations, would seem to be readily available models that can be made consistent with observations most easily. At each location in a basin the four dominant types of observable information requiring satisfaction by a model are (1) observed stratal thicknesses, (2) observed porosity variations with depth, (3) excess fluid pressure variations with depth, and (4) measured permeability variations.

Table 1.4 shows the results of present-day thicknesses using a one-dimensional pseudo-inverse model applied to the Navarin Basin COST No. 1 well (offshore eastern Alaska); and Figs. 1.16(b–e) show the corresponding predicted versus observed behavior for fluid pressure, downhole temperature, permeability, and porosity, respectively. Note that consistency with present-day observations is a necessary requirement, but is not sufficient, because other models (e.g., a two-dimensional compaction/fluid flow model) could equally provide consistency with the observed values. Clearly, other criteria, favoring one class of models over another, are required. We will return to such selection criteria later in this volume.

### 2. Basement Motion

Apart from the generation of variable structural and stratigraphic behavior in a basin by differential compaction and facies variation across a basin, a major factor influencing structural behavior is the evolution of the basement on which lie the sediments. Turcotte and Schubert (1982) provide a compendium of studies varying from isostatic through elastic flexural, thermoelastic, elastic–plastic plate behaviors, to viscoelastic plate models. The important point about all such models is that they individually have several parameters that are to be adjusted until minimal discord is obtained between current predicted and observed basement positions. Again, the most flexible models available to date are those of the pseudo-inverse sort. However, even here care must be exercised to be aware of the implicit and

**a**

| Depth (feet) | Temperature (°F) |
|---|---|
| 1,000 | 63 |
| 2,000 | 89 |
| 3,000 | 109 |
| 4,000 | 129 |
| 5,000 | 146 |
| 6,000 | 164 |
| 7,000 | 181 |
| 8,000 | 199 |
| 9,000 | 216 |
| 10,000 | 235 |
| 11,000 | 253 |
| 12,000 | 271 |
| 13,000 | 289 |
| 14,000 | 306 |
| 15,000 | 325 |
| 16,000 | 341 |
| 16,400 T.D. | 353 |

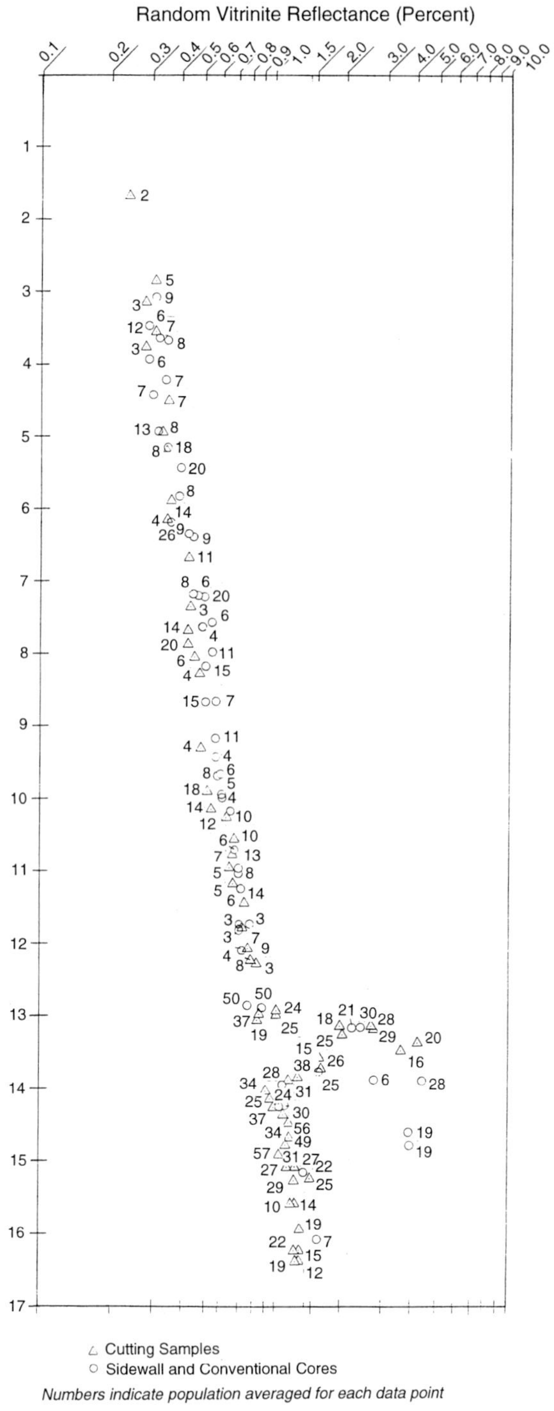

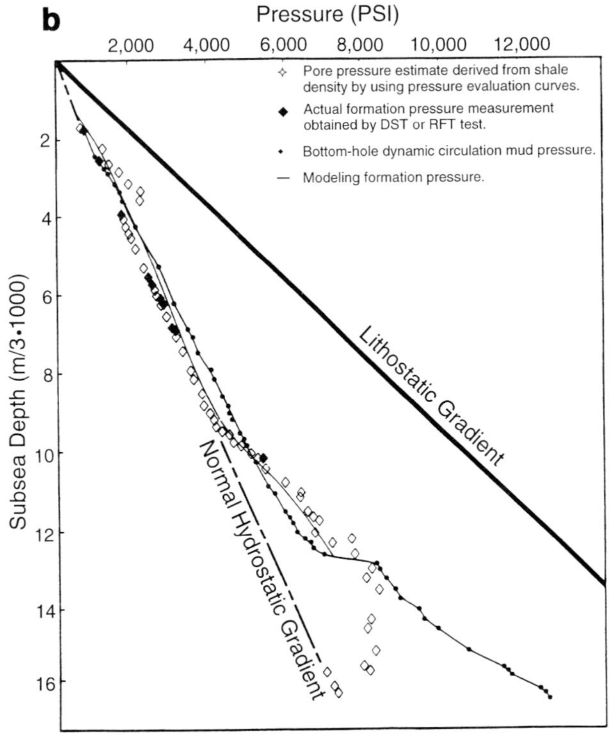


FIGURE 1.16 (a) Plot of vitrinite reflectance observations with depth (symbols). Superimposed is the solid curve representing the predicted vitrinite reflectance obtained by thermal inversion. (b) Pore pressure measurements with depth. The three different symbols represent different methods of estimating pore pressure. The light solid curve is the predicted variation of pore pressure with depth obtained from running a fluid flow/compaction code; (c) Several measured present-day temperature profiles. To obtain a present-day heat flux, the curve shown is used. (d) Permeability variations with depth. Stars denote measured values. Between about 610 to 3000 m, the dominant lithology is sand with shale "stringers." Because the permeability measurements are made in the sand "stringers" below about 3000 m, they seriously misrepresent the bulk permeability of the dominant shale unit. The solid line is the predicted variation of permeability with depth obtained from a fluid flow/compaction code. (e) Porosity measurements with depth in COST No. 1 (dots) and the predicted porosity variation (solid line) obtained from a fluid flow/compaction code.

explicit assumptions and limitations of a model. For instance, application of an elastic plate model intrinsically presupposes that the data will indeed conform to the model. When the parameters have been optimally adjusted to minimize discord between predictions and observations, then a judgment has to be made as to whether the residual mismatch is within some tolerance level deemed acceptable—and it is this last statement that requires a judgment call.

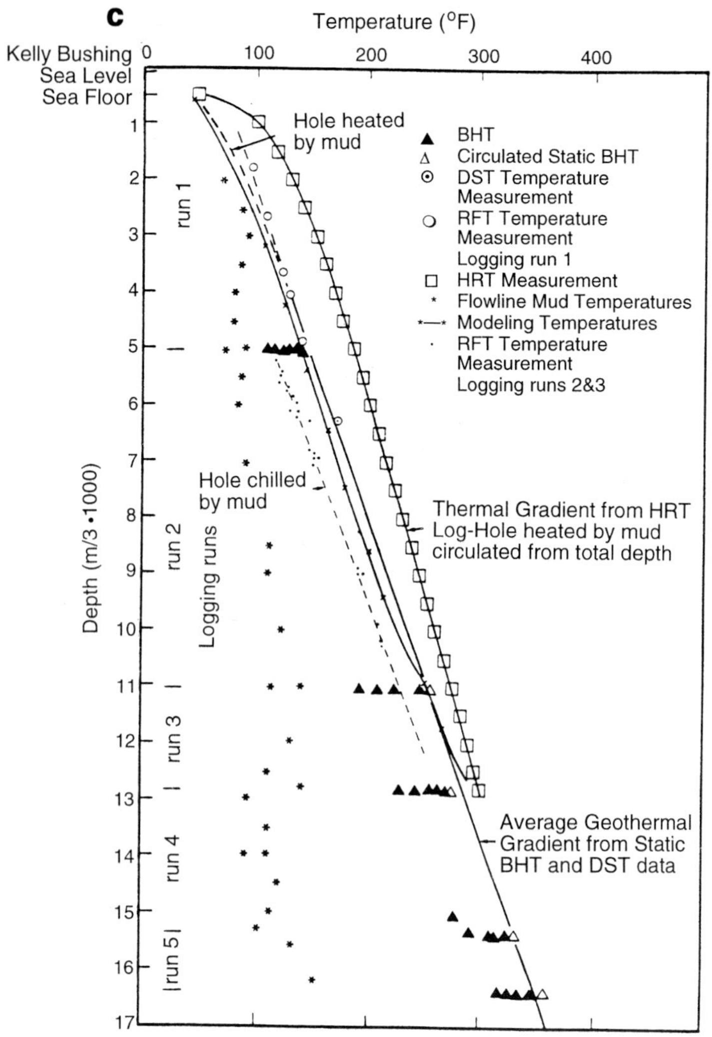


FIGURE 1.16 (*Continued*)

For instance, consider the case exhibited in Fig. 1.17, which reflects the application of a finite-length, elastic plate model to the Carpathian foredeep zone of Czechoslovakia (Cogan *et al.*, 1992). The situation depicted in Fig. 1.17 provides the minimum mismatch between predicted and observed basement deflections, and Fig. 1.18 shows how the minimum mean square residual (MSR) mismatch varies as each of the parameters is, in turn, varied away from the optimal values. In the case of a finite-length elastic plate, of homogeneous properties along its length (intrinsic assumption), there

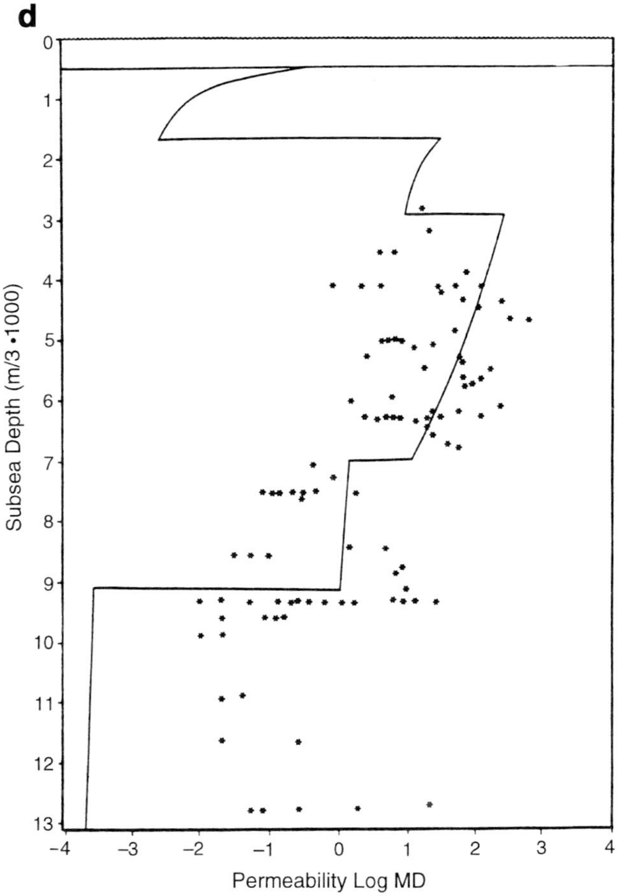


FIGURE 1.16 *(Continued)*

are seven parameters: rigidity, $D$; bending moment, $M_0$, at the free end of the plate in the absence of sediment load; initial hinge angle, $\theta_0$, of the unloaded plate; initial equivalent end point load force, $V_0$, on the free end of the plate, which mimics nonsedimentary force considerations (Turcotte and Schubert, 1982); the length, $L$, of the plate, which can be longer than the observed data segment length but presumably cannot be shorter; and the vector hinge point position (i.e., both lateral and vertical coordinates of the hinge point have to be determined and may themselves vary in time).

A satisfactory job is done by the pseudo-inverse model in matching the observed loaded basin shape, as indicated in Fig. 1.17. Yet the effects of compressive or extensive thrusts, of nonhomogeneous plate properties, of

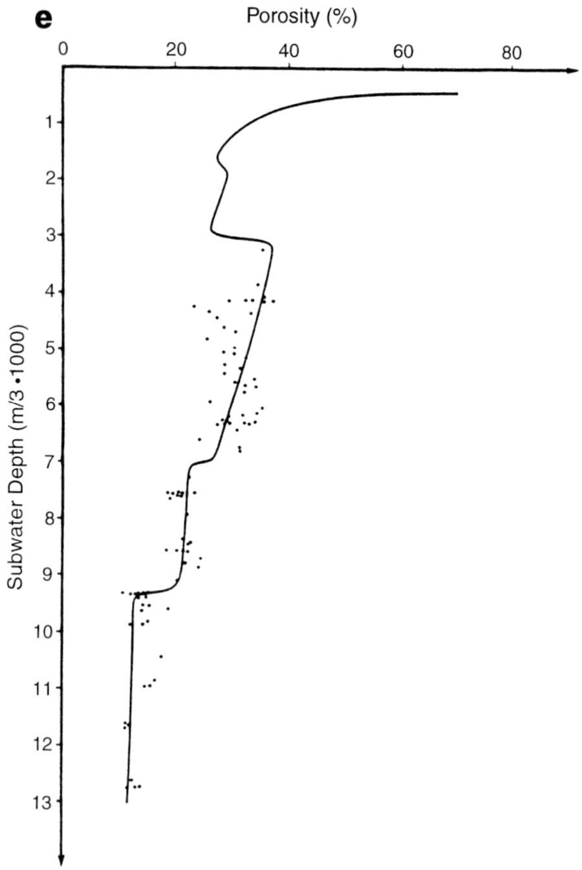


FIGURE 1.16 (*Continued*)

plasticity, of thermal variations of parameters, or of parameters describing relaxation of effective stress with time, are not included—as is also true of the lack of inclusion of any other intrinsic assumption on the behavior of the basement except that the plate is to be considered as acting like a purely elastic plate. The questions of uniqueness, precision, sensitivity, and accuracy of the results to intrinsic model assumptions, to parameters of the model, and to data resolution and finite sampling then arise. A discussion of such problems is deferred to later in this volume.

### 3. Dynamic Salt Effects

In a large number of sedimentary basins, salt plays a major role in controlling structural, stratigraphic, and thermal variations (Lerche and

TABLE 1.4 Thickness Change (m) of Each Modeled Layer with Time (Ma)

| Layer number | 1 shale | 3 shale | 5 shale | 6 sandy mud | 7 sandy | 8 sandy | 9 sandy | 10 sandy | 11 sandy | 12 gray shale | 13 mudstone | Total thickness |
|---|---|---|---|---|---|---|---|---|---|---|---|---|
| Deposition time (Ma) | | | | | | | | | | | | |
| 7 (1) | 230.0 | | | | | | | | | | | |
| 11 (2) | 229.4 | | | | | | | | | | | |
| 13 (3) | 196.8 | 477.2 | | | | | | | | | | |
| 17 (4) | 187.5 | 436.8 | | | | | | | | | | |
| 18.3 (5) | 181.0 | 401.1 | 691.5 | | | | | | | | | |
| 20.2 (6) | 171.3 | 385.0 | 558.1 | 800.1 | | | | | | | | |
| 21.3 (7) | 164.4 | 373.4 | 530.1 | 734.1 | 553.6 | | | | | | | |
| 23.8 (8) | 161.3 | 368.3 | 528.2 | 729.8 | 514.6 | 157.1 | | | | | | |
| 26.4 (9) | 159.1 | 364.8 | 524.6 | 726.8 | 508.7 | 123.3 | 154.0 | | | | | |
| 37.1 (10) | 153.3 | 357.0 | 515.5 | 717.5 | 487.1 | 118.3 | 115.8 | 426.8 | | | | |
| 39.7 (11) | 151.4 | 355.9 | 513.6 | 712.4 | 476.0 | 115.5 | 113.1 | 380.5 | 261.5 | | | |
| 41.6 (12) | 148.0 | 355.1 | 511.3 | 705.0 | 452.9 | 109.6 | 107.3 | 361.8 | 211.3 | 440.5 | | |
| 45.0 (13) | 144.6 | 353.4 | 508.7 | 701.8 | 422.2 | 107.0 | 104.8 | 353.6 | 206.7 | 375.0 | 431.2 | 3729.0 predicted |
| Measured thickness | 152.4 | 359.7 | 502.9 | 707.1 | 434.6 | 104.8 | 106.7 | 350.5 | 207.3 | 384.1 | 427.6 | 3737.7 measured |

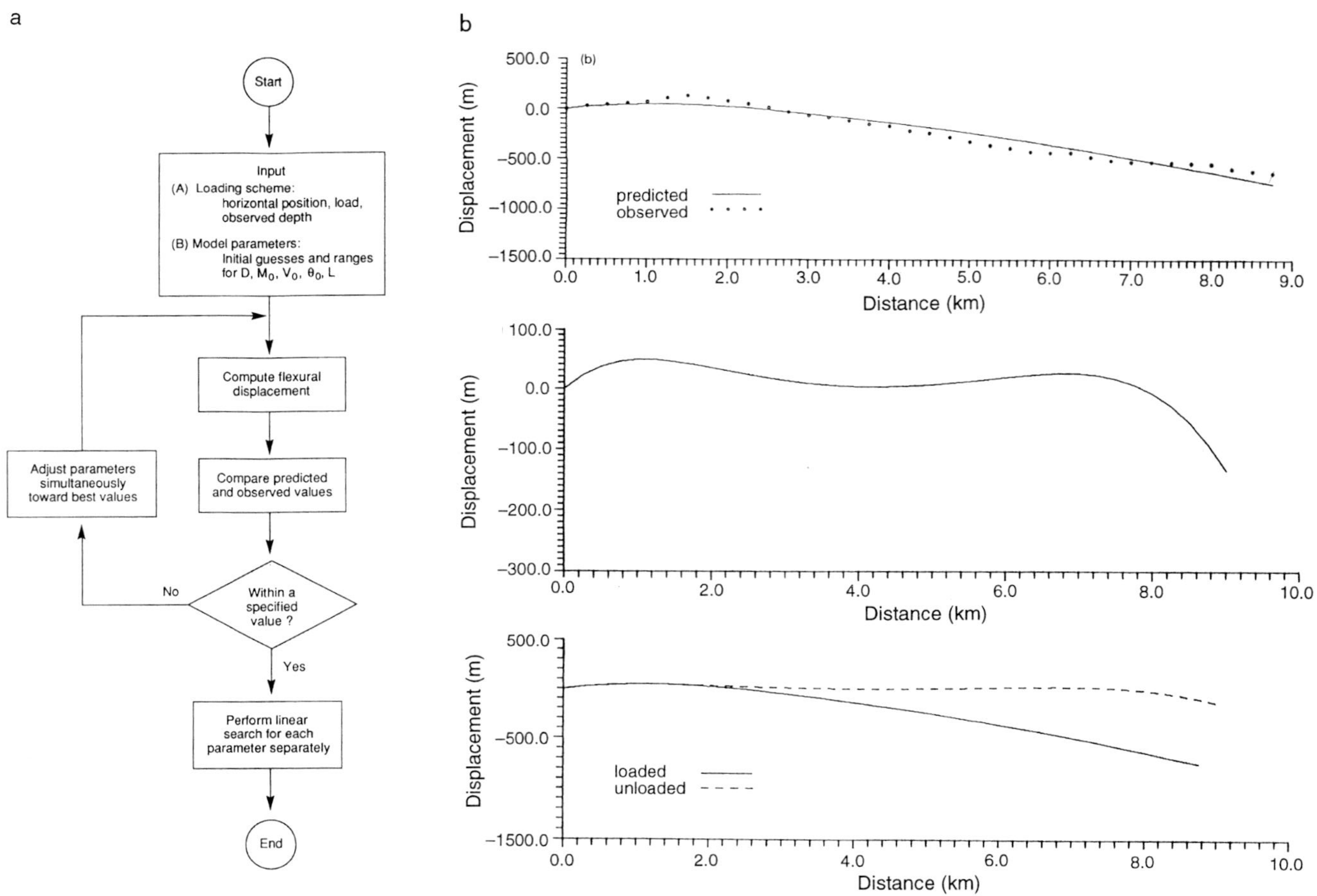


FIGURE 1.17 (a) Flow diagram of the numerical procedure used to determine flexural plate parameters. (b) Predicted versus observed loaded deflections of the basement in the Carpathian foredeep.

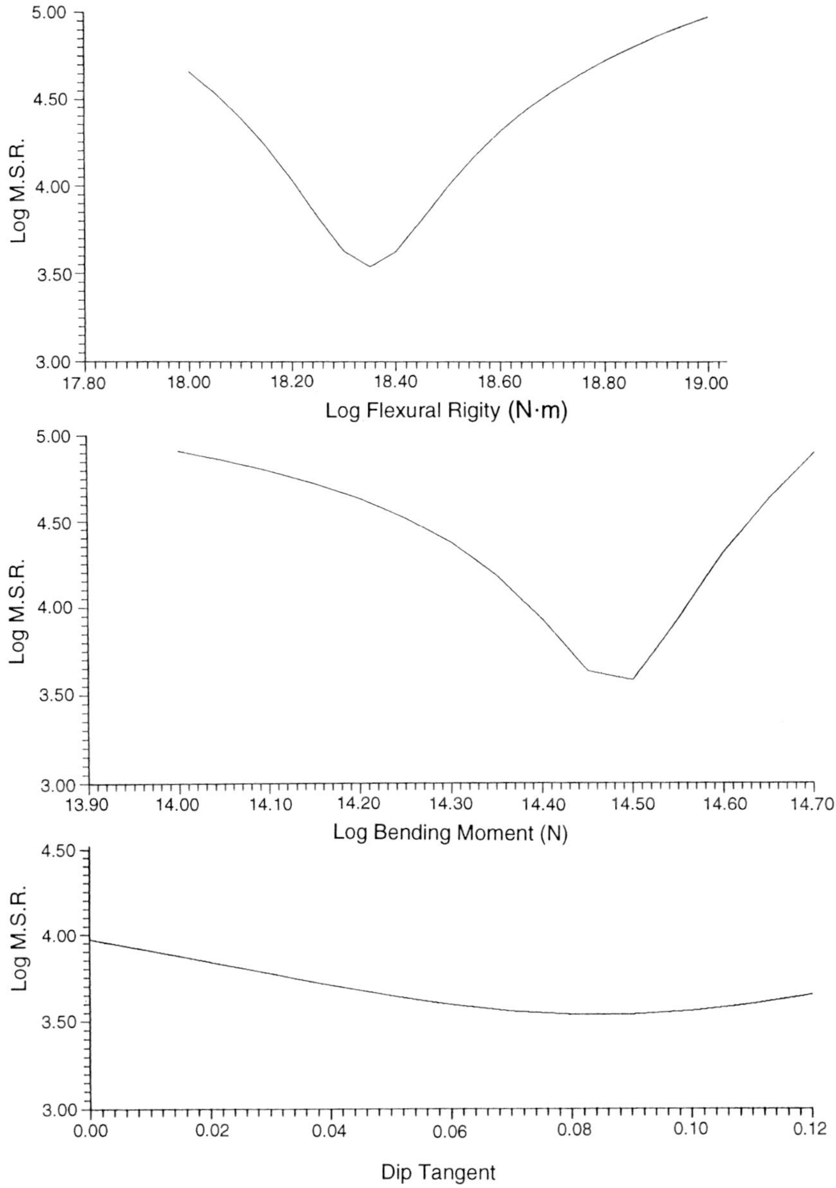


FIGURE 1.18 Sensitivity resolution of the degree of mismatch for each parameter $D$, $\theta_0$, $M_0$, $V_0$, and $L$ as each, in turn, is varied away from the optimal mismatch value. (*Continues*)

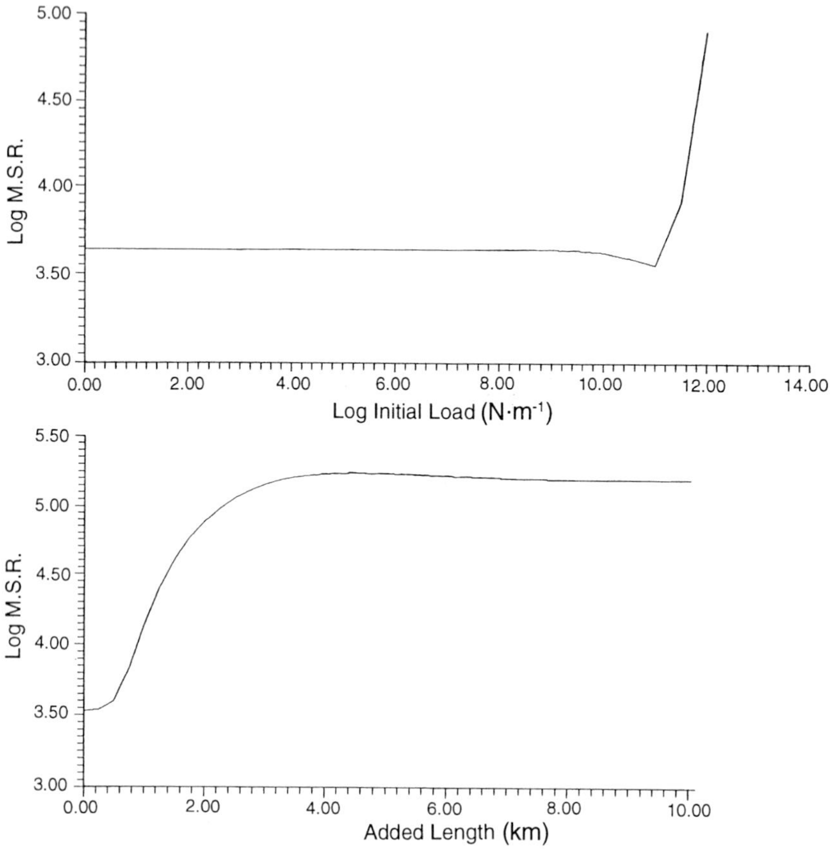


FIGURE 1.18 *(Continued)*

O'Brien, 1987). The reason for this major effect of salt is due to five basic properties of salt:

1. On a geologic timescale salt has little to no internal structural strength and flows like a fluid under applied stress, such as sediment load.

2. Salt has a nearly constant density of between about 2–2.2 $g \cdot cm^{-3}$, a value intermediate between sediment density at deposition (1.6–1.9 $g \cdot cm^{-3}$) and sediment density at full compaction (2.6–2.8 $g \cdot cm^{-3}$). Thus at some point in the evolution of a basin the sediment density overlying the salt will exceed that of the salt. Thereafter the salt will attempt to rise up through the sediments by buoyancy unless impeded by differential sediment strength (such as a competent formation) or by differential buoyancy decrease (such as from development of an overpressured zone). Before and after the neutral buoyancy point is reached, the salt will attempt to flow to locations of minimum applied stress if not otherwise impeded.

Examples of such behavior from the Gabon area of Africa and from the German–Polish salt basins are provided in Thompson *et al.* (1989).

3. Salt has a thermal conductivity of two to three times higher (at 25°C) than sediments. Thus a salt shape acts to distort heat flow in the salt's vicinity by focusing effects, thereby creating spatially variable domains of enhanced and lowered (relative to regional) temperature, which, in turn, affect sediment thermal maturity, sediment draping, and hydrocarbon trapping over and around the mobile salt masses, and which affect thermal fluid convection and salt dissolution (Lerche and O'Brien, 1987a,b).

4. The permeability of salt is variously estimated at $10^{-4}$–$10^{-12}$ mD, making salt sheets and overhangs admirable seals for hydrocarbon entrapment (Petersen and Lerche, 1995).

5. Salt dissolves in water up to about 200 to 300 ppt, promoting collapse features around and above salt masses.

Thus the effects of salt are felt throughout a sedimentary basin both in terms of regional structural and thermal developments and in terms of hydrocarbon generation, migration, and accumulation. The impact of salt dissolution in producing collapse features in sediments has been ably described by Jenyon (1987).

A problem of some concern is to provide quantitative, self-consistent, dynamical models of salt–sediment interactions that allow for all of the foregoing effects and that are constrained by present-day observations or inferences from observations. It is relatively easy to construct models in which a prescribed salt behavior is investigated for its impact on surrounding sediments (Lerche and O'Brien, 1987c), or in which a prescribed sediment behavior forces a corresponding response from a salt mass. But a combined interactive dynamic model that includes the effect of salt on sediments and, equally, the effect of sediments on salt is not so easy to construct. However, recent work (Petersen and Lerche, 1995) has provided at least some capabilities in this arena. This problem is of considerable concern if we are not to overlook potential hydrocarbon accumulations.

## 4. Faults: Open or Closed?

A debate of considerable proportion has occasionally been waged over methods for estimating the timing of faults and their motions (active versus nonactive) in relation to sedimentary strata, and over the question of whether or not faults act or can act as open conduits for fluid flow, and thus hydrocarbon escape, from the basinal system.

From a basin analysis viewpoint, attempts to answer these questions are also of considerable concern. The absolute timing of fault motions throughout the evolution of a basin has an impact on structural blocks (horsts versus grabens), on fault block rotation, on lateral and vertical fluid flow through juxtapositional variations of different lithologies across a fault that are dependent on the throw of the fault, and so on. *Relative* timing of fault motion is often inferred by inspecting the locations and strata where faults peter out and by then arguing that the fault activity ceased at the ages of such strata. Activity of a fault is often estimated by comparing similar age stratal displacements on either side of the fault to see if the displacement is constant or varying with age (in the latter case the fault is assigned to the "growth" fault type, evolving syndepositionally). But procedures for estimating, rather than assuming, absolute age dating of fault motions from observations would still seem to be difficult to obtain.

Even putting aside the question of absolute timing of fault motions, the problem of assessing whether a fault is, or has been, open or shut for geologic time, either along the total length of the fault or along only a part of the fault is, perhaps, of even more pressing concern. Proponents can be found for both points of view, and field evidence (e.g., cement fill versus no fill) suggests that both open and closed faults occur. The problem from basin analysis considerations is not whether a fault is open or closed, but rather whether a fault is sufficiently open that it can have a significant impact on the flow of fluids (and so on both sediment compaction and hydrocarbon loss) throughout the course of a basin's evolution. In addition any dynamic model that ascribes openness or closedness to a fault must have predictable consequences that can be tested by comparison against observations.

Figures 1.19a and 1.19b show the effects of running a two-dimensional dynamic basin analysis model, which allows for fluid flow and compaction, in the Pinedale Anticline area of the Green River Basin, Wyoming (Wei and Lerche, 1988). A major fault cuts the sedimentary strata at the position of the Wagon Wheel No. 1 well. Two extreme end-member cases were run: fault open, fault shut. The open fault case, as shown in Fig. 1.19a, does not permit the presence of any fluid overpressuring today at the Wagon Wheel No. 1 location. By way of contrast the shut fault case permits a major build-up and retention of overpressuring, as shown in Fig. 1.19b.

Observationally, as seen in Fig. 1.19c, there is indeed significant overpressuring today at the Wagon Wheel No. 1 site, implying that the closed fault situation is a better representation of the observational information than the open fault end-member case. However, the question remains: Is it possible that the fault was open at an earlier time, leading to fluid escape, with the shutting of the fault being a phenomenon of fairly recent origin? On the basis of estimates of the recovery time for reoverpressuring (by sealing) of a hydrostatic basin, it would appear that the appropriate answer in the case of the Pinedale anticline is that the fault has been shut since roughly the time of the Wyoming thrust (~60 MYBP) (Wei and Lerche, 1988). However, this argument does not rest directly on observational testing, but rather rests indirectly on the timescale estimate that is based on the low permeability of shale; that is, without a major erosional event

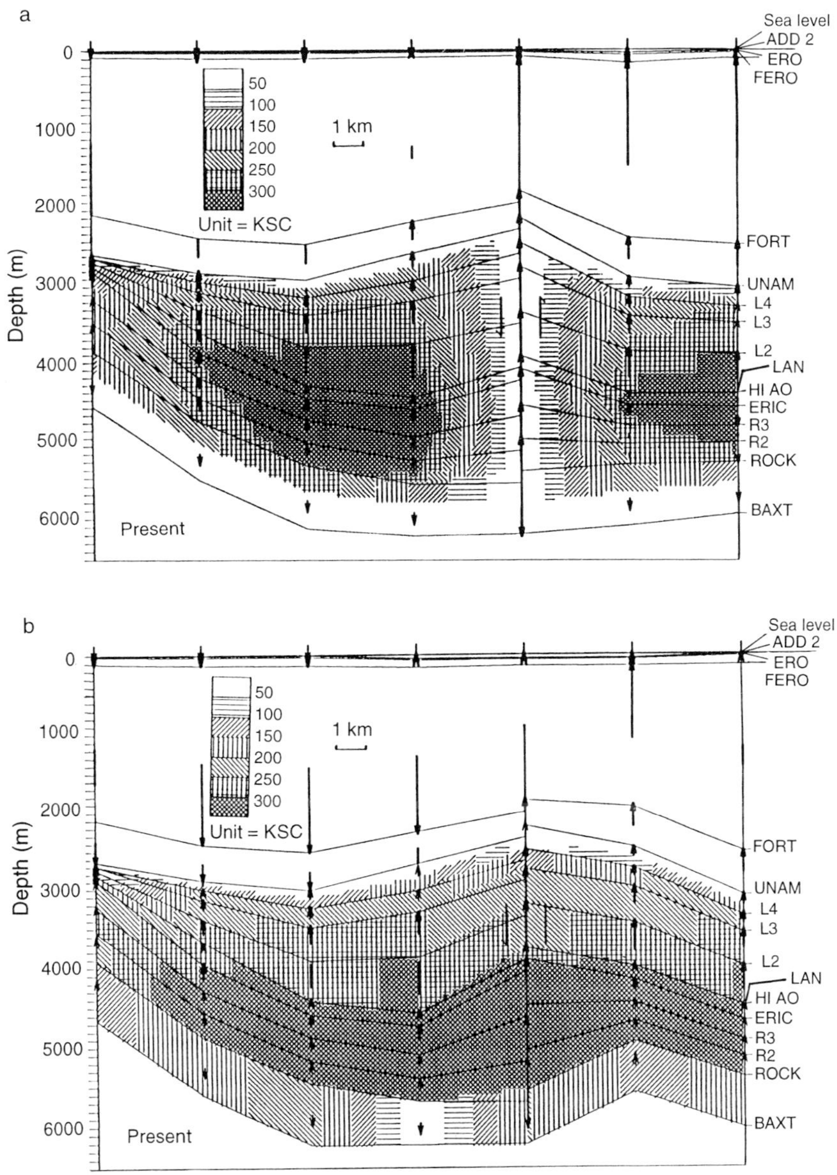


FIGURE 1.19 (a) Basin analysis simulation at present of the Pinedale Anticline, Green River Basin, Wyoming, treating the fault, which intersects the Wagon Wheel No. 1 location, as a completely open sand dike for all time. Note the absence of any predicted overpressure, in opposition to the observations. (b) As for part (a), but with the fault considered closed for all time. Note the high degree of overpressuring. (c) Agreement of predicted and observed overpressure [figure panel (a)] at Wagon Wheel No. 1 location, indicating that a closed fault provides a greater degree of consistency with the observed overpressure than does an open fault (KSC = kg · cm$^{-2}$). Panels (b), (c), and (d) show the equivalent variations of predicted and observed porosity, temperature, and vitrinite reflectance, respectively, at the same location.

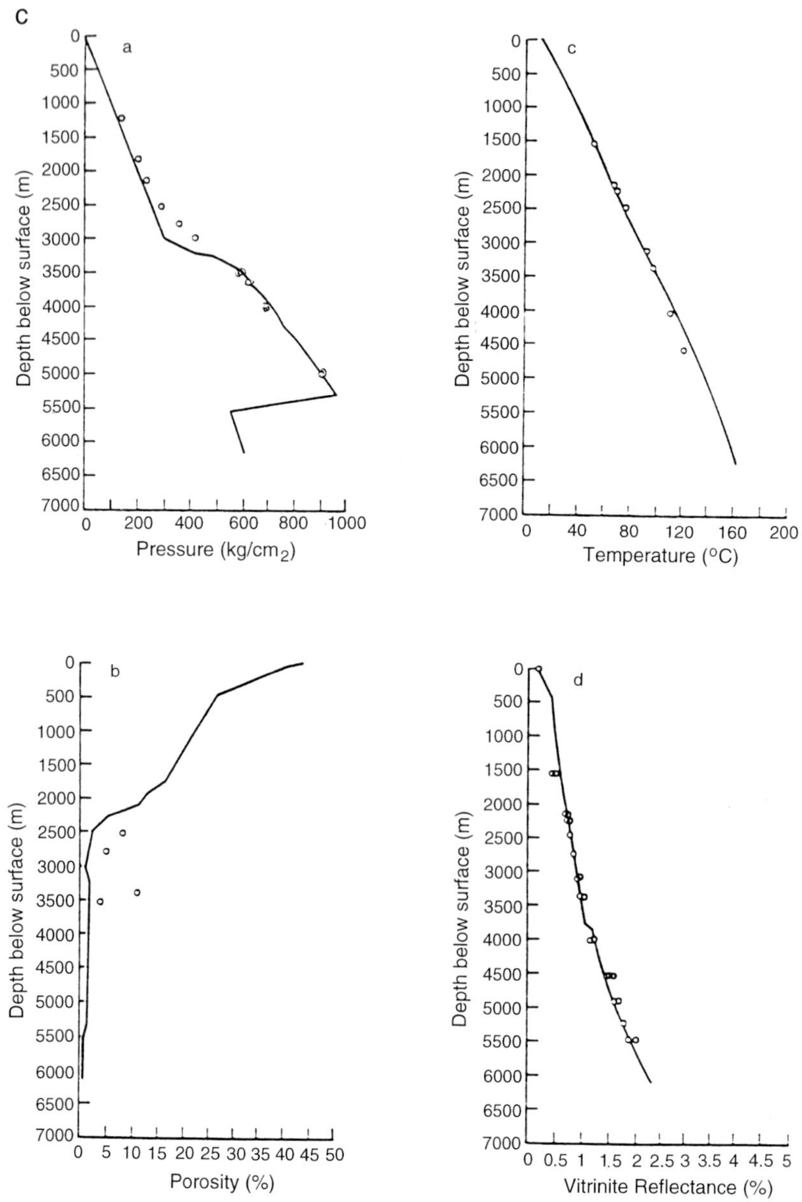


FIGURE 1.19 *(Continued)*

it is a long, slow process to bleed overpressure out of shales because of their low permeability.

In any event the twin requirements of providing pragmatic procedures for determining, rather than assuming, both the absolute timing of fault motions and the openness/shutness of faults within a generic framework are problems that will be addressed later in this volume.

## C. PROBLEMS ASSOCIATED WITH THERMAL HISTORY

Although the structural, stratigraphic, and fluid flow aspects of the dynamic evolution of a basin are of overriding influence in determining the likely migration pathways and potential present-day accumulation sites for hydrocarbons, without the production of hydrocarbons at some time in a basin's history, the geologic evolution is of little economic significance.

Current thought (Hunt, 1979; Waples, 1984; Brooks and Welte, 1984) suggests that thermal degradation of kerogen is the dominant mechanism for producing hydrocarbons (oil and/or gas). Given that viewpoint, the problem of determining hydrocarbon production hinges on two underpinning facets:

1. Can estimates be provided of paleotemperature evolution with such estimates constrained by present-day data?
2. Can quantitative models be provided of temperature-dependent kinetic degradation reactions that are based on an observational framework and can be shown to be extrapolatable to the geologic time frame with a quantifiable degree of resolution?

In the following section we address the first of these two questions.

### 1. Present-Day Temperature and Heat Flux

Most models used to construct thermal history rely on present-day thermal data to act as control constraints on either the relative scaling of paleoheat flux compared to the present-day value or on the temperature evolution relative to present-day values.

One of the major problems that has been put into focus more sharply over the last few years is the difficulty of obtaining accurate formation temperatures. Using several hundred DST (drill stem tests) measurements from many well bores, Hermanrud *et al.* (1990) have shown that, *at best*, it is unlikely that downhole formation temperatures are measured or inferred to a *statistical* (random) accuracy of better than about ±8 to 10°C, whereas the *systematic* uncertainty on true formation temperatures is also typically about 10°C. Both sorts of errors then compound the difficulty of being accurate about the scaling of paleotemperature to present-day values. In addition, even if these problems are ignored, and if present-day formation temperatures are taken as both precise and accurate, one of the problems that is then made manifest is the determination of present-day heat flux, $Q_0$, from the temperature measurements.

In estimating $Q_0$ several implicit or explicit assumptions are normally made: that thermal conduction dominates; that a steady-state treatment is appropriate; that a one-dimensional spatial model (vertical heat conduction only) is acceptable; that the present-day thermal conductivities of the sediments are well determined. Even within these confining, and likely uncertain, assumptions there are still several technical procedures for determining $Q_0$, as have been detailed elsewhere (Lerche, 1990b). Some of these procedures are the following: the use of only the bottomhole and surface temperatures; use of all the downhole temperature data to construct a constant average thermal gradient and to then use this gradient to assess $Q_0$; use of temperature measurements only from similar lithologies, so that effects of lithology dependences of thermal conductivity are minimized; use of a least-squares weighted average procedure to produce a $Q_0$ that is in least discord with *all* of the observed temperature data, irrespective of whether the data are from spatial domains that violate the confining assumptions above or are poorly measured.

Assessments of variations that arise in present-day heat flux as a consequence of using one or more of the procedures mentioned indicate that heat flux values are likely overestimated by about 10%, with a most likely smallest statistical uncertainty also of around 10%, at best (Lerche, 1990b). A problem of some concern is to provide measurement methods and interpolation procedures that will narrow the uncertainties on present-day downhole temperatures and heat flux estimates.

## 2. Paleoheat Flux and Paleotemperature

For the purposes of assessing hydrocarbon production in a basin it is not the paleoheat flux that is of direct concern; rather it is the paleotemperature. A high paleoheat flux does not unequivocally imply a high sediment temperature (e.g., the sediments could have been at shallow subsurface depth at the time of the high heat flux and so would then record only the low temperature of the shallow depth). In addition most modern codes that handle compacting sediments in basins allow for lateral and vertical convective transfer of fluids with the attendant transfer of thermal energy, as well as for focusing and defocusing of heat spatially because of lithology and overpressure variations.

However, if the paleoheat flux variation with time and spatial position could be determined then, of course, that knowledge could be used in basin analysis codes to predict the evolution of temperature with time for each sediment parcel at each location in the basin. The nub of the problem is to devise a procedure to obtain the required information. As we have remarked already, two basic and very different procedures are available. Either one assumes a model connecting paleoheat flux and dynamic basin evolution, and then uses the basinal evolution parameters determined for the particular dynamic model under consideration to connect to the proposed paleoheat flux variation; or one attempts to use measured present-day thermal indicator information, in an inverse or pseudo-inverse sense (Lerche, 1990a), to obtain an assessment of paleoheat flux. Difficulties with the first approach are that a double model connection must be made and different models, although consistent with present-day data, lead to quite different estimates of paleoheat flux. Difficulties with the second procedure are that each thermal indicator has its own dynamic range of sensitivity, is often sparsely or poorly measured, can be influenced in its development by other than purely thermal factors (e.g., mud additives and subsurface migration have significant influences on biomarker isomerization and aromatization ratios, which are then measured with depth in boreholes), and each thermal indicator has physical and chemical kinetic parameters, which often also carry some significant degree of uncertainty.

Even when measurements on multiple thermal indicators are available in the same well, it is often difficult to reduce uncertainty on paleoheat flux determinations because of the foregoing considerations (Lerche, 1990a). A concern is to find a procedure for assessing paleoheat flux and/or paleotemperatures that will be not only consistent with downhole measurements of multiple thermal indicators in many wells, but that will simultaneously provide a capability of separating thermal-basement motion models into classes which are viable from those which are not, based on *both* dynamical and thermal considerations. In resolving this problem, a pressing concern is to provide an estimate of the resolution with which the paleoheat flux is determined in order that minimum, average, and maximum ranges can be accurately categorized.

## D. PROBLEMS ASSOCIATED WITH HYDROCARBON KINETICS AND MIGRATION

Most of the sophisticated basin analysis codes currently in operation have some form of thermally dependent kinetic degradation scheme that permits the breakdown of solid kerogen to liquid and gaseous mobile hydrocarbons of high buoyancy, which are then capable of migrating away from the source rock(s).

Here two problems are exhibited. The first is connected with the form and substance of kinetic schemes employed in kerogen production of hydrocarbons and the second is concerned with hydrocarbon migration problems.

### 1. Hydrocarbon Kinetics

Two basic approaches have been taken to the kerogen breakdown to hydrocarbons. The first approach refers to the kerogen only, without being unduly concerned with the products produced by the kerogen breakdown. This approach is called *residual kerogen kinetics.* It is clear that this form of kinetic scheme is not mass conserving because the analysis considers only the fractional mass of remaining kerogen, which gradually disappears as maturity proceeds.

The second approach keeps track of the products of the kerogen breakdown and the interrelation of one product with another (e.g., oil converting to gas) as maturity proceeds. This approach is a pyrolysis product analysis and, if all products are traced, it is a mass-conserving kinetic framework. Consider each in turn.

### *a. Residual Kerogen Analysis*

The main thrust of such kinetic schemes has changed little since the introduction by Lopatin (1971) of TTI as a measure of kerogen thermal maturity, although technical details have changed significantly in the last few years (Ungerer *et al.*, 1988a,b; Braun and Burnham, 1987). In essence the procedure is to propose a set of kinetic reactions for a pure end-member component of kerogen [normally done pragmatically by declaring type I, II, or III kerogens to be end-members based on the H/C and O/C ratios in a modified van Krevelen diagram (Tissot, 1969; Tissot and Espitalie, 1975; Tissot and Welte, 1978)]. Each reaction is usually taken to be of the temperature-dependent exponential Arrhenius form so that both an activation energy, $E$, and a preexponential frequency factor, $k$, are required for each reaction pathway. (Note the intrinsic assumption that the reactions are Arrhenius in form.) Determination of the parameters $k$, $E$ for each reaction as well as the number, $N$, of reaction pathways needed is usually accomplished using laboratory measurements of residual kerogen under increasing temperature (Braun and Burnham, 1987), and then attempting some least-squares fit to the laboratory data. The problems with this approach are well known. First, there is the potentially serious difficulty that the number of reaction pathways can be increased systematically and steadily until the assumed kinetic process is bound to be closer and closer to any set of observations. Current models use around 26 to 30 channels for reactions, suggesting that between 52 to 60 parameters have to be determined, which seems excessive. Second, even if such a large number of parameters is truly needed to describe residual kerogen analysis, the uncertainty (both statistical and systematic) in the parameters needs to be addressed. Third, the heating rates and temperature regimes of the laboratory analysis are well in excess of those pertaining to geological conditions. How does one prove that high-temperature reactions, done under pristine, laboratory conditions, are really the appropriate representations to extrapolate to the slow-heating, low-temperature, crude conditions of geologic basins? And how are such extrapolations to be constrained by data? A detailed sensitivity analysis investigating such considerations would seem to be a problem of some significance to reducing risk in finding hydrocarbons.

### *b. Pyrolysis Product Analysis*

Apart from the efforts made to describe the thermal maturation of kerogen, efforts have also been undertaken to consider the products of thermal maturation (and of the further conversion of one product to another) with the ultimate goal of providing a self-consistent kinetic description of hydrocarbon generation (Ungerer *et al.,* 1988a,b). Such kinetic schemes rely on a sequence of parallel or series kinetic reactions, again most often of the first-order Arrhenius form, and the reactions are usually taken to be solely temperature dependent. Again the problems here are very close, in a formal sense, to those arising in the residual kerogen analysis scheme. For each reaction pathway both an activation energy and a frequency factor must be specified, and the resolution of each such parameter determined. Again the question arises of extrapolation of kinetic schemes derived from high-heating-rate laboratory experiments to the low-heating rates appropriate to geological conditions. Two further problems arise that are not encountered in the residual analysis kinetics. First, the liquid and gaseous pyrolysis products can migrate in the subsurface so that the thermodynamic mix of materials changes with time, making even more difficult the extrapolation from laboratory conditions; second, the *forward* modeling fits to the laboratory sets of data (Ungerer *et al.,* 1988a,b) systematically undershoot or overshoot (Fig. 1.20) even when the number of reaction pathways is increased and optimal fit parameter adjustments are used. Recent work (Lerche, 1990c; Liu and Lerche, 1990, 1991) using an *exact* inverse of the general kinetic schemes proposed to date, has shown that the reaction kinetics are intrinsically time dependent based on application to the laboratory data (Fig. 1.21). Clearly, a greater degree of effort is needed to uncover a kinetic framework that can be extrapolated to geological conditions outside the laboratory with some sharp and defined degree of confidence in the veracity of the framework. This problem would seem to be of pressing concern as well.

## 2. Hydrocarbon Migration and Accumulation

As remarked earlier, the problems of hydrocarbon migration and accumulation are intimately intertwined with the dynamic and thermal evolution of a basin. Recognizing that a detailed discussion of all problems connected with primary and secondary hydrocarbon migration would make for a very long volume (Lerche, 1990a,b), here we consider just two problems, one of migration and the other of accumulation, in order to underscore some of the problems that need to be addressed if we are to improve on our ability to understand the flow of hydrocarbons in evolving subsurface conditions.

### *a. Capillary Pressure Effects*

In concert with hydrodynamic forces moving hydrocarbons, Hubbert (1953) has also written extensively on the capability of capillary pressure

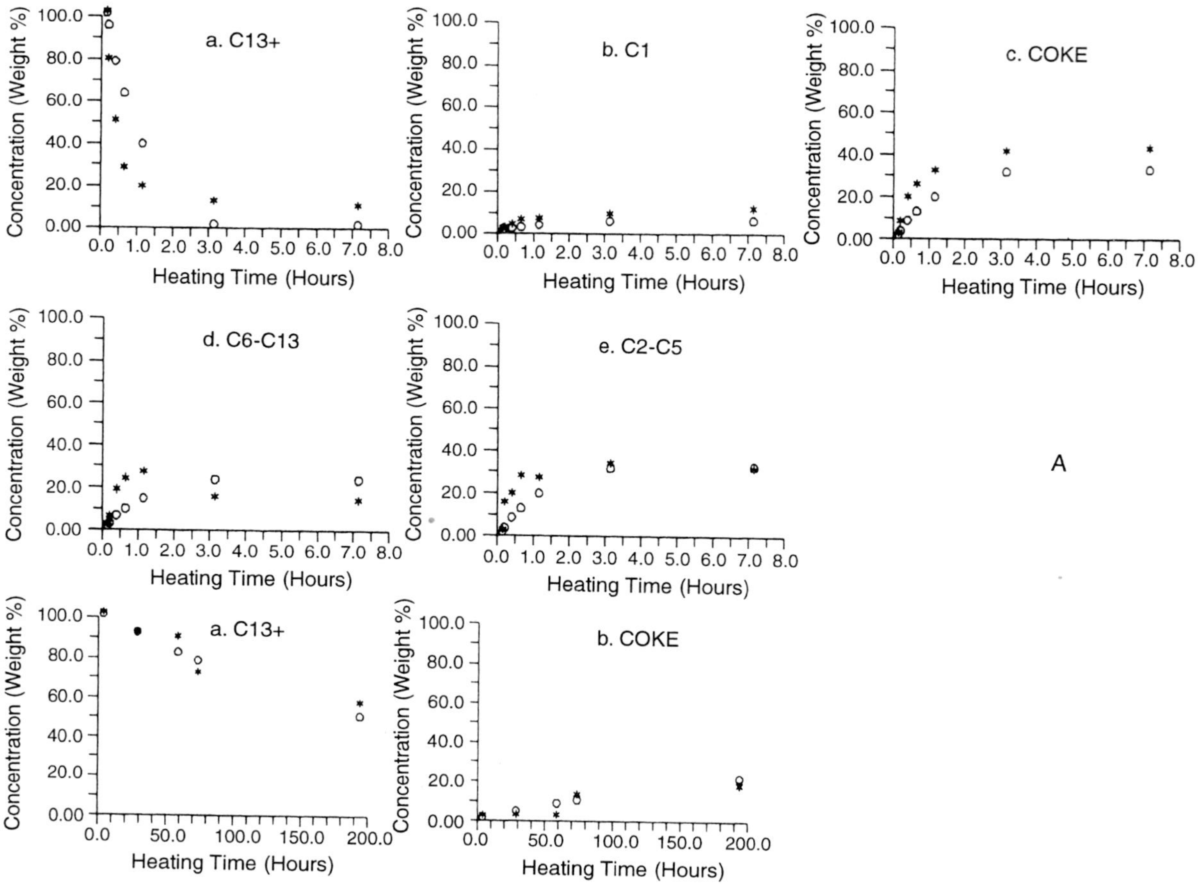

a. C13+
b. C1
c. COKE
d. C6-C13
e. C2-C5
a. C13+
b. COKE
A
Concentration (Weight %)
Heating Time (Hours)

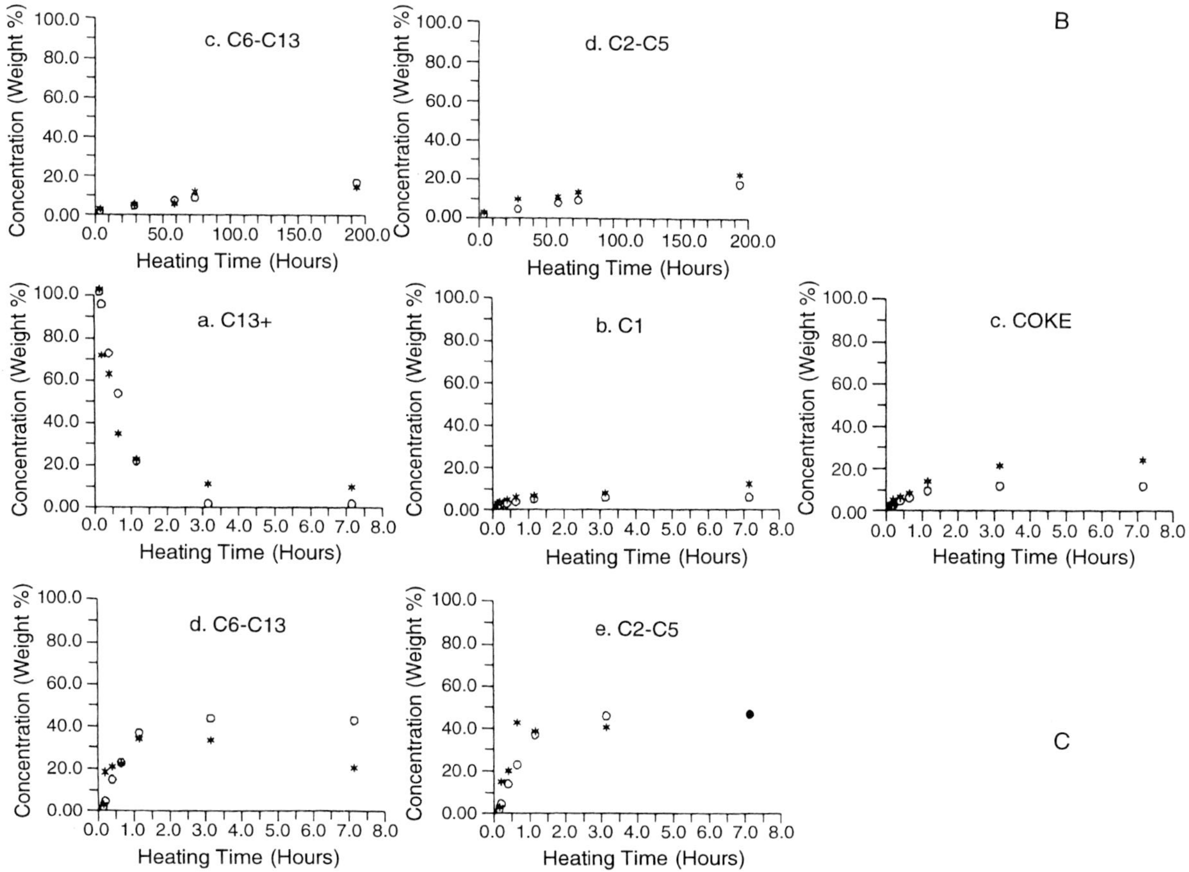


FIGURE 1.20 Forward kinetic model applied to three sets of autoclave product pyrolysis data (from Ungerer *et al.*, 1988a,b). Asterisks denote observations, open circles denote predictions.

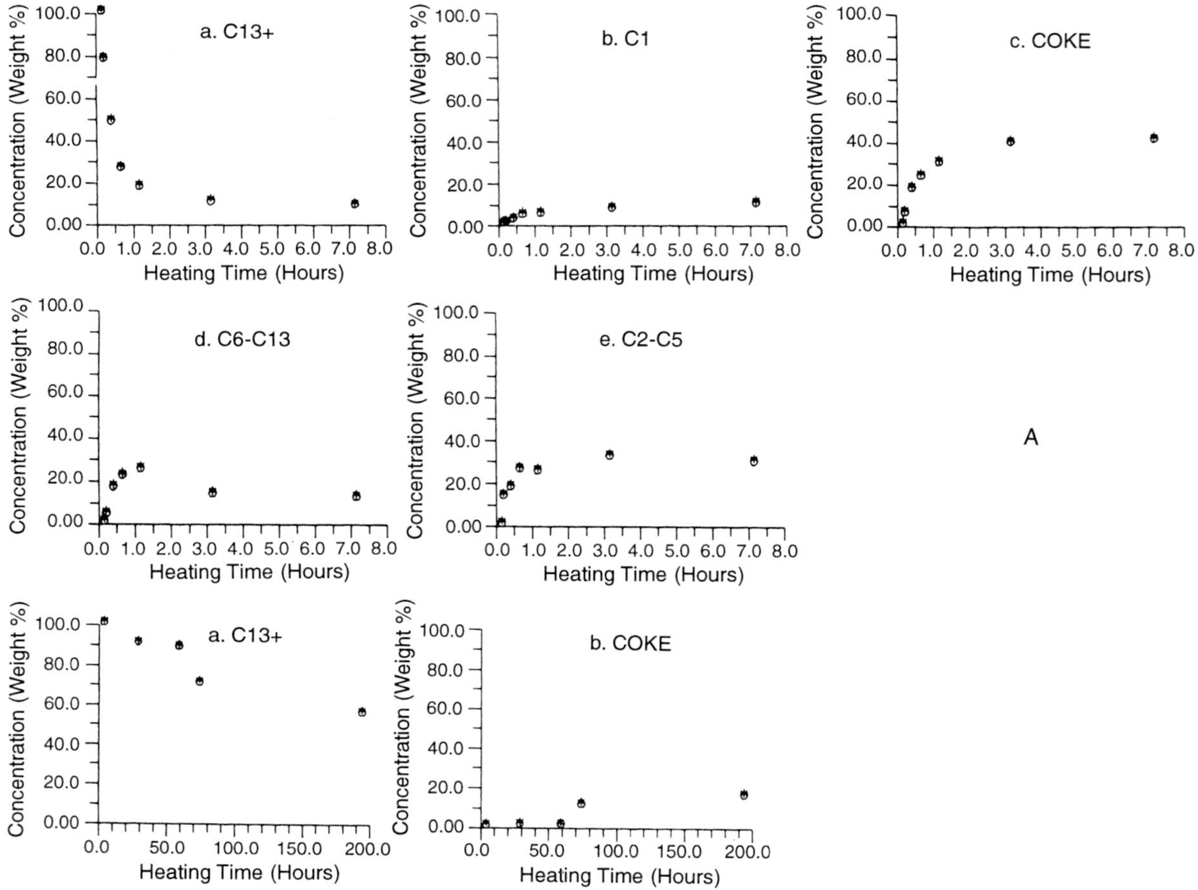
a. C13+
b. C1
c. COKE
d. C6-C13
e. C2-C5
A
a. C13+
b. COKE
Concentration (Weight %)
Heating Time (Hours)

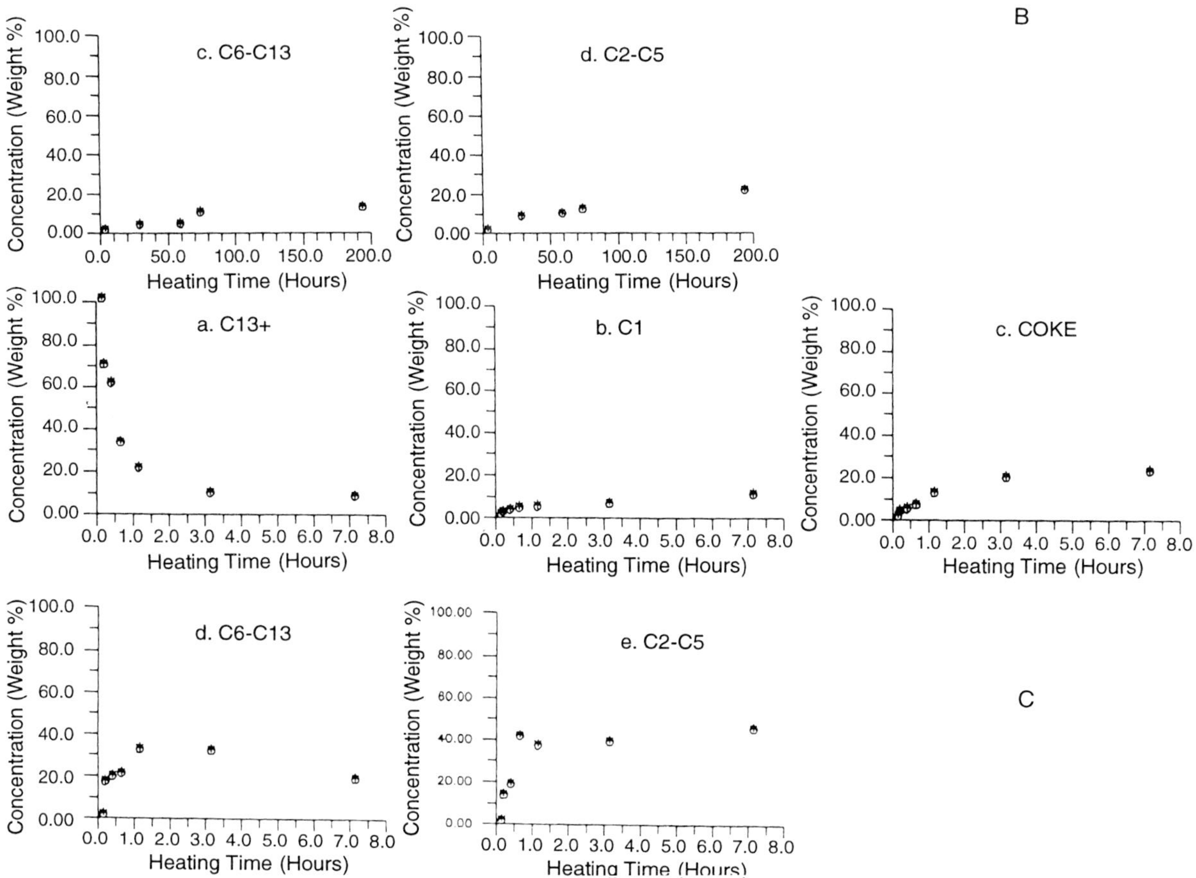


FIGURE 1.21 Kinetic model of Ungerer *et al.* (1988a,b) applied in an inverse sense to the three sets of data of Fig. 1.20. Note the significant improvement in match to the data (from Liu and Lerche, 1991).

forcing liquid hydrocarbons out of hydrocarbon-wet, fine-grained source rocks into water-wet, coarse-grained carrier aquifer beds. Later work by Dahlberg (1982) has acted to underpin the essential correctness of Hubbert's pioneering considerations. And yet some points seem to be still underappreciated despite nearly a half century of effort (Lerche and Thomsen, 1994).

Consider a hydrocarbon-bearing fine-grained shale stratum abutted on both sides by an initially water-wet, coarse-grained sandstone. A capillary pressure of some tens to hundreds of atmospheres exists at the upper and lower interfaces between the shale and the surrounding sands. Accordingly, as Hubbert (1953) has noted, the hydrocarbons are driven by capillary pressure into the sands with water being correspondingly driven into the shale from the sands. Thus *both* the overlying and underlying sands become hydrocarbon-charged. Hydrocarbons in the overlying sand, being buoyant relative to water, will promptly seek to move vertically away from the sand–shale interface; hydrocarbons in the underlying sand, also being buoyant relative to water, will tend to hug the top of the sand and so remain near the underlying shale–sand interface. But now a major question arises as follows: If the sands stay *water-wet,* then capillary pressure will continually force hydrocarbons from the shale into both sands. However, if the sands become *hydrocarbon-wet* (perhaps as a consequence of their initial charging by hydrocarbons), then the capillary force changes its character completely, as Hubbert (1953) has noted. The initial water-wet conditions imply no capillary impedance to the flow of water across a sand–shale boundary, only an impediment to the flow of hydrocarbons. Equally, hydrocarbon-wet conditions imply no capillary impedance to the flow of hydrocarbons across a sand–shale boundary, only an impediment to the flow of water. Thus, if the sands become oil-wet, then not only is there enhancement of the buoyant escape of hydrocarbons from the shale source-rock into the overlying sand, but hydrocarbons previously driven into the underlying sand can now use their buoyancy to transit back across the shale into the overlying sand. Thus the behavior of primary migration of hydrocarbons is particularly and sensitively dependent on the water-to-hydrocarbon wetness of the coarse-grained and fine-grained rocks. Hence a problem of considerable concern is to assess the water-wetness of coarse-grained sand strata as hydrocarbons flow from fine-grained source rocks, because the dynamic evolution of wetness can have a major impact on the ability of sands to act as aquifers for hydrocarbon migration.

### *b. Accumulation Effects*

Suppose that, by whatever mechanism, liquid hydrocarbons, in the form of oils, have reached a structural accumulation site at sometime in the history of a basin. Two factors dictate that knowledge of this site is not adequate to assess a low risk for hydrocarbon exploration today—unless, of course, the oils reached the site in the very recent past. First, as the basinal evolution proceeds, the accumulation site may be exhumed, buried deeper, hydrodynamically flushed by meteoric waters, faulted, and so on. Thus the retention of hydrocarbons to the present day in a putative accumulation site is intimately intertwined with the continued dynamic evolution of the basin. Second, even if the dynamic evolution of the accumulation site is known after hydrocarbon entrapment, the continued thermokinetic evolution of liquid hydrocarbons to gaseous hydrocarbons after emplacement puts its own constraints on the accumulation site. Conversion of oil to gas increases the buoyancy pressure pushing on the overlying seal, and may end up fracturing the trap, thereby releasing hydrocarbons. This effect clearly depends on the total amount of hydrocarbons entrapped, the amount converted to gas, and the evolving strength and permeability of the overlying seal, which may be increased or decreased by compaction, stylotization, diagenetic cementation, dolomitization, and so on.

Clearly, a concern with respect to hydrocarbon accumulations is to provide an integrated picture of hydrocarbon generation and migration in relation to dynamic and thermal evolution of a basin, and with respect to the structural, stratigraphic, thermal, diagenetic, and kinetic considerations particularly and especially relevant to the basin under consideration.

## E. CONTROLS, CONSTRAINTS, AND UNCERTAINTIES

The essence of science is model building to account for known data and observations, and to predict new phenomena that are then measured and compared with model predictions. When a model ceases to predict observed phenomena, it must be modified, abandoned, or replaced such that the perception of science is adjusted to reality. But within this quintessential framework, the philosophies and strategies of model building and model testing are as broad and varied as circumstances and imagination allow.

There is a general understanding that causative agents (sources) are modulated by dynamic processes (filters) to produce products (outputs). In modeling, the specification of the sources and filters determines the outputs. We mortals have a brief lifespan compared to the geological timescale, and it is rare when we can observe a system from beginning to end and thus know unequivocally what relations exist among sources, filters, and outputs. Instead, usually we are restricted to measuring outputs and inferring filters and sources. This limitation imposes severe difficulties and constraints on model development. Each model of sources and/or filters depends on intrinsic assumptions and on the functional relationships that are assumed to connect one set of physical variables to another. In addition, the functional dependencies among intrinsic and extrinsic variables and parameters are often uncertain, yet they must be provided for or adjusted in models. Determination of these dependencies is one of the major goals of models. The uniqueness, resolution, and sensitivity of a model relative to other competing models then must be assessed by comparing their output behaviors. But effective comparisons of outputs are limited by measurements with finite resolution, noise, sampling, uniqueness, and reproducibility.

Pragmatic strategies for assessing model veracity must be tailored both to the proposed sources and filtering behaviors as well as to the quality, quantity, and limitations of data available at the time the model is developed. Various observational and theoretical constraints (for example, requiring that positive variables remain positive or that mass, momentum, and energy be conserved) are often effective aids in designing model strategies, as will be discussed in detail later in this volume, although the arguments advanced apply equally to any scientific discipline.

### 1. Assumptions and Ambiguities

Prediction of the amounts and locations of oil and gas in sedimentary basins is the primary goal of sedimentary basin analysis from an oil industry viewpoint. To accomplish this goal, we require knowledge of three major components of basin evolution: a dynamic burial history of sediments; a thermal history of the basin; and a hydrocarbon generation, migration, and accumulation history. Unfortunately, we can observe only the end result of the interactive processes that contributed to the basin evolution and, even then, only with limited resolution and limited quality and quantity of data. The task is to invert observations made today to obtain the dynamic and thermal evolution of the basin. A better determination can then be made of where and when hydrocarbons were produced in the basin, and where they might be now. Several factors limit the precision of any inversion scheme. Irrespective of either model behavior or intrinsic assumptions from which a model was derived, measurements of input data are limited in their resolution, accuracy, precision, and uniqueness. Even if measurements of input data were perfect, requirements of finite spatial and temporal sampling (discrete samples) imposed by the models further limit resolution.

If we ignore these limitations of data input quality and sampling, three additional sources of ambiguity remain that limit precision of inversion models. The first includes all intrinsic assumptions inherent to a class of behaviors. As examples, a one-dimensional model is inherently a poor aproximation of the real four-dimensional world and may not behave like the real world; an assumed constant, equilibrium-balanced, conductive heat flux ignores likely changes in heat flow or thermal gradients through time; a time-invariant porosity with depth function ignores geographic, stratigraphic, and temporal changes in subsurface fluid pressures, and therefore porosity, through time.

The second source of ambiguity is the type of behavior proposed or assumed within an intrinsic assumption class. For example, it might be assumed that the paleoheat flux of a basin has the particular form of a McKenzie (1978) stretching model with three parameters: time of onset, magnitude "jump" at onset, and thermal decay timescale. Using this assumption, a pseudo-inverse procedure can be employed to determine a value or a family of values for each parameter as precisely as possible. This method is termed *pseudo-inverse* because it posits specific forms of the functional behavior. Other models with different functional behaviors and a similar number of parameters might provide equally good fits to the data, and it would not be possible to discriminate among conflicting model behaviors.

Within an intrinsic assumption class, the third source of ambiguity arises from a true inverse procedure in which the functional behavior of a required quantity is determined (rather than assumed as in the pseudo-inverse situation) by direct inverse techniques applied to the data. True inverse techniques also must determine the degree to which a functional behavior can be resolved relative to "noise" in the system.

In addition to these philosophies two other end-member classes are available in forward models. One could build an absolutely determined model and calculate a quantity of interest at the present day that evolved from assumed initial conditions. To assess the validity of the forward model, the calculated quantity is then compared in one of two ways with measured values. First, the calculated quantity can be compared directly with observed data of comparable type (e.g., stratal thickness, lithology, or porosity). Alternatively, the calculated quantity can be intrinsically unobservable, in which case an extra postulate of the quantity's relation to an observable must be made, which can never be proven (e.g., TTI to vitrinite reflectance). In both cases, the forward model may be judged acceptable or not depending on how close the calculated quantity matches the observed data. However, even if the calculated and observed quantities are close and the model is judged sufficient, the condition of necessity is not also necessarily proven—that is, the model may be inaccurate (a different model might be a better one) and its calculated quantity may be only fortuitously similar or identical to observed data.

## 2. Sensitivity, Noise, Resolution, and Uniqueness

In discussing the forward, pseudo-inverse, and true inverse philosophies for analyzing burial, thermal, and hydrocarbon histories, three underlying considerations were taken for granted, either explicitly or implicitly. First, we assumed that any input data required (e.g., stratigraphic ages, formation thicknesses, porosity variations with depth, thermal indicator measurements, present-day heat flux) were accurate (no error) and continuous with depth (no finite sampling interval problem). Neither assumption is true in general. Second, we assumed that, apart from the parameters or functional forms we were attempting to determine, all other parameters and functional forms were known precisely (e.g., thermal conductivity, permeability scale values, mantle density). This assumption is probably invalid. Third, we assumed that the intrinsic underlying assumptions of a model were accurate enough to describe the real world (e.g., one-, two-, three-dimensional, conductive heat flux only, no diagenesis, hydrocarbon kinetic model forms). But this third assumption is known to be perhaps the most invalid point of any model. Indeed, even the models connecting heat flux variations to basement subsidence vary among themselves quite markedly and all meet constraints of present-day data (Sleep, 1971; Royden *et al.*, 1980a,b; McKenzie, 1978; Falvey and Middleton, 1981; Turcotte and Ahern, 1977). Each of these problems is associated with uniqueness, resolution, sensitivity, precision, and signal versus noise, in forward, pseudo-inverse, and true inverse models (Cao, 1987). This area would seem to be one requiring a more detailed investigation and recognition than it has so far received, and will be pursued in this volume.

Perhaps the best that can be hoped for in our attempts to determine hydrocarbon accumulations is to bracket least probable, average, and most probable situations with confidence intervals so that we can at the least quantify precisely the level of our lack of determinism. And that step forward is precisely what we are about in this book.

In general forward models suffer from being presented independently of the observed data, but have the advantage of guiding thought about dynamic sedimentary behavior. Forward models often conflict one with another and occasionally have internal inconsistencies as well. True inverse procedures are often more difficult to obtain; they usually honor some components of data sets, but may be so restrictive in their underpinning assumptions as to omit the dominant physical and geological behaviors of the system. Pseudo-inverse models would seem to represent a compromise between forward models and true inverse models; on one hand they permit determination of a finite number of parameters for a given functional shape, but on the other contain the capability of allowing an extended set of functional shapes should the data warrant such an extension.

All models suffer from the vagaries of data noise and sampling, from assumptions concerning parameter values and functional behaviors within a model framework, and from intrinsic model assumptions. Measures of determination that can provide resolution, uniqueness, and precision of dynamical evolution are needed.

## IV. SUMMARY

The purpose of these examples from the three areas of predictions (shortfall), discovered and undiscovered resources (gas estimates) and basin analysis models, is to show that each and every procedure one uses to evaluate economically and/or scientifically the values and behaviors of quantities related to hydrocarbon assessment are bound to greater or lesser extents to

1. Model assumptions
2. Model parameters
3. Data quality, quantity, and sampling frequency

Quantities of interest must be evaluated within the framework of these three dependencies for

1. Uniqueness
2. Precision
3. Accuracy
4. Sensitivity

so that one has some idea of the worth and uncertainty of the assessments made.

Then, with quantitative knowledge of the degree of uncertainty, efforts can be focused on narrowing down those components of the system that provide the greatest degree of uncertainty, and of using the degree of uncertainty to assign some quantitative measures of risk and relative importance to the individual assessments of likely hydrocarbon accumulations. These measures can then be used in economic worth calculations, and also in setting up priorities and strategies for evaluating corporate options within a set of plays and prospects to maximize likely fiscal value (Rankey and Watney, 1996; Bornholdt and Westphal, 1996).

Part I of this volume examines how to evaluate quantitatively uncertainties in basin analysis models, and Part II shows how to incorporate the uncertainties into probabilistic and risk evaluations of prospects and plays, together with economic uncertainties.

## APPENDIX A: SOME PROPERTIES OF A LOG-NORMAL DISTRIBUTION

### A.I. EXACT STATEMENTS

Consider the log-normal probability distribution

$$p(x|a, \mu)\, dx \propto \exp[-(\ln x - a)^2/2\mu^2]\, dx/x, \tag{A.1}$$

where $a$ and $\mu$ are fixed parameters, and the variable $x$ ranges in $0 \leq x \leq \infty$; and where $p(x|a, \mu)$ is the differential probability of finding the value $x$ in the range $x$ to $x + dx$.

With the normalization

$$\int_0^\infty p(x|a,\mu)\,dx = 1, \tag{A.2}$$

it follows that

$$p(x|a,\mu)\,dx = \tfrac{1}{2}\pi^{-1/2}\mu^{-1}\exp[-(\ln x - a)^2/2\mu^2]\,dx/x. \tag{A.3}$$

The mean value of $x^n$, $E_n(x)$, is given by

$$E_n(x) = \int_0^\infty x^n p(x|a,\mu)\,dx \equiv \exp(an + n^2\mu^2/2). \tag{A.4}$$

Thus,

$$E_1(x) = \exp(a + \mu^2/2), \tag{A.5a}$$

$$E_2(x) = \exp(2a + 2\mu^2), \tag{A.5b}$$

and hence

$$\mu^2 = \ln(E_2(x)/E_1(x)^2). \tag{A.6}$$

It also follows that the value $x_{1/2}$ at which

$$\int_0^{x_{1/2}} p(x|a,\mu)\,dx = 0.5 \tag{A.7}$$

is given by

$$a = \ln x_{1/2} \tag{A.8}$$

so that

$$E_1(x) = x_{1/2}\exp(\mu^2/2) > x_{1/2}. \tag{A.9}$$

Consider the cumulative probability

$$P(x|a,\mu) \equiv \int_0^x p(u|a,\mu)\,du, \tag{A.10}$$

which provides the probability of obtaining a value less than or equal to $x$. On $x_1 = x_{1/2}\exp(\mu)$, $P(x_1|a,\mu)$ is given through

$$P(x_1|a,\mu) = \frac{1}{2} + (2\pi)^{-1/2}\int_0^{2^{-1/2}} \exp(-s^2)\,ds \cong 0.84 \equiv P(84), \tag{A.11a}$$

which is independent of $a$ and $\mu$; whereas on $x_2 = x_{1/2}\exp(-\mu)$, $P(x_2|a,\mu)$ is given through

$$P(x_2|a,\mu) = 1 - P(x_1|a,\mu) \cong 0.16 \equiv P(16), \tag{A.11b}$$

which is also independent of $a$ and $\mu$. Thus,

$$\text{on } P(16) \qquad \text{one has } x = E_1(x)\exp(-\mu - \mu^2/2) < x_{1/2}, \tag{A.12a}$$

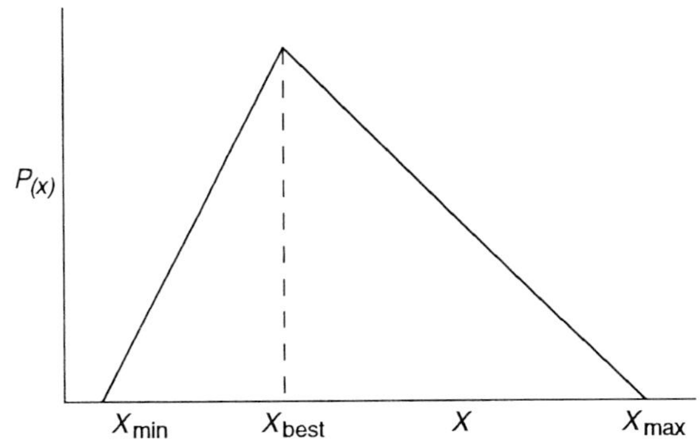


FIGURE A.1 Sketch of the triangular differential probability used with special values labeled.

$$\text{on } P(84) \qquad \text{one has } x = E_1(x)\exp(\mu - \mu^2/2) > x_{1/2}, \tag{A.12b}$$

$$\text{on } P(50) \qquad \text{one has } x = x_{1/2} = E_1(x)\exp(-\mu^2/2) < E_1(x). \tag{A.12c}$$

Note that for $\mu < 2$ [i.e., $E_2(x) < e^4 E_1(x)^2$], $E_1(x)$ occurs between $P(50)$ and $P(84)$, whereas for $\mu > 2$, $E_1(x)$ occurs at greater than $P(84)$. For $\mu \lesssim 2$, a good pragmatic approximation is that on $x = E_1(x)$, $P(E_1(x)|a, \mu)$ occurs at about 68% [i.e., about midway between $P(50)$ and $P(84)$]. A slightly better approximation is that $E_1$ occurs at a cumulative probability of about $(50 + 17\,\mu)\%$.

## A.II. APPROXIMATE STATEMENTS

Empirically it is often difficult, if not impossible, to obtain enough information to determine the precise shape of the frequency distribution of a particular parameter or variable. Indeed, it is considered a good achievement to be able to estimate a likely minimum, $x_{\min}$, a likely maximum, $x_{\max}$, and a likely most probable value, $x_{\text{p}}$, for a parameter. A rough idea of relevant mean and variance can then be obtained from Simpson's triangular rule, as sketched in Fig. A.1. We then have

$$E_1(x) \cong \tfrac{1}{3}(x_{\min} + x_{\text{p}} + x_{\max}), \tag{A.13a}$$

$$E_2(x) = E_1(x)^2 + \sigma^2, \tag{A.13b}$$

with

$$\left.\begin{aligned}\sigma^2 &\cong \tfrac{1}{2}E_1(x)^2 - \tfrac{1}{6}[x_{\min}x_{\max} + x_{\text{p}}(x_{\min} + x_{\max})] > 0\\ &\equiv \tfrac{1}{18}\{[x_p - \tfrac{1}{2}(x_{\min} + x_{\max})]^2 + \tfrac{3}{4}(x_{\max} - x_{\min})^2\}\end{aligned}\right\}. \tag{A.13c}$$

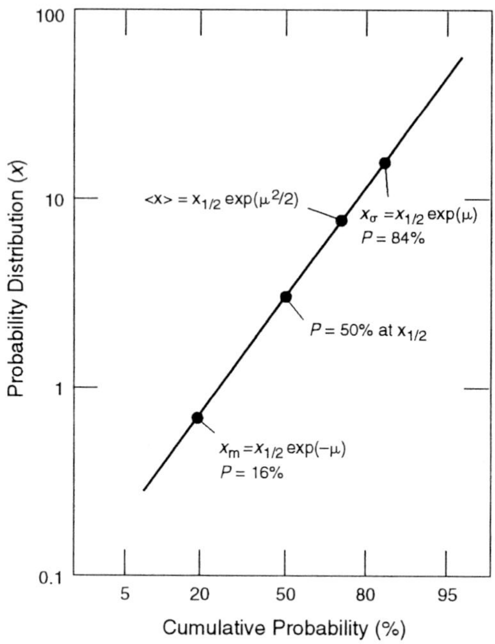


FIGURE A.2 General schematic plot of the cumulative probability versus arbitrary parameter, $x$, with special values labeled.

If it is further assumed that the variable is approximately log-normally distributed, then

$$\mu \cong \{\ln[1 + \sigma^2/E_1(x)^2]\}^{1/2} \tag{A.14}$$

so that an equivalent log-normal approximate distribution can be constructed (Fig. A.2) using Eqs. (A.12) together with the estimates of $E_1(x)$ from Eq. (A.13a) and $\mu$ from Eq. (A.14).

## A.III. MULTIPLE PARAMETER DISTRIBUTIONS

In assessments of exploration economic objectives and scientific basin analysis outputs, many parameters occur, either alone or in combination, and each of the parameters has its own uncertainty. We need practical procedures for estimating the combined effects of uncertainty of parameters on an exploration project.

Two sorts of fundamental parameter combinations seem to be prevalent: sums of parameters and products or rates of parameters. Consider each in turn.

### A. Sums of Parameters

It is well known that two or more independent random variables, $A$ and $B$, both with *normal* distributions, combine to give a sum $(A \pm B)$ that is also precisely normally distributed with mean value

$$E_1(A \pm B) = E_1(A) \pm E_1(B) \tag{A.15a}$$

and variance

$$\sigma(A \pm B)^2 = \sigma(A)^2 + \sigma(B)^2. \tag{A.15b}$$

Empirically it appears that $N$ independent random variables from any frequency distributions (not necessarily normally distributed) add to give a sum $S_N(\equiv x_1 \pm x_2 \pm x_3 \pm \cdots \pm x_N)$, which is approximately normally distributed as $N$ becomes large, with mean value

$$E_1(S_N) \cong E_1(x_1) \pm E_1(x_2) \pm \cdots \pm E_1(x_N) \tag{A.16a}$$

and with variance

$$\sigma(S_N)^2 = \sum_{i=1}^{N} \sigma(x_i)^2. \tag{A.16b}$$

### B. Products of Parameters

It is well known that multiple independent random variables, $X$, $Y$, $Z$, ..., from *log-normal* distributions combine in generic product form $X^aY^bZ^c \cdots$ to give a distribution for the product that is also precisely log-normally distributed with mean value

$$E_1(X^aY^bZ^c \cdots) = E_1(X^a)E_1(Y^b)E_1(Z^c) \cdots, \tag{A.17a}$$

with second moment

$$E_2(X^aY^bZ^c \cdots) = E_2(X^a)E_2(Y^b)E_2(Z^c) \cdots, \tag{A.17b}$$

and with scale factor $\mu$ given through

$$\mu^2 = \ln[E_2(X^aY^bZ^c \cdots)/E_1(X^aY^bZ^c \cdots)^2]. \tag{A.17c}$$

Empirically it appears that $N$ independent random variables from any frequency distributions (not necessarily log-normally distributed) tend to combine to produce a product $P_N$ $(\equiv X_1^a X_2^b X_3^c \cdots X_N^d)$, which is approximately log-normally distributed as $N$ becomes large, with mean value

$$E_1(P_N) \cong E_1(X_1^a)E_1(X_2^b) \cdots, \tag{A.18a}$$

and scale parameter

$$\mu^2 \cong \ln[E_2(P_N)/E_1(P_N)^2] = \sum_{i=1}^{N} \mu_i^2, \tag{A.18b}$$

where

$$\mu_i^2 = \ln[E_2(X_i^p)/E_1(X_i^p)^2]. \tag{A.18c}$$

These results will be used throughout the volume.

# PART I

# Models and Their Uncertainties

As basin modeling techniques become used routinely in hydrocarbon exploration, the uncertainty and sensitivity associated with basin modeling become vital issues in order to avoid decision making based on uncertain outputs. When data availability decreases, the uncertainty involved will increase and so will the exploration risk. Therefore, it is essential to know the uncertainty and associated risk when applying basin modeling techniques to individual basins. For a given model it is necessary to examine how the model results are influenced by the uncertainty in model assumptions and model parameters and by the errors in and finite sampling of the input data.

The purpose of this part of the volume is to provide quantitative measures to outcomes of model calculations in order to obtain objective, reproducible estimates of the likelihood of uncertainty. From such measures strategic exploration decisions can be made, tied also to the probability of finding economic hydrocarbon accumulations (see Part II).

Three groups of variables that influence the results are input data, equation parameters, and intrinsic assumptions.

Input Data. Very commonly errors occur in the measurements of the geological and geochemical data and, in addition, the data are nonuniformly sampled with depth at different locations as well as being of different total number at each location. The variables that need to be examined in the input data are depth and age of each formation base, lithology and paleowater depth of each formation, unconformity time and eroded thickness, temperature at the sediment surface, bottom hole temperature, heat flow at present day, thermal gradient, vitrinite reflectance, kerogen type, and content of each formation.

EQUATION PARAMETERS. Most of the parmeters/constants in the equations used in the models are based on empirical data (for example, a depositional fractional porosity of shale, 0.62, is often used as a default). At the least, the following parameters need to be examined for the sensitivity of output results to variations in the input parameters: depositional porosity, permeability and frame pressure, viscosity of the fluid in the sediments, parameters in void ratio/frame pressure functions and in the void ratio/permeability functions for each lithology, critical temperature and doubling temperature in the vitrinite reflectance equation, thermal conductivity of each lithology, activation energies and frequency factors in the generation model, and so on.

INTRINSIC ASSUMPTIONS. Most models contain intrinsic assumptions such as the 1-D nature versus a 2-D problem, such as the replacement of complete isostatic movement of basement by partial flexural compensation, etc. The results of models have to be evaluated for four factors: (1) the sensitivity of the model results to changes in input assumptions, parameters, or data, that is, how much model results vary with input changes; (2) the resolution of the model results, that is, whether the mean output value of a quantity is well resolved (low variance) or poorly resolved (high variance); (3) the precision of output results (e.g., there is little point in quoting output temperature values with depth to decimal level values if the input controls are uncertain by much larger factors); and (4) the uniqueness of the model results. For example, it can happen that two very different models of heat flux variations with time are consistent with observed thermal indicator data at the present day. In that case the model results are not unique. Thus, on the path to achieving an evaluation of hydrocarbon accumulation site determinations, the four major concerns on model results are (1) resolution, (2) sensitivity, (3) uniqueness, and (4) precision.

These four factors are controlled to a greater or lesser extent by the three overriding inputs of (1) intrinsic assumptions, (2) parameters values, and (3) quality, quantity, and frequency distribution of data.

The appropriateness of any quantitative basin analysis result in relation to the evolution of a basin can be no better than the quality and quantity of the modeled behaviors. At any given stage of knowledge of a sedimentary basin, there are always the questions of determining what can be garnered without equivocation, what can be achieved with a degree of equivocation that is acceptable, what cannot be evaluated without further data, and what measures of "acceptable equivocation" can be provided. In this part of the volume individual components of a basin are addressed for uncertainty, all of which will be used in Part II when dealing with probability and risk.

# 2

# Flexural Plate Motion

## Elastic, Thermo-Elastic, and Plastic-Elastic Models

### I. Overview

Quantitative basin analysis has undergone a revolution in the last 20 years due to new approaches, of which one of the more exciting is computer application to simulate the evolutionary processes of both basement motion and infilling materials of a basin. One often-used assumption is that the lithosphere is in local Airy isostasy, with crustal loads in pressure balance with the upper mantle. Models of compensation by flexure are based on the assumption that the lithosphere has some finite strength. Loads are supported by the mechanical strength of the lithosphere as well as by buoyant forces (Turcotte, 1979).

Four categories of flexural models have been proposed. *Linear elastic models*, based on Hooke's law, such that strain is directly proportional to stress (Turcotte and Schubert, 1982); deformation occurs instantaneously when a given stress is applied and there is complete recovery when the stress is removed (no permanent deformation). *Viscoelastic models* express strain as a function of both stress and time. Following Maxwell's law, strain is taken to decrease exponentially with time (Turcotte and Schubert, 1982); there is a relaxation of stress with time, resulting in permanent deformation of the material. *Thermal flexure models* are based on elastic (or viscoelastic) theory with the additional assumption that the rigidity is related to heat transfer within the material (McMullen *et al.*, 1981; Quinlan and Beaumont, 1984; Haxby *et al.*, 1976; McMullen and Mohraz, 1987). The strength of the material decreases with increasing temperature. *Plastic-elastic models* are based on elastic models with the additional assumption that the plate curvature becomes infinite at some critical bending moment (Turcotte and Schubert, 1982).

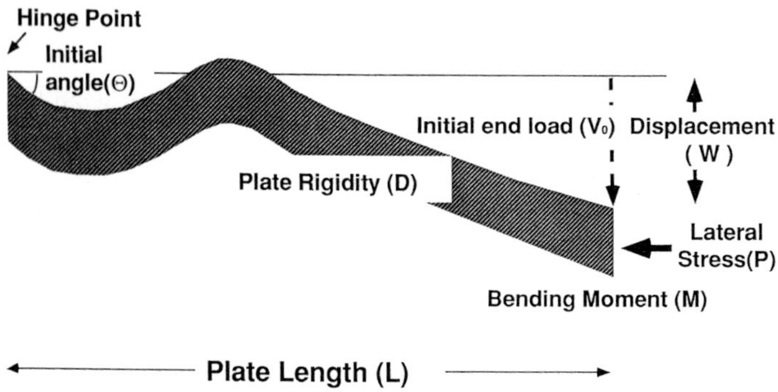


FIGURE 2.1 A flexural model with compressive thrusting. The displacement, $w$, for an elastic plate depends on both flexural parameters (the rigidity of the plate, $D$; the initial bending moment, $M$; the initial end point load, $V_0$; the dip angle of the plate at the hinge point, $\theta$; and the length of the plate, $L$) and nonplate parameter (the lateral stress $P$ applied to the end of the plate.

Often pure elastic flexure cannot account for the observed deflection of the lithosphere. Beaumont (1981) produced poor results initially in the Alberta Basin using an elastic model. However, his results were greatly improved by employing a viscoelastic model. Similarly, Quinlan and Beaumont (1984) had little success modeling the Appalachian Basin using elastic or viscoelastic theory; acceptable results were achieved only after the incorporation of thermal effects. Patton and O'Connor (1988), using an elastic model, determined an "anomalously low" flexural rigidity, $D$, for the Northern Orman Mountain Foredeep ($D = 10^{22}$ N · m) and attributed this finding to compressional stresses and/or mechanical weakness of the lithosphere.

The elastic models employed by the researchers mentioned, and by others (Jordan, 1981; Royden and Karner, 1984; Jacobi, 1981; Karner and Watts, 1983), assume that the lithosphere is an infinite plate. Such a lithosphere must be horizontal prior to loading, so that the plate cannot have an initial dip angle or initial bending moment. Only two parameters then control the bending of an infinite elastic plate: the flexural rigidity ($D$), and the load ($V$), a situation that limits the number of plate geometries that can be produced. A lithospheric plate of *finite* length has four more parameters entering the elastic model: plate length ($L$), initial dip angle ($\theta$), initial bending moment ($M$), and lateral stress ($P$) (Fig. 2.1).

In a two-dimensional situation, one end of a plate is taken as a hinge point, normally representing a geological boundary. Compressive or extensional stress is applied at the other end of the plate. The hinge point position is either specified based on geological requirements (e.g., a bounding fault) or can be adjusted vertically and laterally (see later) when the hinge point position is then a parameter to be determined. The sediment cover is applied on the plate as a vertical force. The deformation of the plate due to the applied stresses is calculated relative to the hinge point as zero displacement (Fig. 2.1). The unloaded plate configuration is an indication of the basement geometry that existed prior to any sediment loading or thrust. The allowed variation in preloaded basement shapes can produce greater variability in the final (loaded) basement geometry than can an initially flat (infinite) plate.

In plastic-elastic and thermo-elastic plate models more parameters enter the model description (see Appendix B at the end of this chapter), because a model response of rigidity to thermal effects and a model response of bending moment to critical curvature must both be incorporated into the flexure equations, and such responses normally contain parameters describing the scale and functional behavior of the response.

The purposes of this chapter are (1) to illustrate how the flexural plate equations depend on the functional behavior assumed for a response mechanism; (2) to provide a general inverse scheme for obtaining parameter values, together with their resolution, for any model behavior from knowledge of observed sedimentary load distribution along a basement horizon; and (3) to provide a case history illustrating how the method works in practice. The detailed mathematical analysis of the flexural plate equations is developed in the appendixes to this chapter.

## II. MODEL DESCRIPTION

The flexural response of basement to sediment loading and to lateral stress, due to compressive thrusting or extensional rifting, depends on the basic parameters: rigidity of the basement, $D$; initial bending moment, $M_0$; initial end point load $V_0$; dip angle of the basement at the hing point, $\theta$; length of the basement, $L$; lateral stress, $P$, applied to the end of the basement; and the density difference between mantle and sediment $\Delta\rho$ (Fig. 2.1). The deformational process is taken to be independent of depositional rate; that is, the plate response is sufficiently rapid that at each instant of time the steady-state response is an accurate representation. Therefore, all of the parameters in the model are taken as constants through geological time. However, the parameters in the model are related to spatial position along the basement. The parameters divide into three groups: the first group sets the overall scales of the basic parameters; the second group determines whether the basic parameters systematically increase or decrease along the basement; the third group is used to simulate a sinuous shape of basement deformation. Simulations then allow for the effects of regional geological events on basement and also for the effects of local events on basement deformation, such as local intrusions of volcanic material, metamorphic phase changes, etc.

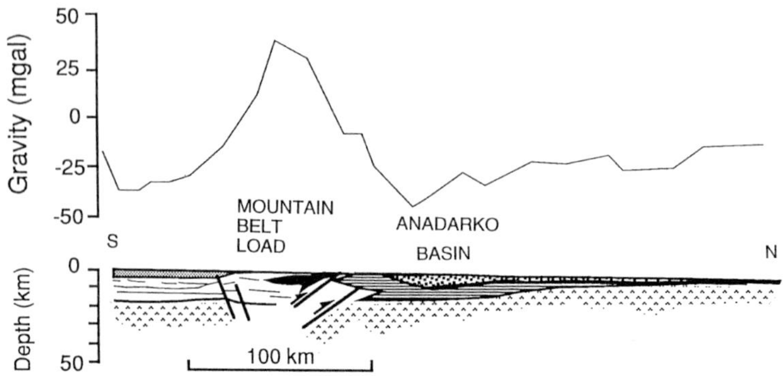


FIGURE 2.2 The deep, flexural, Anadarko Basin was produced by Pennsylvanian–Permian overthrusting of the Wichita Mountains but has no present-day topographic relief (from Steckler, 1990). Black, volcanics; horizontal lined, Wichita Mountain sediments; stippled, post–Pennsylvanian–Permian sediments; light shading, post–Permian sediments; teepee-hatching, crustal material.

An example of a flexural basin generated by subsurface loads is the Anadarko basin of Oklahoma (Fig. 2.2). Steckler (1990) states "this 6 km-deep flexural basin was produced by Pennsylvanian-Permian overthrusting of the Wichita Mountains which has *no present-day topographic relief.* Thus, the entire load creating the Anadarko basin is in the subsurface. The core of the uplift is covered by synrift volcanics of the southern Oklahoma aulacogen. Compression of the Late Precambrian–Cambrian aulacogen resulted in overthrusting of the heavy rift crust. This uplifted block acts as the subsurface load for the Anadarko basin." Gravity data show a high over the mountain-belt load with an accompanying broad low over the Anadarko basin (Steckler, 1990).

Inverse methods determine the parameters in models (Cogan *et al.,* 1989; Tang *et al.,* 1990, 1992; Tang and Lerche, 1992). Parameter values are adjusted in a systematic nonlinear optimization procedure to provide a close approximation to the observed, loaded basement profile determined by the best fit between calculated and observed basement profiles. (Appendix A of this chapter provides a detailed description of the inverse method.) With the best parameters so determined the inverse method removes sediment cover to determine an unloaded basement profile.

To investigate the effects of individual layers on basement deformation, the basement evolution with geological time is modeled as follows: First, the inverse model is used to determine the flexure parameters that best fit the observations; second, by running a forward model with the best values of the parameters (without sediment cover on the basement) we obtain an unloaded profile; third, we add sedimentary layers to the unloaded profile in a sequential fashion, allowing the basement to "bend" and the basement shape to be modified by sedimentation. When all sedimentary units are laid down, the final calculated basement shape will be that having the minimum mismatch to the observed profile. As the load of each layer is added, modification of the basement shape occurs, yielding the deformation due to the change in differential sediment loading over a given geological time interval.

We examine three cases of plate models: elastic, thermo-elastic, and plastic-elastic, with each allowed to occur in a setting where the geologic properties vary; for example, the underlying mantle density on which the plate rests can vary from oceanic to continental (Appendix B of this chapter).

## A. ELASTIC MODELS

### 1. Constant Plate Rigidity

The simplest situation, which illustrates the logic of the procedures, is the case of an elastic plate whose rigidity is constant (termed *elastic model 1*). Flexure of a finite *elastic* beam of length $L$, anchored at one end, due to the force of a single applied end load is dictated by six parameters: flexural rigidity, $D$; bending movement, $M$; dip angle of the anchored end, $\theta$; end load, $V$; length, $L$; and lateral stress, $P$ (Fig. 2.1; also Appendix A at the end of this chapter). Initial application of this model is limited by the assumption that the total load of the basin can be approximated by a point load applied at the end of the plate. The single load model has to be modified to account for differential lateral loading (Cogan *et al.*, 1989).

Once a basement horizon is picked (usually on the basis of seismic and gravity data) and the hinge point chosen (often taken to be the highest point of the basement on the section), the sediment load overlying the basement is then divided into a series of blocks of equal width (Fig. 2.3). The lines between the blocks are drawn from the seismic datum to the top of the basement. If the porosity is not measured then assume that the porosity decreases exponentially with depth:

$$\phi(z) = \phi_0 \exp(-z/l),$$

where $\phi(z)$ is the porosity at a given depth $z$, $\phi_0$ is the porosity at the depositional surface, and $l$ is a scaling length (Sclater and Christie, 1980). The values of the surface porosity and scaling length have to be specified. For default values we use Sclater and Christie's suggested values of $\phi_0 = 0.56$ and $l = 2300$ m for sandy shales (Table 2.1). The total sediment load within each block is then calculated as follows:

$$V_s = gA \int_0^z [\rho_w \phi(z) + \rho_s(1 - \phi(z))]\, dz, \tag{2.1}$$

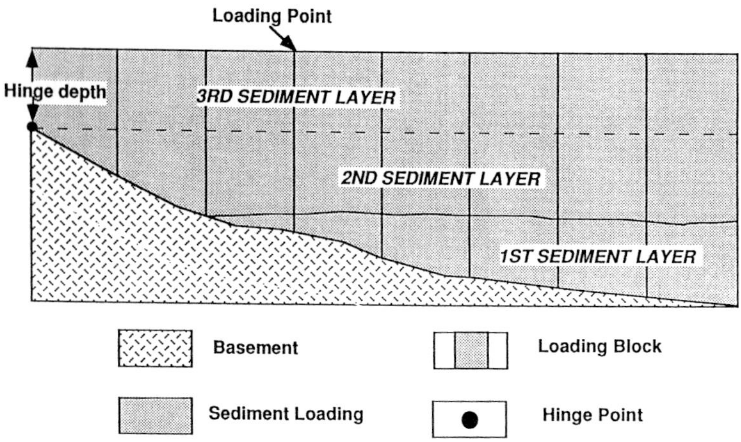


FIGURE 2.3 Sketch of point loading relative to hinge point from geological sedimentary layers (see text for a detailed description).

where $V_s$ is the total sediment load within the block, $g$ is the acceleration due to gravity, $A$ is the cross-sectional area of the block, $\rho_s$ is the density of the sediment matrix material, and $\rho_w$ is the density of water. If a basin is partially filled by water, we have to consider the effect of water loading on deformation of basement. The weight of water of column height $h_w$ in the basin is calculated from

$$V_w = gA\rho_w h_w. \tag{2.2}$$

The total load on the basement within each block is then

$$V_T = V_s + V_w. \tag{2.3}$$

A theoretical basement profile is calculated for the observed loading scheme with the parameter values set at some initial values. The theoretical

TABLE 2.1 Values of Constant Parameters Used to Calculate Sediment Load on the Basement in Elastic Model 1

| Parameter | Description | Units |
|---|---|---|
| $E$ | Young's modulus | $7.0 \times 10^{11}$ (Pa) |
| $\nu$ | Poisson's ratio | 0.25 (dimensionless) |
| $\rho_m$ | Mantle density | $3.4 \times 10^3$ kg $\cdot$ m$^{-3}$ |
| $\rho_w$ | Water density | $1.0 \times 10^3$ kg $\cdot$ m$^{-3}$ |
| $\rho_s$ | Sediment matrix density | $2.7 \times 10^3$ kg $\cdot$ m$^{-3}$ |
| $\phi_0$ | Mean surface porosity of sediment | 56(%) |
| $l$ | Scaling length for porosity calculation | $2.3 \times 10^3$ m |

profile is compared to the observed profile using a standard least-squares fit. The average degree of mismatch is represented by the MSR, or mean square residual. The lower the MSR, the better the fit. (The unit of MSR is length squared, but we take the logarithm of MSR in order to simplify the calculations. Henceforth, all MSR values refer to the logarithm of a number whose unit is meters squared.) After an MSR is calculated for the initial parameter estimates, an inverse method (Menke, 1984) is employed to adjust all the parameter values simultaneously to minimize the MSR (Fig. 2.4) using the procedure given in Appendix A at the end of the chapter. The length, $L$, of the plate is treated differently in the inverse approach than in the forward model. It is possible that the plate extends beyond the limits of the data, but cannot be shorter than the observed end of the plate (Fig. 2.5). Therefore, plate length can be varied giving rise to an "added length," which refers to the segment of plate added onto the end of the observed plate furthest from the hinge. Likewise the initial choice of hinge point position may be mistaken. Accordingly, the hinge point can be adjusted both vertically and laterally using the nonlinear procedure to minimize the MSR. When the best values of the parameters have been determined the model is run for a final time. The theoretical basement profile is determined, the final MSR calculated, and the preloading basement profile obtained (Cogan *et al.*, 1989; Tang *et al.*, 1990).

**2. Variable Plate Rigidity**

Turcotte and Schubert (1982) noted that it is unlikely that all flexural plates have the same rigidity, and even more unlikely that any given plate has the same rigidity at every point along the plate. It is therefore appropriate to ask what is the effect on an elastic flexural plate of a varying rigidity (termed elastic model 2).

In Appendix B of this chapter, the general solution is given to the elastic flexure equations when the rigidity $D(x)$ varies along the plate from the hinge point at $x = 0$ in the following manner:

$$D(x) = D_0 \exp\left(\beta x + \sum_{n=1}^{N} \beta_n \sin(n\pi x/L)\right), \tag{2.4}$$

where $\beta$, and $\beta_1, \beta_2, \beta_3, \ldots, \beta_N$ are parameters to be determined; $D_0$ sets the overall scale of rigidity; $\beta$ determines whether the rigidity is increasing ($\beta > 0$) or decreasing ($\beta < 0$) along the plate in a systematic manner, perhaps indicative of a thickening or thinning plate; and the sinusoidal terms measure departures from the overall trend behavior.

Typical thicknesses of the lithosphere under the oceans vary from 0.5 km at midocean ridges to about 100 km in subduction zones. Across continents, crustal thickness increases near mountain ranges. For example, the crustal thickness varies from 70 km at the Pamir Mountain range of Eurasia to

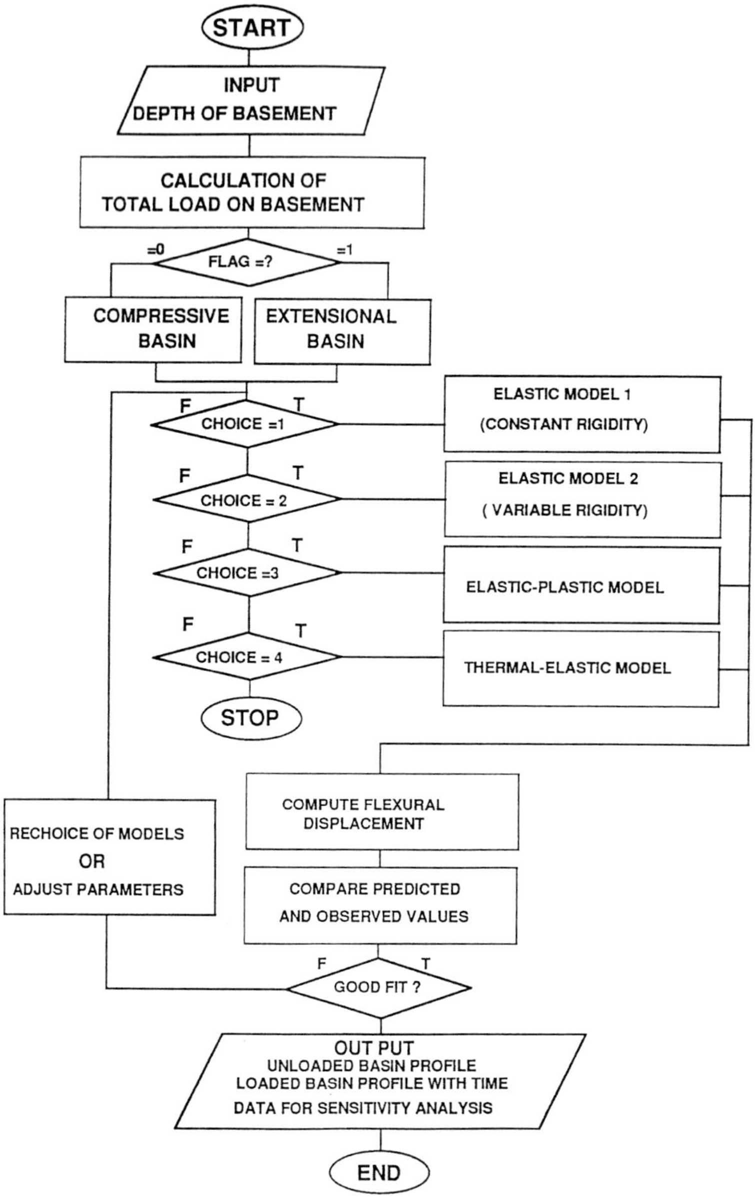


FIGURE 2.4 Flowchart illustrating the inversion technique.

less than 20 km in the back-arc basin of Japan. In the model, $\beta$ is treated as a coefficient determining the slope of the variation in basement rigidity, while the sinusoidally modulated coefficients $\beta_n$ are treated as describing local variations of the rigidity caused by geological events.

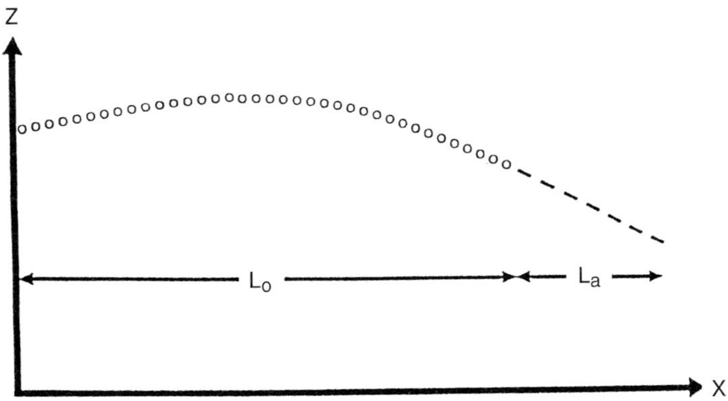


FIGURE 2.5 Representation of the observed length of the plate and the added length (from Cogan *et al.*, 1989).

## B. THERMO-ELASTIC MODELS

Here there is a connection between rigidity and temperature. Because plate displacement implies relocation to a different temperature regime, it is often the case that the local plate displacement is used in place of the temperature (on the assumption that temperature increases roughly linearly with increasing depth). For two model behaviors (Turcotte and Schubert, 1982) of an inverse power law and an exponential, the rigidity, $D$, and the plate displacement, $w$, are taken to be connected by the rules

$$D(x) = D_0(1 + w(x)/w_0)^{-n} \tag{2.5}$$

and

$$D(x) = D_0 \exp(-w(x)/w_0), \tag{2.6}$$

where, $D_0$, $w_0$, and $n$ are positive parameters to be determined by minimizing the mismatch between predicted and observed loaded basement positions.

## C. PLASTIC-ELASTIC MODELS

In this case the model behavior is taken from the curvature dependence of the flexural plate (Turcotte and Schubert, 1982). Inspection of the equations (Appendix B) governing plate flexure indicates a plastic curvature term governed by the dependence $(1 - M/M_*)^{-1/2}$ where $M(x)$ is the plate curvature at any point and $M_*$ is the scaling value at which the plate curves infinitely rapidly (the plastic limit). In the plastic-elastic model the problem is the determination of the parameter $M_*$. The inverse procedure given in Appendix A at the end of the chapter is used to determine the value of $M_*$, which minimizes discord between observed and predicted basement positions including sedimentary cover.

### D. NONPLATE PARAMETERS

The internal plate parameters and the sedimentary load are not the only factors that provide contributions to plate flexure. A compressive (or extensive) force, externally imposed and which may vary with both time and spatial location along the plate, will also influence plate motion. As shown in Appendix B, when the spatial dependence of this external stress component is written

$$P(x) = P_0 \exp\left(\gamma x + \sum_{n=1}^{N} \gamma_n \sin(n\pi x/L)\right), \tag{2.7}$$

then determination of the parameter values $P_0, \gamma, \gamma_1, \ldots, \gamma_N$ is required. This external stress parameter in the flexural model is able to provide an estimate of the tectonic subsidence or uplift throughout a basin related to compressive thrusting or extensional rifting (Tang *et al.*, 1990).

The upthrust, due to *relative* buoyancy of the plate with respect to the mantle on which the plate rests, can also be spatially variable. This effect is modeled by a density difference, $\Delta\rho$, written in the form

$$\Delta\rho(x) = \Delta\rho_0 \exp\left(\delta x + \sum_{n=1}^{N} \delta_n \sin(n\pi x/L)\right). \tag{2.8}$$

Such an effect can occur at a suture zone between oceanic and continental lithosphere, or as a consequence of mineralogy variations along the plate or in the mantle, or as a consequence of a variable plate thickness, etc. This spatially variable density allows simulation of the effects of subplate tectonic variations on basement deformation.

## III. NUMERICAL CONSIDERATIONS

### A. NUMBER OF LOADING BLOCKS

In elastic models 1 and 2 and in the plastic-elastic and thermo-elastic models the observed sedimentary load overlying the basement is used as input to calculate a theoretical basement profile. The observed total load

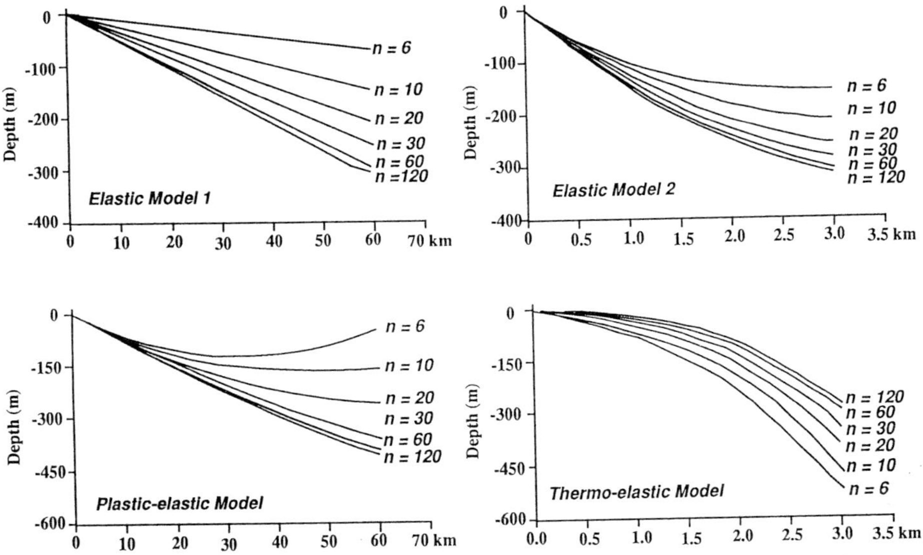


FIGURE 2.6 Effect of loading block size variations on the calculated subsidence of the synthetic basements. The total load on the basement is the same in each case ($n$ = number of loading blocks). The ordinates (in meters) give the displacement of the plate, the abscissas (in kilometers) give the distance along the plate from the hinge point.

along the plate is divided into a series of blocks of equal width, $\Delta x$ (Fig. 2.3). The smaller the width of each loading block, the more precise the calculation of the basement profile; however, more computer time is then needed. Tests for $\Delta x$ values were conducted with different numbers of loading blocks, ranging through 6, 10, 20, 30, and 60 to 120 along the plate. The length of the plate was set ot 60 km for both elastic model 1 and the plastic-elastic model, whereas for elastic model 2 and the thermo-elastic model the length of the plate was 3 km. There is no significant change in the synthetic basement profiles produced when the number of loading blocks is higher than about 60 for all models (Fig. 2.6). The results also demonstrate that the models are considerably more sensitive to the number of loading blocks than to the length scale of the plate. A finely divided load spacing is important for calculation of an accurate theoretical basement profile. Such numerical test results are in accord with the exact analytic results for simple cases presented in Cogan *et al.* (1989).

## B. ACCURACY AND STABILITY

The deformation of the basement depends both on plate parameters (rigidity of plate, sediment/thrust load, plate curvature, plate angle) and

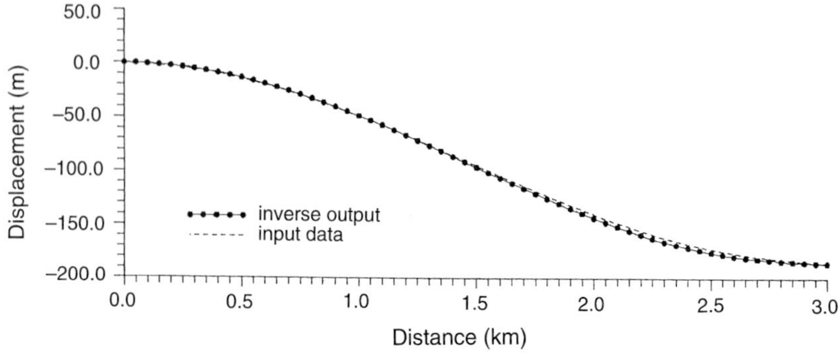


FIGURE 2.7 Comparison between input plate profile and inverse calculated profile in elastic model 2. Mismatch = 2.1 m.

on nonplate parameters (compressive or extensive stress and variations in density along the plate). The problem is to determine the values of these parameters so as to match the available observations. In the flexural models there are many parameters, so a systematic inverse method (Cogan *et al.*, 1989; Tang and Lerche, 1992) is used to determine them simultaneously. A more detailed description of the method can be found in He and Lerche (1989). We take elastic model 2 as an example to illustrate the inverse procedure. First we simulated a synthetic basement profile by running the model in a forward fashion with a loading scheme and parameters specifically chosen. The model was then run in an inverse fashion to recover the parameters. The synthetic basement profile produced by forward modeling is 3 km long, divided into 60 loading blocks with a spacing of 50 m between blocks. After running the inverse procedures with 10 nonlinear iterations, a basement geometry was obtained (Fig. 2.7) determined by a combination of the parameter values (Table 2.2). The mismatch between forward-modeled basement and the calculated profiles was less than 1.0%, that is, the average mismatch is about 2 m. The synthetic test demonstrates that the inverse method is successful in obtaining the input parameter values

TABLE 2.2 Comparison of Input Parameters from Forward Model to Output Parameters from Inversion for Elastic Model 2

| | Input (forward) | Output (inverse) |
|---|---|---|
| Rigidity (N · m) | 10E19.0 | 10E18.7 |
| Lateral stress ($N \cdot m^{-1}$) | 10E13.0 | 10E12.6 |
| Bending moment (N) | 10E15.0 | 10E14.7 |
| Dip tangent | 0.0 | 0.03 |

(Table 2.2) and also produces a basement profile in close accord with observations.

## C. SENSITIVITY ANALYSIS OF PARAMETERS

After the best parameter values have been determined by the inverse method, one parameter is varied linearly around its best value while the other parameters are held fixed at their best values. At each increment of the parameter the MSR is calculated between predicted and observed basement, and a curve of parameter value versus MSR is generated. If the curve is relatively flat, small departures from the minimum do not affect the fit to any great extent; the value of the parameter can lie within a broad range and produce essentially the same final result. If the curve is steep, and there is a sharply defined minimum, small departures from the minimum will cause a significant change in the MSR. The value of the parameter is then tightly constrained and can be assigned a sharply determined value with greater confidence. Similarly, if two parameters are allowed to vary around their best values (while the remaining parameters are held fixed at their best values) then a contour map of MSR will show how sensitive the system is to the two parameters as well as showing any relationship between the two parameters. Random "noise" can also be added to test the ability of the inverse method to recover parameter values. Such tests indicate a high degree of stability of the inverse method (Cogan *et al.*, 1989; Tang *et al.*, 1992).

We now examine why the inverse procedure may not recover the original values of the parameters but, instead, gives values close to the input values (Table 2.2). We use elastic model 2 to illustrate the point. Figure 2.8 is a map of iso-log (MSR) values, representing the mismatch between calculated and synthetic basement profiles, showing the existence of several minima. For example, one minimum occurs at a rigidity of $10^{19}$ N · m and dip tangent = 0.0, which is the forward model result. Another minimum occurs at a rigidity of $10^{18.7}$ N · m and a range ($\pm 0.03$) of dip tangent. Figure 2.9 shows iso-log MSR values when lateral stress and dip tangent are varied. Note the presence of multiple minima in this case, as well. Thus, because of both a finite number of iterations (we used 10 iterations) and the interactions of the parameters, the inversion did not recover precisely the known values. Instead a combination of parameter values was determined (Table 2.2) that gave a "best fit" between calculated and synthetic basement profiles.

If the number of iterations is increased and/or the initial estimates of parameters are changed, a best fit between predictions and observations can be obtained without any error for the synthetic test. For real case histories, because of noise in the geological data and uneven sampling distribution of the data, predictions from the inverse method will mis-

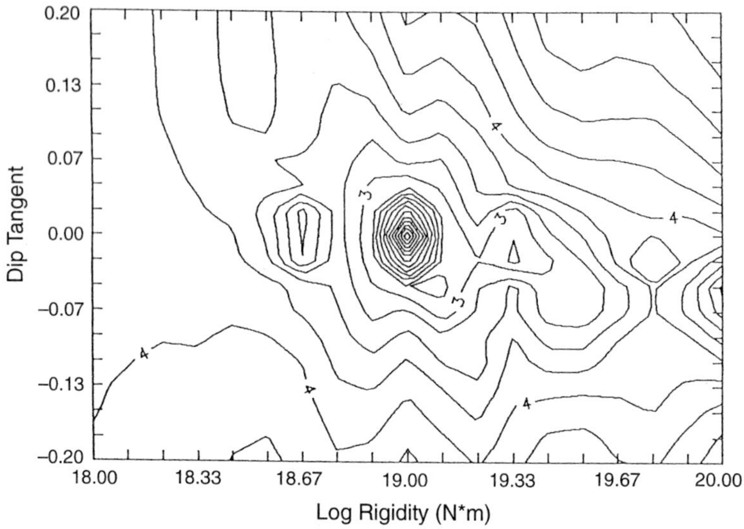


FIGURE 2.8 Contour map of log MSR between calculated and input basement profiles for variations of two parameters (rigidity and dip tangent).

match against the observations. The inverse method guarantees to provide a closer approximation to the observations at each iteration, whereas sensitivity analysis provides information on the degree of resolution of each parameter.

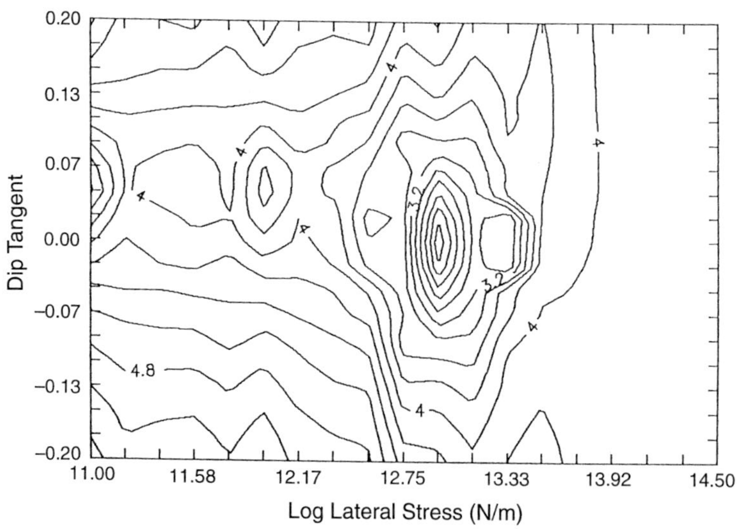


FIGURE 2.9 Contour map of log MSR between calculated and input basement profiles for variations of lateral stress and dip tangent.

# IV. APPLICATION TO THE CENTRAL CHUKCHI SEA BASIN, ALASKA

The modeling results depend on two aspects: (1) assumptions of the model and (2) input data. If one model yields a fit to measured data more accurately than do the other models, then the basin evolution may have been dominated by the tectonic activity incorporated in that particular model. If the results from different models give roughly equal fits to observations, then there is no special reason (without independent evidence) to favor any one model. This fact is of importance in the example presented here (see later). The flexural model is used here to simulate the *basement* evolution of the Central Chukchi Basin. A quantitative basin analysis for the sediments has been given elsewhere (Cao *et al.,* 1993).

## A. REGIONAL GEOLOGY

The Central Chukchi Basin is located in the Chukchi Sea planning area, which covers nearly 49,000 square miles of the northwestern Alaska Continental shelf. To the west, the Central Chukchi Basin is bounded by the Chukchi Platform, a structurally positive block during deposition of the Cretaceous strata (Fig. 2.10). The Herald Arch, in the southwestern part of the Central Chukchi Basin, was a structurally positive area from Early Mississippian to Early Cretaceous time and was actively uplifted in the Early Cretaceous (Thurston and Theiss, 1987). The Paleozoic structural trend is in a north–south direction, whereas the Early Cretaceous to Tertiary structural trend is northwest–southeast (Thurston and Theiss, 1987, their Figs. 22 and 23).

The tectonic evolution of the Central Chukchi Basin and surrounding area has been divided into three stages (Grantz *et al.,* 1975, 1979; Grantz and May, 1983; Coakley and Watts, 1991): (1) Cambrian to Devonian eugeosynclinal and miogeosynclinal facies; (2) Permian to latest Jurassic cratonic facies; and (3) Cretaceous flysch facies and Tertiary molasse facies. Figure 2.11 shows the stratigraphic framework of northern Alaska, which has been correlated and extrapolated to seismic data of the Central Chukchi Basin by Thurston and Theiss (1987). Three regional stratigraphic sequences have been distinguished: (1) the Franklinian sequence (Precambrian to Middle Devonian); (2) the Ellesmerian sequence (Late Devonian to Late Jurassic); and (3) the Brooks sequence (latest Jurassic to present day), corresponding to the tectonic evolution of the Central Chukchi Basin.

## B. MODELING

Four seismic cross-sections have been used to investigate the basement motion of the Central Chukchi Basin (lines 1, 2, 4, and 5, Fig. 2.10). By

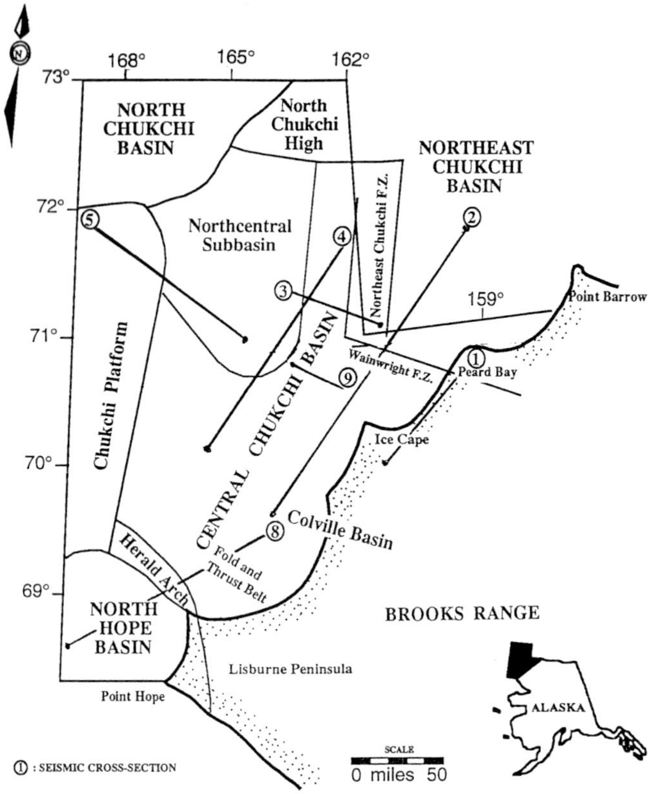


FIGURE 2.10 Study area map showing major structural provinces and sedimentary basins of the Chukchi Sea planning area (after Thurston and Theiss, 1987).

adding in three other seismic cross-sections (Lines 3, 8, and 9, Fig. 2.10), a generalized isopach map of the Upper Ellesmerian sediments in the Central Chukchi Basin was constructed as shown later in Fig. 2.13. The seismic sections were interpreted by Thurston and Theiss (1987), as were seismic sequences descriptions and depth conversions. From the point of view of the modeling, the Lower Ellesmerian sequence has been defined as basement. Theoretically, the hinge point represents the fixed end of the flexural basement and all displacements are calculated relative to this point. The hinge points were chosen approximately where the PU unconformity reflector appears at the shallowest position of each seismic section (Fig. 2.12). The high points of lines 1, 2, and 4 were chosen to be at the northeastern margin of the Central Chukchi Basin (Fig. 2.13), (in the Wainwright Fault Zone for line 1; near the northeast Chukchi Basin for line 2, and near the northeast Chukchi Fault Zone for line 4, Fig. 2.10). For line 5 the hinge point was chosen at the east flank of the Chukchi Platform (Fig.

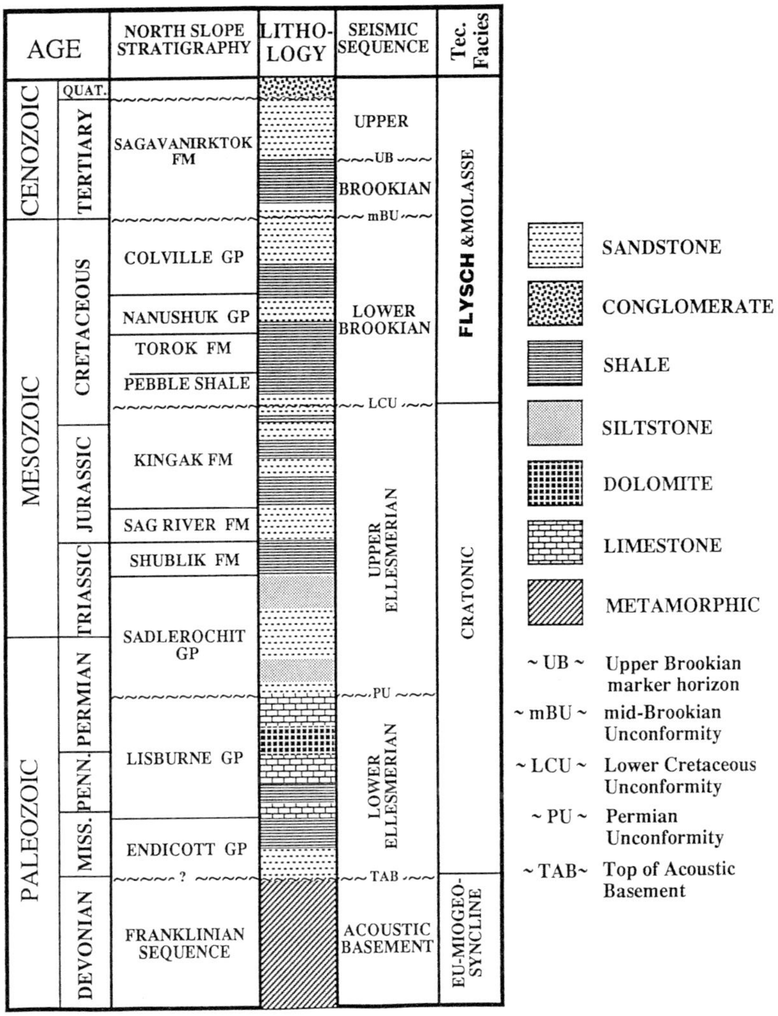


FIGURE 2.11 Stratigraphy of northern Alaska and seismic sequences recognized in the Chukchi Sea planning area (after Thurston and Theiss, 1987).

2.12). The sediment cover overlying basement for the seismic sections was divided into 85 blocks on line 1, 152 on line 2, 103 on line 4, and 54 on line 5, with 1-mile block width. Figure 2.13 provides the generalized isopach map.

The inverse procedure was used to determine the best flexural parameters for the different sections. The results with the best fit between observations and model calculations came from the first group of flexural parameters for the four seismic sections (Table 2.3). Mismatch (Fig. 2.14),

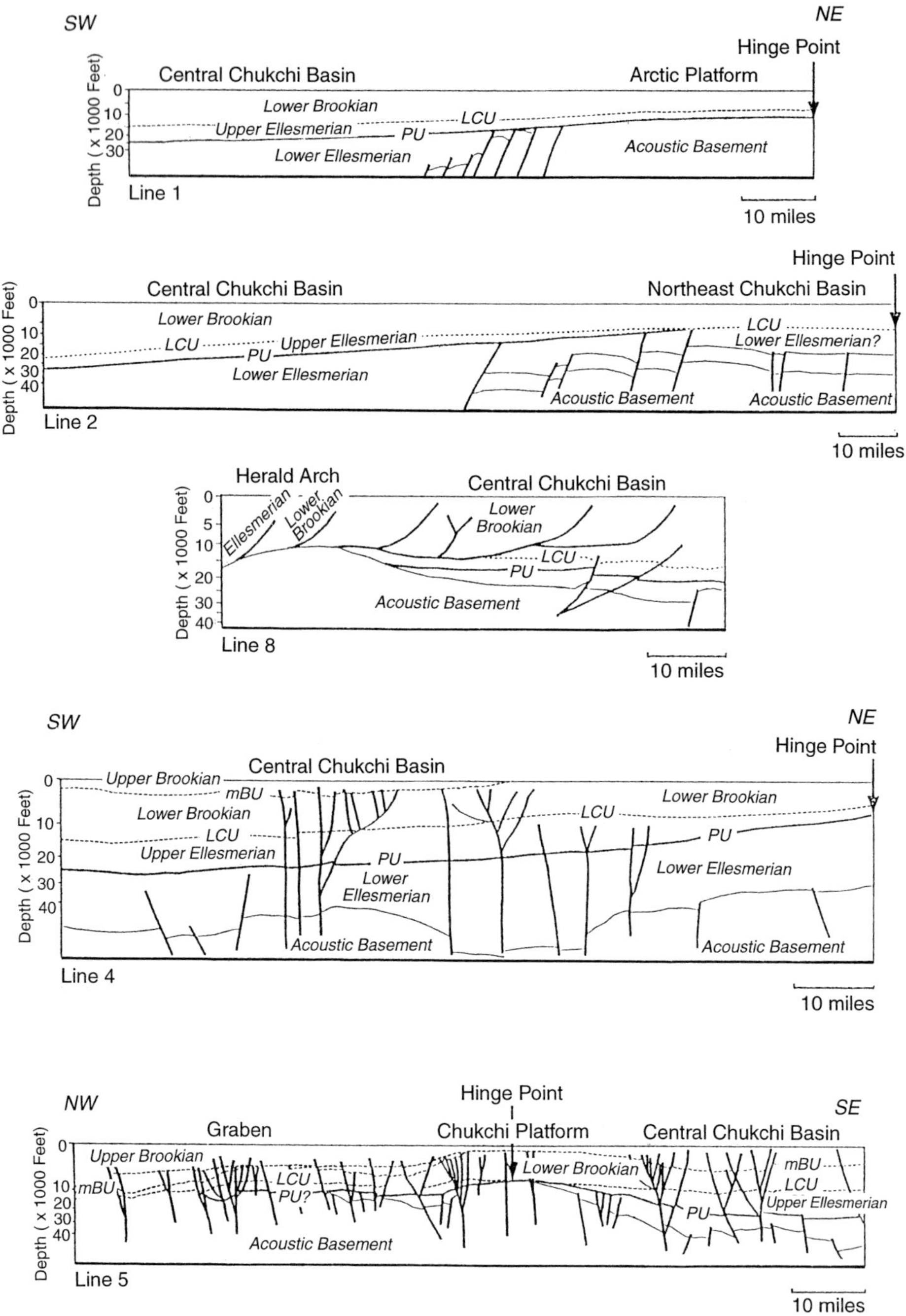


FIGURE 2.12 Seismic-reflection profiles used to model basement motion (for location of the profiles see Fig. 2.10): lines 1, 2, and 8; and lines 4 and 5 (modified from Thurston and Theiss, 1987).

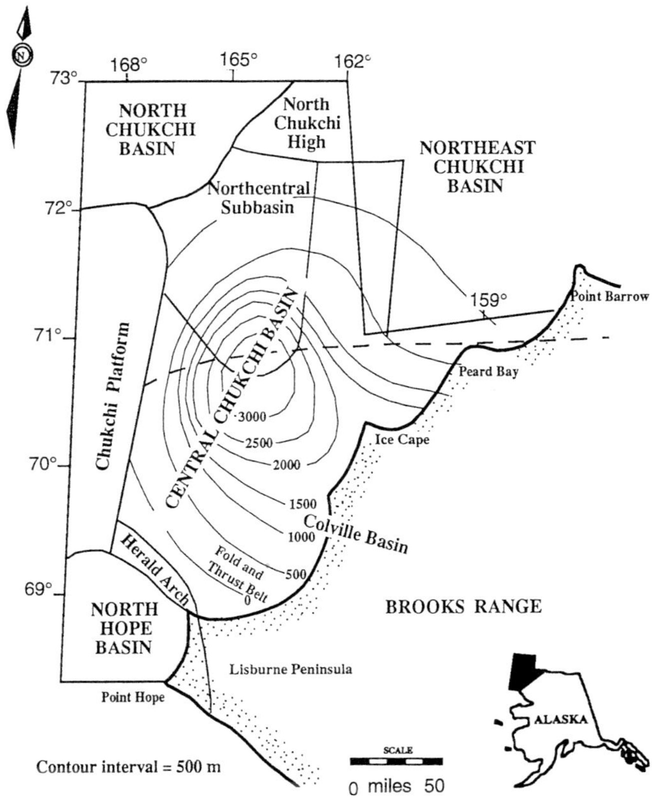


FIGURE 2.13 Generalized isopach map of Upper Ellesmerian sediment distribution in the Central Chukchi Basin.

magnitude of the flexural parameters determined by inversion (Table 2.3), and sensitivity analysis results (Fig. 2.15) from all three models are roughly equal in degree of fit to the data, so that basement deformation of the Central Chukchi Basin may be due to pure elastic flexure, flexure involving plastic bending, or thermal effects. There is no reason to prefer the results of one model over the others if *only* the degree of fit between observations and calculations is considered. Sediment cover is now removed from the basement to reconstruct the unloaded basement geometry for the three different models (Fig. 2.16).

## C. INITIAL BASEMENT PATTERNS

Figure 2.17 provides isopach maps derived from the three models showing the initial depth contours before the Upper Ellesmerian sediments

TABLE 2.3 Best Inverse Method Parameters for Each Model

| Model | Parameter | Seismic Cross-Section 1 | 2 | 4 | 5 |
|---|---|---|---|---|---|
| Elastic | Rigidity (N · m) | $10^{22.9}$ | $10^{23.6}$ | $10^{23.9}$ | $10^{23.9}$ |
| | Lateral stress (N/m) | $10^{14}$ | $10^{14}$ | $10^{14.3}$ | $<10^{12}$ |
| | Density difference ($kg \cdot m^{-3}$) | 1600 | 1700 | 1600 | 1600 |
| | Bending moment (N) | $10^{17}$ | $10^{17.2}$ | $10^{15.6}$ | $10^{15.7}$ |
| | Initial load ($N \cdot m^{-1}$) | $10^{11}$ | $10^{11}$ | $10^{12}$ | $10^{10}$ |
| | Initial angle (rad) | −0.0001 | −0.003 | −0.05 | −0.02 |
| Plastic-Elastic | Rigidity (N · m) | $10^{23.1}$ | $10^{23.5}$ | $10^{23.4}$ | $10^{23}$ |
| | Lateral stress (N/m) | $10^{14}$ | $10^{14}$ | $10^{14.2}$ | $<10^{12}$ |
| | Density difference ($kg \cdot m^{-3}$) | 1800 | 1600 | 1600 | 1700 |
| | Bending moment (N) | $10^{17.2}$ | $10^{17.2}$ | $10^{17.2}$ | $10^{16.9}$ |
| | Initial load ($N \cdot m^{-1}$) | $10^{11}$ | $10^{11}$ | $10^{11}$ | $10^{10}$ |
| | Initial angle (rad) | 0.01 | 0.01 | −0.05 | −0.002 |
| Thermo-Elastic | Rigidity (N · m) | $10^{23.1}$ | $10^{23.5}$ | $10^{23.6}$ | $10^{22.7}$ |
| | Lateral stress (N/m) | $10^{14}$ | $10^{14}$ | $10^{14.2}$ | $<10^{11}$ |
| | Density difference ($kg \cdot m^{-3}$) | 1700 | 1800 | 1600 | 1600 |
| | Bending moment (N) | $10^{17.2}$ | $10^{17.2}$ | $10^{17.2}$ | $10^{16.8}$ |
| | Initial load ($N \cdot m^{-1}$) | $10^{11}$ | $10^{10}$ | $10^{10}$ | $10^{10}$ |
| | Initial angle (rad) | 0.0 | 0.015 | −0.05 | −0.009 |
| | Thermal gradient (°C/100 m) | 3.0 | 3.6 | 3.0 | 2.2 |

began to fill the basin. A deep subsidence center is well defined, located on line 4, with a shallowing toward the margins. However, the subsidence contour trends from the pure elastic model (Fig. 2.17c) are different from those of the plastic-elastic and thermo-elastic models (Figs. 2.17a and 2.17b). The former shows a broad northeast-trending, trough-shaped pattern, whereas the unloaded basin geometries constructed from the other two models both yield a northwest trend. Comparing the basement geometries with the trends of the Paleozoic fault system (Thurston and Theiss, 1987, their Fig. 23), the unloaded basin simulated from the pure elastic model generally parallels the trends of the fault system.

Note that the unloaded basin geometry from the pure elastic model (Fig. 2.17c) is consistent with the thickness distribution of the Upper Ellesmerian sediments (Fig. 2.13). In contrast, the unloaded basin geometries from the other two models (Figs. 2.17a and 2.17b) suggest that there was a mountain range near the south end of lines 2 and 4 in Late Permian time. A difficulty then arises because the reconstructed unloaded geometries from the two models do not conform to either the Paleozoic fault distribution (Thurston

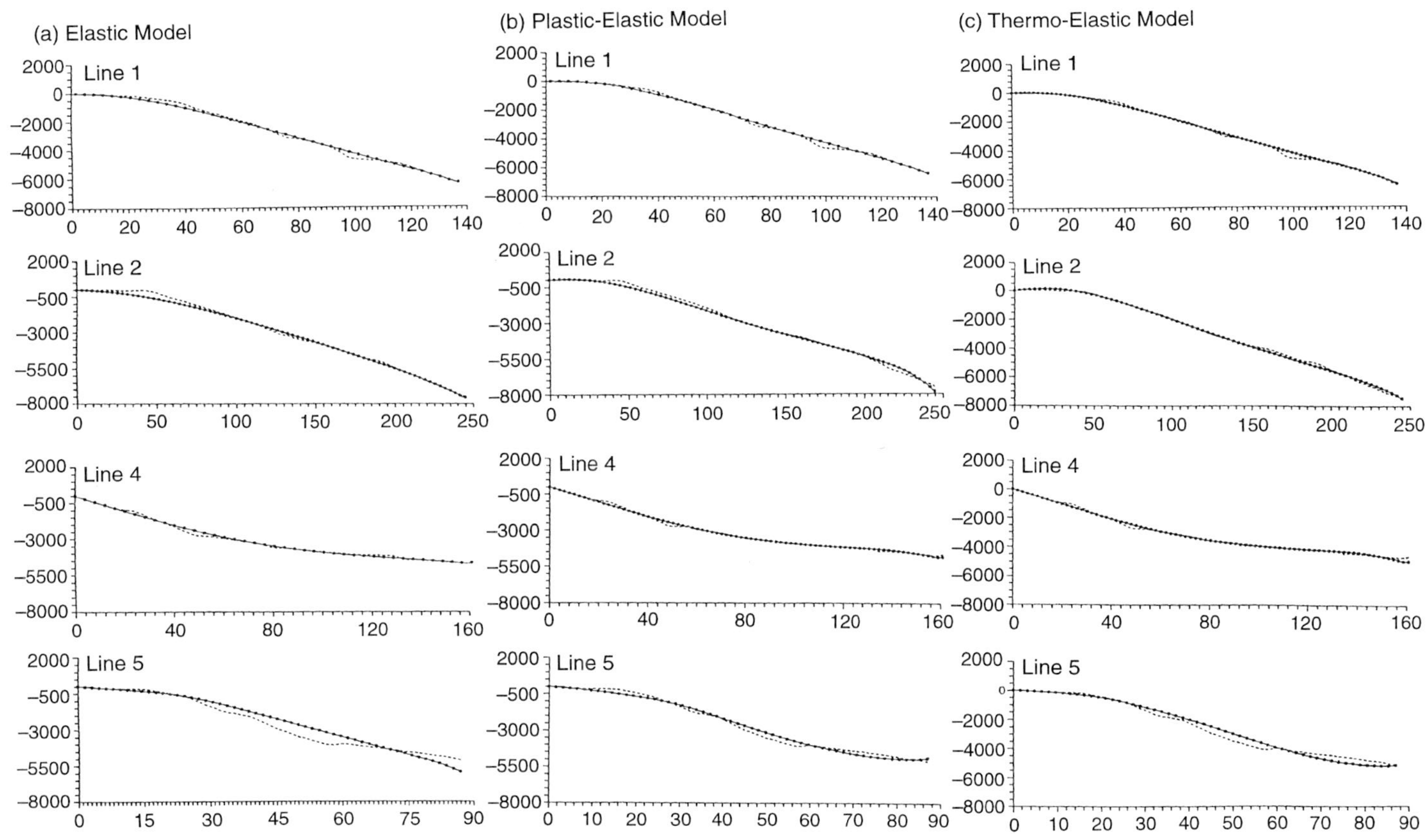


FIGURE 2.14 Comparison between observed basement geometries (dashed line) and modeled basement profiles (dotted line) for four seismic sections. Ordinates in meters; abscissas in kilometers.

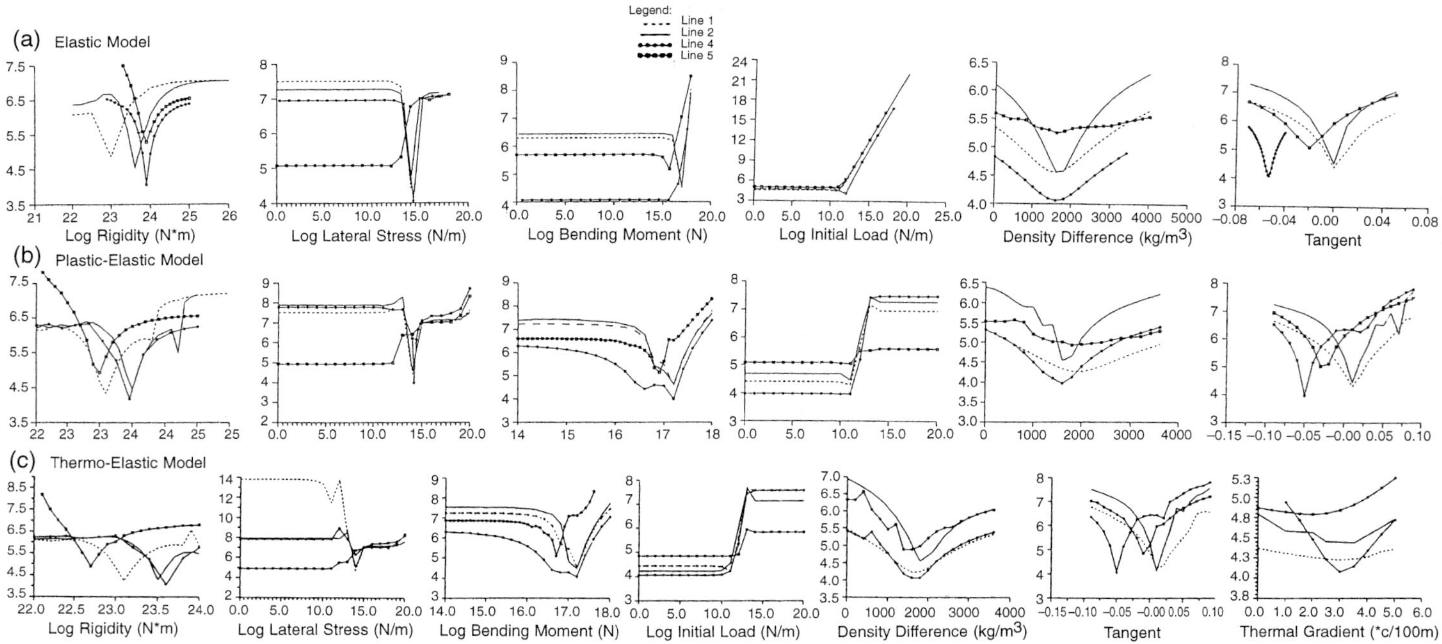


FIGURE 2.15 Sensitivity of basement subsidence for the Central Chukchi Basin to variations of parameters (vertical scale, logarithmic units for MSR).

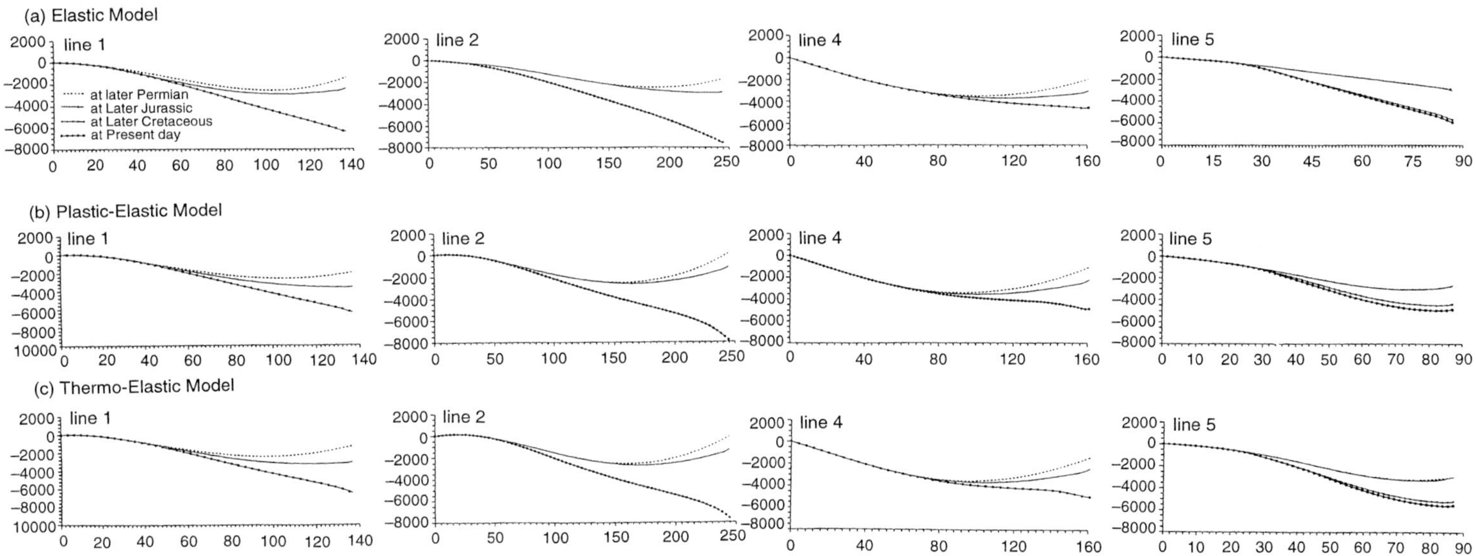


FIGURE 2.16 Two-dimensional basement evolution of the Central Chukchi Basin from Late Permian to present day, reconstructed from four seismic sections: (a) elastic model; (b) plastic-elastic model; (c) thermo-elastic model. Ordinates in meters; abscissas in kilometers.

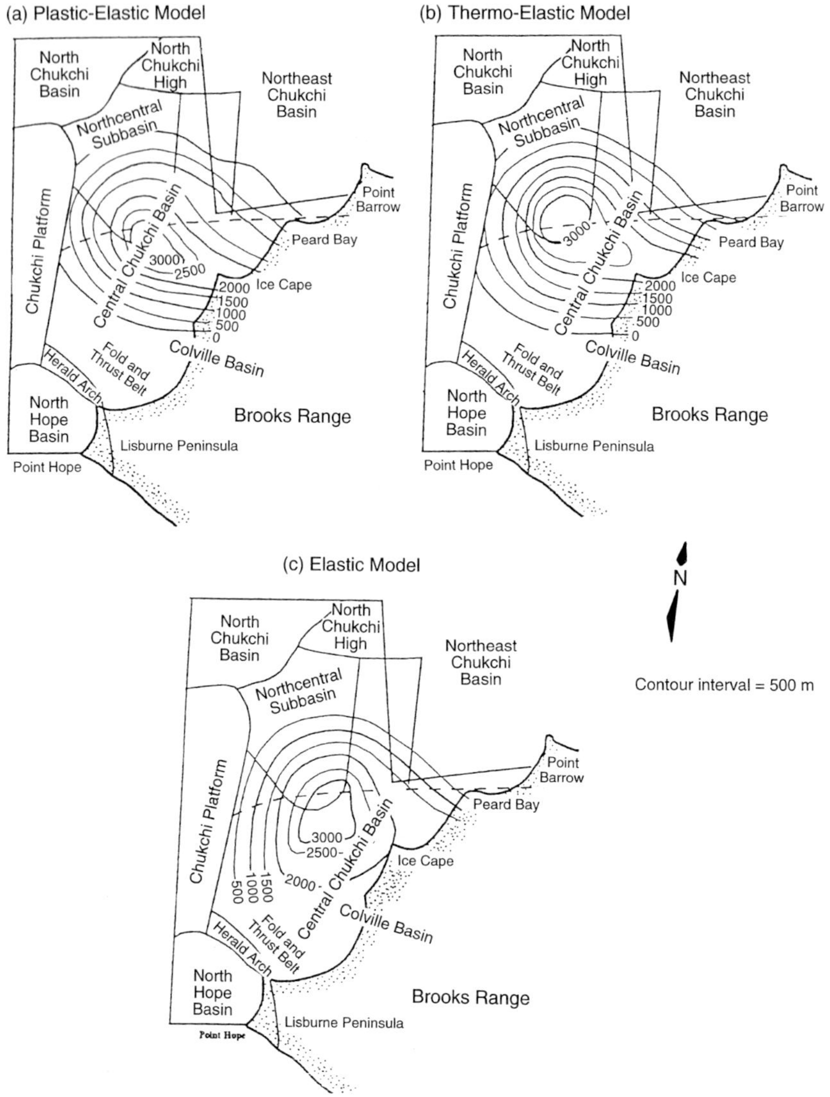


FIGURE 2.17 Initial depth contours of basement in Late Permian times before the Upper Ellesmerian sediments began to fill the Central Chukchi Basin: (a) plastic-elastic model; (b) thermo-elastic model; (c) elastic model. The dashed line represents the northern limit of the Colville Basin according to Thurston and Theiss (1987).

and Theiss, 1987, their Fig. 23) or the isopach map of the Upper Ellesmerian sediments (Fig. 2.13) in the Central Chukchi Basin. Therefore, the pure elastic flexural model results better characterize the evolution of the Central Chukchi Basin, not because of a better degree of fit to the observed data, but because the model results conform better to regional geological data.

The origin of the unloaded basin geometry can be studied by sensitivity analysis of flexural parameters (Fig. 2.15). For example, without the lateral stress parameter the mismatches between observations and calculations are larger than 700 m for the present-day observed flexural basement profiles. On the other hand, the lateral stress value is sensitive to the observed basement profiles of lines 1, 2, and 4, but is not sensitive to that from line 5, suggesting that a compressive stress was applied to the south margin of the basin, while the west and east margins of the basin were under extensional stress. Structural characteristics of the Chukchi Sea planning area indicate that the compressive stress came from the Herald Arch region because this high structural element existed to the southwestern margin of the Central Chukchi Basin during Mississippian to Early Permian. The structural activity and development allowed compressive stress to be applied to the southwest margin of the basin, leading to the raising of the basement at the margin (Fig. 2.16). Evidence from seismic data shows that the extensional force in the west–east direction was active before Permian time. On the basis of the basement configuration, the Chukchi Sea planning area can be divided into three structural elements in the west to east direction: (1) the west flank of the Chukchi Platform, characterized by southeast dipping normal faults; (2) the structural high Chukchi Platform; and (3) the Central Chukchi Basin on the east flank of the Chukchi Platform (Fig. 2.12b, line 5). The diapir evolution model of Thurston and Lothamer (1991, their Fig. 7) indicates that extensional faults were active at the beginning of the Late Mississippian and lasted to the Tertiary.

Rigidity, a sensitive flexural parameter, is determined to be in the range from $10^{23}$ to $10^{24}$ N·m for the four seismic sections (Table 2.3). For a Young's modulus of $E = 10^{11}$ Pa and a Poisson's ratio of $\nu = 0.25$, the flexural rigidity is equivalent to an elastic thickness greater than about 20 km. The regional variation in flexural rigidity may represent thickness changes of the lithosphere. The bending moment curves from Fig. 2.15 show well-defined minima for lines 1 and 2, unlike lines 4 and 5.

From the seismic data, together with numerical results and ranges of the elastic model flexural parameters, we conclude the following: (1) The coincidence of late Permian basement subsidence contour trends (Fig. 2.17c) with Permian fault trends indicates that, to first order, the fault system was controlled by both compressive stress in a north–south direction and extensional stresses in a west–east direction. These stresses are probably the key factor in development of the Upper Permian basement geometry. (2) The similarity between basement geometry (Fig. 2.17c) and the thickness distribution of the Upper Ellesmerian sediments (Fig. 2.13) suggests that basement tectonic subsidence is dominated by flexural bending due to sediment loading.

## APPENDIX A: ELASTIC MODEL WITH CONSTANT PLATE RIGIDITY

The flexural response of a uniform elastic lithosphere at a position $x$ to an applied end lateral stress and sediment load is given by (Turcotte and Schubert, 1982)

$$D\frac{d^4w}{dx^4} + p\frac{d^2w}{dx^2} + (\rho_m - \rho_o)gw = q_1(x), \tag{A.1}$$

where $w$ is the displacement of the lithosphere and $D$ the flexural rigidity $[D = Eh^3/12(1-\nu^2)]$, with $E$ being Young's modulus, $h$ the plate thickness, $\nu$ is Poisson's ratio, $\rho_m$ and $\rho_o$ the densities of the mantle and the overlying material, respectively, $g$ the acceleration due to gravity, $q_1(x)$ the applied sedimentary load, and $P$ the lateral stress applied to the end of the plate (Fig. 2.1). The displacement of the plate depends on six parameters: the rigidity of the plate, $D$; the initial bending moment, $M_0$; the initial load, $V_0$; the dip angle of the plate at the hinge point, $\theta$; the length of the plate, $L$; and the lateral stress, $P$, applied to the end of the plate.

The analytical solution of Eq. (A.1) in the absence of $q_1(x)$ is:

$$w = c_1e^{\delta_1 x} + c_2e^{\delta_2 x} + c_3e^{\delta_3 x} + c_4e^{\delta_4 x}, \tag{A.2}$$

where

$$\delta_1 = \{-P + [P^2 - 4gD(\rho_m - \rho_o)]^{1/2}\}^{1/2}(2D)^{-1/2},$$
$$\delta_2 = \{-P - [P^2 - 4gD(\rho_m - \rho_o)]^{1/2}\}^{1/2}(2D)^{-1/2},$$
$$\delta_3 = -\delta_1, \qquad \delta_4 = -\delta_2.$$

The constants $c_1, c_2, c_3$, and $c_4$ are determined by four boundary conditions. Because there is no displacement at $x = 0$, then

$$c_1 + c_2 + c_3 + c_4 = 0, \tag{A.3}$$

while the initial angle on $x = 0$ is $\theta$ so that

$$c_1\delta_1 + c_2\delta_2 + c_3\delta_3 + c_4\delta_4 = \tan\theta. \tag{A.4}$$

On $x = L$ (at the end of the plate), the bending moment $M = -Dd^2w/dx^2$ is given by

$$M = -D[c_1\delta_1^2e^{\delta_1 L} + c_2\delta_2^2e^{2\delta_2 L} + c_3\delta_2^2e^{\delta_3 L} + c_4\delta_4^2e^{\delta_4 L}]; \tag{A.5}$$

while the end point load $V_0 = -dM/dx - P\,dw/dx$ on $x = L$ is given by

$$V_0 = c_1e^{\delta_1 L}(D\delta_1^3 - P\delta_1) + c_2e^{\delta_2 L}(D\delta_2^3 - P\delta_2) + c_3e^{\delta_3 L}(D\delta_3^3 - P\delta_3) + c_4e^{\delta_4 L}(D\delta_4^3 - P\delta_4). \tag{A.6}$$

Equations (A.3) through (A.6) provide expressions for $c_1$, $c_2$, $c_3$, and $c_4$, and then the deflection of lithosphere is given by Eq. (A.2). If all six parameters are known, vertical displacement as a function of horizontal distance from the hinge point can be calculated and a theoretical basement profile plotted. However, the values of the flexural rigidity $D$, lateral stress $P$, bending moment $M_0$, initial end point load $V_0$, dip angle $\theta$, and the length of the plate $L$ have to be inferred from present-day geological or geophysical data. Denote by $\mathbf{p}$ the vector of all parameters describing the system. For a given choice of vector $\mathbf{p}$, the predicted behavior, $w_M(x, \mathbf{p})$ will be determined at present day. However direct measurements are available for $w(x_i)$ ($i = 1, \ldots, R$), where the vertical coordinate reference is the same as for the forward run model calculation. Clearly, with arbitrary choices of the vector $\mathbf{p}$ there is no reason why model predictions $w_M(x_i; \mathbf{p})$ and observations $w(x_i)$ at each location $x_i$ should be similar. The aim is to use the difference between observed and predicted behaviors to provide a procedure that iteratively corrects the model parameters toward values leading to the least discord between predictions and observations. Such a procedure operates as follows. The parameters in the vector $\mathbf{p}$ have different dimensions. Let the $j$th component of $\mathbf{p}$ have a maximum value $\max_j$ and a minimum value $\min_j$. Then set

$$a_j = (p_j - \min_j)/(\max_j - \min_j) \tag{A.7a}$$

so that $0 \leq a_j \leq 1$. Regard the vector $\mathbf{a}$ as fundamental with $\mathbf{p}$ being a dependent vector given through

$$p_j = \min_j + a_j(\max_j - \min_j). \tag{A.7b}$$

In this way all components of $\mathbf{a}$ are dimensionless and are required to be in the range of zero to unity.

Introduce the least-squares control function:

$$X^2(\mathbf{a}) = R^{-1} \sum_{i=1}^{R} [w_M(x_i, \mathbf{a}) - w(x_i)]^2. \tag{A.8}$$

An iteration scheme that *guarantees* to produce a closer correspondence of predicted and observed bed locations at each iteration and that *guarantees* to keep each component of $a$ in the bounds $0 \leq a_j \leq 1$ is as follows: At the $n$th iteration represent the updated value $a_j(n + 1)$ to $a_j$ by

$$\begin{aligned} a_j(n) &= \sin^2\theta_j(n), \\ \theta_j(n+1) &= \theta_j(n) \exp\left[-\tanh\left\{\alpha_j\delta_j(n)\frac{\partial X^2(\mathbf{a}(n))}{\partial a_j(n)}\right\}\right], \\ a_j(n+1) &= \sin^2\theta_j(n+1), \end{aligned} \tag{A.9a}$$

where

$$\alpha_j = \left|\frac{\partial X^2(\mathbf{a}(0))}{\partial a_j(0)}\right|^{-1} \ln\{1 + (Ma_j(0))^{-1}\}, \tag{A.9b}$$

$$\delta_j(n) = q_j(n) \Big/ \left[ \frac{1}{P} \sum_{p=1}^{P} q_p(n) \right], \tag{A.9c}$$

with

$$q_j(n) = (|a_j(n) - a_j(n-1)|/a_j(n)) + \beta^2, \tag{A.9d}$$

where $M$ is the number of times the nonlinear iteration is to be undertaken, $a_j(0)$ is the initial choice for $a_j$ [with $0 \leq a_j(0) \leq 1$], $P$ is the number of parameters being varied, and $\beta$ is the scale choice for calculating numerically the derivative $\partial X^2/\partial a_j$ (see later). The factor $\delta_j(n)$ chooses the most sensitive parameters for more rapid variation after the first iteration ($n \geq 2$) while for $n = 1$, $\delta_j(1)$ is set to unity.

This procedure guarantees to increase (decrease) $a_j$ accordingly as $\partial X^2/\partial a_j$ is negative (positive), guarantees to keep each component of $a$ in the range $0 \leq a_j \leq 1$, and guarantees to make the largest changes in those parameters that are furthest from satisfying the least-squares control.

Two caveats are (1) the numerical accuracy with which a minimum in least-squares mismatch can be achieved is limited by the accuracy with which the partial derivatives, $\partial X^2/\partial a$, are calculated and (2) the fact that the nonlinear iteration scheme guarantees to find a minimum least-squares mismatch is no guarantee that the procedure has found a global minimum.

The first caveat can be handled to an extent by decreasing the range of a particular component of the vector $a$ used to calculate a derivative. Thus, with the approximation

$$\begin{aligned} \partial X^2/\partial a_i = (\beta a_i)^{-1}\{&X^2(a_1, a_2, \ldots, a_i + \beta a_i, a_{i+1}, \ldots) \\ &- X^2(a_1, a_2, \ldots, a_i, a_{i+1}, \ldots)\} \end{aligned} \tag{A.10}$$

and the default value of $\beta = 0.1$, we can in turn replace $\beta$ by $10^{-2}$, $10^{-3}$, ..., until some preset criterion is reached for accuracy of the derivative calculation. By and large this problem is not severe. As a by-product, if numerical resolution allows $a_i(n)$ to exceed unity, then it is an appropriate strategy to immediately set $a_i(n) = 0.9$, which forces $a_i(n)$ to be in the preset domain $\{0, 1\}$, and which is also a valid value from which to calculate the partial derivative because, then, $a_i + 0.1a_i = 0.99$, which is also in the required domain.

The second caveat is of greater concern because some procedure must be found to decide whether multiple minima exist, and to decide which minimum is the appropriate value to use in parameter determination, from which one constructs the combined paleo-evolution of the basement and associated sedimentary beds. A strategy that has proven useful in other similar problems is as follows.

First make a rough assessment of the minimum and maximum ranges that one anticipates will encompass each parameter. Then do a linear search over pairs of parameter values while holding the remaining parameters at their midpoint values. Two-dimensional plots of mean square mismatch contours in each two-parameter space then provide a rough idea of where minima lie, as well as suggesting how to modify the initial estimated ranges of parameter values in order to improve the ability to focus on the minima uncovered.

Then a multidimensional nonlinear iterative search, as outlined earlier, can be done to improve the degree of fit between predicted and observed behaviors. The point here is that *all* parameters are allowed to vary simultaneously (rather than just two), thereby improving the overall parameter values. A linear search can be redone after the nonlinear iterations, so that we are even more certain that a global minimum in the search ranges specified has been obtained. The pair of intertwined procedures (linear search, nonlinear iteration) can be repeated as many times as desired. The improvement in least-squares mismatch using this type of strategy is considerable, and also helps to ensure that the search procedure does not become "stuck" in a local minimum.

# APPENDIX B: GOVERNING EQUATIONS FOR MODELS

## B.I. THE GOVERNING EQUATIONS FOR ELASTIC FLEXURE WITH VARIABLE PLATE RIGIDITY

For a two-dimensional elastic plate the vertical displacement, $w(x)$, is controlled by the downward load per unit length, $q(x)\,dx$, acting between $x$ and $x + dx$ (where $x$ is the horizontal coordinate), by the bending moment $M(x)$, and by the applied horizontal force $P(x)$ through (Turcotte and Schubert, 1982) the equation

$$\frac{d^2M}{dx^2} = -q(x) + \frac{d}{dx}\left\{P(x)\frac{dw}{dx}\right\}. \tag{B.1}$$

For a plate of local thickness $h(x)$, with a Young's modulus $E(x)$, and a local Poisson's ratio $\nu(x)$, the connection between the bending moment $M(x)$ and the displacement $w(x)$ is given through (Turcotte and Schubert, 1982)

$$M(x) = -D(x)\frac{d^2w}{dx^2}, \tag{B.2}$$

where the rigidity $D(x)$ is given by

$$D(x) = E(x)h(x)^3/[12(1 - \nu(x)^2)]. \tag{B.3}$$

Substituting Eq. (B.2) into Eq. (B.1) gives the equation governing displacement of the plate in the form

$$\frac{d^2}{dx^2}\left[D\frac{d^2w}{dx^2}\right]+\frac{d}{dx}\left[P(x)\frac{dw}{dx}\right]-q(x)=0. \tag{B.4}$$

For lithospheric material of density $\rho_m$, overlain by material of average density $\rho_o$, Archimedes' principle provides an expression for the net downward load per unit length of (Turcotte and Schubert, 1982)

$$q(x) = -(\rho_m - \rho_0)gw, \tag{B.5}$$

where $g$ is the acceleration due to gravity.

Equation (B.5) when substituted in Eq. (B.4) yields

$$\frac{d^2}{dx^2}\left[D\frac{d^2w}{dx^2}\right]+\frac{d}{dx}\left[P(x)\frac{dw}{dx}\right]+(\rho_m-\rho_o)wg=q_1(x), \tag{B.6}$$

where $q_1(x)$ is the applied sedimentary load on the plate.

There is no *a priori* reason to suppose that either the rigidity, $D$, the compressive thrust, $P$, or the anomalous density contrast, $(\rho_m - \rho_o)$, is independent of spatial position along the plate. For instance, for a plate that extends under both oceanic and continental domains the overlying density, $\rho_o$, could range from that of water ($\rho_o \approx 1\ \text{g}\cdot\text{cm}^{-3}$) to that of continental crust material ($\rho_o \approx 2.7$–$3\ \text{g}\cdot\text{cm}^{-3}$) with position along the plate (Turcotte and Schubert, 1982). Equally, for a plate of variable thickness, but constant Young's modulus, the plate rigidity would vary as the cube of the plate thickness, and so on. The response is required of an elastic plate when the "parameters" describing the plate's physical characteristic behavior [namely, $D$, $P$, and $\Delta\rho \equiv (\rho_m - \rho_o)$] are location dependent.

An avenue of approach is available. We can attempt numerically to find the general solution to

$$\frac{d^2}{dx^2}\left[D\frac{d^2w}{dx^2}\right]+\frac{d}{dx}\left[P(x)\frac{dw}{dx}\right]+g\,\Delta\rho(x)w=q_1(x), \tag{B.7}$$

directly with $D$, $P$, and $\Delta\rho$ varying, and then compute the effects of the spatially variable sediment loading.

The boundary conditions on the plate are those employed in the body of Chapter 2. At the hinge point ($x = 0$) the plate displacement is zero, that is,

$$w = 0 \qquad \text{for } x = 0; \tag{B.8a}$$

and the plate hinge angle is $\theta$, that is,

$$\frac{dw}{dx} = \tan\theta \qquad \text{for } x = 0. \tag{B.8b}$$

On the end of the plate, $x = L$, the conditions are that the bending moment $M_L$ is given, that is,

$$M_L = -D\frac{d^2w}{dx^2} \qquad \text{for } x = L; \tag{B.8c}$$

and the line load $V_0$ at the end of the plate is given as follows:

$$\frac{d}{dx}\left(D\frac{d^2w}{dx^2}\right) + P\frac{dw}{dx} = -V_0 \qquad \text{for } x = L. \tag{B.8d}$$

The problem of numerical integration is best handled by constructing the coupled implicit equations

$$w(x) = x\tan\theta - \int_0^x (x - x')M(x')/D(x')\,dx' \tag{B.9a}$$

and

$$\begin{aligned} M(x) = {} & M_0 + xP(x = L)\tan\theta + xR \\ & + \int_0^x P(x')\left[\int_0^{x'} M(x'')/D(x'')\,dx''\right]dx' \\ & + \int_0^x (x - x')[(\rho_m - \rho_o)gw(x') - q_1(x')]\,dx' \end{aligned} \tag{B.9b}$$

where the constant $R$ is related to the end point load $V_0$ through

$$R = V_0 - \int_0^L [(\rho_m - \rho_o)gw(x') - q_1(x')]\,dx', \tag{B.10}$$

where $M_0$ is the bending moment at the hinge point ($x = 0$), and $P(x = L)$ is the value of $P(x)$ at the end of the plate on $x = L$. The bending moment, $M_L$, on the end of the plate is related to $M_0$ through Eq. (B.9b) evaluated on $x = L$:

$$\begin{aligned} M_L = {} & M_0 + L\tan\theta P(x = L) + LR \\ & - \int_0^L P(x)\left[\int_0^x M(x')/D(x')\,dx'\right]dx \\ & + \int_0^L (L - x)[(\rho_m - \rho_0)gw(x) - q_1(x)]\,dx. \end{aligned} \tag{B.11}$$

Equations (B.9a) and (B.9b) are integral representations of Eq. (B.7) and automatically satisfy the boundary conditions. They can also be easily integrated numerically once the general functional forms of $D(x)$, $P(x)$, and $\Delta\rho(x)$ ($\equiv \rho_m - \rho_0$) are specified.

To allow for spatial variations, write $D$, $P$, and $\Delta\rho$ in the following forms:

$$\begin{aligned} D(x) &= D_0\exp\{\beta x + b_1\sin\pi x/L + \beta_2\sin 2\pi x/L + \ldots\}, \\ P(x) &= P_0\exp\{\gamma x + \gamma_1\sin\pi x/L + \gamma_2\sin 2\pi x/L + \ldots\}, \\ \Delta\rho(x) &= \Delta\rho_0\exp\{\delta x + \delta_1\sin\pi x/L + \delta_2\sin 2\pi x/L + \ldots\}, \end{aligned} \tag{B.12}$$

where $D_0$, $P_0$, $\Delta\rho_0$ are scaling constants and $\beta$, $\gamma$, and $\delta$ measure the overall trend of change in elastic parameters along the plate; $\beta_1, \beta_2, \ldots, \gamma_1, \gamma_2, \ldots, \delta_1, \delta_2, \ldots$, measure local scale fluctuations relative to the trend values. The advantages of the forms (B.12) is that they keep $D$, $P$, and $\Delta\rho$ manifestly positive and, by the addition of oscillatory factors, the degree of resolution of each parameter can be determined as well as allowing for a plate to start rigid on $x = 0$, become less (more) rigid in the middle, and more (less) rigid at its end $x = L$, as the need dictates.

## B.II. GOVERNING EQUATIONS FOR ELASTIC-PLASTIC FLEXURE

The general equation, Eq. (B.1), for elastic plate flexure is valid regardless of the constitutive relation between bending moment, $M$, and displacement, $w$, as has been emphasized by Turcotte and Schubert (1982) and Lliboutry (1987). For a plate bent beyond the elastic recovery point, Turcotte and Schubert (1982) proposed that the connection between bending moment and displacement be written in the following form:

$$\frac{d^2w}{dx^2} = -\frac{M_*}{D}(1 - M/M_*)^{-1/2} \qquad \text{in } 2/3M_* \lesssim M \lesssim M_*, \tag{B.13}$$

where $M_*$ is the critical flexure at which the curvature of the plate becomes infinite.

A functional form that combines the low elastic limit ($M \leq 2/3M_*$) behavior described by Eq. (B.2) and the plastic limit, described by Eq. (B.13), is

$$\frac{d^2w}{dx^2} = -\frac{M}{D}(1 - M/M_*)^{-1/2}. \tag{B.14}$$

For an elastic plate under a constant applied horizontal force $P$, Eq. (B.1), in combination with Eq. (B.5), reads

$$\frac{d^2M}{dx^2} = P\frac{d^2w}{dx^2} + (\rho_m - \rho_o)wg - q_1(x). \tag{B.15}$$

For numerical purposes of integration, it is best to use boundary conditions as follows. On the hinge point ($x = 0$) of the elastic-plastic plate we have

$$w = 0, \tag{B.16a}$$

$$dw/dx = \tan\theta, \tag{B.16b}$$

$$M(0) = M_0, \tag{B.16c}$$

while on the end ($x = L$) of the plate we have the general condition

$$\frac{dM}{dx} - P\frac{dw}{dx} = V_0 \qquad x = L,$$

where $V_0$ is the end point load.

Then rewrite Eqs. (B.14) and (B.15) as

$$w = x\tan\theta - D^{-1}\int_0^x M(x')[1 - M(x')/M_*]^{-1/2}(x - x')\,dx' \tag{B.17}$$

and

$$\begin{aligned} M(x) &= M_0 + Sx + Pw(x) \\ &\quad + \int_0^x (x - x')[(\rho_{\rm m} - \rho_{\rm o})gw(x') - q_1(x')]\,dx', \end{aligned} \tag{B.18}$$

with

$$V_0 = S + \int_0^L [(\rho_{\rm m} - \rho_{\rm o})gw(x') - q_1(x')]\,dx' \tag{B.19}$$

and with the requirement $M(x) \le M_*$ at all $L \ge x \ge 0$.

Inspection of Eqs. (B.17) and (B.18) shows that they can be easily integrated numerically when values are specified for the parameters $S$, $\tan\theta$, $P$, $D$, $M_0$, and $M_*$, yielding a solution for the displacement along the plate in this form:

$$w = w(x; D, P, M_0, S, \theta, M_*). \tag{B.20}$$

Determination of the unknown parameters is then a relatively straightforward matter. Observations yield the sediment load $q_1(x)$ at positions along the plate, as well as the observed plate displacements $\{w_i\}$ at a set of positions $x_i$ ($i = 1, \ldots, N$). Then construct

$$X^2(a) = \frac{1}{N}\sum_{i=1}^{N} [w_i - w(x_i, a)]^2, \tag{B.21}$$

where $a$ is the parameter vector $a \equiv (D, P, M_0, S, \tan\theta, M_*)$; and use the nonlinear search procedure given in Appendix A for each component of $a$.

## B.III. GOVERNING EQUATIONS FOR THERMO-ELASTIC FLEXURE

Turcotte and Schubert (1982) and Lliboutry (1987) both point out that the rigidity, $D$, of elastic plates is temperature, $T$, dependent. Various models have been proposed in which the rigidity decreases with increasing temperature, occasionally as an inverse power law, $D\alpha T^{-n}$, where $n$ is a fixed index, occasionally as an exponential, $D\alpha\,\exp(-T/T_{\rm c})$, where $T_{\rm c}$ is a fixed scaling temperature, and occasionally as more complex functions of increasing temperature. The link of temperature to elastic displacement is then conventionally made by assuming a temperature increasing linearly with depth, of gradient $G$, so that $T = T_0 + wG$, where $T_0$ is the temperature at the hinge point of the plate (where $w = 0$). For an exponential behavior of $D$ with temperature we would then have

$$D = D_0 \exp[-w/w_0], \tag{B.22}$$

where $w_0 = T_c/G$, and $D_0$ is the rigidity at the hinge point $w = 0$.

For an inverse power law the equivalent relation is

$$D = D_0(1 + w/w_0)^{-n}. \tag{B.23}$$

In any event, for any prescribed dependence of rigidity on temperature, the equivalence is $D = D(w, \mathbf{p})$ where $\mathbf{p}$ is the parameter vector describing the modeled behavior of $D$ on $w$. Thus $\mathbf{p} = (D_0, w_0)$ for the exponential case, while $\mathbf{p} = (D_0, w_0, n)$ for the inverse power law case. Thus the elastic plate response equation under a constant horizontal force, $P$, takes the form

$$\frac{d^2M}{dx^2} = P\frac{d^2w}{dx^2} + q_1(x) - (\rho_m - \rho_0)gw, \tag{B.24}$$

$$\frac{d^2w}{dx^2} = -\frac{M}{D(w, \mathbf{p})}(1 - M/M_*)^{-1/2}. \tag{B.25}$$

For any prescribed dependence of $D$ on $T$ (or $w$) the procedure for obtaining parameters follows a line similar to Section B.II of this appendix. Written out, implicitly we have

$$w = x \tan\theta - \int_0^x D(w(x'), \mathbf{p})^{-1} M(x')[1 - M(x')/M_*]^{-1/2}(x - x')\, dx', \tag{B.26a}$$

$$M(x) = M_0 + Sx + Pw(x) + \int_0^x (x - x')[(\rho_m - \rho_0)gw(x') - q_1(x')]\, dx', \tag{B.26b}$$

where the boundary conditions on the plate are those employed in Section B.II, as is the symbol notation. Thus we see that $w(x)$ now depends on the parameter vector $a \equiv (P, M_0, M_*, S)$ as well as on $\mathbf{p} \equiv (D_0, w_0)$ for the exponential case; $\mathbf{p} \equiv (D_0, w_0, n)$ for the inverse power law case. The procedure is as before. For $\mathbf{a}$ and $\mathbf{p}$ given, we have a theoretically calculated displacement $w(x; a, p)$. Construct

$$X^2(\mathbf{a}, \mathbf{p}) = \frac{1}{N}\sum_{i=1}^{N} [w_i - w(x_i; \mathbf{a}, \mathbf{p})]^2 \tag{B.27}$$

and follow the nonlinear minimization technique given in Appendix A with respect to the individual components of $\mathbf{a}$ and $\mathbf{p}$.

# 3

# Dynamic Models of Sediment Evolution and Their Uncertainties

## I. Introduction

The dynamic evolution of sediments, and their compaction, is the underpinning keystone on which rests the ability to model thermal history of a basin and also hydrocarbon generation, migration, and accumulation histories. In the last decade or so, numerical frameworks have been constructed that permit data-driven models to be analyzed for fluid flow and compaction in sediments (Yukler, 1979; Welte and Yukler, 1981; Bethke, 1985; Cao *et al.*, 1986; Lerche, 1990a,b). These models, by and large, include auxiliary procedures for handling other phenomena allied to the evolution of fluids and compaction in sediments, such as chemical cementation and dissolution (Bethke, 1985; Cao *et al.*, 1986), faulting and fracturing (Cao *et al.*, 1989; Wei and Lerche, 1988), erosion events (Armagnac, 1985; Armagnac *et al.*, 1989), thermal and hydrocarbon histories, and so on. But the dominant support to analysis of these subsidiary processes is provided by the burial history of the sediments.

Elsewhere (Lerche, 1988a,b; He and Lerche, 1989; Pantano and Lerche, 1990) we have provided a scheme such that multiple thermal indicators, which carry the impressed signature of a thermal record, can be inverted to yield information concerning paleoheat flux variations with time, unconformity events, physical and chemical parameters of the individual thermal indicators, stratigraphic ages, and overthrust events (Kuckelkorn *et al.*, 1990). The essence of the method used to invert present-day downhole thermal indicator data was based on thermal indicator tomography, which is itself an offshoot of seismic tomography.

The tomographic procedure can be generalized to allow the burial history of sediments at single- and multiple-well sites to be subject to a similar procedure. The rationale for wishing to obtain such a scheme is the plethora of seemingly freely adjustable parameters and functions that can enter into basin analysis burial history calculations. For instance, the quasi-empirical relation between permeability, $\kappa$, and void ratio, $e$, for shales is often written in the form

$$\kappa = \kappa_*(e/e_*)^B, \tag{3.1}$$

where $\kappa_*$ is the permeability on $e = e_*$, $e_*$ is the surface void ratio [void ratio, $e$, is related to porosity $\phi$, through $e = \phi/(1 - \phi)$], and $B$ is a number, usually of order $3 \pm 3$ for most shales (Cao *et al.*, 1986; Dutta, 1987). This equation of state for shales therefore contains three freely adjustable parameters. Equally, the equation of state relating frame pressure (effective stress), $P_f$, and void ratio for shales is often written as the quasi-empirical relation

$$P_f = P_{f_*}(e/e_*)^{-A}, \tag{3.2}$$

where, again, $P_{f_*}$ and A are freely adjustable parameters.

Each lithology has its own equations of state relating permeability and frame pressure to void ratio, and each of these equations has its own adjustable parameters. In addition, different equations of state can be included for the same lithology under different conditions. For instance, a shale may have an anisotropic permeability, different parallel and perpendicular to bedding. Thus, from the point of view of using observed data to provide constraints on parameters in the equations of state, the advantages are clear in having available a dynamic inversion scheme. Equally, the burial histories of sediments are influenced by dynamic events that occurred in the past. Thus the development of a fault can have a major impact on the ongoing evolution of the basin. If the throw of the fault is significant, then different lithologic units can be brought into juxtapositional butting, leading to a change in the paths allowed thereafter for migration of fluids. In turn, this change will impact the evolution of the dynamic compaction of the sediments. Likewise, whether a fault is open to hydrodynamic fluid flow or closed to such flow has a significant impact on the loss of fluid and consequent fluid overpressure in the basin, as has been shown for the Pinedale Anticline, Green River Basin, Wyoming, by Wei and Lerche (1988). The development of fracturing and the critical pressure at which fracturing can take place also have roles to play in controlling the evolution of sediment compaction and fluid loss in a basin (Cao and Lerche, 1989). Similarly, the presence or absence of diagenetic cementation and dissolution, the timing of such events, the degree of cementation or dissolution, and the formations subject to such effects all influence the rate, amount, and timing of fluid loss in sediments.

In addition, the presence of erosional events and whether they are depositional hiatae or represent deposition followed by erosion (in which case the thickness and lithology of the eroded units and the timing of such erosion events are the critical unknown parameters influencing the burial history) are of concern in attempting to assess the consequences to sedimentary burial history of such events. We see then two sorts of parameters entering burial history calculations:

1. Parameters associated with equations of state of lithologic units
2. Parameters associated with geologic events

The determination from observational data of the two sorts of parameters is not independent in the sense that one cannot first determine, say, the parameters relevant to the equations of state and then the geological event parameters. The reason is that one set of parameter values impacts the other. Thus, if the shale scaling permeability, $\kappa_*$ is arbitrarily chosen to be very small, then the loss of fluids from shale layers is extremely slow. An erosional event may then have little consequence for the shale. On the other hand, if $\kappa_*$ is larger, so that fluids can escape relatively quickly, then a depositional/erosional event may have a significant impact on the loss of fluid from the underlying shale. Clearly the problem is one of determining *simultaneously,* for both equations of state and for geological events, the sets of parameters that are most consistent with present-day data. In addition, the resolution of each parameter must be determinable. Supposing such a scheme to have been devised, it would then appear that burial histories most consistent with observed data, and with known resolution and uncertainty, are constructed. However, as has been noted elsewhere (Lerche, 1990a), thermal conductivities, fluid viscosities, chemical diagenesis, and conversion of kerogen to mobile and soluble oil and gas phases are all temperature dependent, as is the thermal expandability of fluids, the pressure of gases, and the solubility of methane and ethane in water and oil. Thus the burial history evolution is dependent to a greater or lesser extent on the thermal evolution of the sediments and on the hydrocarbon evolution within the sediments.

It would appear that the thermal history of the system must be known simultaneously with the dynamic burial history of the sediments. This point is reinforced when it is also recognized that thermal indicators contained information concerning the impressed signature of the temperature history to which they have been exposed, which, in turn, owes some allegiance to the burial paths of sedimentary units as well as to intrinsic variations in paleoheat flux. The effects of unconformities (Armagnac, 1985; Armagnac *et al.,* 1989), of stratigraphic age uncertainties (Pantano and Lerche, 1990), and so on are recorded with some resolution in thermal indicators. Conversely, observed present-day thermal indicator values can be used to determine the degree to which such geological events can be determined. But then these events, whether determined from dynamic burial path information or from thermal indicators, should have the same parameter values.

Hence, it should be possible to use both dynamic information and thermal indicator information simultaneously to better determine event parameters. Such a procedure for multiple thermal indicators has already been used (He and Lerche, 1989) to better constrain not only paleoheat flux variations but also physical and chemical parameters for each thermal indicator, because all thermal indicators must have been exposed to the same paleoheat flux, which information therefore acts as a constraint on allowed variations of other parameters.

This chapter first sets up the formal quantitative development for handling all of these problems on an equal footing in order to illustrate the underlying mathematical framework of the procedure. Case histories are then considered.

First we present the methodology for handling the dynamic inversion scheme when only single-well information is available. Then a procedure is provided for generalizing the method to allow for multiple-well information. As is then shown, the dynamic inverse tomographic scheme can be combined with the thermal indicator tomographic procedure already developed (Lerche, 1988a,b).

## II. DYNAMIC TOMOGRAPHY IN ONE DIMENSION

The arguments put forward by He and Lerche (1989) for dealing with thermal indicator tomography are valid in like manner when dealing with dynamic tomography. Indeed both the nod to seismic tomography and the general procedures developed in He and Lerche (1989) are transportable almost wholesale to the dynamic tomography problem. Despite the closeness in general development, it is nevertheless appropriate to spell out some of the detailed reasons underlying the rationale, as well as the appropriate generalizations necessary to carry through the machinations for dynamic tomography. These developments seem appropriate for two main reasons: first, so that one does not have to have available the earlier papers on thermal indicator tomography and so this chapter stands on its own in complete form; second, because there are interesting technical differences of procedures and control functions for dynamic tomography which, when combined with thermal indicator tomography, lead to a general procedure for handling both dynamic and thermal parameters on the same footing. For these reasons we believe it is useful to construct the dynamic tomography procedure from first principles.

### A. GENERAL REMARKS

As with thermal indicator tomography, dynamic tomography is a nonlinear inverse problem. The nonlinearity arises because the parameters and/

or functions to be determined influence the burial paths of the sediments, the thicknesses of stratigraphic units, the porosity of formations, and the fluid paleo-overpressure. The essence of most tomographic procedures devised to date from many disciplines has been to construct "almost linear" solutions. Thus, the true nonlinear behavior is linearized around initial estimates for the parameters. The linearized equations are then solved to obtain linearized corrections to the parameters, the new parameters are then used to linearize again around the new nonlinear estimate, and the process iterated (Aki and Lee, 1976; Hawley *et al.*, 1981; Menke, 1984; Lerche, 1988a,b). In constructing a thermal indicator tomography inverse scheme, we originally (Lerche, 1988a,b) followed this tried and proven path. However, several difficulties arise that make it desirable to devise a more general procedure. First, as noted by Menke (1984), a damping constant has to be introduced to stabilize inversion of matrices in the linearized equations. The value of this damping constant is arbitrary, but tests against many synthetic data sets have shown that a value of about 5–10% of the main diagonal elements of the relevant matrices produces acceptable inverse solutions. But in any particular problem there is no way of knowing, short of trial and error, the sensitivity of the inverse solutions to the value chosen for the damping constant. (We would prefer to avoid the introduction of an arbitrary constant; we have enough parameters to determine already.) Second, it can—and does—happen that the linearized corrections to the initial parameters can cause the updated (initial plus linearized correction) parameters to land in nonphysical space (e.g., a correction to a small initial erosional unconformity may give a negative amount of erosion). Such facts can cause rapid numerical instability of the linearized iteration scheme. It is, of course, possible to rewrite parameters to enforce the physical requirements, but it would be preferable if the tomographic procedure would ensure such correct physical behaviors automatically. Third, the initial estimates of parameter values can be very far from the best parameter values so that many iterations are necessary to approach the minimum mismatch between predicted and observed data. Such effects can cause massive computer run times. We would prefer to have a procedure that would automatically find initial parameter values close to the "best" values. Fourth, there is no guarantee that, once a local minimum mismatch is found, there is not some other, more global, minimum within the physically acceptable domain. We would prefer to have a procedure that could readily be used to search quickly for such a potential minimum. Fifth, it can happen that the system is not sensitive to the precise value of a parameter, or a set of parameters, in some range, so that the system cannot resolve such parameters and attempts to find values should then be discontinued. We would prefer a procedure that would rapidly sort out such insensitive parameters from sensitive parameters that can be determined.

As a consequence of such difficulties with the conventional procedure, we have gradually abandoned the linearized procedure during the last few years in favor of a nonlinear scheme that does not suffer from the mentioned drawbacks and that is remarkably fast numerically. We consider this framework in the next subsection.

## B. CONTROL FUNCTIONS AND NONLINEAR TOMOGRAPHY

In a single well the information usually available is (1) the depth, $Z_T$, measured from the sediment–water interface to the total depth (TD) drilled; (2) the thicknesses, $\tau_i$, and lithologies and ages of each formation in an ordered sequence $i = 1, 2, \ldots, I$. The sum of the thicknesses should equal the total depth drilled,

$$Z_T = \sum_{i=1}^{I} \tau_i; \tag{3.3}$$

(3) a set of porosity measurements $\phi_j$ measured at depths $z_j$, with $j = 1, 2, \ldots, J$; (4) a set of total fluid pressure measurements, $P_k$, measured at depths $z_k$, with $k = 1, 2, \ldots, K$; (5) occasionally, but not always, there will also be a set of permeability measurements, $k_l$, measured at depths $z_l$, with $l = 1, 2, \ldots, L$; and (6) if unconformities have occurred, their depth positions, $Z_n$, $(n = 1, 2, \ldots, U)$ are also noted. One of the purposes of dynamic burial history modeling is to devise time-dependent, fluid flow compaction models that will honor these data. As we have remarked already, in the construction of burial history models, a suite of parameters is specified relating both to equations of state and to geological events. Let the vector, $\mathbf{p}$, denote the complete set of all such parameters. We recognize that the parameters are often of different dimensions, which, in the nearly linear procedure discussed earlier, requires a renormalization to dimensionless form prior to initiating a tomographic procedure (Menke, 1984). As we shall see, the same is true with the nonlinear tomographic procedure to be exhibited here.

The argument is now relatively straightforward. A quantitative burial history model constructs dynamically evolving values for total depth $Z_M(\mathbf{p}, t)$, for porosity $\phi_M(z, \mathbf{p}, t)$, for total fluid pressure $P_M(z, \mathbf{p}, t)$, for formation thicknesses $t_M(z, \mathbf{p}, t)$, for permeability $k_M(z, \mathbf{p}, t)$, and will also insert unconformities with depth and time so that the modeled unconformities occupy positions $Z_M(z, \mathbf{p}, t)$. Here $t$ is time, $z$ is depth below the water base, and $\mathbf{p}$ is the parameter vector to show explicitly that all modeled quantities depend on the parameters invoked.

At the present day ($t = 0$), the modeled quantities take on the values $Z_M(\mathbf{p})$, $\phi_M(z, \mathbf{p})$, $P_M(z, \mathbf{p})$, $t_M(z, \mathbf{p})$, $k_M(z, \mathbf{p})$, and $Z_M(z, \mathbf{p})$. For consistency all of the modeled values should be in agreement with the present-day measured values. Criteria used to assess the degree of mismatch between predicted and observed values are least-squares control functions as follows:

1. For TD,

$$X^2_{\mathrm{TD}}(\mathbf{p}) = (Z_{\mathrm{M}}(\mathbf{p}) - Z_{\mathrm{T}})^2. \tag{3.4a}$$

2. For formation thicknesses,

$$X^2_{\mathrm{T}}(\mathbf{p}) = I^{-1} \sum_{i=1}^{I} (\tau_i - \tau_{\mathrm{M}}(z_i, \mathbf{p}))^2. \tag{3.4b}$$

3. For porosity variations,

$$X^2_{\phi}(\mathbf{p}) = J^{-1} \sum_{j=1}^{J} (\phi_j - \phi_{\mathrm{M}}(z_j, \mathbf{p}))^2. \tag{3.4c}$$

4. For fluid pressure variations,

$$X^2_{\mathrm{p}}(\mathbf{p}) = K^{-1} \sum_{k=1}^{K} (P_k - P_{\mathrm{M}}(z_k, \mathbf{p}))^2. \tag{3.4d}$$

5. For permeability variations,

$$X^2_{\kappa}(\mathbf{p}) = L^{-1} \sum_{l=1}^{L} (\ln \kappa_l - \ln \kappa_l(z_l, \mathbf{p}))^2. \tag{3.4e}$$

6. For unconformity positions,

$$X^2_{\mathrm{u}}(\mathbf{p}) = U^{-1} \sum_{n=1}^{U} (Z_n - Z_{\mathrm{M}}(z_n, \mathbf{p}))^2. \tag{3.4f}$$

The reason for using the natural logarithm of the permeability in the least-squares control function (3.4e) is that the observed permeabilities of shales are typically several orders of magnitude smaller than those of sands (Glezen and Lerche, 1985; Dutta, 1987), with the lower shale permeabilities having dominant control on the escape of fluids from the subsurface (Cao and Lerche, 1990; Lerche, 1990a,b). The logarithmic dependence is appropriate to describe this dominance of shales, as has also been noted by Warren and Price (1961), Freeze (1975), Palciauskas and Domenico (1980), and others. Note that the individual control functions in Eqs. (3.4a) through (3.4f) have different dimensions so that they are not immediately comparable.

Suppose, for the moment, that some initial estimate has been chosen for the parameter vector, $\mathbf{p}$. Let this value be $\mathbf{p}_0$. (We shall return later to a procedure for determining the initial parameter vector $\mathbf{p}_0$). Then construct the dimensionless quantities:

$$y_1^2(\mathbf{p}) = X_{\mathrm{TD}}^2(\mathbf{p})/X_{\mathrm{TD}}^2(\mathbf{p}_0); \qquad y_2^2(\mathbf{p}) = X_{\mathrm{T}}^2(\mathbf{p})/X_{\mathrm{T}}^2(\mathbf{p}_0),$$
$$y_3^2(\mathbf{p}) = X_{\phi}^2(\mathbf{p})/X_{\phi}^2(\mathbf{p}_0); \qquad y_4^2(\mathbf{p}) = X_{\mathrm{p}}^2(\mathbf{p})/X_{\mathrm{p}}^2(\mathbf{p}_0), \tag{3.5}$$
$$y_5^2(\mathbf{p}) = X_{\kappa}^2(\mathbf{p})/X_{\kappa}^2(\mathbf{p}_0); \qquad y_6^2(\mathbf{p}) = X_{\mathrm{u}}^2(\mathbf{p})/X_{\mathrm{u}}^2(\mathbf{p}_0).$$

On $\mathbf{p} = \mathbf{p}_0$, each of the $y$'s is unity. Clearly a smaller mismatch to the observations occurs when each of the $y$'s tends to zero. Now construct a global control function $Y^2(\mathbf{p})$ as follows:

$$Y^2(\mathbf{p}) = \sum_{r=1}^{6} a_r y_r^2(\mathbf{p}), \tag{3.6}$$

where the $a_r$ are positive fractional weight coefficients whose sum is unity. The weight coefficients can be chosen to emphasize one aspect of the model (and agreement between observations and predictions) over others. For instance, if no permeability measurements are available, it would be inappropriate to compare model predictions of log permeability against nonexistent observations. In that case one might set $a_5 = 0$. Or again it might be thought that porosity measurements are less reliable than, say, formation thickness measurements in which case $a_3$ might be made smaller than the remaining weight coefficients.

Before we can construct the nonlinear tomographic procedure one further factor needs to be addressed. The parameter vector, $\mathbf{p}$, is a conglomerate of scalar parameters all of which have different dimensions. To construct a simple nonlinear tomographic procedure, it is best to normalize parameters. To this end suppose that a scalar parameter is considered to be physically constrained to lie between a minimum $p_{\min}$, and a maximum $p_{\max}$. Then write

$$a = (p - p_{\min})/(p_{\max} - p_{\min}) \tag{3.7a}$$

so that $a = 0$ on $p = p_{\min}$ and $a = 1$ on $p = p_{\max}$. Perform a similar normalization on each component of the parameter vector, $\mathbf{p}$, so that for each component we have

$$p = p_{\min} + a(p_{\max} - p_{\min}), \qquad 0 \le a \le 1. \tag{3.7b}$$

Regard a vector $\mathbf{a}$, of identical dimensionality to the vector $\mathbf{p}$, as being fundamental with the connection between each component of $\mathbf{a}$ and each component of $\mathbf{p}$ being given by Eq. (3.7b). The vector $\mathbf{a}$ then has entries that are dimensionless for each component and are in the range $\{0, 1\}$. Regard $Y^2$ as a direct function of the vector $\mathbf{a}$, with $\mathbf{p}$ being determined parametrically through the component representations as in Eq. (3.7b). Thus an initial estimate for each component $\mathbf{p}$, in the range $\{p_{\min}, p_{\max}\}$, provides an initial estimate of $a_0$, through Eq. (3.7a) and so an initial estimate for the vector $\mathbf{a}_0$, directly related to the initial estimate of the vector $\mathbf{p}_0$.

Consider then the following nonlinear iteration scheme for each scalar component, $a_j$ of $\mathbf{a}$, to be run $N$ times:

$$a_j(n) = \sin^2\theta_j(n),$$

$$\theta_j(n+1) = \theta_j(n)\exp\left[-\tanh\left\{\delta_j(n)\alpha_i\frac{\partial Y^2(\mathbf{a}(n))}{\partial a_i(n)}\right\}\right], \tag{3.8}$$

$$a_j(n+1) = \sin^2\theta_j(n+1),$$

where

$$\alpha_i = \left|\frac{\partial Y^2(\mathbf{a}_0)}{\partial a_i(0)}\right|^{-1} \ln[1 + (Na_i(0))^{-1}] \tag{3.9a}$$

and

$$\Gamma_j(n) = |a_j(n+1) - a_j(n)|/a_j(n) + \beta^2, \tag{3.9b}$$

with

$$\delta_j(n) = \Gamma_j(n)/\left\{J^{-1}\sum_{j=1}^{J}\Gamma_j(n)\right\} \tag{3.9c}$$

and with $a_i(0) = a_0$, and the derivative calculated numerically from

$$\begin{aligned}\frac{\partial Y^2(\mathbf{a}(n))}{\partial a_i(n)} = {} & [Y^2(a_i(n), a_2(n), \ldots, a_i(n) + \beta a_i(n), a_{i+1}(n), \ldots) \\ & - Y^2(\mathbf{a}(n))]/(\beta a_i(n)),\end{aligned} \tag{3.9d}$$

where $\beta$ (default) is 0.1. The nonlinear iteration scheme given by Eq. (3.8) has several interesting characteristics. First, it guarantees that if $a_i$ is positive at any iteration (including the initial estimate), then $a_i(n)$ will be positive at every iteration. Second, it guarantees that $Y^2(\mathbf{a})$ will always be smaller or the same at each iteration if the partial derivatives are calculated exactly. [If the derivatives are only calculated numerically approximately, as by Eq. (3.9d), then small localized oscillations around a minimum can occur.] Thus $Y^2$ is always decreased away from its initial value so the procedure guarantees that a minimum will be found. (The questions of resolution, precision, uniqueness, and sensitivity are deferred to later.) Third, it is a relatively simple matter, once a minimum is achieved, to "freeze" all parameters except one at the values corresponding to the minimum, and to then perform a linear or random search between minima and maxima. The procedure is repeated for each and all parameters in turn. In this way a relatively confident position can be taken that no minima have been overlooked. Fourth, with respect to the choice of initial estimates for the parameters, it now becomes clear how to ensure that they will be close to a minimum position. Choose the search range of each parameter (i.e., $p_{min}$, $p_{max}$). Then perform a linear or random search on each within the parameter range to find the minimum of $Y(\mathbf{p}_0)^2$, thereby identifying the rough value of the parameter. Choose that value as the estimate of $\mathbf{p}_0$. In practice, this procedure works extremely well. Indeed, when a rough estimate of the initial parameter vector has been found, the linear search procedure can be iterated to provide closer approximations to the minimum value of $Y(\mathbf{p}_0)^2$ and hence also a correspondingly better estimate of the parameter vector. Fifth, as a tactical maneuver, a more rapid approach to a minimum value is obtained if $N$ nonlinear iterations are done, the updated parameter vector, $\mathbf{p}_N$, at the end of the iterations is used to replace $\mathbf{p}_0$, and then $N$ more nonlinear iterations performed. This tactic is far superior to doing $2N$ nonlinear iterations around the original estimate, $\mathbf{p}_0$.

To date, numerical experience on single-well studies and on synthetic data has shown the following:

1. The nonlinear inversion generally converges rather rapidly in 5 to 10 iterations.
2. The pragmatic procedure (linearly estimate $\mathbf{p}_0$, nonlinearly invert, linearly estimate, nonlinearly invert) is usually more than enough to exhaust the precision of the input data, so that the model arrived at after the inversion procedure is a significantly better fit to data than the results of the first linear search estimate.

## III. MULTIPLE WELLS IN A BASIN AND 2-D DYNAMIC TOMOGRAPHY

For a single-well study there is little need to discriminate between parameters that are associated with equations of state from parameters that are, or may be, associated with geological events. However, such is not the case in multiple-well studies because geological events can change from spatial location to spatial location while, presumably, parameters in equations of state should be invariant. Exceptions have to be provided for; thus, the anisotropy in shale permeability parallel and perpendicular to bedding can be influenced by the local depositional and compactional history of a system, and so can be location dependent. But frame pressure–void ratio and permeability–void ratio rules should, presumably, not be dependent on location within a given basin. Two fundamentally different nonlinear inverse procedures are available depending on the assumptions made concerning the parameter vector. On the other hand, there is the extreme end-member decision that every component of the parameter vector may be location dependent for reasons not completely understood. On the other hand, there is the extreme end-member strategy which regards only geological event parameters as location dependent, and any and all equation-of-state parameters are considered independent of location.

In either event there exists a procedure for determining the validity of the extreme end-member scenarios. First consider the situation in which *all* components of the parameter vector **p** are considered to be dependent on location $x_i$ ($i = 1, 2, \ldots, X$) as well as on depth, $z$. Write $\mathbf{p} = \mathbf{p}(x)$ to express this dependence. Then suppose that at position $x_i$, thicknesses, TD, porosity, fluid pressure, permeability, and unconformity depths have been measured. Construct, as in Section II, $Y_r(\mathbf{p}(x_i))^2$ for each quantity that now depends on the parameter vector, **p**, at lateral location $x_i$. The evolution and compaction of sediments in the basin are now laterally connected because fluids can flow horizontally as well as vertically. Thus consistency of predicted and observed behaviors at a given location does not guarantee a minimum discord at other positions in the basin. To obtain a minimum mismatch at all positions in the basin, introduce the equation:

$$y_r(\mathbf{p}(x_l), \mathbf{p}(x_2), \ldots)^2 = X^{-1} \sum_{i=1}^{X} y_r(\mathbf{p}(x_i))^2. \tag{3.10}$$

Then proceed as noted earlier. Construct

$$Y^2 = \sum_{r=1}^{6} a_r y_r(\mathbf{p}(x_l), \mathbf{p}(x_2), \ldots)^2 \tag{3.11}$$

and minimize $Y^2$ with respect to the total set of parameter values, both geological event and equation of state, each of which is considered location specific. Then perform the linear or random search routine of Section II for each parameter, thereby constructing a range of sensitivity for each parameter around the minimum value of $Y^2$.

Now repeat the analysis but hold parameters thought to be independent of geological events at common, location-independent values. Again minimize $Y^2$ and again construct the sensitivity range for each parameter around the minimum value of $Y^2$. Compare the degrees of resolution and sensitivity of this end-member situation with the previous analysis. If, to within some imposed degree of tolerance either with respect to intrinsic data resolution or with respect to predicted and observed mismatch weighted or unweighted, there is no difference in the parameter values, then justification can be given for arguing that parameters representing equations of state are indeed location independent. A significant departure of the sensitivity curves one from the other in the two cases implies that, to within the resolution limit set, the parameter under investigation *is* location dependent. Thus the procedure in the 2-D case is a simple generalization of the 1-D procedure with only minor variations in technique—but providing the potential for assessing the validity of intrinsic assumptions of different extreme end-member behaviors.

## IV. COMBINED DYNAMIC AND THERMAL TOMOGRAPHY

Information on the dynamic evolution of sedimentary basins is imparted not only by present-day downhole data such as TD, formation thicknesses, porosity, permeability, total fluid pressure, and unconformity depths, but some information related to dynamic evolution has also been impressed on thermal indicators throughout their evolution (Lerche, 1990a,b). Indeed, elsewhere we have used multiple present-day thermal indicator data in an inverse manner to determine not only paleoheat flux variations and physical and chemical parameters associated with the thermal indicators, but also parameters associated with geological events (such as unconformity thicknesses, stratigraphic ages, emplacement time of overthrusts, insertion temperature, and time of igneous overthrusts) (Lerche, 1988a,b, 1990a,b; Pantano and Lerche, 1990; He and Lerche, 1989).

Clearly, parameters for geological events and equations of state are making their effects felt on both dynamic tomography (in a direct sense) and on thermal indicator tomography (in an indirect sense through the effects on temperature experienced by thermal indicators). Thus a combined dynamic and thermal indicator tomography should better resolve, or constrain, common parameters influencing both types of indicators.

To allow for such a situation, simply introduce the $\nu$th thermal indicator, $\mathrm{TI}_\nu(z_s)$, measured at the set of depths $z_s$, $s = 1, 2, \ldots, S$. Then, following the procedure laid down elsewhere (Lerche, 1988a,b) for a thermal indicator with a given kinetic description, we can write

$$U_\nu^2 = S^{-1} \sum_{s=1}^{S} (\mathrm{TI}_\nu(z_s, \mathbf{p}) - \mathrm{TI}_\nu(t_s, \mathbf{p}))^2, \tag{3.12}$$

where $\mathrm{TI}_\nu(t_s, \mathbf{p})$ is the predicted value of the $\nu$th thermal indicator with time $t_s$ corresponding to present-day depth $z_s$. To combine a set of different multiple thermal indicators, in a manner similar to that set out in Section 2, we introduce

$$u_\nu^2(\mathbf{p}) = U_\nu(\mathbf{p})^2 / U_\nu(\mathbf{p}_0)^2 \tag{3.13}$$

and

$$U^2(\mathbf{p}) = V^{-1} \sum_{\nu=1}^{V} b_r u_\nu(\mathbf{p})^2, \tag{3.14}$$

where $V$ is the total number of thermal indicators, and the $b_r$ are positive, fractional weight factors. In the case of thermal indicators not only is the parameter vector $\mathbf{p}$ made up of components due to geological events and equations of state, but also components due to paleoheat flux and physical/chemical parameters associated with each thermal indicator. Absorbing

TABLE 3.1 Synthetic Input Data

| Layer number | Depth (m) | Age (Ma) | Lithology |
|---|---|---|---|
| 1 | 300 | 3.4 | Shale |
| 2 | 700 | 5.2 | Shale |
| 3 | 1200 | 7.9 | Shale |
| 4 | 1510 | 10.6 | Shale |

these "extra" parameters into the definition of **p**, we can then combine the dynamic and thermal indicator control functions to write

$$W(\mathbf{p})^2 = Y(\mathbf{p})^2 + U(\mathbf{p})^2 \tag{3.15}$$

and perform the nonlinear inversion procedure of Section II on the combined control function $W(\mathbf{p})^2$.

The formal mathematical analysis is now complete because, with both the thermal and dynamic components successfully mated, we have available a procedure for assessing parameters of geological processes that affect the evolution of sedimentary basins in both a local and a regional sense.

The combined determination of structural and dynamic properties of petroliferous basins and of quantitative paleoheat flux variations, as well as physical/chemical parameters of equations of state and thermal indicators, together with quantitative measures of their uncertainties consistent with downhole data, offers the greatest probability of accurately assessing not only the timing of hydrocarbon production and migration, but also the relationship to the timing and development of structural and stratigraphic traps that serve as oil reservoirs in a basin.

## V. SYNTHETIC TESTS OF DYNAMIC INDICATOR TOMOGRAPHY

In a nonlinear sense, dynamic indicator tomography (DIT) provides a tool to do inversion of the multiple dimension parameters with control indicators. In this model we can simultaneously decide whether more than one parameter is important in practical applications and we can determine the sensitivity of the system to each parameter. Here we use a synthetic data set to do an inversion in order to demonstrate the variation of each MSR and the precision reachable with the multiparameter DIT.

Table 3.1 shows the input synthetic data for the forward model. Because of the low permeability of shale and its rapid variation with porosity, four

TABLE 3.2 Tomography Results for Synthetic Data

| | $A$ | $B$ | $\kappa_*$ (md) | $P_{f_*}$ (atm) | $e_*$ | MSR |
|---|---|---|---|---|---|---|
| Input | 2.8 | 3.0 | 10.0 | 1.0 | 1.85 | Default |
| | 2.74 | 2.91 | 9.2 | — | — | 1.75 |
| | 2.76 | 2.89 | 9.12 | 0.96 | 1.88 | 1.59 |

shale layers with default parameters (see Table 3.2) are used to calculate the present-day porosity, permeability, fluid pressure with depth, and formation thickness. These present-day values are then used as four dynamic indicators. With this synthetic data set we can use different parameters for individual or combined indicators to do an inversion and determine if the default input parameters are recovered by the inverse procedure.

The testing with synthetic data involved (1) choosing one or two parameters with one or more indicators to do DIT and then to compare the MSR variation with that of a linear search (see later) with two parameter limits; (2) supposing there is no error in the data and no resolution limit from the data, using all of the parameters with all of the indicators to check precision and resolution of the DIT; and (3) determining the sensitivity of parameters, the physical domain of parameters relative to the initial estimated values for the parameters.

Figure 3.1 shows the MSR variation with each single parameter using all of the indicators: pressure, porosity, permeability, and formation thickness, respectively. The other parameters are "frozen" at their default values (midpoint of the user-specified range). All the curves of Fig. 3.1 demonstrate that there is at least one minimum MSR value in the chosen range of each parameter, but the curves have different shapes, representing the sensitivity variation of the parameters relative to each individual indicator. Figure 3.2 shows the total MSR variation through each nonlinear iteration, illustrating the approach to a minimum. Once the minimum value is reached, there is no further variation of a parameter around the local minimum. In this synthetic case there is only one minimum in the given parameter range, so that the minimum value found by DIT must be the best value. If there are multiple minima in a given parameter range, another minimum MSR and different best values of parameters could be obtained by different initial starting values for the parameters.

The direction and speed of the MSR variation is controlled by the size of the various partial derivatives. Figure 3.3 shows that the inverse tomography procedure, when started at three different points labeled A, B, and C, reached the minimum area (closure by contour value 1.5) after 30 iterations. For this simple case the method guarantees to find the best

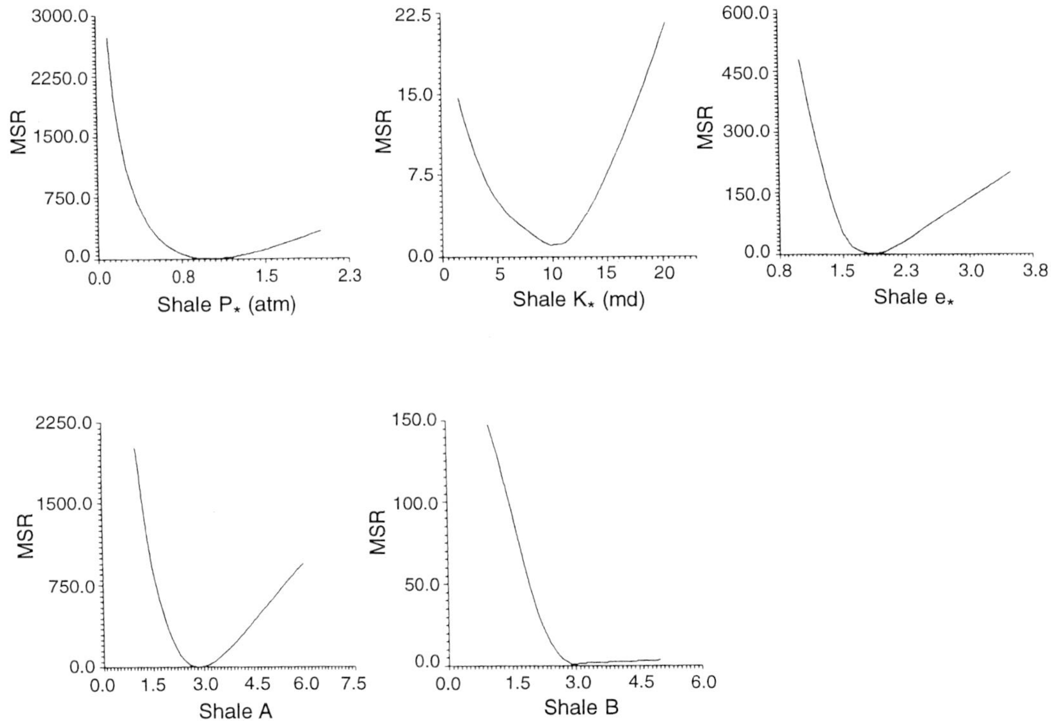


FIGURE 3.1 Sensitivity of each parameter for the synthetic shale case. Parameters are $A$, $B$, $\kappa_*$, $P_*$, and $e_*$.

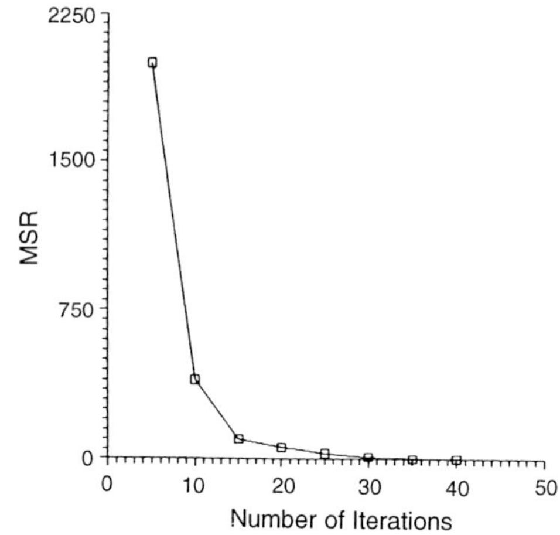


FIGURE 3.2 MSR decreases with each nonlinear iteration and becomes constant after reaching the minimum.

values. In more complicated cases, several MSR minimum values could occur. The minimum arrived at finally by the nonlinear procedure depends on the location of the initial parameter values. Table 3.2 lists the results of tomography with more than two parameters. Using three parameters ($A$, $B$, and $\kappa_*$ for shale), a minimum MSR was reached at 1.75 and then, using all five shale parameters ($P_{f_*}$, $A$, $B$, $\kappa_*$, and $e_*$), the minimum MSR

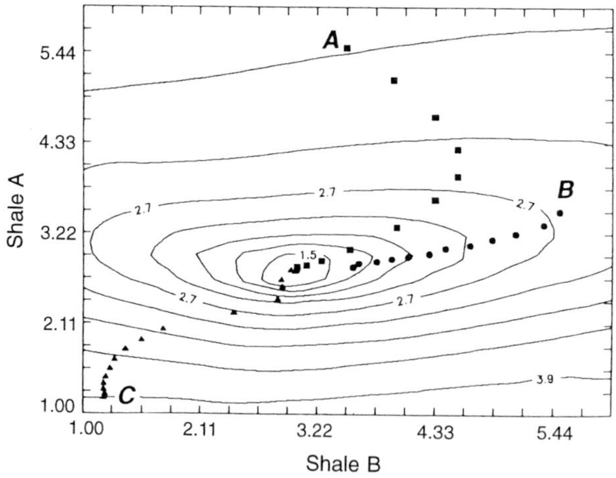


FIGURE 3.3 Log MSR contours in space of shale $A$ and shale $B$ with all indicators. Symbols show MSR paths of tomography, starting from three initial points $A$ (solid squares), $B$ (solid circles), and $C$ (solid triangles) and reaching the minimum value of the contour domain by 30 iterations.

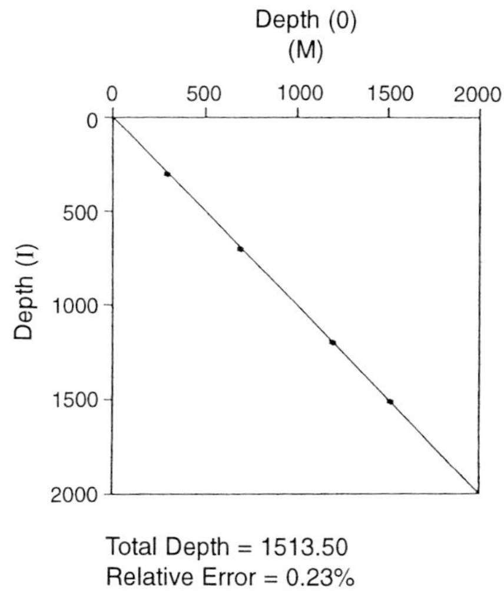


FIGURE 3.4 Comparison of the input and output formation thicknesses. If both thicknesses are equal, points will land on the 45° diagonal line.

reached is 1.59, which is about a 10% improvement. The tomography results for the synthetic tests are shown in Figs. 3.4, 3.5, and 3.6. On the profiles of pore pressure, porosity, permeability, and depth, the present-day ($t$ = 0) indicators fit well and, in the inverse sense, the history of sediment compaction and the evolution of pressure, porosity, permeability, and formation thickness are constrained more precisely. Based on the synthetic

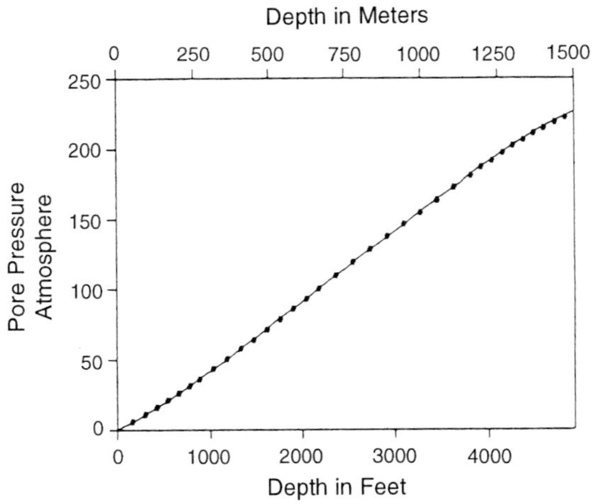


FIGURE 3.5 Pressure profiles. The solid line represents calculated values; starred points are synthetic data from the forward model.

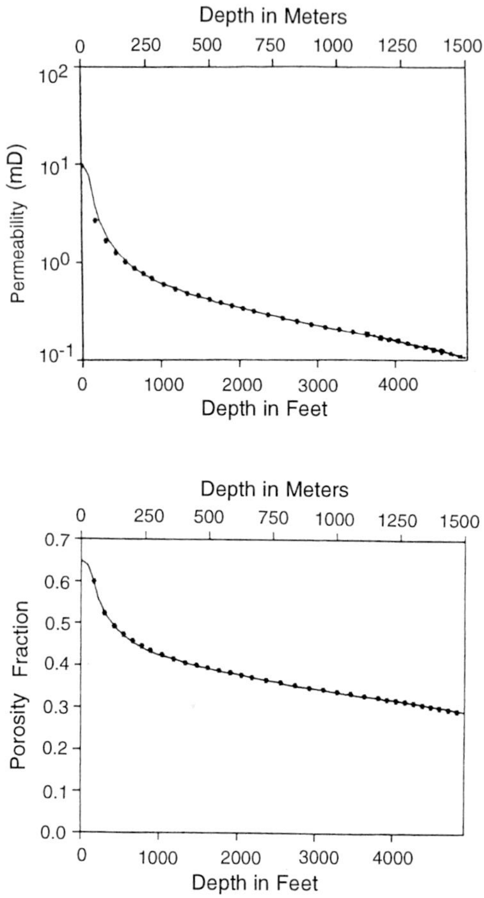


FIGURE 3.6 Permeability (top) and porosity (bottom) profiles. Starred points are the synthetic data obtained from the forward model with default parameters. Dynamic tomography reduces the mismatch between the synthetic data and calculated values (solid line).

data, the known best parameters, and no measurement error, DIT is an effective way to do inversion. The code speed is fast, requiring about 2–5 minutes of CPU time for completion of all the calculations on a Sun SparcIII.

## VI. A SINGLE-WELL CASE HISTORY: NAVARIN BASIN COST NO. 1 WELL, ALASKA

For the Navarin COST well, most of the detailed geological, geophysical, and geochemical data are reported in *Geological and Operational Summary,*

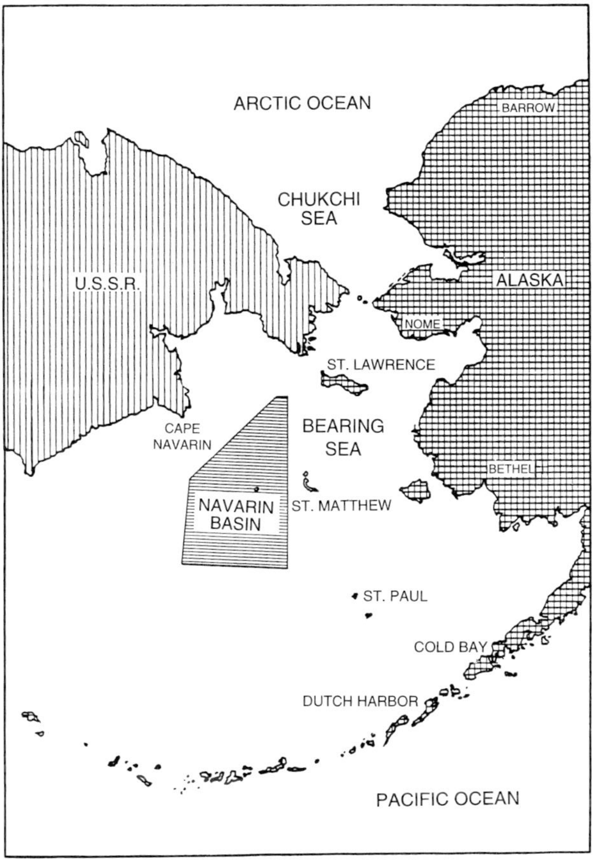


FIGURE 3.7 Location of the Navarin Basin COST No. 1 well.

*Navarin Basin COST No. 1 Well, Bering Sea, Alaska* (Turner *et al.*, 1984) (OCS Report MMS 84-0031). Navarin Basin COST No. 1 well was the first well drilled in the remote Navarin Basin (Fig. 3.7), which is a Tertiary sedimentary basin on the outer continental shelf on the Bering Sea, and covers an area of about 83,400 km$^2$.

The Navarin Basin COST No. 1 well was located in the east-central part of the basin, reaching a depth of 4999 m. The sediments penetrated by the COST well are composed of Neogene and Paleogene clastic sedimentary rocks deposited in a marine environment, and Cretaceous clastic sediments deposited in fluvial, floodplain, and deltaic environments. The Cretaceous and Tertiary sedimentary rocks are separated by a diachronous angular unconformity. In the interval between 3901 and 4587 m, Late Cretaceous rocks are intruded by numerous basaltic and diabasic sills (Turner *et al.*, 1984).

TABLE 3.3 Input Data for Navarin Basin COST No. 1 Well

| Layer number | Depth (ft) | Age (Ma) | Lithology | Clay content | Paleowater depth (ft) |
|---|---|---|---|---|---|
| 1 | 1,920 | 3.4 | Shale | 1.0 | 400 |
| 2 | 3,180 | 5.3 | Sandy shale | 0.5 | 300 |
| 3 | 3,860 | 7.9 | Sandy shale | 0.5 | 600 |
| 4 | 5,010 | 18.6 | Sand | 0.0 | 600 |
| 5 | 5,360 | 21.1 | Sand | 0.0 | 600 |
| 6 | 5,704 | 23.7 | Sand | 0.0 | 600 |
| 7 | 7,130 | 24.8 | Sandy shale | 0.5 | 300 |
| 8 | 9,450 | 26.7 | Sandy shale | 0.5 | 600 |
| 9 | 11,000 | 28.0 | Shale | 1.0 | 1500 |
| 10 | 12,280 | 34.0 | Shale | 1.0 | 1500 |
| 11 | 12,780 | 45.0 | Shale | 1.0 | 600 |

*Note:* Depth and age are at the base of each layer.

Table 3.3 lists the basic input data of the Navarin COST No. 1 well. The Cenozoic sequence is split into 11 model layers from bottom to top, based on lithology, paleontology, biostratigraphy, and depositional environments. Measured porosity data from sidewall and conventional cores are available in the interval from 914 to 3962 m. Figure 3.8 shows the pressure profile for COST No. 1 well. The existence of abnormally high pore pressure in the Navarin COST No. 1 well has been shown by wireline logs, formation drill-stem tests, and the drilling exponent method (Turner *et al.,* 1984). The wireline log and formation test data clearly identify a major zone of abnormal formation pore pressure in the interval below 2874 m in this well. Pore pressure (11 points from RFT) data are used as an indicator of total fluid pressure.

Because the measured permeability data vary randomly (in the sandy shale layer), a considerable amount of noise would be introduced into the inversion procedure for parameter determination if the permeability were to be included as a dynamic indicator. On these grounds we treat only thickness, pressure, and porosity as indicators, and confirm the appropriateness of determined parameters by seeing how well the predicted permeability honors the main trend of the observed, noisy, permeability data. Using the default parameters (Table 3.4) for shale and sandy-shale, the first forward run ("freeze" all parameters without using any dynamic indicators) gave relatively large differences between measurements and predictions. For instance, the relative error of thickness is above 10% for shale. In this case, 15 parameters are involved for the lithologies of sand, shale, and

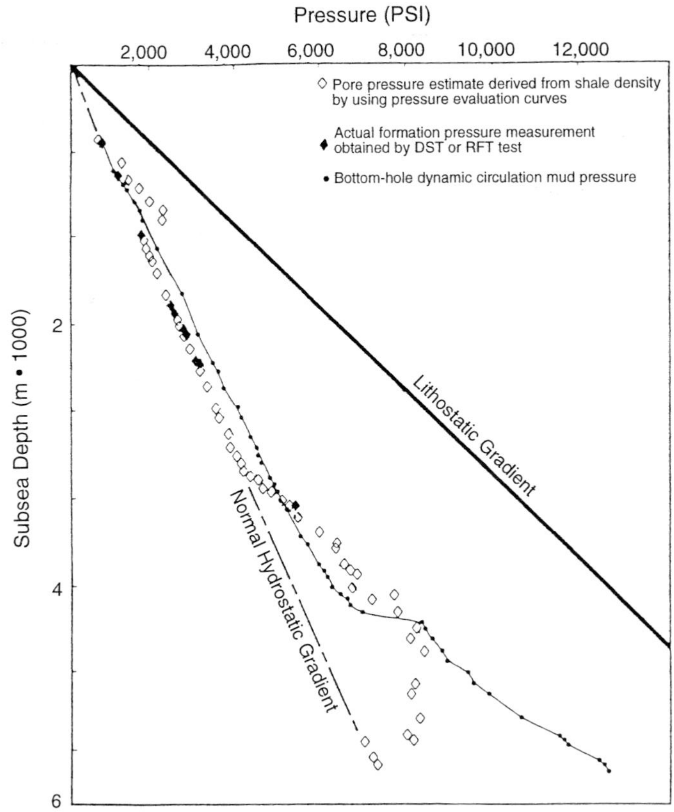


FIGURE 3.8 Pore pressure measurements with depth. The three different symbols represent different methods of estimating pore pressure as explained in Turner *et al.* (1984).

sandy-shale. It is then necessary to adjust the parameters with the inversion method in order to construct a more precise model based on the measured indicators.

Figure 3.9 shows the lithology model for this well. The top and bottom layers are shale. Because the 1-D fluid flow/compaction model is concerned only with vertical fluid flow, the low permeability shale at the top of the sequence overlying the sandier units has a significant effect on evolution of the fluid pressure, porosity, and thickness. To first order in sensitivity, then, parameters related to shales must be adjusted first and then step-by-step changes in the other parameters made until the improvement in mismatch of all indicators is exhausted.

Table 3.4 shows the final results from DIT. Table 3.4 lists three parameters for shale and five parameters for sandy-shale, with five sand parameter values and two for shale held fixed (i.e., no further resolution capability

TABLE 3.4 Parameter Values

1. Parameter values used by Bour and Lerche (1994) and taken as first initial guesses for this study.

| | $A$ | $B$ | $\kappa_*$ | $e_*$ |
|---|---|---|---|---|
| Shale | −0.48 | 3.2 | 3 (mD) | 1.75 |
| Sandy-shale | −0.285 | 2.5 | 2.0 (Darcy) | 1.75 |
| Global MSR (normalized) = 1.0 | | | | |

2. Parameter values determined by using the tomography method yielding 10% improvement of fit.

| | $A$ | $B$ | $\kappa_*$ | $e_*$ |
|---|---|---|---|---|
| Shale | −0.42 | 1.5 | 2 (mD) | 1.40 |
| Sandy-shale | −0.20 | 3.0 | 1.5 (Darcy) | 2.50 |
| Global MSR (normalized) = 0.90 | | | | |

was available in the data). Comparison of input and output formation thicknesses for this well is shown in Fig. 3.10; relative error is 0.31%. Figure 3.11 shows present-day predicted porosity versus depth for the Navarin COST No. 1 well, where the points are measured porosity data. The minimum MSR of predicted and measured porosity corresponds to average fractional relative error of only 0.15. Figure 3.12 shows the total fluid pressure profile. The solid line crosses the 11 points of measured pore pressure (with a minimum MSR equal to an average relative error of 5%).

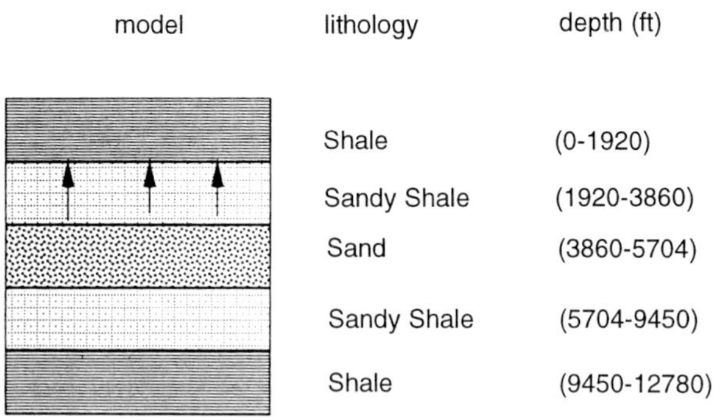


FIGURE 3.9 Lithology model of COST No. 1 well, Navarin Basin. Arrows denote vertical fluid flow direction.

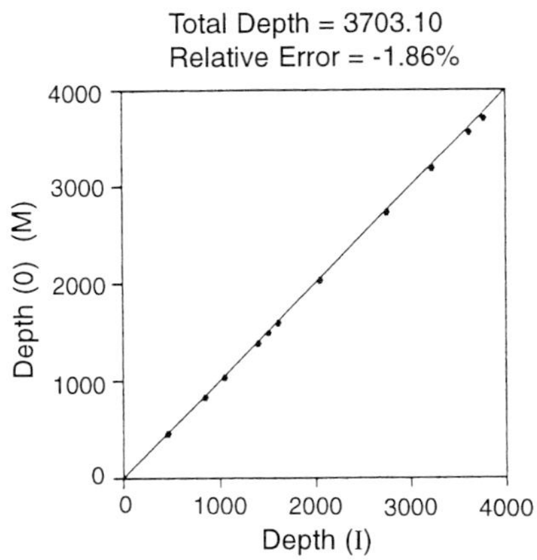


FIGURE 3.10 Comparison of input and output formation thicknesses for COST No. 1 well. All of the points are sited on the 45° diagonal line, implying minor error between input and output data. In total, the relative error of depth is 0.31%.

The abnormal pore pressure zone starts at a depth of about 2804 m and extends to the bottom layer. An RFT test at 3108 m (subsea depth) within the zone yielded 357 atm (5250 psi) on pressure build-up; the predicted value at this point is 353 atm (5191 psi). The permeability profile predicted is also shown in Fig. 3.11 as the solid line with the data points given, even though this indicator was not used in the tomography procedure.

To determine the uncertainty associated with each parameter, a linear search procedure was used in which one variable is allowed to vary, while holding all others fixed at their optimal values determined from DIT, to see how much the mismatch (MSR) will change when a single parameter deviates from its best value. Results of such an investigation are shown for COST No. 1 in Fig. 3.13a. For the three parameters related to shale, $A$, $B$, and $\kappa_*$, only $\kappa_*$ has a unique minimum value in the given range. The other two parameters have several minima but the smallest minimum was found with DIT. Among the five sandy-shale parameters, $A$, $P_{f_*}$, and $e_*$ show minima in curves for MSR versus each of the three parameters, allowing the best parameter corresponding to the minimum MSR to be determined (Fig. 3.13b). The MSR valley is broad and flat for $\kappa_*$ and $B$ for sandy-shale, implying a relatively large uncertainty for these two parameters. The uncertainty of each parameter can be affected by the error distribution and resolution of data points. In the present case, the majority of data points for porosity are located in sandy-shale and sand lithologies. There are no data points in shale to control shale parameters. But, because of the control of the low-permeability shales on the fluid flow/compaction processes, there is an indirect control on shale parameter values.

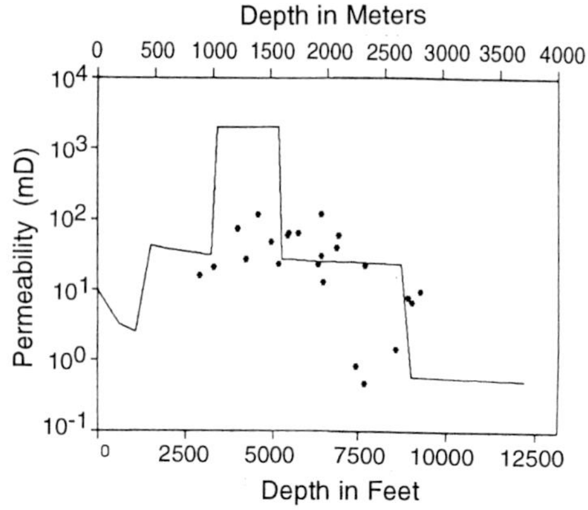


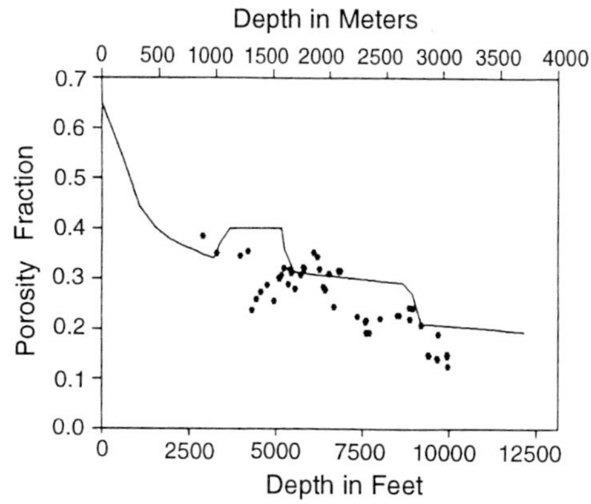


FIGURE 3.11 Permeability (top) and porosity (bottom) profiles. The solid line represents the prediction and starred points are measured values. MSR of porosity corresponds to an average relative error of 14.7%.

## VII. A MULTIWELL CASE HISTORY: NAVARIN BASIN, ALASKA

For single-well studies the general use of the inverse procedure is clear from both the synthetic well study and a real case history. However, two major problems remain, as discussed next.

First, it is not obvious that the general procedure is going to be of practical use in more than one dimension, because there is no *a priori* guarantee that parameters describing empirical equations of state for lithological end-members will remain at the same values at different locations in a basin. With multiwell information in a basin it is also not obvious that the general inverse procedure for parameter determination will converge

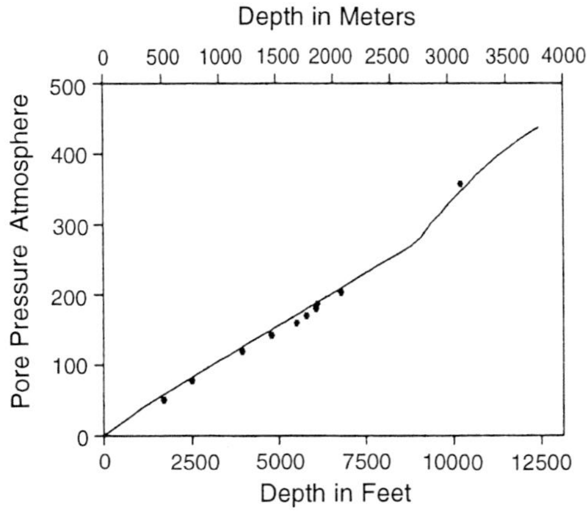


FIGURE 3.12 Pore pressure profile. MSR between measured values (starred points) and calculated values represents an average relative error of 5.1%.

at all—there may not be a minimum mismatch between predicted and observed multiwell values. For example, one well may have its pressure profile well matched but not its porosity profile; a second well may have exactly the opposite match. It is then possible for an oscillating state to develop in an inverse iteration loop. It is also not obvious how rapidly the inverse procedure will converge in general. Clearly, a systematic procedure for assessing parameter values, and their uncertainties, is to be preferred over random trial-and-error searches, particularly when many parameters can be varied. However, if such a systematic procedure is too slow numerically to be integrated into other aspects of hydrocarbon exploration, then the procedure, while correct, will be of use only in extreme geologic problems.

Second, it is not obvious that the total amount of information available is always being used consistently. For instance, it is already known that downhole single and multiple thermal maturation indicators can be used to determine not only paleoheat flux but also estimates of erosion amounts and their timing, as well as stratigraphic ages, and also parameters influencing the burial history of the sediments (Lerche, 1990a,b). The argument here is that the development of, say, paleo-overpressure influences porosity, which, in turn, impacts thermal conductivity, which leads to higher temperature values in overpressured regions (Hunt, 1979), thereby influencing thermal maturation indicators. The extent and timing of such overpressure are dependent on parameters influencing the dynamic evolution of the sediments. Thus, present-day thermal indicator profiles possess some knowledge of dynamic parameters. Clearly, if only dynamic indicators (formation thicknesses, porosity, pressure, permeability, TD, unconformity depths) are used to determine the dynamic parameters, it is not obvious

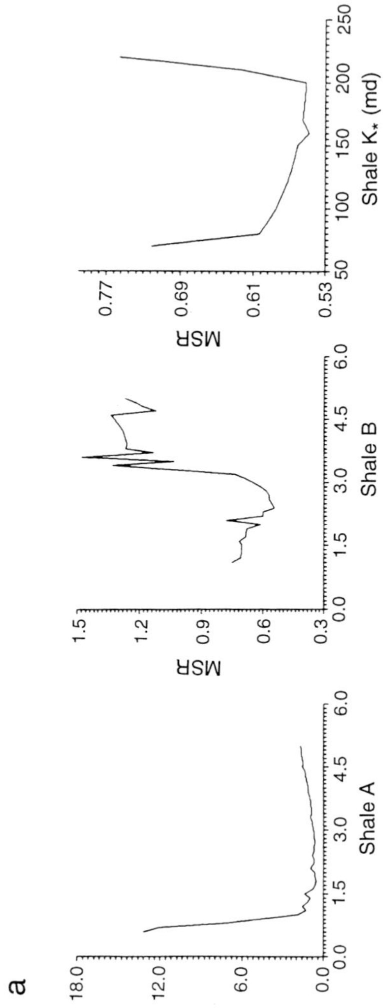
a
MSR
18.0
12.0
6.0
0.0
0.0
1.5
3.0
4.5
6.0
Shale A
MSR
1.5
1.2
0.9
0.6
0.3
0.0
1.5
3.0
4.5
6.0
Shale B
MSR
0.77
0.69
0.61
0.53
50
100
150
200
250
Shale K* (md)

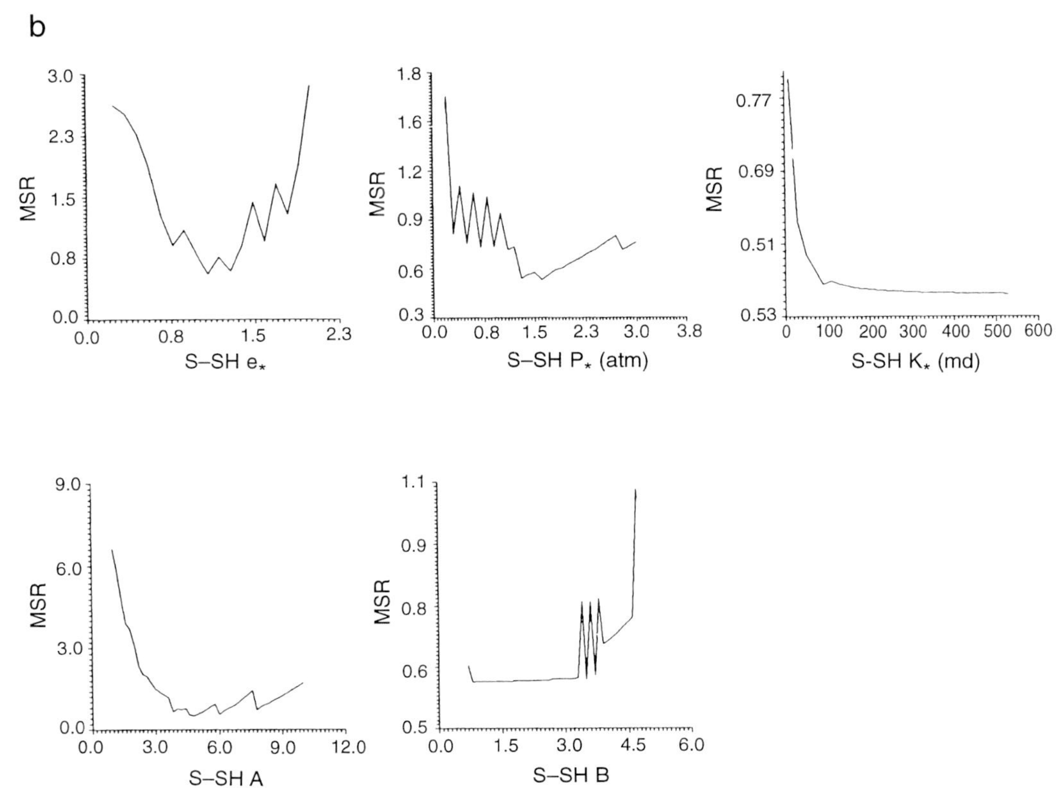


FIGURE 3.13 MSR variation with parameters related to lithologies for (a) shale and (b) sandy-shale using the linear search method with pore pressure, porosity, and formation thickness as dynamic indicators.

that the parameters found will be maximally resolved without the inclusion of thermal maturation indicators (vitrinite, sterane and hopane isomers, sterane aromatization, pollen translucency, apatite fission tracks, hydrogen index, etc.), nor is it obvious that parameters that influence only thermal indicators will then be resolved correctly. The point here is that if the thermal indicators have *not* been involved in determining dynamic parameters, but are taken to be merely passive receptors of dynamic parameters, then the sensitivity of a thermal indicator to determining a set of parameters is restricted to those parameters which are *not* part of the dynamic set—something which forces all uncertainty into paleoheat flux evolution and kinetic parameters, a less than desirable option.

There is a point worth emphasizing here, as has been done elsewhere (Lerche, 1990a,b) in great detail. The outputs of any basin analysis model are beholden to three intrinsic input factors: (1) intrinsic assumptions of the model (e.g., 1-D, 2-D, isostatic basement motion, etc.); (2) input parameters in equations of state (rock density, water density, permeability/void ratio relations, etc.); and (3) the quality, quantity, and sampling frequency of data used as constraint information.

Although the influence of different intrinsic assumptions on output results can be easily tested by using different models, and the parameters in equations of state can be resolved (as done in this chapter) and analyzed for the sensitivity of output results to changes in input values, the control exerted by the constraint data is often one of the crucial ingredients. Indeed, it can, and does, happen that a particular method is, *per se,* highly regarded as an accurate device; but it can also happen that the data available from which to extract estimates are sufficiently sparse, or of poor quality, or sampled so nonuniformly that the ability of the method is compromised by the vagaries of the data. In such a case it is not the method which is at fault, but rather the sharpness of resolution attainable with a particular set of data. In most cases one cannot obtain a better set of (or more) geological data, so the resolution is then whatever it is; one can, of course, use the resolution obtained to argue that the next set of data obtained in a different area should include the required improvements. While such a learning curve situation is useful, it does not obviate the fact that with any given set of data some output results are required, and the resolution, sensitivity, precision, and uniqueness of the outputs must all be assessed in some quantitative, objective, reproducible manner.

This section concentrates on the first major problem: the effect of multiple-well information on the ability of the general inverse procedure to resolve dynamic parameters. Perhaps the point here has not so much to do with the quantitative development of the formal mathematical methods, but rather how well they operate in real situations to delineate or determine dynamic parameters.

The computer time needed to perform the calculations can be estimated as follows. Suppose it takes a computer time of CPU(2-D) to run through the two-dimensional basin model code once. Then to calculate the derivative of $Y(\mathbf{a})^2$ [Eq. (3.9b)] for $Q$ components of $\mathbf{a}$ requires a computer time $(Q + 1)$CPU(2-D). If the parameter vector $\mathbf{a}$ is updated once, so that $\mathbf{a}_1$ replaces $\mathbf{a}_0$ throughout, then the process is repeated again. Thus for $N$ nonlinear iterations the total computer time is $(N + 1)(Q + 1)$CPU(2-D). For instance, if 10 nonlinear iterations of just five parameters are done on a computer requiring 1 hour of CPU time to run a 2-D basin modeling code once, then a total of 66 CPU hours will be required.

Although one has an absolute guarantee that the set of parameter values after the $N$ iterations provides a mismatch to all the observations which is smaller than that using the original parameter values, and although this systematic procedure is guaranteed to be a significant improvement over random trial-and-error searches in $Q$ dimensions, nevertheless the total CPU time involved can be considerable. Two strategies are available to speed up the process: (1) Use the fastest machine possible to cut down the basic CPU time per run and (2) use different strategies to determine which parameters are likely to dominate the system and concentrate on obtaining a minimum mismatch with them first. For instance, a 1-D basin analysis code will often run in seconds to minutes. Thus if the same number of iterations is performed on the same number of parameters at $X$ well locations but using a 1-D code, then total CPU time will be $X(N + 1)(Q + 1)$CPU(1-D). This strategy will produce $X$ values for each of the $Q$ parameters, each of which fits a *single* well best, but not all wells. The average of all the values of each parameter is then a likely good starting value for a 2-D run so that, often, only minor changes are needed to obtain a good fit. The requirement for such a strategy is that

$$X\ \mathrm{CPU(1\text{-}D)} \ll \mathrm{CPU(2\text{-}D)}.$$

Thus if the 1-D code runs in 1 minute and the 2-D code in 2 hours, then the $Q$ parameters should be evaluated for less than 60 wells; otherwise, no saving in time is achieved over a direct run of the 2-D code. Many other strategies are available, of course. But there does not yet seem to be available a unique royal road to success. Welte (1994) has suggested that a different strategy is likely to emphasize a better geological data set.

In the same vein Dunbar (1994) has pointed out that "In other more complexly structured basins, lateral flow through sands and up faults dominates the dewatering process. In such cases the 1-D solution would be far from the correct solution and may lead to convergence to an incorrect minimum." If such a difficulty is suspected then, of course, one runs the 2-D inverse model *de novo* to investigate the likelihood of such a problem having occurred. The price that is paid is more computer time, but such a price is well worthwhile to alleviate suspicions, and it is a very cheap price to pay relative to drilling a well at a probably incorrect site.

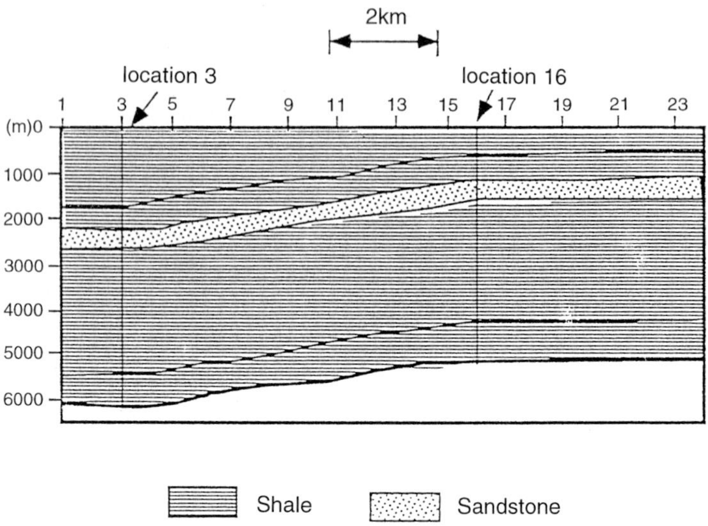


FIGURE 3.14 Synthetic 2-D cross-section with five layers and 20.5 km in lateral extent. At locations 3 and 16, porosity, permeability, and fluid pressure variations with depth, as well as the formation thickness, are checked at each inversion iteration.

The question, as always, is this: Does the inverse procedure produce a result that is practically useful? If so, then the computer time spent is worthwhile. To answer this question, we consider here both a synthetic case history and also a real case history to illustrate the behaviors that result.

### A. A SYNTHETIC CASE HISTORY

The simple synthetic 2-D cross-section displayed in Fig. 3.14 consists of a thin sheet sand immersed in a shale matrix. The present-day section varies between 6000 to 5500 m in total thickness, progressing left to right across the section. The synthetic data were constructed by doing forward calculations using the 2-D fluid flow/compaction code GEOPETII (developed at the University of South Carolina), which has been described in detail elsewhere (Nakayama, 1987; Yu, 1992; Lerche, 1990a,b). The forward calculations were done with a fixed set of parameters for each lithology. At both locations 3 and 16 of figure 3.14 information is available on porosity, thickness, permeability variations with depth, as well as fluid pressure profiles, as depicted in figures 3.15 and 3.16. Of interest here are two parameters: the parameter $A$ in the void rate–frame pressure equation of state for shales $(e/e_*) = (P_f/P_{f_*})^A$, and the parameter $B$ in the permeability—void ratio equation of state for shales $(\kappa/\kappa_*) = (e/e_*)^B$. All other parameters are retained at their forward calculational input values; thus, $e_* = 3$ (deposi-

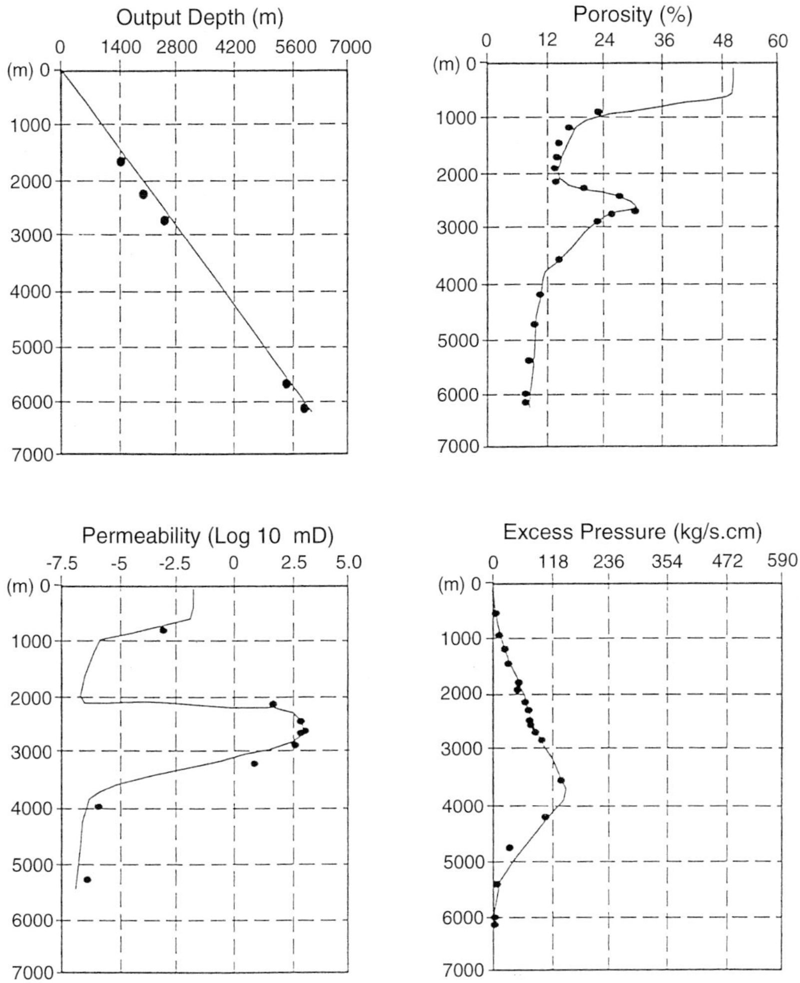


FIGURE 3.15 Comparison of the input and output dynamic indicators of formation depth, porosity, permeability, and fluid pressure respectively at location 3. Solid lines show the tomographic results and the dots indicate input measurement data.

tional surface void ratio), $P_{f_*} = 1\ \text{kg} \cdot \text{cm}^{-2}$ (scaling value for frame pressure), and $\kappa_* = 1$ mD (depositional surface value). The input values of $A$ and $B$ are $A = -0.40$ and $B = 4.3$, which were used to produce Fig. 3.14 and also the porosity, permeability, and pressure variations with depth of Figs. 3.15 and 3.16.

For this synthetic illustration, the inverse tomographic procedure uses the formation thicknesses at locations 3 and 16, as well as the fluid pressure, permeability, and porosity with depth profiles at the same two locations.

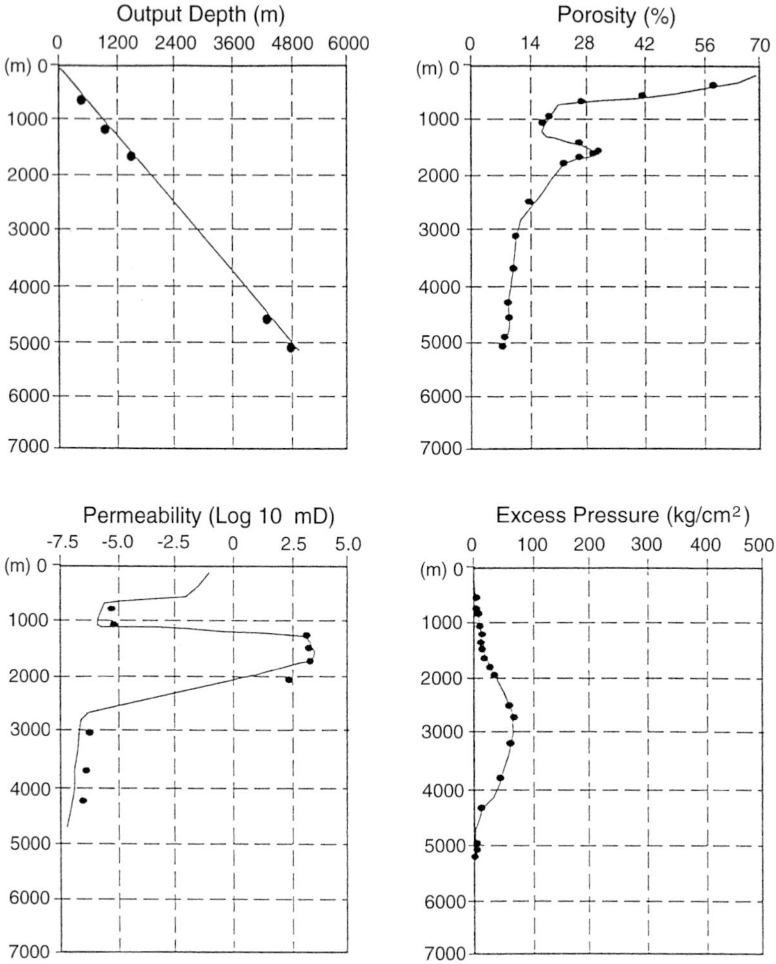


FIGURE 3.16 Comparison of the input and output dynamic indicators of formation depth, porosity, permeability, and fluid pressure respectively at location 16. Solid lines show the tomographic results and the dots indicate input measurement data.

Initial estimates for the starting values of $A$ and $B$ to use in the full 2-D dynamic inverse procedure were provided by first running a 1-D dynamic inverse code at each of locations 3 and 16, as though there were only vertical flow effects present. Thus each well was initially considered in isolation. The corresponding $A$ and $B$ values that best fit each well individually were, of course, different because of the omission of two-dimensional effects. The average of each was taken as a first input estimate to the full 2-D code yielding the estimates $A \approx -0.55$ and $B \approx 4.6$. The 2-D dynamic inverse code was then run through 10 nonlinear iterations, when the mis-

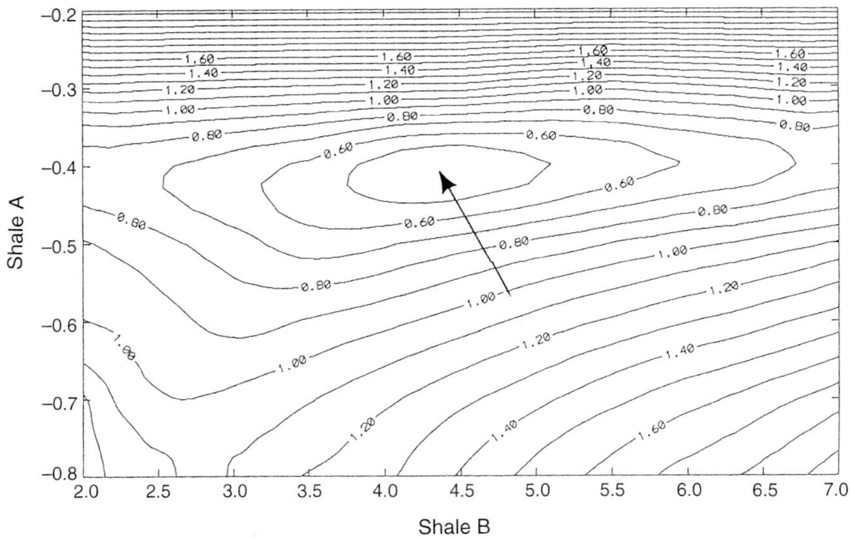


FIGURE 3.17 A contour map of normalized mismatch of dynamic indicators in the parameter space of shale $A$ and shale $B$ obtained using the systematic linear search method. The tomographic search path (starting with $A = -0.55$ and $B = 4.6$) is superposed, showing the minimum value reached after 10 iterations.

match between predicted and observed quantities at locations 3 and 16 had reached a numerical minimum in MSR. To ensure that a global minimum had been reached, two procedures were then followed. First, a linear search along lines of constant $A$ (constant $B$) was performed around the minimum, and the contours drawn of normalized constant MSR mismatch. The path followed in $A$, $B$ space by the nonlinear tomographic inversion procedure was then superposed on this contour diagram (solid arrow line), as shown as Fig. 3.17. The tomographic variation path uniformly heads to the minimum, and the double linear search shows that it is the only minimum. Second, holding parameter $A(B)$ at its best value (i.e., lowest MSR), the $B(A)$ parameter was then scanned in a range around the best $A(B)$ value in small steps to ensure that no minimum had been overlooked. The results of these two linear searches are shown in Fig. 3.18 with MSR versus $A(B)$ plots. The final minimum found corresponds precisely (to within numerical resolution) to the forward model input values of $A = -0.40$ and $B = 4.3$.

The corresponding predicted pressure, permeability, and porosity profiles with depth at locations 3 and 16 are shown by the solid lines on Figs. 3.15 and 3.16, demonstrating high-quality matches to observed values.

To further demonstrate that the tomographic method can be useful in bracketing any parameters in the GEOPETII system [in which there are about 378 possible adjustable parameters (Yu, 1992)], an example is given,

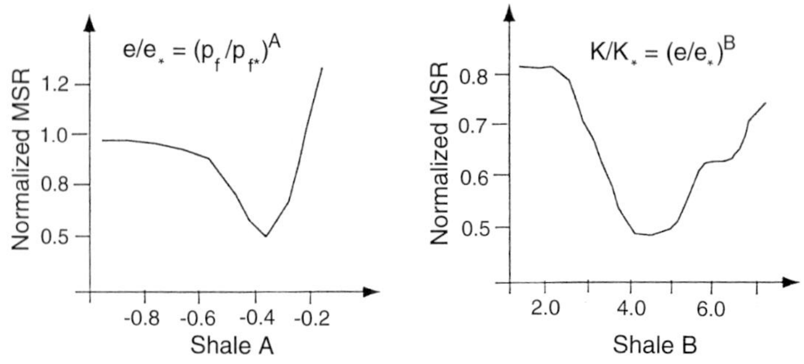


FIGURE 3.18 Fine-scale linear search results for parameters for shale $A$ and shale $B$, obtained by varying one parameter at a time with the other fixed.

in the space of shale $A$ and rock fracturing coefficient, $C_f$ (rocks are taken to fracture when the total fluid pressure reaches the fraction $C_f$ of the vertical sedimentary load). The MSR contour map of dynamic indicators is plotted in Fig. 3.19 for the same synthetic section of Fig. 3.14 by running the linear search method first with $C_f$ and $A$. In this case an important

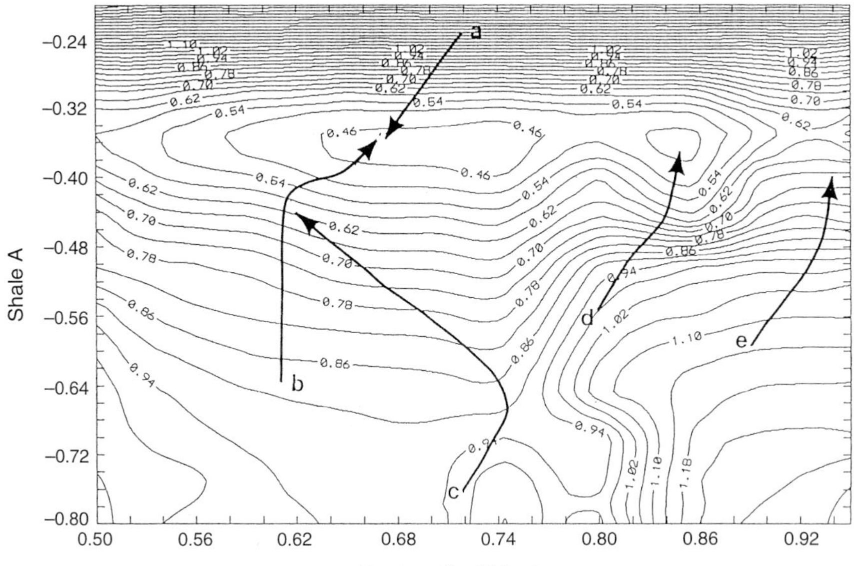


FIGURE 3.19 Mismatch contour map in the parameter space of the fracture coefficient $C_f$ and shale $A$ for the synthetic model of Fig. 3.14 and with the same input data controls shown on Figs. 3.15 and 3.16. Multiple minima are clearly shown. The starting point of a tomographic search determines which minimum will be approached as shown by the five paths labeled a through e.

point is uncovered by the linear search. There are multiple minima in the system and so the final tomographic results can depend on where the initial values start (Lerche, 1991). As shown in Fig. 3.19, five different $A$, $C_f$ parameter value pairs were chosen as initial places from which to start the tomographic runs. Each run was done with 20 iterations, and the tomographic paths taken are superposed on the MSR contours of Fig. 3.19. All paths head toward a minimum, but not always the same minimum. In practice, therefore, it is important to evaluate which of the minima provides the best degree of fit to the observations and, in the case of almost equal mismatches, to find other criteria to determine the sensitivity, resolution, and precision of different dynamic behaviors that would result for the basin, depending on which sets of parameters are used to describe the evolution.

The points of these synthetic examples are several-fold: First, the examples illustrate that the dynamic inversion procedure does recover parameter values correctly. Second, the strategy of first using a fast 1-D dynamic inversion code on each well in turn, and then using the average parameter values so obtained as first estimates for the full 2-D code, is a much faster procedure than using the 2-D code itself to search over a broad range for best starting estimates. Third, the ability to contour mismatch, and then to vary each parameter in turn while holding the remaining parameters at their best values, provides a useful strategy for not only assessing how many nonlinear tomographic iterations are needed, but also provides a powerful method to ensure that no other minima have been overlooked in the search ranges undertaken.

The consequences are that there is now available a fairly rapid and extremely powerful strategic procedure for addressing simultaneously the concerns of computer speed requirements and of maximal use of all available dynamic indicator information at multiple well locations.

## B. NAVARIN BASIN, ALASKA

Data for this field case were taken from Turner (1984), Steffi (1991), and Worrall (1991). Figure 3.20 presents the location of the cross-section A-A′ used in the investigation. The A-A′ section is intersected by the four wells: Amoco Misha No. 1, Exxon Redwood No. 1, Exxon Redwood No. 2, and Amoco Danielle. The nearest seismic cross-section available is that of A-B (on the line of cross-section) and B-B′ which is at a slight angular offset from B-A′, the line of the wells. Projection onto the line of section A-A′ was used to provide information at locations other than the four wells. Figure 3.21 shows the interpreted seismic profile.

The cross-section modeled, together with the four well locations and their TDs, is shown in Fig. 3.22. The total cross-section is about 240 km wide, with pseudo-well spacing at 5 km. A major wrench fault, nearly vertical (Worrall, 1991), intersects the cross-section at location 24 (120 km),

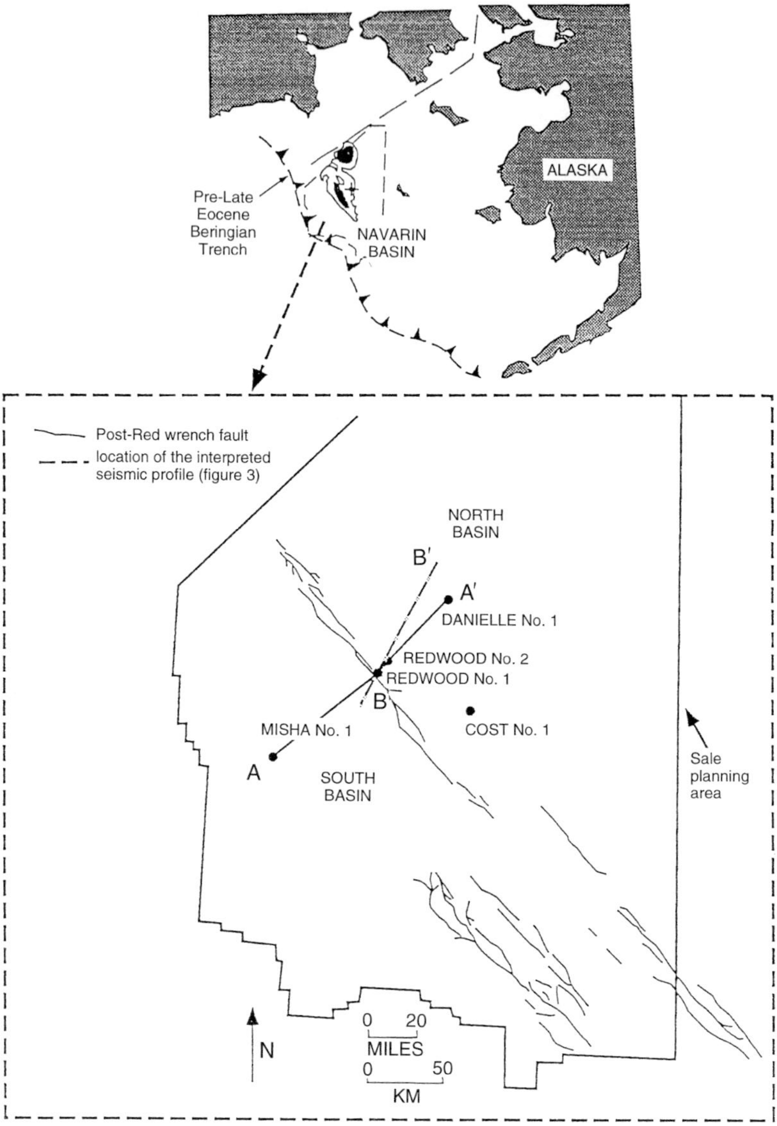


FIGURE 3.20 Location of cross-section A-A′ in the Navarin Basin with the four control wells: Amoco Misha No. 1, Exxon Redwood No. 1, Exxon Redwood No. 2, and Amoco Danielle. Line B-B′ shows the location of the seismic profile (Fig. 3.21) (modified from Worral, 1991).

causing a throw of about 4000 m in the lowermost formation overlying interpreted basement. The seismic expression of this part of the section is reproduced here as Fig. 3.21 (taken from Worrall, 1991) and corresponds to line B-B′ of Fig. 3.20. The Red, Ochre, Green, and Blue horizons of

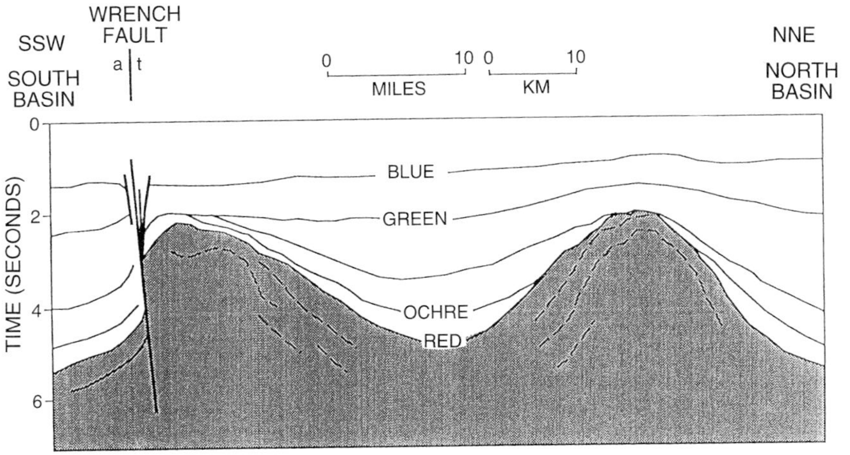


FIGURE 3.21 Cross-section interpreted from seismic section (modified from Worral, 1991). See Fig. 3.20 for location.

Fig. 3.21 have three obvious counterparts on the right side of Fig. 3.22, as does the unnamed horizon of Fig. 3.21 lying between Ochre and Green. The dominant geology of the area is well described elsewhere (Turner, 1984; Steffi, 1991; Worrall, 1991) as is the lithologic distribution throughout

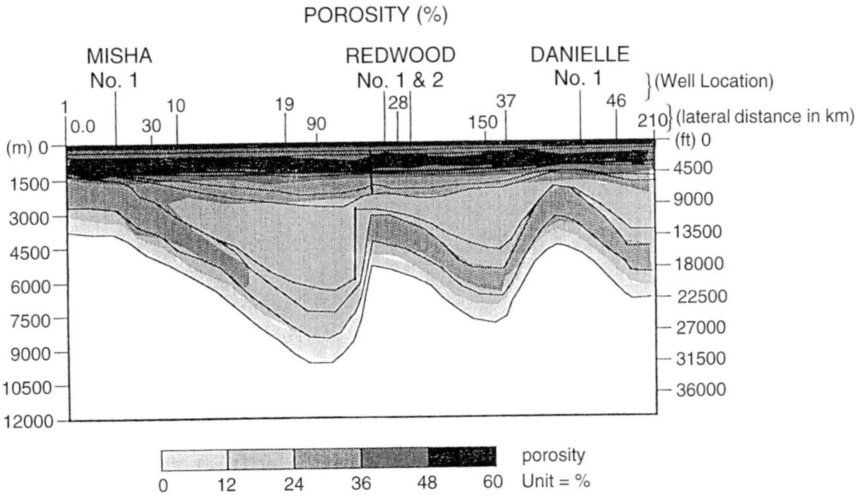


FIGURE 3.22 Modeled cross-section (A-A′) with predicted present-day porosity. Four control wells with porosity, fluid pressure, and formation thicknesses with depth provide dynamic indicators at locations 5 (Amoco Misha No. 1), 27 (Exxon Redwood 1), 29 (Exxon Redwood 2), and 43 (Amoco Danielle), respectively.

the basin. For the purposes of assessing the use of the 2-D dynamic inversion process in a real basin, an abbreviated description will suffice here.

**1. Geologic Description**

The Navarin Basin, a Tertiary sedimentary basin, is located on the outer continental shelf of the Bering Sea about 480 km west of mainland Alaska (Fig. 3.20). After the end of the Beringian trench activity, ~45 my BP, the Navarin Basin sustained a large sedimentary fill from Late Eocene through to the present (Worrall, 1991). The sedimentary section penetrated by the wells is composed mainly of clastic sedimentary rocks deposited in a marine environment. Locations of depocenters are related to the wrench tectonic activity of the basin, but have been better interpreted as the result of an elastic dislocation model (Rodgers, 1980), rather than a basic pull-apart basin model (Worrall, 1991). Indeed, no pull-apart normal faults are present, and the smoothly warped, prebasin surface of the Navarin Basin suggests that deformation occurs more by elastic-plastic stretching than by brittle deformation (Worrall, 1991).

Sediments pre–Eocene are considered as basement in the model because of the lack of data control for earlier rocks and the complex of erosion events that developed pre–Eocene. The Late Eocene organic-rich shales are of special petroleum interest because of their deposition in an anoxic environment during rapid subsidence.

Large excess pressure was developed for the shale/shale-sand section in the Navarin Basin due to undercompaction, fluid thermal expansion, and dehydration of smectites (Steffi, 1991). Measurements from RFT and well logs, such as shale velocities, show directly or can be used to infer the existence of high geopressure.

**2. Dynamic Tomographic Parameter Determination**

The procedure for determining parameter values basically follows the method developed for the synthetic case discussed previously. For each of the four wells, information to be matched consists of formation thicknesses, porosity profiles, and pressure profiles, as shown in Figs. 3.23 through 3.26. Bour and Lerche (1994) did a basin analysis study of the Navarin Basin in both one and two dimensions using a forward model only, without the dynamic inverse procedure. For our purposes here, the range and first estimates of parameter values can be chosen based on that previous study. The solid lines in Figs. 3.23 through 3.26 show the final, tomographically determined, predicted patterns. Table 3.4, given earlier, gives the parameter values used by Bour and Lerche (1994) and also the values determined in this study using the inverse tomographic method. Although the general geologic results agree in both studies, an improvement of 10% in the MSR mismatch between predicted and observed quantities is obtained by the tomographic method.

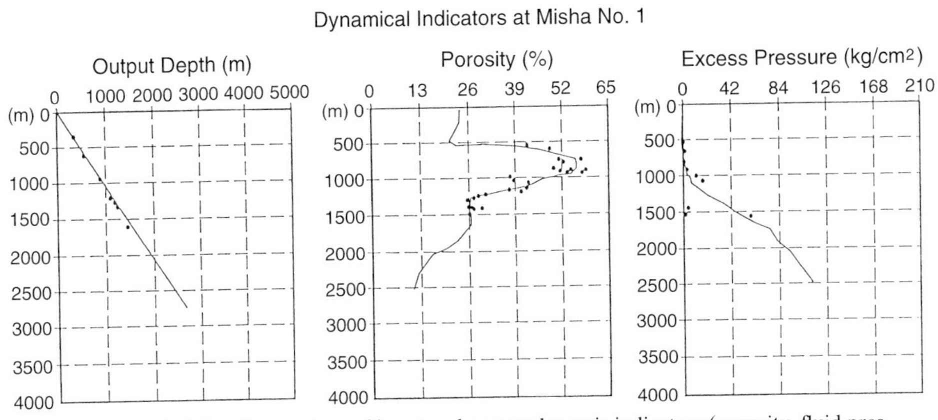


FIGURE 3.23 Comparison of input and output dynamic indicators (porosity, fluid pressure, and formation thickness) at location 5, Amoco Misha No. 1 well. Solid lines show the final tomographic prediction and the dots indicate input data.

For a sensitivity analysis four parameters were examined ($A$, $B$, $\kappa_*$, and $e_*$) for shales (see Fig. 3.27), although many other parameters were also included in the study to accommodate the various lithologies in the Navarin Basin. A 2-D linear search was performed with parameters $A$ and $B$ around their best values in order to draw the contours of mismatch of predicted and observed quantities (Fig. 3.28) with the remaining parameters fixed at their best values (i.e., those that give minimum MSR). Figure 3.28 indicates that only a single minimum is present for variations of the $A$ and $B$ values for shale, with the minimum located at about $A = -0.40 \pm 0.05$ and $B = 1.3 \pm 0.5$. Guided by this fact, based on MSR contours, and the rough

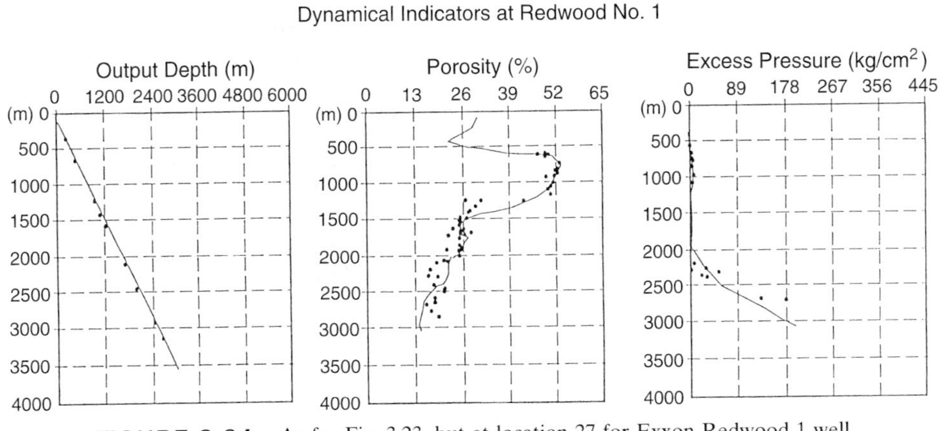


FIGURE 3.24 As for Fig. 3.23, but at location 27 for Exxon Redwood 1 well.

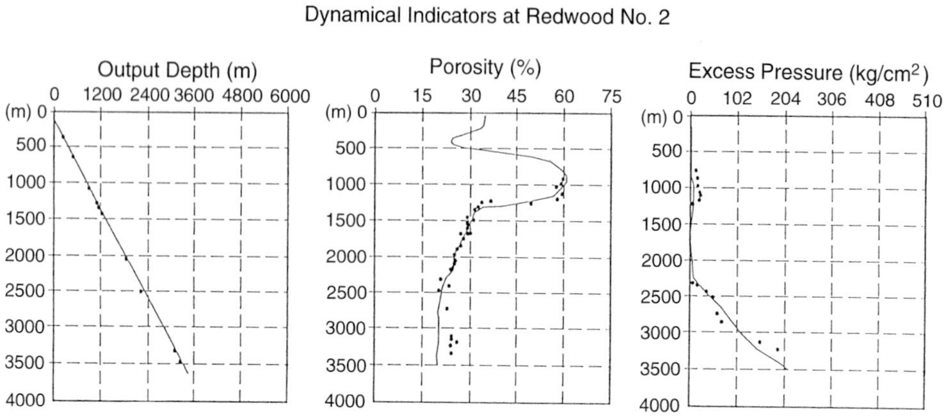


FIGURE 3.25 As for Fig. 3.23, but at location 29 for Exxon Redwood 2 well.

ranges of parameter values from the linear search procedure, more accurate values of the parameters with better matches of predictions to observations are obtained by the tomographic method, as given in Table 3.4, with a run of 10 nonlinear iterations.

Table 3.4 also provides the relative normalized global MSR for formation thickness, porosity, and fluid pressure at the four wells. With an MSR of unity set for the parameter values used in Bour and Lerche (1994), an improvement in MSR to a minimum value of 0.90 was obtained by the tomographic procedure, indicating at least a 10% improvement of fit to the data and also a better control of basinal burial history reconstruction in two dimensions, as well as of the concomitant thermal history and hydrocarbon generation behavior. The parameters for each of the *multiple* lithologies

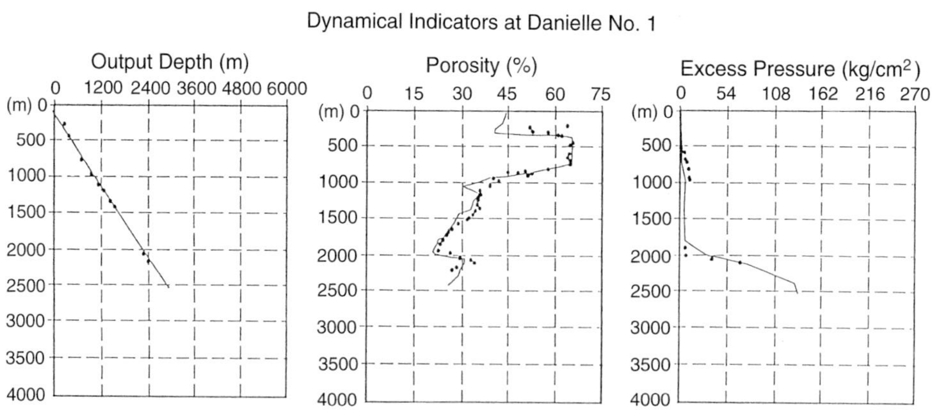


FIGURE 3.26 As for Fig. 3.23, but at location 43 for Amoco Danielle well.

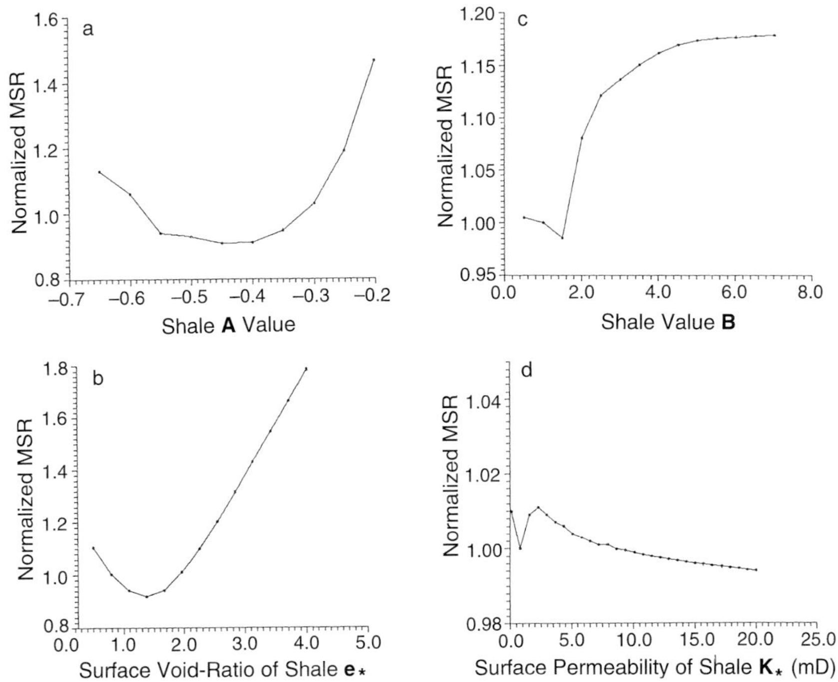


FIGURE 3.27 Sensitivity analysis of shale $A$, shale $B$, surface void ratio $e_*$, and surface permeability of shale $\kappa_*$, showing MSR variation in the given space, with the other parameters fixed at their best values.

for the Navarin Basin can also be evaluated by the tomographic method. An exhaustive exhibition of all the parameters for other lithologies in the Navarin Basin would be a major piece of work, without adding to the point that such things can be done. However, a "snapshot" of one linear search, with the two parameters $A$ for both sandy-shale and shale, respectively, is presented in Fig. 3.29 to ensure that the parameter sets determined here from minimal MSR consideration are not overlooked by other minimal solutions when the influence of various lithologies is involved. Fortunately, in this case, only a single minimum is uncovered and the values of parameter $A$ for both sandy-shale and shale, respectively, as constrained from the linear search map given in Fig. 3.29, are in accord with the values determined (and used) by the tomography method (see Table 3.4).

The corresponding predicted variations of porosity and fluid pressure across cross-section A-A′ are then as given in Figs. 3.22 and 3.30, respectively, and are in maximal accord with the observed variations at the four well sites. In brief, the main overpressure, deeper than about 1500 ± 200 m and reaching about 800 atm in the central graben, probably is the dominant factor driving fluid movement in the basin toward the shallower and mar-

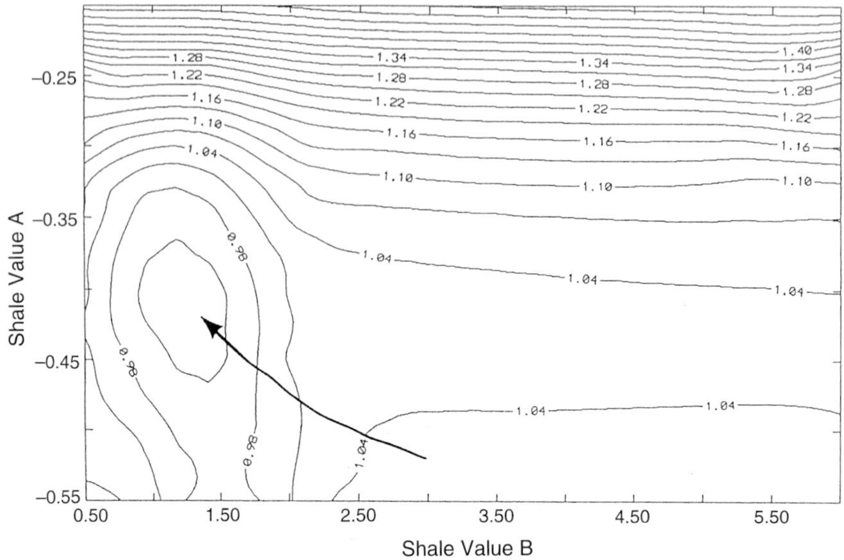


FIGURE 3.28 Normalized MSR contours in space of shale *A* and shale *B* with dynamic indicators of formation thicknesses, porosity, and fluid pressure at four wells, Navarin Basin, Bering Sea. Solid arrow line shows the path of the tomography search.

ginal-ward directions, which could lead to accumulations when traps, such as up-dip lithologic sandy layers, faulted blocks, and syndepositional structures overlying horsts, are encountered. The rapid deposition of low-permeability shale materials and/or thermal expansion are the main causes for the extensive overpressure in the Navarin Basin. Porosity retention within the overpressured zone (Fig. 3.22) provides a favorable condition for hydrocarbon accumulations in the basin. Further discussion of the general basin evolution, and potential exploration strategies for hydrocarbon accumulations are given in Bour and Lerche (1994).

## VIII. DISCUSSION AND SUMMARY

Although more than 300 parameters can be adjusted in the 2-D system, it is impractical for any single run to determine all of these parameters, considering the computer CPU time required and the specifics of a particular basin. Therefore, a question here is how to select the parameters to be determined.

Generally, the parameters in the 2-D model fall into three groups with respect to their influence on the dynamic and thermal evolution of the basin. The first group of parameters is dynamical parameters, such as the

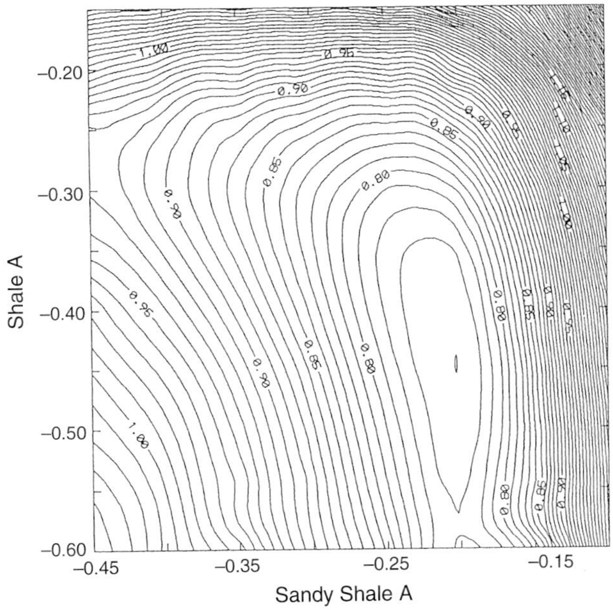


FIGURE 3.29 Normalized MSR contours in space of shale $A$ and sandy-shale $A$ with dynamic indicators of formation thickness, porosity, and fluid pressure at four wells, Navarin Basin, Bering Sea.

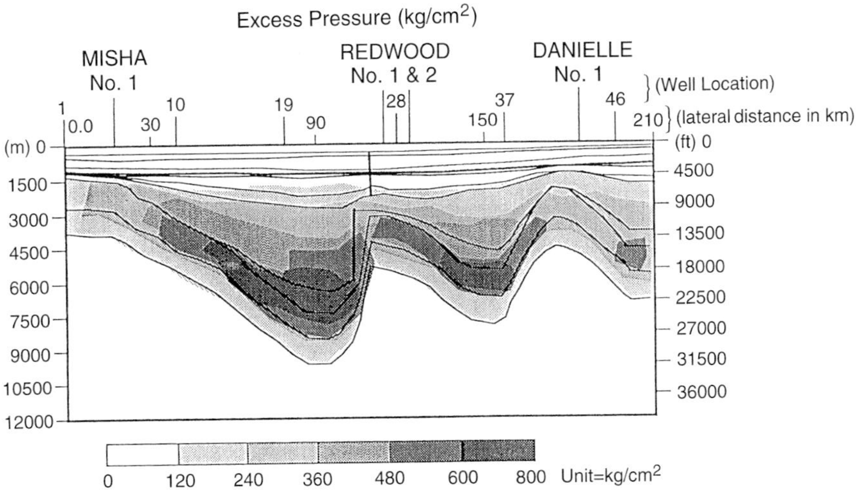


FIGURE 3.30 Two-dimensional predicted pressure distribution on the cross-section A-A′, Navarin Basin, showing overpressure at present day.

$A$, $B$, $C$, $e_*$, and $P_{f_*}$ values for each lithology and the fracture coefficient $C_f$, which dominantly control the dynamic evolution as the formation pressure varies with the rate of fluid escape from the sediments, and as rock porosity as well as permeability evolve with compaction. The formation thickness (or total depth) is also directly related to the dynamic parameters. Therefore, dynamic indicator behaviors depend primarily on the variations of parameters in this group, but not exclusively. The second group of parameters are dominantly thermal controls, such as thermal conductivities of each rock type, which control the thermal character, maturity, and hydrocarbon generation in the basin. This group of parameters indirectly influences the dynamic indicator evolution. For example, a lower thermal conductivity value of shale will lead to higher temperature and so produce greater hydrocarbon generation and thermal expansion pressure, which comprise part of the dynamic fluid pressure in the basin. The third group of parameters are "neutral" parameters, which are relatively insensitive to dynamic indicators. This group consists of more or less basin-dependent parameters, such as densities of rock and fluid. Therefore, the question now devolves to the ability of a model to determine sensitivity, precision, resolution, and uniqueness (Lerche, 1990a,b).

Based on experience with dynamic and thermal tomography procedures in the single-well cases, in general a good recommendation is to choose the most sensitive parameters of the first group when applying dynamic tomography in the multiple-well case. It is also recommended that any single run should *not* exceed 20 parameters, and should also *not* be iterated more than 10 times in order to be efficient. Indeed one can always update the parameters after 10 iterations (two lots of 10 iterations each is more accurate than one lot of 20 iterations).

The assumption of invariance to spatial location of parameters entering the equations of state might weaken the ability of the tomography model. For example, the same lithological shale could have different dynamic parameters at the basin margin relative to the basin center. To compensate for this drawback in the model assumptions, spatial dependent lithologies must be specified. The GEOPETII system accommodates for this problem through user-defined lithologies.

The development of dynamic indicator tomography closely parallels the corresponding development for thermal indicator tomography. In essence, the idea is to use observed, or inferred, present-day data to constrain parameters or functions occurring in dynamic models of paleo-evolution of sediments. The logic for assessing the degree to which model parameters can be constrained, or determined, by present-day data is founded on a nonlinear inversion of least-squares control functions in which total depth, formation thicknesses, porosity, fluid pressure, permeability variations with depth, and depths to unconformities, act to constrain model behaviors.

This chapter has spelled out the main theoretical procedures for enabling quantitative numerical codes to be constructed in order to examine the resolution, precision, sensitivity, and degree of uniqueness with which information can be extracted concerning the dynamic aspects of compaction and evolution of sedimentary basins. The main factor that controls the evolution of compaction is the rate at which fluid can escape from the sediments. That loss is dominated by both the permeability variation and the excess fluid pressure gradient of shales, although other factors such as overburden load, unconformities, cementation, and so on play roles.

The inverse, nonlinear, dynamic tomography scheme set up here is designed to address directly the questions of which parameters control the detailed, dynamic development, the degree to which parameters can be determined or constrained, and the degree to which different hypotheses (e.g., open/closed faults; erosion/hiatus; equation-of-state parameters dependent/independent of spatial location) can be evaluated for their veracity. In addition to demonstrating that the design and implementation of a dynamic indicator tomography procedure are eminently practical goals for a single well (one-dimensional system), we also showed that both a two-dimensional generalization and a combined dynamic/thermal indicator tomographic procedure were implementable.

The advantage to having the preceding method available is that trial-and-error searches do not have to be undertaken for each and every parameter, something that is both time consuming and very difficult to implement systematically. The tomographic nonlinear inversion automatically incorporates a systematic method for determining multiple parameters and their sensitivity, while guaranteeing that parameters will be kept within any preset ranges.

Underlying the development of the multiple dynamic and/or combined thermal indicator methodology is the idea that multiple types and quantities of dynamic downhole information act as powerful constraints to provide checks and balances of one dynamic indicator against another, such as formation thickness, porosity with depth, and fluid pressure with depth. The point here is that different dynamic indicators spanning the same depth range in the same well must have been exposed to precisely the same burial path behavior and to the same variations of fluid pressure or fluid flow with time. Each dynamic indicator (TD, formation thickness, porosity, permeability, fluid pressure, unconformity depth) must then produce the same value for a parameter common to all indicators. By demanding such agreement, and because each dynamic indicator has its own particular dependence on time and fluid flow, the inverse procedure can extract those parameters which are most consistent with the downhole data of a particular dynamical indicator, and also with the geological parameters determined by inversion from other downhole dynamic indicators. Potential applications of the dynamic tomography procedure on its own, in one or two dimensions, and in combination with the thermal indicator tomography procedure developed previously (He and Lerche, 1989) are legion. For example, as we have remarked already, the determination of parameters associated with permeability is a crucial ingredient if we are to determine more accurately the temporal loss of fluid from a sedimentary basin. In turn, the loss of fluid impacts the evolution of paleo-overpressure, which is linked to both rock fracturing and to increased temperature (through retention of fluid), thereby influencing thermal indicator evolution and hydrocarbon generation. Or again, the presence of an erosional unconformity, and the determination of both the amount of sediment eroded as well as the timing of erosion, influences the dynamic evolution of sediments in a basin, the thermal history of the sediments, and the potential for hydrocarbon retention. Many such cases can be investigated for the ability of present-day dynamic and thermal indicator information to provide knowledge of the paleo-evolution of a basin. Indeed, it could be argued that *all* quantitative basin analysis requires such a treatment, leading to the exciting possibility of treating every sedimentary basin by such techniques.

In terms of the ability to determine the sensitivity, resolution, and precision of all parameters being considered, a thorough examination with both combined dynamic and thermal tomographic models is still a pressing concern, although the dynamic parameters are generally dominantly well determined by the dynamic indicators.

In the case of the synthetic model for the shale-sand section, with two pseudo-wells, the known parameter values are recovered by the dynamic inverse procedure. The procedure provides a way to better determine the parameters that control the basin evolution, and therefore to establish more precise burial history models, which are the basis of quantitative basin models.

Application of the method to the Navarin Basin, Bering Sea, Alaska, yielded improvement of predicted fit to the observed data at all well locations relative to the parameters used previously without the availability of the tomographic inverse model.

# 4

# Thermal Models of Sediment Maturation and Their Uncertainties

To assess the thermal history of sediments we need to determine how their temperatures changed with time. Four factors influence the temperature history of sediment layers: conductive transport, convective transport, focusing and defocusing by heterogeneous thermal conductivity variations, and basement heat flux, $Q(t, x)$ variations with time, $t$, and spatial location, $x$. In principle, if basement heat flux through time is known, then the remaining factors also are known if the burial history is accurate. It is possible to construct two end-member models: forward models with undetermined parameters and pseudo-inverse models based on present-day observations.

## I. Forward Models of Paleoheat Flux

McKenzie (1978) argued that rapid stretching of continental lithosphere produces thinning and passive upwelling of hot asthenosphere, associated with block faulting and subsidence. After the heating event, the lithosphere thickens by cooling, and the surface slowly subsides, typically without faulting. The exponentially decreasing rate of subsidence and heat flow depend only on the amount of stretching, which can be estimated numerically. In the McKenzie model, the initial subsidence ($S_i$) is given by

$$S_i = a\left((\rho_0 - \rho_c)\left(\frac{t_c}{a}\right)\left(1 - \alpha T_1\left[\frac{t_c}{2a}\right]\right) - \frac{\alpha T_1 \rho_0}{2}\right)\left(\frac{1 - 1/\beta}{[\rho_0(1 - \alpha T_1) - \rho_w]}\right), \tag{4.1}$$

where $a$ is the thickness of the lithosphere, $t_c$ is the initial thickness of the continental crust, $\rho_0$ is the density of the mantle, $\rho_c$ is the density of the continental crust, $\rho_w$ is the density of seawater, $\alpha$ is the thermal expansion coefficient of the mantle, $T_1$ is the temperature of the asthenosphere, and $\beta$ is the crustal stretching factor.

For large values of time, $t$, and small values of $\beta$, the surface elevation, $E(t)$, is given by

$$E(t) = \frac{4a\rho_0\alpha T_1}{\pi^2(\rho_0 - \rho_w)} \exp(-t/\tau), \tag{4.2}$$

where

$$\tau = a^2/\pi^2\kappa \tag{4.3}$$

and $\kappa$ is the thermal diffusivity. The heat flow, $Q(t)$, is given by

$$Q(t) = \left(\frac{KT_1}{a}\right)\left[1 + \left(\frac{2\beta}{\pi}\right)\sin(\pi/\beta)\exp(-t/\tau)\right], \tag{4.4}$$

where $K$ is the thermal conductivity.

For small values of $\beta$ and large values of time, a straight-line fit that best approximates Eq. (4.2) for the subsidence is given by

$$Y = E_0 r\left[\left(\frac{2t}{e\tau}\right)^{1/2} - 2e^{-1/2} + 1\right], \tag{4.5}$$

where $E_0 = 4a\rho_0\alpha T_1/\pi^2(\rho_0 - \rho_w)$, $r = (\beta/\pi)\sin(\pi/\beta)$, and $\tau = 62.8$ My. The subsidence predicted by the McKenzie model is similar in form to that predicted by the Turcotte–Ahern model (1977).

In the McKenzie model, prior to the time of stretching, the heat flow is given by the leading coefficient $KT_1/a$ in Eq. (4.4). Sclater and Christie (1980) noted that the present-day heat flow (approximately $0.8 \times 10^{-2}$ cal/m$^2 \cdot$s) predicted by the McKenzie model is lower than present measurements of heat flow in the North Sea (approximately $1.5 \times 10^{-2}$ cal/m$^2 \cdot$s). Sclater and Christie modified the McKenzie model by adding another heat flow component of $0.8 \times 10^{-2}$ cal/m$^2 \cdot$s from the decay of radioactive elements within the crust. Another modification was proposed by Keen *et al.* (1981), in which the effect of thermal blanketing by low thermal conductivity sediments is included, slowing the cooling of the lithosphere. This effect is claimed to be particularly important to studies of older rifted margins, such as the North Sea, where sediment thicknesses on the order of 15 km (49,000 ft) occur. Royden and Keen (1980) and Hellinger and Sclater (1983) proposed a modified stretching model encompassing two layers to explain uplift and erosion on the flanks of a rift basin.

Royden and Keen (1980) proposed another model that provides an exponential subsidence compatible with observational data. This model assumes fracturing of the continental lithosphere and its intrusion by dikes from the mantle. Replacement of light crustal rocks by denser ultramafic materials results in initial subsidence and avoids the general problem of uplift. If $\gamma_d$ is the fraction of lithosphere composed of dike material intruded from the asthenosphere, then the heat flow $Q(t)$, and surface elevation $E(t)$, respectively, are given by

$$Q(t) = \left(\frac{T_m K}{l}\right)\left(1 + 2\gamma_d \sum_{n=1}^{\infty} \exp(-n^2\pi^2\kappa t/l^2)\right) \tag{4.6}$$

and

$$\begin{aligned} E(t) = {} & \left(\frac{\alpha l \rho_m T_m}{(\rho_m - \rho_w)}\right)\left(\frac{4\gamma_d}{\pi^2}\right) \\ & \times \sum_{n=0}^{\infty}\left[\frac{1}{(2n+1)^2}\right] \exp[-(2n+1)^2\pi^2\kappa t/l^2], \end{aligned} \tag{4.7}$$

where $T_m$ is the temperature at the base of the mantle, $l$ is the thickness of the lithosphere, $\kappa$ is the thermal diffusivity of the lithosphere, $\alpha$ is the thermal expansion coefficient of the mantle, $\rho_m$ is the density of the mantle, $\rho_w$ is the density of water, and $K$ is the thermal conductivity.

Subsidence may be computed from the surface elevation and, because subsidence also is *inferred* from the burial history, the discordance between model predictions and observations can be minimized by parameter adjustment. These adjusted parameters are then used directly in the posited heat flux models. These examples show how model strategies may be mixed for increased effectiveness; the subsidence components are treated as pseudo-inverse models, but the heat flux components are treated as forward models.

These theories of basin formation rely on thermal contraction of the crust to produce subsidence. An alternative explanation, proposed by Falvey (1974), and expanded by Middleton (1980) and Falvey and Middleton (1981), suggests that the initial stage of subsidence is caused by deep crustal metamorphism in basins of relatively young, or unaltered, crust. Falvey (1974) demonstrated that elevation of the geotherm during continental rifting may produce subsidence along the margins due to deep crustal metamorphism. He also suggested that some intracratonic basins may have similar origins. The deep crustal metamorphism mechanism implies a subsidence of

$$\frac{\rho_m(\rho_2 - \rho_1)}{\rho_1(\rho_m - \rho_s)} L(t), \tag{4.8}$$

where $L(t)$ is the thickness of greenschist facies rocks metamorphosed to amphibolite facies at time $t$, $\rho_1$ is the density of the greenschist facies, $\rho_2$ is the density of the amphibolite facies, $\rho_s$ is the density of sediment, and

$\rho_m$ is the density of the upper mantle. Falvey (1974) suggested that the metamorphism driving basin subsidence occurs predominantly at the greenschist–amphibolite facies boundary. The model entails a period of heating followed by a period of cooling, with subsidence occurring in two stages. The initial subsidence, which occurs during the latter part of the heating period is attributed to deep-crustal metamorphism and continues as long as crustal temperatures increase. As temperatures decline during the subsequent cooling period, subsidence continues due to thermal contraction of the lithosphere. The time of transition between the heating phase and the cooling phase is $t_1$.

The downward movement, $l(t)$, of the earth's surface due to metamorphism is offset by the tendency for uplift, $h(t)$. Total displacement, taking these two factors into account and neglecting erosion, is given by

$$S(t) = (l(t) - h(t)) \tag{4.9}$$

for $l(t) < h(t)$, and by

$$S(t) = \left(\frac{\rho_m}{\rho_m - \rho_s}\right)(l(t) - h(t)) \tag{4.10}$$

for $l(t) > h(t)$, where $-S(t)$ is uplift, $\rho_m$ is the density of mantle, $\rho_s$ is the density of the sediment, and $\rho_m/(\rho_m - \rho_s)$ is the isostatic loading factor. Subsidence due to thermal contraction is given by

$$S(t) = \left(\frac{\rho_m}{\rho_m - \rho_s}\right)(h(t_1) - h(t) + h(t - t_1)), \tag{4.11}$$

where $t_1$ is the time of transition between the heating and cooling phases.

The distinguishing features of the Falvey–Middleton metamorphism model are that heat flow *increases* to the point of breakup and thereafter *decreases.* This behavior contrasts with other models in which heat flow decreases after the time of stretching. The heat flow from the metamorphism model is on the order of half that of the crustal extension model.

These geologic models for basin formation predict different heat flows through time. If we are to use these models quantitatively to predict heat flow, it is necessary to determine the optimum set of parameters for a given model for specific wells. Even then there are major differences among predicted histories, all consistent with inferences based on present-day data. Figure 4.1 compares the McKenzie (1978) and Falvey–Middleton (1981) models of heat flux tied to subsidence. The models have a discordance in heat flux that varies by a factor of 2. Subsidence histories described by these models may be evaluated by comparison with inferred subsidence from a burial history model. The large discordances in reconstructing estimates of paleoheat flux are caused by the different model *postulates* that

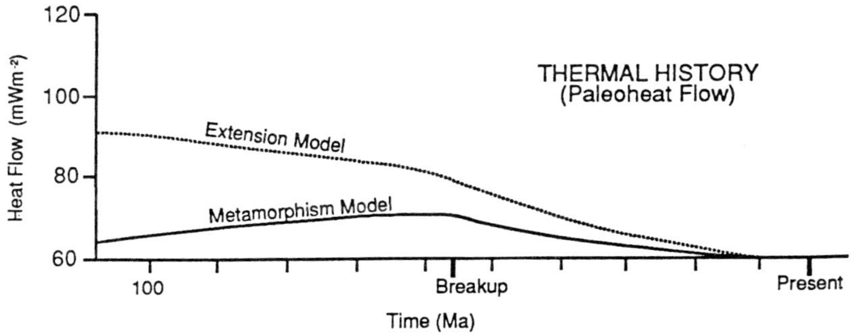


FIGURE 4.1 Comparison of heat flow resulting from McKenzie crustal extension model, derived using stretching factor of $\beta$ = 1.8, and low heat flow from Falvey deep crustal metamorphism model (see Falvey and Middleton, 1981).

connect modeled basement heat flux changes with the dependent subsidence (elevation).

## II. INVERSE AND PSEUDO-INVERSE MODELS OF PALEOHEAT FLUX

### A. GENERAL OBSERVATIONS

Dissatisfaction with forward heat flux models has focused attention on inverse methods applied to measured thermal indicators. If the chemistry and/or physics of the kinetic evolution of a thermal indicator is known, and if the burial history also is known, then the evolution of a thermal indicator from deposition to the present day depends solely on the temperature variation along the burial path of the sediments. By connecting the temperature to basement heat flux through a conductive or conductive–convective model, and by measuring the thermal indicator today at a large suite of depths, paleoheat flux through time may be recovered by inversion. Detailed mathematical developments of inversion procedures for several different thermal indicators have been presented elsewhere (e.g., vitrinite reflectance: Toth *et al.*, 1981, Lerche *et al.*, 1984; sterane and hopane isomers and sterane aromatization: MacKenzie and McKenzie, 1983; Armagnac *et al.*, 1988; Lerche, 1987; $Ar^{39}/Ar^{40}$ variations: Harrison and Be, 1983; Lerche, 1988a; pollen translucency: Grayson, 1975; McKenna and Lerche, 1991; apatite fission tracks: Gleadow *et al.*, 1983; Huntsberger and Lerche, 1987; Lerche, 1988a,b).

Thermal indicator inversion schemes for one-dimensional systems are constructed with the following logic. In a one-dimensional burial history, variations of *all* quantities can occur only with depth and time. Therefore the temperature, $T(z, t)$, varies only with depth, $z$, below the sediment–water interface and with time, $t$. Because the position of a sedimentary stratum changes with time we reference the $i$th layer by its present position, $z_i$. If the sedimentation, subsidence, and compaction history of the basin are known, the position of any stratum through time is given by the function $z = \zeta(t; z_i)$ and the time of sedimentation of any stratum is given by the function $\tau_i = \tau(z_i)$. Because pressure, $p(z, t)$, also affects the physical and chemical evolution of the sediments, we will assume that it, too, is known. A thermal indicator, $\mathrm{TI}(z)$, is then related to the temperature history of the stratum in which it is measured:

$$\mathrm{TI}(z) = \int_{\tau(z_i)}^{t_{\text{present}}} F[(\zeta(t; z_i), t), p(\zeta(t; z_i), t)]\, dt, \tag{4.12}$$

where $F$ is a known function.

If several different thermal indicators, $\mathrm{TI}_j(z_i), j = 1, \ldots, N$, are measured at several depths, $z_i = 1, \ldots, M$, can we reconstruct the temperature history, $T(z, t)$? The temperature field is a function of position and time, and because temperature must obey the heat flow equation, the problem may be transformed into one with a one-dimensional unknown. The heat flow equation may be written

$$LT(z, t) = 0, \tag{4.13}$$

where $L$, the linear differential operator of the heat flow equation, is assumed known. That is, the thermal conductivity, $K$, heat capacity, and source terms are known. Then the temperature is completely determined by an initial condition (e.g., sediment is added to the basin with a given temperature) and two boundary conditions [e.g., the temperature at the top of the basin is at the known value, $T_s(t)$, and that the heat flux through the bottom of the basin is $Q(t) = -K(\partial T/\partial z)$]. Thus we can replace the problem of determining the entire temperature field with the problem of determining the heat flux, $Q(t)$, at one depth in the basin. The unknowns are reduced from a two-dimensional function to a one-dimensional function. At any depth in the basin the temperature history $T(z, t)$ is described by:

$$T(z, t) = T_s + Q(t) \int_0^z \frac{dz'}{K(z')}. \tag{4.14}$$

This equation applies if the thermal recovery time through the sedimentary overburden is shorter than the timescale marking changes in the basement heat flux, and if convective heat transfer is either minimal or accounted for in dealing with real basins.

Inserting $T(z, t)$ into Eq. (4.12) expresses the thermal indicator $\mathrm{TI}(z)$ in terms of the one unknown, $Q(t)$. By measuring $\mathrm{TI}(z)$ at numerous depths, it is possible to invert to obtain directly the unknown heat flux variation

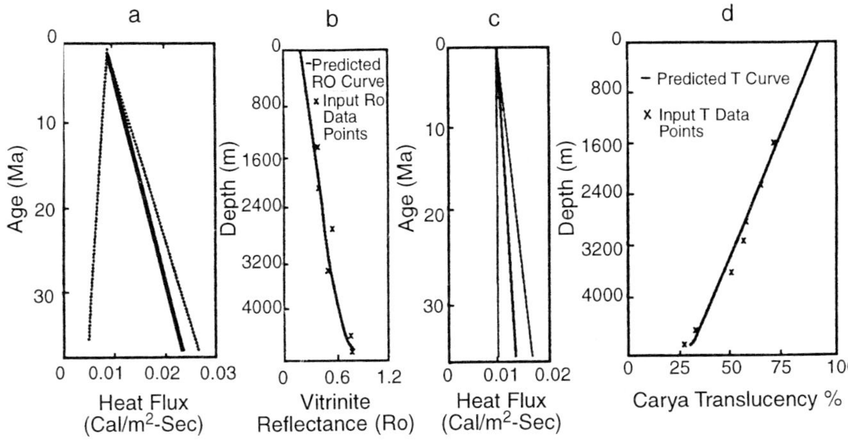


FIGURE 4.2 Heat flux variations with time in well Pan Am No. 27 based on pseudo-inversion of vitrinite reflectance data and on pollen translucency data. (a) Heat flux calculated from the vitrinite reflectance data shown in (b). (c) Heat flux calculated from the Carya translucency data shown in (d). Note the common domain of overlap of both indicators in predicted heat flux behavior. Also note the high degree of fit of observed versus predicted behavior with depth for both indicators (from McKenna and Lerche, 1987).

$Q(t)$. As shown elsewhere (Lerche, 1987a,b), the details change depending on the individual thermal indicators used, but the logic stands. Figure 4.2, from McKenna and Lerche (1991), shows the results of such an inversion procedure using both pollen translucency and vitrinite reflectance measurements.

With the exception of fission track measurements, all other thermal indicators known are *cumulative* recorders of the thermal history undergone by a particular formation. Fission tracks in a single crystal of apatite differ in this respect because, after deposition, there is continuous generation of new tracks with time in the buried apatite as uranium fissions. Thus, at a given depth of burial, a track length distribution measured today records the *differential* thermal history along a sedimentary burial path.

Each measured thermal indicator is dependent in some manner (usually nonlinear) on a cumulative thermal history along a sedimentary formation burial path. As a consequence there is some time–temperature integral, specific to each indicator, that is related to an observable quantity through a model description for the maturation of the indicator. For all thermal indicators, except fission tracks, the integral extends over all time since deposition of the formation through to the present day. Because fission tracks are produced at an essentially constant rate with time, the corresponding time–temperature integrals extend over a uniform distribution of times in the interval between deposition time of the sediment formation and the present day.

Each thermal indicator has its own characteristic physicochemical parameters (such as activation energies, frequency factors, power law dependencies, etc.) which are specific to the modeled functional behavior taken to connect the observable quantity to the relevant time–temperature integral. Each thermal indicator is also responding to geological factors connected to both the burial path and the thermal history, such as the paleoheat flux variations.

The ability to determine geological event parameters, and their associated resolution, from a set (or multiple sets) of thermal indicator values with depth then depends on (1) the intrinsic validity of the model connecting the measurable thermal indicator quantity with the geological parameter being sought; (2) having available quantitative values for the physicochemical parameters in the model (or of determining them from the same set of data), together with the ranges of uncertainty of the parameters; and (3) having data of high quality, sufficient quantity, and sampled at enough depths that one is sure the output results are statistically sharp within some confidence interval, and are not sensitively beholden to either the quality, quantity, or sampling distribution of the data used.

Failure to allow for such factors can give the perception that a particular thermal indicator is much better (or much worse) at resolving particular parameters (be they geological or physicochemical) than is, in fact, the case. We illustrate these points by considering examples of different thermal indicators from four wells, one in Alaska, two in Australia and one in the Barents Sea.

We first focus on the problem of determining geological parameters under a prescribed model of thermal indicator behavior with physicochemical parameters set at precisely fixed values; then we consider the problem of determining geological *and* physicochemical parameters under a prescribed model of thermal indicator behavior; finally we relax the assumption of a prescribed model behavior and investigate the determination of geological parameters under different models. In this way effects due to each of the three points raised earlier are illuminated by their impact on the determination of parameters of interest.

The general method used to determine parameter values is the thermal indicator tomography procedure, which has been described in technical detail elsewhere (Lerche, 1990a, 1991), so that only a précis of the logic of the method is presented here.

## B. MATHEMATICAL LOGIC AND PROGRAM OUTLINE

Any present-day indicator of thermal history in a basin records some cumulative measure of temperature experienced along its burial path. Thus if the burial path of sedimentary units is known precisely, the paleoheat flux history can be determined. Conversely, if the burial path is imprecisely known (say, because of an unconformity in the past or because the formation age is not known) but the thermal history is prescribed, then parameters associated with such geological events can be determined, because their variation influences the thermal history.

The basic logic runs as follows: Some particular thermal indicator, TI, has values $TI_i$ at increasing depths $z_i$ ($i = 1, 2, \ldots, n$) at the present day. The physicochemical laws that govern evolution of the thermal indicator allow us to derive an expression for a predicted value at the present day, $TI(\tau_i)$, of the thermal indicator, of age $\tau_i$. The predicted value depends on the evolution of temperature, that is, on the burial path at each instant of time and the heat flux variation with time. Hence, if one or several of these variables are not known, the variables can be determined by demanding agreement between predicted values of the thermal indicator and the measured values. This logic applies to any geological variable if the temperature history experienced by the thermal indicator is altered through:

1. Changing the burial path
2. Alteration of the fluid flow/compaction process and hence the burial path and thermal conductivity
3. Adding heat sources to the system
4. Any combination of the above

In other words, any geological parameter can be determined if its variation causes a change in the predicted value $TI(\tau_i)$ of the thermal indicator.

If $m$ different thermal indicators are measured in a well, a measure of the degree of agreement between the observations and the theoretical predictions is provided by constructing

$$X^2 = \sum_{j=1}^{m} \frac{1}{(n_j - 2)} \sum_{i=1}^{n_j} [TI_j(\tau_i) - TI_j(z_i)]^2, \tag{4.15}$$

where $n_j$ is the number of measurements of the $j$th indicator.

Thus, the task is to find a set of parameters which gives the minimum mismatch (smallest $X^2$) between observations and predictions (Lerche, 1988a,b). The method to be used is precisely that of Chapter 3 and does not need to be repeated here. The essence of the thermal indicator tomography is to (1) construct a model which approximates past geological and thermal behavior of a well in the subsurface, (2) compute theoretical thermal indicators for each sedimentary layer by forward tracing in time through that model, (3) compare the predicted present-day thermal indicator variations with depth with the actual maturation profile, and then (4) modify the model to minimize the difference between computed and measured maturations. The process is nonlinearly updated until a preassigned degree of mismatch is obtained (Lerche, 1988a,b).

The geothermal model just presented is based on a one-dimensional fluid flow/compaction model (Cao *et al.*, 1986) (appropriate for single-well studies) which contains three basic types of parameters, with the paleoheat flux function, $Q(t)$, varying with time $t$ as

$$Q(t) = Q_0 \exp[\beta t + a_1 \sin(\pi t/t_{\max}) + a_2 \sin(2\pi t/t_{\max}) + a_3 \sin(3\pi t/t_{\max})]. \tag{4.16}$$

Through the inversion of multiple thermal indicators for many wells to determine a large number of different parameters, the preceding tomography scheme has been found to be remarkably stable and successful. The same tomographic method also has been tested by determining the true formation temperature from downhole temperature recovery measurements with time. The accuracy and resolution of the method are testable directly in some such cases, where formation temperature measurements taken long after shut-in provide a "ground truth" (Cao *et al.*, 1988).

In the last few years, a deeper understanding of the chemical and physical response of several thermal indicators to temperature and time has enabled quantitative estimation of thermal history and geological history parameters. In particular, the inversion of present-day downhole thermal indicator information has enabled assessment of the effective paleoheat flux (Lerche *et al.*, 1984), of the thickness of sediments removed at an unconformity (Armagnac *et al.*, 1989), and of unknown stratigraphic ages (Pantano and Lerche, 1990).

The present version of the computer code, called 1D-DTIT (dynamical and thermal indicator tomography) operates in three modules:

PART A. One-dimensional fluid-flow/compaction uses well information to reconstruct 1-D burial history allowing for (1) paleo-overpressure, (2) present-day overpressure, (3) unconformities (erosional and hiatae), (4) fracturing, (5) cementation and dissolution, (6) lithology changes, (7) basement subsidence, and (8) paleowater depth. The program also includes two different hydrocarbon generation and migration models, and determines the evolution of permeability and porosity of formations with time. The code can be run in both forward and inverse modes.

PART B. Thermal indicators are used in conjunction with Part A to inverst observed variations with depth in order to determine paleoheat flux and paleotemperature variations with time. Currently available in operating mode are:

1. Vitrinite reflectance
2. Sterane isomerization
3. Sterane aromatization
4. Hopane isomerization

5. Apatite fission track distributions
6. Pollen translucency
7. S2/TOC variations with depth

The program allows the *simultaneous* self-consistent determination by inversion of thermal indicator data for (1) *multiple* unconformities (timing and amount of eroded material), (2) variable paleoheat flux in time (both increasing, decreasing, and multiple heat pulses), (3) multiple unknown stratigraphic ages, (4) the onset and termination times of paleo-overpressuring as well as the depth range of such overpressuring, (5) the "best" chemical and kinetic parameters consistent with the observational data, and (6) the code will also handle all thermal indicators at the same time such as vitrinite, steranes, and hopanes (both isomeric and aromatic), apatite fission tracks, pollen translucency and S2/TOC data. The code uses the 1-D fluid flow/compaction code as its base dynamics routine.

Part C. The dynamic indicator part of the code takes the 1-D fluid flow/compaction code of Part A and uses an inverse tomographic procedure to determine dynamic input parameters (e.g., surface depositional values for porosity, permeability, frame pressure) for each lithology, and also determines parameters in equations of state for each lithology connecting permeability/porosity rules or frame pressure/porosity rules. The inverse procedure uses present-day formation thicknesses, porosity with depth, fluid pressure with depth, permeability with depth, position of unconformities, and TD to determine those parameters which permit minimum mismatch of predicted and observed behaviors. The code can also determine (in an inverse or forward manner) parameters in geological events which influence burial histories such as unconformity erosion times and amounts removed, as well as stratigraphic ages, and critical fracturing values for release of overpressure. The dynamic and thermal indicators can be used together to determine parameter values by inverse methods, or they can be run separately, or a default forward run can be done with input values for paleoheat flux, and for dynamic parameters in equations of state.

## C. APPLICATION TO INIGOK-1 WELL, ALASKA

Here we consider a *fixed* model for increase in vitrinite reflectance, $R$, maturity in the form (van Krevelen, 1981; Toth *et al.,* 1981; Lerche *et al.,* 1984)

$$R(z_i) = [R_s^{1/2} + \alpha \mathrm{TTI}(t_i)]^2, \tag{4.17}$$

where $R_s$ is the surface depositional value of the reflectance, $z_i$ is the current depth at which the value $R(z_i)$ is measured, $t_i$ is the age of the formation currently at depth $z_i$, $\mathrm{TTI}(t_i)$ is a time–temperature integral taken in the form

$$\mathrm{TTI}(t_i) = \int_0^{t_i} \exp((T(t) - T_c)/T_D)\, dt, \tag{4.18}$$

where the integral is taken along the burial path of the sedimentary layer, and the integrand is zero in $T(t) < T_c$; $T_c$ and $T_D$ are fixed scaling temperatures. The temperature, $T(t)$, and heat flow, $Q(t)$, are taken to be connected through

$$T(z(t)) = T_s(t) + Q(t) \int_0^{z(t)} dz'/k(z', t), \tag{4.19}$$

where $T_s(t)$ is the paleosurface temperature, $k(z(t), t)$ is the thermal conductivity at burial depth $z(t)$ at time $t$, and the thermal resistance integral in Eq. (4.19) is taken from the surface to depth $z(t)$. The coefficient $\alpha$ is determined from a set of measurements of $R_i$ at depths $z_i$ ($i = 1, \ldots, N$) as

$$\alpha = \sum_{i=1}^{N} [R_i^{1/2} - R_s^{1/2}]\mathrm{TTI}(t_i) / \sum_{i=1}^{N} (\mathrm{TTI}(t_i)^2). \tag{4.20}$$

The parameters $R_s$ (0.2%), $T_c$ (295 K), and $T_D$ (200 K) are fixed in this first example. The point here is to see how well geological parameters can be determined with a fixed model and fixed parameters. It is within that constraining framework that the results of this subsection should be viewed.

The well Inigok-1 is in the Ikpikpuk Basin of the National Petroleum Reserve, Alaska (NPRA). The well was drilled through a thick sequence of coastal to shallow marine sediments which have undergone several episodes of uplifting and erosion (Armagnac, 1985). Vitrinite was sampled from 125 to 6060 m and reflectance values range from approximately 0.42 to 4.80%. Three unconformities in the well were dated at 1.8–69, 131–140, and 278–285 MYBP, respectively, with unknown amounts of erosion. As first-order estimates of paleoheat flux and the surface unconformity, two-dimensional grid searching (in an unconformity amount, $\beta$-space with the other two unconformities modeled as hiatuses) yielded an erosion amount of 2900 m for the surface unconformity and $\beta = 0.6 \times 10^{-3}\ \mathrm{MY}^{-1}$ (Figs. 4.3 and 4.4). The "best" $X^2$ is 0.092. Inspection of the number of vitrinite data points and the systematic offset between predictions and measurements in the shallow section (approximately between 400 and 3000 m) (Fig. 4.4), led to the suspicion that the assumptions of (1) a linear variation of heat flux with time and/or (2) a fixed time of 30 MYBP to start the erosion were not adequate. Therefore, the next step taken was to include the age when the erosion began, and Eq. (4.19) for the heat flux, both to be varied within the 1D-DTIT framework. An $X^2$ of 0.074 was obtained (Fig. 4.5), which is a 20% improvement over the "best" values from the linear search. The final heat flux variation (Fig. 4.6) suggests that heat flow was greater during the Late Cretaceous. This timing for the heat pulse corresponds to

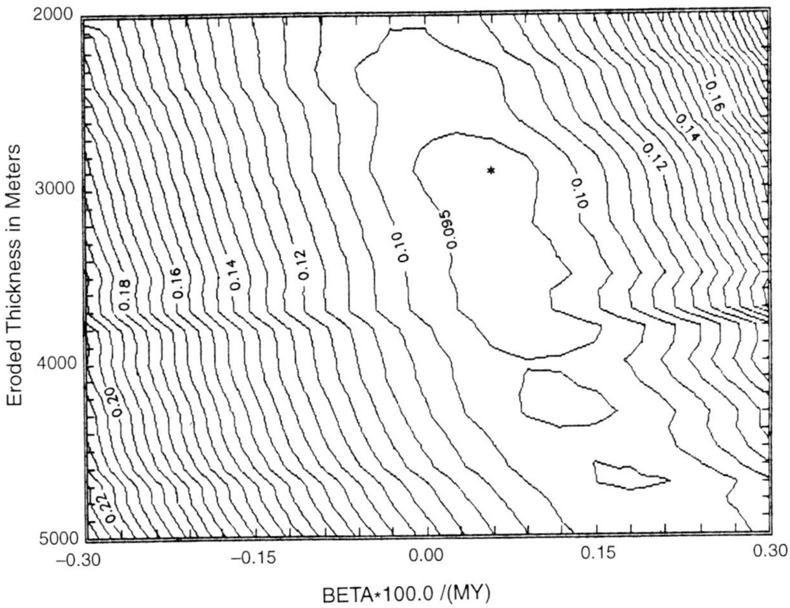


FIGURE 4.3 $X^2$ contoured in a $\beta$-eroded thickness parameter space for Inigok-1, NPRA. Beta defines the fractional change of heat flow with time with the absence of oscillatory terms. Star denotes minimum of $X^2$. [This figure and Figures 4.4–4.32 are reprinted from Van Gijzel *et al.* (1992), *Applied Geochemistry,* Vol. 7, pp. 135–143, with permission from Elsevier Science Ltd.]

the period of rifting which separated the NPRA from its northern source and formed the Canada Basin (Grantz and May, 1983). The increase in deposition rate and basement depth, seen on the geohistory plot (Fig. 4.7)

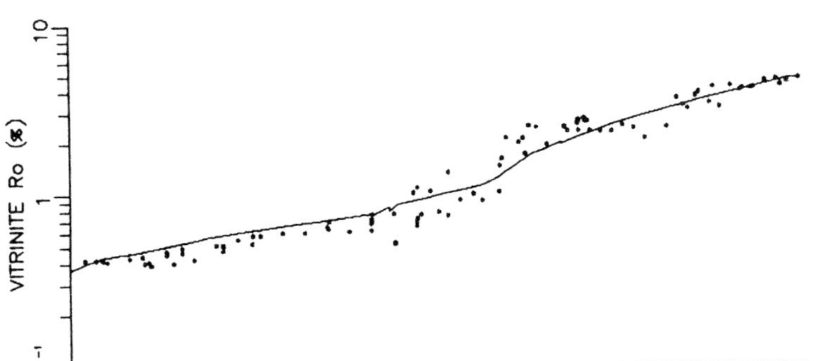


FIGURE 4.4 Reflectance versus depth, including predicted reflectance line, for Inigok-1, Alaska. Asterisks are measured vitrinite values; dark line is predicted vitrinite reflectance derived using $\beta = 0.6 \times 10^{-3}$ MY$^{-1}$ and a surface erosion of 2900 m derived from grid searching shown in Fig. 4.3. Note the systematically greater predicted values between 400 and 3000 m. Compare with Fig. 4.5. (Reprinted with permission from Van Gijzel *et al.*, 1992.)

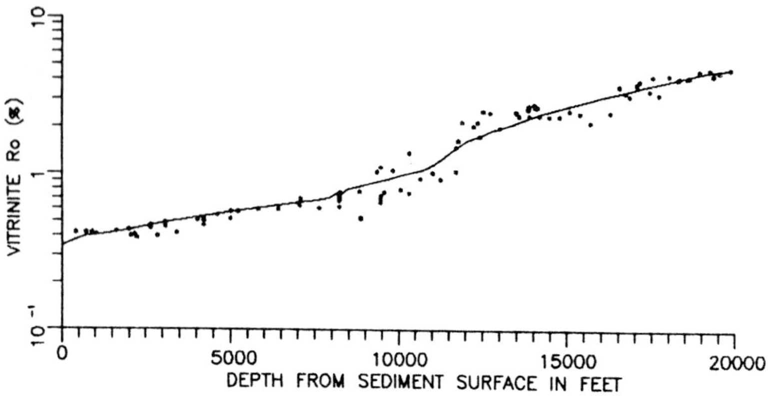


FIGURE 4.5 Same as Fig. 4.4, except predicted reflectance line is derived using the parameters listed in Table 4.1. (Reprinted with permission from Van Gijzel *et al.*, 1992.)

from approximately 120 to 90 MYBP, may be the result of both rifting and an increase of sedimentary influx from the Brooks Range. Continued subsidence during the early Tertiary may be caused by postrift cooling (Armagnac, 1985). The "best" parameters determined by 1D-DTIT (Table 4.1) indicate an age of 30 MYBP for the onset of erosion. This timing of uplift is younger than the youngest dated sediments (37 MYBP) immediately below the same unconformity in a neighboring well, whereas the youngest rocks immediately below the unconformity in Inigok-1 are dated at about 69 MY in age. When the amounts of erosion for the two other possible unconformities (at depths of 9040 and 13,890 ft) were included as parameters to be determined, the tomography favored a hiatus for both unconformities. The resolution limit from the data (Figs. 4.4 and 4.5) has been reached for the determination of these unconformities. If any erosion occurred, the "noise" in the vitrinite data is obscuring the amount of erosion.

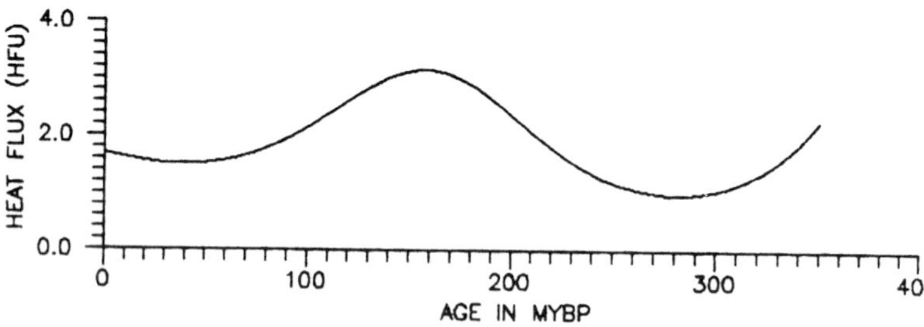


FIGURE 4.6 Heat flow versus geological age for Inigok-1, Alaska, depicted by 1D-DTIT code. (Reprinted with permission from Van Gijzel *et al.*, 1992.)

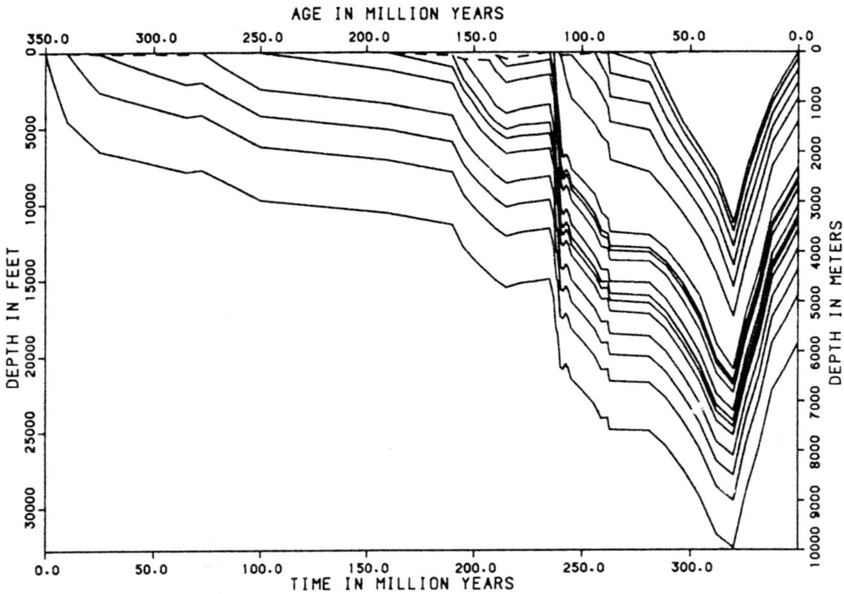


FIGURE 4.7 Geohistory plot for Inigok-1, Alaska. Erosion amount and timing are determined using vitrinite reflectance as a thermal indicator in 1D-DTIT. (Reprinted with permission from Van Gijzel *et al.*, 1992.)

## D. APPLICATION TO TWO AUSTRALIAN WELLS

Here we consider a model of *fixed* functional form for shortening of fission track lengths in apatite, but in which *both* the geological parameters (namely, paleoheat flux) *and* the physical parameters describing the track shortening are to be determined. The point here is that if the proposed model is to be a valid descriptor of track shortening per se, then the physical parameters determined from several wells should be the same to within resolution, even if geological parameters (describing different conditions) differ from well to well. Of course, such consistency of physical parameters does not prove that the model is correct, only that it is consistent, but inconsistency of physical parameters, outside of resolution ability,

TABLE 4.1 Tomography Results for Inigok-1, NPRA

| $X^2$ | $T_D$ (K) | $\beta$ (MY$^{-1}$) | $a_1$ | $a_2$ | $a_3$ | Eroded thickness (m) | Age (MY) |
|---|---|---|---|---|---|---|---|
| 0.078 | 192 | $-0.84 \times 10^{-3}$ | $-0.015$ | 0.36 | $-0.44$ | 2500 | 12 |

*does* indicate that the model needs modification, all other factors being equal.

The model we use here to illustrate the point is that in which tracks, of length $l$, shorten with time and temperature at a rate

$$\frac{dl}{dt} = -lk_1 \exp(-T_c/T(t)), \tag{4.21}$$

where $k_1$ and $T_c$ are parameters to be determined, as is the paleoheat flux $Q(t)$. The assumption is also made that, when produced, each generated track is of initial length $L$, with a normal probability distribution proportional to $\exp(-(L_{av} - L)^2/\Delta^2)$, where $L_{av} \gg \Delta$, with $L_{av}$ the average track length at production and $\Delta$ the half-width spread of $L$ around $L_{av}$. The assumption is also made that tracks are produced at an essentially constant rate with time because the decay half-life ($\sim 10^{10}$ years) of uranium is much longer than that of typical sedimentary basins, and is certainly long compared to both the cases of Curringa-1 and Kambara-1 studied here, as well as to the Barents Sea example given in the next subsection.

For the distribution of track length method, at each depth $z_i$ one measures the number of tracks $N(l_j, z_i)$ in $0 \leq l_j \leq L_{max}$ with $l_j$ the track length (in microns) at the center of the histogram "bin" (for 1-$\mu$m bins, $l_1 = 0.5$, $l_2 = 1.5, \ldots$). What is usually plotted at each depth is the probability distribution of track lengths observed in each range of $l$ (frequency of occurrence):

$$P(l_j, z_i) = \frac{N(l_j, z_i)}{\sum_{j=1}^{J} N(l_j, z_i)}, \tag{4.22}$$

where $J$ is the total number of bins. Based on the track shortening formula (4.21), the theoretical expression directly comparable to $P(l_j, z_i)$ is given by (see next subsection)

$$\begin{aligned} p(l, t_i) = [\pi^{1/2} t_i \Delta]^{-1} \int_0^{t_i} dt' \exp\left[\int_{t'}^{t_i} \alpha(t'')\, dt''\right] \\ \cdot \exp\left\{-\Delta^{-2}\left[L_{av} - l \exp \int_{t'}^{t_i} \alpha(t'')\, dt''\right]^2\right\}, \end{aligned} \tag{4.23}$$

where $\alpha(t) = k_1 \exp[-T_c/T(t)]$, with $k_1$ being the rate constant and $T_c$ the critical temperature corresponding to an activation energy of $E = T_c R$ ($R$ is the gas constant). The predicted distribution $p(l, t_i)$ for an apatite sample deposited at time $t_i$ and currently at depth $z_i$ depends not only on the temperature along the burial path $T(t)$, but also on two "chemical" constants $T_c$ and $k_1$.

Using the tomographic inversion method, an attempt was made to determine simultaneously the heat flow history together with $k_1$, $T_c$, and other variables, such as the original track length distribution parameters $\Delta$ and $L_{av}$. [For ease in numerical computation $k_1 \exp(-T_c/T)$ is rewritten in the form $k_* \exp[T_c/273 - T_c/T)]$ with $k_* = k_1 \exp(-T_c/273)$. This procedure minimized numerical overflow and allowed $T_c$ and $k_*$ to be determined accurately.] The test was carried out on four wells from the NW Canning Basin and four other wells in two other basins. To retain the emphasis on tomography in this chapter, here only the results for wells Curringa-1 and Kambara-1 are given, which were studied by Huntsberger and Lerche (1987) and He and Lerche (1989).

Igneous intrusions, dated Late Carboniferous to Jurassic but thought to be Permian in age, have been intersected by six wells in the northwestern Canning Basin. Both seismic and magnetic interpretations suggest that these intrusions are common over a large area. Regionally, mafic intrusions are restricted to the northwest edge of the Canning Basin adjacent to the continental margin of northwest Australia. Ingeous intrusions of a similar age are known from the northern Carnarvon Basin and these, together with the Canning Basin intrusions, are thought to be related to the initiation of tensional tectonics at the beginning of rifting (Reeckmann and Mebberson, 1984).

The final best values of parameters determined using tomography (Table 4.2) show that, despite the similarity in burial histories of the two wells, heat flow variations are somewhat different (Fig. 4.8). A heat pulse is needed between 250 and 350 MYBP for Kambara-1, which is coincident with the dating of the igneous intrusions. Although Kambara-1 is 3 km away from the nearest mapped igneous intrusion, the existence of a widespread heating event was suggested by Reeckmann and Mebberson (1984). For Curringa-1, which is 26 km from the nearest mapped intrusion (Gleadow and Duddy, 1984), this thermal event is not evident in apatite tracks even with the use of 1D-DTIT. Whether the thermal event was not felt by Curringa-1 or whether it is beyond the resolution of the technique is unknown. Results from two other wells in the basin, which are penetrated by intrusive bodies, have shown pronounced effects of the heating event (He and Lerche, 1989).

Nevertheless, the combination of parameters (Table 4.2) determined from 1D-DTIT gave good fits to the data (Figs. 4.9 and 4.10). The values of $X^2$ for Curringa-1 and Kambara-1 are $0.335 \times 10^{-2}$ and $0.80 \times 10^{-2}$, which are improvements by about 2 to 3 orders of magnitude over results obtained by Huntsberger and Lerche (1987). Much of the improvement is due largely to the better constrained chemical constants that describe the thermal behavior of fission tracks. As a result, the heat flow history determined using better defined parameters is certainly more reliable as described later in the section on resolution.

TABLE 4.2 Tomography Results for Curringa-1 and Kambara-1, NW Canning Basin

| Well | $X^2$ | $T_c \times 10^{-3}$ (K) | $k_* \times 10^5$ (MY$^{-1}$) | $\beta$ (MY$^{-1}$) | $a_1$ | $a_2$ | $a_3$ |
|---|---|---|---|---|---|---|---|
| Curringa-1 | $0.349 \times 10^{-2}$ | 7.3 | 0.14 | $0.66 \times 10^{-3}$ | 0.091 | 0.007 | −0.020 |
| Kambara-1 | $0.799 \times 10^{-2}$ | 7.0 | 0.13 | $0.35 \times 10^{-5}$ | $0.69 \times 10^{-2}$ | −0.20 | $0.59 \times 10^{-1}$ |

*Note:* $k_* = k_1 \exp(-T_c/273)$.

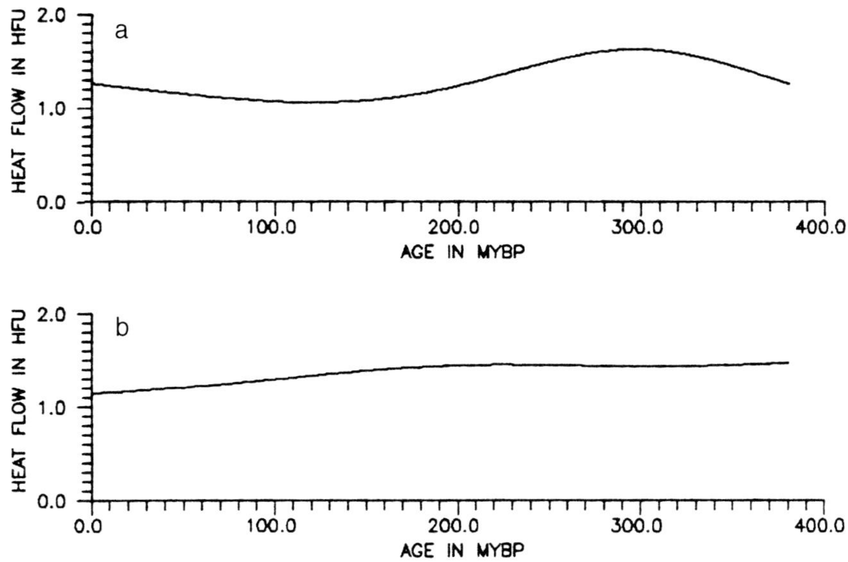


FIGURE 4.8 Heat flow versus geological age for (a) Curringa-1 and (b) Kambara-1, determined using length distribution of fission scar tracks as a thermal indicator in 1D-DTIT. (Reprinted with permission from Van Gijzel *et al.*, 1992.)

## E. APPLICATION TO BARENTS SEA WELL 7323/12

Here we consider the geological and physical parameters obtained by using *different* functional forms for the shortening of fission tracks in order to illustrate the dependence on assumed functional behaviors. To provide a quantitative method for describing the evolution of track length distributions in a given apatite sample, and for relating the thermal history of the sample to the measured track length distribution, a relation is needed describing the rate of track length shortening to the track length and to temperature. Two alternatives are available:

*Alternative A.* For a track length $l$, the rate or reduction of track length with time follows the law:

$$\frac{dl}{dt} = -k_1 l \exp(-T_c/T) \equiv -l\alpha(T), \tag{4.24}$$

which is deducible from the experimental results reported by Gleadow *et al.* (1983), where $k_1$ is a fixed rate constant and $T_c$ is a fixed critical temperature.

*Alternative B.* More recently (Green and Duddy, 1988) it has been suggested that a more appropriate track length shortening formula is[1]

[1] Green (1995) informs me that this track shortening formula has been changed once more because it is not completely error free.

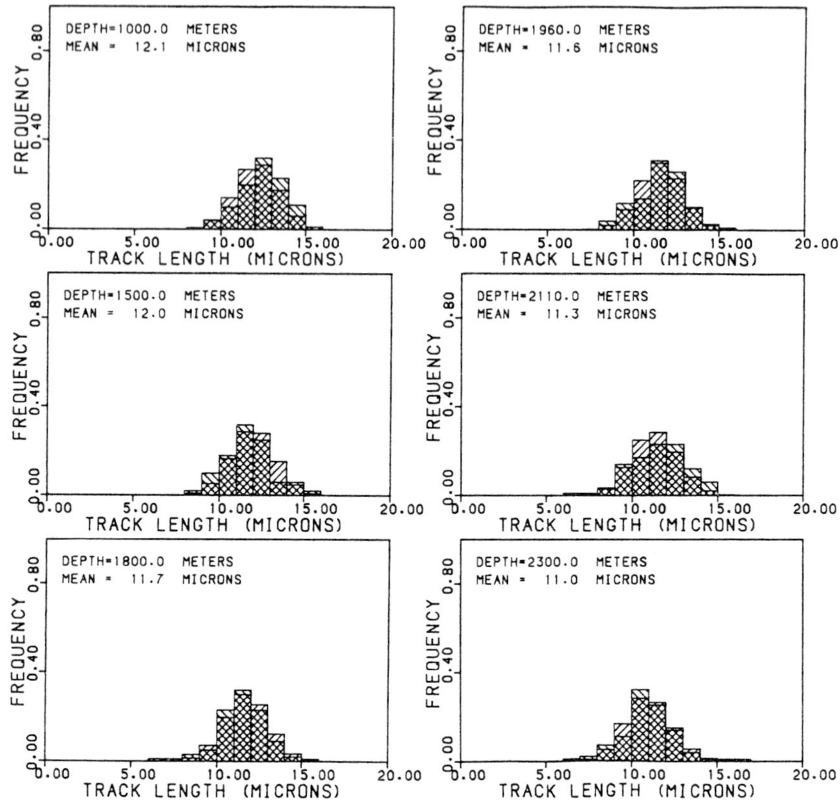


FIGURE 4.9 Theoretically predicted (hashed at 45°) and observed (hashed at −45°) track length distribution with depth at the present day for Curringa-1. The heat flux variation is shown in Fig. 4.8a. (Reprinted with permission from Van Gijzel *et al.*, 1992.)

$$\frac{dl}{dt} = -L(1 - l/L)^{-a}k(T), \tag{4.25}$$

where $k(T) = k_1 \exp(-T_c/T)$ with $a$, $k_1$, and $T_c$ again positive constants, and $L$ the fixed track length at production. In the next section we investigate the formal theoretical consequences for the frequency of occurrence of length distributions using the two formulas of Eqs. (4.24) and (4.25). Note that formula (4.25) has $dl/dt$ infinite on $l = L$ so that freshly formed tracks start to anneal at an infinite speed, formula (4.25) also has $dl/dt$ *finite* on $l = 0$ so that tracks of *zero* length are still shortening!

The important point in either case is that the parameters $a$, $k_1$, and $T_c$, as well as the variation of paleoheat flux with time and the amount of eroded material, must be deducible from downhole data. To accomplish

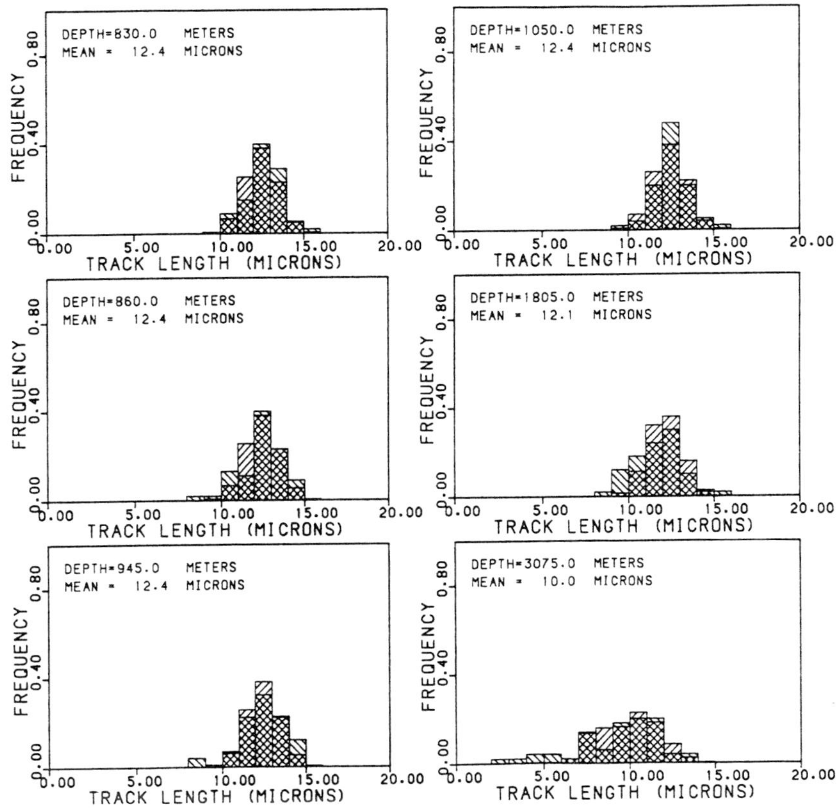


FIGURE 4.10 As for Figure 4.9 but for track length distribution data for Kambara-1. The heat flux variation is shown in Fig. 4.8b. (Reprinted with permission from Van Gijzel *et al.*, 1992.)

these objectives we need to be able to provide a theoretical expression for the track length distribution.

### 1. General Considerations

Let $k_0$ be the rate at which uranium produces tracks, $L$ the original track length, and $U$ the uranium concentration. Let $n(l, t)$ be the number density of tracks in the length range $l$ to $l + dl$ at time $t$ when all tracks were created of length $L$ at time $t = 0$.

The rate of production of tracks of length $l$ is $k_0U\delta(l - L)$. The rate of loss of tracks of length $l$ is $\partial/\partial l(n\, dl/dt)$. Thus the total number density of tracks, $n(l, t, L)$, of length $l$ at time $t$ (created of length $L$ at time $t = 0$) follows the Langevin conservation equation:

$$\frac{\partial n}{\partial t} = k_0U\delta(l - L) - \frac{\partial}{\partial l}\left(n\frac{dl}{dt}\right). \tag{4.26}$$

Because the original track length distribution is a narrow Gaussian centered on $L_{av}$, with an rms spread, $\Delta = 0.9$ $\mu$m (Gleadow *et al.*, 1983), we can write the number of tracks in the length range $l$ to $l + dl$ at time $t$, independent of the precise value that the original track length had, as[2]

$$n(l, t) = \pi^{-1/2}\Delta^{-1} \int_{-\infty}^{\infty} n(l, t; L) \exp[-(L - L_{av})^2/\Delta^2]\, dL, \qquad (4.27)$$

The frequency of occurrence, $P(l, t)$, at time $t$ is given by

$$P(l, t) = n(l, t) \Big/ \int_0^{\infty} n(l, t)\, dl, \qquad (4.28a)$$

with the normalization

$$\int_0^{\infty} P(l, t)\, dl = 1. \qquad (4.28b)$$

Clearly, then, the first task before us is to solve the Langevin conservation equation for the two generic functional forms of $dl/dt$ given through Eqs. (4.24) and (4.25), respectively. Armed with those solutions, construction of the frequency of occurrence is then a matter of quadratures. We first provide the solutions to the Langevin conservation equation for both cases, and then discuss the inversion procedure for obtaining the paleoheat flux evolution with geologic time as well as the parameters for each case.

In alternative A the Langevin equation takes on the form

$$\frac{\partial n}{\partial t} = k_0 U \delta(l - L) + k_1 \exp[-T_c/T(t)] \frac{\partial}{\partial l}(nl). \qquad (4.29)$$

For a sample of apatite deposited on the sedimentary surface at time $t = 0$ with no track density, the resulting number density at time $t$ is given by

$$n(l, t; L) = k_0 U \int_0^t dt' \exp\left[\int_{t'}^{t} \alpha(t'')\, dt''\right] \delta\left(L - l \int_{t'}^{t} \alpha(t'')\, dt''\right), \qquad (4.30)$$

where

$$\alpha(t) = k_1 \exp[-T_c/T(t)].$$

[For a sample of apatite already containing fission tracks at deposition the unfolding of later-produced tracks and those at deposition can be accom-

[2] Because $\Delta \ll L_{av}$ we can extend the range of $L$-integration to $-\infty$ (instead of its more proper value of zero) without introducing significant error.

plished using a variant of the inverse argument given elsewhere (Lerche, 1988a, 1990a)].

Using Eq. (4.30) in Eq. (4.27) we have

$$n(l,t) = (\pi^{1/2}\Delta)^{-1} k_0 U \int_0^t dt' \exp\left[\int_{t'}^t \alpha(t'')\, dt''\right] \times \exp\left[-\Delta^{-2}\left[L_{av} - l \exp\int_{t'}^t \alpha(t'')\, dt''\right]^2\right]. \tag{4.31}$$

It then follows that (see footnote 2)

$$\int_0^\infty n(l,t)\, dl = k_0 U t, \tag{4.32}$$

so that the frequency of occurrence is given by

$$P_A(l,t) = (\pi^{1/2} t\Delta)^{-1} \int_0^t dt' \exp\left[\int_{t'}^t \alpha(t'')\, dt''\right] \times \exp\left[-\Delta^{-2}\left[L_{av} - l \exp\int_{t'}^t \alpha(t'')\, dt''\right]^2\right]. \tag{4.33}$$

We shall return to this functional form in a while. For the moment let us leave Eq. (4.33) and concentrate on obtaining an equivalent result for the competing form of track length loss given by Eq. (4.25).

In alternative B the Langevin conservation equation takes on the form

$$\frac{\partial n}{\partial t} = k_0 U \delta(l - L) + L\alpha(t) \frac{\partial}{\partial l}[n(1 - l/L)^{-a}], \tag{4.34}$$

where $\alpha(t) = k_1 \exp[-T_c/T(t)]$; with solution in parametric form

$$n(l,t;L) = k_0 U \int_0^t dt'\, \delta[L - l(\zeta, t')] \left\{ \frac{\left[\zeta + (a+1)\int^t \alpha(t'')\, dt''\right]}{\left[\zeta + (a+1)\int^{t'} \alpha(t'')\, dt''\right]} \right\} \tag{4.35}$$

with

$$l(\zeta, t') = L\left[l - \left[\zeta + (a+1)\int^{t'} \alpha(t'')\, dt''\right]^{1/(a+1)}\right] \tag{4.36}$$

and

$$\zeta = (1 - l/L)^{a+1} - (a+1)\int^t \alpha(t'')\, dt''. \tag{4.37}$$

Using Eq. (4.35) in Eq. (4.27) we have

$$n(l,t) = k_0 U(\pi^{1/2}\Delta)^{-1} \int_0^t dt' \left\{1 - \left[(a+1)\int_{t'}^t \alpha(t'')\,dt''\right]^{1/(a+1)}\right\}^{-1} \times \exp\left[-\Delta^{-2}\left[L_{\text{av}} - l\left\{1 - \left[(a+1)\int_{t'}^t \alpha(t'')\,dt''\right]^{1/(a+1)}\right\}^{-1}\right]^2\right]. \tag{4.38}$$

It then follows (see footnote 2) that

$$\int_0^\infty n(l,t)\,dl = k_0 U t, \tag{4.39}$$

so that the frequency of occurrence is given by

$$P_{\text{B}}(l,t) = (\pi^{1/2} t\Delta)^{-1} \int_0^t \left\{1 - \left[(a+1)\int_{t'}^t \alpha(t'')\,dt''\right]^{1/(a+1)}\right\}^{-1} \times \exp\left[-\Delta^{-2}\left[L_{\text{av}} - l\left\{1 - \left[(a+1)\int_{t'}^t \alpha(t'')\,dt''\right]^{1/(a+1)}\right\}^{-1}\right]^2\right] dt'. \tag{4.40}$$

The tomographic inversion method (Lerche, 1990a) allows determination of the heat flow history [i.e., the parameters $\beta$, $a_1$, $a_2$, $a_3$ in Eq. (4.16)] simultaneously with $k_*$, $a$, $T_c$ as well as other variables, such as the original track length distribution parameters $L_{\text{av}}$ and $\Delta$, and erosion amounts, by demanding that the predicted track length distribution $P_{\text{A}}(l, t)$ (alternative A) or $P_{\text{B}}(l, t)$ (alternative B) be as close to the observed distribution $P(l_j, z_i)$ for tracks at depth $z_i$ (corresponding to sedimentary bed age $t_i$), and in the length range $l_j \pm \Delta l_j$ (where $2\Delta l_j$ is the histogram bin width of observed tracks, typically 1 $\mu$m).

If $M$ distributions of tracks are measured in a well at the depths $z_i$ ($i = 1, 2, \ldots, M$), a measure of the degree of agreement between the observations and theoretical predictions from alternative A is provided by constructing

$$X^2 = \sum_{i=1}^{M}\sum_{j=1}^{J} (P_{\text{A}}(l_j, t_i) - P(l_j, z_i))^2, \tag{4.41}$$

and $P_{\text{A}}$ can be replaced by $P_{\text{B}}(l_j, t_i)$ for alternative B. As usual, the task is to find the set of parameters yielding a minimum mismatch (smallest $X^2$) between predictions and observations.

Fission track data for Barents Sea well 7323/12 have been published by Løseth *et al.* (1993), as has the stratigraphic column. We have used these data together with the 1D-DTIT code for each of alternatives A and B to obtain the best set of parameters satisfying the observed track length

TABLE 4.3 Tomography Results for Barents Sea Well 7323/12

| | $k_*$ (1/MY) | $T_c$ (K) | $L_{av}$ ($\mu$m) | $\Delta$ ($\mu$m) | $\beta$ (1/MY) | $a_1$ | $a_2$ | $a_3$ |
|---|---|---|---|---|---|---|---|---|
| Alternative A | $0.27 \times 10^{-6}$ | 1900 | 12.6 | 1.90 | −0.0010 | 0.00 | −0.05 | −0.15 |
| Alternative B | $0.41 \times 10^{-9}$ | 3400 | 12.6 | 1.39 | −0.0017 | −0.05 | −0.10 | −0.05 |

*Note:* $a = 0.7$ for alternative B.

distributions in the well. Table 4.3 presents the tomographic results for Barents Sea well 7323/12 indicating that, while the best fitting parameters are different for the two alternative kinetic formulas, the best uplift onset time is the same at 5 MaBP and the best estimate of eroded material is also the same at 2200 m. The heat flux parameters for each alternative are different, but the heat flux variation in time is remarkably similar as shown in Fig. 4.11. The degrees of fit of each of the alternatives, measured through $X^2$ in Table 4.3, are identical to four figures, implying that either kinetic model is acceptable. The only discordant note is in the power index ($a = 0.7$) for alternative B; laboratory data have been used (Green and Duddy, 1988) to suggest an index of a $\approx$ 3 to 4 for tracks produced artificially. Such a discordance is likely due to the long conditions of geological burial relative to the short-time conditions in the laboratory.

The burial history path for well 7323/12 using the best parameter values from Table 4.3 is given in Fig. 4.12, whereas the fits to the observed distributions of fission tracks for alternatives A and B are given in Figs. 4.13a and 4.13b, respectively.

The point about this illustration is to note that different model assumptions, leading to different functional forms containing different scaling parameters, must each be tested against observational data to determine the ability of each model to honor the available data. In this case both models provide identical fits to the data, both provide the same best parameters for uplift time and erosion amount, and both provide virtually identical paleoheat flux variations with time. On the basis of their best fits to the data, either kinetic model appears acceptable.

**2. Resolution and Uncertainty**

The ability of a model or models to provide one or more sets of parameters representing best fits to a set of data is not the end of the analysis by any means. The resolution of each parameter in providing changes in the minimum mismatch to a set of data is, perhaps, one of the more important factors to investigate; for only by having an appreciation of the sensitivity and rugged stability of such variations can one obtain an insight into the degree to which one can determine with known precision the model, or its

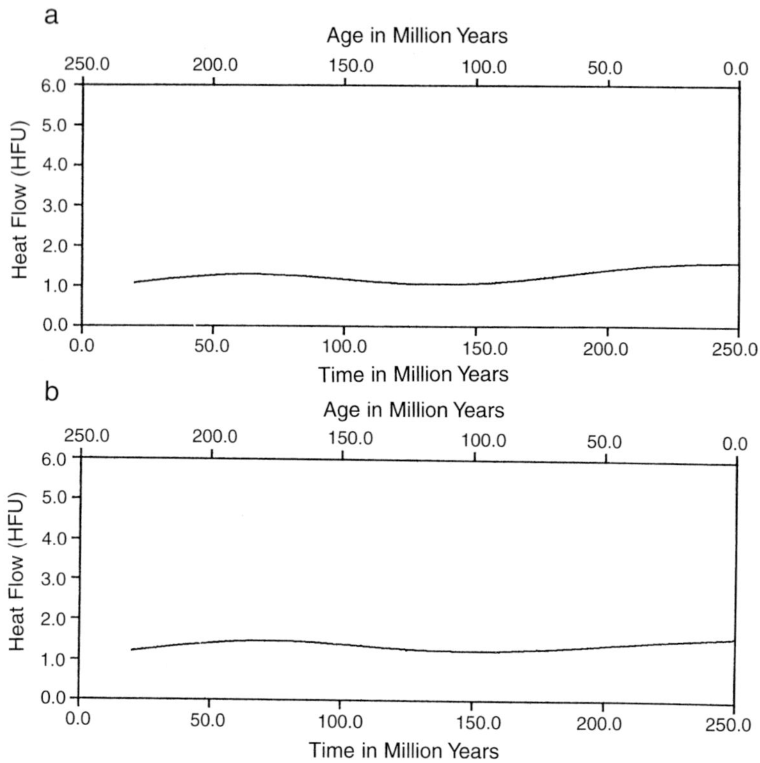


FIGURE 4.11 Paleoheat flux variations for Barents Sea Well 7323/12 derived from minimum mismatch to observed fission track distributions: (a) alternative A for fission track shortening; (b) alternative B for fission track shortening. (Reprinted with permission from Van Gijzel *et al.*, 1992.)

parameter values, or geological values of interest (such as the amount of an erosion). Here, we examine resolution and uncertainty for the three cases examined in the last subsection.

Obviously, the resolution limit from the data has not been reached when a systematic mismatch is observed. As discussed in the case of Inigok-1, varying only the unconformity size and the $\beta$ value could not make up the mismatch. Some other parameters must have caused the offset (unconformity age and/or a heat pulse) and could be determined. Also clearly, the data do not provide enough information for a parameter to be examined when an expected change (such as a "kick" in the trend of vitrinite data caused by an unconformity) is not observed or is buried in the scatter of the data. But the question still exists: When the "best" parameters are found through tomography and no further improvement is possibe by adding more variables, what is the uncertainty associated with each parameter? To examine this problem, one variable is allowed to vary step by step while

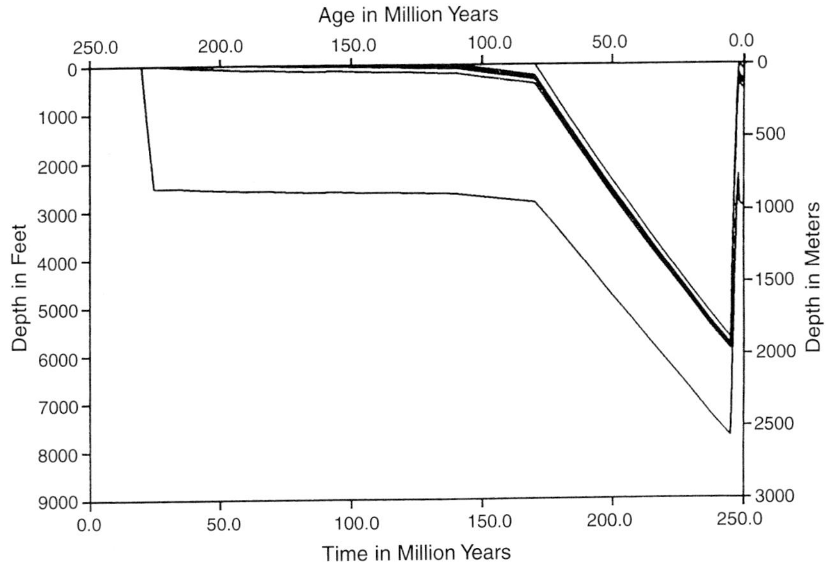


FIGURE 4.12 Burial history path for Barents Sea well 7323/12 showing the major erosional event of about 2000 m commencing at 4 MYBP, representing values derived from minimum mismatch to observed fission track distributions. (Reprinted with permission from Van Gijzel *et al.*, 1992.)

holding all others fixed at their optimal values to see how much the mismatch ($X^2$) will increase when a single parameter deviates from its best value. Results of such an investigation are shown for Inigok-1 where vitrinite reflectance was the thermal indicator (Figs. 4.14, 4.15, and 4.16). The $X^2$ versus $T_D$ curve gives a best $T_D$ of 204 K, consistent with the findings of Toth *et al.* (1984) on about 400 wells which yielded a $T_D$ of 200 $\pm$ 50 K. However, the $T_D$ value is less (more) confined to the higher (lower) side of 200 K in this well. All the heat flow parameters appear to be well confined (Fig. 4.15), except that a broad range of $a_2$ is observed where $X^2$ has a "flat" valley. But an examination of the heat flow equation with different values of $a_2$ within the valley does not change the overall shape of the heat flow curve (Fig. 4.7). Therefore, the resolution of the data is high enough to define the heat pulse. As for the burial history parameters, the curve of $X^2$ versus unconformity size (Fig. 4.16a) indicates that a larger or smaller amount of erosion will not only increase $X^2$ but will also influence the stability of the fluid flow/compaction process. By way of contrast, when the time of onset of erosion is varied around its best value (Fig. 4.16b), only an age of about 20–50 MY could be assessed. Because an age of 30 MY was chosen when applying the linear search, the improvement of fit was due mostly to a more precise depiction of the paleoheat flux variation compared to a linear varying pattern.

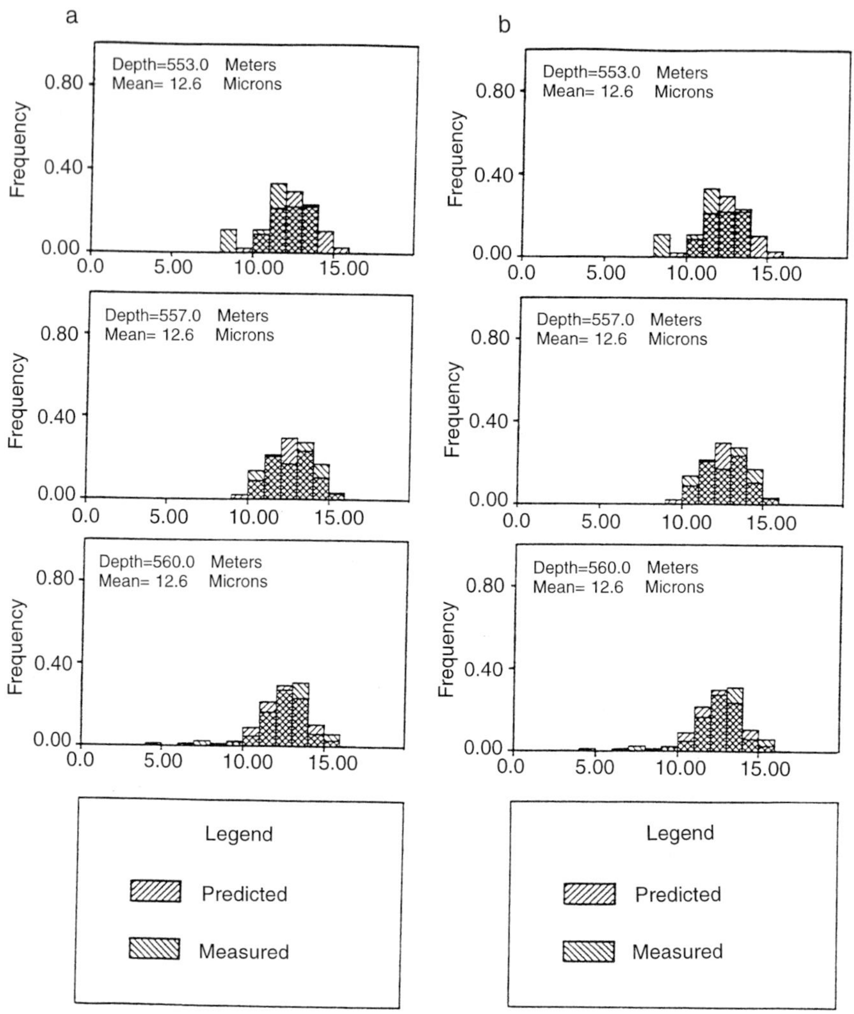


FIGURE 4.13 Minimum mismatch fits of predicted and observed fission track distributions in Barents Sea well 7323/12: (a) alternative A for fission track shortening; (b) alternative B for fission track shortening. (Reprinted with permission from Van Gijzel *et al.*, 1992.)

When this uncertainty analysis was applied to the apatite distribution method for the Australian wells, uncertainties in the heat flow parameters were even smaller (Figs. 4.17 and 4.20; note the finer scales of the abscissa as compared to those in Fig. 4.15) compared to those for Inigok-1, and almost every parameter had a well-developed $X^2$ minimum (Fig. 4.18), suggesting a greater sensitivity of the apatite track distribution as a thermal indicator. Furthermore, significant resolution on chemical parameters $k_1$

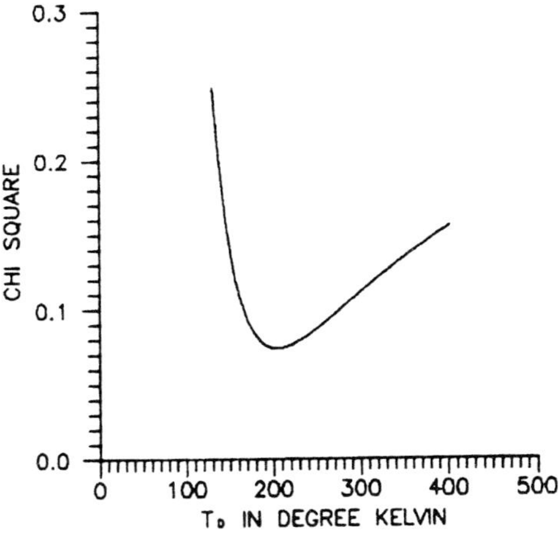


FIGURE 4.14 $X^2$ versus $T_D$ for Inigok-1, Alaska. (Reprinted with permission from Van Gijzel *et al.*, 1992.)

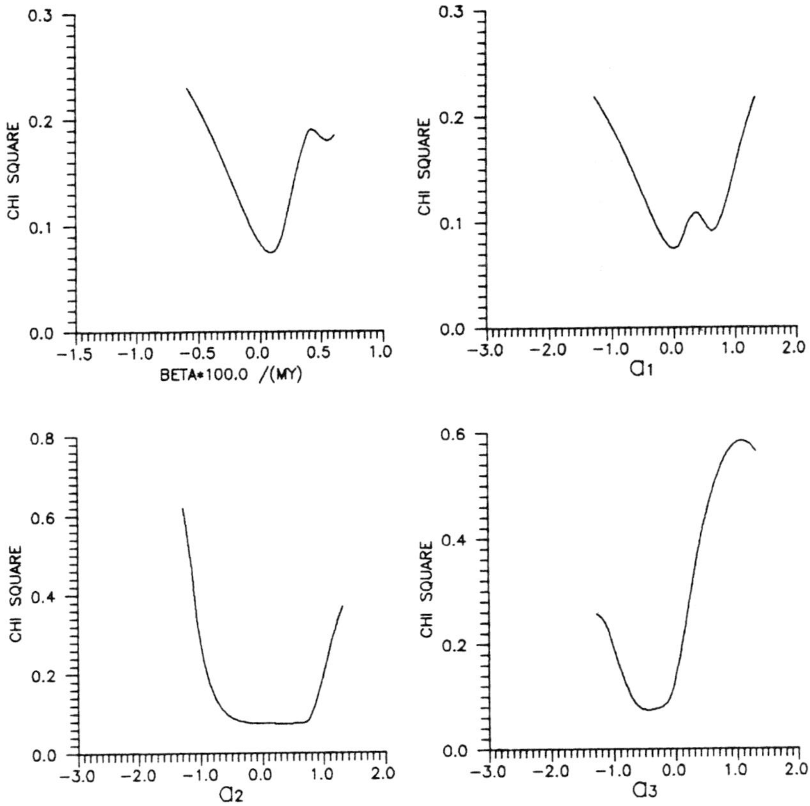


FIGURE 4.15 $X^2$ versus heat flow parameters for Inigok-1, Alaska. (Reprinted with permission from Van Gijzel *et al.*, 1992.)

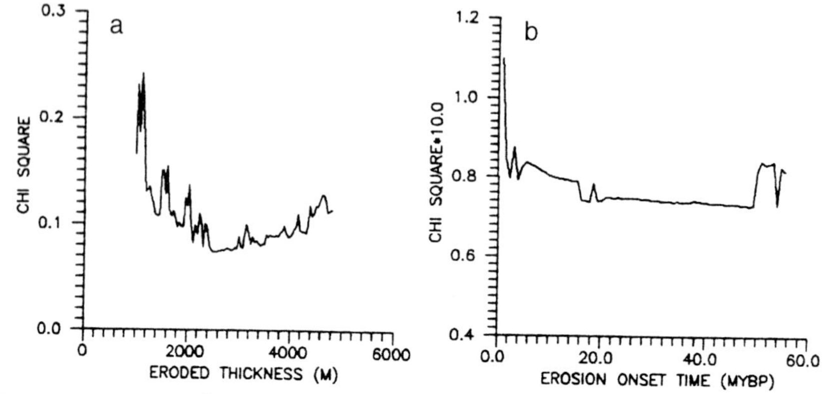


FIGURE 4.16 $X^2$ versus (a) amount of surface erosion and (b) onset time of erosion for Inigok-1, Alaska. (Reprinted with permission from Van Gijzel *et al.*, 1992.)

and $T_c$ (Figs. 4.17 and 4.19), and their consistency, implies a smaller uncertainty in both the paleoheat flow and geological parameters determined by the 1D-DTIT tomography method.

Similar sensitivity and resolution studies have been carried out for both kinetic model alternatives for fission track shortening used on Barents Sea well 7323/12, in order to determine the degree of uncertainty of each parameter (geological or physicochemical) in each model. Figure 4.21 shows the comparisons between kinetic parameters for the two models. Note that there is an extra parameter, $a$, in alternative B which has no comparison counterpart in alternative A. Note from Figs. 4.21a and 4.21b that the best average track length, $L_{av}$, at track production has a well-resolved minimum to within about $\pm 1$ $\mu$m and is similar for both alternatives; equally the spread in uncertainty, $\Delta$, of track lengths at production (shown in Figs.

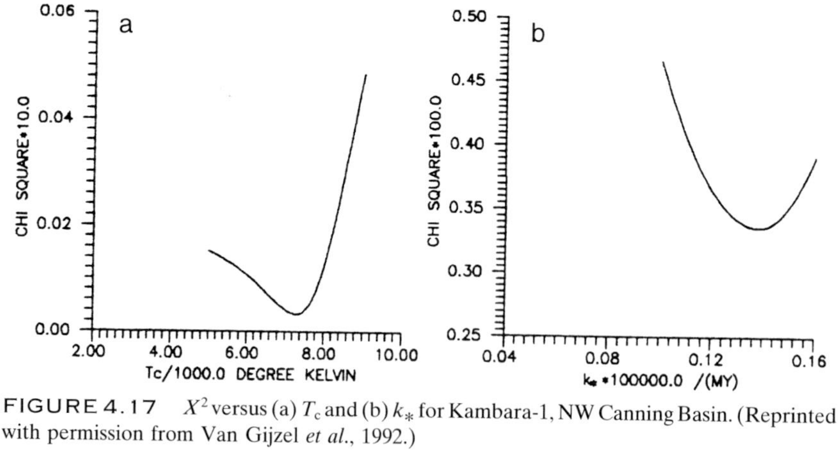


FIGURE 4.17 $X^2$ versus (a) $T_c$ and (b) $k_*$ for Kambara-1, NW Canning Basin. (Reprinted with permission from Van Gijzel *et al.*, 1992.)

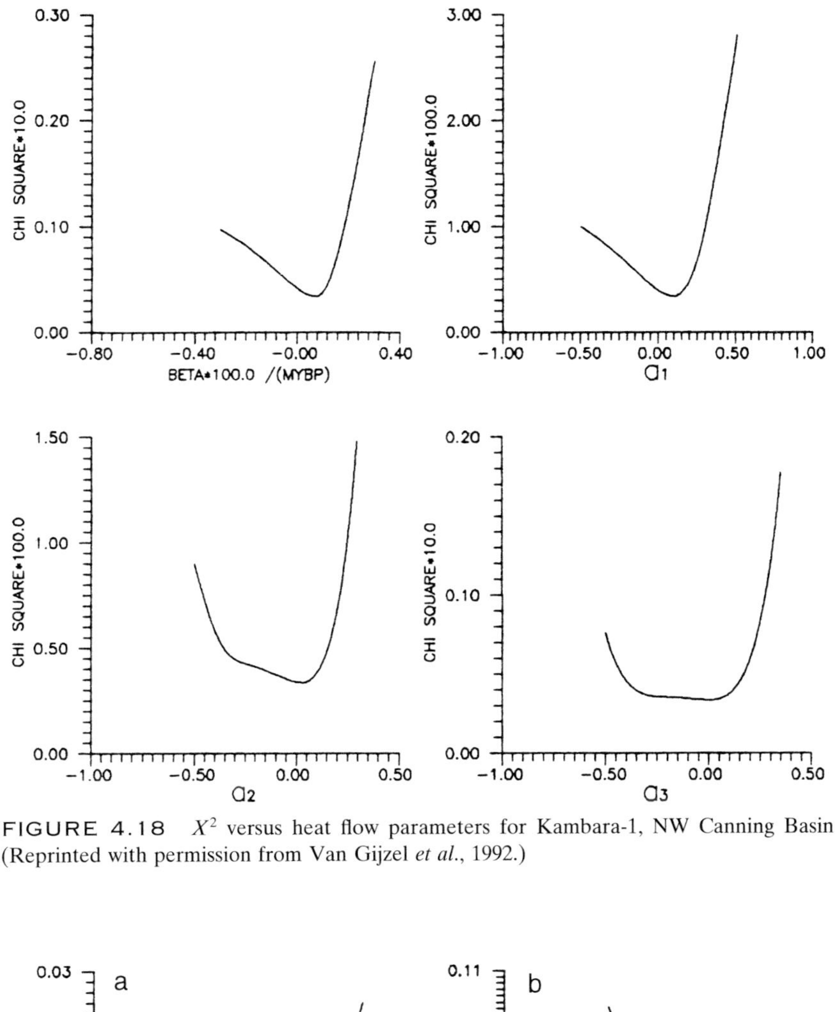


FIGURE 4.18 $X^2$ versus heat flow parameters for Kambara-1, NW Canning Basin. (Reprinted with permission from Van Gijzel *et al.*, 1992.)

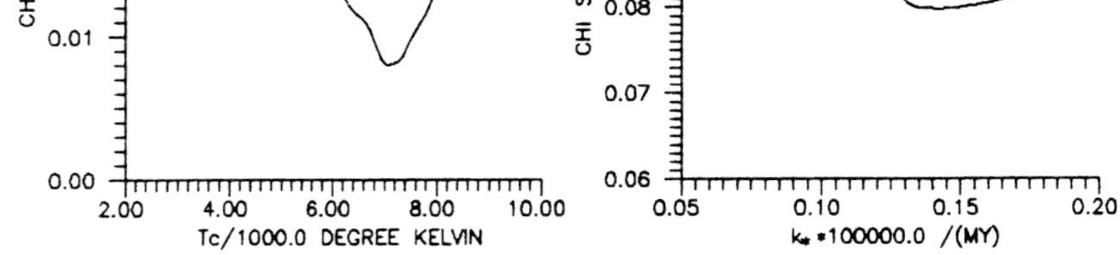


FIGURE 4.19 $X^2$ versus (a) $T_c$ and (b) $k_*$ for Curringa-1, NW Canning Basin. (Reprinted with permission from Van Gijzel *et al.*, 1992.)

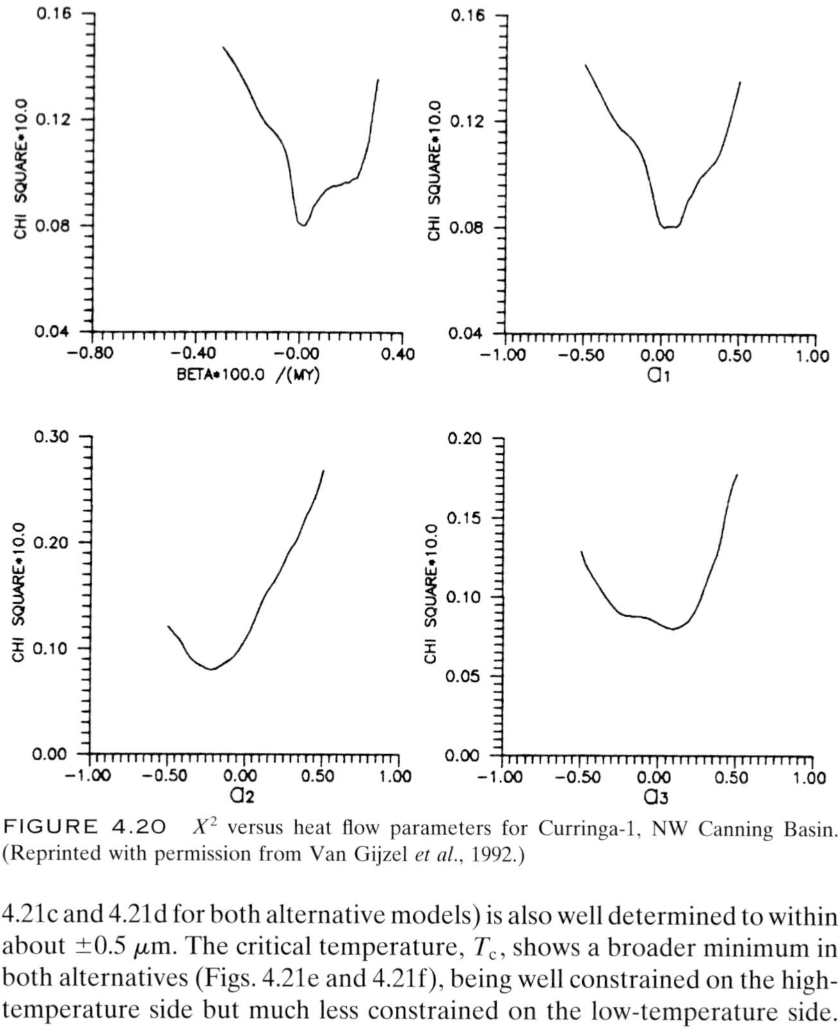


FIGURE 4.20 $X^2$ versus heat flow parameters for Curringa-1, NW Canning Basin. (Reprinted with permission from Van Gijzel *et al.*, 1992.)

4.21c and 4.21d for both alternative models) is also well determined to within about $\pm 0.5\ \mu$m. The critical temperature, $T_c$, shows a broader minimum in both alternatives (Figs. 4.21e and 4.21f), being well constrained on the high-temperature side but much less constrained on the low-temperature side. Thus the best tomographic values of Table 4.3 for $T_c$ are uncertain by about 1000 K on the low-temperature side of the best values and about 500 K uncertain on the high-temperature side. Thus, for alternative A one should write $T_c = 3400^{+500}_{-1000}$ K. Within this error range, values of $T_c$ from both alternatives can be made coincident. Because the frequency factor, $k_1$, always occurs in combination with $\exp(-T_c/T)$ in either alternative, it follows that a slight uncertainty in the high value of $T_c$ will cause a large uncertainty in $k_1$. For instance, with $T \approx 300$ K, a $T_c$ of 3300 K yields $\exp(-T_c/T) = \exp(-11)$, whereas at $T_c = 2300$ K, one has $\exp(-T_c/T) \approx \exp(-7.7)$. Correspondingly the relevant values of $k_1$ will be in the ratio $\exp(3.3)$ $(\exp(11 - 7.7))$, which is a factor of 3000 different. The point is that while $k_1$ may be extremely variable, as shown in Table 4.3, the quantity

$k_1 \exp(-T_c/273)$ is much better constrained than either of the individual components $k_1$ or $T_c$. This factor is the one used throughout both the NW Australia wells and Barents Sea well 7323/12, accounting for the well-determined degrees of fit to the observational data in all cases.

For alternatives A and B, the range of the frequency factors ($k_1$) is presented in Figs. 4.21g and 4.21h, respectively, showing well-defined minima for each. The parameter $a$ in alternative B, shown in Fig. 4.21i, has a threshold type of behavior—below a value of about $a \cong 1$, any value of $a$ will fit the data almost equally well (there is a nominal numerical minimum at $a = 0.7$), indicating little resolution to the parameter $a$, provided it is chosen such that it is below the threshold value of $a \cong 1.0$.

Similar plots can be drawn for resolution assessment of the paleoheat flux parameters, as depicted in Fig. 4.22 for alternatives A and B. Both alternatives prefer a negative $\beta$ (cooler in the past) but the resolution is low below a threshold of $\beta \approx 0$, as shown in Figs. 4.22a and 4.22b.

The oscillatory term, $a_1$, has a broad range roughly centered on zero (Figs. 4.22c and 4.22d) and, when it is remembered that this term enters the paleoheat flux assessment as $\exp(a_1 \sin(\pi t/t_{max}))$, the range of uncertainty around the best value presents only about a 20% maximum uncertainty at any time in the paleoheat flux. Similar results are obtained for the second oscillatory coefficient $a_2$ in both alternatives (Figs. 4.22e and 4.22f), again representing a worst uncertainty of about 20% on the paleoheat flux. The coefficient $a_3$ exhibits a similar behavior (Figs. 4.22g and 4.22h). The combined result of all of these factors is to make the trend term $\exp(\beta t)$ in the paleoheat flux dominant but, in all situations, the range of uncertainty on variations of paleoheat flux is so broad that a constant heat flux will equally well satisfy the fission track data. There is little resolution in the fission track data to determining any departure of paleoheat flux from its present-day value.

Varying the amount of eroded material and the time of uplift around the nominal best values of 2200 m and 4 MYBP are valid exercises if we are to be able to assess the accuracy of results in relation to the dynamic history of the Barents Sea. Shown in Figs. 4.23a and 4.23b are the changes in fit to the fission track data as both amount and onset time are varied for both alternatives. The resolution of the amount eroded can obviously not be determined to better than $\pm 1$ km, while the uplift time can hardly be constrained at all from the apatite fission track data. There is, apparently, little resolution of the data to the geological events.

## F. TRANSMITTANCE COLOR INDEX AND VITRINITE REFLECTANCE

To improve on our current capability for determining the thermal history of a basin and to better bracket the uncertainties in determination of

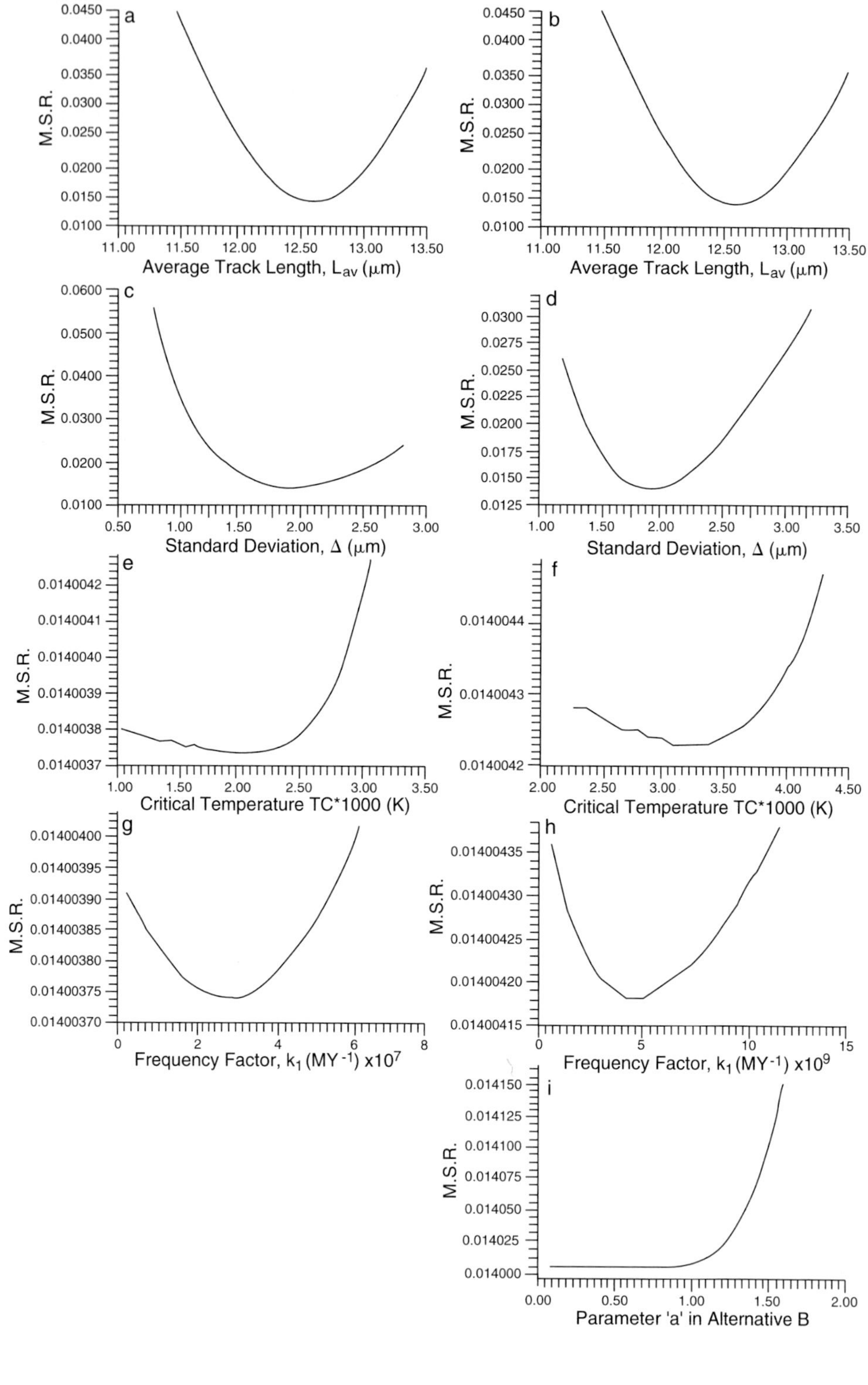
a
M.S.R.
Average Track Length, $L_{av}$ (μm)
b
M.S.R.
Average Track Length, $L_{av}$ (μm)
c
M.S.R.
Standard Deviation, Δ (μm)
d
M.S.R.
Standard Deviation, Δ (μm)
e
M.S.R.
Critical Temperature TC*1000 (K)
f
M.S.R.
Critical Temperature TC*1000 (K)
g
M.S.R.
Frequency Factor, $k_1$ (MY$^{-1}$) x10$^7$
h
M.S.R.
Frequency Factor, $k_1$ (MY$^{-1}$) x10$^9$
i
M.S.R.
Parameter 'a' in Alternative B

paleoheat flux and sedimentary thermal maturity, we need to expand our ability in two major respects: (1) Quantify more thermal indicators for use both independently and in a cross-check mode in diverse geological situations and (2) improve the methodology in processing such thermal indicators quantitatively and with greater precision, so that statistical data uncertainties are reduced.

The transmittance color index (TCI) (Van Gijzel, 1989; Smith and Van Gijzel, 1989) is one of the newly developed thermal indicators. In comparison with some other direct thermal indicators, such as the widely used vitrinite reflectance (VR) and thermal alteration index (TAI), TCI has significant advantages. First, measurements of TCI are made on physically amorphous organic matter (AOM), facilitating direct assessment of thermal maturity of oil-prone source rocks that lack vitrinite and spores or pollen. Thus, TCI is especially well suited to applications involving carbonate sequences, siliceous black shales, and so on. Second, unlike vitrinite, which is unknown in rocks older than Late Silurian, AOM occurs in pre-Devonian rocks; therefore TCI has wider applicability in thermal history studies of basins. Third, TCI is measured by quantitative microscope spectophotometry, making it easy to standardize between different laboratories. For these reasons, it is expected that TCI will become a commonly applied thermal indicator in the study of the thermal history of sedimentary basins.

This section first describes briefly how TCI measurements are made and then focuses on the relationship between TCI and VR. As a case history study, we apply both TCI and VR data from the COST G-2 well (offshore northeastern United States) in a one-dimensional basin analysis model to demonstrate the use of TCI as a quantitative thermal indicator. The relation of TCI to VR is established in three stages:

1. Determine both a theoretical and an empirical relation between TCI and vitrinite reflectance (VR).
2. Determine how to use TCI directly as a thermal indicator and, in an inverse sense, determine quantitatively sedimentary thermal history, and provide an algorithm for use in a one-dimensional fluid flow/compaction model.
3. Confirm the relation between TCI and thermal history by "ground truth" control in a well in which an estimate of thermal history is already available from inversion of another thermal indicator. Vit-

---

FIGURE 4.21 Mean square residual (MSR) mismatch for parameters $L_{av}$, $\Delta$, $T_c$, $k_1$, and $a$ (alternative B only) for fission track data versus predictions in Barents Sea well 7323/12. The left-hand column depicts the resolution of the four parameters using alternative A for fission track shortening, whereas the right-hand column depicts the resolution of the five parameters using alternative B. (Reprinted with permission from Van Gijzel *et al.*, 1992.)

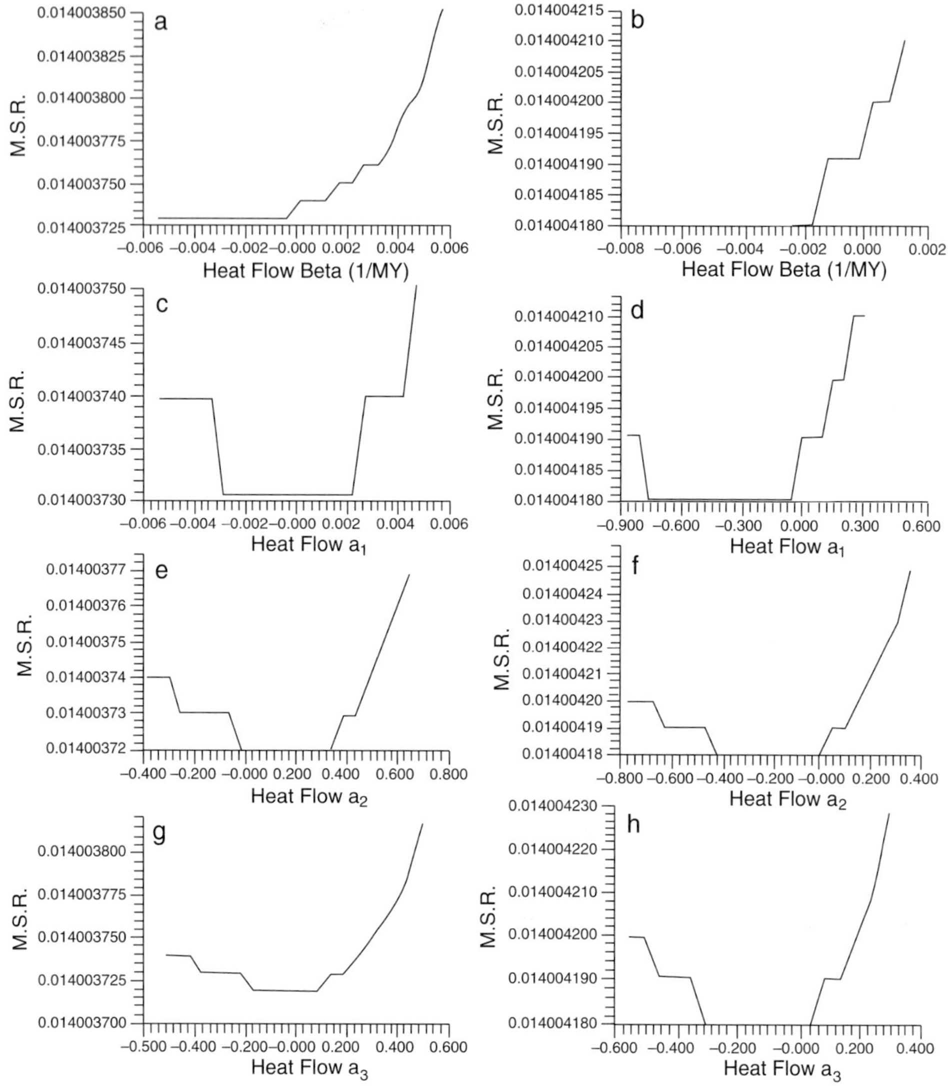


FIGURE 4.22 MSR mismatch for parameters for paleoheat flux determination given in Eq. (4.18). The left column of the figures shows the resolution obtained using alternative A for fission track shortening and the right side shows that for alternative B. (Reprinted with permission from Van Gijzel *et al.*, 1992.)

rinite reflectance and TCI data from well COST G-2, Georges Bank, are used here for this purpose.

### 1. The Transmittance Color Index Technique

TCI measurements are obtained from a rapid spectral scanning system (developed by Texaco, patent pending), which provides a fast, stable means

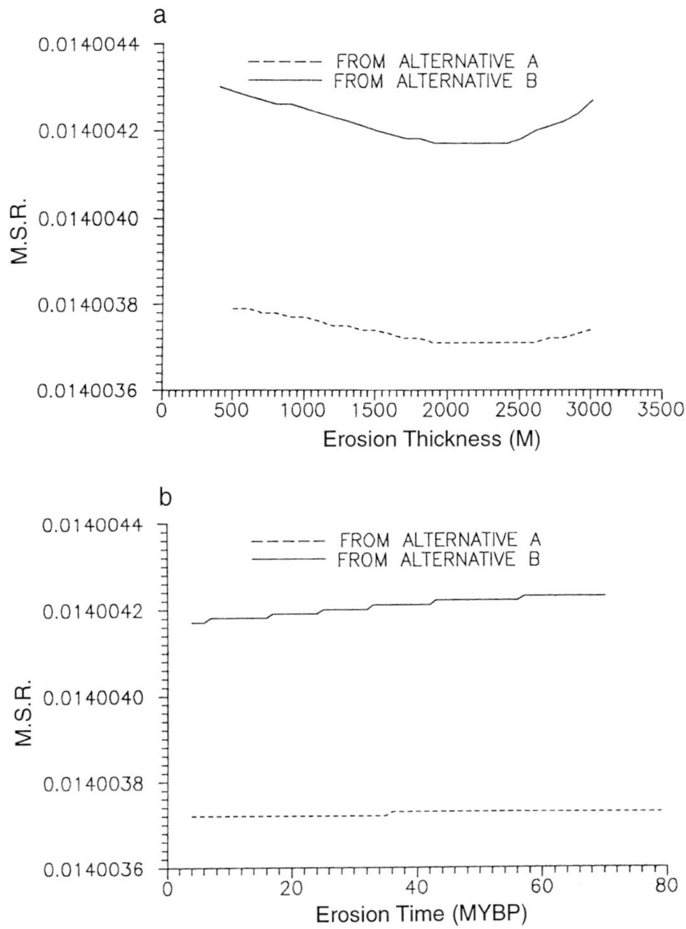


FIGURE 4.23 MSR mismatch against fission track measurements as (a) erosion thickness and (b) erosion onset time are varied for Barents Sea well 7323/12 for both alternatives A and B for fission track shortening. While nominal minima exist in both cases, the resolution is virtually nonexistent. (Reprinted with permission from Van Gijzel *et al.*, 1992.)

of spectral analysis. TCI analysis uses the tungsten spectrum at a given radiation temperature (2850 K) and power (6 volt, 12 amp) for standard spectral analysis. A personal computer serves as both controller and data recorder. TCI utilizes the wavelength ($\lambda_{1/2}$) at which the transmitted intensity drops to half the incident value. A software package provides statistical data treatment.

TCI measurements are made on the same kerogen slides used for visual kerogen description. If possible, TCI analysis is conducted only on particles of amorphous material of uniform thickness and color. Because measure-

TABLE 4.4 Variation in Average $\lambda_{1/2}$ (nm) Values as a Function of Number of Measured Spectra

| Sample number | $R_0$ | Number of measured spectra 20 | 30 | 40 | 50 | 60 | 70 | 80 |
|---|---|---|---|---|---|---|---|---|
| SAH1 | 0.22 | 581.1 | 581.9 | 581.7 | 582.0 | — | — | — |
| HMS5 | 0.24 | 583.0 | 583.6 | 583.2 | 583.1 | 582.6 | — | — |
| MSN | 0.28 | 583.2 | 583.5 | 583.9 | 584.0 | 584.3 | 584.0 | — |
| HMS4 | 0.30 | 583.2 | 584.6 | 584.0 | 584.9 | — | — | — |
| VIL3 | 0.34 | 586.3 | 587.5 | 586.4 | 587.3 | — | — | — |
| MOD12 | 0.41 | 587.3 | 586.3 | 586.8 | 586.8 | 587.1 | — | — |
| MCU | 0.47 | 594.6 | 594.6 | 593.4 | 593.1 | 593.0 | 593.4 | — |
| MOD16 | 0.50 | 595.2 | 594.2 | 594.3 | 594.7 | 594.7 | 594.3 | — |
| ASTA | 0.60 | 598.1 | 597.1 | 597.0 | 596.5 | 596.2 | 596.2 | — |
| MOD7 | 0.68 | 601.8 | 602.3 | 602.3 | 601.3 | — | — | — |
| MOD6 | 0.70 | 601.1 | 602.5 | 602.6 | 602.2 | — | — | — |
| MOD2 | 0.69 | 602.0 | 601.3 | 601.7 | 602.6 | 603.1 | 602.5 | 602.6 |
| MOD8 | 0.76 | 605.7 | 604.7 | 604.2 | 603.9 | — | — | — |
| MOD15 | 0.79 | 604.8 | 605.6 | 605.9 | 606.4 | 605.7 | 606.2 | — |
| 312 | 0.86 | 609.8 | 608.7 | 608.3 | 607.5 | 607.2 | 607.8 | — |
| M/10 | 0.95 | 610.1 | 611.2 | 609.7 | 609.6 | 610.2 | 610.6 | — |
| M/12 | 0.99 | 610.2 | 611.6 | 611.1 | 611.5 | 611.4 | 611.2 | — |
| MOD14 | 1.19 | 623.7 | 621.1 | 621.6 | 621.1 | 621.3 | — | — |
| M/5(811) | 1.26 | 622.7 | 623.7 | 622.7 | 623.1 | 622.6 | 623.0 | — |
| M/8(808) | 1.37 | 628.5 | 628.4 | 627.5 | 627.9 | 628.5 | 629.1 | — |

ments are made on amorphous kerogen, the level of thermal maturity is directly established for potentially oil-prone organic matter. This procedure also allows for a quantitative determination of thermal maturity on vitrinite-poor (e.g., carbonates) or devoid (e.g., pre-Late Silurian) rocks. The principle of spectral transmittance measurements is based on the absorption of light from the tungsten spectrum by kerogen particles. Consequently, a measured spectral curve for kerogen will always be shifted toward longer wavelengths than the standard. With increasing thermal maturation the transmittance spectra gradually change and, consequently, $\lambda_{1/2}$ increases from ~570 nm for immature material up to 670 nm or more for very mature kerogen. When the number of measured spectra is large enough, the average $\lambda_{1/2}$ tends to be constant (Table 4.4). The TCI value of a given rock sample is defined as the average value of $\lambda_{1/2}$ of at least 30 amorphous kerogen measurements (Table 4.5).

TABLE 4.5 Summary of Measured Results of Transmittance Spectra

| Sample number | Type of organic matter | $R_0$ | Number of spectra measured | $\lambda_{1/2}$ (nm) | | | |
|---|---|---|---|---|---|---|---|
| | | | | Average | Minimum | Maximum | Standard deviation |
| SAH1 | LD+HD | 0.22 | 50 | 582.1 | 570.8 | 595.0 | 7.5 |
| HMS5 | LD | 0.24 | 50 | 582.8 | 563.6 | 603.3 | 9.5 |
| MSN | LD+HD | 0.28 | 70 | 584.2 | 566.9 | 605.5 | 10.1 |
| HMS4 | LD | 0.30 | 40 | 585.0 | 569.7 | 603.8 | 8.8 |
| VIL3 | LD | 0.34 | 50 | 587.5 | 569.7 | 609.3 | 10.5 |
| MOD19 | AOM | 0.38 | 50 | 588.7 | 570.2 | 609.3 | 10.5 |
| MOD12 | AOM | 0.41 | 60 | 587.3 | 568.8 | 609.3 | 12.2 |
| MOD1 | AOM | 0.42 | 80 | 589.9 | 563.1 | 608.2 | 9.4 |
| 792 | AOM | 0.45 | 40 | 592.2 | 573.5 | 608.2 | 7.9 |
| MCU | LD+HD | 0.47 | 70 | 593.5 | 571.9 | 616.5 | 11.6 |
| MOD16 | AOM±SM | 0.50 | 70 | 594.5 | 575.7 | 618.1 | 11.2 |
| ASTA | LD | 0.60 | 70 | 596.7 | 573.3 | 619.8 | 12.2 |
| MOD7 | AOM | 0.68 | 50 | 601.5 | 578.5 | 623.6 | 11.2 |
| MOD2 | AOM | 0.69 | 80 | 602.8 | 572.4 | 623.6 | 11.3 |
| MOD6 | AOM | 0.70 | 50 | 602.4 | 583.4 | 619.2 | 10.5 |
| MOD8 | AOM | 0.76 | 50 | 604.1 | 576.8 | 623.1 | 10.9 |
| M/11 | LD | 0.76 | 45 | 601.5 | 580.7 | 623.6 | 10.1 |
| MOD15 | AOM | 0.79 | 70 | 606.4 | 578.5 | 626.9 | 11.4 |
| MOD9 | AOM | 0.85 | 30 | 606.6 | 581.8 | 629.7 | 14.9 |
| 312 | AOM | 0.86 | 70 | 608.0 | 581.8 | 626.4 | 9.3 |
| M/10 | LD# | 0.95 | 70 | 610.8 | 586.7 | 638.5 | 12.6 |
| M/12 | LD# | 0.99 | 70 | 611.4 | 590.6 | 640.1 | 11.6 |
| MOD14 | AOM | 1.19 | 60 | 621.5 | 599.4 | 650.1 | 12.4 |
| M/5 | LD | 1.26 | 70 | 623.1 | 600.5 | 649.0 | 14.3 |
| M/8 | LD* | 1.37 | 70 | 629.2 | 603.3 | 659.4 | 14.6 |
| MOD13 | AOM | 1.57 | 70 | 635.8 | 609.9 | 663.8 | 14.7 |
| M/4 | LD* | 1.76 | 60 | 641.1 | 629.7 | 667.1 | 8.2 |

AOM, amorphous organic matter; HD, humodetrinite; LD, liptodetrinite; LD*, liptodetrinite from bituminous coal; SM, structured material.

To calibrate, TCI analyses were performed only on kerogen slides of samples for which reliable VR data were available. These samples were selected, in part, from material made available by the International Committee on Coal Petrology (ICCP). Samples were also selected from a collection of reliable published data (Van Gijzel, 1967) and are listed in Tables 4.4 and 4.5.

**2. Empirical Connections between TCI and Vitrinite Reflectance**

A detailed rationale is given in the appendix at the end of this chapter for expecting a unique relation between the quantity, $\lambda_{1/2}$, the TCI indicator (Table 4.5), and $R_0$, percent mean vitrinite reflectance. The predicted relationship is

$$\lambda_{1/2} = AR_0^{1/2} + B, \tag{4.42}$$

where $A$ and $B$ are constants. We have tested the theoretically predicted relationship in two ways: (1) with the general data provided in Table 4.5 and (2) with the specific TCI and reflectance data for the COST G-2 well (Tables 4.6 and 4.7).

To be sure that $\lambda_{1/2} \alpha R_0^{1/2}$ gives the best linear relation between TCI and VR, we construct:

$$X^2 = \frac{1}{N} \sum_i \left\{ \frac{\lambda_i - \lambda_s}{\Sigma_j (\lambda_j - \lambda_s)^2} - \frac{(R_{0i}^b - R_s^b)}{\Sigma_j (\lambda_j - \lambda_s)(R_{0j}^b - R_s^b)} \right\}^2, \tag{4.43}$$

where $X^2$ is a measure of mismatch of TCI and VR data, $N$ is the number of samples, $\lambda_i$ is the TCI value of $i$th sample, $\lambda_s$ is the surface value for TCI, $R_{0i}$ is the VR value of the $i$th sample, $R_s$ is the VR surface value, and $b$ is the parameter to be determined in the VR–TCI relationship. By varying $b$, we obtain a series of $X^2$ values. The best $b$ is chosen when $X^2$ reaches its minimum (Figs. 4.24a and 4.24b). The minimum of $X^2$ for data from Table 4.4 is at $b = 0.5 \pm 0.1$ as shown in Fig. 4.24a. Data from the COST G-2 well show the same range of $b$ values (Fig. 4.24b). Both cases show a linear relationship between TCI and $R_0^{1/2}$. Figure 4.25 shows a plot of $\lambda_{1/2}$ versus $R_0^{1/2}$ for the data presented in Table 4.5. Superimposed on the data (with associated error bars on TCI quoted in Table 4.5) are two "eyeball" straight lines. No attempt was made to improve the fit beyond the "eyeball" level because of the error bars on TCI. What is apparent, however, is that even at this level the relationship is indeed linear as predicted theoretically.

Although the general relationship $\lambda_{1/2} \alpha R_0^{1/2}$ appears to be broadly confirmed from Table 4.5 data, it seemed that the coefficients $A$ and $B$ in Eq. (4.42) might be specific to either kerogen type, correct vitrinite identification, or basinal setting. Accordingly, we examined the behavior for data from a specific well. Figure 4.26 shows $\lambda_{1/2}$ versus $R_0^{1/2}$ for data (Tables 4.6 and 4.7) from COST G-2 well, Georges Bank, where both TCI and VR data are available. Again we see that a straight-line plot occurs (to within error assessments). To obtain an estimate of degree of fit we proceed as follows. At the set of depths in COST G-2, $z_1, z_2, \ldots, z_n$ where both vitrinite

TABLE 4.6 TCI Data from COST G-2 Well, Georges Bank Basin

| Depth (m) | TCI (nm) | Depth (m) | TCI (nm) |
|---|---|---|---|
| 807 | 589 | 4892 | 629 |
| 1198 | 585 | 4947 | 626 |
| 1353 | 587 | 4984 | 634 |
| 1505 | 590 | 4999 | 644 |
| 1701 | 590 | 5066 | 634 |
| 1874 | 594 | 5097 | 648 |
| 2056 | 594 | 5142 | 641 |
| 2216 | 595 | 5151 | 627 |
| 2374 | 599 | 5209 | 631 |
| 2522 | 603 | 5212 | 643 |
| 2659 | 610 | 5295 | 630 |
| 2664 | 613 | 5337 | 633 |
| 2676 | 614 | 5340 | 647 |
| 2819 | 615 | 5423 | 654 |
| 2987 | 615 | 5459 | 641 |
| 3121 | 616 | 5517 | 654 |
| 3289 | 614 | 5572 | 651 |
| 3344 | 614 | 5642 | 644 |
| 3356 | 614 | 5715 | 640 |
| 3420 | 616 | 5786 | 637 |
| 3572 | 615 | 5831 | 641 |
| 3609 | 616 | 5883 | 642 |
| 3728 | 617 | 5972 | 644 |
| 3886 | 619 | 6115 | 645 |
| 4042 | 629 | 6212 | 652 |
| 4435 | 622 | 6289 | 652 |
| 4450 | 622 | 6343 | 654 |
| 4487 | 626 | 6429 | 655 |
| 4770 | 625 | 6487 | 657 |
| 4834 | 626 | 6557 | 659 |

data and TCI data were available, we computed $A$ and $B$ from a least-squares fit model to Eq. (4.42), that is, from the two equations

$$\sum \lambda_i = A \sum R_{0i}^{1/2} + nB, \tag{4.44a}$$

$$\sum \lambda_i R_{0i}^{1/2} = A \sum R_{0i} + B \sum R_{0i}^{1/2}. \tag{4.44b}$$

TABLE 4.7 Vitrinite Reflectance Data from COST G-2 Well, Georges Bank Basin

| Depth (m) | $R_0$ (m) | Depth (m) | $R_0$ (m) |
|---|---|---|---|
| 488 | 0.386 | 2523 | 0.647 |
| 808 | 0.380 | 2820 | 0.689 |
| 1006 | 0.400 | 3122 | 0.721 |
| 1200 | 0.432 | 3420 | 0.720 |
| 1280 | 0.480 | 3887 | 0.946 |
| 1353 | 0.407 | 4042 | 0.997 |
| 1554 | 0.472 | 4771 | 1.394 |
| 1701 | 0.497 | 4984 | 1.487 |
| 1801 | 0.456 | 5786 | 1.969 |
| 1905 | 0.465 | 6115 | 2.028 |
| 2056 | 0.579 | 6344 | 1.965 |
| 2092 | 0.598 | 6487 | 2.134 |
| 2374 | 0.575 | 6557 | 2.066 |

Then TCI was written in a vitrinite equivalent, $R_e$, form as

$$R_e = (\lambda - B)^2/A^2, \quad (4.45)$$

and $R_e$ is plotted with depth on top of the observed VR measurements with depth. The result is shown in Fig. 4.27, from which it can be seen that the behavior of $R_e$ with depth closely parallels the measured VR with depth behavior, thus confirming the appropriateness of the square root relationship between TCI and VR.

### 3. Thermal Inversion Procedures

To examine the applicability of TCI as an effective thermal indicator in the quantitative modeling of paleoheat flux in a basin, we use both VR data and TCI data in a one-dimensional fluid flow/compaction model (Cao and Lerche, 1989). This procedure permits us to see if TCI can replace VR in the inversion process. The COST G-2 well in the Georges Bank Basin provided sufficient data for this analysis.

We first reconstruct the burial history, then reconstruct thermal history based on the burial history, because basement subsidence, sediment fill, compaction, and diagenesis processes directly influence the reconstructed thermal history of the basin.

#### *a. Burial History of the COST G-2 Well*

*Geological Setting.* Georges Bank Basin is a collection of sub-basins and grabens that lie in the North Atlantic outer continental shelf of the

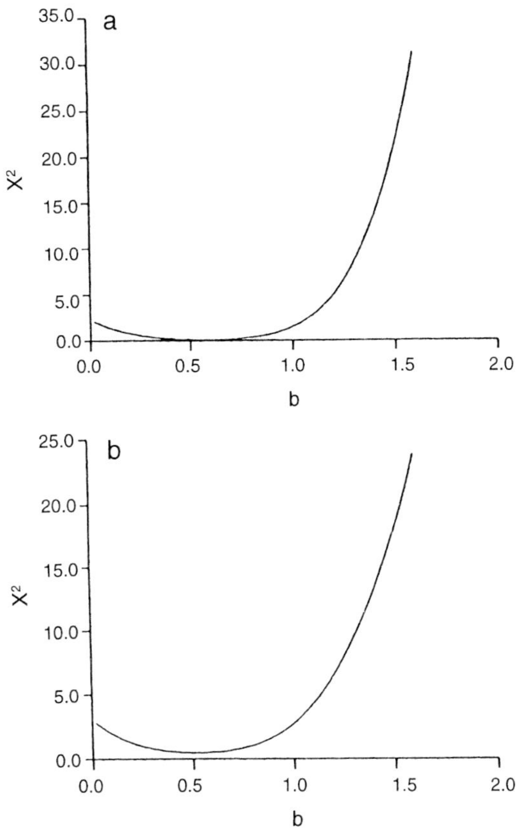


FIGURE 4.24 Plots of $X^2$ versus $b$ show that $\lambda_{1/2}$ and $R_0^b$ have a linear relationship when $b \cong 0.5$ where $X^2$ has its minimum value. (a) Data from Table 4.4; (b) data from COST G-2 well. (Reprinted with permission from Van Gijzel *et al.*, 1992.)

United States (Fig. 4.28). They are filled by a wedge of Triassic and younger sedimentary rocks that overlie rifted basement. The formation and development of the basin have been closely associated with the rifting and separating of the African plate from the North American plate. Before the Early Jurassic late rifting phase, northwest Africa was adjacent to the Georges Bank region (Klitgord *et al.*, 1988). The development of the strata in the basin started with Late Triassic synrift carbonates and evaporites (Manspeizer and Cousminer, 1988). The stratigraphy of the basin provided by the G-2 well (Cousminer and Steinkraus, 1988) is shown in Fig. 4.29. The total section encountered in the G-2 well is ~6700 m thick and is composed of limestone, dolomite, anhydrite, shale, sandstone, and mudstone. The Bathonian (Middle Jurassic) and younger section contains abundant sandstone and mudstone, whereas the older rocks are dominantly carbonates

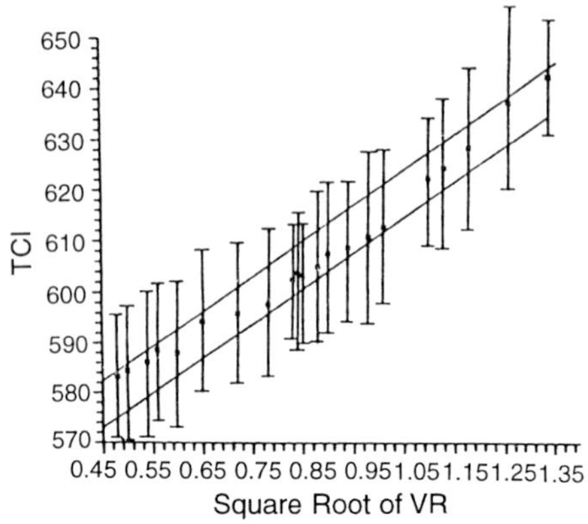


FIGURE 4.25 Plot of $\lambda_{1/2}$ versus $R_0^{1/2}$ for the data presented in Table 4.4. Superimposed on the data (with associated error bars on TCI quoted in Table 4.4) are two "eyeball" straight lines. (Reprinted with permission from Van Gijzel *et al.*, 1992.)

and evaporites. Analysis of the stratigraphic and structural evolution of the basin indicates that the basin developed from a rifting environment in which restricted oceanic circulation and intermittent sabkha conditions resulted in evaporite and carbonate deposition (Smith and Van Gijzel, 1989) during the Late Triassic and Early Jurassic. Then, along with the separation of the plates, the basin became wider and had more continental detrital sediments deposited, indicating the formation of a "normal" passive margin basin on the shelf along the East Coast of the United States.

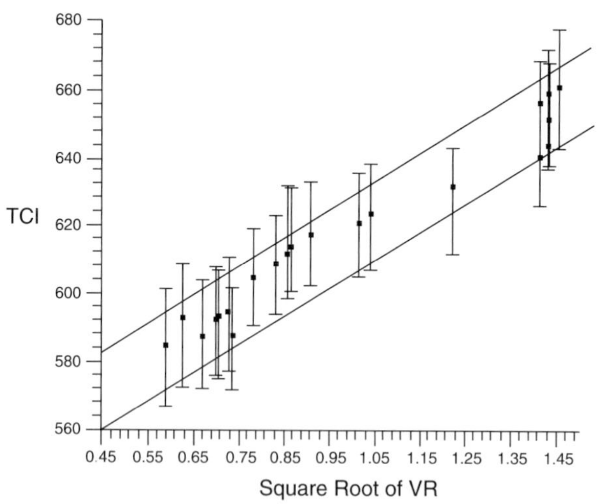


FIGURE 4.26 Plot of $\lambda_{1/2}$ versus $R_0^{1/2}$ for the data from COST G-2 well presented in Table 4.6. The meanings of error bars and straight lines are the same as for Fig. 4.25. (Reprinted with permission from Van Gijzel *et al.*, 1992.)

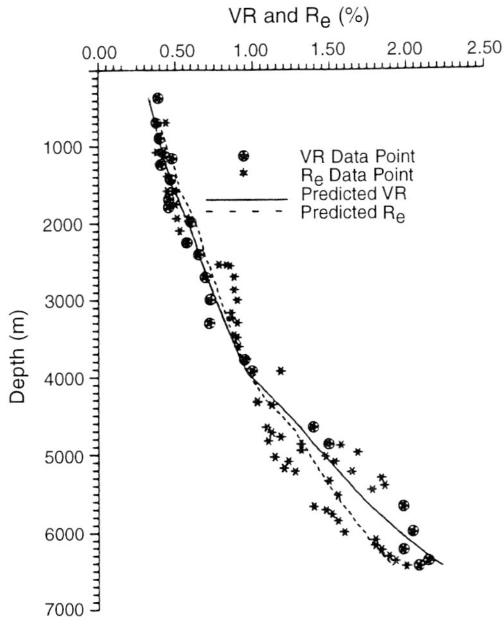


FIGURE 4.27 Comparison of $R_e$ and VR measurements with depth. $R_e$ with depth closely follows the trend of measured VR with depth. The predicted VR (solid line) and $R_e$ (dashed line) are also compared with VR and $R_e$ measurements. (Reprinted with permission from Van Gijzel *et al.*, 1992.)

*Modeling of the Well.* Figure 4.30 exhibits the basement subsidence and tectonic subsidence variation with time at the G-2 well site. The simulation of tectonism-caused basement subsidence follows Steckler and Watts (1978). Based on the total basement subsidence and tectonic subsidence, variation of basement subsidence rate and tectonic subsidence rate, and using prior studies of the lithology, stratigraphy, and geophysics in this area (Schlee and Klitgord, 1982), five stages, A, B, C, D and E, are examined (shown on Fig. 4.30).

Among these stages, the A and B stages are two cycles in which the basement subsided very rapidly. In stage A particularly (prior to 100 MYBP) the rapid subsidence of the basement led to the formation of strata ~3600 m thick. The average deposition rate is 66 m/Ma. Stage A corresponds to the rift to postrift stage. In stage B (100–125 MYBP) the thickness of the strata is 1900 m; deposition rate is 54 m/Ma. Stage B corresponds to a basement uplifting and erosion event. Stages C and D represent gentle development stages in the basinal history. In stage C (125–37 MYBP), the average deposition rate is 3 m/Ma. In stage D, basement uplift and sediment erosion occurred at about 5.2 MYBP. The formation of a stable basement accompanied these stages. Stage E is the beginning

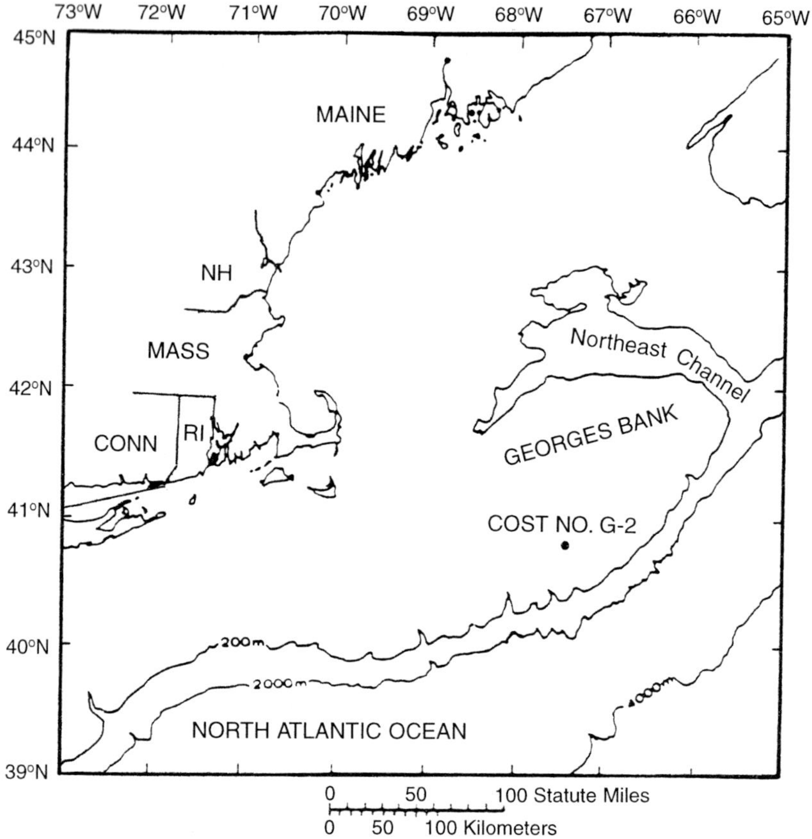


FIGURE 4.28 Location of COST G-2 well in the Georges Bank Basin, East Coast of the United States. (Reprinted with permission from Van Gijzel *et al.*, 1992.)

of a new basement subsidence/sediment fill cycle of the basin, with almost 300 m of sediment deposited withinthe last 5 Ma.

A one-dimensional fluid flow/compaction model (Cao and Lerche, 1989; He and Lerche, 1989) is applied to simulate the dynamic processes of the burial history. We optimize the modeling by minimizing the mismatch between predicted values and geological measurements such as porosity variation with depth, thickness of the strata, and so on, so as to make the simulation of thermal history more reliable (Fig. 4.31).

Figure 4.32 presents the modeled burial history of the well site, which shows the variation of sediment compaction with time. The following discussion of TCI and VR, as well as the thermal history, is based on this dynamic burial history. Having reconstructed the burial history of the basin, it is then possible to reconstruct the thermal history. Moreover, different thermal indicators such as TCI and VR can be used in the model, and comparisons

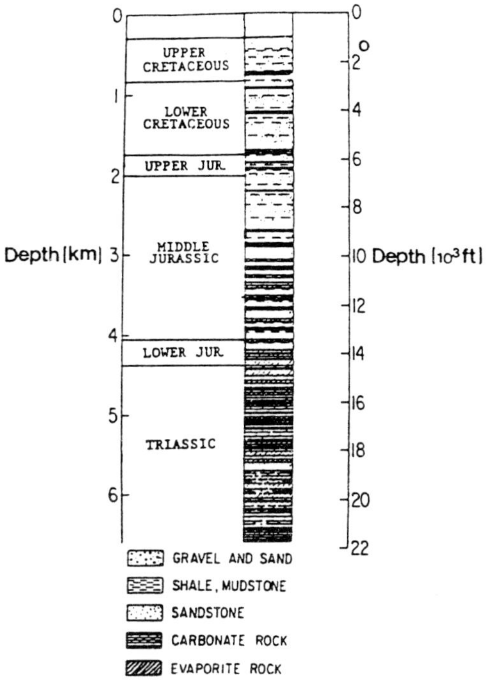


FIGURE 4.29 Stratigraphic column of the COST G-2 well. (Reprinted with permission from Van Gijzel *et al.*, 1992.)

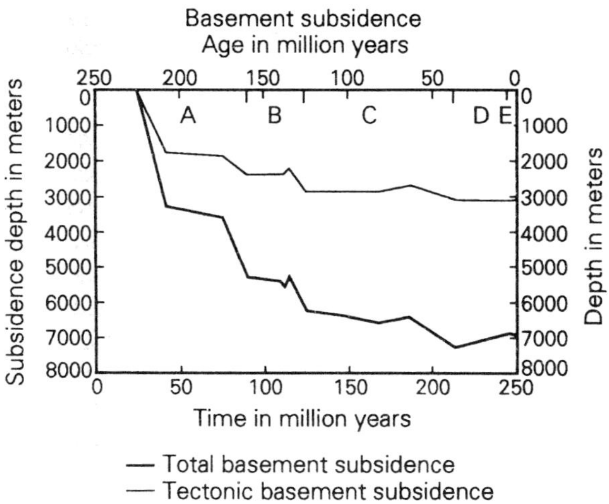


FIGURE 4.30 Basement subsidence and tectonic subsidence variation with time at the COST G-2 well site. The capitals A, B, C, and so on represent stage A, stage B, stage C, etc. (Reprinted with permission from Van Gijzel *et al.*, 1992.)

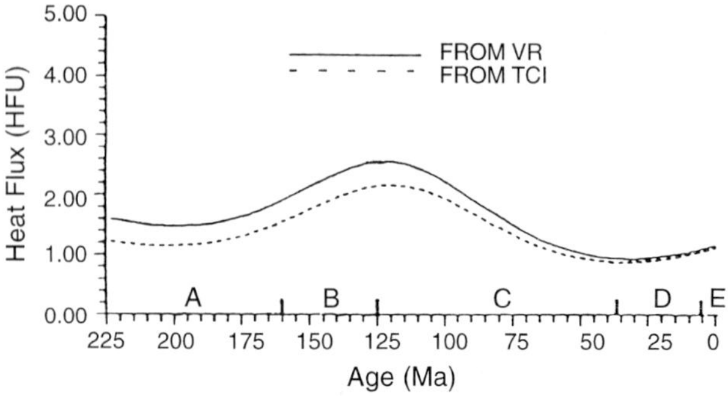


FIGURE 4.31 Variations of paleoheat flow determined from VR data (solid line) and $R_e$ (dashed line). The stages are also labeled in the figure. (Reprinted with permission from Van Gijzel *et al.*, 1992.)

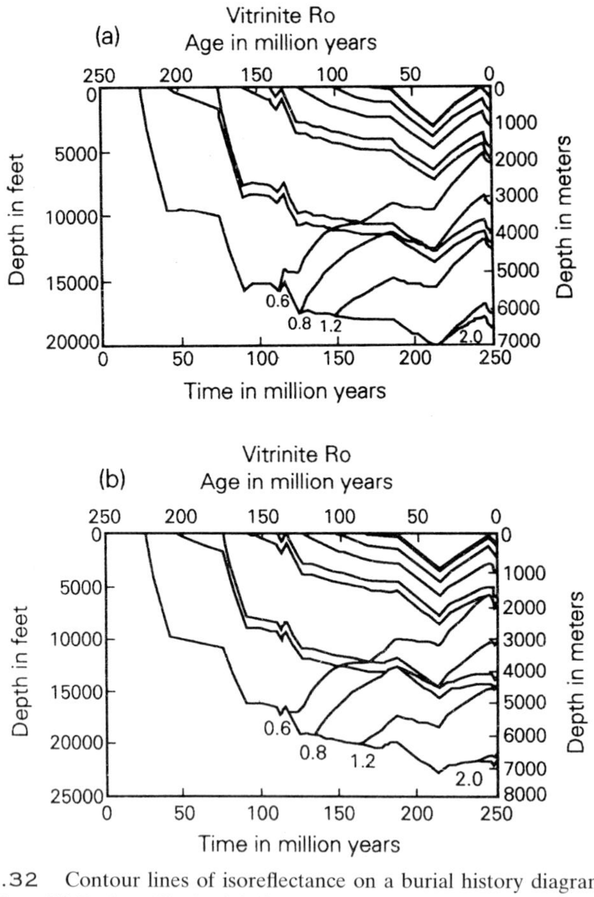


FIGURE 4.32 Contour lines of isoreflectance on a burial history diagram determined from (a) VR data; (b) $R_e$ data. The burial history diagram shows sedimentary fill and compaction with time at the COST G-2 well site. (Reprinted with permission from Van Gijzel *et al.*, 1992.)

given of the effect of using such different thermal indicators in the inverse model for determining paleoheat flux variations.

### *b. Thermal History Modeling*

Armed with the quantitative deposition history, in which the thermal conductivity, porosity, permeability, and fluid pressure have been simulated with time, thermal history reconstruction is then relatively simple. Applying inversion methods (Lerche *et al.,* 1984) to the measured and equivalent ($R_e$, converted from TCI) VR data, the thermal history, as well as the effects of the thermal indicators in the thermal inversion procedure, are obtained and compared.

The paleoheat flux function used here is:

$$Q(t) = Q_0 \exp[\beta t + a_1 \sin(\pi t/t_{\max}) + a_2 \sin(2\pi t/t_{\max}) + a_3 \sin(3\pi t/t_{\max})], \tag{4.46}$$

where $Q(t)$ is the heat flux at time $t$, $Q_0$ is the present-day heat flux, $t$ is the time, $t_{\max}$ is the bottom hole age, and $\beta$, $a_1$, $a_2$, and $a_3$ are parameters to be determined. The inverse optimization tomographic procedure is employed to search out the $\beta$, $a_1$, $a_2$, and $a_3$ assemblage of values by minimizing the mismatch of both predicted VR and TCI indicator values against VR and TCI measurements from the G-2 well.

In the modeling, the formation temperature at time $t$ and depth $z(t)$ below the sediment–water interface is

$$T(t, z(t)) = T_s + Q(t) \int_0^{z(t)} dz/k(z), \tag{4.47}$$

where $T_s$ is the surface temperature and $k(z)$ is the sedimentary thermal conductivity at depth $z(t)$ described through

$$k(z) = (k_{fl})^{\phi(z)}(k_s)^{(1-\phi(z))}, \tag{4.48}$$

where $k_{fl}$ is the thermal conductivity of the pore fluid, $k_s$ is the thermal conductivity of the solid rock, and $\phi(z)$ is the porosity at depth $z$ (Lewis and Rose, 1970; Andrews-Speed *et al.,* 1984).

A comparison between predicted VR values and measured VR values using both VR data and $R_e$ data (from TCI), respectively, is shown in Fig. 4.27. The variations of the paleoheat flow determined from VR data and $R_e$ data are shown in Fig. 4.31. There are only slight differences between the inverse results using either VR data or $R_e$ data (Fig. 4.32). Figure 4.27 shows the distribution behaviors of VR and $R_e$ data with depth. Although the VR and $R_e$ data have very similar distributions, the broadly scattered $R_e$ data deeper than ~4000 m cause slight differences in the heat flow determined using the different thermal indicators VR and $R_e$. The heat flux determined from $R_e$ data is slightly lower than that determined from the VR data, but to within data resolution almost identical thermal histories for the basin are obtained.

The paleoheat flux shows a peak at ~125 MYBP, tied in time to stage B tectonics of uplift and erosion. The nominal heat flux peak is several million years later than the erosion event, but the width of the peak is about ±25 Ma, which encompasses the time of stage B tectonics. In concert with the heat flux variations of the basin, the other thermal and maturation indices (such as paleotemperature, TTI, and VR distributions in a burial history plot) also show that the modeled results are almost identical from both VR and $R_e$ data, as shown in Figs. 4.32a and 4.32b.

**4. Pragmatic Procedure for TCI Thermal Inversion**

Based on the theoretical/empirical relation that TCI is proportional to the square root of vitrinite reflectance, and to the close correspondence of the paleoheat flux results in the COST G-2 well for both measured and equivalent vitrinite data, a pragmatic procedure for using TCI as a thermal indicator on its own is recommended as follows. Set $\lambda_s = 575 \pm 5$ (nm) as a surface depositional value based on both the fit of TCI to VR from Table 4.4 and from COST G-2—both of which yield essentially the same value for $\lambda_s$. Then at each depth $z_i$ ($i = 1, \ldots, n$) where TCI is measured construct the equivalent vitrinite reflectance value

$$R_e^{(i)} = (\lambda_i/\lambda_s - 1 + R_s^{1/2})^2, \qquad \text{for } i = 1, \ldots, n, \tag{4.49}$$

where $R_s$ is the surface depositional value of vitrinite reflectance, estimated at the fraction $R_s = 0.2 \times 10^{-2}$ based on several hundred wells (Lerche, 1990a). Then follow *precisely* the inversion method for vitrinite reflectance as a thermal indicator provided elsewhere (Lerche, 1990a) using $R_e^{(i)}$ as the equivalent reflectance value at depth $z_i$.

This is a pragmatic procedure for using TCI as an inverse thermal indicator on its own in depositional and lithologic regions lacking vitrinite.

**5. Discussion**

From the preceding analysis, we conclude the following:

1. TCI shows a close relation with VR data for the same sedimentation series. Thus TCI is directly convertible to equivalent VR data. The empirical connection of TCI to VR is in accord with the theoretically predicted behavior.
2. The COST G-2 well case history, where both TCI and VR data are available, shows that TCI and VR yield essentially identical inverse thermal histories for the well, corroborating that TCI is a useful general-purpose thermal indicator.

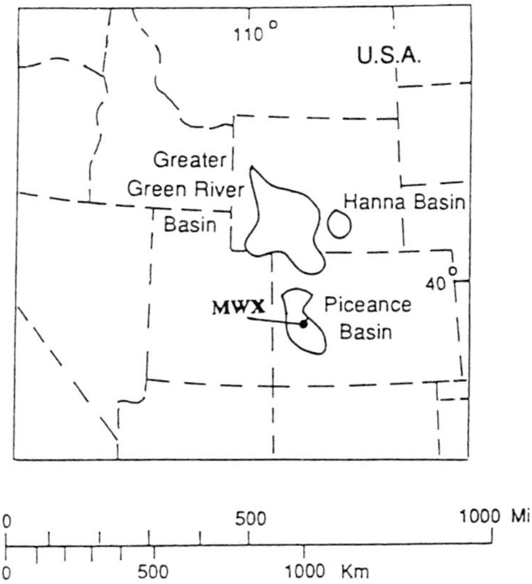


FIGURE 4.33 Index map showing the location of the MWX site in the Piceance Creek Basin, Colorado, USA (from Law *et al.,* 1989).

## G. APPLICATION TO THE MULTIWELL EXPERIMENT SITE, PICEANCE CREEK BASIN, COLORADO

The purposes of this section are (1) to apply the 1-D model to an area in which there are excellent thermal indicators and a well-defined geological setting; (2) to examine the uncertainty associated with each parameter determined, in particular those parameters connected with paleoheat flux and erosion events; and (3) to explain or extrapolate the geohistory and thermal history in the area. The multiwell experiment (MWX) site (Fig. 4.33), Piceance Creek Basin, has been selected for the study.

The Piceance Creek Basin contains major resources of natural gas (derived from coal formations) in low permeability (tight) reservoirs. The National Petroleum Council (Spencer and Keighin, 1984) estimated a maximum recoverable gas resource in this area of about 32 TCF (= $10^9$ $m^3$). Major gas reserves occur in fluvial and marine sandstones in the Upper Cretaceous Mesaverde Formation and in fluvial sandstones in the Lower Tertiary Wasatch Formation. Geological and geothermal data are available from USGS Open File Report 84-757 (Spencer and Keighin, 1984) and vitrinite reflectance well profiles (Law *et al.,* 1989).

For the MWX site, most of the detailed geological, geophysical, and geochemical data are reported in geological studies of the U.S. Department of Energy Multiwell Experiment, Garfield County, Colorado, by Spencer and Keighin (USGS Open File Report 84-754).

Figure 4.33 shows the location of the MWX site. The wells are designated CER Nos. 1, 2, and 3 MWX and are all located in the northwest quarter of Sec. 34, T.6.S, R.94.W. Garfield County, Colorado. The wells were spudded in the upper part of the Tertiary Wasatch Formation and reached a total depth of 8000 ft (2500 m) at or near the base of the Upper Cretaceous Mesaverde Formation. The three wells deviated only slightly from vertical with only minor lateral drift with depth. Because all three wells were in proximity and had similar downhole data, they are treated as a single composite well in the discussion that follows. Numerous commercial and research geophysical logs were run in all three wells on multiple logging runs. More than 4000 ft (1250 m) of core was cut in the three wells with nearly complete recovery. These cores and their subsequent analyses provide the basic data for a number of research studies, including origin of gases, thermal maturation of source rock, well-log interpretation, and vitrinite reflectance from the three wells (Law *et al.,* 1989), while Fig. 4.34 (taken from Bostick and Freeman, 1984) shows present-day temperature with depth in each borehole.

**1. Geological Setting**

The Piceance Creek Basin is a latest Cretaceous through Eocene age sedimentary and structural basin (Johnson and Nuccio, 1986). Prior to the Laramide uplift, which broke up the much larger Cretaceous epiric sea, the basin was not a separate intermountain basin as it is today; rather it was part of a large, rapidly subsiding trough that covered much of the central part of the United States and Canada, and which formed to the east of the actively rising area. The orogenic belt was occupied by an epicontinental seaway for much of the Cretaceous. As much as 5000 ft (1520 m) of marine sediments (Mancos shale and marine portion of the Mesaverde Group) were deposited in the Piceance Creek Basin before the sea retreated to the east in the late Cretaceous. The eastward retreat of the seaway resulted in regressive sequences including marginal marine, blanket sandstone, coal swamp, and fluvial deposits. A basin-wide unconformity occurs at the top of the Mesaverde, and the upper part of the Mesaverde was deeply weathered. The thickness of Cretaceous rocks removed by the unconformity is unknown, but possibly thousands of feet of section were lost. When sedimentation resumed in the early or middle Paleocene, the unconformity was onlapped by sediments of the Wasatch Formation. The oldest sediments of the Wasatch were coarse clastic, but by late Paleocene a poorly drained lowland or swamp covered most of the basin.

A permanent lake formed in the northcentral part of the basin near the beginning of the Eocene. The lake expanded to cover a larger area (Johnson and Nuccio, 1986) and then formed Lake Uinta in which the widespread Mahogany oil shale zone of the Green River Formation was deposited.

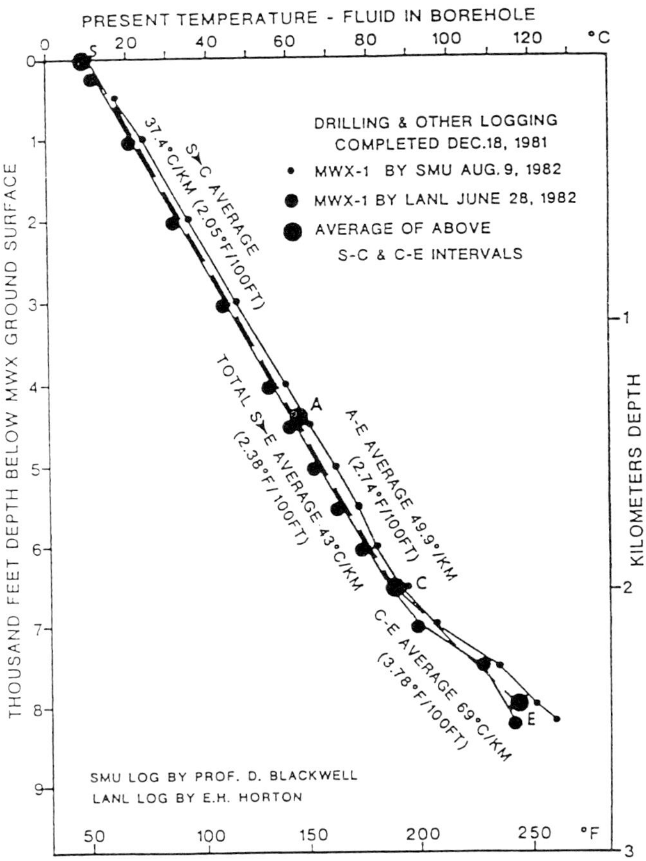


FIGURE 4.34 Present rock temperatures based on temperatures of fluid in borehole 6 and 8 months after drilling (from Bostick and Freeman, 1984).

Oligocene magmatism affected the southeastern corner of the Piceance Creek basin with shallow intrusions throughout the area from about 34 to 29 Ma. Several of the intrusions formed an extensive surface sheet of volcanic breccia. Basalts covered much of the central part of the Piceance Creek Basin. About 10 Ma ago major tectonism resumed throughout the Rocky Mountain region. Old Laramide structures were reactivated and some new folds and faults were formed as the entire region was uplifted vertically. Downcutting of the Colorado River Canyon across the middle of the basin began about 9 Ma. The Colorado River system has removed as much as 5000 ft (1525 m) of overburden from the central part of the Piceance Creek Basin.

At the MWX site (Fig. 4.34) the preserved formations, from oldest to youngest, are the Mancos shale, Lies Formation, Williams Fork Formation,

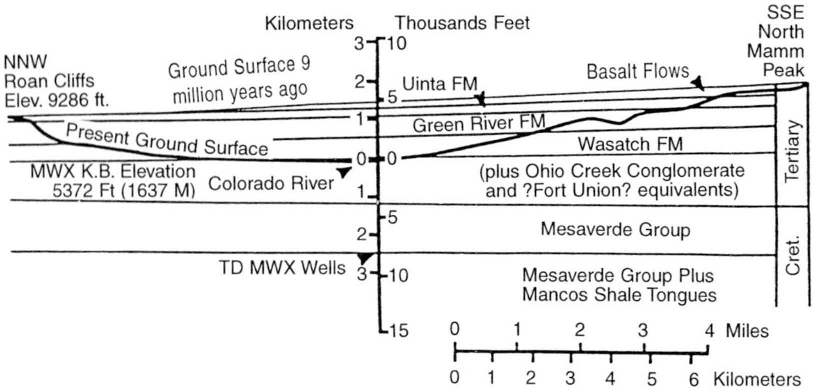


FIGURE 4.35 Schematic stratigraphic cross-section through the MWX boreholes and extending to the valley walls on either side. Formation contacts are extended across the present valley approximately where they were prior to erosion by the Colorado River (from Bostick and Freeman, 1984).

and Wasatch Formation. In the Mesaverde Formation the sequence is composed predominantly of carbonaceous shale, coal, and lenticular sandstone. The lower part, about 900 ft (274 m) of sandy shale, grades up into a more sandy section of overlapping lenticular sandstones interbedded with gray shale and mudstone. The interpreted depositional environment is that of medium to low sinuosity fluvial environment. This upper part is about 2500 ft (762 m) thick and is truncated at the top by a regional unconformity.

Overlying the upper Cretaceous unconformity at the MWX site is about 3900 ft (1189 m) of Late Paleocene through early Eocene age variegated mudstone and lenticular sandstone of the Wasatch Formation. The MWX site is in the canyon of the Colorado River, and about 4700 ft (1433 m) of Eocene age sediments are still preserved on the nearby canyon walls (Fig. 4.35, taken from Bostick and Freeman, 1984). Thus a reasonably good estimate of the type and thickness of sediments removed by erosion at the MWX site can be made. This estimate is necessary in order to construct a burial curve for the MWX site for use in the thermal maturation studies. Approximately 600 ft (183 m) of early Eocene variegated mudstone and lenticular sandstone of the Wasatch Formation, and 4100 ft (1250 m) of early to late Eocene marlstone, oil shale, and sandstone of the Green River Formation were removed.

Between 10 and 40 Ma volcanic dikes were emplaced about 40 miles southeast of MWX, causing some local heating adjacent to each dike. The MWX site was probably covered by a thin basalt flow, which was subsequently cut through by the Colorado River (see Fig. 4.35). All of these events provide clues to regional heat flux "pulsations" in this period (Figs. 4.36 and 4.37).

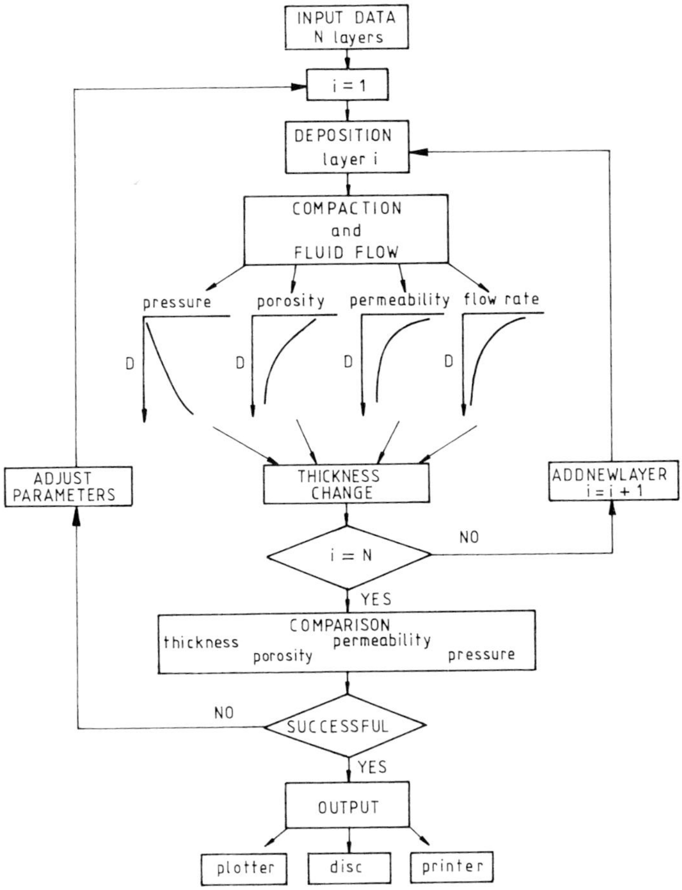


FIGURE 4.36 Flow diagram for 1-D compaction model, step 1 (from Cao, 1985).

## 2. Thermal Maturation and Burial History

Figure 4.38 shows the burial history in the Piceance Creek Basin at the MWX Site. At about 75 Ma the area began to subside. Three major deposition stages occurred in the Mesaverde Group (Upper Cretaceous) and Wasatch Formation (Tertiary). Major uplift, starting in the late Miocene, resulted in the present-day surface unconformity (Table 4.8).

A long hiatus, of about 25 million years duration, predated the uplifting and a short time unconformity (3 Ma) occurred between two major depositional phases. The greatest basement depth is at about 13,000 ft (4500 m) at about 35 Ma. The present day total depth is 8400 ft (2560 m). Table 4.9 shows the thicknesses of individual layers comparing input and output data results.

Examining the history of compaction at the MWX site, we find that the

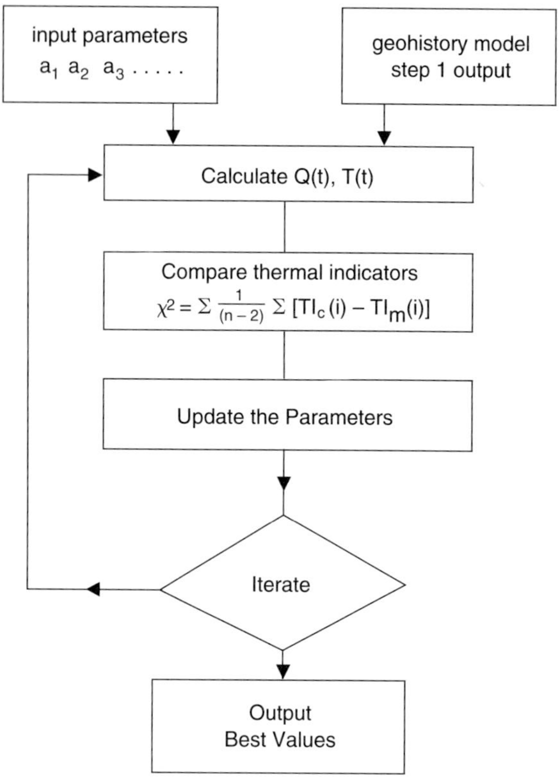


FIGURE 4.37 Flow diagram for 1-D thermal indicator tomography, step 2.

Upper Cretaceous and Tertiary shales and sandy-shales were overpressured soon after their deposition and retained overpressure until the present day (Fig. 4.39). The profiles of pressure with depth at present are shown in Fig. 4.40.

In general, overpressuring is considered to be the result of compaction caused by the combined effects of three factors: (1) lack of a permeable bed, (2) a rapid rate of sedimentation, and (3) thick accumulation of sediments (Magara, 1978). In the MWX site, the burial processes of shale and sandy-shale at the early stage of sedimentation could be the reason for overpressure. Alternatively, Law *et al.* (1989) have emphasized that overpressuring in low-permeability *sandstone* reservoirs in the Rocky Mountain region is due to the accumulation of thermogenic gas. In the thermal indicator tomography procedure, if we consider the thermogenic gas to be completely trapped, predicted fluid pressure will be larger than the presently observed values. For instance the interpreted pressure for MWX site well at about 8000 ft (2430 m) (Spencer, 1987) is 6200 psi (420 atm), which is 1800 psi (120 atm) higher than the calculated fluid pressure (see Fig. 4.40).

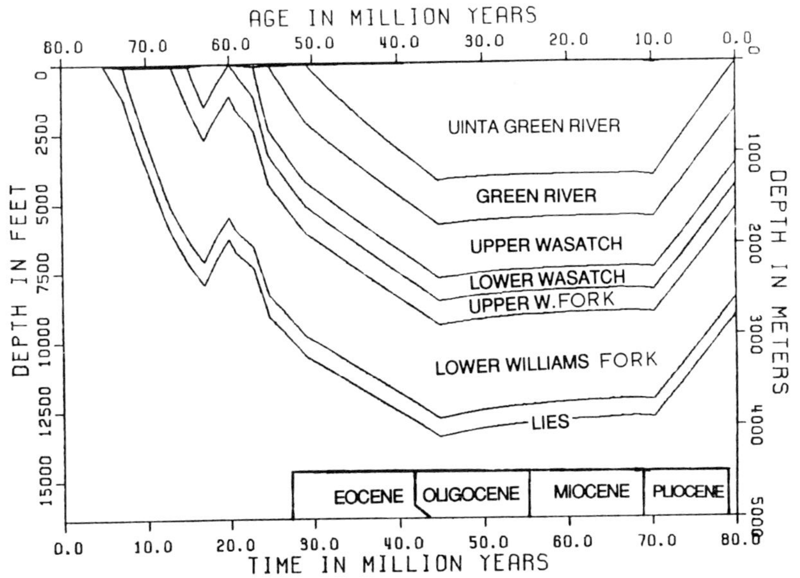


FIGURE 4.38 Geohistory profile at MWX site, Piceance Creek Basin. Horizontal line marks a hiatus; tilted lines represent subsidence or uplift.

TABLE 4.8 Basic Input Data for Piceance Basin, MWX Well

| Layer number | Depth at base of formation (m) | Age at base (Ma) | Lithology | Thermal conductivity ($cal \cdot cm^{-1} \cdot s^{-1} \cdot °C^{-1}$) |
|---|---|---|---|---|
| 12 | −1220.0 | 9.8 | Unconformity | |
| 11 | 165.0 | 11.8 | Shale | $3.87 \times 10^{-3}$ |
| 10 | 165.0 | 35.2 | Hiatus | |
| 9 | −1220.0 | 50.8 | Unconformity | |
| 8 | 457.0 | 55.2 | Sandy-shale | $6.5 \times 10^{-3}$ |
| 7 | 990.0 | 57.0 | Sandy-shale | |
| 6 | 1220.0 | 60.0 | Shale | |
| 5 | −457.0 | 63.0 | Unconformity | |
| 4 | −457.0 | 65.0 | Unconformity | |
| 3 | 1463.0 | 67.0 | Shale | |
| 2 | 2377.0 | 72.7 | Shale | |
| 1 | 2560.0 | 75.0 | Shale | |

*Note:* Depth is measured from the surface and negative means erosion. Age is based on GSA 1983 geologic timescale. Lithology data from Spencer and Keighin (1984). Input thermal conductivity for the same lithology is identical.

TABLE 4.9 Comparison of Input and Output Thickness and Porosity of Each Layer

| Layer | Depth (I) (m) | Depth (O) (m) | Thickness (I) (m) | Thickness (O) (m) | Porosity (I) (fraction) | Porosity (O) (fraction) |
|---|---|---|---|---|---|---|
| Uinte-Green River | 61.0 | 51.4 | 45.7 | 45.8 | — | 0.26 |
| Green River | 457.2 | 448.9 | 396.2 | 397.4 | — | 0.24 |
| Upper Wasatch | 990.6 | 987.5 | 533.4 | 538.6 | 0.23 | 0.23 |
| Middle Wasatch | 1219.1 | 1216.5 | 228.6 | 229.0 | — | 0.13 |
| Lower Wasatch | 1371.5 | 1368.8 | 152.4 | 152.3 | — | 0.14 |
| Upper Williams Fork | 1463.0 | 1451.3 | 91.4 | 82.4 | — | 0.18 |
| Lower Williams Fork | 2316.4 | 2315.5 | 853.4 | 864.3 | — | 0.16 |
| Cameo Coal Zone | 2377.3 | 2377.1 | 61.0 | 61.5 | — | 0.16 |
| Lies | 2560.2 | 2561.8 | 182.9 | 184.7 | 0.15 | 0.13 |

*Key:* I, input; O, calculated output; –, no data.

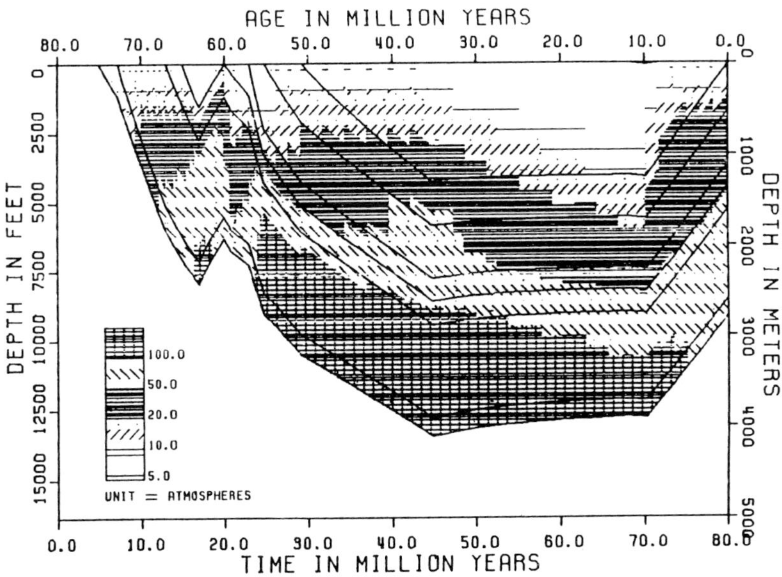


FIGURE 4.39 Excess pressure variation with geohistory.

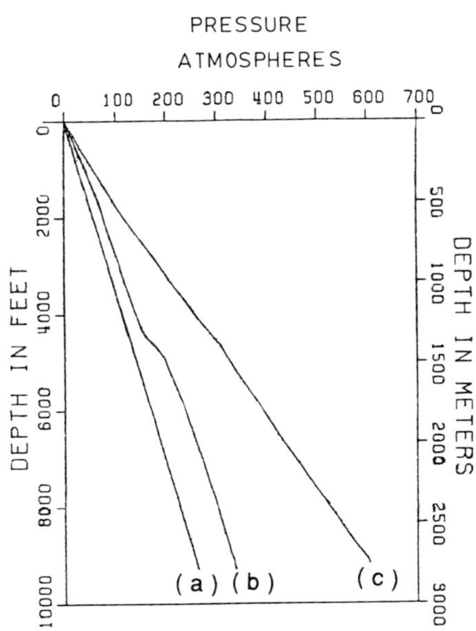


FIGURE 4.40 Three kinds of calculated pressure: (a) hydrostatic pressure; (b) fluid pressure; (c) overburden pressure.

TABLE 4.10 Tomography Results and Error Bar of Parameters for MWX Site, Colorado

| MSR | $\beta$ (MY$^{-1}$) | $a_1$ | $T_c$ (K) | $T_D$ (K) | Eroded thickness (m) | Age erosion begins (Ma) |
|---|---|---|---|---|---|---|
| 0.011 | $(-0.83 \pm 0.1) \times 10^{-3}$ | $0.55 \pm 0.2$ | $260 \pm 40$ | $160 \pm 30$ | $1250 \pm 1000$ | $10.5 \pm 2$ |

*Note:* The error bars were determined by 10% increase of MSR from its minimum value.

But the general pervasive increase in overpressure with depth through the stratigraphic sequences of shales, sandy-shales, etc., extending over more than 1000 m, is probably not due to the thermogenic gas component, which can only serve to balance fluid overpressure in a sandstone. Rather, there is a systematic increase of water overpressuring with depth, in concert with the restricted gas component in sandstone reservoirs.

Because an unconformity represents erosional removal, the trend of measured vitrinite reflectance values with depth shows a discontinuity at the unconformity depth (Armagnac *et al.,* 1989). For the MWX site the surface measured vitrinite reflectance is 0.6, higher than the general surface depositional value of 0.2, suggesting that erosion happened in the relatively recent past. To fit the vitrinite reflectance data, we adjusted both the geohistory model and geothermal model parameters (i.e., the amount of thickness eroded and paleoheat flux) to obtain the best fit values determined by the MSR between measured and predicted vitrinite reflectance. Table 4.10 shows the tomography result for the MWX site. Predicted present-day vitrinite reflectance is compared with the measured vitrinite reflectance in Fig. 4.41. The degree of fit is measured by MSR, the mean square mismatch per point between predicted and measured values. The paleoheat flux pattern and erosion thickness and erosion timing were all determined by the tomography method. Figure 4.42 shows the heat flux variation with time. The peak heat flux is about 2.74 HFU at about 24 Ma, attributed to the effect of the volcanic activity in Oligocene-Miocene time (Bostick and Freeman, 1984). Table 4.10 shows the values of eroded thickness and erosion timing, corresponding to the least mismatch, which are basically consistent with the known geologic data and events. Calculated eroded thickness in the range of $1000 \pm 500$ m and erosion commenced about 10 Ma.

Uncertainty of the chemical parameters $T_c$ and $T_D$, which relate to the calculation of vitrinite reflectance as a thermal indicator (Lerche *et al.,* 1984), causes a variation and uncertainty of the heat flux pattern. As shown later in Fig. 4.47 for $T_c < 300$ K and $T_D > 90$ K the MSR has only minor changes. But with the best default values of $T_c = 295$ K and $T_D = 200$ K, the paleotemperature reached 450°C at the base of the Mesaverde Group

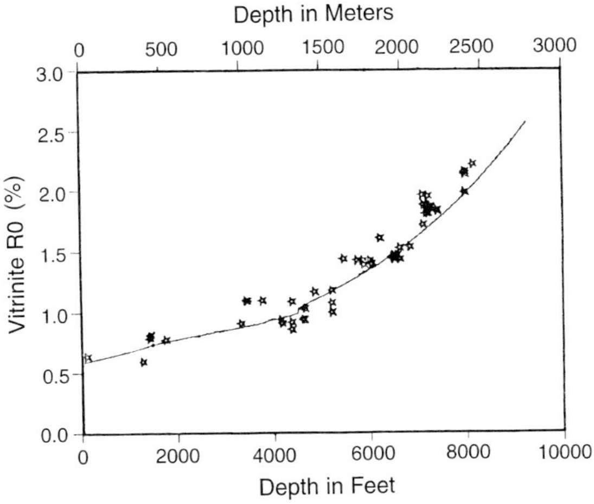


FIGURE 4.41 Vitrinite reflectance profile. The stars represent measured values (Law *et al.*, 1989), while the line shows calculated values.

and the peak heat flow value is 4 HFU at 22 Ma. The range of acceptable $T_c$ and $T_D$ values based on several hundred wells is $T_c = 295 \pm 30$ K and $T_D = 200 \pm 50$ K (Lerche *et al.*, 1984). If one chooses $T_c = 260$ K and $T_D = 160$ K (within the statistical range of uncertainty) the peak heat flow is then only about 2.7 HFU at 24 Ma and the maximum temperature will be 340°C, well below a metamorphic regime, implying that lower $T_c$ and $T_D$ values are more appropriate than the statistical average values for

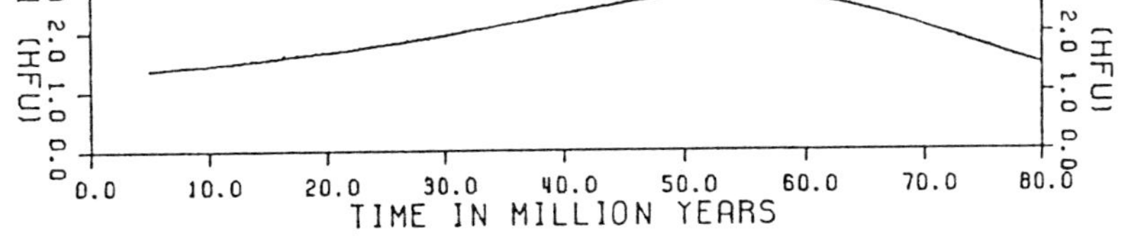


FIGURE 4.42 Heat flow variation with time. Peak heat flux of 2.74 HFU occurred at 24 Ma.

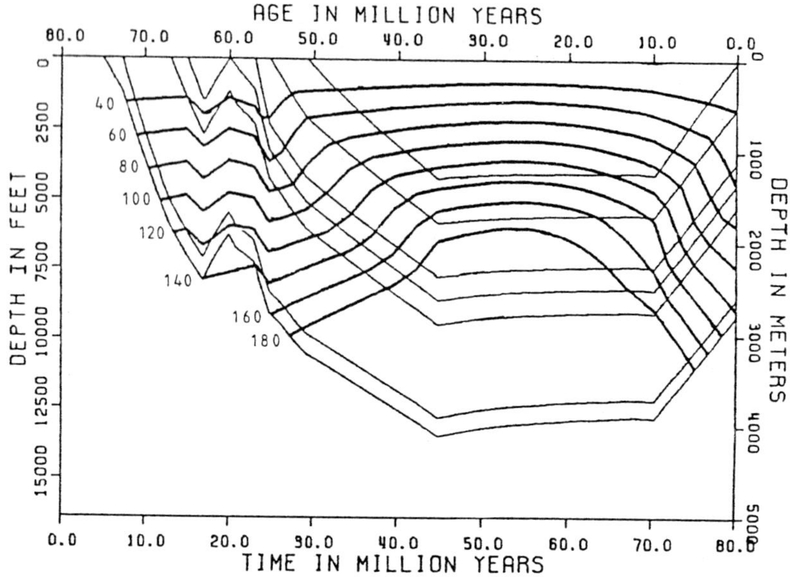


FIGURE 4.43 Temperature variation with geohistory at MWX site, Piceance Creek Basin, which is a function of heat flux and thermal conductivity.

this well. Figures 4.43, 4.44, and 4.45 show the evolution of temperature, Lopatin's TTI, and vitrinite reflectance, respectively. The present-day temperature is calculated from Eq. (4.50), and Fig. 4.44 shows the profile of measured (asterisks) and predicted (solid line) temperature values.

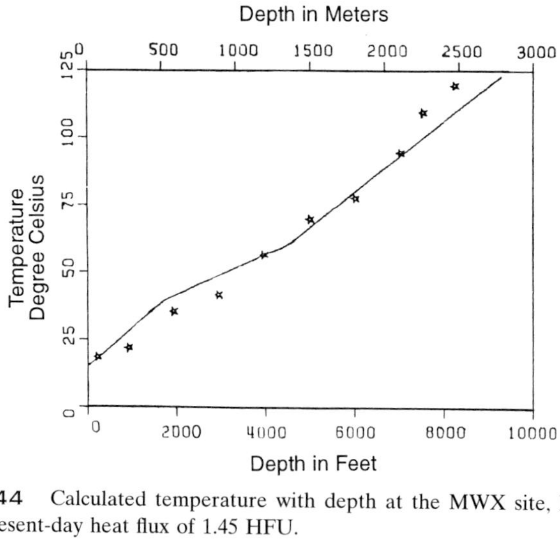


FIGURE 4.44 Calculated temperature with depth at the MWX site, Piceance Creek Basin, using present-day heat flux of 1.45 HFU.

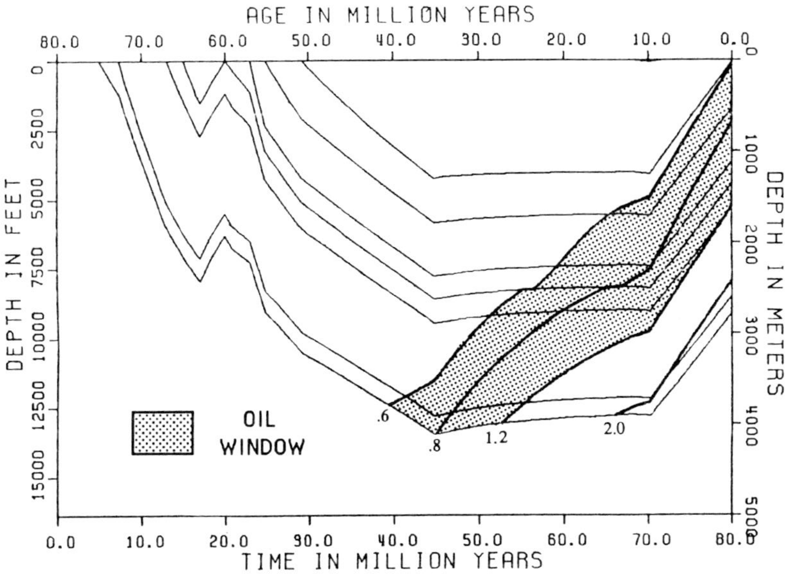


FIGURE 4.45 Predicted vitrinite reflectance values as a function of geologic time and temperature. Iso-vitrinite reflectance lines are labeled.

Three measures are used to depict the level of maturation:

1. The predicted iso-vitrinite reflectance lines, plotted in Fig. 4.45, show the thermal maturation of each layer. If a vitrinite reflectance value of 0.6 is chosen as defining the onset of the "oil window" (Waples, 1980), the depth and time at which each layer entered the "oil window" indicate hydrocarbon generation started within the Mesaverde Formation at 45 Ma. The present-day "oil window" range of depths is from the surface to 5250 ft (1600 m), corresponding to a vitrinite reflectance range from 0.6 to 1.2, respectively.
2. Lopatin's TTI (1980) is plotted in Fig. 4.46. If a TTI value of 15 is defined as the top of the "oil window," hydrocarbon generation at the MWX site started at 58 Ma and the present-day "oil window" reaches to the surface, with corresponding TTI value from 15 to 150.
3. Temperature variation with geohistory is plotted in Fig. 4.43. If we assume a threshold of 80°C for hydrocarbon generation onset, such would occur at about 4000 ft (1220 m) at about 72 Ma.

A naive average of the parameters from these three methods suggests that hydrocarbon generation started at about 45 ± 25 Ma and at a depth of about 7500 ± 2500 ft (2290 ± 760 m). The present-day average depth of the "oil window" is then from the surface to about 5000 ft (1500 m).

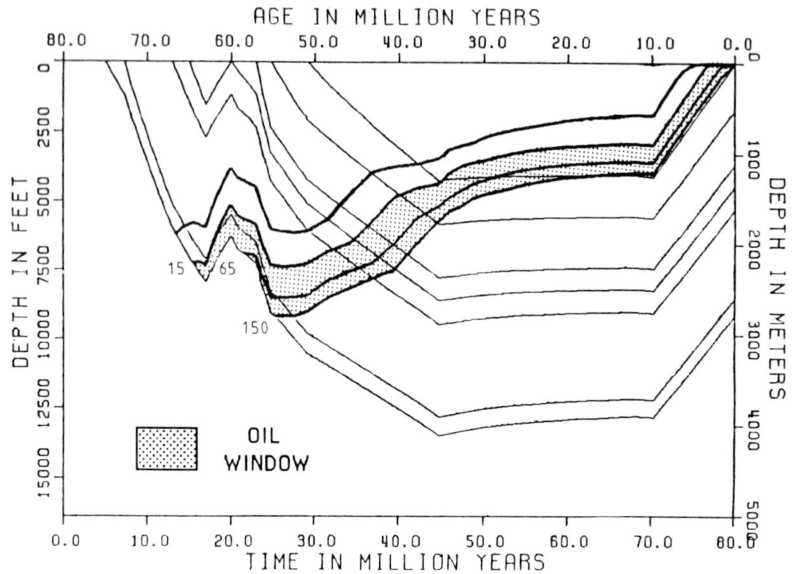


FIGURE 4.46 Lopatin's TTI variation with geohistory. Iso-TTI lines are labeled.

### 3. Discussion

Parameters resolvable with the available data appear to be limited by the resolution of the data. What is the uncertainty associated with each parameter? To examine this problem, one variable is allowed to vary systematically while all other parameters are held fixed at their optimal values. The least-squares residual mismatch, MSR, increases when a single parameter deviates from its best value. Results of such an investigation are shown for the MWX site (see Fig. 4.47 for the heat flux parameters $\beta$, $a_1$, $T_c$, and $T_D$). All the heat flow parameters appear to be well constrained. For the burial history parameters, the curve of MSR versus unconformity size (Fig. 4.47) indicates that the amount of erosion is limited to less than about 2250 m when the time of onset of erosion is varied around its best value (Fig. 4.47).

The MWX site, Piceance Creek Basin, Colorado, went through three stages: subsidence, hiatus, and uplift. Both a high sedimentation rate, which resulted in overpressure in the sequence, and high paleotemperature and rapid uplift are beneficial to hydrocarbon generation, migration, and accumulation. Regional volcanic activity caused a high heat flux around Oligocene-Miocene time. Major source rocks, the Williams Fork and Wasatch Formations, reached maturity about 45 million years ago.

The fluid flow/compaction model provides a combined determination of quantitative paleoheat flux variation with time, and associated geological processes of the MWX site in the Piceance Creek Basin, Colorado, together

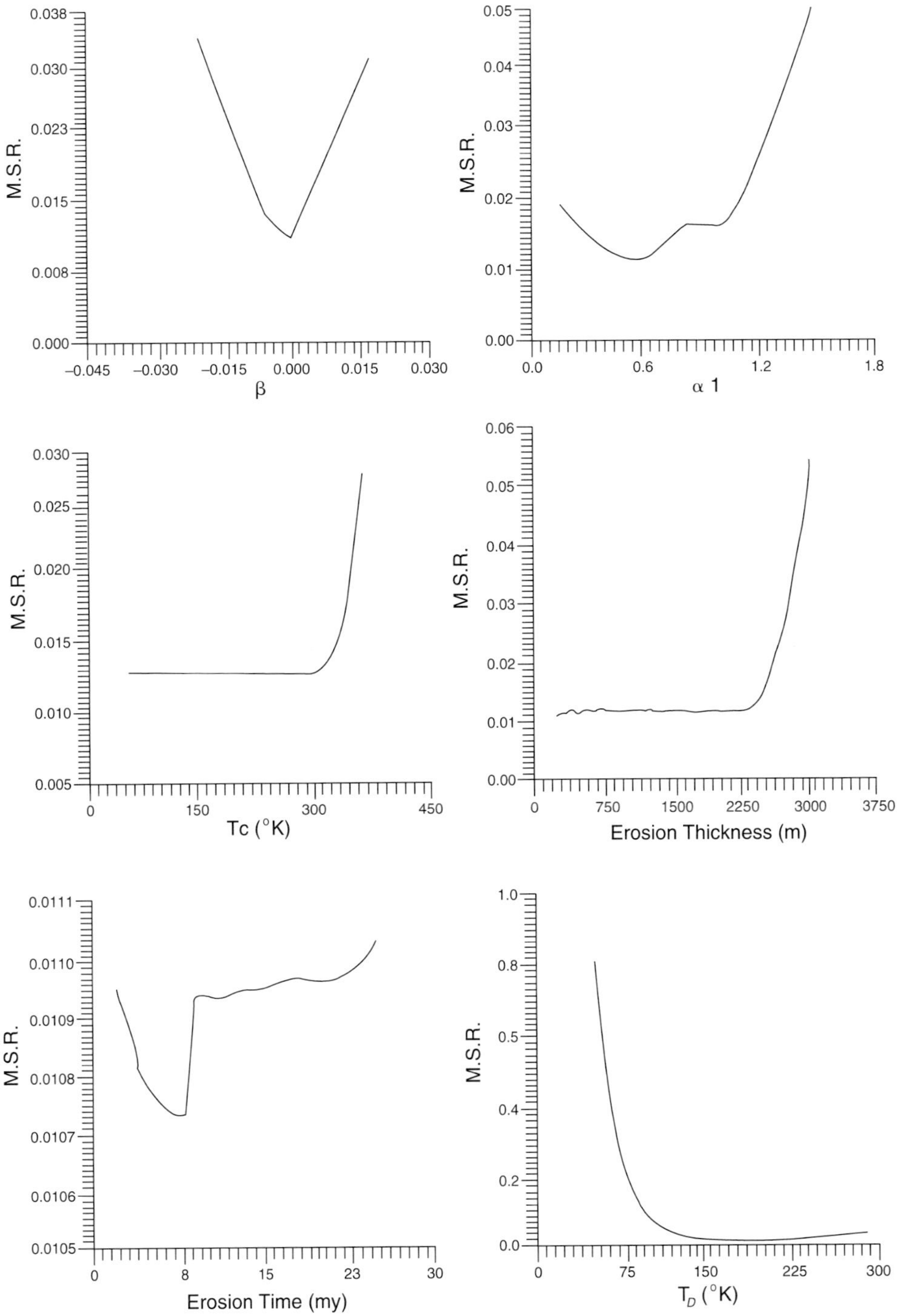


FIGURE 4.47 Sensitivity analysis results of parameters $\beta$, $a_1$, $T_c$, and $T_D$, erosion thickness and erosion time, which show the MSR variation with each parameter.

with quantitative measures of their uncertainties consistent with downhole data. This sort of analysis offers the greatest probability for accurately assessing the petroleum potential of a basin. The hydrocarbon generation model was not used because of lack of downhole kerogen data.

## APPENDIX A: TRANSMITTANCE COLOR INDEX DERIVATION

The propagation of light at angular frequency $\omega$ through a solid of thickness $L$, and complex refractive index $n = n_0 + i\omega\alpha$ is such that the original intensity, $I_0$, is diminished with the emergent intensity, $I(L)$, given by (Jackson, 1975)

$$I(L) = I_0 \left|\exp(i\omega n L)\right|^2 \tag{A.1a}$$

[where (Jackson, 1975) the absorptive part of the refractive index, $\omega\alpha$, is such that $\alpha$, the absorption per unit frequency, is proportional to the number density $N$ and a cross-section $\sigma$, which is slowly varying with frequency]. That is,

$$I(L) = I_0 \exp(-2\omega^2\alpha L). \tag{A.1b}$$

Inserting $\alpha = N\sigma$, $\omega = c2\pi/n_0\lambda$, where $\lambda$ is the wavelength, we can write

$$I(L) = I_0 \exp[-\gamma N L\sigma/\lambda^2], \tag{A.2a}$$

where $\gamma$ is the scaling constant $8\,\pi^2/n_0^2$. Now refer the intensity emerging at wavelength $\lambda$ to the intensity (defined as unity) at a template wavelength $\lambda_0$ (this template wavelength is 700 nm for the TCI work reported in text). Then

$$I(\lambda, L) = \exp\left[-\gamma N L\sigma\left(\frac{1}{\lambda^2} - \frac{1}{\lambda_0^2}\right)\right]. \tag{A.2b}$$

What we are measuring is the wavelength, $\lambda_{1/2}$, at which $I(\lambda, L) = 1/2$, that is,

$$\lambda_{1/2}^2 = \frac{\lambda_0^2}{[1 + \lambda_0^2(\ln 2)/(\gamma N L\sigma)]} \tag{A.3}$$

and $\lambda_{1/2}$ is referred to as the TCI value.

Observationally, $\lambda_{1/2}$ is ~570 nm at low maturity and tends towards $\lambda_0$ (700 nm) at high maturity. Thus $|\lambda_{1/2} - \lambda_0|/\lambda_0 \ll 1$.

Expanding Eq. (A.3) for $\lambda_{1/2} \approx \lambda_0$ we have

$$\lambda_{1/2} = \lambda_0\left[1 - \frac{\ln 2}{2\gamma L\sigma}\lambda_0^2\frac{1}{N}\right]. \tag{A.4}$$

Note that for a given thickness of sample, as the number density $N$ increases, then $\lambda_{1/2}$ tends towards $\lambda_0$, and that $\lambda_{1/2}$ is smaller than $\lambda_0$, as observed. Now the dipole density, $N$, of a solid increases with increasing maturity such that (Lerche *et al.*, 1984)

$$N = N_0 \exp(k\mathrm{TTI}) \cong N_0(1 + k\mathrm{TTI}), \tag{A.5}$$

where $k$ is a constant, and TTI is a time-temperature integral. Thus, using Eq. (A.5) in Eq. (A.4) we have

$$\lambda_{1/2} \cong \lambda_0 \left[ 1 - \frac{\ln 2\lambda_0^2}{2\gamma L\sigma N_0}(1 - k\mathrm{TTI}) \right]. \tag{A.6}$$

Set $\lambda_s = \lambda_0[1 - \lambda_0^2 \ln 2/2\gamma L\sigma N_0]$, which is the TCI value at surface deposition, when we have

$$\lambda_{1/2} \cong \lambda_s + \varepsilon\mathrm{TTI}, \tag{A.7}$$

where $\varepsilon = k\lambda_0^3 \ln 2/2\gamma L\sigma N_0$.

Now we already know (Lerche *et al.*, 1984) that vitrinite reflectance maturity obeys

$$R^{1/2} = R_s^{1/2} + \beta\mathrm{TTI}, \tag{A.8}$$

where $R_s$ is the surface depositional value of vitrinite reflectance, and $\beta$ is a constant. Hence, eliminating the thermal maturity integral, TTI, between Eqs. (A.7) and (A.8) we obtain the theoretical prediction for TCI:

$$\lambda_{1/2} - \lambda_s = \mathrm{constant}(R^{1/2} - R_s^{1/2}). \tag{A.9}$$

Equation (A.9) is examined empirically in this chapter and found to be an accurate enough representation of the general data in Tables 4.4 and 4.5, and of specific data from COST G-2, Georges Bank, given in Tables 4.6 and 4.7.

# 5

# Hydrocarbon Kinetic Models and Their Uncertainties

## I. INTRODUCTION

The production of liquid and gaseous hydrocarbons from organic matter by thermal degradation processes was put on a firm footing in the 1960s [see, e.g., Philippi (1965)]. Much earlier it was already known that biogenic processes could also produce degradation of organic material [see, e.g., Hunt (1979) and references therein]. But knowing that processes exist which lead to hydrocarbon generation and quantifying those processes are two very different matters. Two major difficulties are apparent at the outset of any attempts to provide quantitative kinetic models.

The first difficulty is the sheer magnitude of the task: Organic matter (mainly insoluble kerogen) is composed of biological polymers (lipids, proteins, amino acids, etc.) with molecular weights in the multimillion range; by way of contrast, methane ($CH_4$) has a molecular weight of 16. We are invited to construct a complete kinetic reaction scheme which accounts correctly for the detailed breakdown of organic matter to methane.

The second difficulty is the sheer time scope of the task. We know from isotopic measurements, stereochemistry, and gas chromatography/mass spectrometry, that the production of hydrocarbons in the Earth occurs on timescales ranging from fractions of a megayear to hundreds of megayears, in a temperature regime in the rough range of 0–200°C, and in an evolving environmental "smog" of sedimentary pores containing water laced with varying amounts of ions, molecules, and minerals. We are invited to construct a chemical kinetic framework which allows correctly for all of these factors, as well as for surficial reaction chemistry on grain surfaces.

Faced with these two major problems, as well as a host of important, but subsidiary, difficulties, two major thrusts of research have evolved during the last 20 years or so in attempts to bypass, or at least constrain and bracket, the major problems.

Recognizing that it will be a very long time indeed before a complete thermokinetic behavior is available (if ever) pyrolysis experiments in the laboratory have been used, in conjunction with theoretical concepts, to try to determine at least the dominant controlling reaction kinetics in two areas: first, in attempts to discover the order of the chemical reactions (first-order, second-order, etc.) and, second, in attempts to discover the sequence of chemical reactions (series, parallel, catalytic, etc.).

## A. SOME SIMPLE KINETIC MODELS

The simplest model assumes that oil generation can be modeled by a single first-order reaction of the Arrhenius type:

$$\frac{dM}{dt} = k(1 - M), \qquad \text{where } k = A\exp(-E/RT), \tag{5.1}$$

with $M$ = the mass of oil produced per unit mass of kerogen, $t$ = time, $E$ = activation energy, $R$ = gas constant, $T$ = absolute temperature, and $A$ = frequency factor (occasionally we also write $E/R = T_0$, the activation temperature). [As an aside, note that if temperature is expressed in degrees Celsius, $T = 273 + C$, then for $|C| \ll 273$ we have $k = A\exp(-T_0/(273 + C)) \approx A\exp(-T_0/273) \times \exp(CT_0/(273)^2)$. Write $T_D = (273)^2/T_0$, when $k$ takes the form

$$k = A'\exp[(C - T_c)/T_D], \tag{5.2}$$

where $A' = A\exp(-T_0/273 + T_c/T_D)$ = constant, and $T_c$ is arbitrary. Then the solution for $M$ is in the form

$$\ln(1 - M) = \{-A'\mathrm{TTI}(t)\}, \tag{5.3}$$

where

$$\mathrm{TTI}(t) = \int_0^t \exp((C(t') - T_c)/T_D)\,dt', \tag{5.4}$$

which we recognize as providing a basis for the Lopatin (1971) rule-of-thumb estimate for hydrocarbon maturity popularized by Waples (1980)].

On the other hand such simplistic models [and the determination of the best values of $A$ and $E$ from matching predicted behavior to observations; see, e.g., van Heek and Juntgen (1968) and Connan (1974)] are known not to be very relevant. Oil is not a single molecular substance, no gas production is allowed for, and different types of kerogen, from provenances ranging from terrestrial to marine, produce different oils. In addition, there has persisted an annoying irreconcilable difference between the activation energy of

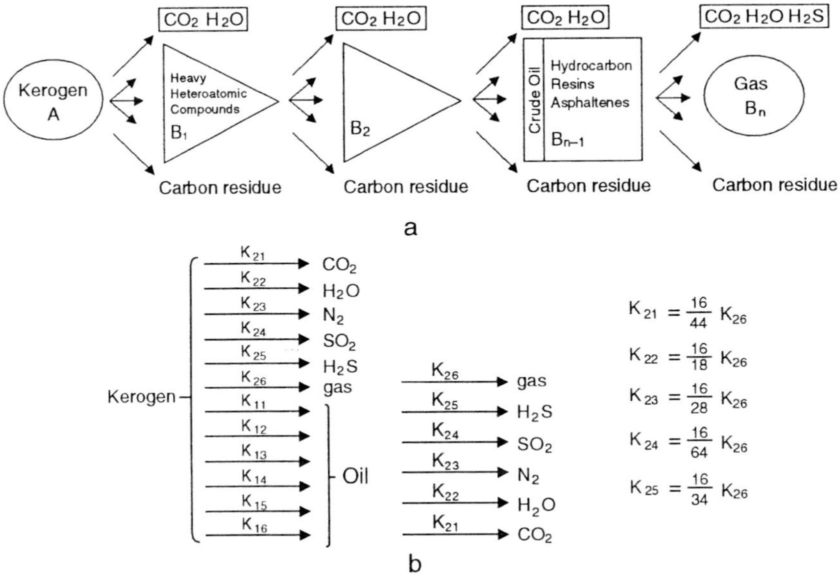


FIGURE 5.1 (a) General scheme of kerogen degradation under thermocatalytic cracking. (b) Two detailed kinetic frameworks of oil and gas production. The Tissot (1969) scheme is produced by setting $K_{21}$ through $K_{25}$ to zero, and by ignoring $K_{26}$ in the direct transformation of kerogen to oil and gas.

40–70 kcal/mol required from laboratory pyrolysis and the range of 10–20 kcal/mol derived from geological data alone (Sweeney *et al.,* 1987).

Only a quantitative approach, allowing computation of the amount of oil and gas generated in any place in the basin as a function of time, can provide information on (1) amount of oil and gas generated per square kilometer in each source rock and in every part of the basin; (2) timing of hydrocarbon formation, timing of the development of impermeable seals, and timing of folding and faulting, so traps can be identified; (3) amounts of oil and gas that have been expelled from the source rock into the porous reservoirs and how much is still retained in the source rock; and (4) evaluation of the ultimate oil and gas reserves of a sedimentary basin. However, because many of the intermediate steps in petroleum and gas generation from kerogen degradation are unknown, such a scheme has to be phenomenological; that is, parameters have to be determined by least-squares fitting to basin and laboratory data. Such a mathematical model of petroleum generation accounting explicitly for geologic time was first introduced by Tissot (1969) and is discussed fully by Tissot and Espitalié (1975). As sketched in Fig. 5.1 the idea is to simulate the formation of oil from kerogen in six parallel, first-order reactions (first stage), and the formation of gas ($CH_4$) from oil in a single reaction (second stage). The model is based on kerogen degradation and uses the general scheme of first-order chemical kinetics. Kerogen is a macromolecule composed of polycondensed nuclei bearing alkyl chains and functional groups—the links between nuclei being heteroatomic bonds or carbon chains. As the burial depth and temperature increase, the bonds are successively broken, roughly in order of increasing rupture energy. The products generated are heavy heteroatomic compounds, carbon dioxide, and water, then progressively smaller molecules, and finally hydrocarbons. At the same time, the residual kerogen becomes progressively more aromatic and evolves toward a carbon residue.

To apply the phenomenological model of Tissot and Espitalié (1975), as popularized in Tissot and Welte (1978), to the prediction of change in hydrocarbon content of rocks with respect to time and temperature, it is necessary to assume first-order chemical reaction kinetics for the breakdown of kerogen to hydrocarbons as governed by

$$-dx_i/dt = k_i x_i, \tag{5.5a}$$

$$k_i = A_i \exp(-E_i/k_B T(t)), \tag{5.5b}$$

where $x_i$ is the fraction of kerogen (see Appendix A) reacting with the rate constant $k_i$, $k_B$ is Boltzmann's constant, $A_i$ is the reaction rate (frequency factor) in the limit of high temperature, $E_i$ is the activation energy, and $T(t)$ is the temperature as a function of time acting on the component $x_i$. In the Tissot and Espitalié scheme, three types of initial kerogens are considered. The three types differ in their hydrogen-to-carbon ratio. Loosely speaking, these three types can be considered type I (algal) with a high petroleum potential, type II (amorphous) with an intermediate petroleum potential, and type III (woody) with a low petroleum potential. Tables of the activation energies ($E_i$), the reaction constants ($A_i$), and the initial values ($x_i$) for each kerogen fraction have been given by Tissot and Welte (1978, Table V.4.1).

To calculate the quantity of hydrocarbons produced by potential source rocks, it is necessary to use the preceding calculations and measurements or estimates of the organic material percentage in those source rocks. Nakayama and Van Siclen (1981) detailed the methodology of this system to work out a hydrocarbon budget for a basin. To make predictions about the hydrocarbon history of a basin, it is also necessary to determine the temperature as a function of time. Tissot and Welte (1978) assumed a constant geothermal gradient in their models. However, by allowing the heat flux to vary with time, their model is improved, provided the variation of heat flux with time can be determined.

Presumably what should be done at the present stage of understanding is to let the frequency factors and activation energies be "free" parameters, and determine the values most consistent with observational data in a globally optimized least-squares sense, using the best burial, fluid flow, and thermal histories currently available. In addition a sensitivity study should be undertaken to determine the degrees of resolution, precision, uniqueness, and stability of any chemical parameters so obtained. On the other hand, two major problems still remain: First, there is no guarantee that the kinetic framework proposed is the correct phenomenological description. (For example, why have only six channels for oil production? Why have only one channel for gas production?) Two extreme end-member lines of development can be carried forward to provide different quasi-phenomenological hydrocarbon generation schemes: (1) Modify the Tissot-type of sequence of reaction kinetics and (2) modify the number of channels of the reaction kinetics.

Yet one can question the sense in retaining a fixed number of reactions degrading kerogen (6 in the Tissot-type of model). The charge can be leveled that the number of chemical parameters to be determined can be increased systematically until the reaction kinetics are sure to agree with any set of measurements. Arguments for simplicity (Occam's razor) suggest that other types of reaction kinetics (with fewer parameters but greater flexibility) might be more appropriate. Braun and Burnham (1987), Sweeney *et al.* (1987), and Burnham *et al.* (1987) have provided both laboratory pyrolysis data and an application to the Uinta Basin which, they suggest, are better understood in terms of a distributed activation energy model (an infinite number of channels for hydrocarbon production but with an energy weighting per channel).

In its simplest form the distributed activation energy model writes a kerogen mass fraction $K$ loss equation under thermochemical kinetics in the form

$$\frac{dK}{dt} = -\alpha(T, E_0, \sigma)K, \tag{5.6}$$

but, instead of $\alpha$ being the rate constant, $k$, appropriate to a single kinetic reaction, what we write is

$$\alpha(T, E_0, \sigma) = A \int_0^\infty dE \exp(-E/RT)P(E; E_0, \sigma); \tag{5.7}$$

that is, the reaction kinetics are distributed over energy with a channel probability $P(E; E_0, \sigma)$. The most common assumption (Braun and Burnham, 1987) on the channel probability is that

$$P\alpha \exp(-(E - E_0)^2/\sigma^2), \tag{5.8}$$

where $E_0$ and $\sigma$ are constants such that $\alpha$ takes the form

$$\alpha = A \int_0^\infty dE \exp\{-E/RT - (E - E_0)^2/\sigma^2\}. \tag{5.9}$$

The distributed activation energy kinetics then depend on a frequency factor, $A$, a mean activation energy, $E_0$, and a spread in activation energies, $\sigma$, together with the assumption that the probability is Gaussian distributed in activation energy. Braun and Burnham (1987), Sweeney *et al.* (1987) and Burnham *et al.* (1987) report good fits to laboratory pyrolysis data for petroleum source rocks with such a phenomenological model and further suggest that the chemical parameters derived from such laboratory fitting are also appropriate in geological applications. This model has also been used by Quigley *et al.* (1987) for petroleum formation. Burnham *et al.* (1987) have shown that the Gaussian model gives essentially the same predictions as the Ungerer–Pelet (1987) discrete model when calibrated on laboratory data. A kinetic modeling of oil and gas formation has recently been suggested (Ungerer *et al.*, 1988a) as an improvement over earlier versions (Ungerer and Pelet, 1987; Espitalié *et al.*, 1987; Ungerer *et al.*, 1988b). The essence of this model is also beholden to the Tissot–Espitalié (1975) framework. The sense of the model splits into three parts: kerogen degradation kinetics (i.e., primary hydrocarbon production kinetics), secondary hydrocarbon cracking, and source-expulsion efficiency kinetics.

Thus: "The kinetic model of primary cracking is obtained by detailing the bulk distribution of activation energies. ... The kinetic scheme is represented by the following formulation:

$$\frac{dq}{dt} = Cx, \tag{5.10}$$

where $q$ is the vector of the amounts $q_j(t)$ of products of the fraction $j$ generated at time $t$, $x$ is the vector of the residual potential $x_j(t)$ of the $j$th parallel reaction, solution of

$$\frac{dx_j}{dt} = -A\exp(-E_j/RT)x_j(t), \tag{5.11}$$

where $A$, $E_j$ are the pre-exponential factor and activation energy attached to reaction $j$, $R$ is the gas constant and $T$ is the temperature. $C$ is a $4 \times n$ matrix with coefficients given by $C_{ij} = a_{ij}A\exp(-E_j/RT)$ where the relative yields $a_{ji}$ are" defined in Ungerer *et al.* (1988a).

The model of secondary cracking is similar to the model of oil cracking except that Ungerer *et al.* (1988a) add a reaction of degradation of the $C_2$–$C_5$ fraction. "The coupling with primary cracking yields the following equation for a *closed* system:

$$\frac{dq}{dt} = Cx + Bq, \tag{5.12}$$

where $q$, $C$, $x$ have the same meaning as above, $B$ is a $4 \times 4$ matrix with coefficients

$$b_{ij} = a_{ji}k_j' \qquad \text{for } i = 1\text{–}4, j = 1\text{–}3, i \neq j,$$
$$b_{jj} = -k_j' \qquad \text{for } j = 1\text{–}3,$$
$$b_{ij} = 0 \qquad \text{for } i = 1\text{–}4, j \neq 4, \text{ and}$$

$k_j' = A_j' \exp(-E_j'/RT)$ are the first-order rate constants of the secondary reactions, the stoichiometric coefficients $\alpha_{ij}$ being defined" in Ungerer *et al.* (1988a). This formulation makes the ad hoc assumption that the $k_j$ are, indeed, required to be of the Arrhenius form. The free parameters of this model ($A_j'$, $E_j'$, $\alpha_{ij}$) are optimized versus autoclave pyrolysis data, accounting for conservation of carbon and hydrogen. The model is said to reproduce correctly the successive stages of light oil, gas condensate, and dry gas formation. The maximum yield is lower than in an open system, because of the partial transformation of hydrocarbons into coke (i.e., pyrobitumen). Ungerer *et al.* (1988a) add an expulsion term as follows: "Expulsion is considered to occur only after a certain saturation threshold of the hydrocarbons in the porous medium is reached.... The products expelled during a small time step have the same composition as the residual hydrocarbons in the source rock at the same time, and we assume that the excess volume of generated hydrocarbons above the saturation threshold is totally expelled. Once the expulsion threshold is reached, the mathematical formulation is:

$$\frac{dq}{dt} = Cx + Bq - \lambda q, \tag{5.13a}$$

$$\frac{dq'}{dt} = \lambda q, \tag{5.13b}$$

where the scalar expulsion rate $\lambda$ is determined so that:

$$\frac{q_1 + q_2 + q_3 + q_4}{\rho} = VS_{\max}, \tag{5.13c}$$

with $C$, $x$, and $B$ the same as previously defined, $\rho$ is the density of the hydrocarbon phase (kg/m$^3$), $V$ is the pore volume, expressed in cubic meters per ton of initial rock, $q$ and $q'$ are the compositions of the residual and expelled hydrocarbons (kg/ton of initial rock).

This model predicts a composition of hydrocarbons expelled which is intermediate between the open system and the closed system. The residual hydrocarbons have a composition similar to the closed system (i.e., tending to dry gas) but their total amount decreases once the expulsion has started

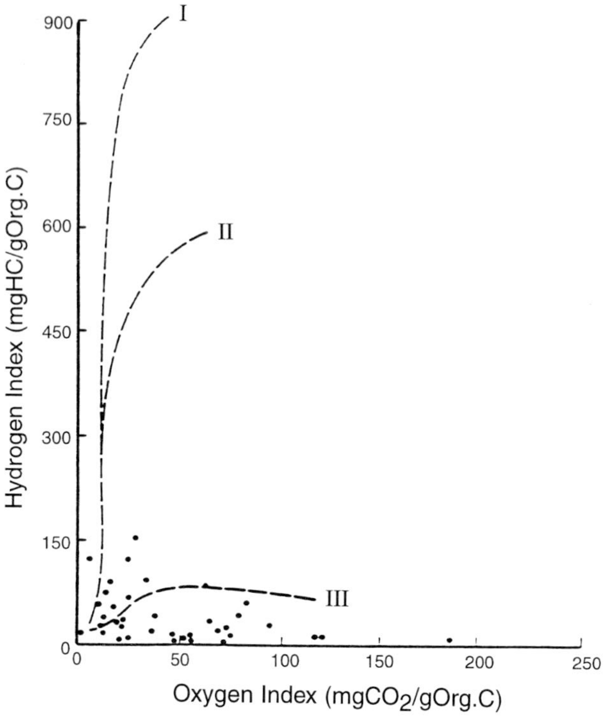


FIGURE 5.2 A diagram of hydrogen index to oxygen index for organic samples taken from the Pinedale Anticline area, Wyoming, USA. The dashed lines provide the general trend paths expected for type I, II, and III kerogens.

due to the combined influence of the decrease in porosity and in hydrocarbon density."

The appearance of several models for hydrocarbon generation may seem disquieting, for it suggests that there is still some distance to go to unravel the correct quantitative scheme for hydrocarbon production. But this very variation can be used to advantage in assessing hydrocarbon proneness of a basin. The obvious strategy is to invoke *all* of the models of hydrocarbon generation in putting together the development of a sedimentary basin. Each will predict differing amounts and rates of hydrocarbon production (both oil and gas) at different stages in the basin's evolution. But the important point is that these individual assessments of hydrocarbon generation provide a quantitative measure of the uncertainty and resolution of the timing and amount of hydrocarbon production. This uncertainty can then be invoked in an exploration economic sense to assign ranking probabilities (Nind, 1981).

In addition, some qualitative controls are available to limit the degree of uncertainty of hydrocarbon models. For example, Fig. 5.2 indicates a systematic loss of hydrogen and oxygen in maturing kerogen, while measurements of the location of a partially matured kerogen sample set a qualitative limit on its total maturity, which can be compared, in a rough sense, with modeled hydrocarbon generation. Thus if measurements indicate low H/C and O/C ratios, indicating high maturity, but the particular hydrocarbon generation model has hardly started to produce, then either the model is inappropriate or the model may be appropriate but some of the empirically determined model parameters may be out of adjustment. Again, gas chromatography/mass spectrometry measurements of the composition of oils indicate whether they were produced early in the basin's history or are more recent (Hunt, 1979; Waples, 1984). A generation model not conforming to these rough rule of thumb restrictions is unlikely to be a highly probable reaction kinetic pathway. Other models would then be preferable.

Regardless of which technique is invoked, three factors are apparent. First, any model of *total* amount of hydrocarbon *generation* depends on (1) the amount of kerogen per unit volume of rock (the "richness" or "leanness" of a potential source rock), (2) the burial path of the sedimentary units, and (3) the thermal history of the sediments. Second, the generation of hydrocarbons does not have to take place solely *in situ* with the kerogen. Migration of hydrocarbons in solution, in suspension, or as a separate phase means that generation to lighter hydrocarbons and to gas can carry on during migration, or even after emplacement in a reservoir accumulation site. Third, while kerogen typically makes up less than a few percent of rock mass (Hunt, 1979), generated hydrocarbons influence the fluid flow and fluid pressure, because they form a mobile phase invading pore volume, with a different capillary surface tension than water, and can act to microfracture their very generation and/or accumulation sites thereby being lost to the system. Some of the components of these three factors are roughly insensitive to changes in the absolute amounts of hydrocarbons generated, being dominated by other considerations (e.g., burial history is surely dominated by total deposition rate), whereas others (e.g., a rock close to its fracture limit) can be impacted significantly by a smal change in the amount of hydrocarbon generation. Thus there is a sensitive interplay, as well as a nonsensitive component, between dynamic and thermal evolution of a sedimentary basin, and hydrocarbon production. There is an urgent need to improve on the ability to model hydrocarbon generation in a more fundamental manner than is currently available.

There is a common underlying thread to all of the thermokinetic models developed to date on which we concentrate first. Note that all the models developed consider breakdown of kerogen as the prime mover. The models appear to differ in their technical approaches to the kerogen degradation. But having removed kerogen the question of its conversion to other products becomes the next technical question. Here the models proposed again appear to differ widely, with some models including specific compounds at production (oil, water, $CH_4$, $CO_2$, $N_2$, etc.), other models concentrate more on the specifics of "oil" regarding it as a multichannel process of varying composition (Ungerer *et al.*, 1988a,b) with apportioned channel productivity rates. In addition, some models include an expulsion efficiency from the source rock (whether it is in a natural setting or in the laboratory), while other models do not.

It would seem that it might be possible to combine all of these models into a few classes of generic model and to use data directly in an inverse method sense to determine the degree to which resolution, uniqueness, and precision are contained within any posited kinetic description. The inverse problem is considered in two main stages: (1) autoclave product analysis and (2) residual kerogen analysis.

## II. AUTOCLAVE PRODUCT ANALYSIS

In controlled autoclave product pyrolysis experiments, where the products can be measured *in situ* as temperature increases, we can examine proposed kinetic models for thermal evolution of hydrocarbons and ascertain if the experimental data provide enough information for determination of the correctness of either a kinetic model or of parameters within a model. In this section we focus on (1) the numerical implementation of a proposed inverse method (Lerche, 1990a) and (2) the application of this inverse model to autoclave pyrolysis products.

### A. MATHEMATICAL MODEL

For a closed system, a kinetic model of primary cracking and secondary cracking of hydrocarbons is written in the vector form (Ungerer *et al.*, 1988a):

$$\frac{d\mathbf{q}}{dt} = \underline{\underline{\mathbf{C}}}\mathbf{x} + \underline{\underline{\mathbf{B}}}\mathbf{q}, \tag{5.14}$$

where $\mathbf{q}$ is the vector of the amounts $q_j(t)$ of products of the fraction $j$ generated at time $t$, and $\mathbf{x}$ is the vector of the residual potential $x_j(t)$ of the $j$th parallel reaction, given through the solution of

$$\frac{dx_j}{dt} = -A \exp(-E_j/RT)x_j(t). \tag{5.15}$$

The general scheme of this kinetic model is shown in Fig. 5.3. The general form of the equations is as follows. Suppose the amount of the $i$th product

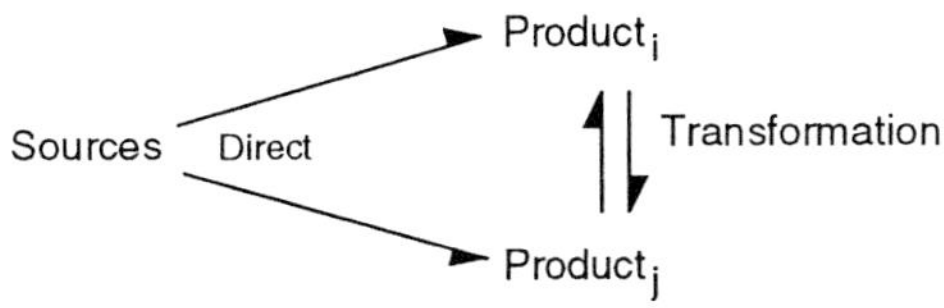


FIGURE 5.3 General scheme of the model which shows hydrocarbon products arising in two ways: (1) from direct production from sources and (2) from transformation from other products.

is $q_i(t)$ at time $t$. This amount may arise because of transformation from other products or source degradation (such as kerogen degradation or oil cracking). Then

$$\left(\frac{dq_i}{dt}\right)_{\text{transformation}} = B_{ij}q_j, \tag{5.16}$$

where the Einstein summation convention over repeated Latin indices is used. Here $B_{ij}$ is a transformation matrix, which is *nominally* solely temperature dependent (Ungerer *et al.*, 1988a).

The amount of the $i$th product produced can also be generated directly from the source or sources:

$$\left(\frac{dq_i}{dt}\right)_{\text{direct}} = C_{i\alpha}x_\alpha(t), \tag{5.17}$$

where the Einstein summation convention over repeated Greek indices is used, where $C_{i\alpha}$ is a generation matrix, and where $x_\alpha(t)$ $(\equiv x(\alpha, t))$ is the source, which can be the amount of kerogen of type $\alpha$ or the amount of oil of composition $\alpha$ at time $t$. In the experimental data we use, taken from Ungerer *et al.* (1988b), the source material is oil.

Thus the total generation with time of product $i$ is written

$$\frac{dq_i(t)}{dt} = B_{ij}q_j(t) + C_{i\alpha}x_\alpha(t), \tag{5.18}$$

which has the same meaning and structure as Eq. (5.14).

Direct inspection of Eq. (5.18) tells us that the matrices $B_{ij}$ and $C_{i\alpha}$ are *not* uniquely determinable. For suppose that $B_{ij}$ was of the form

$$B_{ij} = B(T)[\delta_{ij} - q_iq_j/q \cdot q] + \varepsilon_{ijk}(q_k/|q|)C(T), \tag{5.19}$$

where $q \cdot q = q_iq_i$, $|q| = (q_iq_i)^{1/2}$, $\delta_{ij}$ is the Kronecker $\delta$ function (unity if $i = j$, zero otherwise), and $\varepsilon_{ijk}$ is the alternating Levi–Civita symbol (+1 if $ijk$ are cyclic, $-1$ if $ijk$ are anticyclic; zero otherwise). Then

$$B_{ij}q_j = 0. \tag{5.20}$$

Hence no components of Eq. (5.19) survive projection with $\mathbf{q}$. (A similar argument holds for $C_{i\alpha}$ with $j$ replaced by $\alpha$, and $\mathbf{q}$ replaced by $\mathbf{x}$).

Hence the only components of $B_{ij}$ which are determinable are those that project along $q_i$ or $q_j$. Likewise the only components of $C_{i\alpha}$ which are determinable are those that project along $x_\alpha$. Fortunately these are precisely and solely the components which are needed for use in Eq. (5.18). A determination of the components of $B_{ij}$ and $C_{i\alpha}$ can be made as follows. Introduce the vector $\mathbf{U}(T) = (\mathbf{q}(T), \mathbf{x}(T))$ and the matrix, $D_{ij}$, so that

$$B_{ij}q_j + C_{i\alpha}x_\alpha = D_{ij}U_j, \tag{5.21}$$

where $D_{ij}$ is the bulk rate matrix, which should be temperature dependent only if the kinetic scheme proposed by Ungerer *et al.* (1988a,b) is to hold.

Then Eq. (5.18) takes on the form

$$\frac{dq_i}{dt} = D_{ij}(T)U_j(T). \tag{5.22}$$

An exact solution for Eq. (5.22) is provided by

$$D_{ij}(T) = dq_i/dt\ U_j(T)/|U_s^{\mathrm{P}}(T)U_s^{\mathrm{P}}(T)|, \tag{5.23}$$

where uppercase P means $U_s^{\mathrm{P}}$ is the sum of all $U_j$ except the $U_i$ term when calculating $D_{ij}$.

Equation (5.23) completes the inverse solution to the kinetic problem because from a given set of pyrolysis experiments $D_{ij}(T)$ is determined. Then in any other set of experiments [with values $u_j(T)$ replacing the test set $U_j$; and $Q_i$ replacing $q_i$], and on the supposition that $D_{ij}(T)$ is an absolute, kinetic process determined, fixed functional form so that the same $D_{ij}(T)$ applied to all pyrolysis product measurements, we can write:

$$\frac{dQ_i}{dt} = D_{ij}(T)u_j(T), \tag{5.24}$$

which provides a predictive measure for autoclave product measurements—*if* the kinetic description is accurate. By running many such experiments, a best set of values can be determined by averaging the results of $D_{ij}$ determinations from many experiments and the resolution thus obtained.

The questions we address here are basically simple: (1) Using the exact inverse solution given, does it follow that the matrix $D_{ij}$ is really only temperature dependent? (2) Even if the matrix $D_{ij}$ is time dependent, is it of the same behavior for all experiments and/or does some average value serve to satisfy the data fields?

## B. INPUT DATA

The inverse procedure has been applied to three sets of pyrolysis product analysis experimental data taken from Ungerer *et al.* (1988b). The data are

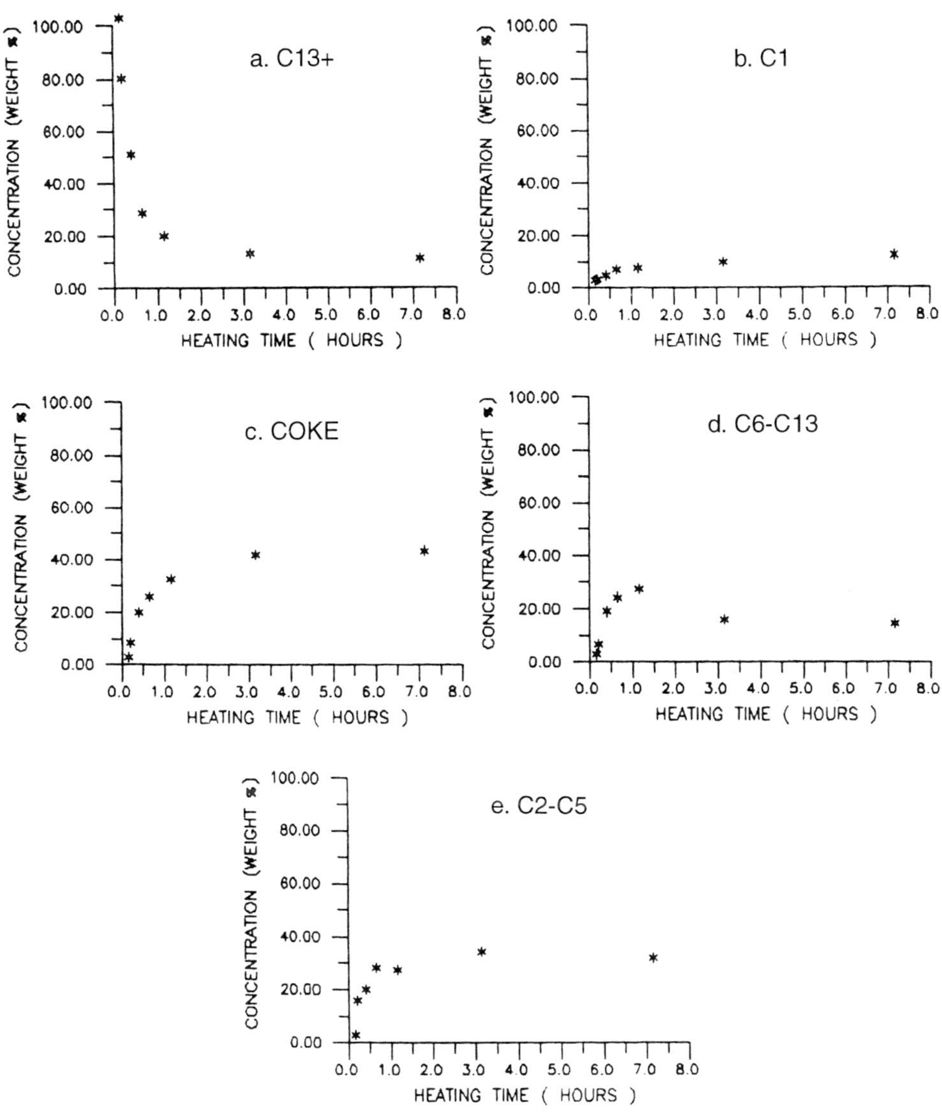


FIGURE 5.4 Measured concentrations of the various fractions from the pyrolysis of Boscan oil at 450°C (from Ungerer *et al.*, 1988a,b). (a) $C_{13+}$; (b) $C_1$; (c) coke; (d) $C_6$–$C_{13}$; (e) $C_2$–$C_5$.

taken to be true reflections of the kinetic process of oil cracking (Figs. 5.4, 5.5, and 5.6). In the experiments reported by Ungerer *et al.* (1988b), the compositions of the hydrocarbon products are defined by the molecular weight fractions. There are five classes of hydrocarbon fractions in their

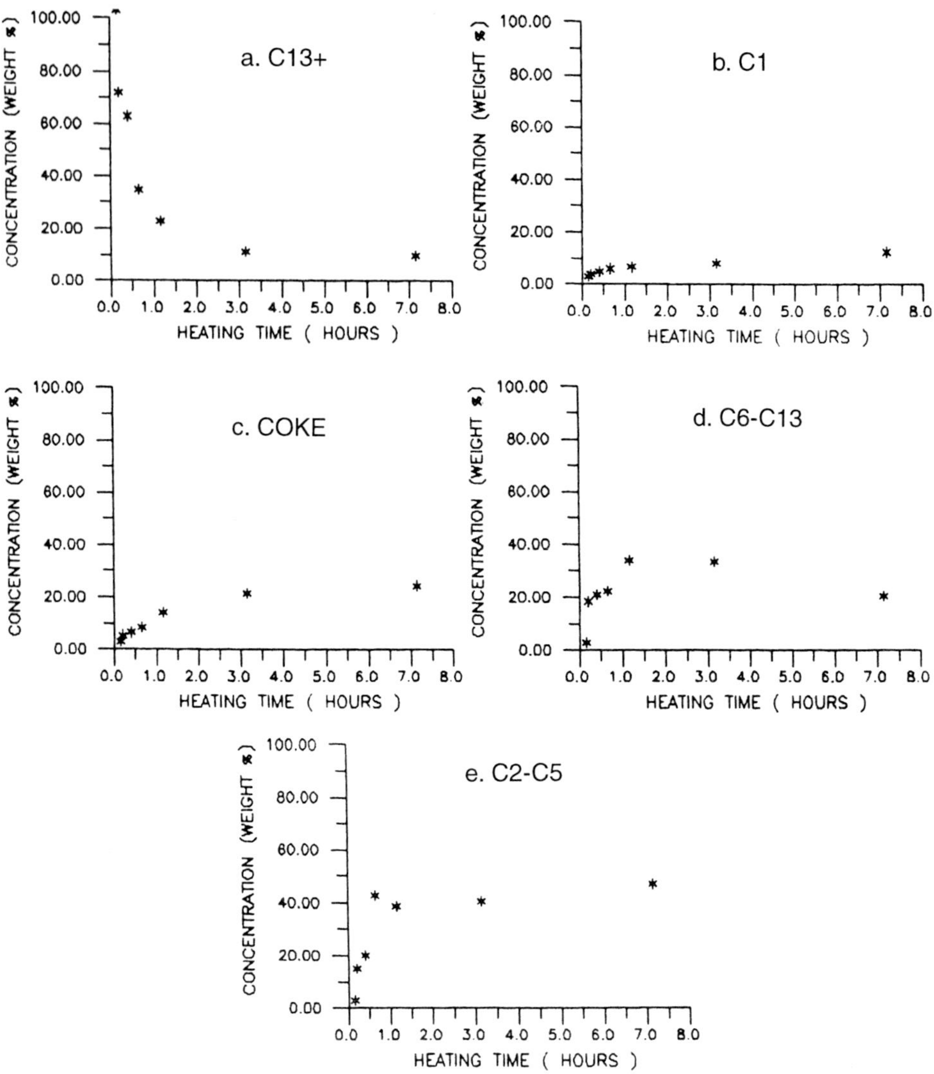


FIGURE 5.5 Measured concentrations of the various fractions from the pyrolysis of Pematang oil at 450°C (from Ungerer *et al.*, 1988a,b). (a) $C_{13+}$; (b) $C_1$; (c) coke; (d) $C_6$–$C_{13}$; (e) $C_2$–$C_5$.

experimental results ($C_{13+}$, $C_6$–$C_{13}$, $C_2$–$C_5$, $C_1$, coke); $C_{13+}$ is taken as the source material in their experiments. Among the three data sets, data sets 1 and 2 were obtained for the same temperature (450°C) and pressure (≥9 MPa) and heating time (7 hr) with different source materials [data set 1: Boscan (aromatic oil); data set 2: (Pematang (paraffinic oil)]. Data set 3 is

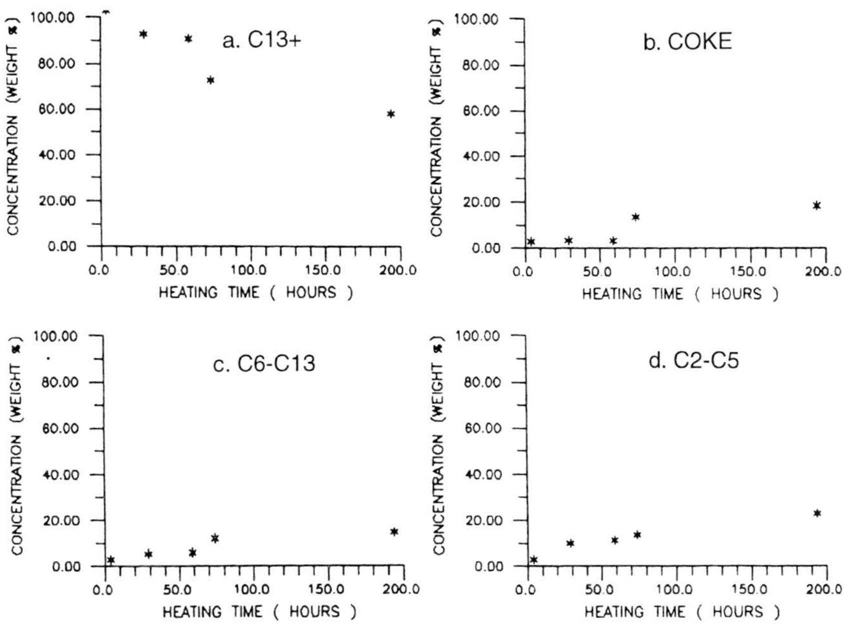


FIGURE 5.6 Measured concentrations of the various fractions from the pyrolysis of Boscan oil at 357°C (from Ungerer *et al.*, 1988a,b). (a) $C_{13+}$; (b) coke; (c) $C_6$–$C_{13}$; (d) $C_2$–$C_5$.

derived from the same source material as data set 1 (Boscan), but with a different temperature (357°C) and heating time (190 hr).

By inverting these data sets, we learn about (1) the dependence of the experimental results on temperature and heating time and (2) the influence of different source material on the experimental results, and the influences of these two factors on the modeling results.

## C. MODELING RESULTS

An inverse code was constructed along the lines given in the mathematical section of this chapter, and applied to the three data sets. The kinetic matrix, $D_{ij}$, is supposedly only temperature dependent so that once $D_{ij}$ is determined, the prediction model for oil cracking is also determined. Therefore we first discuss the behavior of $D_{ij}$ for each data set and then consider the possibility of using an average $D_{ij}$ in the prediction equation.

### 1. Behavior of $D_{ij}$

Data sets 1 and 2 are derived from different source materials but under the same experimental thermodynamic condition (both at $P \geq 9$ MPa, $T = 450$°C, and time = 7 hr), whereas data sets 1 and 3 are from the

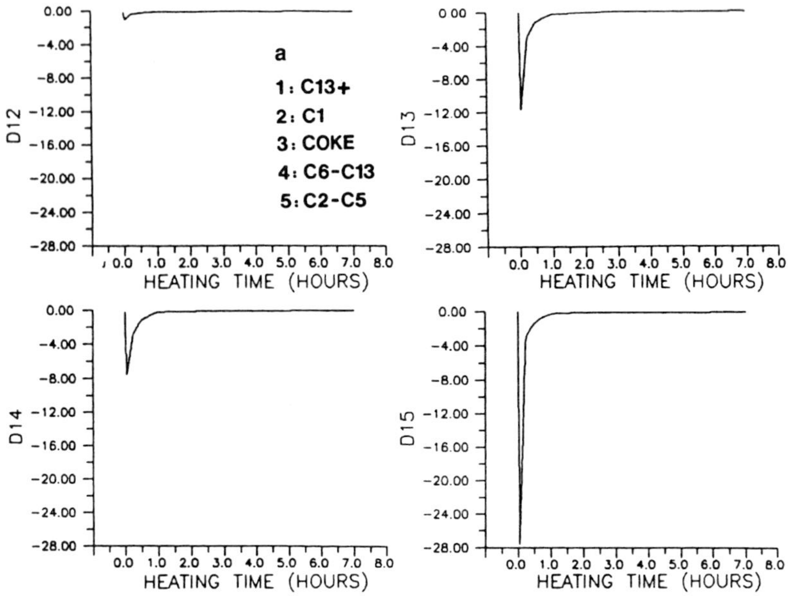


FIGURE 5.7 The variations of the elements of the transformation matrix $D_{ij}$ with heating time for Boscan oil at 450°C. (a) $D_{1j}$; (b) $D_{2j}$; (c) $D_{3j}$; (d) $D_{4j}$; (e) $D_{5j}$.

same source material but different thermodynamic conditions. Therefore, a comparison of $D_{ij}$ between data sets 1 and 2, as well as between data sets 1 and 3, provides measures of the sensitivity of $D_{ij}$ both to the source material and to the thermodynamic conditions.

Figures 5.7a–e (data set 1), Figs. 5.8a–e (data set 2), and Figs. 5.9a–e (data set 3) show that the $D_{ij}$ distributions vary with time for the three cases for each component of $D_{ij}$. From Figs. 5.7 and 5.8, four points are apparent:

1. Despite the difference in source material, $D_{ij}$ values determined for data sets 1 and 2 show some similar variations with increasing time. The two source materials appear to cause differences mainly in the magnitude of $D_{ij}$.
2. In the first hour, $D_{ij}$ increases and decreases very sharply. After the first hour, all $D_{ij}$ values tend toward zero. The values of most $D_{ij}$'s in the first hour of the reaction are apparently greater ($D_{ij} > 0$) or smaller ($D_{ij} < 0$) than the $D_{ij}$'s after the first hour of reaction, indicating that products are not only derived from the degradation of source material but also transformed from other products.
3. All the $D_{i1}$ are highest (where $j = 1$ refers to the source material), indicating that products are generated mainly from the degradation of source material.

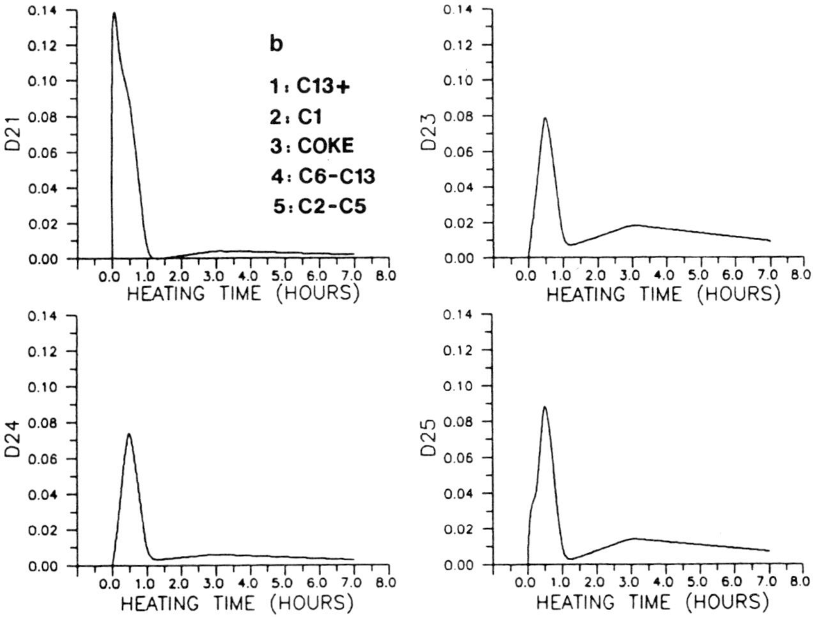


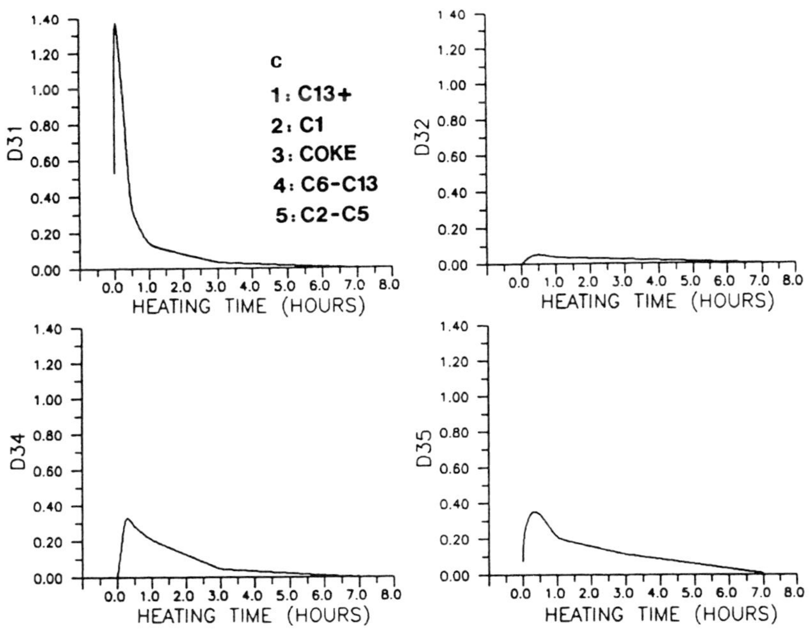


FIGURE 5.7 (*Continued*)

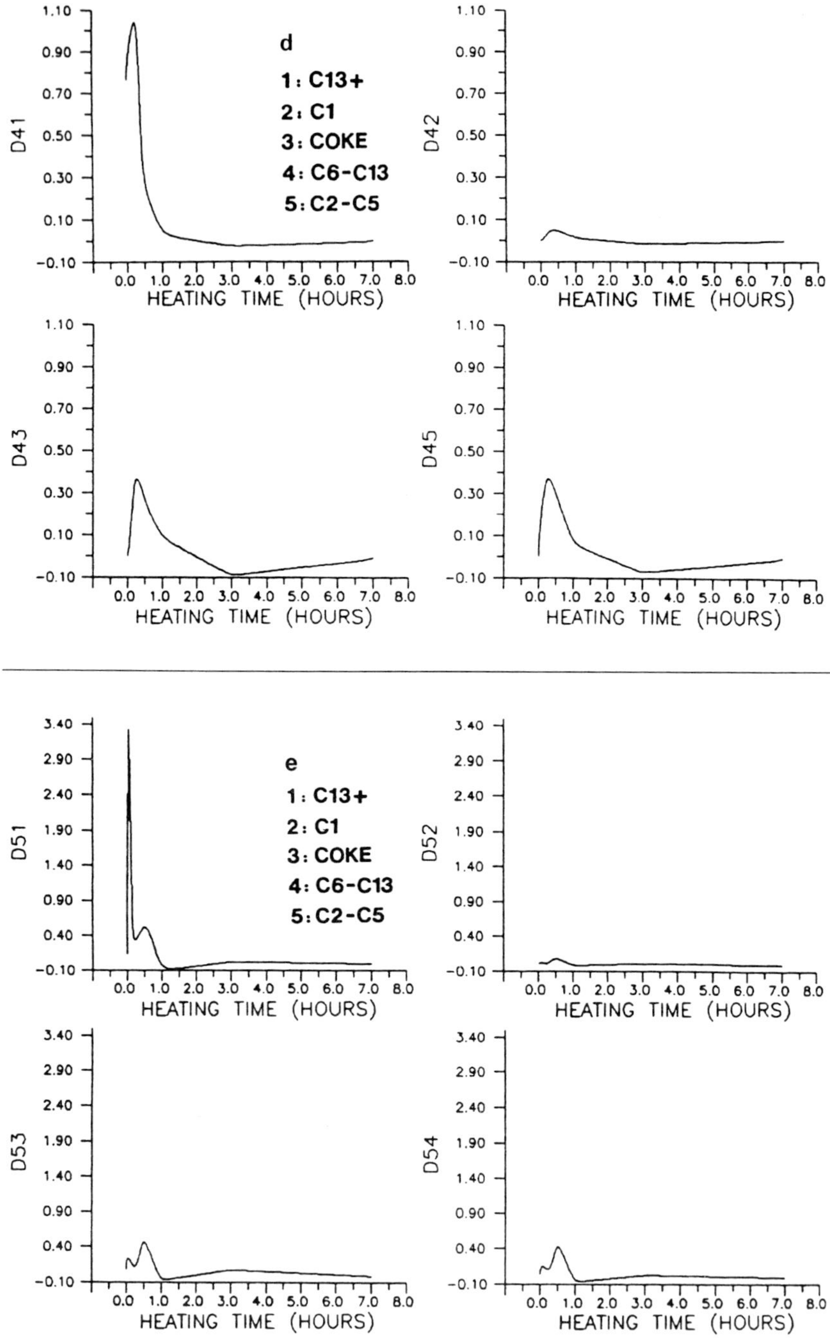


FIGURE 5.7 *(Continued)*

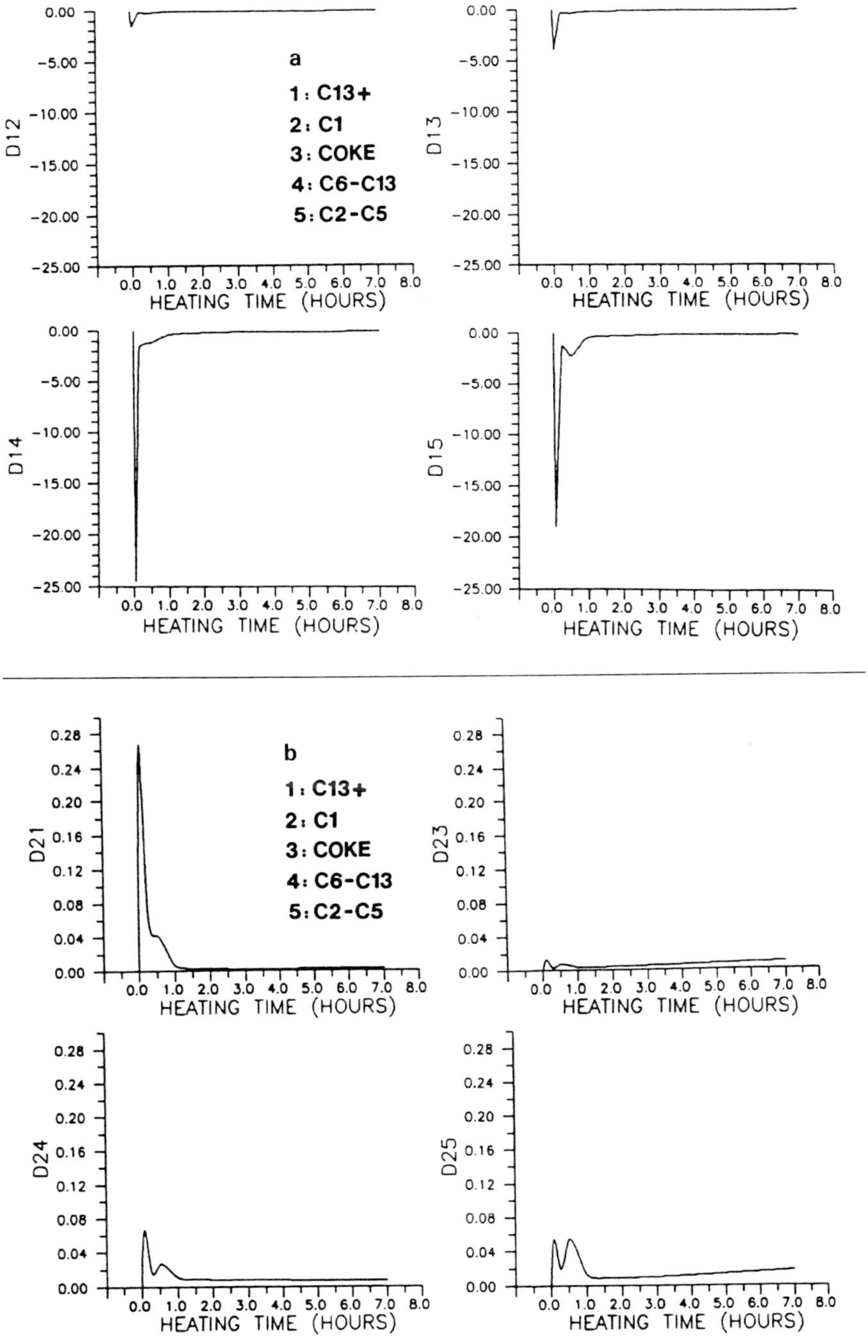


FIGURE 5.8 The variations of the elements of the transformation matrix $D_{ij}$ with heating time for Pematang oil at 450°C. (a) $D_{1j}$; (b) $D_{2j}$; (c) $D_{3j}$; (d) $D_{4j}$; (e) $D_{5j}$.

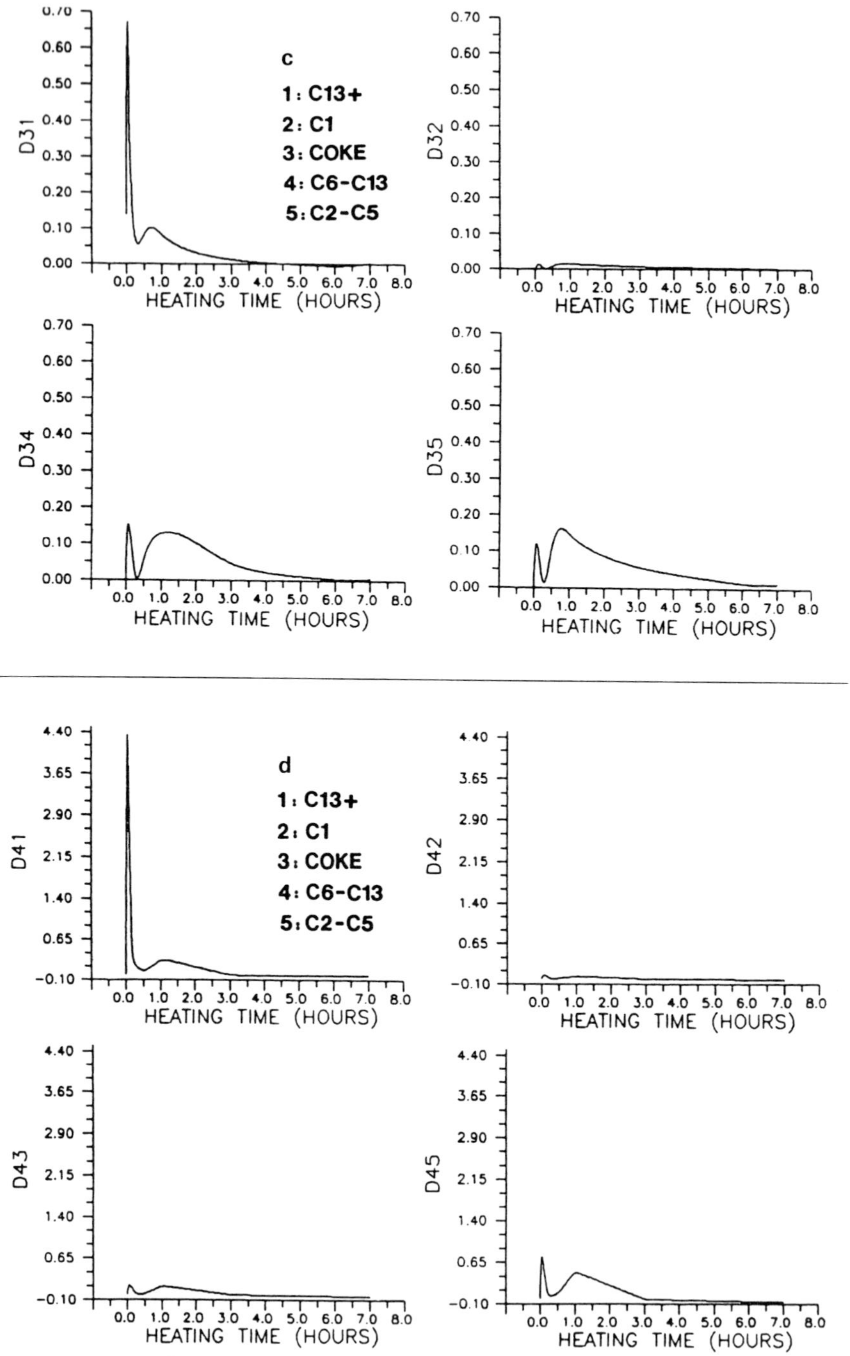


FIGURE 5.8 *(Continued)*

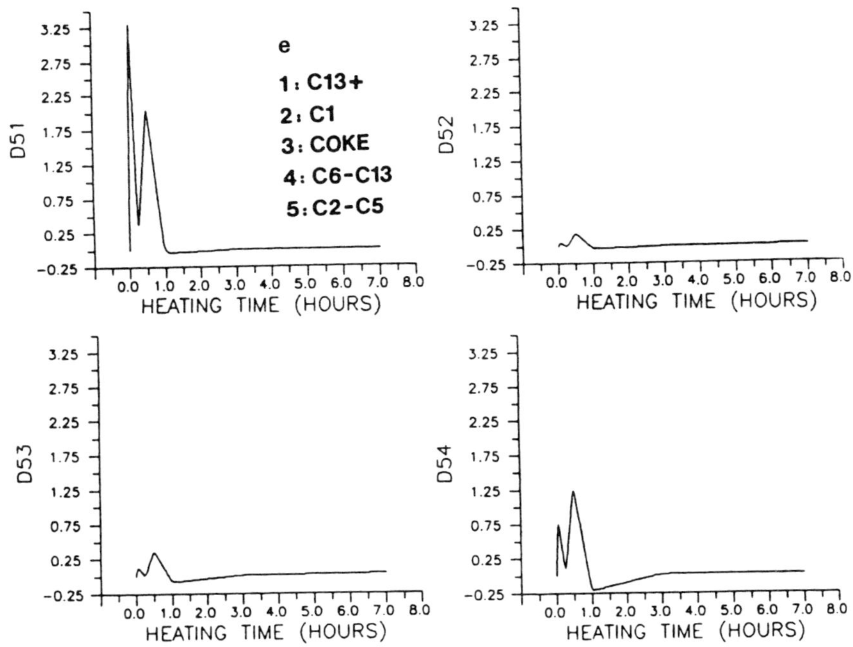


FIGURE 5.8 *(Continued)*

4. In the first hour of reaction, $D_{4j}$ shows an increasing trend, but a decreasing trend thereafter, indicating that the $C_6$–$C_{13}$ product was generated from the degradation of $C_{13+}$ in the first hour, and then cracking to other products caused a decrease in the amount of $C_6$–$C_{13}$.

By comparing Figs. 5.7 and 5.9 we find the following:

1. Although they have the same source material, data sets 1 and 3 show very different behaviors for $D_{ij}$ with time. Variations of $D_{ij}$ in data set 3 are relatively smooth, there are no sharp changes between the highest $D_{ij}$ values and lower $D_{ij}$ values. This behavior is in contrast to the variation of $D_{ij}$ for data set 1.
2. Transformation between different products is relatively weak. $D_{i1}$ tends to be higher than other $D_{ij}$ values, so that source production dominates.

Comparison of the $D_{ij}$ for the three data sets shows that (1) the thermodynamic conditions (rather than source materials) have more control on oil cracking kinetics and (2) there are also apparent differences in the magnitude of $D_{ij}$ when the source material is different.

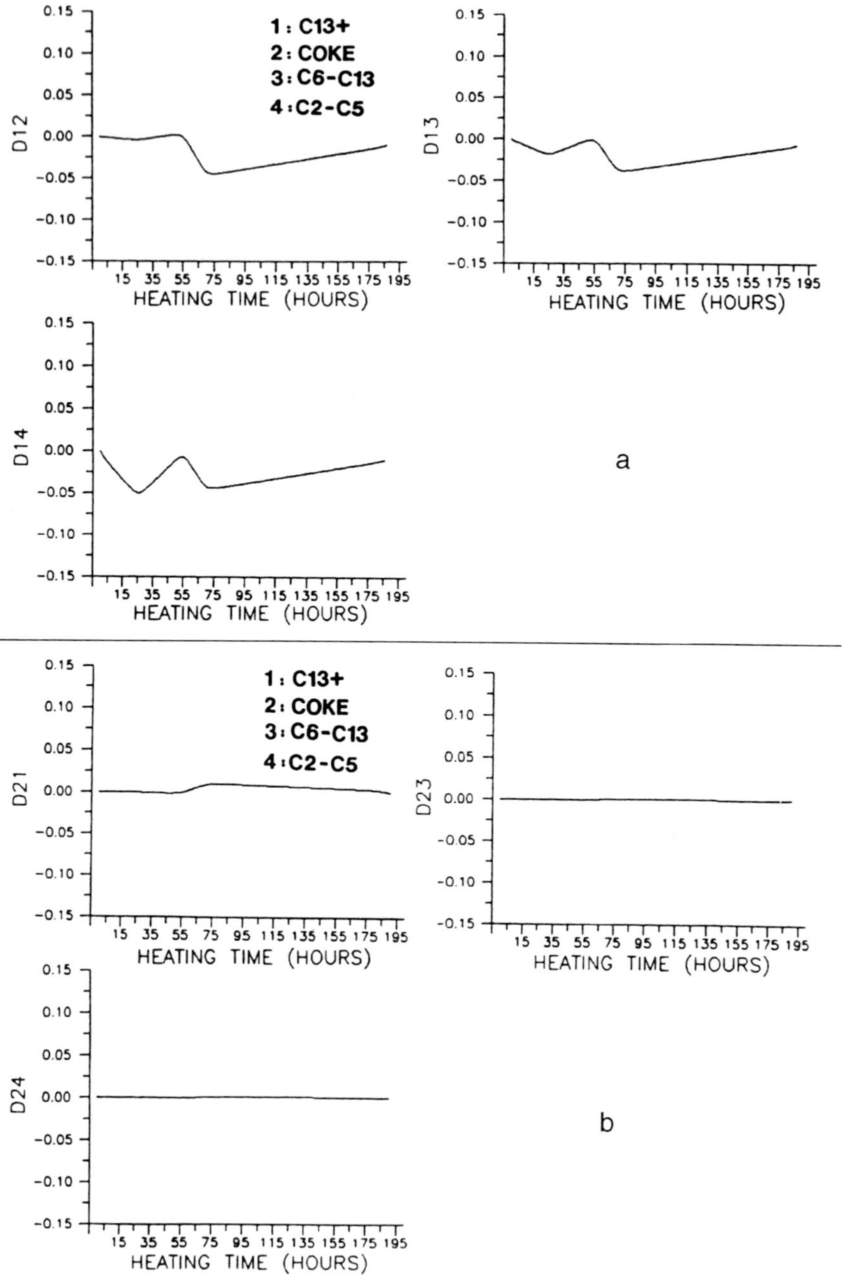


FIGURE 5.9 The variations of the elements of the transformation matrix $D_{ij}$ with heating time for Boscan oil at 357°C. (a) $D_{1j}$; (b) $D_{2j}$; (c) $D_{3j}$; (d) $D_{4j}$.

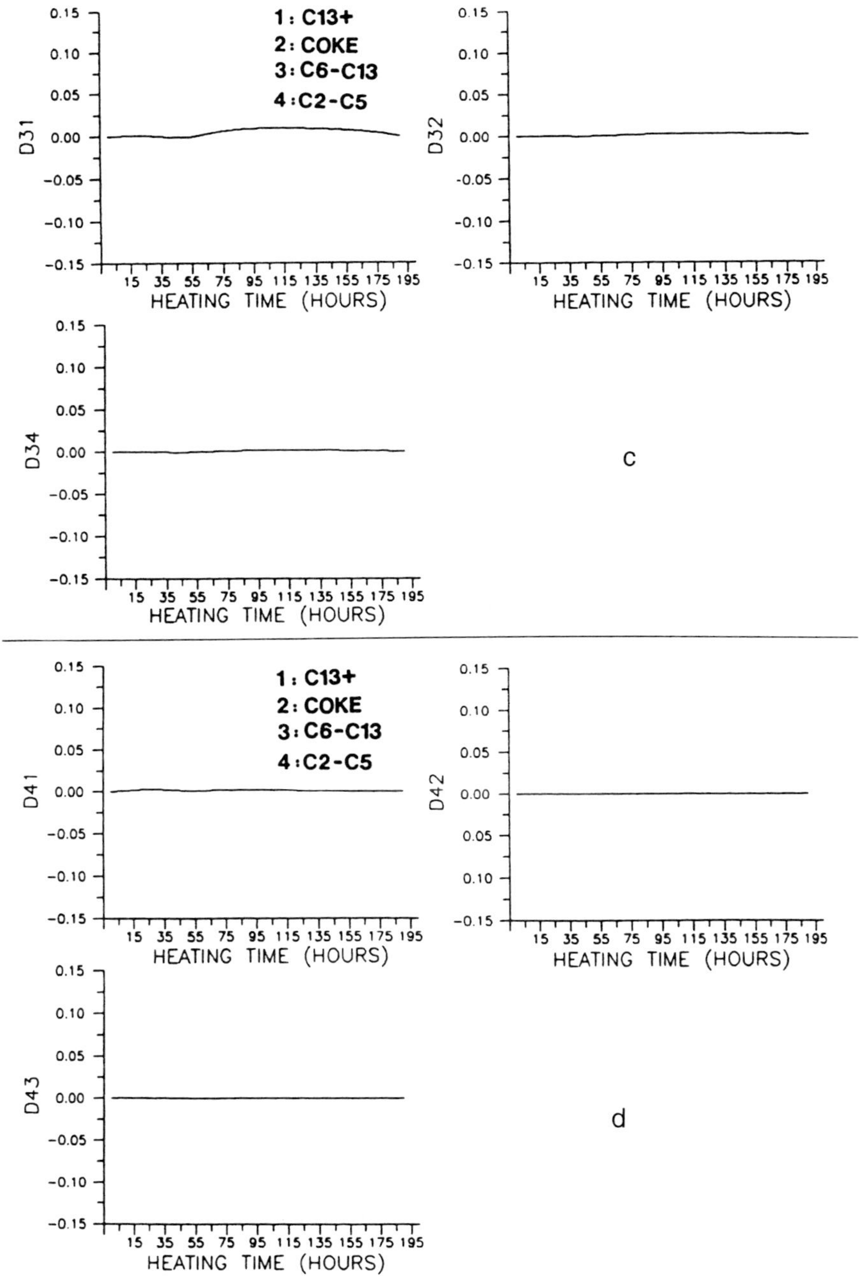


FIGURE 5.9 *(Continued)*

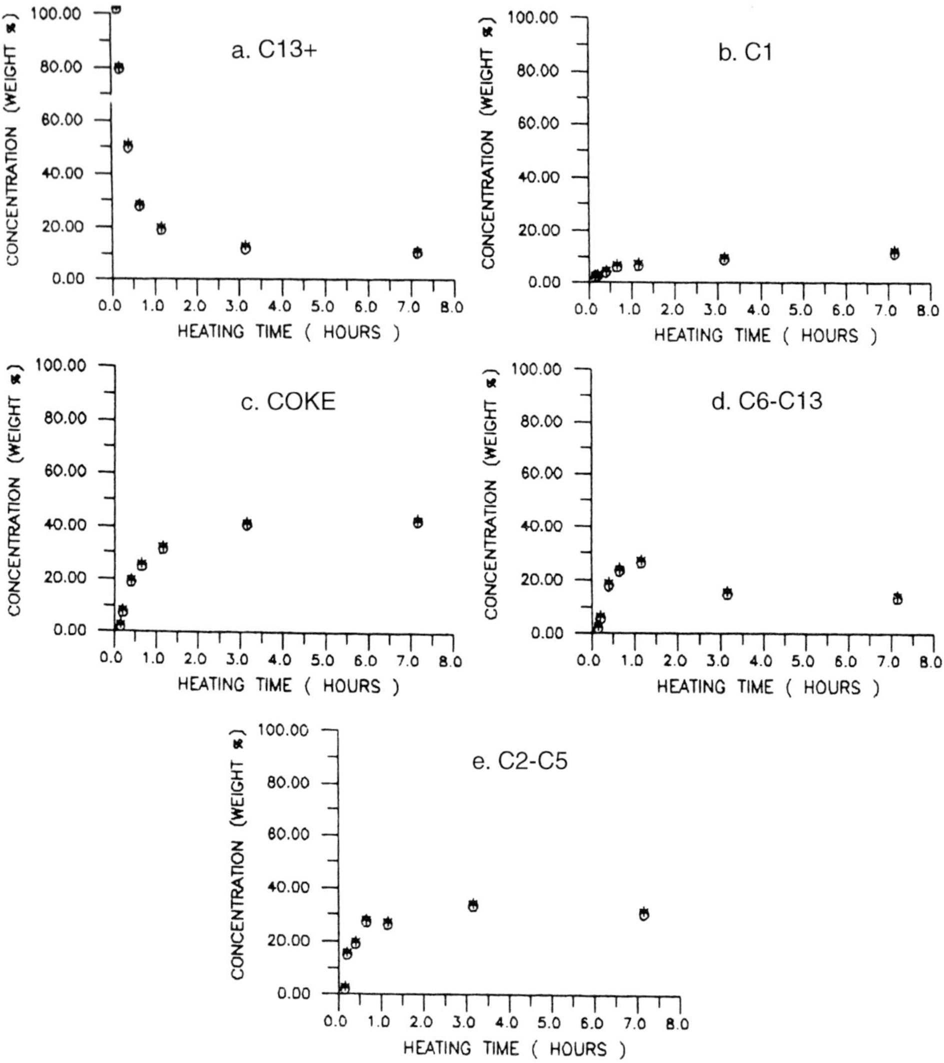


FIGURE 5.10 Measured concentrations of the various fractions from pyrolysis of Boscan oil (stars) at 450°C and for the values computed from the inverse model (circles) plotted versus heating time at 450°C using $D_{ij}$ as a function of heating time. (a) $C_{13+}$; (b) $C_1$; (c) coke; (d) $C_6$–$C_{13}$; (e) $C_2$–$C_5$.

## 2. Model Results Using $D_{ij}$ as a Function of Time

The preceding section has shown that the $D_{ij}$ obtained from the three data sets vary with the heating time. Especially in the high-temperature and short heating time experiments (data sets 1 and 2), $D_{ij}$ shows a strong dependence on the heating time. Indeed Figs. 5.10a–e (data set 1), Figs.

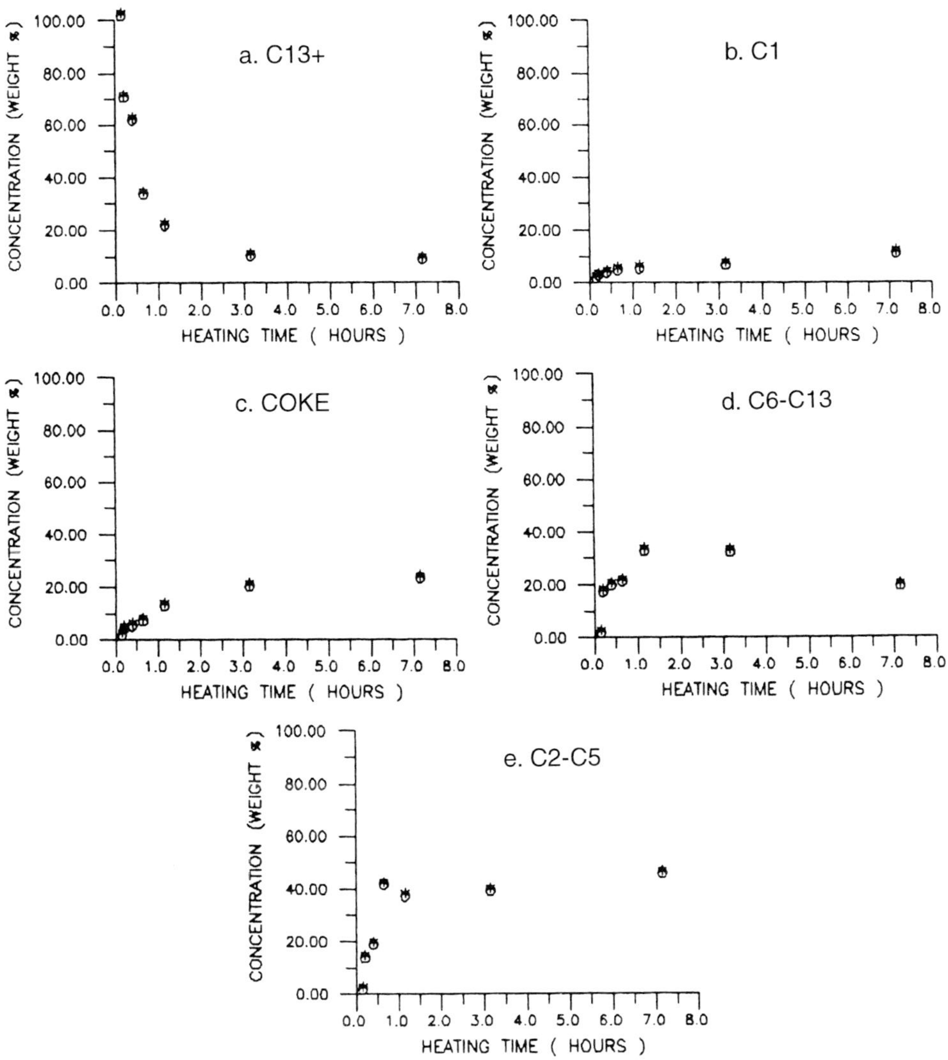


FIGURE 5.11 Measured concentrations of the various fractions from pyrolysis of Pematang oil (stars) at 450°C and from values computed from the inverse model (circles) plotted versus heating time using $D_{ij}$ as a function of heating time. (a) $C_{13+}$; (b) $C_1$; (c) coke; (d) $C_6$–$C_{13}$; (e) $C_2$–$C_5$.

5.11a–e (data set 2), and Figs. 5.12a–e (data set 3) show the fit of predicted values and the input data sets. The figures all show that the predicted values and data sets are virtually identical, indicating that $D_{ij}$ is indeed time dependent. Thus, the reaction mechanisms, and therefore the kinetics in these experiments, are different with time. This result accounts for the

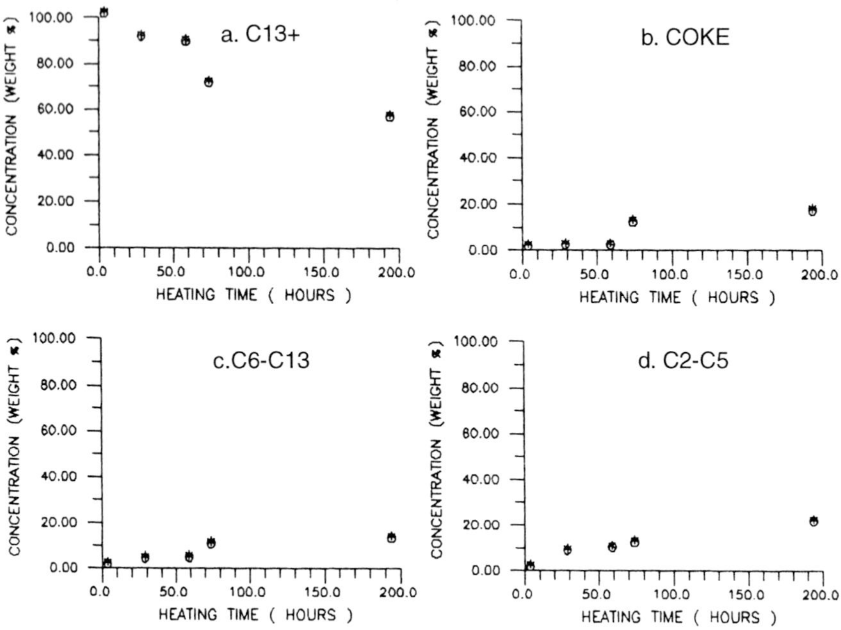


FIGURE 5.12 Measured concentrations of the various fractions from pyrolysis of Boscan oil (stars) plotted versus heating time at 357°C and from values computed from the inverse model (circles) using $D_{ij}$ as a function of heating time. (a) $C_{13+}$; (b) coke; (c) $C_6$–$C_{13}$; (d) $C_2$–$C_5$.

relatively large mismatch between the predictions and input data based on other modeling with these three experimental data sets (Ungerer *et al.*, 1988b). In turn, the applicability of the experimental reaction kinetics to geological situations needs to be rethought.

### 3. An Average Matrix

Although the $D_{ij}$ varied with time in these three cases, it is still interesting to know whether an average $D_{ij}$, $\overline{D_{ij}}$, will provide rough predictions for the products. We obtained a least-squares estimate of $\overline{D_{ij}}$ by averaging over time, with

$$\overline{D_{ij}} = \sum^{N} \left\{ \frac{(dq_i/dt)U_j}{N} \Big/ |U_s^P U_s^P| \right\}, \tag{5.25}$$

where $N$ is the number of time points. Figures 5.13a–e (data set 1), Figs. 5.14a–e (data set 2), and Figs. 5.15a–e (data set 3) compare input data and predicted values for all three data sets using the best least-squares average

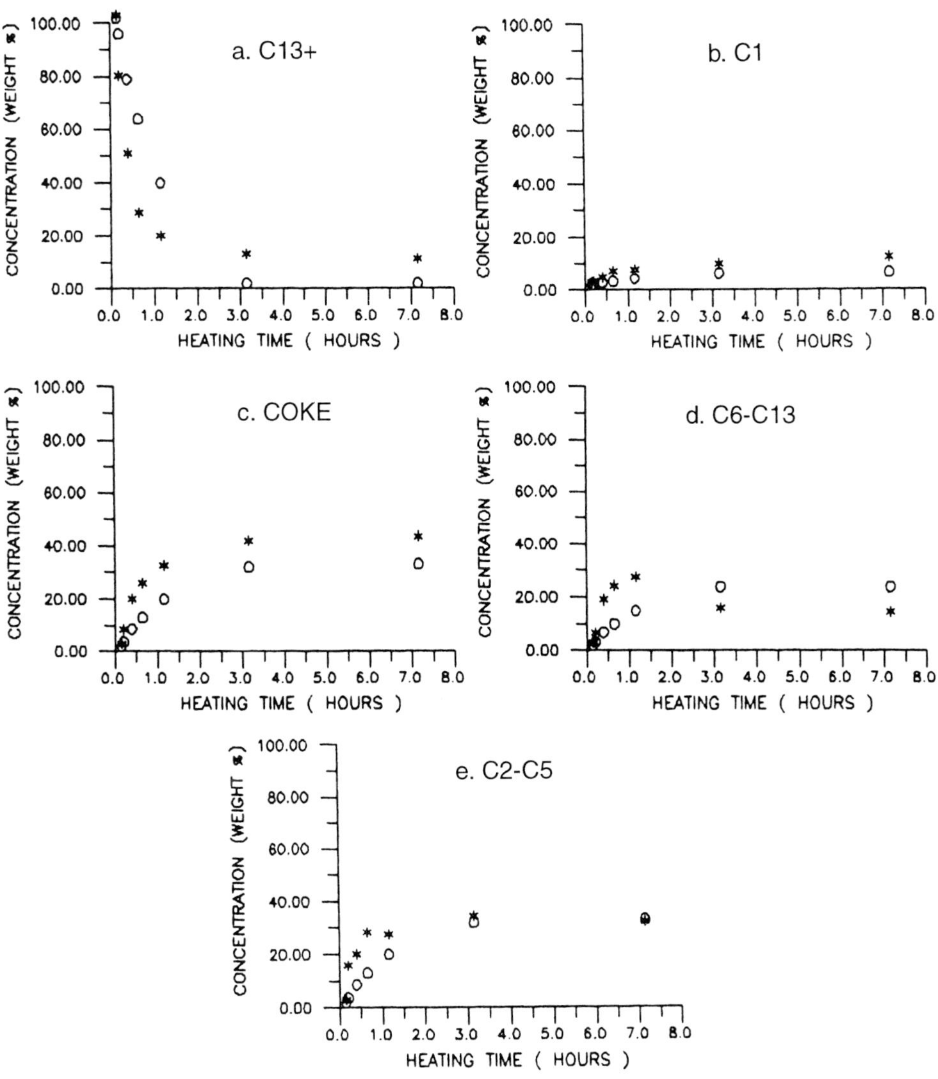


FIGURE 5.13 Measured concentrations of the various fractions from pyrolysis of Boscan oil (stars) at 450°C and from values computed from the inverse model (circles) using an average $\overline{D}_{ij}$. (a) $C_{13+}$; (b) $C_1$; (c) coke; (d) $C_6$–$C_{13}$; (e) $C_2$–$C_5$.

$\overline{D_{ij}}$. Because the $D_{ij}$ for data sets 1 and 2 vary sharply with time, the fits between input data and predicted values are poor. A 30–50% mismatch arises in most of the products for these two cases, especially in the first hour of reaction, which includes the majority of the reaction information in these two cases. For data set 3, $D_{ij}$ varies slowly with time so that the

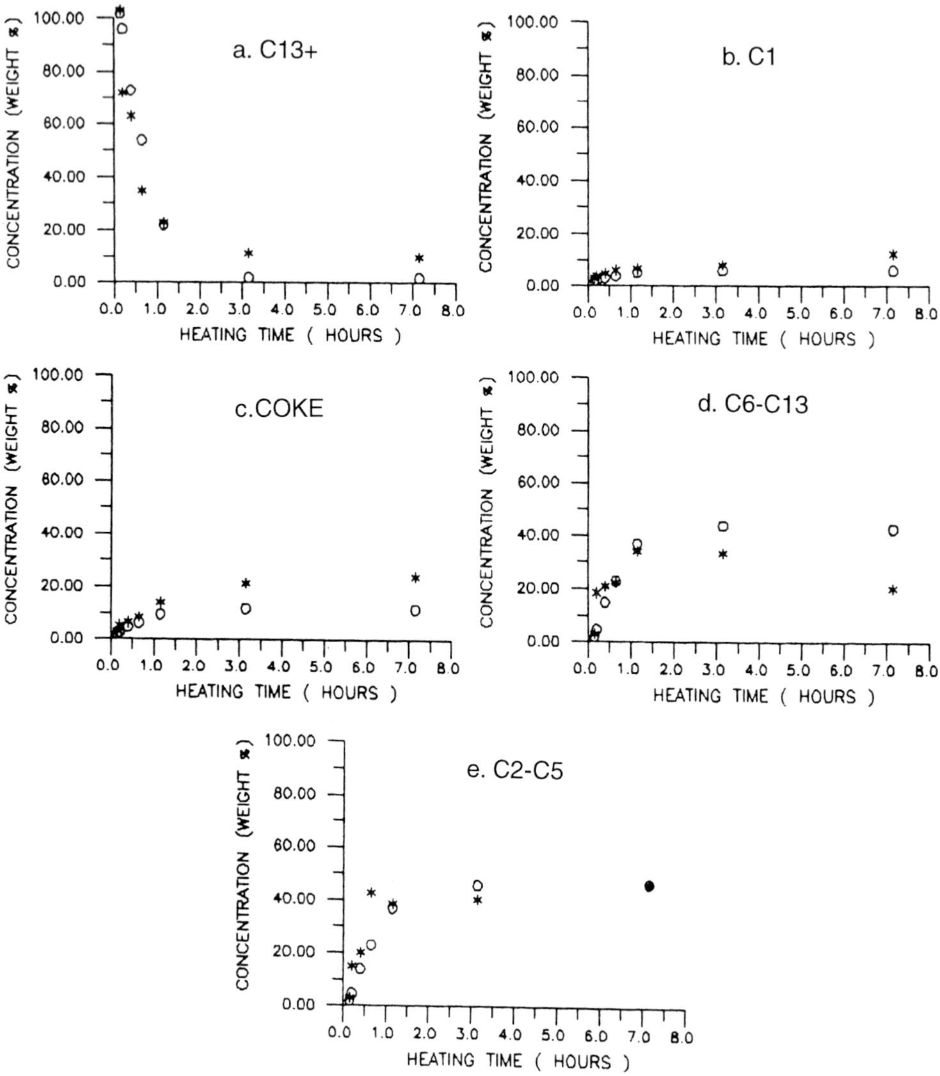


FIGURE 5.14 Measured concentrations of the various fractions from pyrolysis of Pematang oil (stars) at 450°C and from values computed from the inverse model (circles) using an average $\overline{D}_{ij}$. (a) $C_{13+}$; (b) $C_1$; (c) coke; (d) $C_6$–$C_{13}$; (e) $C_2$–$C_5$.

predicted behaviors are closer to the observations using the best least-squares estimate of $D_{ij}$ than in the other two cases.

It appears that with longer heating time and lower temperature, the experimental results may provide reaction rate "constants" closer to constant values, which would be more reproducible and thus useful in assessing hydrocarbon kinetic processes on a geologic timescale.

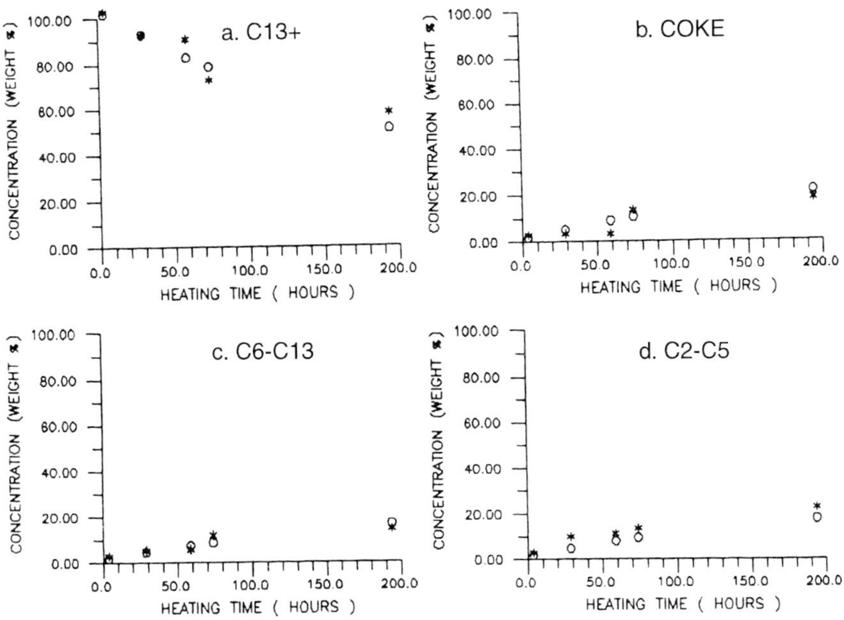


FIGURE 5.15 Measured concentrations of the various fractions from pyrolysis of Boscan oil (stars) at 357°C and from values computed from the inverse model (circles) using an average $\overline{D}_{ij}$. (a) $C_{13+}$; (b) coke; (c) $C_6$–$C_{13}$; (d) $C_2$–$C_5$.

## D. SUMMARY

Based on the results obtained above, it appears that we can conclude the following:

1. Checking of the model results using $D_{ij}$'s as functions of time shows accurate fits between predictions and experimental data. This fact indicates that the model is appropriate both in terms of the mathematical procedure and numerical implementation.
2. The $D_{ij}$ provide quantitative information about the thermal changes of the different hydrocarbon products. The $D_{ij}$ show a time dependence for experimental data obtained under high temperatures and short time periods.
3. $D_{ij}$ versus time can be used to indicate if laboratory experimental data can be applied to a prediction model. Usually this step is very important when extrapolating the experimental results to geological timescales because if $D_{ij}$ is time dependent, as appears to be the case, then a useful hydrocarbon prediction equation of generic validity is not available.
4. The experimental data from lower temperature and longer time period experiments may be more pertinent for constructing a kinetic

model for oil cracking and thermal changes of hydrocarbon products.

From this discussion, it seems necessary (1) to classify different source materials according to their behaviors in the oil cracking processes to determine whether a few sets of $D_{ij}$ values can be obtained to construct different prediction models for different groups of source materials; and (2) to have more laboratory experiments performed at low temperatures and over longer time periods, in order to determine whether it is possible to construct stable values of $D_{ij}$ which are independent, in a statistical sense, from the experiments, and which are therefore of generic validity.

## III. RESIDUAL KEROGEN ANALYSIS (S2)

### A. MATHEMATICAL MODEL

Here a method is developed to use S2 data from routine Rock–Eval pyrolysis, from which the thermal history can be reconstructed by an inverse scheme and, simultaneously, hydrocarbon generation can be simulated as kerogen degradation by kinetic modeling. The inverse procedure for S2 measurement was built into the existing dynamic and thermal indicator tomography (DTIT) model, developed by the USC Basin Analysis Group (Lerche, 1990a), which is a comprehensive one-dimensional fluid flow and compaction model with the ability to (1) reconstruct burial history from a simulation of sediment compaction and fluid flow; (2) reconstruct thermal history from inversion of multiple thermal indicators, such as vitrinite reflectance, apatite fission tracks, both hopane and sterane isomerization and aromatization, and pollen translucency; and (3) simulate hydrocarbon generation through kinetic models. S2 data were used as one of the thermal indicators in the DTIT model to determine paleoheat flux following the same inversion logic provided in the DTIT model. Several synthetic data sets were used to test the stability and sensitivity of the S2 inverse procedure. After the synthetic tests, two real well data sets were used to examine field case applications of the S2 inverse procedure. These two real data sets were (1) well 30/6-5 in the Norwegian sector of the North Sea (Dahl and Yukler, 1991) and (2) well Alaska State D1, in the Arctic National Wildlife Refuge, Alaska (Magoon *et al.*, 1987).

This section of the chapter first describes the basic theory and numerical procedure to invert S2 data as a thermal indicator for reconstructing thermal history and simulating hydrocarbon generation, and then demonstrates the sensitivity, accuracy, and resolution of the S2 inverse model through synthetic tests and case studies.

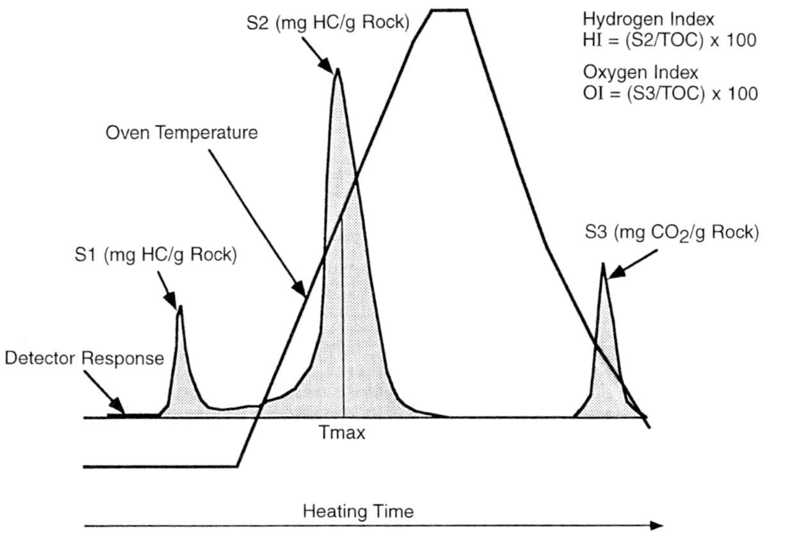


FIGURE 5.16 Schematic showing evolution of organic compounds from rock sample during heating (modified from Peters, 1986). [This figure and Figures 5.17–5.41 are reprinted from Cao and Lerche (1994), *Applied Geochemistry,* Vol. 9, pp. 93–117, with permission from Elsevier Science Ltd.]

## 1. Basic Theory of S2 Kinetic Inversion

Rock–Eval pyrolysis is widely used in the petroleum industry to evaluate rapidly the petroleum generation potential and thermal maturity of rocks. Pyrolysis is defined as the heating of organic matter, in the absence of oxygen, to yield organic compounds. In Rock–Eval pyrolysis, pulverized samples are gradually heated under an inert atmosphere. This heating distills the free organic compounds, then cracks pyrolytic products from the nonvolatile organic matter (Espitalié *et al.,* 1985; Peters, 1986). Figure 5.16 shows the evolution of organic compounds from a rock sample during heating. The first peak (S1) represents milligrams of hydrocarbons that can be thermally distilled from one gram of the rock. The second peak (S2) represents milligrams of hydrocarbons generated by pyrolytic degradation of the kerogen in 1 g of rock. The third peak (S3) represents milligrams of carbon dioxide generated from 1-g rock during heating up to 390°C. The temperature at which the maximum amount of S2 hydrocarbons is generated is called $T_{max}$. A value of S2 normalized relative to the total organic carbon percentage (TOC) in the sample is called the hydrocarbon index (HI) with units of mg hydrocarbon/g TOC (mg HC/g TOC). Similarly, the normalized S3 is called the oxygen index (OI) with units of mg $CO_2$/g TOC.

### *a. Inversion of a Particular Kinetic Model*

Because the TOC or kerogen in the sample is the residual kerogen, that is, the remaining part of the original (depositional) kerogen after thermal maturation, S2 represents the residual potential of hydrocarbons which can be generated from the rock, and HI represents the residual (or remaining) potential of hydrocarbons from 1 g of TOC (kerogen).

Consider a given type kerogen of amount $x_0$ per gram of rock at deposition time ($t = 0$). Suppose that at time $t$, when the temperature is $T(t)$, the rate of loss of material is given by

$$\frac{dx(t)}{dt} = -\alpha(T(t), t)x(t), \tag{5.26}$$

where $\alpha(T(t), t)$ represents the time- and temperature-dependent rate "constant" for thermal degradation. It is often *assumed* that $\alpha(T(t), t)$ is an explicit function of temperature only, and not of time. A close correspondence in functional form with an Arrhenius-type of kinetic degradation can be drawn if one writes

$$\alpha(T) = k_0 \int_0^\infty p(E) \exp(-E/RT)\, dE, \tag{5.27}$$

with $k_0$ being an overall rate constant, $R$ the gas constant, and $p(E)$ a differential probability of activating a degradation channel of energy $E$. The normalization on $p(E)$ is often taken to be in the form

$$\int_0^\infty p(E)\, dE = 1, \tag{5.28}$$

although any normalization will do because $k_0$ will automatically compensate. For later convenience note that if one were to rescale $p(E)$ by $\exp(-E/RT_0)$, with $T_0$ a fixed scaling temperature, then one can also write $\alpha(T)$ in the form

$$\alpha(T) = k_1 \int_0^\infty \theta(E) \exp[-E/R(1/T - 1/T_0)]\, dE, \tag{5.29}$$

with $\theta(E) = p(E) \exp(-E/RT_0)$. The normalization on $\theta(E)$ can also be chosen in any relevant manner. For instance, one could choose

$$\int_0^\infty \theta(E)\, dE = 1 \tag{5.30}$$

or one could choose any other representation. It is *not* necessary for $p(E)$ ($\theta(E)$) to be normalized so that it has unit integral. The constant $k_0(k_1)$ will accommodate for any chosen departure of the normalization from the conventional values of unity given earlier.

In any event, with the functional forms of Eqs. (5.26) and (5.27) it follows that the kerogen degradation with time can be described by the integral form

$$x(t) = x_0 \exp\left\{-k_0 \int_0^t dt' \int_0^\infty p(E) \exp(-E/RT(t'))\, dE\right\}, \tag{5.31}$$

where $x(t)$ is the amount of kerogen at time $t$ since deposition, $x_0$ is the amount of kerogen at deposition ($t = 0$), $k_0$ is a fixed kinetic scaling constant with dimensions $(\text{time})^{-1}$, $p(E)$ is a channeling probability with dimension $(\text{energy})^{-1}$, $R$ is the gas constant, $E$ is activation energy, $T(t')$ is the temperature (K) at time $t'$ along the burial path of the kerogen, and $t$ is present-day time after deposition.

The integral over time in Eq. (5.31) is taken along the burial path. The amount of kerogen at present day after burial history pyrolysis is taken to be the S2 laboratory measurement from pyrolysis of a sample recovered from depth $z$ ($z$ is the present-day depth of the kerogen). Therefore, for a set of S2 measurements from samples at different depths in a well, Eq. (5.31) can be rewritten, *assuming all S2 measurements share the same initial amount of kerogen,* $x_0$, *and kinetic parameters* $k_0$, p(E). Then one obtains

$$x(z_i) = x_0 \exp\left\{-k_0 \int_0^{t_i} dt' \int_0^{\infty} p(E) \exp(-E/RT_i(t'))\, dE\right\}, \quad (5.32)$$

where $x(z_i)$ is the S2 measurement at depth $z_i$, $T_i(t')$ is the temperature along the burial path currently at depth $z_i$, and $t_i$ is the age of the S2 sample currently at depth $z_i$.

For a given burial history, $T_i(t')$ in Eq. (5.32) can be calculated from

$$T_i(t') = T_s(t') + Q(t') \int_0^{z_i(t')} \frac{dt''}{K(z_i(t''))}, \quad (5.33)$$

where $T_s(t')$ is the sediment surface temperature at time $t'$, $Q(t')$ is the heat flow through the basement at time $t'$, $z_i(t')$ is the burial depth at time $t'$, and $K(z_i(t'))$ is the sediment thermal conductivity at time $t'$.

Following the thermal inversion scheme of Lerche (1990a), we can construct

$$X^2 = \frac{1}{N}\sum_{i=1}^{N}\left[x(z_i) - x_0 \exp\left\{-k_0 \int_0^{t_i} dt' \int_0^{\infty} p(E) \exp(-E/RT_i(t'))\, dE\right\}\right]^2, \quad (5.34)$$

where $N$ is the number of S2 measurements, and then minimize Eq. (5.34) to obtain the unknown values $Q(t)$, $x_0$, $k_0$, and $p(E)$.

### *b. Comparison of Two Kinetic Models*

Lerche (1991) pointed out that many kinetic models proposed to date for kerogen degradation can be put in the general form described by Eqs. (5.26) and (5.31). The major difference is in the kinetic parameters used. The original Tissot (1969) and Tissot and Espitalié (1975) models assume $p(E)$ to be proportional to a sum of $\delta$ functions at prescribed energies, that is,

$$p(E) = \sum_{i=1}^{N} p_i \delta(E - E_i), \tag{5.35}$$

with $p_i$ a set of constants. Then Eq. (5.31) becomes

$$x(t) = x_0 \exp\left\{-\sum_{i=1}^{N} A_i \int_0^t dt' \exp(-E_i/RT(t'))\right\} \tag{5.36}$$

with $A_i = k_0 p_i$. The parallel channel kinetic models of Braun and Burnham (1987) assume $p(E)$ is Gaussian in structure, whereas Sweeney and Burnham (1990) used a discrete model for $p(E)$ [see Eq. (5.42) for the parallel kinetic model].

Another difference that can arise is in the functional forms. The S2 kinetic model uses a bulk set of reactions that influences only the departure of the decay "constant" $\alpha(T)$ from a simple Arrhenius form ($\exp(-E/RT)$) but which otherwise consider the total decay of a bulk mass of kerogen. Other kinetic models use parallel kinetic reactions. The difference between these two functional forms is demonstrated as follows: For the S2 kinetic model of Eq. (5.26) we have

$$\left(\frac{x(t)}{x_0}\right)_{\mathrm{b}} = \exp\left\{-k_0 \int_0^t dt' \int_0^\infty p(E) \exp(-E/RT(t'))\, dE\right\}, \tag{5.37}$$

where b denotes the bulk set and we have chosen the normalization $\int_0^\infty p(E)\, dE = 1$.

Parallel channel kinetic models (Ungerer and Pelet, 1987; Sweeney, 1990) have the basic form

$$\frac{dx_i(t)}{dt} = -k_0 x_i \exp(-E_i/RT(t)) \tag{5.38}$$

with

$$\frac{x(t)}{x_0} = \sum_{i=1}^{M} f_i \frac{x_i(t)}{x_{i0}} \tag{5.39}$$

so that $\Sigma_{i=1}^{M} f_i = 1$, where $f_i$ are stoichiometric factors (all positive) and the number of channels is $M$. Rewrite Eq. (5.38) as

$$x_i(t) = x_{i0} \exp\left\{-k_0 \int_0^t \exp[-E_i/RT(t')]\, dt'\right\}. \tag{5.40}$$

Then

$$\frac{x(t)}{x_0} = \sum_{i=1}^{M} f_i \exp\left\{-k_0 \int_0^t \exp[-E_i/RT(t')]\, dt'\right\}. \tag{5.41}$$

Converting Eq. (5.41) to a continuous representation we have

$$\left(\frac{x(t)}{x_0}\right)_{\mathrm{p}} = \int_0^\infty p(E) \exp\left[-k_0 \int_0^t \exp(-E/RT(t'))\, dt'\right] dE, \qquad (5.42)$$

where p denotes the parallel form and we have again chosen the normalization

$$\int_0^\infty p(E)\, dE = 1 = \sum_{i=1}^{M} f_i.$$

Set

$$k_0 \int_0^t \exp[-E/RT(t')]\, dt' = U(E, t) > 0. \qquad (5.43)$$

Then for the bulk set of reaction kinetics we have

$$\left(\frac{x(t)}{x_0}\right)_{\mathrm{b}} = \exp\left\{-\int_0^\infty p(E)U(E, t)\, dE\right\}, \qquad (5.44)$$

while the parallel kinetic reactions yield

$$\left(\frac{x(t)}{x_0}\right)_{\mathrm{p}} = \int_0^\infty p(E) \exp[-U(E, t)]\, dE. \qquad (5.45)$$

Now we consider $U(E, t) \leq 1$. The only regime of interest is $U(E, t) \leq 1$ because if $U(E, t) \gg 1$, then, because $\int_0^\infty p(E)\, dE = 1$, both Eqs. (5.44) and (5.45) would yield values of $x(t)/x_0$ close to zero.

For $U(E, t) \leq 1$, the first-order approximations for Eqs. (5.44) and (5.45) are

$$\left(\frac{x(t)}{x_0}\right)_{\mathrm{b}} \cong 1 - \int_0^\infty p(E)U(E, t)\, dE, \qquad (5.46)$$

$$\left(\frac{x(t)}{x_0}\right)_{\mathrm{p}} \cong 1 - \int_0^\infty p(E)U(E, t)\, dE, \qquad (5.47)$$

which are *identical*. The second-order approximations yield

$$\left(\frac{x(t)}{x_0}\right)_{\mathrm{b}} \cong 1 - \int_0^\infty p(E)U(E, t)\, dE + \frac{1}{2}\left[\int_0^\infty p(E)U(E, t)\, dE\right]^2, \qquad (5.48)$$

$$\left(\frac{x(t)}{x_0}\right)_{\mathrm{p}} \cong 1 - \int_0^\infty p(E)U(E, t)\, dE + \frac{1}{2}\int_0^\infty p(E)(U(E, t))^2\, dE, \qquad (5.49)$$

and, if $U(E, t) = 1$, both Eqs. (5.44) and (5.45) yield $x(t)/x_0 = e^{-1} = 0.368$. Therefore for $U(E, t) < 1$, the difference between a bulk set and a parallel set of reaction kinetics is extremely small to the second order of approximation.

Next we consider $U(E, t) < k_0 t$. From Eq. (5.43) we have $U(E, t) < k_0 t$ at *any* finite values of $E$ and $T(t')$. Then, from Eqs. (5.44) and (5.45), we have

$$\left(\frac{x(t)}{x_0}\right)_{\rm b} \geq \exp(-k_0 t), \tag{5.50}$$

$$\left(\frac{x(t)}{x_0}\right)_{\rm p} \geq \exp(-k_0 t). \tag{5.51}$$

Introduce

$$V(E, t) = k_0 \int_0^t [1 - \exp(-E/RT(t'))]\, dt' \equiv k_0 t - U(E, t) \geq 0. \tag{5.52}$$

We have

$$\left(\frac{x(t)}{x_0}\right)_{\rm b} = \exp\left\{-k_0 t + \int_0^\infty p(E) V(E, t)\, dE\right\}, \tag{5.53}$$

$$\left(\frac{x(t)}{x_0}\right)_{\rm p} = \exp(-k_0 t) \int_0^\infty p(E) \exp(V(E, t))\, dE. \tag{5.54}$$

Both Eqs. (5.53) and (5.54) are greater than or equal to $\exp(-k_0 t)$. An expansion in powers of $V(E, t)$ then gives

$$\left(\frac{x(t)}{x_0}\right)_{\rm b} \cong \exp(-k_0 t)\left\{1 + \int_0^\infty p(E) V(E, t)\, dE + \frac{1}{2}\left(\int_0^\infty p(E) V(E, t)\, dE\right)^2 + \ldots\right\}, \tag{5.55}$$

$$\left(\frac{x(t)}{x_0}\right)_{\rm p} \cong \exp(-k_0 t)\left\{1 + \int_0^\infty p(E) V(E, t)\, dE + \frac{1}{2}\int_0^\infty p(E)[V(E, t)]^2\, dE + \ldots\right\}. \tag{5.56}$$

Again both models are identical to the first-order expansion and have only a slight difference at the second-order expansion level.

Therefore, any differences between the S2 bulk kinetic model developed in this section and other kinetic models *cannot* be from the functional forms. Any such differences must come from different numerical choices for $k_0$ and $p(E)$ in the kinetic models that have been suggested to date. In this connection it is important to note that a determined set of numerical choices for $k_0$ and $p(E)$ for, say, the parallel channel kinetic model would not then necessarily be the relevant choices for the bulk kinetic model; those parameters would themselves have to be determined. The purpose of this section of the chapter is to use the functional forms with downhole data to *determine*, rather than assume, the kinetic parameters.

For the case of constant heating rate experiments which start at temperature $T_0$ at time $t = 0$ an accurate analytic approximation for $U(E, t)$ is provided in Appendix B, Section B.I, at the end of this chapter. In addition, for heating, cooling, and reheating experiments, *exact* analytic expressions for $U(E, t)$ are given in Appendix B, Section B.II, provided that temperature increases (decreases) in a manner with $dT/dt \; \alpha T^2$.

Further, the theoretical behavior for the bulk and parallel kinetic equations developed here assumes (in both cases) that only one end-member type of pyrolyzable kerogen is present. The data we have available in the downhole case history examples (see later) manifestly indicate the presence of more than one kerogen type (reversal of hydrogen index with increasing burial depth from decreasing to increasing and back to decreasing). Accordingly in Appendix B, Section B.III, we illustrate, using a mix of two kerogen types, how both the mix of the kerogens and the maturation path (heating, cooling, reheating) have impacts on the *total* amount of pyrolyzable kerogen at a given time, and also how both influence the temperature, $T_{max}$, at which the maximum S2 will be reached.

## B. NUMERICAL PROCEDURE FOR S2 KINETIC INVERSION

The logic of the inverse procedure used is most easily exhibited for the bulk kinetic model, but applies equally well to the parallel kinetic model. In Appendix A at the end of this chapter, we develop the details of the corresponding inverse procedure for the parallel kinetic model in two ways: first, independently of the technical details to be exhibited here for the bulk kinetic model; second, by using the functional forms of the bulk and parallel kinetic models and then by providing an inverse procedure that uses the *difference* between the two models.

We now concentrate on exhibiting the logic procedure. Before executing the thermal inversion scheme to obtain heat flux with time, S2 kinetic parameters must be determined for a given burial history, temperature history, and a set of S2 measurements for samples from different depths. In the general form [Eq. (5.31)] for the S2 kinetic model, the kinetic parameters that need to be determined are $x_0$ (initial value of kerogen), $k_0$ (kinetic scaling constant), and $p(E)$ (channeling probability of activation energy).

The general form of Eq. (5.31) can be rewritten in a discrete representation as

$$x_j = x_0 \exp\left\{-k_0 \sum_{i=1}^{M} P_i F_{ji}\right\}, \tag{5.57}$$

with $F_{ji} = \int_0^{t_j} \exp(-E_i/RT_j(t'))\, dt'$, $P_i = p_i(E_i)\, \Delta E$, and where $M$ is the number of probability channels for the activation energy. Note that $F_{ji} \geq$

0, and that $\Sigma_{i=1}^{M} P_i = 1$. Suppose an approximation $\hat{P}_i$ to $P_i$ is available and an improvement is required. Then we can write exactly that

$$x_j = x_0 \exp\left\{-k_0 \sum_{i=1}^{M} \hat{P}_i F_{ji} - k_0 \sum_{i=1}^{M} (P_i - \hat{P}_i) F_{ji}\right\}. \tag{5.58}$$

Assume $|k_0 \Sigma_{i=1}^{M} (P_i - \hat{P}_i) F_{ji}| \ll 1$, that is, $\hat{P}_i$ is an accurate approximation. Then we have

$$\frac{x_j}{x_0} \exp\left\{k_0 \sum_{i=1}^{M} \hat{P}_i F_{ji}\right\} \cong 1 + k_0 \sum_{i=1}^{M} \hat{P}_i F_{ji} - k_0 \sum_{i=1}^{M} P_i F_{ji}. \tag{5.59}$$

Hence by rearrangement we have

$$k_0 \sum_{i=1}^{M} \hat{P}_i F_{ji} = -\frac{x_j}{x_0} \exp\left\{k_0 \sum_{i=1}^{M} \hat{P}_i F_{ji}\right\} + 1 + k_0 \sum_{i=1}^{M} \hat{P}_i F_{ji}. \tag{5.60}$$

Because $F_{ji} \geq 0$, and $P_i$ is required to be positive, $x_0$ must be chosen so that, in Eq. (5.60), we have

$$x_0 \geq \max\left\{\frac{x_j \exp\left[k_0 \sum_{i=1}^{M} \hat{P}_i F_{ji}\right]}{\left(1 + k_0 \sum_{i=1}^{M} \hat{P}_i F_{ji}\right)}\right\}. \tag{5.61}$$

Assign

$$Y_j = 1 + k_0 \sum_{i=1}^{M} \hat{P}_i F_{ji} - \frac{x_j}{x_0} \exp\left\{k_0 \sum_{i=1}^{M} \hat{P}_i F_{ji}\right\}; \tag{5.62}$$

then Eq. (5.60) can be rewritten as

$$k_0 \sum_{i=1}^{M} P_i F_{ji} = Y_j, \tag{5.63}$$

which can be solved by standard least-squares procedures to yield $P_i$ (Lerche, 1991).

Thus with this $P_i$, the requirement is to fit the S2 data better, that is,

$$X^2 = \frac{1}{N} \sum_{j=1}^{N} \left[x_j - x_0 \exp\left(-k_0 \sum_{i=1}^{M} P_i F_{ji}\right)\right]^2 \tag{5.64}$$

should be a minimum. For a given value of $k_0$, the minimum occurs when

$$x_0 = \frac{\sum_{j=1}^{N} x_j \exp\left(-k_0 \sum_{i=1}^{M} P_i F_{ji}\right)}{\sum_{j=1}^{N} \exp\left(-2k_0 \sum_{i=1}^{M} P_{ij} F_{ji}\right)}. \tag{5.65}$$

In this way $x_0$ is updated.

It also follows from Eq. (5.63) that a least-squares solution for $k_0$ is given by

$$k_0 = \frac{\sum_{j=1}^{N} Y_j \left( \sum_{i=1}^{M} P_i F_{ji} \right)}{\sum_{j=1}^{N} \left[ \left( \sum_{i=1}^{M} P_i F_{ji} \right)^2 \right]}. \tag{5.66}$$

Then use the updated $P_i$, $k_0$, and $x_0$ to start the next iteration. The S2 kinetic optimization procedure is summarized as follows:

1. Choose an initial $P_i$ = constant = $1/M$, where $M$ is the number of probability channels and $P_j$ is normalized over the $M$ channels so that $\Sigma_{i=1}^{M} P_i = 1$.
2. Calculate $F_{ji}$ from the temperature history which is obtained during the thermal inversion.
3. Use Eqs. (5.65) and (5.66) to get $x_0$ and $k_0$.
4. Calculate the $X^2$ value from Eq. (5.64).
5. Solve the least-squares equation [Eq. (5.63)] to get the new $P_j$.
6. Repeat steps 3, 4, and 5 until the $X^2$ values converge to a minimum. Usually it takes five to eight iterations to achieve a minimum.
7. Then update the thermal history using parameters in the paleoheat flux function of the DTIT model (see later) and repeat steps 1 through 6. Retain the minimum of all minima of $X^2$ as the best value. This procedure provides both residual kerogen kinetics parameters and paleoheat flux parameters consistent with available data. Figure 5.17 shows the flowchart of S2 kinetics inversion. A similar inverse procedure is also available for the parallel model of Eq. (5.43) and is given in Appendix A of this chapter.

## C. SYNTHETIC TESTS

The synthetic tests of this section are designed to ensure that the inverse procedure will truly recover known input parameters. To this end, a series of forward models was generated with given parameters [both kinetic, e.g., $k_0$ and $p(E)$, and geological, e.g., heat flux variation with time]. Calculations were then carried forward in time thereby generating synthetic hydrogen index (S2) profiles with depth at the present day. These profiles were then presented to the inverse procedure and the questions asked whether the inverse method could recover the known parameters, and to what degree of resolution. Thus this section tests the convergence, accuracy, resolution, precision, and stability of the inverse method per se using a particular kinetic functional framework. The emphasis here is on testing the inverse method.

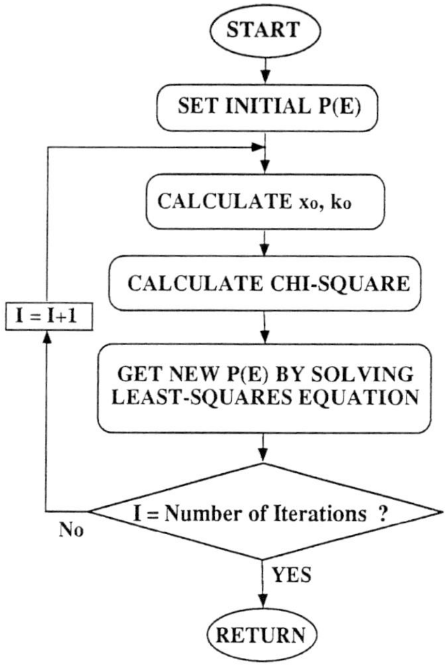


FIGURE 5.17 Flowchart of S2 kinetic inversion. (Reprinted with permission from Cao and Lerche, 1994.)

### 1. Sensitivity Tests on $M$, $Q_0$, BETA, and $A_1$

A simple burial history, shown in Fig. 5.18, is used to create synthetic S2 data sets. Heat flow change with time ($Q(t)$) is calculated from

$$Q(t) = Q_0 \exp[\text{BETA}^* t + A_1 \sin(\pi t/t_{\max})], \tag{5.67}$$

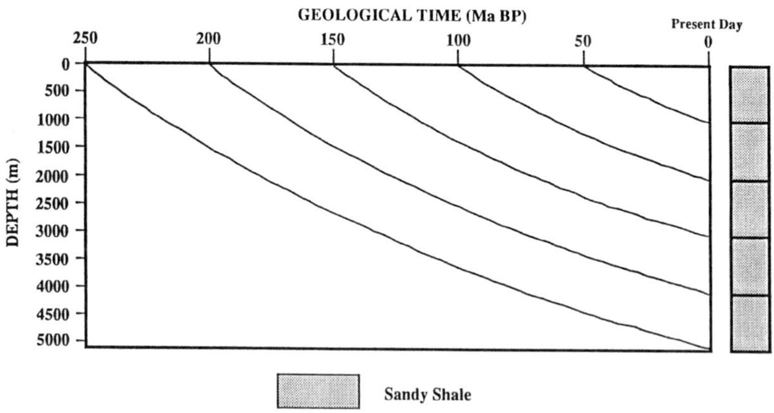


FIGURE 5.18 Burial history for synthetic test data. (Reprinted with permission from Cao and Lerche, 1994.)

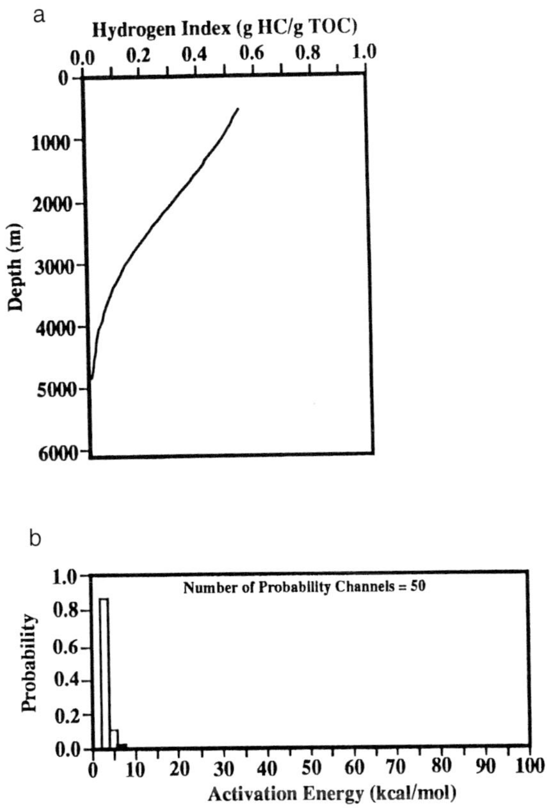


FIGURE 5.19 (a) Synthetic S2 data set 1 with (b) input kinetic parameters. (Reprinted with permission from Cao and Lerche, 1994.)

where $Q_0$ is the present-day heat flow value, BETA and $A_1$ are heat flow constants to be determined, $t$ is the geological time, and $t_{max}$ is the maximum time in the burial history. A constant heat flow of 1.5 HFU is assumed through geological time, that is, $Q_0 = 1.5$ HFU, all other heat flow parameters (BETA, $A_1$) were set to zero, Thus, the burial history and thermal history were given. Figure 5.19a shows the first set of S2 synthetic data generated from the given burial history and thermal history with the input probability distribution of activation energy as shown in Fig. 5.19b. The input values of $x_0$ and $k_0$ are set to 0.6 g HC/g TOC and 2.0 $MY^{-1}$, respectively. The range of the activation energy is from 0.0 to 100 kcal/mol with 50 channels ($M = 50$). This S2 synthetic data set is used to test the sensitivity of the S2 inversion model on $M$, $Q_0$, BETA, and $A_1$ parameters.

1. *Parameter M. M* is the number of probability channels for activation energy. In this test, the thermal history was kept at the input (default) value, the values of $M$ searched linearly, and the kinetic parameters, $x_0$, $k_0$, and $p(E)$, optimized simultaneously with a fixed

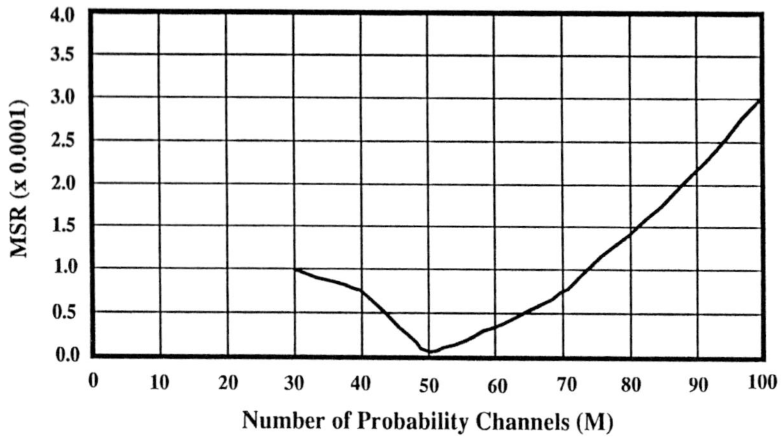


FIGURE 5.20 MSR plot for parameter $M$ (synthetic test). (Reprinted with permission from Cao and Lerche, 1994.)

range of activation energy as given in Fig. 5.19. Figure 5.20 shows the linear search results for $M$, indicating that $M$ is a sensitive parameter in S2 inversion. The best value of $M$ is 50 and the corresponding mean square residual (MSR) between calculated and synthetic S2 values is $0.794 \times 10^{-8}$. The optimized kinetic parameters are basically the same as the input forward kinetic parameters, with optimized $x_0 = 0.59976$ g HC/g TOC and $k_0 = 1.98172$ MY$^{-1}$ (Fig. 5.21).

2. *Parameter* $Q_0$. $Q_0$ is the present-day heat flux, which should be a sensitive parameter in the S2 inversion. When $Q_0$ changes, the thermal history will change. In this test, $M$ is fixed at 50, the activation energy range is unchanged, the kinetic parameters, $x_0$, $k_0$, and $p(E)$, are optimized through the S2 kinetic inversion when $Q_0$ is changed linearly. Figure 5.22 shows the MSR plot for $Q_0$, indicating that $Q_0$ is a sensitive parameter. The best value of $Q_0$ is 1.50 HFU, the same as the forward value, with an MSR value of $0.813 \times 10^{-8}$. The optimized kinetic parameters are basically the same as the forward kinetic parameters with optimized $x_0 = 0.59975$ g HC/g TOC and $k_0 = 1.98162$ MY$^{-1}$.
3. *Parameters M and BETA*. BETA is the constant in the heat flow function equation (5.67). When BETA changes, the thermal history will change. In this test, $M$ and BETA are varied linearly when the kinetic parameters are optimized, and other parameters are fixed at the forward values. Figure 5.23 shows the MSR plot in two dimensions (the best values of $M$ and BETA are 50.0 and 0.0, respectively) with an MSR value of $0.805 \times 10^{-8}$. The optimized kinetic

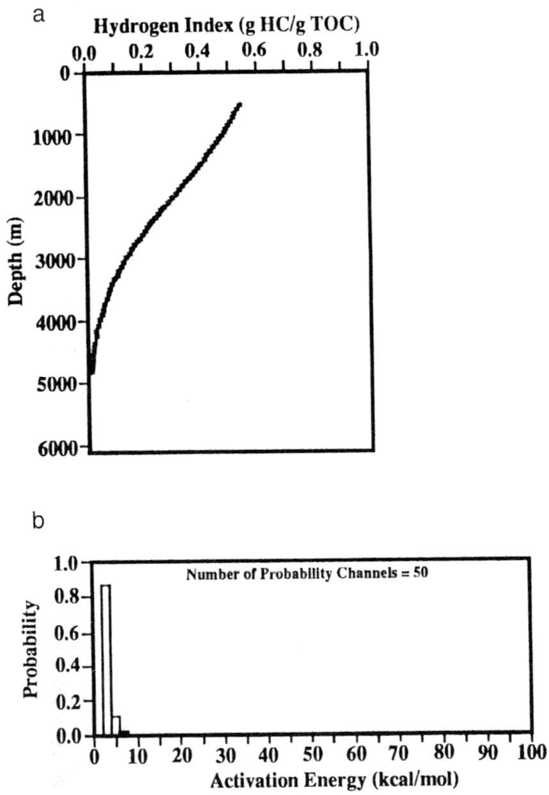


FIGURE 5.21 S2 plot showing (a) the fit of S2 data and (b) determined kinetic parameters for parameter $M$. (Reprinted with permission from Cao and Lerche, 1994.)

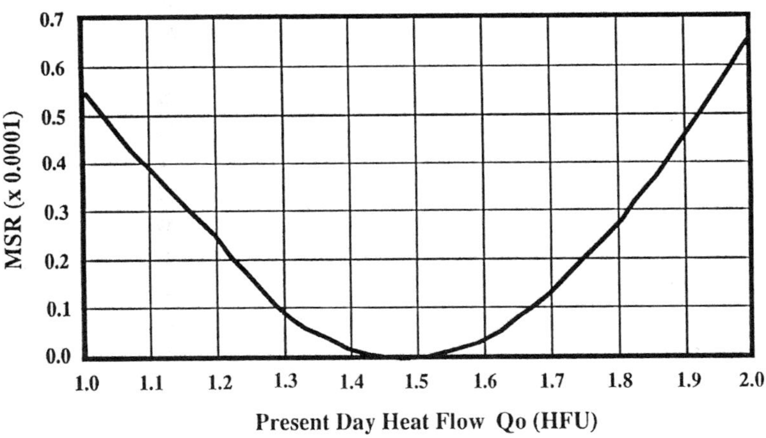


FIGURE 5.22 MSR plot for parameter $Q_0$ (synthetic test). (Reprinted with permission from Cao and Lerche, 1994.)

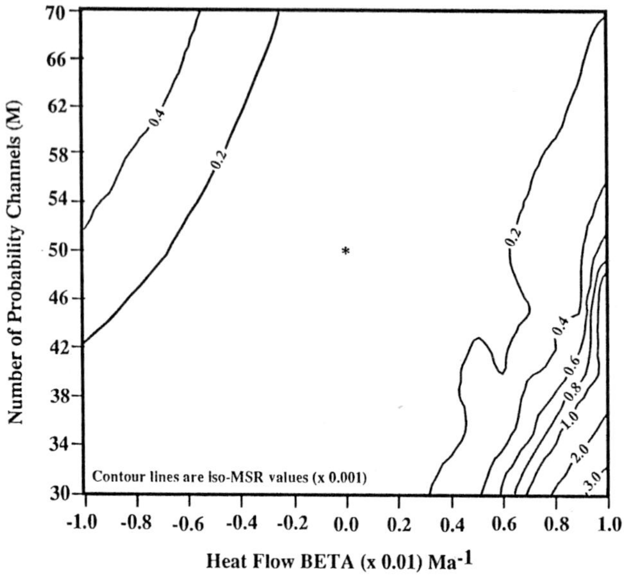


FIGURE 5.23 MSR plot for parameters $M$ and BETA (synthetic test). (Reprinted with permission from Cao and Lerche, 1994.)

parameters are basically the same as the forward ones with optimized $x_0 = 0.59975$ g HC/g TOC and $k_0 = 1.98166$ MY$^{-1}$.

4. *Parameters BETA and $A_1$*. BETA and $A_1$ are constants in Eq. (5.67) that control the heat flow pattern in geological time. In this test, BETA and $A_1$ are linearly varied when the kinetic parameters are optimized, with other parameters fixed at the forward values. Figure 5.24 shows the MSR plot in the two dimensions (BETA and $A_1$). The best values for both BETA and $A_1$ are 0.0, the same as the forward values, with a MSR value of $0.816 \times 10^{-8}$. The optimized kinetic parameters are basically the same as the forward ones with optimized $x_0 = 0.9974$ g HC/g TOC and $k_0 = 1.9816$ MY$^{-1}$.

## 2. Sensitivity Test for Activation Energy

Based on a constant heat flux history, we created a second synthetic S2 data set using the activation energy distribution given by Sweeney and Burnham (1990) with $k_0$ and $x_0$ fixed at $3.2 \times 10^{18}$ MY$^{-1}$ and 0.85 g HC/g TOC, respectively. Figure 5.25 shows the synthetic S2 data profile and activation energy distribution with $E_{min} = 30$ kcal/mol, $E_{max} = 70$ kcal/mol, and $M = 22$. Figure 5.26 shows the calculated HI variation during the burial path. We first used the linear search method to fit the synthetic S2 data given in Fig. 5.25 by varying $E_{min}$ from 0 to 40 kcal/mol and $M$ from 20 to 100 with the fixed temperature history. Figure 5.27 shows the fit

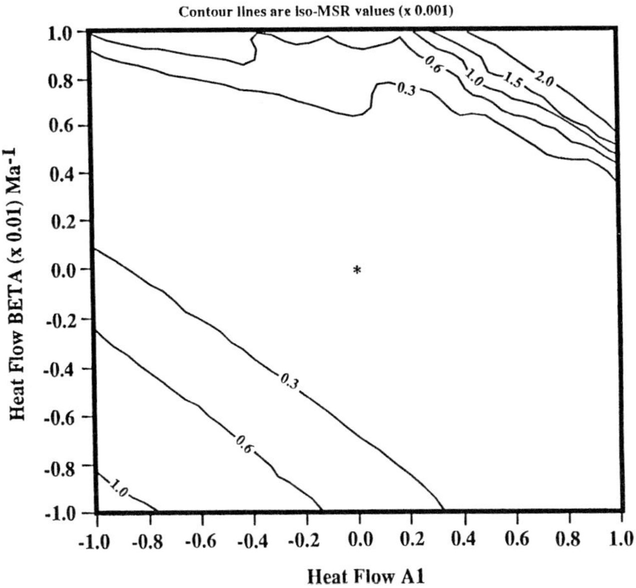


FIGURE 5.24 MSR plot for parameters BETA and $A_1$ in heat flow function (synthetic test). (Reprinted with permission from Cao and Lerche, 1994.)

between calculated (solid curve) and synthetic (points) S2 data, and the kinetic parameters determined. A good fit to the synthetic S2 data is achieved with an MSR value of $0.5405 \times 10^{-3}$. However, the activation energy distribution of Fig. 5.27b from the inverse method is different from the forward energy distribution (Fig. 5.25b), with only a single channel probability at 26 kcal/mol. The determined $E_{min}$ and $M$ are 25 kcal/mol and 90, respectively, with $k_0 = 7.6 \times 10^{12}$ $MY^{-1}$ and $x_0 = 0.8613$ g HC/g TOC. Figure 5.28 shows the calculated HI variation along the burial path, indicating a very similar pattern to that from forward modeling (Fig. 5.26). This result suggests that different activation energy distributions, that is, different kinetic parameters, can produce similar kerogen degradation results for a given temperature history. In other words, for a given set of downhole S2 data with a fixed temperature history, there can be different numerical solutions in terms of kinetic parameters.

To confirm this point, we introduce the scaling temperature, $T_0$, given in Eq. (5.29) (Liu and Lerche, 1990) and linearly searched both $T_0$ and $M$ with $E_{min}$ fixed at 20 kcal/mol. Figure 5.29 shows the fit between calculated (solid curve) and synthetic (points) S2 values, with an MSR value of $0.8135 \times 10^{-3}$—the same level of fit as the previous case (Fig. 5.27). The kinetic parameters are totally different from the previous ones. The activation energy distribution with the scaling temperature factor ranges from

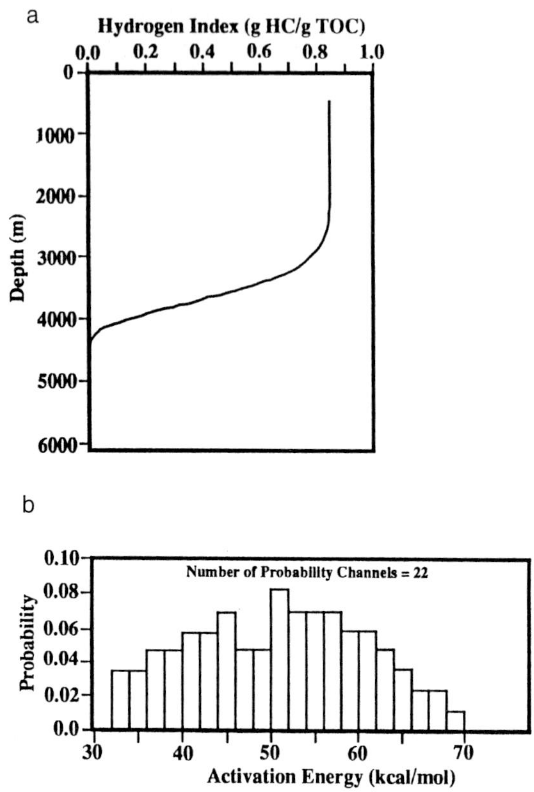


FIGURE 5.25 (a) Synthetic S2 data set 2 with (b) input kinetic parameters. (Reprinted with permission from Cao and Lerche, 1994.)

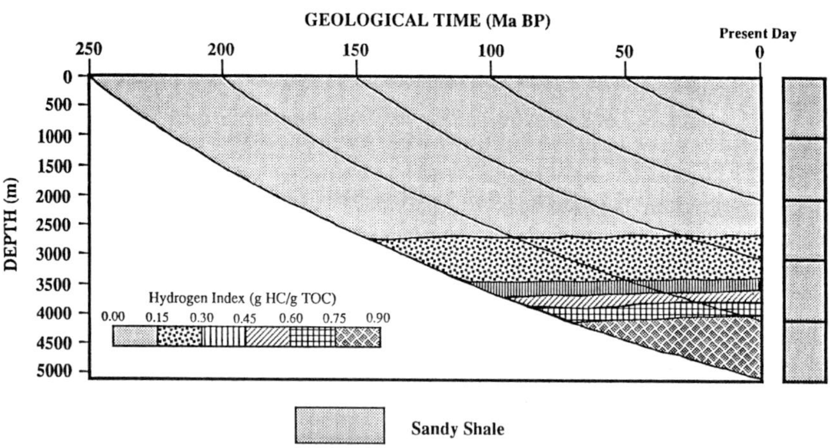


FIGURE 5.26 Calculated hydrogen index variation with burial path from forward S2 model. (Reprinted with permission from Cao and Lerche, 1994.)

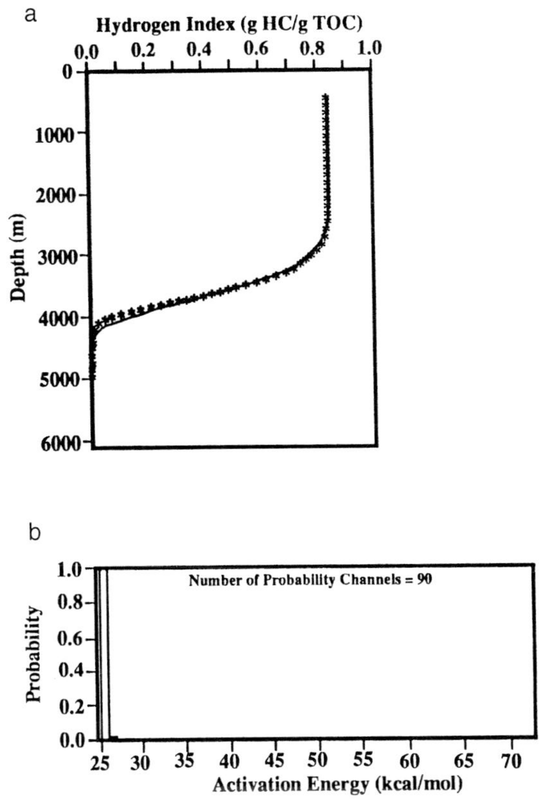


FIGURE 5.27 S2 plot showing (a) the fit of S2 data and (b) determined kinetic parameters with a search done on parameters $E_{min}$ and $M$. (Reprinted with permission from Cao and Lerche, 1994.)

22 to about 40 kcal/mol (dashed lines in Fig. 5.29b) with a high probability around 25 kcal/mol; that is $\theta(E)$ is plotted rather than $p(E)$, with $\theta(E) = p(E)\exp(-E/RT_0)$. The activation energy distribution $p(E)$ [*not* $\theta(E)$] without the scaling temperature factor (solid lines in Fig. 5.29b) ranges from 62 to 70 kcal/mol with a high probability at 68 kcal/mol. The corresponding $x_0$ and $k_0$ values are 0.8531 g HC/g TOC and 417.2960 $MY^{-1}$. Even though there is a large difference in the activation energy distributions (Figs. 5.25b, 5.27b, and 5.29b), the calculated HI variation with burial path (Fig. 5.30) has a very similar pattern to the previous two cases shown in Figs. 5.26 and 5.28, respectively. Therefore, for the same synthetic S2 data shown in Fig. 5.25a, there are at least three sets of kinetic parameters which fit the S2 data numerically equal well and *yield the same HI evolution history*.

Note that to keep $U \lesssim 1$, for both the parallel and bulk kinetic models, a very rough estimate (accurate to within a couple of orders of magnitude)

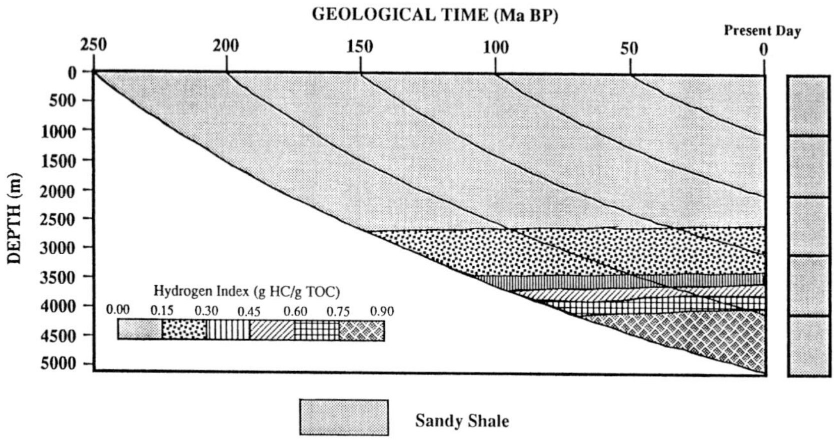


FIGURE 5.28 Calculated hydrogen index variation with burial path with the best fit parameters $E_{min}$ and $M$. (Reprinted with permission from Cao and Lerche, 1994.)

can be given of the interdependence of $k_0$, activation energy, and burial time, $t$, because

$$U = O(k_0 t \exp(-E/RT))$$

where $E$ is likely the lowest activation energy with a significant probability, $t$ the age of the oldest formation with an S2 measurement, and $T$ a rough estimate of maximum temperature. For instance, with a 100-MY old layer at about 127°C (400 K) and an activation energy of around 50 kcal/mol it follows that

$$\begin{aligned} k_0 &= O(t^{-1} \exp(E/RT)) \\ &\cong O(10^{-2} \exp(50.10^3/2 \times 400))\ \text{MY}^{-1} \\ &\cong O(10^{-2} \exp(62.5))\ \text{MY}^{-1} \\ &\cong O(10^{25})\ \text{MY}^{-1}, \end{aligned}$$

while at 40 kcal/mol, the corresponding value is $k_0 = O(10^{19})$ MY$^{-1}$. Thus the value of $k_0$ is highly sensitive to the thermal and burial histories and to the activation energy; conversely, the thermal and burial histories and the activation energy distribution are not sensitive to the value of $k_0$. Unfortunately, when all three ($k_0$, activation energy distribution, and thermal history) have to be determined from downhole data, the lion's share of data quality, quantity, and sampling frequency with depth is needed to get some degree of resolution first on $k_0$. It would seem that this reason is the dominant cause of virtually identical hydrogen index histories being achieved by different kinetic distribution models and associated parameters.

We have performed the inverse procedure on the parallel kinetic model as well as on the bulk model, using the method described in Appendix A,

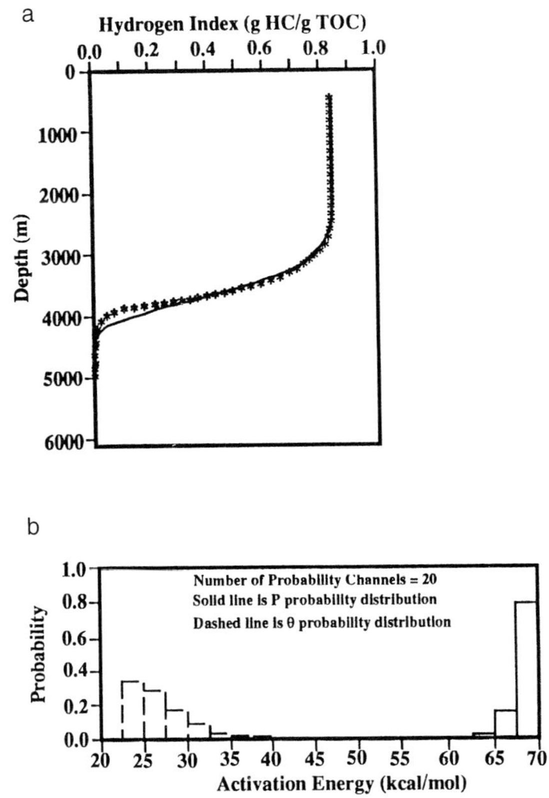


FIGURE 5.29 (a) S2 plot showing the fit of S2 data and (b) determined kinetic parameters for $M$ and $T_0$ with $E_{\min} = 20$ kcal/mol. (Reprinted with permission from Cao and Lerche, 1994.)

Eqs. (A.7) through (A.9), achieving identical consequences (of multiple kinetic activation energy distributions satisfying the HI data and producing the same HI evolution history). Either model would seem to be difficult to constrain from a kinetic perspective with a set of downhole data, but either model yields the same HI evolution history so that, from a geothermal perspective, the production of hydrocarbons is a ruggedly stable result unaffected by this lack of kinetic parameter or kinetic model resolution. In addition, as we shall now show with real downhole data, the problems of mixed kerogen types and of quality of data make these kinetic uncertainties pale in comparison to the vagaries of real situations.

## D. CASE STUDIES

The S2 inversion model was applied to borehole data sets from two different basins in order to demonstrate how to use the S2 inversion code in temperature history reconstruction.

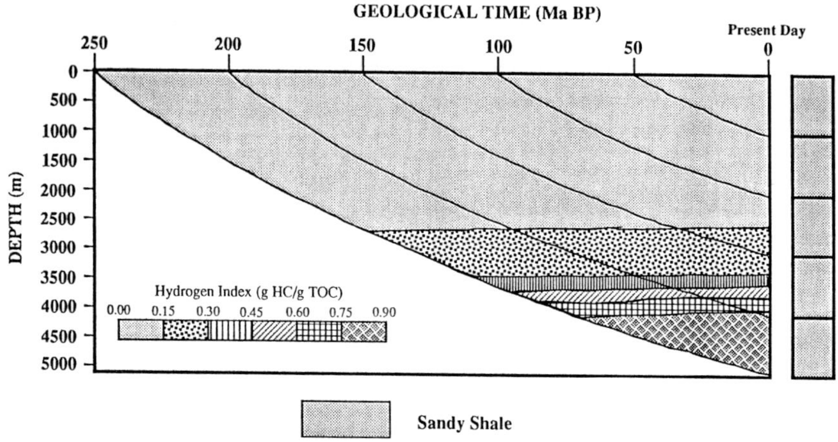


FIGURE 5.30 Calculated hydrogen index variation with burial path with best fit parameters $M$ and $T_0$ when $E_{min} = 20$ kcal/mol. (Reprinted with permission from Cao and Lerche, 1994.)

### 1. Well 30/6-5 in the Norwegian Sector, North Sea

The data for well 30/6-5 in the Norwegian sector, North Sea, are taken from Dahl and Yukler (1991). Well 30/6-5 has three kinds of thermal indicators: vitrinite reflectance ($R_0$), sterane isomerization ratios (SI), and hydrogen index (HI). The $R_0$ and SI data were sampled from the Draupne Formation (Lower Cretaceous) down to the Statfjord Formation (Lower Jurassic), and the HI data were sampled from the Draupne and Heather Formations. The burial history for the well was reconstructed as shown in Fig. 5.31. The heat flow function with time is taken in the form

$$Q(t) = Q_0 \exp[\text{BETA}^*t + A_1 \sin(\pi t/t_{max}) + A_2 \sin(2\pi t/t_{max}) + A_3 \sin(3\pi t/t_{max})] \quad (5.68)$$

where $Q(t)$ is the heat flow at time $t$, $Q_0$ is the present-day heat flow, BETA, $A_1$, $A_2$, and $A_3$ are parameters to be determined by minimizing the difference between calculated and measured thermal indicator values, and $t_{max}$ is the maximum time spanning the data (250 million years). For the thermal history reconstruction of the well, first we used a linear search method to determine (1) the present day heat flow value ($Q_0$), which is not very certain (Dahl and Yukler, 1991), and (2) the parameter BETA which controls the general trend of heat flow in the past. All three thermal indicators were used with parameters $A_1$, $A_2$, and $A_3$ fixed at zero. Figures 5.32a–c are the results from the $R_0$ data, with Fig. 5.32a showing the fit

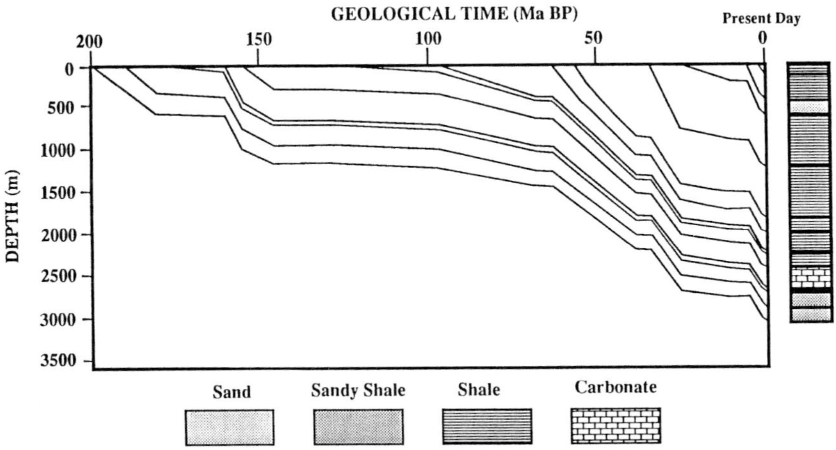


FIGURE 5.31 Burial history for well 30/6-5, North Sea. (Reprinted with permission from Cao and Lerche, 1994.)

between calculated[1] and measured $R_0$ values, Fig. 5.32b showing the determined heat flow variation with time, and Fig. 5.32c showing the reconstructed temperature history. The best values for $Q_0$ and BETA are 1.2 HFU and 0.007 $MY^{-1}$, respectively, with an MSR value of $0.1645 \times 10^{-2}$. Figures 5.33a–c are the results for the SI data, with Fig. 5.33a showing the fit between calculated[1] and measured SI values, Fig. 5.33b showing the heat flow variation with time, and Fig. 5.33c showing the temperature history. The best values for $Q_0$ and BETA are 1.2 HFU and 0.009 $MY^{-1}$, respectively, with an MSR value of $0.1545 \times 10^{-2}$. Figures 5.34a–c are the results from the HI data, with the best $Q_0$ = 1.2 HFU, BETA = 0.010 $MY^{-1}$, and MSR = $0.3269 \times 10^{-2}$. Figure 5.34a shows the HI fit and optimization kinetic parameters with $x_0$ = 0.488 g HC/g TOC and $k_0 = 1.0 \times 10^{23}$ $MY^{-1}$. The activation energy ranges from 41 to 46 kcal/mol with a probability high at 42 kcal/mol. Figures 5.34b and 5.34c show the heat flow variation with time and reconstructed temperature history, respectively. All three thermal indicators give the same $Q_0$ value of 1.2 HFU, and all three indicate a higher paleoheat flow in the past with different heating rates (BETA values are different), but with an average value of BETA from all three thermal indicators of $(8.7 \pm 1) \times 10^{-3}$ $MY^{-1}$.

Next we use the tomography method to determine all heat flow parameters ($Q_0$, BETA, $A_1$, $A_2$, $A_3$) from these three thermal indicators. Table

[1] The calculated vitrinite reflectance values at depth $z_i$ today are given by formula (4.47) of Lerche (1991) while the corresponding sterane isomerization calculated values are given by formula (16) of He and Lerche (1992).

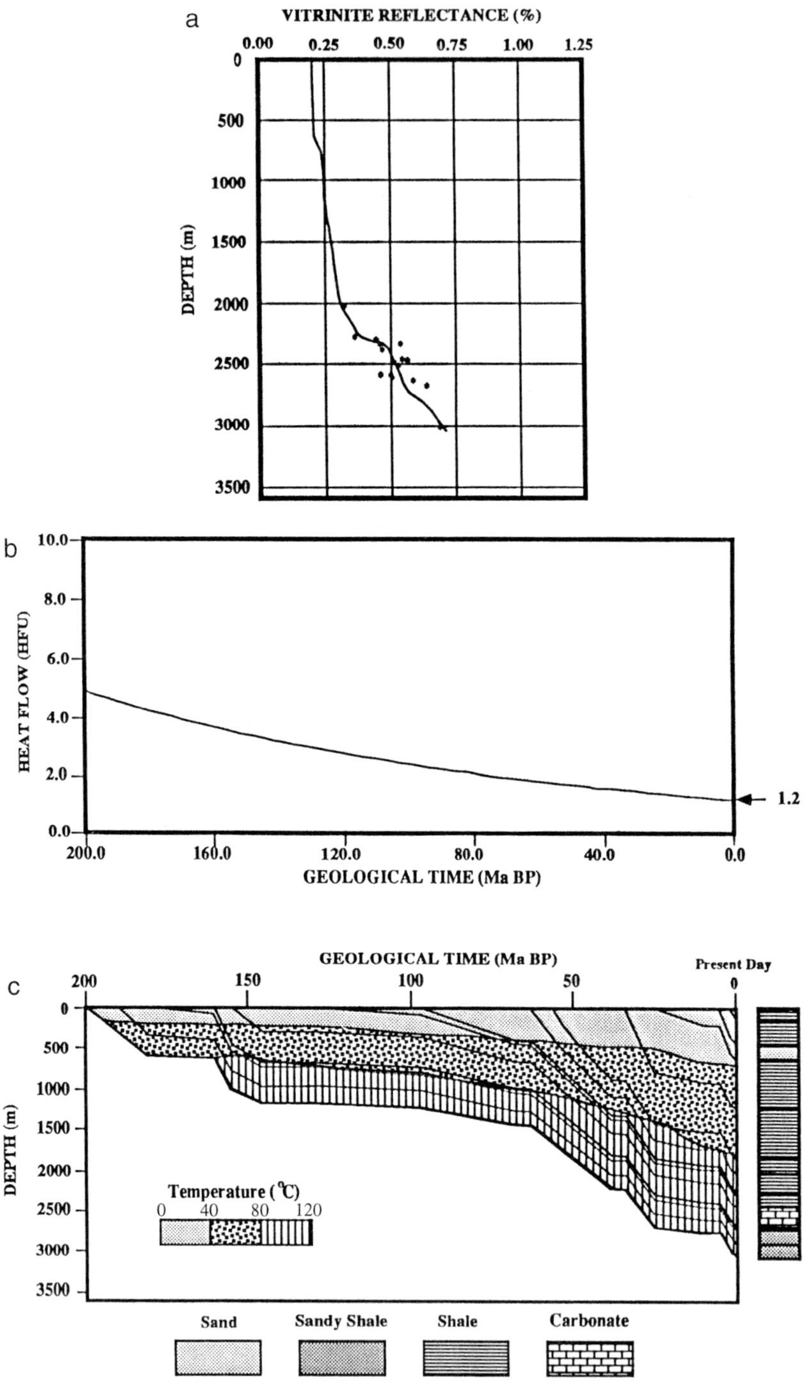
a
VITRINITE REFLECTANCE (%)
0.00 0.25 0.50 0.75 1.00 1.25
DEPTH (m)
0 500 1000 1500 2000 2500 3000 3500
b
HEAT FLOW (HFU)
10.0 8.0 6.0 4.0 2.0 0.0
1.2
200.0 160.0 120.0 80.0 40.0 0.0
GEOLOGICAL TIME (Ma BP)
c
GEOLOGICAL TIME (Ma BP)
Present Day
200 150 100 50 0
DEPTH (m)
0 500 1000 1500 2000 2500 3000 3500
Temperature (°C)
0 40 80 120
Sand
Sandy Shale
Shale
Carbonate

5.1 gives a summary of the results. Figures 5.35a and b, 5.36a and b, and 5.37a and b are the determined heat flow variation with time and corresponding temperature history for all three thermal indicators, showing a similar heat flow pattern with a relative high during the early Tertiary and a relative low value during the early Cretaceous. There is not much difference between the tomography method and previous linear search methods in terms of the fit between calculated and measured thermal indicator values.

This case study shows that S2 data can be used as a thermal indicator to reconstruct the thermal history including paleoheat flow and temperature variation with time, due concern, and allowance being made for determinations of resolution, uniqueness, sensitivity, and precision in the manner outlined.

**2. Well Alaska State D1**

Well Alaska State D1 is located offshore in the 1002 Area in the Arctic National Wildlife Refuge, Alaska (Magoon *et al.*, 1987). Figure 5.38 shows the burial history from this well based on the data from Magoon *et al.* (1987). There are two thermal indicators available from this well. Figure 5.39a shows the vitrinite reflectance ($R_0$) data points and Fig. 5.39b shows the hydrogen index profile. It is obvious from Fig. 5.39a that the $R_0$ data are poor in this well, because the $R_0$ data show little increasing trend with depth (from 500 to 3500 m) and most of the samples were contaminated or reworked. Basically, it is impossible to determine a reliable paleoheat flow from this $R_0$ data set. The HI data clearly show a mixture of at least two types of kerogen in this well. For modeling purposes, we separated the HI data into two subsets arbitrarily, although we recognize that unmixing codes, in the sense of Williams *et al.* (1988), would do a considerably better job of identifying individual end-member kerogen contributions with depth. (For the influence of a mix of different kerogen fractions on the S2 behavior, see Appendix B, Section B.III.) Subset 1 is the HI data above 2500 m and subset 2 is the HI data below 2500 m. We used both subsets independently to determine the paleoheat flow, with the present-day heat flow ($Q_0$) of 0.82 HFU calculated form the bottom hole temperature data in this well.

Figure 5.40a shows the S2 kinetic inversion results for subset 1 with $x_0 = 0.1513$ g HC/g TOC and $k_0 = 2.5 \times 10^{14}$ MY$^{-1}$. The activation energy ranges from 21 to 25 kcal/mol with a probability high at 22 kcal/mol. The fit of HI data is not good because of the poor quality of the data. Figure

FIGURE 5.32 (a) Fit of vitrinite reflectance ($R_0$) to data (linear search on BETA) for well 30/6-5, North Sea; (b) corresponding heat flow with time; (c) corresponding temperature history. (Reprinted with permission from Cao and Lerche, 1994.)

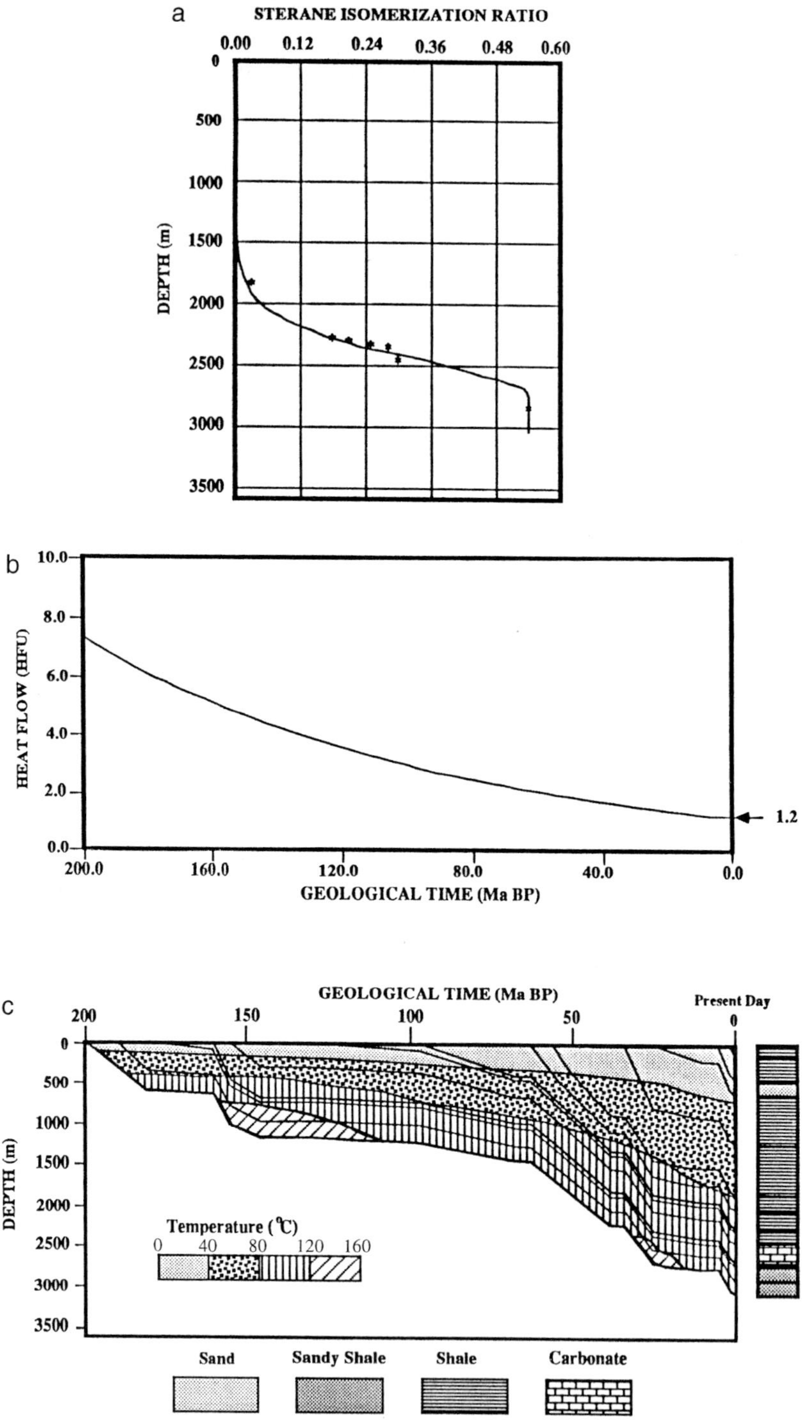
a
STERANE ISOMERIZATION RATIO
0.00 0.12 0.24 0.36 0.48 0.60
DEPTH (m)
0 500 1000 1500 2000 2500 3000 3500
b
HEAT FLOW (HFU)
10.0 8.0 6.0 4.0 2.0 0.0
1.2
200.0 160.0 120.0 80.0 40.0 0.0
GEOLOGICAL TIME (Ma BP)
c
GEOLOGICAL TIME (Ma BP)
Present Day
200 150 100 50 0
DEPTH (m)
0 500 1000 1500 2000 2500 3000 3500
Temperature (°C)
0 40 80 120 160
Sand
Sandy Shale
Shale
Carbonate

5.40b shows the corresponding paleoheat flow with time. Figure 5.41a shows the S2 kinetic results for subset 2 with $x_0$ = 0.2678 g HC/g TOC and $k_0$ = $1.8 \times 10^{12}$ $MY^{-1}$. The activation energy distribution is not resolvable with this poor data set except to note that one channel at 21 kcal/mol is acceptable. Figure 5.41b shows the corresponding paleoheat flow with time.

Table 5.2 gives a summary of determined heat flow parameters from both subsets. The two paleoheat flow patterns determined from both HI subsets are similar, showing a relatively lower heat flow in the past, even though the HI data are of poor quality. Yu and Lerche (1992) have done a basin modeling study in the area. Based on $R_0$ data from five wells, their study indicates a similar pattern trend of decreasing paleoheat flux in the past, but with a relatively high heat flow (1.5–2.0 HFU) at about 20–25 million years ago. The paleoheat flow determined for HI data in this study (Figure 5.41b) did not show this relative high value at about 20–25 million years ago, due possibly to the poor quality of the HI data.

## E. SUMMARY

The computer model to inverse S2 pyrolysis data from Rock–Eval as a thermal indicator was developed and tested against both synthetic data and measured downhole data from three wells. The synthetic tests show that the S2 inverse model is sensitive to both kinetic parameters and to heat flow parameters. From the heat flow parameters, we can determine the variation of heat flow with time and reconstruct the temperature history. From the kinetic parameters, we can simulate hydrocarbon generation from kerogen degradation. Thus, S2 inversion models make S2 pyrolysis data more useful in petroleum basin analysis. Application of the S2 inversion model to the two downhole S2 data sets confirmed the applicability and usefulness of the S2 data as a thermal indicator in determination of paleoheat flow variations and reconstruction of temperature history.

In this study, we also find that for a given set of downhole S2 data, there can be different sets of kinetic parameters which give the same level of fit between calculated and measured S2 values. In other words, there are no unique numerical solutions of kinetic parameters for a given set of downhole S2 data. Therefore, different kinetic models can fit a given S2 data set equally well with different kinetic parameters. The modeling results of the S2 inversion are sensitive to the quality of downhole S2 data. The S2 inversion model assumes degradation of a single type of kerogen. Any S2 data set with multiple-type kerogen can produce unrealistic results.

FIGURE 5.33 (a) Fit of sterane isomerization ratio (SI) data (linear search on BETA) for well 30/6-5, North Sea; (b) corresponding heat flow with time; (c) corresponding temperature history. (Reprinted with permission from Cao and Lerche, 1994.)

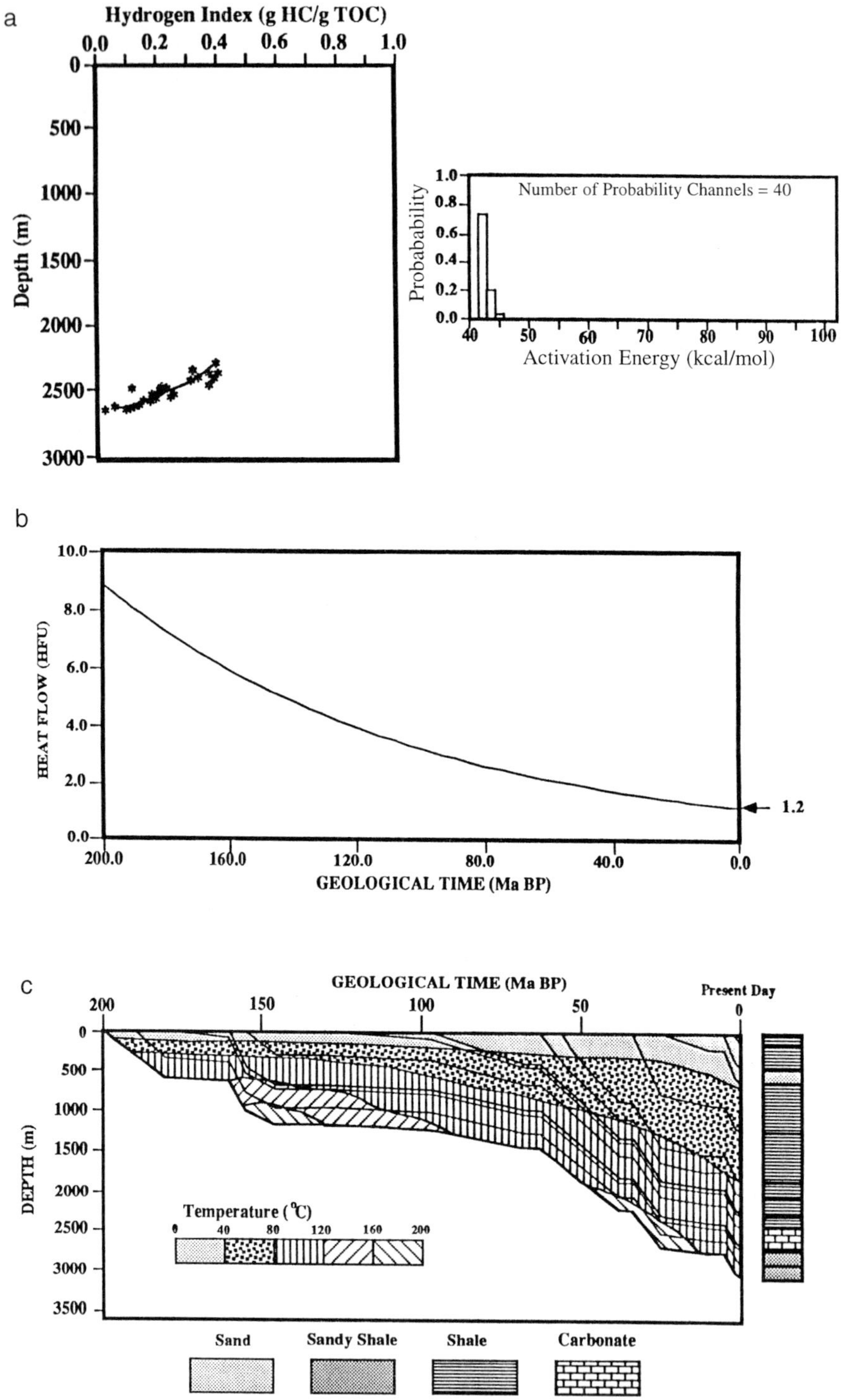
a
Hydrogen Index (g HC/g TOC)
0.0 0.2 0.4 0.6 0.8 1.0
Depth (m)
0 500 1000 1500 2000 2500 3000
Number of Probability Channels = 40
Probabability
1.0 0.8 0.6 0.4 0.2 0.0
40 50 60 70 80 90 100
Activation Energy (kcal/mol)
b
HEAT FLOW (HFU)
10.0 8.0 6.0 4.0 2.0 0.0
200.0 160.0 120.0 80.0 40.0 0.0
GEOLOGICAL TIME (Ma BP)
1.2
c
GEOLOGICAL TIME (Ma BP)
Present Day
200 150 100 50 0
DEPTH (m)
0 500 1000 1500 2000 2500 3000 3500
Temperature (°C)
0 40 80 120 160 200
Sand
Sandy Shale
Shale
Carbonate

TABLE 5.1 Determined Heat Flow Parameters from Three Thermal Indicators for Well 30/6-5, North Sea

| Thermal indicators | $Q_0$ (HFU) | BETA (1/MY) | $A_1$ | $A_2$ | $A_3$ | MSR ($\times$ 0.01) |
|---|---|---|---|---|---|---|
| | Parameters | | | | | |
| Vitrinite reflectance | 1.203 | 0.007 | −0.180 | 0.975 | 0.000 | 0.159 |
| Sterane isomerization | 1.231 | 0.005 | 0.448 | 0.381 | 0.264 | 0.645 |
| Hydrogen index | 1.224 | 0.003 | 0.300 | 0.333 | 0.211 | 0.317 |

Therefore, geochemical studies on kerogen characterization and classification and application of unmixing procedures are necessary before the S2 inversion model is used.

*Note Added February 15, 1996.* Recently C. C. Lakshmanan and N. White (A New Distributed Activation Energy Model Using Weibull Distribution for the Representation of Complex Kinetics, *Energy and Fuels* **8,** 1158–1167, 1994) have made the points that (1) a Weibull distribution, containing only three parameters, does a more than adequate job of representing the kinetics of complex thermal transformations; one does not need massive numbers of parameters; (2) there is multiplicity—a number of sets of kinetic parameters describe the observed data within the limits of experimental accuracy; and the parameter values vary substantively from set to set; and (3) both isothermal and nonisothermal data can be described by the Weibull distribution, and both have multiplicity problems.

## APPENDIX A: INVERSE PROCEDURE FOR EQUATION (5.45)

The general form of Eq. (5.45) can be rewritten in a discrete representation as

$$x_j = x_0 \sum_{i=1}^{M} P_i \exp(-k_0 F_{ji}), \tag{A.1}$$

FIGURE 5.34 (a) S2 plot showing the fit of S2 data and determined kinetic parameters for well 30/6-5, North Sea; (b) corresponding heat flow with time; (c) corresponding temperature history. (Reprinted with permission from Cao and Lerche, 1994.)

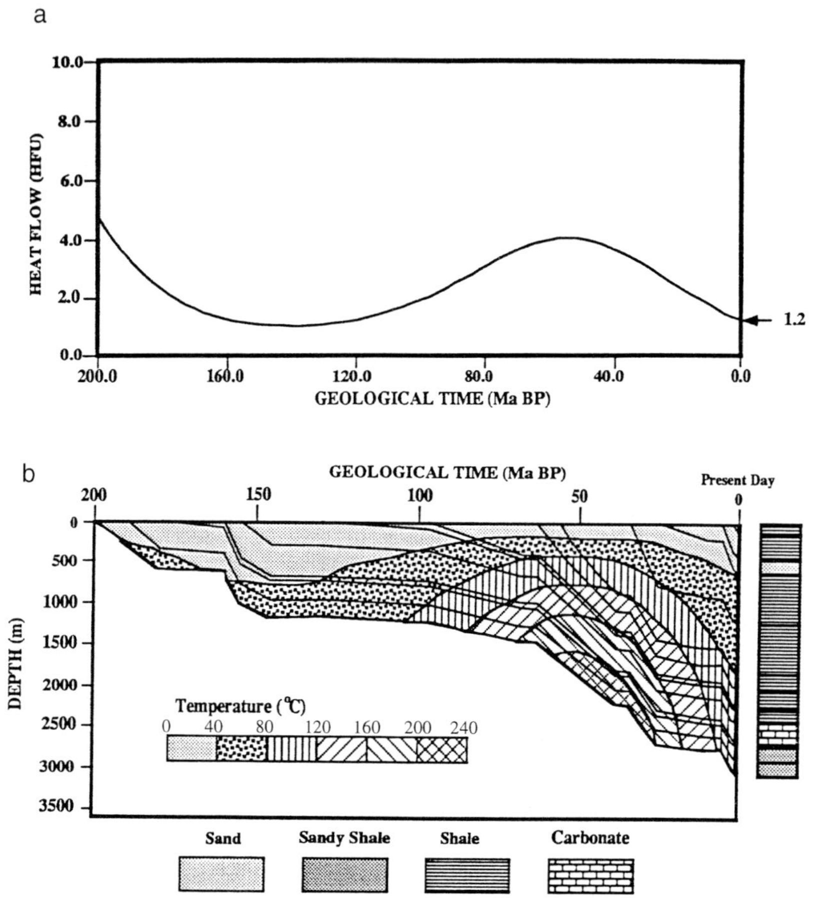


FIGURE 5.35 (a) Calculated heat flow with time from $R_0$ data (tomography search on all parameters) for well 30/6-5, North Sea; (b) corresponding temperature history. (Reprinted with permission from Cao and Lerche, 1994.)

with $F_{ji} = \int_0^{t_j} \exp(-E_i/RT_j(t'))\, dt'$, $P_i = p_i(E_i)\, \Delta E$, and where $M$ is the number of probability channels for the activation energy. Note that $F_{ji} \geq 0$ and that $\sum_{i=1}^{M} P_i = 1$.

In this case, because the factors $P_i$ enter Eq. (A.1) linearly, a direct least-squares solution for $P_i$ is available immediately for given values of $x_0$ and $k_0$ from the linear set of equations

$$x_0 \sum_{r=1}^{M} P_r \sum_{j=1}^{N} M_{jr}(k_0) M_{ji}(k_0) = Z_i(k_0), \tag{A.2}$$

where

$$M_{ji}(k_0) = \exp(-2k_0 F_{ji}) \geq 0 \tag{A.3a}$$

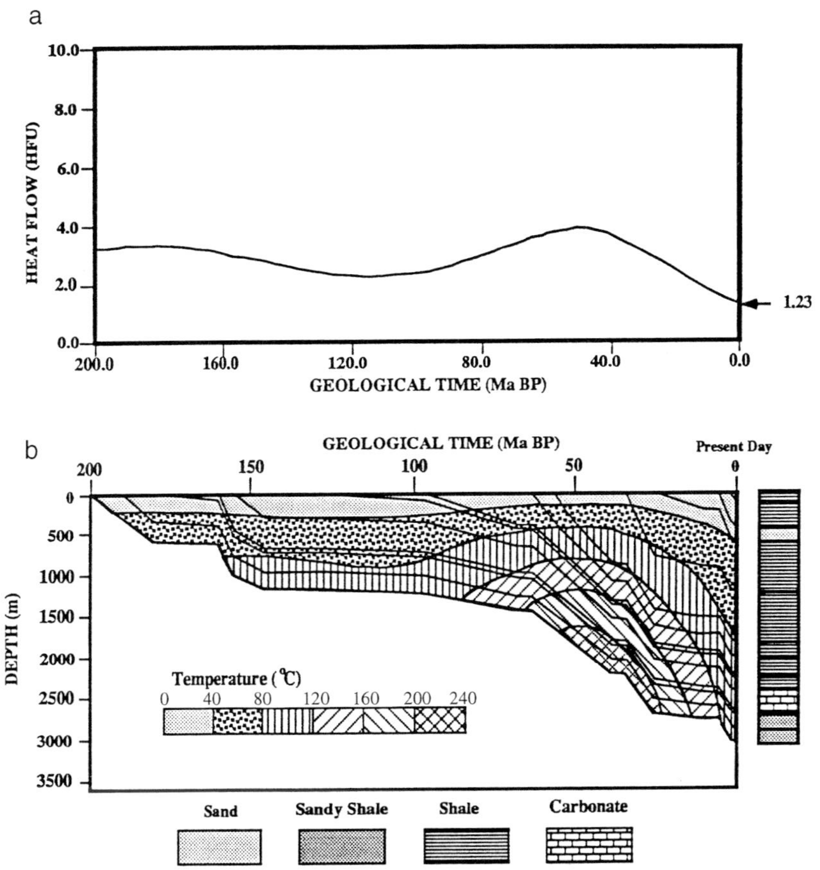


FIGURE 5.36 (a) Calculated heat flow with time from SI data (tomography search on all parameters) for well 30/6-5, North Sea; (b) corresponding temperature history. (Reprinted with permission from Cao and Lerche, 1994.)

and

$$Z_i(k_0) = \sum_{j=1}^{N} x_j \exp(-k_0 F_{ji}) \geq 0. \tag{A.3b}$$

The matrix equation, Eq. (A.2) can be solved by standard least-squares techniques to yield $P_i$.

With this set of $P_i$ (for given $x_0$ and $k_0$) the requirement is to fit the S2 data better, that is,

$$X^2 = \frac{1}{N} \sum_{j=1}^{N} \left[ x_j - x_0 \sum_{i=1}^{M} P_i \exp(-k_0 F_{ji}) \right]^2 \tag{A.4}$$

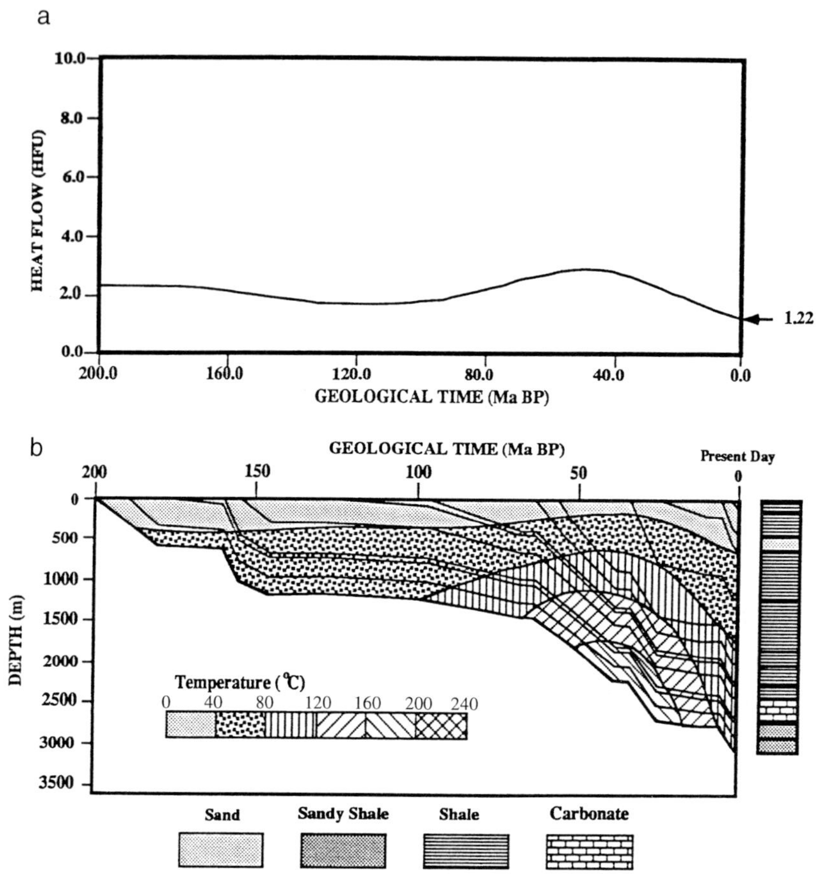


FIGURE 5.37 (a) Calculated heat flow with time from S2 data (tomography search on all parameters) for well 30/6-5, North Sea; (b) corresponding temperature history. (Reprinted with permission from Cao and Lerche, 1994.)

should be a minimum. For a given value of $k_0$ the minimum occurs when

$$x_0 = \frac{\sum_{j=1}^{N} x_j \left( \sum_{i=1}^{M} P_i \exp(-k_0 F_{ji}) \right)}{\sum_{j=1}^{N} \left( \sum_{i=1}^{M} P_i \exp(-k_0 F_{ji}) \right)^2}. \tag{A.5}$$

It also follows from Eq. (A.4) that a least-squares solution for $k_0$ is then given through

$$\begin{aligned} 0 = & \sum_{j=1}^{N} \sum_{i=1}^{M} x_j P_i F_{ji} \exp(-k_0 F_{ji}) \\ & - x_0 \sum_{j=1}^{N} \left( \sum_{i=1}^{M} P_i \exp(-k_0 F_{ji}) \right) \times \left( \sum_{i=1}^{M} P_i F_{ji} \exp(-k_0 F_{ji}) \right). \end{aligned} \tag{A.6}$$

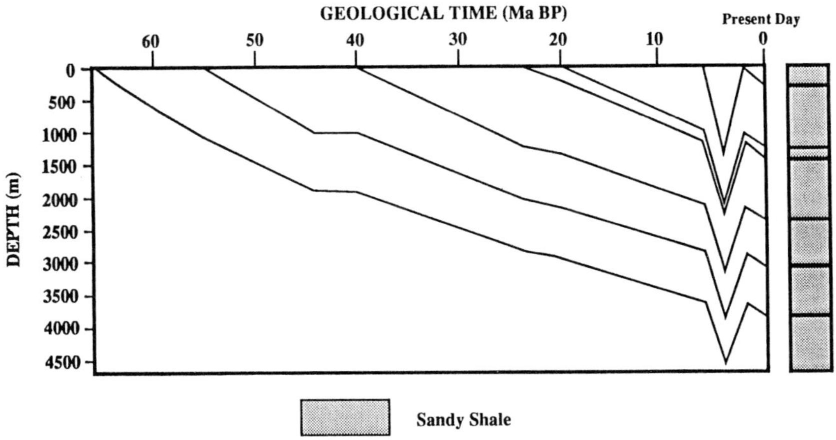


FIGURE 5.38 Burial history for well Alaska State D1, Alaska. (Reprinted with permission from Cao and Lerche, 1994.)

The parallel channel kinetic optimization procedure is then summarized as follows:

1. Choose initial $P_i \equiv$ constant $= 1/M$, where $M$ is the number of probability channels, and $P_i$ is normalized over the $M$ channels so that $\sum_{i=1}^{M} P_i = 1$.

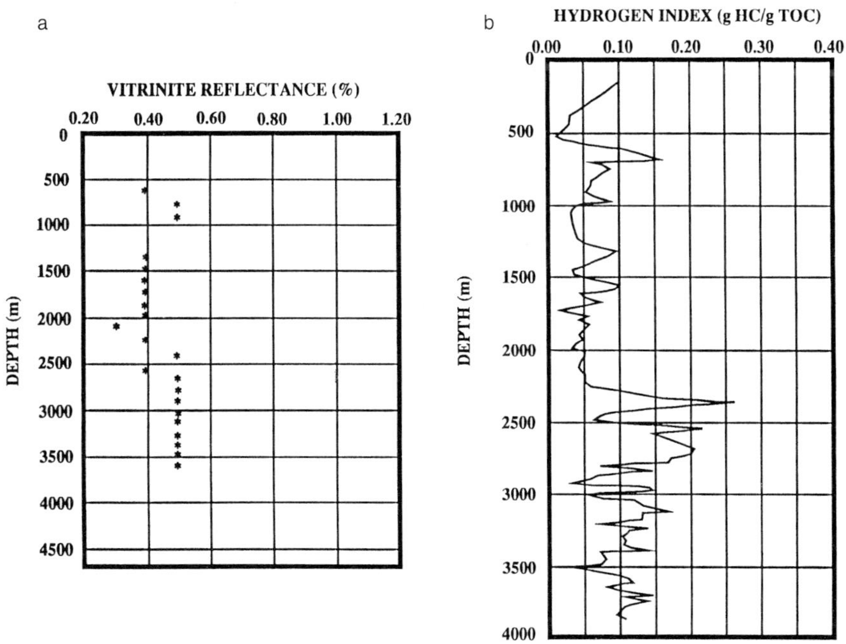


FIGURE 5.39 (a) Profile of measured $R_0$ data (points) for well Alaska State D1, Alaska; (b) profile of measured S2. (Reprinted with permission from Cao and Lerche, 1994.)

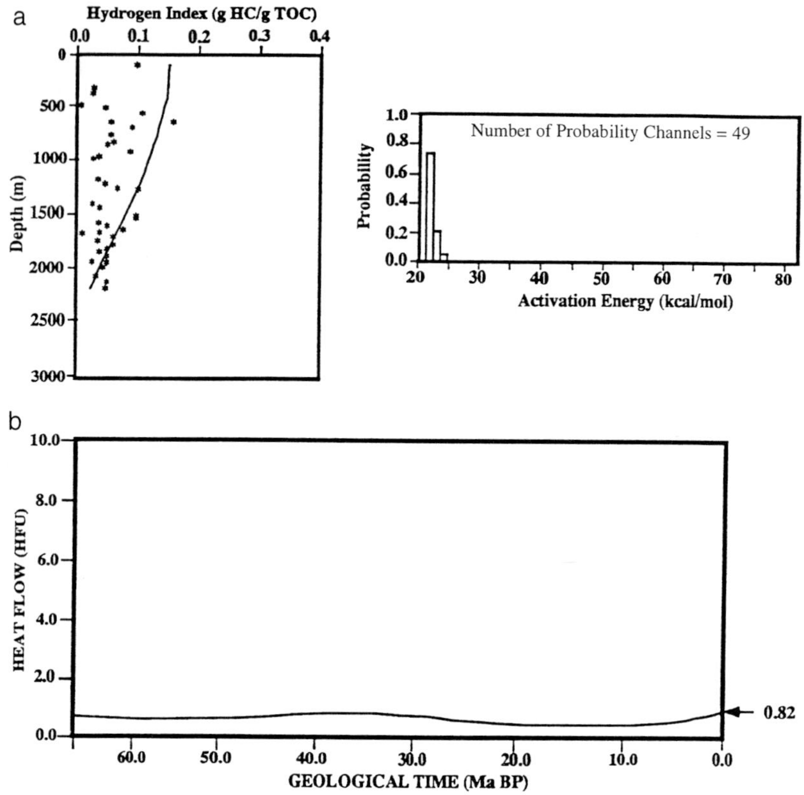


FIGURE 5.40 (a) S2 plot showing the fit of S2 data set 1 and determined kinetic parameters for well Alaska State D1, Alaska; (b) corresponding heat flow with time. (Reprinted with permission from Cao and Lerche, 1994.)

2. Calculate $F_{ji}$ from the temperature history obtained during the thermal inversion.
3. Use Eqs. (A.5) and (A.6) to obtain estimates of $x_0$ and $k_0$.
4. Use Eq. (A.2) to obtain updated values for $P_i$. Then recalculate $X^2$ from Eq. (A.4).
5. Repeat steps 3 and 4 until the $X^2$ values converge to a minimum.
6. Then update the thermal history using parameters in the paleoheat flux function of the DTIT model and repeat steps 1 through 5. Retain the minimum of all minima of $X^2$ as the best value.

An alternative procedure is available. Rewrite Eq. (A.1) in the form

$$x_j = x_0 \left\{ \exp\left[ -k_0 \sum_i P_i F_{ji} \right] + \sum_i P_i \exp(-k_0 F_{ji}) - \exp\left[ -k_0 \sum_i P_i F_{ji} \right] \right\} \tag{A.7}$$

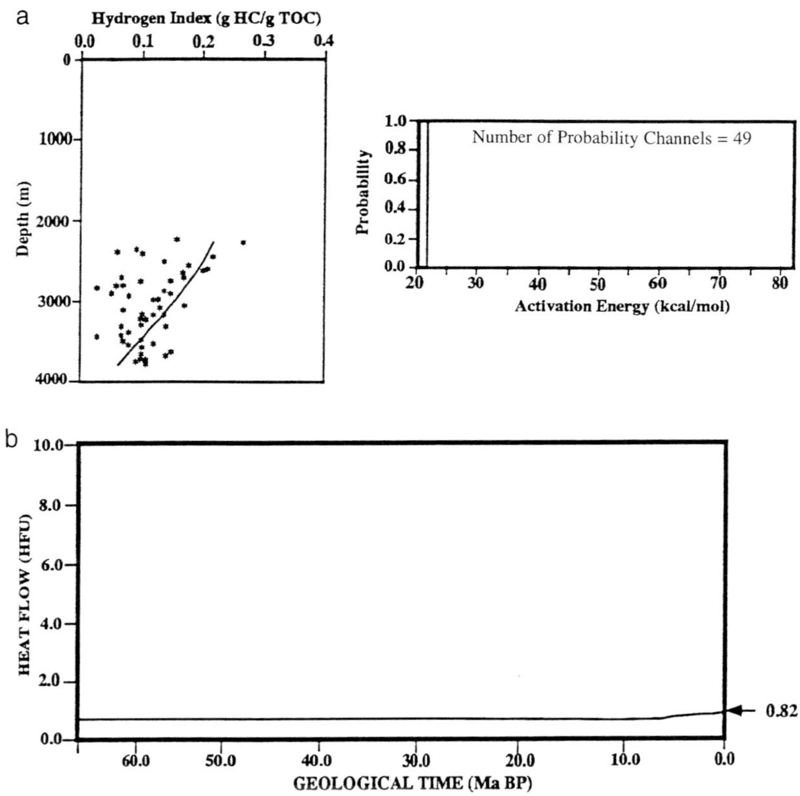


FIGURE 5.41 (a) S2 plot showing the fit of S2 data set 2 and determined kinetic parameters for well Alaska State D1, Alaska; (b) corresponding heat flow with time. (Reprinted with permission from Cao and Lerche, 1994.)

and then rearrange terms to read

$$q_j = x_0 \exp\left[-k_0 \sum_i P_i F_{ji}\right], \tag{A.8}$$

where

$$q_j = x_j - x_0\left[\sum_i P_i \exp(-k_0 F_{ji}) - \exp\left(-k_0 \sum_i P_i F_{ji}\right)\right], \tag{A.9}$$

with the difference $(q_j - x_j)/x_0$ known to be small from the body of the chapter.

The argument for constructing the inverse procedure for the parallel kinetic then operates *precisely* as for the bulk kinetic model with the replace-

TABLE 5.2 Determined Heat Flow Parameters from the S2 Data for Well Alaska State D1

| | Parameters | | | | |
|---|---|---|---|---|---|
| HI data sets | BETA (1/MY) | $A_1$ | $A_2$ | $A_3$ | MSR (× 0.01) |
| Data set 1 | −0.0051 | −0.388 | −0.368 | −0.354 | 0.308 |
| Data set 2 | −0.0049 | −0.229 | −0.139 | −0.112 | 0.363 |

ment throughout of $x_j$ by $q_j$. At each iteration $q_j$ is updated by replacing $x_0$, $k_0$, and $P_i$ with their newly updated values and proceeding through the iteration steps once more.

# APPENDIX B: MATHEMATICAL CONSIDERATIONS

## B.I. APPROXIMATE EVALUATION OF AN INTEGRAL

The integral that occurs throughout both the bulk and parallel kinetic models is

$$U = k_0 I \tag{B.1}$$

where

$$I = \int_0^t \exp(-E/RT)\, dt. \tag{B.2}$$

Consider the evaluation of the integral in Eq. (B.1) starting at time $t = 0$ at temperature $T_0$ and with temperature $T$ at time $t$ under uniform heating rate $dT/dt = H =$ constant. Then write

$$\frac{HU}{k_0} = \int_{T_0}^{T} \exp(-E/RT)\, dT. \tag{B.3}$$

Set $w = (E/R)1/T$ when

$$\frac{HRU}{Ek_0} = \int_{\varepsilon}^{\varepsilon_0} \frac{dw}{w^2} \exp(-w), \tag{B.4}$$

where $\varepsilon = E/RT$, $\varepsilon_0 = E/RT_0$. Change variables in Eq. (B.4) with $w = \varepsilon + \varepsilon x$ to obtain

$$\frac{\varepsilon HRUe^{\varepsilon}}{Ek_0} = \int_0^{\mu} e^{-\varepsilon x} \frac{dx}{(1+x)^2} \equiv J, \tag{B.5}$$

with $\mu = (T - T_0)/T_0 > 0$. Integration by parts in Eq. (B.5) yields exactly

$$J = 1 - \frac{e^{-\varepsilon\mu}}{(1+\mu)} - \varepsilon \ln(1+\mu)\, e^{-\varepsilon\mu} - \varepsilon^2 \int_0^{\mu} e^{-\varepsilon x} \ln(1+x)\, dx. \tag{B.6}$$

Then

$$\begin{aligned} J = e^{-\varepsilon\mu} &\left[\frac{\mu}{(1+\mu)} - \varepsilon \ln(1+\mu) + \varepsilon\mu\right] \\ &- \varepsilon^2 \int_0^{\mu} [\ln(1+x) - x]\, e^{-\varepsilon x}\, dx. \end{aligned} \tag{B.7}$$

For $\varepsilon \gg 1$ ($E/RT \gg 1$) the integrand in Eq. (B.7) drops rapidly to zero over a scale of $x \leq 1/\varepsilon \ll 1$. Then an accurate enough expansion of the last term for $\mu \leq 0(1)$ can be achieved by writing

$$[\ln(1+x) - x] \cong -\frac{x^2}{2}$$

because $|\ln(1 + x) - x| \leq (\mu + \ln(1 + \mu)) = 0(1)$. Then

$$J \cong e^{-\varepsilon\mu}\left[\frac{\mu}{1+\mu} - \varepsilon \ln(1+\mu) + \varepsilon\mu\right] + \frac{\varepsilon^2}{2}\int_0^{\mu} x^2 e^{-\varepsilon x}\, dx, \tag{B.8}$$

which can be integrated exactly in closed form.

Reexpressing factors in terms of $E$, $R$, $T$, and $T_0$ we then have

$$\begin{aligned} U \cong \frac{k_0 E}{RH} &[\exp(-E/RT_0) \\ &\times \left\{\left(\frac{T-T_0}{T_0}\right)\left[1 + \left(\frac{RT}{E}\right)\left(\frac{T}{T_0}\right) + \frac{1}{2}\left(\frac{T}{T_0}\right)\right]\right. \\ &\left. + \frac{E}{RT}\left[\left(\frac{T-T_0}{T_0}\right) - \ln T/T_0\right]\right\} \\ &+ (RT/E)^2\{\exp(-E/RT) - \exp(-E/RT_0)\}] \end{aligned} \tag{B.9}$$

and expression (B.9) is a uniformly accurate approximation as long as

(a) $T_0 \leq T \leq 2T_0$ (K),

(b) $E/RT \gg 1$.

Note that for $T = T_0 + Ht$, in the limits as $H \to 0$, the approximate form of expression (B.9) reduces exactly to

$$U = k_0 \frac{(T - T_0)}{H} \exp(-E/RT_0) \tag{B.10}$$

as it should. Note also that on $T = T_0$ we have $U = 0$ as it should.

For $E \gg RT$, and $E \gg RT_0$, and $(E/RT)/[(T - T_0)/T_0] \gg 1$ a further approximation to expression (B.9) is appropriate, namely,

$$U \approx \frac{k_0 E}{RH} \left[ \exp(-E/RT_0) \frac{E}{RT} \left[ \left( \frac{T - T_0}{T_0} \right) - \ln T/T_0 \right] \right. \\ \left. + (RT/E)^2 \{ \exp(-E/RT) - \exp(-E/RT_0) \} \right]. \tag{B.11}$$

Equation (B.11) can be rewritten

$$U \approx \frac{k_0 RT^2}{EH} \left\{ \left( \frac{E}{RT} \right)^3 \exp(-E/RT_0) \left[ \frac{T - T_0}{T_0} \right. \right. \\ \left. \left. - \ln T/T_0 \right] + \exp(E/RT) - \exp(-E/RT_0) \right\}. \tag{B.12}$$

Now $E/RT \gg 1$ so that provided $(T - T_0)/T_0$ is larger than zero by the very small approximate amount

$$\frac{T - T_0}{T_0} \geq (RT_0/E)^{3/2},$$

one can further approximate expression (B.12) by

$$U \approx \frac{k_0 RT^2}{EH} \left[ \theta \left( \frac{E}{RT} \right)^3 \exp(-E/RT_0) + \exp(-E/RT) \right], \tag{B.13}$$

where $\theta \equiv (T - T_0/T_0) - \ln T/T_0$ is then a number of order unity. For instance on $T = 2T_0$ ($\mu \equiv 1$), $\theta = 1 - \ln 2 \approx 0.3$.

One approximation in the past for $U$ has consisted of ignoring the term factored by $\theta$ in Eq. (B.13). Such an approximation is valid when

$$\exp\left[ -\frac{E}{R} \left( \frac{1}{T} - \frac{1}{T_0} \right) \right] \gg \theta (E/RT)^3;$$

that is, when

$$T - T_0 \geq \frac{RTT_0}{E} \ln[\theta (E/RT)^3], \tag{B.14}$$

where $T$ and $T_0$ are in degrees Kelvin.

For instance, for the case of $E \approx 50$ kcal/mol, $R = 2$ cal/mol and $T_0 =$ 573 K (300°C), inequality (B.14) requires

$$T \gtrsim T_0 \left[ 1 - \frac{RT_0}{E} \ln(\theta(E/RT)^3) \right]^{-1}. \tag{B.15}$$

Now $RT_0/E \cong 2.3 \times 10^{-2}$ so that $\ln(E/RT) \cong 2 \ln_e 10 - \ln 2.3 \approx 3.8$. Hence $RT_0/E \ln(\theta(E/RT)^3) \cong 2.3 \times 10^{-2} [11.4 + \ln \theta]$. With $\theta \approx 0.1$, $\ln \theta \cong -2.3$, then $RT_0/E \ln(\theta(E/RT)^3) \approx 0.29$. Hence, requirement (B.15) is

$$T \geq T_0/(1 - 0.29) \cong \frac{T_0}{0.71} = \frac{573 \text{ K}}{0.71},$$

that is, $T \geq 800 \text{ K} \equiv 527°\text{C}$.

For any given temperature range in a constant heating rate set of experiments approximation (B.9) is generally in force for the $U$ integral, not that approximation which includes only the terms factored by the form $T^2 \exp(-E/RT)$ in Eq. (B.9); one has to show for each set of experiments that neglect of any set of terms in Eq. (B.9) is appropriate.

## B.II. A SET OF EXACT SOLUTIONS

Suppose the rate of increase of temperature with time is described by

$$dT/dt = H(T/T_0)^2, \tag{B.16}$$

with $T = T_0$ on $t = 0$ and $T = T_m$ on $t = t_m$. Then the integral of Eq. (B.2) is given analytically *exactly* by

$$I = \frac{RT_0^2}{EH}(\exp(-E/RT) - \exp(-E/RT_0)); \qquad \text{for } T_m \geq T \geq T_0, \tag{B.17}$$

whereas the temperature with time behavior is given *exactly* by

$$T = T_0/(1 - Ht/T_0) \tag{B.18a}$$

and approximated by

$$T \cong T_0 + Ht(1 + (Ht)/T_0) + \ldots. \tag{B.18b}$$

If the temperature varies with time through (1) a heating cycle (from $T = T_0$ on $t = 0$ to $T = T_m$ on $t = t_m$), then (2) a cooling cycle (from $T = T_m$ on $t = t_m$ to $T = T_*$ on $t = t_m + t_*$), and then (3) a heating cycle again (from $T = T_*$ on $t = t_m + t_*$ to $T = T$ on $t > t_* + t_m$) with the heating (cooling) rates given by

$$dT/dt = H_a(T/T_0)^2 \text{ (first, heating cycle)}, \tag{B.19a}$$

$$dT/dt = -C(T/T_m)^2 \text{ (second, cooling cycle)}, \tag{B.19b}$$

$$dT/dt = H_c(T/T_*)^2 \text{ (third, heating cycle)}, \tag{B.19c}$$

then the integral of Eq. (A.1) at time $t > t_m + t_*$ is given analytically exactly by

$$I = I_a + I_b + I_c, \tag{B.20}$$

where

$$I_a = \frac{RT_0^2}{EH_a}(\exp(-E/RT_m) - \exp(-E/RT_0)); \tag{B.21a}$$

$$I_b = \frac{RT_m^2}{EC}(\exp(-E/RT_m) - \exp(-E/RT_*)); \tag{B.21b}$$

and

$$I_c = \frac{RT_*^2}{EH_c}(\exp(-E/RT) - \exp(-E/RT_*)). \tag{B.21c}$$

## B.III. MAXIMUM TEMPERATURE PEAK FOR A MIX OF TWO KEROGENS

Consider a mix of two kerogens (A and B) with a total amount $x_0$ of pyrolyzable material at time $t = 0$ when the initial temperature is $T = T_0$. Let the fractional weight of type A be $\alpha x_0$, so that the fractional weight of type B is $(1 - \alpha)x_0$.

Let each kerogen have its own frequency factor $k_0$, probability distribution $p$, and activation energies $E$, so that for the bulk kinetic model one can write, in an obvious notation,

$$x_A(t) = \alpha x_0 \exp\left[-\int p_A(E)U_A(E,t)\,dE\right], \tag{B.22a}$$

$$x_B(t) = (1-\alpha)x_0 \exp\left[-\int p_B(E)U_B(E,t)\,dE\right], \tag{B.22b}$$

with correspondingly similar expressions for the parallel channel kinetic model. The total amount of pyrolyzable material at time $t$ is then

$$x(t) = x_0\left\{\left(\alpha \exp\left[-\int p_A(E)U_A(E,t)\,dE\right]\right.\right.$$
$$\left.\left. + (1-\alpha)\exp\left[-\int p_B(E)U_B(E,t)\,dE\right]\right)\right\}. \tag{B.23}$$

The rate of loss of the total amount of kerogen with time is

$$\begin{aligned}\frac{dx(t)}{dt} = -x_0 \Big\{ \alpha \exp\left[-\int p_A(E) U_A(E,t)\, dE\right] \\ \times \int p_A(E) \frac{dU_A(E,t)}{dt}\, dE + (1-\alpha) \\ \times \exp\left[-\int p_B(E) U_B(E,t)\, dE\right] \\ \times \int p_B(E) \frac{dU_B(E,t)}{dt}\, dE \Big\}.\end{aligned} \tag{B.24}$$

If this total rate of loss is taken to be identical to the total rate of appearance of S2 then

$$\frac{dS2}{dt} = -\frac{dx(t)}{dt}. \tag{B.25}$$

Consider then a heating cycle, followed by a cooling cycle and then a reheating cycle as was discussed in Section B.II. The temperature, $T_{max}$, at which $dS2/dt$ has its peak value is then given by the solution to

$$\frac{d}{dT}(dS2/dt) = 0; \tag{B.26}$$

that is, implicitly by the formula

$$\begin{aligned}T_{max}^2 = \frac{H_c}{Rk_0} \Big\{ \alpha \exp\left[-\int p_A U_A(E, t_{max})\, dE\right] \times \int p_A E \exp(-E/RT_{max})\, dE \\ + (1-\alpha) \exp\left[-\int p_B U_B(E, t_{max})\, dE\right] \\ \times \int p_B E \exp(-E/RT_{max})\, dE \Big\} \\ \times \Big\{ \alpha \exp\left[-\int p_A U_A(E, t_{max})\, dE\right] \int p_A E \exp(-E/RT_{max})\, dE \\ + (1-\alpha) \exp\left[-\int p_B U_B(E, t_{max})\, dE\right] \int p_B\, exp(-E/RT_{max})\, dE \Big\}^{-2},\end{aligned} \tag{B.27}$$

where $t_{max}$ is the time at which the temperature $T = T_{max}$ on the reheating cycle; that is,

$$t_{max} = t_* + t_m + (1 - T_*/T_{max}) T_*/H_c \tag{B.28}$$

and where

$$U_A(E, t_{max}) = k_A(I_a + I_b + I_c), \tag{B.29a}$$

$$U_B(E, t_{max}) = (k_B/k_A) U_A. \tag{B.29b}$$

It is clear from Eq. (B.27) that $T_{max}$ depends on the total maturation path (heating, cooling, reheating) of the sample containing the mix of kerogens, as well as on the fractional mix of the two kerogens. Thus the laboratory equivalent Rock–Eval $T_{max}$ for a rock sample containing a mix of kerogens is going to change depending on (1) the maturation path imposed in the laboratory and (2) the fractional mix of pyrolyzable kerogen types included in the rock sample.

Note that the inverse kinetic procedure developed in the body of the chapter and in Appendix A assumes only one type of kerogen. An unmixing procedure must be applied to real geologic data to ensure that only single kerogen types are dealt with—as is abundantly clear from the downhole hydrogen index data presented in the two real case histories, which show *reversals* in the change of hydrogen index with depth.

## B.IV. RESOLUTION OF NUMERICAL INTEGRALS AND PHYSICAL CONDITIONS

A problem can arise in numerical implementation of inverse kinetic procedures unless some care is taken. It seems worthwhile to spell out the problem here so that future confusion is minimized. The difficulty can occur either in computing numerically

$$I_1 = \int_{T_0}^{T} \exp(-E/(Rx))\, dx \tag{B.30a}$$

or

$$I_2 = \int_0^{\infty} p(E) \exp(-E/RT)\, dE \tag{B.30b}$$

over ranges of $\{T_0, T\}$ or $E$ which allow the respective integrands to vary by orders of magnitude.

To put the problem on a common basis, in Eq. (B.30a) set $x = 1/y$, when

$$\begin{aligned} I_1 &= \int_{v}^{v_0} y^{-2} \exp(-Ey/R)\, dy \\ &\equiv (v_0 - v) \exp(-vE/R) \int_0^1 [v + (v_0 - v)w]^{-2} \\ &\times \exp[-E(v_0 - v)w/R]\, dw, \end{aligned} \tag{B.31a}$$

with $v = 1/T$, $v_0 = 1/T_0$; while in Eq. (B.30b), set $p = p_n$ = constant in $n\Delta E \le E \le (n+1)\,\Delta E$, where $\Delta E$ is fixed, when

$$\begin{aligned} I_2 &= \sum_{n=0}^{\infty} p_n \int_{n\Delta E}^{(n+1)\Delta E} \exp(-E/RT)\, dE \\ &= J \sum_{n=0}^{\infty} p_n \exp(-n\,\Delta E/RT)\,\Delta E \end{aligned} \tag{B.31b}$$

with

$$J = \int_0^1 \exp(-\Delta E\, u/RT)\, du. \tag{B.31c}$$

It can happen that $E(v_0 - v)/R$ and $\Delta E/RT$ are both extremely large compared to unity, when the *numerical* evaluation of Eqs. (B.31a) or (B.31c) has to proceed with extreme care.

In the case of Eq. (B.31c) two options are available: (1) One can note that the integral can be done analytically exactly, yielding

$$J = [1 - \exp(-\Delta E/RT)]\, RT/\Delta E, \tag{B.31d}$$

or (2) one can choose to break the integrand into smaller pieces, with the pieces chosen so that $\exp(-\Delta E u/RT)$ does not vary by large factors over the range of each piece. Then write $u = \sin\theta$ with

$$J = \sum_{p=0}^{P} \int_{\theta_p}^{\theta_{p+1}} \cos\theta \exp[-\Delta E \sin\theta/RT]\, d\theta \tag{B.31e}$$

with $\theta_{P+1} = \pi/2$.

If $\theta$ is written $(\theta_p + z)$ so that $\sin(\theta_p + z) \cong \sin\theta_p + \cos\theta_p\, \Delta z$ for $\Delta z \ll 1$, then one requires also that $\Delta z \ll RT \sec\theta_p/\Delta E \ll 1$ in order to write

$$J = \sum_{p=0}^{P} \int_0^{(\theta_{p+1}-\theta_p)} \cos(\theta_p + \Delta z) \exp[-\Delta E \sin\theta_p/RT] \times \exp(-(\Delta E \cos\theta_p/RT)\, \Delta z)\, d\Delta z. \tag{B.31f}$$

Then, with $\theta_{p+1} - \theta_p \lesssim RT \sec\theta_p/\Delta E$ the integral in Eq. (B.31f) can be performed numerically, to an accurate enough approximation, to yield numerical results indistinguishable from those arising from the exact form of Eq. (B.31d). A similar argument holds for Eq. (B.31a).

However, a numerical evaluation of the respective integrals does *not* mean that the physical channel probability, $p(E)$ has to be cut up into small enough pieces in energy space to satisfy the dictates of numerical precision. Rather, methods for obtaining numerical precision of integral evaluation have to be tailored to the imposed requirements of physical determination of the channel probability. This fact seems to be underappreciated in that there are inverse calculations of residual kerogen kinetics in the literature (Sundararaman *et al.*, 1992) which slave the physical channel probability resolution to the numerical needs of integral precision, thereby creating unnecessary and unresolvable fine-scale detail in the probability distribution for $p(E)$.

# 6

# Variable Total Organic Carbon and Variable Solubility Considerations

## I. VARIABLE TOTAL ORGANIC CARBON AND HYDROCARBON GENERATION

### A. INTRODUCTION

Assessing hydrocarbon generation is an essential element in evaluating the petroleum potential of a sedimentary basin. Temperature and time, as well as the organic richness and kerogen types in the source rocks, are important factors controlling hydrocarbon generation. In most generation studies, the first-order Arrhenius chemical reaction law has been applied widely to hydrocarbon generation and degradation. Many generation models have been proposed and a number of publications have addressed generation and accumulation problems (see Tissot and Welte, 1978; Cao *et al.,* 1986; Sweeney and Burnham, 1990; Leythaeuser *et al.,* 1982; Ungerer *et al.,* 1990; Lerche, 1990b; Petersen *et al.,* 1996; Magoon and Sanchez, 1995). One question, however, that needs to be considered is this: How do the various possible total organic carbon (TOC) distributions (with a fixed average value of TOC in a source rock formation) affect the hydrocarbon generation ability and consequent accumulations? For a given average value of TOC, many TOC distributions are possible that influence the overall generation from the source rock. Unfortunately, only the average value, but not the distributions of TOC, is considered in most studies of hydrocarbon generation potential. Therefore, we seek to determine the degree of bias and the amount of uncertainty that is present when only the average values of TOC are used without regard to the total variation of the TOC distributions. Figure 6.1 shows the theme of this study.

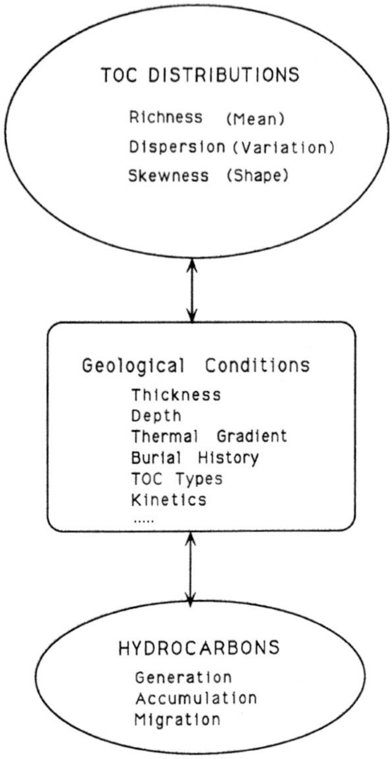


FIGURE 6.1 Impact of various TOC distributions on hydrocarbon generation and accumulation.

## B. METHODOLOGY

Using the two-dimensional fluid flow/compaction model GEOPETII with the options of Tissot's (Tissot and Welte, 1978) and the University of South Carolina's (USC's) hydrocarbon generation models (Nakayama, 1987; Yu, 1992; and Cao, 1985), oil/gas generation rates are examined with various TOC distributions under given geological conditions. A key question here is to define parameters to describe the TOC distributions in a source layer (rather than just using the arithmetic average TOC value for the layer), and then to relate these parameters to the hydrocarbon generation. The statistics most commonly used to represent the properties of a distribution include the following (Davies and Goldsmith, 1972): (1) measures of location, or central value, giving the location of some central or typical values—an example is the arithmetic mean; (2) measures of dispersion, showing the degree of spread around the central values—an example is the standard deviation ($\sigma$); and (3) measures of skewness. Measures of skewness show the extent to which the distribution departs from symmetry around the central value.

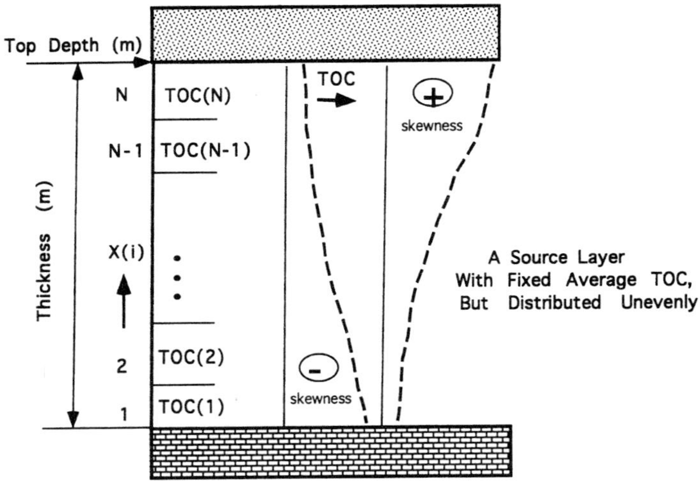


FIGURE 6.2 Schematic diagram showing a source rock layer with a fixed average TOC. The TOC distributions are various in the source layer and are described by skewness (see Appendix A).

A review of definitions of the parameters used here is given in Appendix A at the end of this chapter. Figure 6.2 shows schematically a source formation buried at depth with a *fixed* average TOC value, but the *distribution* of TOC in the formation is variable.

To cover a large range of geologic and geochemical conditions, the following aspects are considered in our investigation:

- Richness of TOC in the source rock (TOC %)
- Various standard deviations of TOC ($\sigma$)
- Thickness of the source rock (m)
- Burial depth of the top of the source rock (m)
- Thermal gradient of the basin through time (°C/km)
- Different kerogen Types (I, II, III)
- Different kinetic models (USC's and Tissot's)

With each setting of these conditions, a number of cases of TOC distributions are examined, which cover a wide range of skewness values. The oil/gas generation rates (in mg/g-rock/Ma) versus the skewness of the TOC distributions are presented for each specific condition.

The kinetic models (e.g., Tissot's and USC's) are based on first-order chemical reaction kinetics for the breakdown of kerogen to hydrocarbons as governed by

$$-dx_i/dt = k_i x_i$$

and

$$k_i = A_i \exp[-E_i/k_B T(t)],$$

where $x_i$ is the number density of component $i$, $k_i$ is the reaction rate for this reaction, $k_B$ is Boltzmann's constant, $A_i$ is the reaction constant (frequency factor), $E_i$ is the activation energy, and $T(t)$ is the temperature as a function of time acting on the element $x_i$. Three types of kerogen are considered, each of which differs in its hydrogen-to-carbon ratios. Activation energies ($E_i$), reaction constants ($A_i$), and the initial values ($x_{i0}$) for each kerogen type have been given in Tissot's (Tissot and Welte, 1978) and USC's models (Cao, 1985). USC's kinetic model has six oil and six gas producing pathways directly from kerogen and, in addition, six pathways for producing gas from oil. The Tissot-type of model has six producing pathways of oil from kerogen and one for producing gas from oil.

## C. SENSITIVITY OF HYDROCARBON GENERATION

In this section, we illustrate a number of test results on the variations of oil and gas generation rates with variable TOC distributions in a source rock layer. The results from two kinetic models are presented under specific geological conditions. In all of the following tests, a Tertiary basin (63 MY old) with a simple structure of flat layers is used as the template. The input parameters are intentionally chosen with wide domains so as to cover various basin situations.

### 1. Various Average Richness

The overall generation ability of a source rock formation is directly related to the availability of organic material. First we take a fixed source rock thickness of 1000 m buried at depth of 2500 m with a thermal gradient of 33°C/km. The temperature within the source rock is in the range of 92–125°C (assuming 10°C at the surface). With different arithmetic average TOC values at 2, 5, 10, and 15% in the source rock and a fixed standard deviation of 5.1%, we then illustrate how much the overall generation ability of the formation varies with different TOC distributions. Figures 6.3a–d show oil and gas generation rates versus the coefficient of skewness of the TOC distributions for two kinetic models for type II kerogen. The richer the source rock, the greater the generation rate, as expected. However, the generation rate with a negative skewness distribution is much higher than that with positive skewness. At the extreme cases, the oil and gas generation rates are nearly doubled at the most negative skewness case relative to the most positive skewness case. This difference, due solely to different TOC distributions, is significant despite the fact that the average TOC value is the same in the source rock layer.

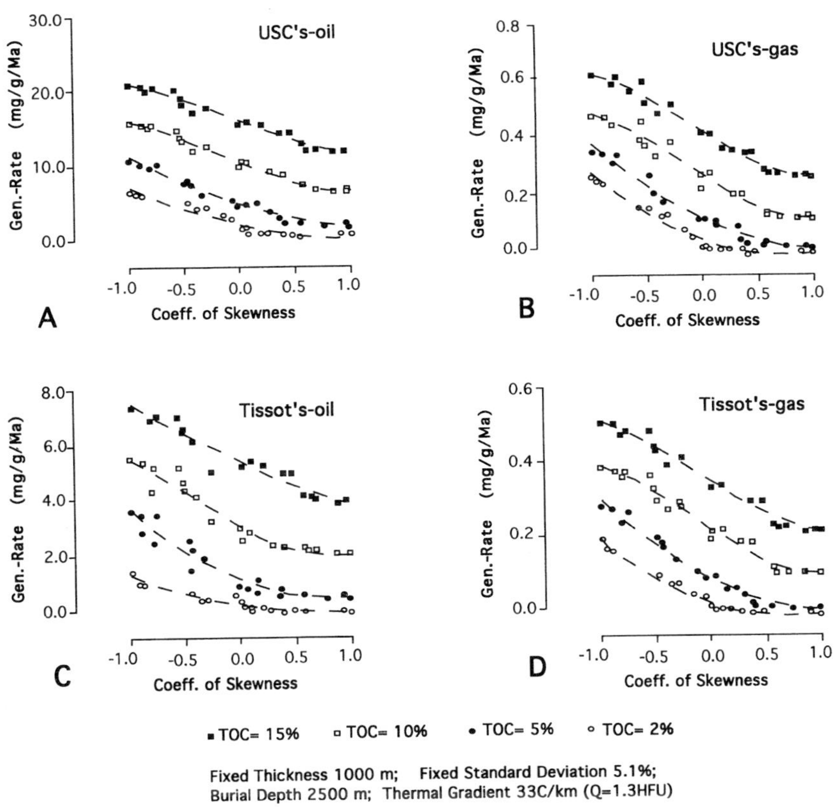


FIGURE 6.3 Hydrocarbon generation rate versus skewness of TOC distributions with varying richness of TOC in the source rock layer showing increasing generation rates with increasing TOC: (a) USC's model for oil; (b) USC's model for gas; (c) Tissot's model for oil; (d) Tissot's model for gas.

## 2. Various Degrees of Spread

Now fix the average TOC in the formation in order to examine solely the effects of spread of the TOC distribution. Take an average 10% TOC in a 1000-m-thick formation, buried at 2500 m with a thermal gradient of 33°C/km. Figures 6.4a–d give the overall hydrocarbon generation rate of source rock versus the coefficient of skewness of the TOC distribution with different standard deviations ($\sigma$) at 2.8, 5.1, 14.2, and 25.2%. When the deviation ($\sigma$) is 2.8%, the generation curves are relatively gentle. The oil rates are 10–14 mg/g/Ma from the USC model (Fig. 6.4a), and 3.6–5 mg/g/Ma from Tissot's model (Fig. 6.4c). Both are 40% higher for the most negative skewness case than for the most positive skewness case. The gas generation rates have the same trend as those for oil generation, with a variation of about 20% at a 2.8% ($\sigma$) spread for both the USC and the

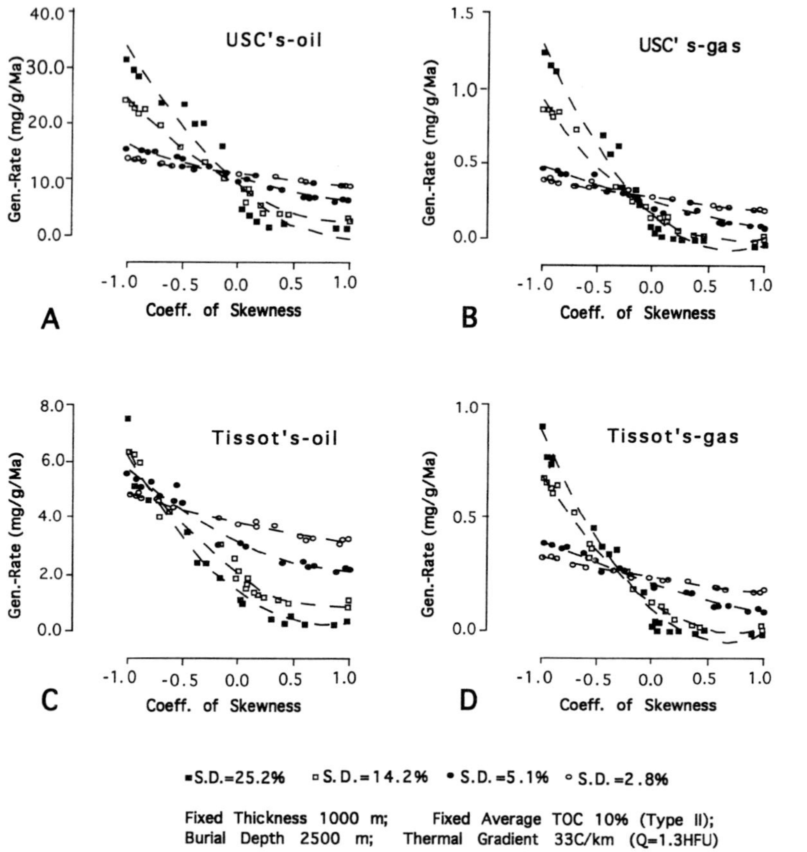


FIGURE 6.4 Hydrocarbon generation rate versus skewness of TOC distributions with fixed average TOC and various standard deviations ($\sigma$) (written as S.D. on the figure) showing that the larger the deviation, the more extreme the variation of generation rate. (a) USC's model for oil; (b) USC's model for gas; (c) Tissot's model for oil; (d) Tissot's model for gas.

Tissot model. As the degree of spread of the TOC distribution ($\sigma$) increases, the generation curves show significantly more variation with skewness. For instance at $\sigma$ = 25.2%, the oil generation rates are in the range of 2–30 mg/g/Ma from the USC model and 0.5–7 mg/g/Ma from Tissot's model, with a factor of 14–15 difference between the most negative and positive skewness distributions. The gas rates are 0.2–1.4 mg/g/Ma from the USC model and 0.1–1.0 mg/g/Ma from Tissot's model, with a factor of 7–10 difference from negative to positive skewness comparisons. Therefore, an increase of spread of TOC distributions in a source layer will result in an increase of the variation of the generation rate due to the different distributions—at the above geological and geothermal conditions.

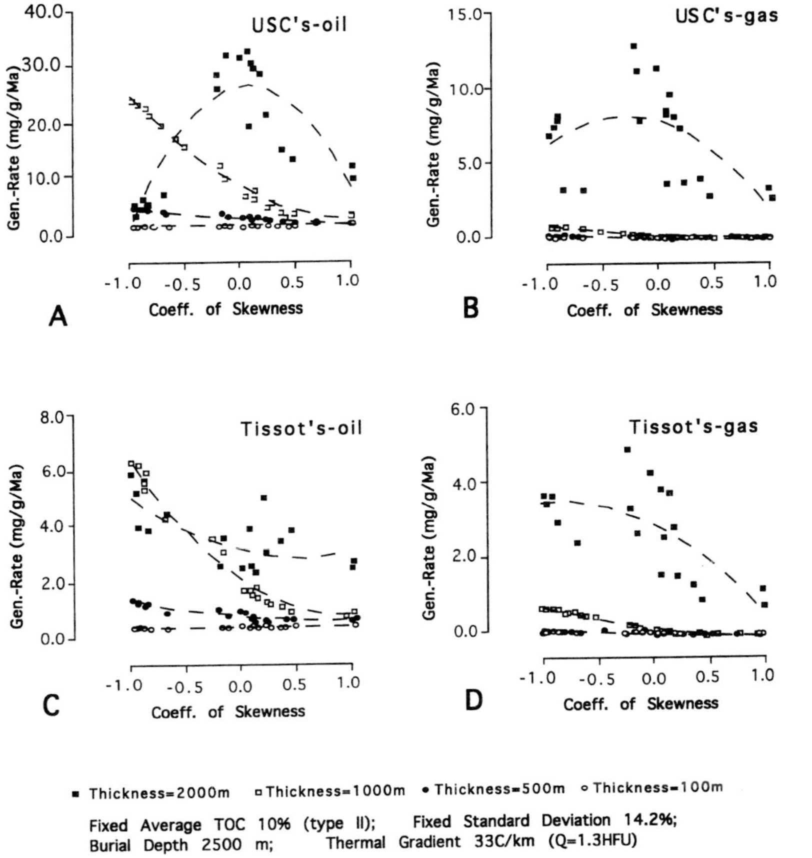


FIGURE 6.5 Hydrocarbon generation rate versus skewness of TOC distributions with varying thickness of source rock layer. Average TOC is 10% and standard deviation is 14.2%. (a) USC's model for oil; (b) USC's model for gas; (c) Tissot's model for oil; (d) Tissot's model for gas.

### 3. Various Thicknesses of the Source Layer

In the previous section, we examined the variation of TOC richness and the spread of TOC distributions while geological conditions such as thickness and burial depth were held fixed. Here the TOC richness is fixed at 10% and standard deviation at 14.2%, and we examine how the hydrocarbon generation varies with different TOC distributions when the top of the layer of variable thickness is buried at a 2500-m depth with a thermal gradient of 33°C/km. Figures 6.5a–d show the oil and gas generation rates versus the coefficient of the skewness of the TOC distribution from both the USC and Tissot models when the thickness of the source rock layer is chosen at 100, 500, 1000, and 2000 m, respectively.

With a 100-m thickness, the generation rates are almost constant due to the small change of the temperature over the 100-m thickness. With a 500-m thickness, the oil generation rates have some increase with negative skewness distributions, but the gas rates are stil unchanged because of the low temperature. When the thickness is 1000 m, the oil generation rates from both models have a significant change with the different distributions. The oil generation rates with extreme positive and negative skewness distributions vary from 3 to 30 mg/g/Ma with the USC model and from 1 to 7 mg/g/Ma with Tissot's model, with a factor of 7–10 variation over the different distributions. The variations of gas generation rate also show a significant change for each kinetic model. When the thickness of the source layer increases to 2000 m (the corresponding temperature is then between about 90 and 156°C), there is a dramatic change in the oil generation rate pattern as shown in Figs. 6.5a and 6.5c. The reason for the lower oil generation rate for the negative skewness cases is because the oil has converted to gas, as supported by the gas generation rates shown in Figs. 6.5b and 6.5c. According to both the USC and Tissot models, the gas generation rates are much greater for negative skewness distributions than positive skewness cases with a 2000-m-thick source layer.

Thus the thickness is an important factor determining the influence of the TOC distributions on the oil and gas generation. When the thickness is small (say, less than 100 m) the temperature variations over the formation depth are also small, and variations in the TOC distribution do not have a strong influence on either oil or gas generation. However, when the source rock formation becomes thicker (approximately greater than 500–1000 m thick), the influence of variations in TOC becomes stronger. Notice that when the thickness of the source formation is large ($\geq$1000 m) and the formation is buried at a depth such that the gas generation window has been reached, the oil generation rate decreases with negative skewness distributions and, as a result, the gas generation rate will be increased because the oil is cracked to gas, as shown by the extreme scatter of points in Fig. 6.5 for the case of a 2000-m-thick source rock. Conversely, positive skewness distributions will retain their oil generating capability for a longer period.

**4. Various Burial Depths**

Now vary the burial depth (top boundary) of the source rock with a formation thickness of 500 m, a fixed TOC richness of 10%, and a standard deviation ($\sigma$) of 14.2%. The thermal gradient is 33°C/km. We examine the effects of the different TOC distributions on the hydrocarbon generation when the source formation is located at different depths and temperatures.

The oil and gas generation rates versus the coefficient of skewness of the TOC distribution are shown in Figs. 6.6a–d for both kinetic models at

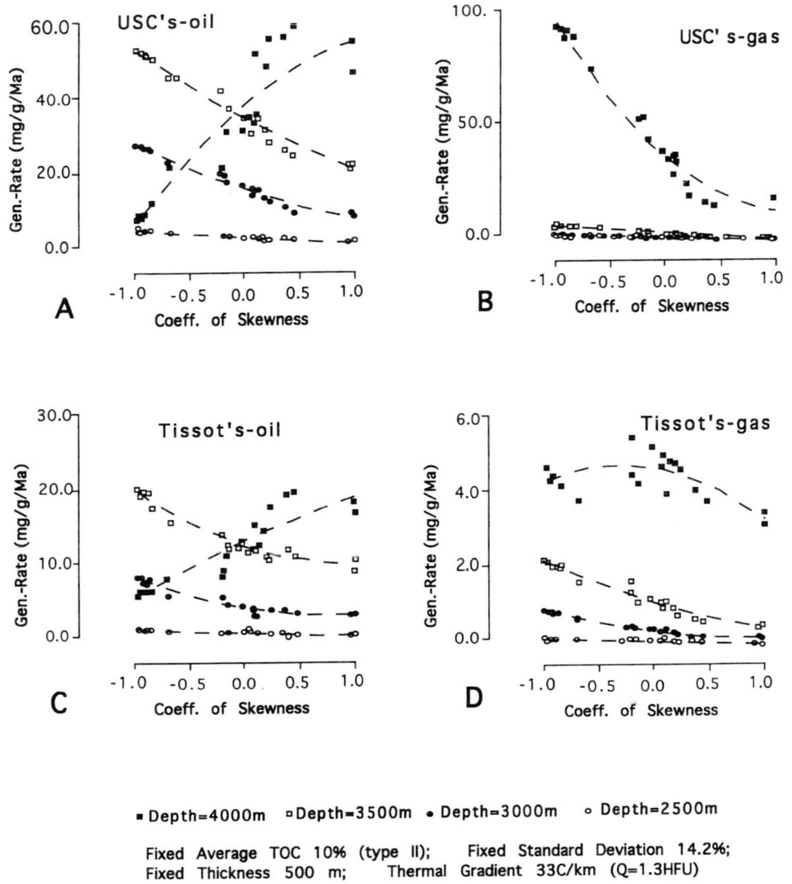


FIGURE 6.6 Hydrocarbon generation rate versus skewness of TOC distributions with varying burial depths of a 500-m-thick source rock. Average TOC is 10% and standard deviation is 14.2%. (a) USC's model for oil; (b) USC's model for gas; (c) Tissot's model for oil; (d) Tissot's model for gas.

burial depths of 2500, 3000, 3500, and 4000 m, respectively. At shallow depth (2500 m), the oil and gas generation rates are nearly constant and at very low rates because of the low temperature. When the burial depths are 3000 and 3500 m, with corresponding temperature ranges of 110–127°C and 128–145°C, respectively, the oil generation rates have a significant variation with various TOC distributions. The oil rate has been doubled with the most negative skewness distribution relative to the most positive skewness distribution from both models. The gas generation rate at the burial depths of 3000 and 3500 m shows some change in Tissot's model and a lesser change in USC's model. When the burial depth is 4000 m, which is in the 142–158°C gas generation stage, the oil generation rates have decreased with the negative skewness distribution and the gas generation rates have increased dramatically according to both kinetic models. Thus, the effects of TOC distributions on the oil and gas generation rate are very significant when the source formation is in the oil and gas generation windows.

**5. Different Kerogen Types**

Only type II kerogen has been used in the preceding discussions. Now examine the effects of the TOC distributions for different types of kerogen on oil and gas generations. Taking a 1000-m-thick source rock formation with 10% average TOC, 5.1% standard deviation ($\sigma$), 2500-m burial depth, and a 33°C/km thermal gradient, the oil and gas generation rates from the two kinetic models are given in Figs. 6.7a–d for three types of kerogen. Notice that the results shown are the generation rates at the present day of a Tertiary basin with the temperature in the source formation between 92–125°C. For kerogen types I and III, the variation of oil generation with the most negative and positive skewnesses is about 100%, while for type II kerogen the variation is about a factor of 4 different with negative and positive skewness distributions according to both kinetic models (Figs. 6.7a and 6.7c). For gas generation, little effect is seen for type I, due to the lower gas potential of type I kerogen. The effects of the TOC distribution on gas generation from types II and III kerogen are very significant, as shown in Figs. 6.7b and 6.7d. Therefore, for all types of kerogen, the variations of the TOC distributions have significant effects on both oil and gas generation, but the magnitude of the influence is variable among different kerogen types.

**6. Different Thermal Gradients**

The thermal gradient varies from basin to basin and, indeed, often from well to well within a basin. Here we examine the effect of various TOC distributions with different thermal gradients, which will move the source rock at a fixed depth into different ranges of temperature. We take a 1000-m-thick source rock layer with an average TOC of 10%, standard deviation of 5.1%, and burial depth of 2500 m. Figures 6.8a–d show the oil and gas generation rates versus the coefficient of skewness of the TOC distributions from the two kinetic models for thermal gradients of 26, 33, and 40°C/km, respectively. At the lowest gradient, 26°C/km, there is no distribution effect because of the low temperature level and the fact that oil and gas generation have not yet started (Figs. 6.8a–d). With a gradient of 33°C/km, the source rock is in the major oil generation stage and the variations of oil generation

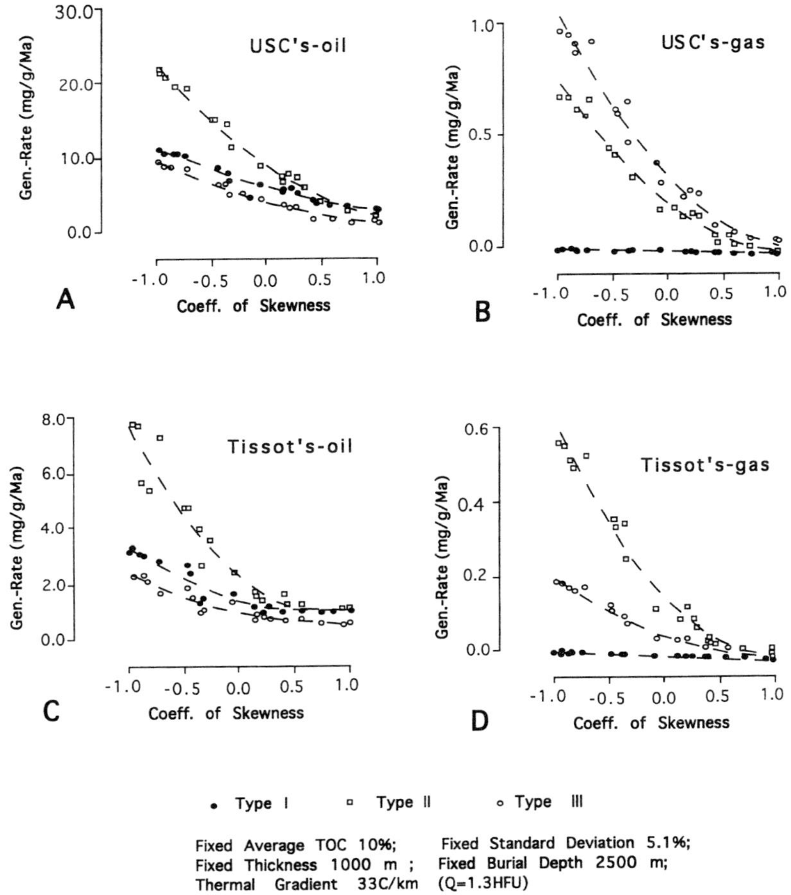


FIGURE 6.7 Hydrocarbon generation rate versus skewness of TOC distributions with three types of kerogen. (a) USC's model for oil; (b) USC's model for gas; (c) Tissot's model for oil; (d) Tissot's model for gas.

from the two kinetic models (Figs. 6.8a and 6.8c) indicate a factor of 2–3 difference between most negative and positive skewness distributions. The gas rates also show significant differences with different TOC distributions. At the highest thermal gradient of 40°C/km, because the source rock formation has entered the gas window, the generated oil has been converted into gas; the gas rates increase rapidly and oil rates decrease with negative skewness distributions. According to this test, the thermal gradient is also an important factor that influences the effect of TOC distributions on generation.

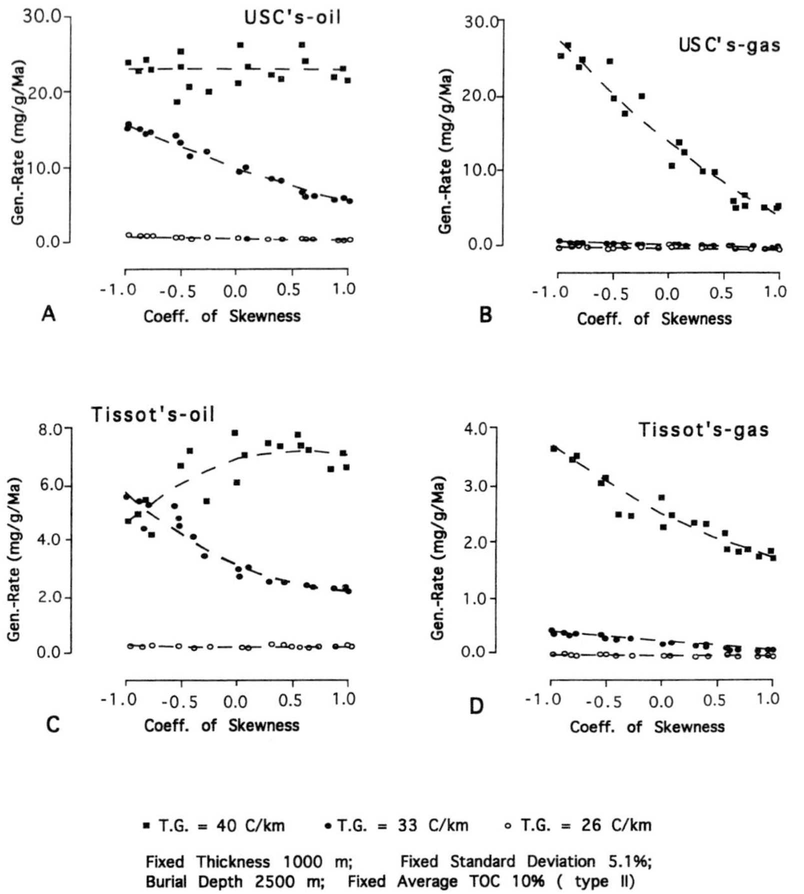


FIGURE 6.8 Hydrocarbon generation rate versus skewness of TOC distributions with different geothermal gradients. (a) USC's model for oil; (b) USC's model for gas; (c) Tissot's model for oil; (d) Tissot's model for gas.

## D. DISCUSSION

From two kinetic models (e.g., USC's and Tissot's models), results show the same variation patterns except for magnitude of hydrocarbon yields. In general, the USC model (Cao, 1985) predicts higher generation rates and more gas generated than Tissot's model under the same conditions.

The dominant factor that controls hydrocarbon generation ability is, of course, temperature (Price, 1982; Waples, 1980) although elapsed burial time cannot be ignored. Therefore, whether a TOC distribution is favorable or unfavorable to oil and gas generation depends on where the majority of the organic carbon is located. If the high-TOC parts of a source layer are in the peak oil/gas generation temperature window, a high overall generation potential of the source formation would be expected. The position of a source rock formation relative to the oil/gas generation window in a basin is determined by the burial depth, thickness of the formation, and the thermal gradient of the basin. Once the source formation is in the relevant generation windows, the TOC distributions within the formation have an important influence on the generation ability, as indicated in this study.

Hydrocarbon accumulation requires a favorable combination of several controlling factors such as generation timing and rate, structural formation timing, expulsion and migration of hydrocarbons, and sealing capability of the accumulation site. In terms of amount and timing, the various TOC distributions influence oil and gas generation, which then influence hydrocarbon accumulation, as well as expulsion and migration of hydrocarbons.

According to the results presented here, the variations of oil and gas generation due to different TOC distributions are characterized as follows:

1. There is a significant influence of variations in TOC distribution when the source formation is positioned in or around the hydrocarbon generation windows (oil/gas). The differences of overall generation rate between the most negative skewness distribution ($b_1 = -1$) and the most positive skewness distribution ($b_1 = +1$) are often about a factor of 5–10 (or more) for a typical formation with 3–15% average TOC and 2–10% standard deviation ($\sigma$). The greater the dispersion of TOC in the formation, the larger the difference of generation ability between favorable and unfavorable distributions. The closer the concentrated TOC portion of a source layer is to the peak generation window conditions, the more favorable the generation.
2. The thickness and burial depth of a formation, as well as the thermal gradient of the basin, are important factors controlling the influence of TOC distributions on hydrocarbon generation. The thicker the source formation or the larger the thermal gradient, the greater the variations of overall oil/gas generation rates. The variation of temperature across a formation is determined by the thickness and thermal gradient. When the thickness is small (say, $\leq$100 m) and the temperature gradient is low (say, $<$30°C/km), the variation of hydrocarbon generation is small and can be ignored. When the thickness is around 500–1500 m, the generation difference must be considered. However, when the thickness is large ($>$2000 m), it is better to divide the source rock into two or more layers because the variation of generation over such a thickness is extreme.
3. The effect of varying TOC distributions on generation may be ignored when the source rock is at a very low temperature ($<$80°C) and so has not yet begun significant generation.

## II. THREE-PHASE FLOW AND VARIABLE SOLUBILITY

### A. INTRODUCTION

The increasing difficulty of successful hydrocarbon exploration for giant oil fields demands more sophisticated techniques. Basin modeling is one of those techniques and has been widely applied in recent years. Computer simulation models integrate all available geological, geochemical, and geothermal data, and provide dynamic basin evolution histories with information of geopressures, rock properties, thermal behaviors, and hydrocarbon generation and accumulations to be expected at the present day (Hunt, 1979; Welte and Yukler, 1981; Nakayama and Van Siclen, 1981; Ungerer *et al.,* 1986; Cao, 1985; Nakayama and Lerche, 1987; Bethke, 1985).

Fluid migration is an important part in basin modeling and has been addressed in various models (Ungerer *et al.,* 1990; Bethke, 1985; Lerche, 1990b). However, the examination of a three-phase fluid system (water, oil, gas), with solubilities of the different phases as functions of temperature and pressure, has been underinvestigated to date despite the fact that three phases of fluids are likely to be the case in most hydrocarbon basins.

Here we study fluid migration for three separate phases (water, oil, gas), applying Darcy's law to each phase. As a new facet of the computer code GEOPETII (developed at the University of South Carolina), solubilities of gas in oil and water, and of oil in water, with temperature and pressure dependencies, are considered in the processes of hydrocarbon generation, migration, and accumulation. User-defined relative permeabilities for each phase are allowed for, with permeabilities taken to be dependent on power law behaviors for partial saturation of a pore space with the separate, free-phase components of the gas/oil/water mixture. Synthetic cases are investigated for hydrocarbon accumulations under various geologic and geochemical conditions. This simulation study is conducted using the two-dimensional model GEOPETII, which allows lateral and vertical flows with horizontal and vertical permeabilities for each lithology, and includes different hydrocarbon generation models (Yu, 1992), each with separate oil and gas flows. Paleo-overpressuring, as well as a fracturing model, are also included. In addition both vertical and slant faults can be included and the faults can be open or shut to fluid flow. In general, this two-dimensional fluid flow/compaction model uses present-day geological data such as lithology, age, depth, paleowater depth, surface temperature, and geochemical data to constrain the basin evolution. The model predicts the evolution with time and basinal location of porosity, permeability, formation depth, excess pressure, fluid pressure, generation pressure, as well as the thermal maturity of the basin, temperature, generation rates of hydrocarbons, kerogen transformation ratio, accumulation amount, and locations of hydrocarbons. Oil and gas saturations in the pore space and fluid movement directions are also given.

## B. MIGRATION MODEL

### 1. Flow, Pressures, and Viscosities

Hydrocarbon migration is the least known aspect in oil/gas accumulation processes, particularly the primary migration. Despite the many theories of primary migration that have been reported (Tissot and Welte, 1978; Bower, 1951; Leythaeuser *et al.*, 1982; Price *et al.*, 1983; Hunt, 1979), none, as yet, provides a satisfactory explanation for the movement of hydrocarbons out of low-permeability source rocks (primary migration) and into carrier beds for further transport (secondary migration). Secondary migration is commonly taken to occur as separate phase hydrocarbon migration (Tissot and Welte, 1978). A prevalent proposal for the primary migration mechanism seems to be expulsion in the pure hydrocarbon phase (Dickey, 1975; McAuliffe, 1980; Durand, 1988), possibly with the help of microfracturing.

In the simulations, separate phase expulsion and migration of each phase are allowed for. Thus, for that component (the free-phase component) of a given phase which is *not* in solution in other phases at a given location at a given time, there must be a flow field for the free phase. Thus we have three separate phase flow systems in a hydrocarbon generating basin, and fluid flow must be calculated for each. Applying Darcy's law to multiple flow in porous media, we have the flow velocities for water ($V_w$), oil ($V_o$), and gas ($V_g$) as:

$$V_w = -\frac{K\,k_{rw}}{\mu_w}\nabla(\Delta p_w), \tag{6.1}$$

$$V_o = -\frac{K\,k_{ro}}{\mu_o}\nabla(\Delta p_o), \tag{6.2}$$

$$V_g = -\frac{K\,k_{rg}}{\mu_g}\nabla(\Delta p_g), \tag{6.3}$$

where $V$ denotes flow rate of each phase with $\Delta p$ = driving pressure, $K$ = intrinsic permeability, $k_r$ = relative permeability (see discussion later), $\mu$ = viscosity, and subscripts w, o, and g denote water, oil, and gas, respectively.

The driving pressures considered here are the excess pressure from disequilibrium compaction, buoyancy from the density differences between oil, gas, and water, and capillary pressure arising from various grain sizes of different lithologies. Generation pressure and aquathermal pressure are accounted for in the model and considered together with excess pressure so that the free-phase densities depend on pressure and temperature (see Appendix B in this chapter). Therefore, the driving pressures for oil and gas at any time between two neighboring grid nodes separated vertically by a distance $\Delta z$ are, respectively,

$$\Delta p_o = \Delta p + p_{co} + g(\rho_w - \rho_o)\,\Delta z, \tag{6.4}$$

$$\Delta p_g = \Delta p + p_{cg} + g(\rho_w - \rho_g)\,\Delta z, \tag{6.5}$$

and, if oil and gas phases are buoyant in the element $\Delta z$, then water is antibuoyant so that the net pressure on the water in the element is

$$\Delta p_w = \Delta p + p_{cw} - g(\rho_w - \rho_o)\,\Delta z - g(\rho_w - \rho_g)\,\Delta z, \tag{6.6}$$

where $g$ = acceleration of gravity, $\rho$ = density, $\Delta p$ = difference in water excess pressure across the cell of height $\Delta z$, $\Delta z$ = difference in physical depth, $p_c$ = capillary pressure, defined as

$$p_c = 2\tau_s(1/R_{c1} - 1/R_{c2}),$$

where $\tau_s$ = interfacial tension is variable between phases (default values are $\tau_s$ = 25 dyn/cm for gas against oil or water; $\tau_s$ = 40 dyn/cm for oil against water) and $R_{c1}$, $R_{c2}$ = smaller and larger pore throat diameters, respectively. Densities ($\rho_w$, $\rho_o$, and $\rho_g$) are taken to be constants in each interval of $\Delta z$ in Eqs. (6.4), (6.5), and (6.6), which is appropriate when the difference in physical depth ($\Delta z$) is small (less than about 20 m).

The viscosities of water and oil are functions of temperature and pressure, respectively, and for simplicity are expressed as

$$1/\mu_w = A_w T^2 + B_w T + C_w, \tag{6.7}$$

$$1/\mu_o = A_o(T)P^2 + B_o(T)P + C_o(T), \tag{6.8}$$

with user-determined values of the coefficients $A$, $B$, and $C$, and where $T$ is temperature (°C) and $P$ is pressure in $kg \cdot cm^{-2}$ (KSC). Gas viscosity is taken to be constant at 0.15 cP.

### 2. Relative Permeability

When multiple fluids move through porous rocks, the flow velocities are related not only to the rock intrinsic permeability, but also to the relative permeabilities which depend on the fraction of each fluid available in the

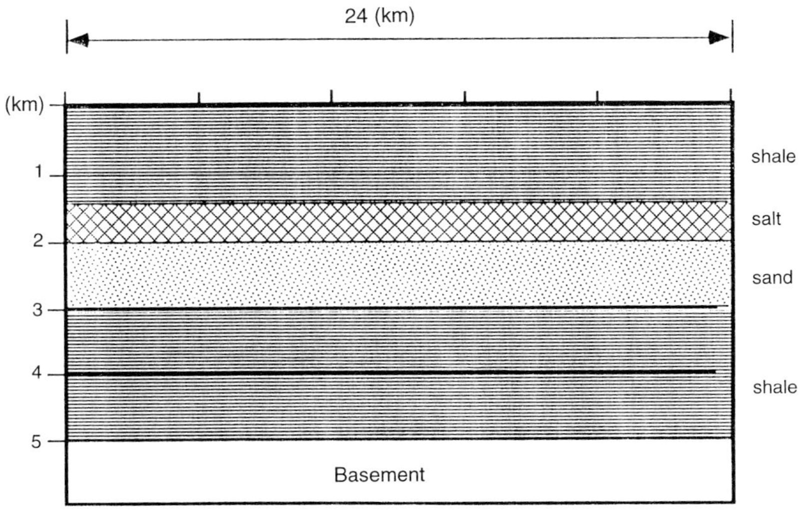


FIGURE 6.9 Stratigraphic section of a synthetic Tertiary basin with simple flat sandstone and shale formations and an overlying salt layer as seal.

free phase and the distribution of each fluid in the pore space. In the simulations, relative permeabilities are defined through

$$k_{rw} = S_w^a(S_w^a + S_o^b + S_g^c)^{-1}, \tag{6.9}$$

$$k_{ro} = S_o^b(S_w^a + S_o^b + S_g^c)^{-1}, \tag{6.10}$$

$$k_{rg} = S_g^c(S_w^a + S_o^b + S_g^c)^{-1}, \tag{6.11}$$

where $S_w$ = effective water saturation, defined as the fraction of movable water in the pore; $S_o$ = oil saturation; $S_g$ = gas saturation, with $S_w + S_o + S_g = 1$; $a$, $b$, $c$ are power indices for water, oil, and gas for each free phase, respectively, and are user adjustable. Thus primary migration of different components of hydrocarbons can be delayed (or accelerated) by the power law indices chosen. Note that the relative permeability is a value within the range 0.0–1.0, and the sum of the partial saturations of the three phases is unity. Empirically, the power indices $a$, $b$, $c$ appear to be in the range $0 \leq a, b, c \leq 5$–6 (Nind, 1981). An example shows the sensitivity of the oil power index ($b$ value) in controlling oil and gas accumulations in the simple stratigraphic situation outlined in Fig. 6.9. Figure 6.10a is drawn for a $b$ value of 1.0 with both $a$ and $c$ equal to 3.0, implying a higher oil permeability and lower gas permeability for equal saturations. Compared with the case drawn in Fig. 6.10a, Figs. 6.10b and 6.10c show

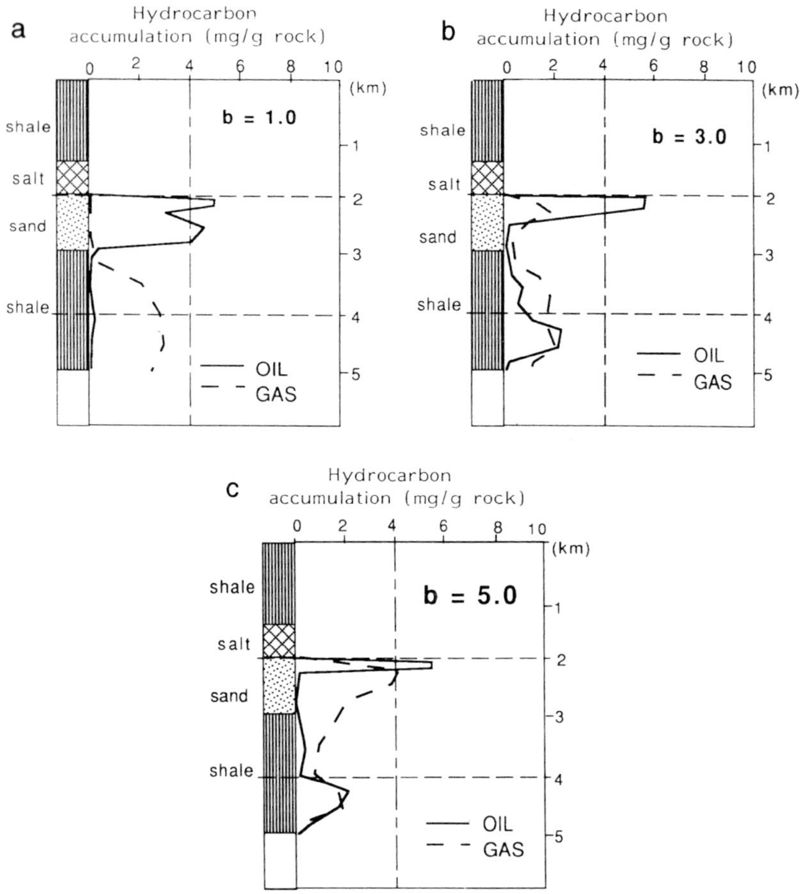


FIGURE 6.10 Hydrocarbon accumulation with various values of the oil power index in the relative permeability formula, showing increased gas migration with *decreasing* oil relative permeability (higher $b$ value) (i.e., an increased gas relative permeability and also a decreased water relative permeability). (a) $b = 1.0$; (b) $b = 3.0$; (c) $b = 5.0$.

the situations for $b$ values of 3.0 and 5.0, respectively, which decrease the oil and increase the gas relative permeabilities. The amounts of oil and gas accumulated in the reservoir show that oil decreases and gas increases with the variation of the relative permeabilities changing from the case of Fig. 6.10a through Fig. 6.10b to the case of Fig. 6.10c. These accumulation patterns indicate the migration response to the relative permeability dependence of oil/gas on saturation and behave as expected, in that higher permeability is associated with higher rates of motion of hydrocarbons.

**3. Solubility Function**

Solubility controls the amount of one phase that can be dissolved in another phase; consequently solubility impacts the migration and accumulation of hydrocarbons because of the relative permeability and fluid density modifications. The solubilities of gas in oil/water and of oil in water are functions of temperature, pressure, and hydrocarbon composition as well as of water salinity. Empirically, the variation of solubility with temperature and pressure is expressed well by the formula

$$S = S_{\min} + (S_{\max} - S_{\min}) \tanh\left(\frac{P - P_c}{P_d}\right) \exp\left(\frac{T_c - T}{T_d}\right), \tag{6.12}$$

where $S$ = solubility (mg/g) with temperature, $T$ (K) and pressure, $P$ (KSC); $S_{\min}(S_{\max})$ = minimum (maximum) solubility of oil/gas (mg/g); $P_c$, $P_d$ = critical and scaling pressure (KSC); $T_c$, $T_d$ = critical and scaling temperature (K). Note that $T_d$ generally takes on negative values for aqueous solubilities and a positive value for gas solubility in oil. Figure 6.11 shows a schematic diagram of gas solubility variation with temperature and pressure, indicating that the gas solubility in oil increases with higher pressure and lower temperature (Nind, 1981).

The aqueous solubilities of hydrocarbons differ for different basins, but generally increase with increasing temperature as shown in Table 6.1 for aqueous solubility of oils as a function of temperature. According to Price (1976), the values of aqueous solubilities of oil in hydrocarbon-bearing basins are in the range of 0–500 ppm at temperatures <200°C, and vary considerably depending on the different oils even in the same temperature ranges.

**4. Effects of Solubility on Hydrocarbon Migration**

The effects of solubility in a three-phase fluid system start at the moment of hydrocarbon generation and continue until the present day in the total process of hydrocarbon generation, migration, and accumulation. Appendix B at the end of this chapter gives the details of calculation descriptions, and Fig. 6.12a shows how solubilities influence the GEOPETII dynamic system. Not all phases are available at each time step for each volume element. The amounts of each phase available depend on the generation of hydrocarbons (see Appendix B).

First, solubilities of gas in oil/water change the amount of free gas in the three-phase system and, consequently, influence the relative permeabilities of each phase because relative permeabilities are related to the saturation of each free phase. The flowchart in Fig. 6.12b shows how the solubility influences hydrocarbon accumulation. For instance, higher gas solubility enables a greater amount of gas to dissolve in both oil and water and so

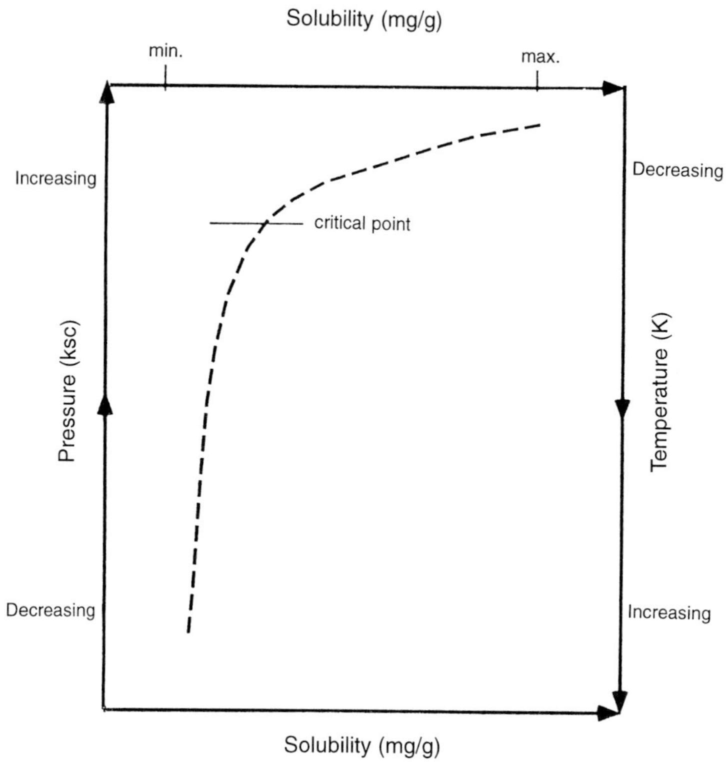


FIGURE 6.11 A sketch of gas solubility in oil with temperature and pressure. The critical point is where the solubility has sharply varied at the corresponding critical pressure and temperature.

less (or no) gas remains as a free phase. In turn, this effect reduces the gas saturation and consequently the gas relative permeability. The free-phase component of the gas is then less mobile and, as a result, gas accumulation will be enhanced *in situ* (or in source rocks) compared to the situation of low gas solubility. Note that this argument applies only to the free-phase gas accumulation, not to the gas migrated in oil or water solution, which is discussed in the following section. In contrast, a lower solubility situation has more gas in the free phase than would have been the case with higher gas solubility. Therefore the free-phase gas motion is enhanced because of the increase of relative permeability; for this situation there would be less gas left in a source rock, that is, less accumulation in the source rock.

In addition the solubility mixes separate phases, thereby altering the densities of each fluid, and thus the buoyancy for each phase is modified. Migrations of both oil and gas in solution are then different from migration in the free phase. Placing more gas in oil will decrease the density of oil

TABLE 6.1 Aqueous Solubility of Whole Oils as Functions of Temperature[a]

| Wyoming Farmers whole crude | | Flower comm. whole crude | | Louisiana State whole oil | |
|---|---|---|---|---|---|
| Solubility (ppm) | Temperature (°C) | Solubility (ppm) | Temperature (°C) | Solubility (ppm) | Temperature (°C) |
| 19.6 ± 0.9 | 25.0 | 51.5 ± 3.2 | 25.0 | 2.43 ± 0.09 | 25.0 |
| 25.0 ± 0.7 | 68.0 | 77.3 ± 8.8 | 86.5 | 5.47 ± 0.49 | 106.3 |
| 42.7 ± 0.9 | 102.6 | 111.0 ± 5.0 | 106.5 | 8.73 ± 0.37 | 123.0 |
| 56.3 ± 2.8 | 114.4 | 123.0 ± 7.0 | 122.6 | 18.4 ± 1.3 | 148.8 |
| 68.1 ± 3.7 | 128.8 | 154.0 ± 11.0 | 133.5 | 25.7 ± 0.8 | 159.6 |
| 101.0 ± 5.0 | 142.5 | 200.0 ± 4.0 | 145.2 | 49.1 ± 3.1 | 179.7 |
| 114.0 ± 5.0 | 151.5 | 269.0 ± 12.0 | 158.1 | | |
| 143.0 ± 10.0 | 161.8 | 446.0 ± 18.0 | 181.7 | | |

[a] As reported by Price (1976).

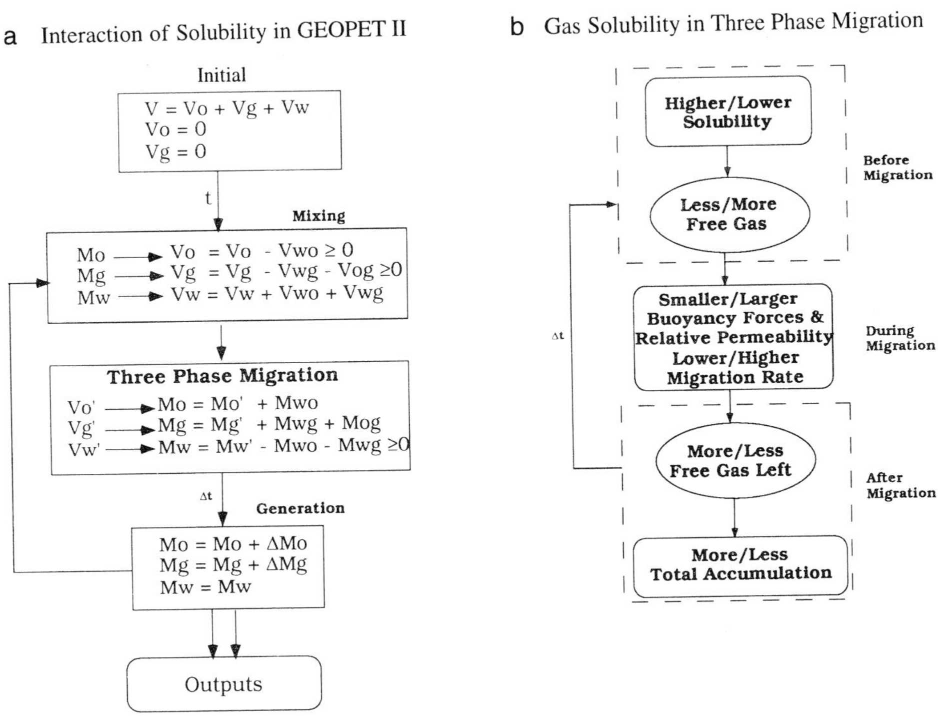


FIGURE 6.12 Flowchart showing the interaction of oil and/or gas generation, mixing, relative migration, and accumulation used in the dynamic simulation model GEOPETII.

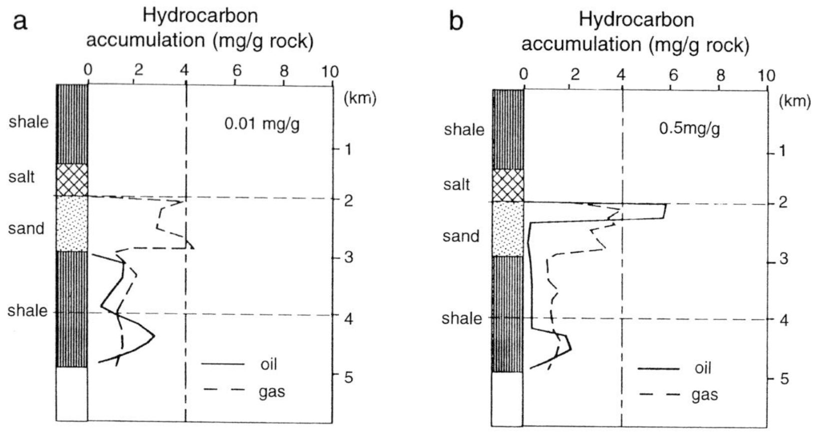


FIGURE 6.13 Effects of aqueous solubilities on hydrocarbon accumulations using the GEOPETII model. (a) 0.01 mg/g; (b) 0.5 mg/g.

and so increase the buoyancy of oil, which accelerates oil motion. As a result, the fraction of gas migrated in oil solution is increased. This character of gas movement is opposite to the free-phase gas motion discussed earlier. Thus the overall gas migration and accumulation do not follow a single trend as the solubility increases or decreases, but rather are complex phenomena depending on which factor prevails under particular conditions, which themselves vary with basinal evolution.

To illustrate simply the solubility effects on hydrocarbon migration and accumulation with a simple stratigraphy, consider a synthetic Tertiary basin comprised of horizontal layers, with source and reservoir layers underlying a salt bed, which acts as a seal (see Fig. 6.9). All conditions are kept invariant except for solubility variations. For the two cases of both oil and gas solubilities in water set at constant values of 0.01 and 0.5 mg/g, respectively, we examine the differences produced in hydrocarbon accumulation.

Figure 6.13 shows the oil/gas accumulations in milligrams per gram of rock with depth at the present day with the solubility for both oil and gas in water at (a) 0.01 mg/g and (b) 0.5 mg/g, respectively, showing the importance of the aqueous solubilities to the hydrocarbon accumulations (gas solubility in oil is set to zero). Figures 6.13a and 6.13b show significant differences of oil and gas accumulation patterns, with more oil moved upward in the case of 0.5 mg/g aqueous solubility. These two cases effectively represent minimum and maximum aqueous solubilities for oil (Price, 1976), and thus indicate the maximum effects of solubility on liquid hydrocarbon accumulation for a fixed functional behavior of relative permeabilities.

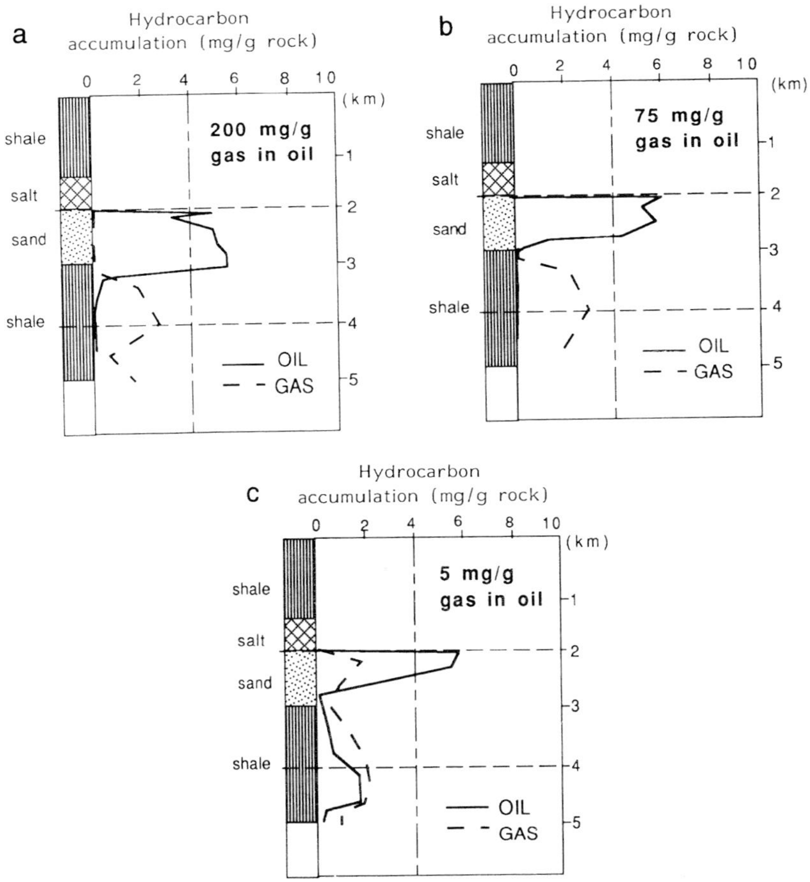


FIGURE 6.14 Effects of gas solubility in oil on hydrocarbon accumulations using the GEOPETII model. (a) 200 mg/g; (b) 75 mg/g; (c) 5 mg/g.

Gas solubility in oil varies depending on the temperature, pressure, and composition of hydrocarbons (Price *et al.*, 1983). Figure 6.14 shows oil/gas accumulation in milligrams per gram of rock with depth at present day when gas solubility in oil is constant at 5, 75, and 500 mg/g, respectively, with the solubilities of gas in water and oil in water set to zero for this illustration. The hydrocarbons are mainly accumulated in the sandstone reservoir beneath the salt sheet, which is expected according to the geological setting. The amount of hydrocarbons accumulated varies significantly with the different values of solubility. However, there seems not to be a simple trend of hydrocarbon accumulation variations with increasing or decreasing gas solubilities because of the effects on relative permeability, buoyancy, saturation, and phase density as already discussed.

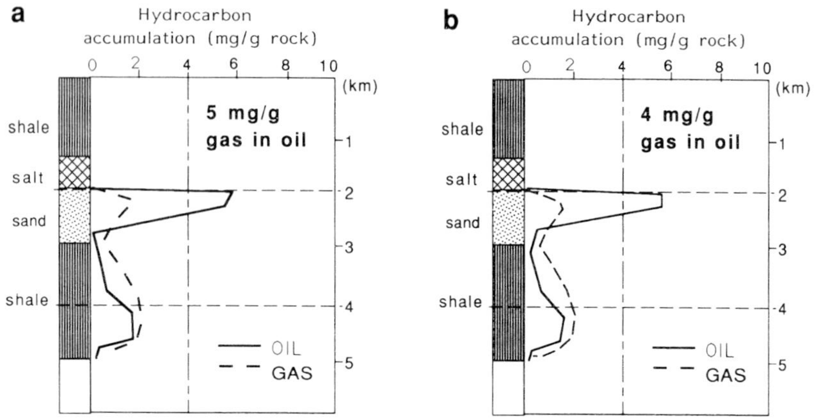


FIGURE 6.15 Oil and gas accumulations with depth for the cases of gas solubility in oil. (a) 5 mg/g; (b) 4 mg/g. Note there is almost no difference in hydrocarbon accumulation with such a small change of gas solubility.

It would seem that the precise and specific geological, geothermal, and geochemical conditions, and their variations with time and location in a basin, dominate how solubility influences hydrocarbon migration and accumulation. Small changes of gas solubility in particular ranges often have insignificant effects on oil/gas accumulations, indicating that threshold values exist that control the importance of solubility effects. For example, Fig. 6.15 shows oil and gas accumulations with the gas solubility set at 4 and 5 mg/g. It is hard to see any difference in the oil and gas accumulations. It would appear that each and every situation has to be carefully examined to evaluate dominance of solubility and saturation effects in impacting oil and gas migration and accumulation.

## C. SIMULATIONS OF THE HAMMERFEST BASIN, BARENTS SEA

Observations in the Hammerfest Basin, Barents Sea, indicate that the hydrocarbon potential in this area has been significantly affected by uplift and erosion (Nyland *et al.,* 1992). To date hydrocarbon exploration on the Barents Sea shelf has been disappointing but, in almost all wells drilled, good shows of residual oils have been found, indicating that oil was generated in the area. In wells where both gas and oil have been tested, high concentrations of extractable hydrocarbons are observed both below and above the present contacts of oil and gas with water (Fig. 6.16). The paleo oil/water contact (PALEO-OWC on Fig. 6.16) at maximum burial is marked, on the assumption that the major drop in the $C_{15+}$ concentration represents this paleocontact. The low concentration of heavier hydrocarbons ($>C_{15+}$) in

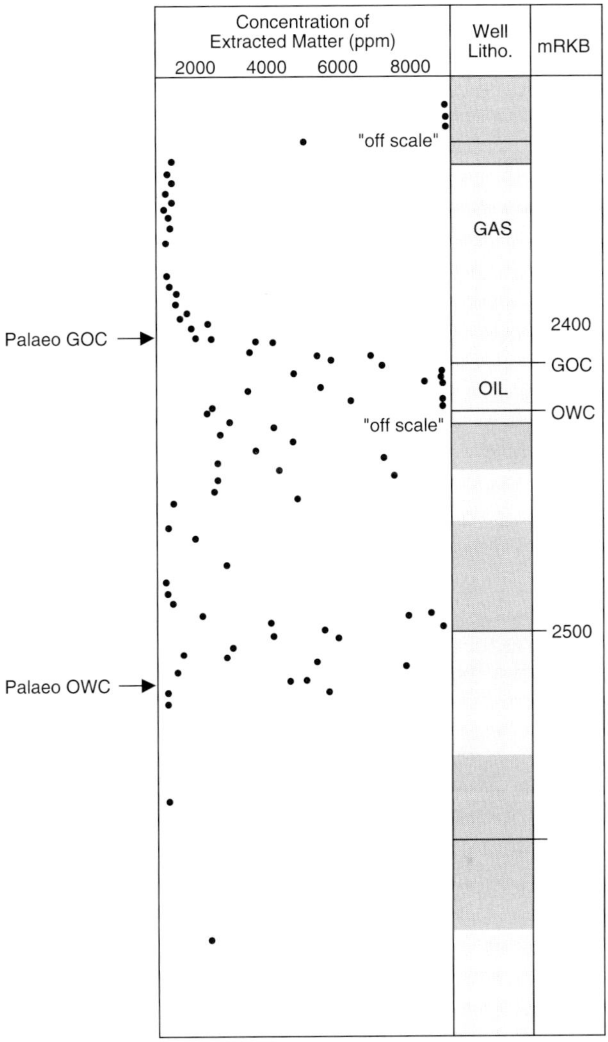


FIGURE 6.16 Paleo oil/water and oil/gas contacts in a Snøhvit well, Barents Sea, estimated from concentrations of extracted organic matter (modified from Nyland *et al.*, 1992).

the upper part of the reservoir may be an indication of a former paleo gas/oil contact (PALEO-GOC), which is shallower than the contact seen today (GOC) (Nyland *et al.,* 1992).

Nyland *et al.* (1992) also suggest that gas expansion during uplift caused oil to spill out of the structures as a result of pressure reduction, as sketched in Fig. 6.17. In the present simulation model, both gas expansion and gas solubility are taken into account as well, of course, as allowing the gas density to depend on pressure and temperature as explained in Appendix

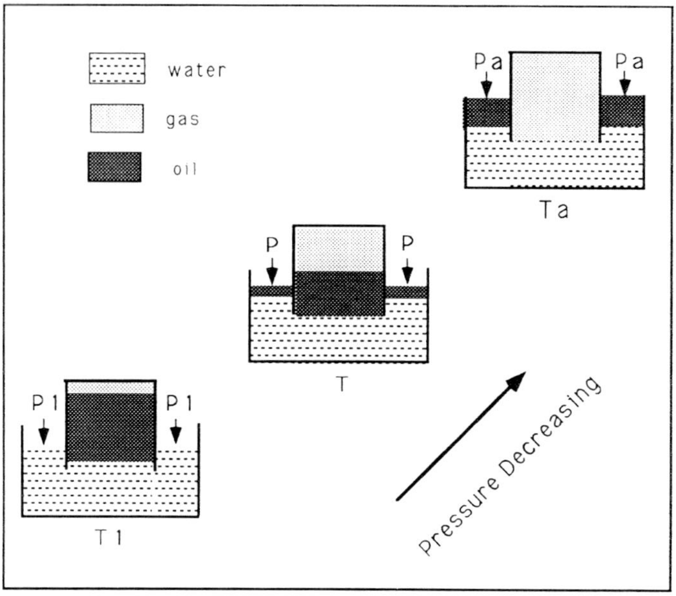


FIGURE 6.17 Phase change caused by pressure and temperature reduction. $P_2$, $T_2$ = initial pressure and temperature; $P$, $T$ = reservoir pressure and temperature; $P_a$, $T_a$ = actual pressure and temperature today (modified from Nyland *et al.,* 1992).

B. Reconstruction estimates are made of the hydrocarbon migration and accumulation histories to understand better the Hammerfest Basin. Figure 6.18a shows a synthetic section about 60 km long with 7000–5000 m thickness of sediments, which broadly represents the conditions of the Hammerfest Basin with respect to age, lithology, and burial history. A Tertiary erosion event of 1500 m, which is about the average magnitude of uplift in the region (Nyland *et al.,* 1992), is included in the simulation. The present-day closure of the top reservoir sand is about 200 m near the structural high. Lateral boundaries at both edges of the section are closed to fluid flow with low-permeability shales. The sole source rock, with 5% type II TOC, is located about 20 km laterally distant from the present-day structural high. With a present-day thermal gradient of 30°C/km, and with a constant heat flow history in the area, we examine the hydrocarbon generation, migration, and accumulation evolutions with respect to relative permeabilities and solubilities in the three-phase system. Figure 6.18b shows the rock permeability at the present day. A Tissot-type of hydrocarbon kinetic generation model is used here for illustrative purposes. For simplicity the aqueous solubilities of oil and gas are taken to be fixed at minimum and maximum values of 0.01 and 0.5 mg/g in this test. Various gas solubilities in oil are used, and the

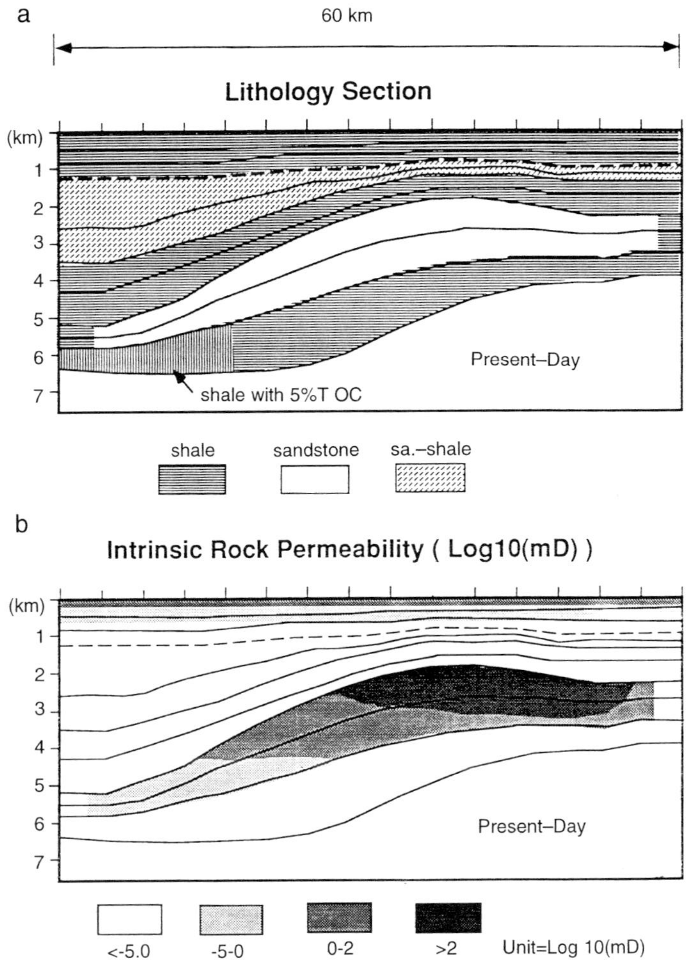


FIGURE 6.18 The present-day two-dimensional synthetic section, broadly constructed with parameters, lithologies, structure, burial history, and formation ages appropriate for the Hammerfest Basin. The lateral scale of the figure is 60 km. (a) lithology section; (b) present-day rock permeability.

hydrocarbon migration and accumulation patterns with time are presented for each case studied.

### 1. Set 1 Examples

Figures 6.19 through 6.29 show oil and gas accumulations with time, for a gas solubility in oil of 5 mg/g and permeability power indices of 3.0 for oil and water, and 1.0 for gas [i.e., $a = b = 3$, and $c = 1.0$ in Eqs. (6.9),

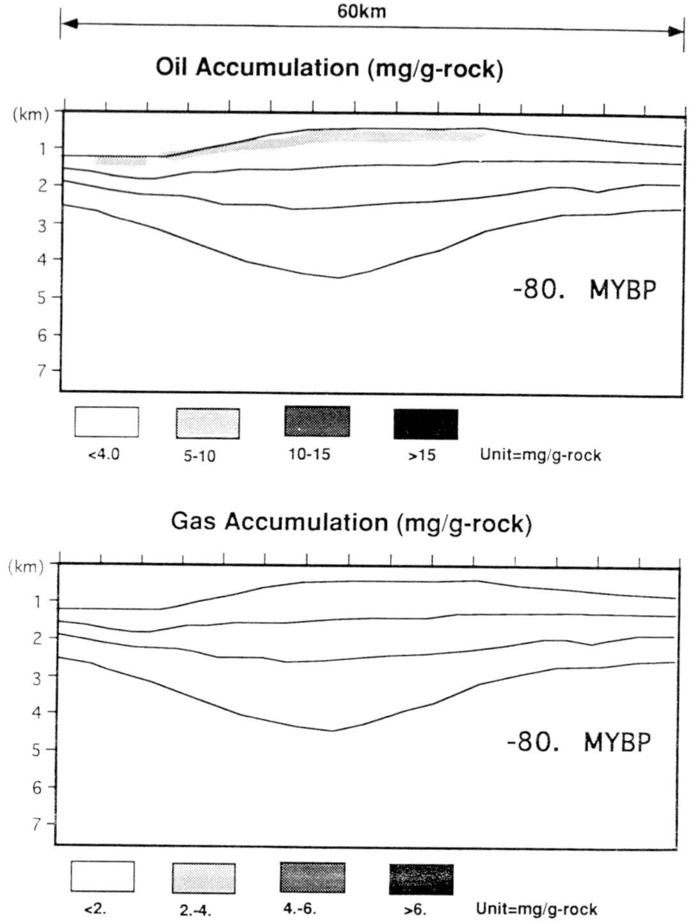


FIGURE 6.19 Oil and gas accumulation section at 80 MYBP, showing the start of oil migration and accumulation in the reservoir layer with the gas solubility set at 5 mg/g and relative permeability indices of 3.0 for water and oil and 1.0 for gas.

(6.10), and (6.11)]. Generally, oil starts to accumulate at 80 MYBP and has a significant amount of about 5–10 mg/g rock at 65 MYBP in the lower part of the reservoir when the source rock is then buried at 2500–3500 m depth (Figs. 6.19–6.29). Gas accumulation starts later, at about 55 MYBP (Fig. 6.21), and has a range of 2–6 mg/g rock in the reservoir. With further burial of the source rock, oil and gas generation continue and, simultaneously, move along the sandstone and reservoir formations (Figs. 6.22 and 6.23). Meanwhile, oil and gas are locally concentrated in the reservoir formation and oil leakage occurs at this time due to microfracturing. Figure 6.24 shows the major oil and gas accumulations at 30 MYBP, before the

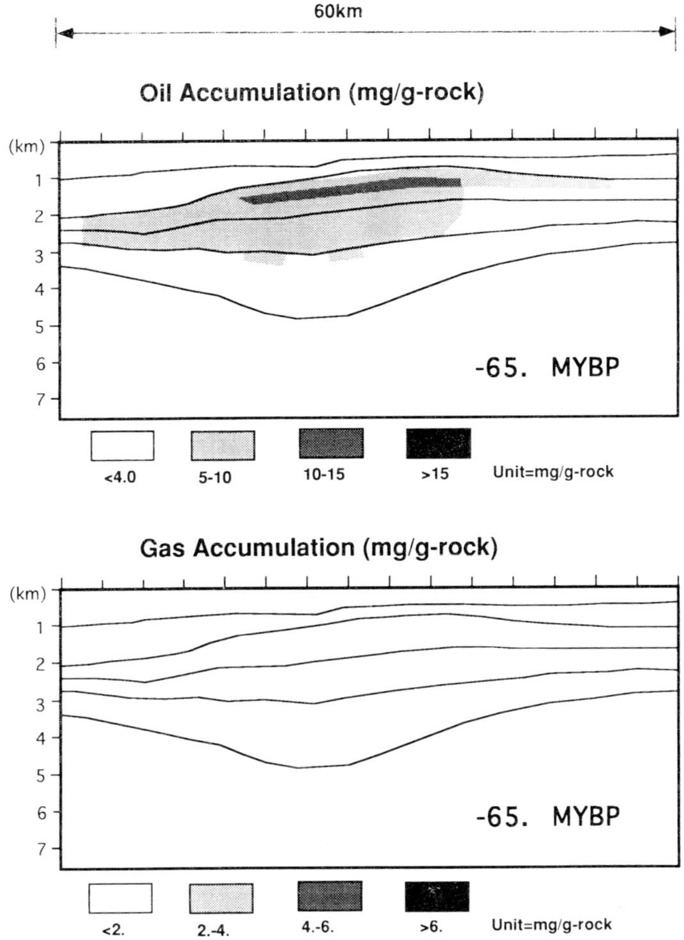


FIGURE 6.20 Oil and gas accumulation section at 65 MYBP, showing oil accumulation, but no gas in the reservoir with the gas solubility set at 5 mg/g.

uplift, with more than 15 mg/g rock of oil in most reservoir areas and 4–6 mg of gas/g rock. Tertiary uplift and erosion, starting at 30 MYBP, has a major impact on prior oil and gas accumulations. Most of the oil has been lost from the sandstone reservoir, possibly due to the pressure decrease and gas expansion, as well as to weakening of the seal capacity in overlying formations (Figs. 6.25 and 6.26). At 15 MYBP, the basin slowly subsides and receives new deposits. Continuation of oil and gas generation from source rock kerogen, as well as oil-cracking to gas, aids in the retention of gas in the reservoir and provide "spotty" oil accumulations. The present-day generation of oil and gas from the source rock provides a further

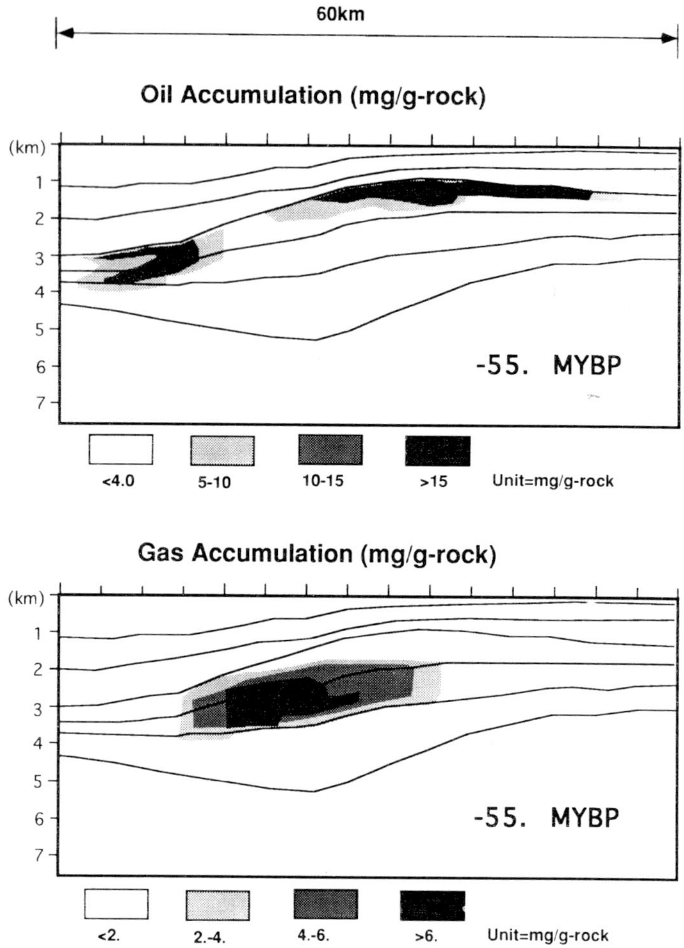


FIGURE 6.21 As for Fig. 6.19, but at 55 MYBP. Note that gas has accumulated.

accumulation in the deep part of the sandstone reservoir (Figs. 6.27, 6.28, and 6.29).

## 2. Set 2 Examples

To examine the oil and gas accumulation patterns due to gas solubility variations under the same geological conditions, Figs. 6.30 through 6.34 shows the oil and gas migration and accumulation in the 2-D section for a gas solubility of 50 mg/g, instead of the 5 mg/g of the preceding case. To avoid exhaustive illustrations, we present only a few time steps to point

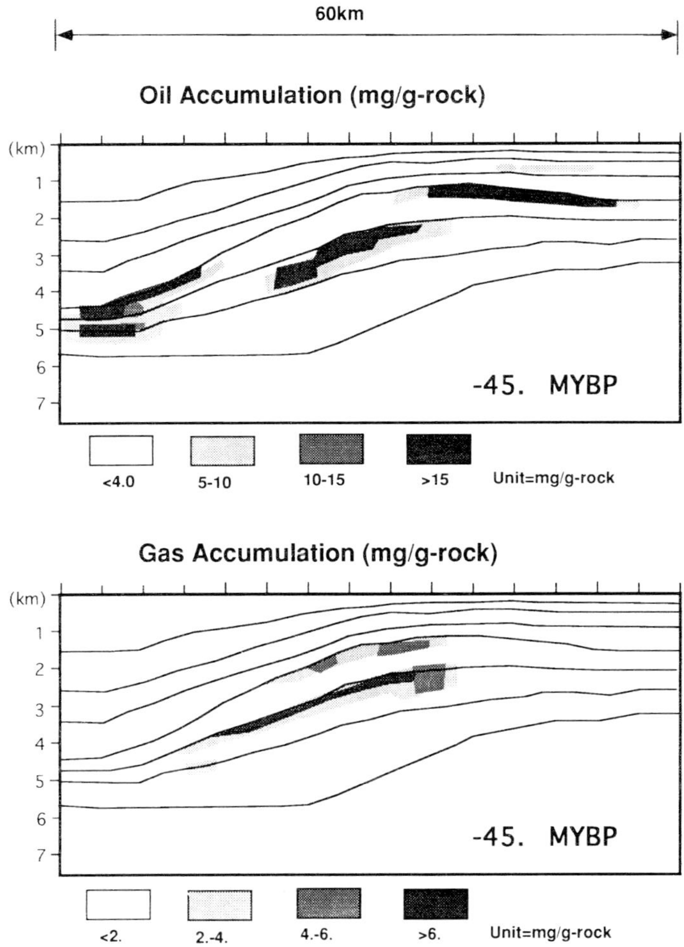


FIGURE 6.22 As for Fig. 6.19, but at 45 MYBP, showing oil and gas migration and accumulation.

out the differences. As expected, the oil starts accumulating with the same pattern as set 1. However, gas does not accumulate at 55 MYBP, unlike the 5 mg/g solubility case. This effect is due to the higher gas solubility, in which more gas is dissolved in oil and consequently the free-phase gas is reduced, therefore the relative permeability of gas is decreased (Figs. 6.30 and 6.20). Comparing the oil and gas migrating upward in the reservoir formations at 45 MYBP (Figs. 6.31 and 6.22), the positions and amounts of oil and gas accumulated are different. Before the uplift at 30 MYBP (Figs. 6.32 and 6.24), however, major oil and gas accumulations occur

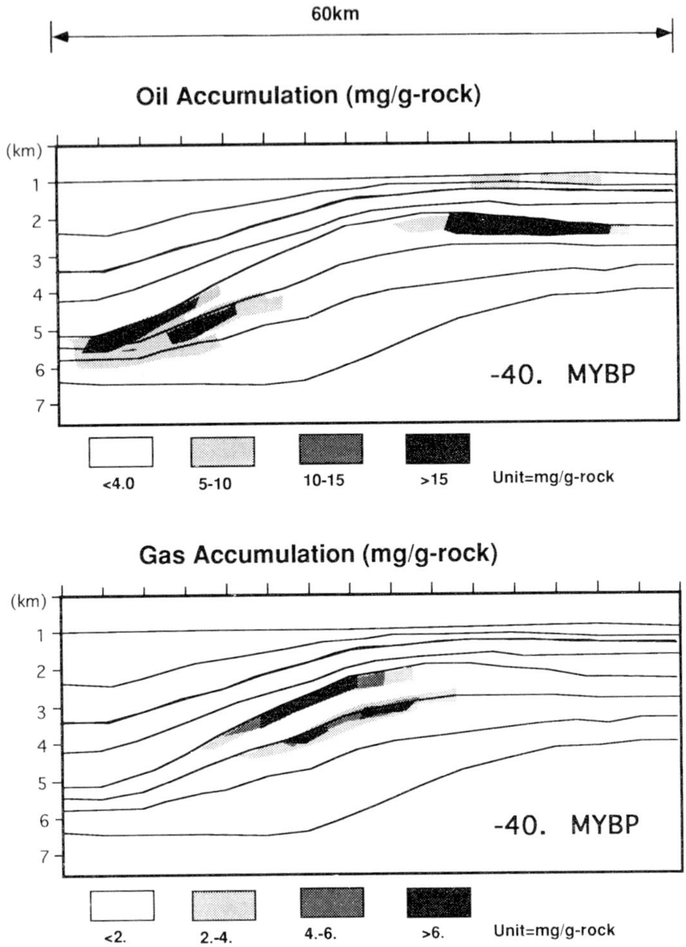


FIGURE 6.23 As for Fig. 6.19, but at 40 MYBP, showing continuation of oil and gas migration and accumulation along the sandstone formations.

despite the fact that the relative amounts are different. At 10 MYBP (Fig. 6.33), the various oil and gas accumulations in terms of amount and position are obvious. At the present day (Fig. 6.34), there is "spotty" oil accumulation and relatively more gas in the reservoir formation. Leakage of oil is also predicted in this case.

### 3. Set 3 Examples

Consider the effects of various relative permeability indices on oil and gas accumulations. Setting the permeability indices at 3.0 for water, oil, and gas (e.g., $a = b = c = 3.0$) in this set, instead of $a = b = 3.0$, $c =$

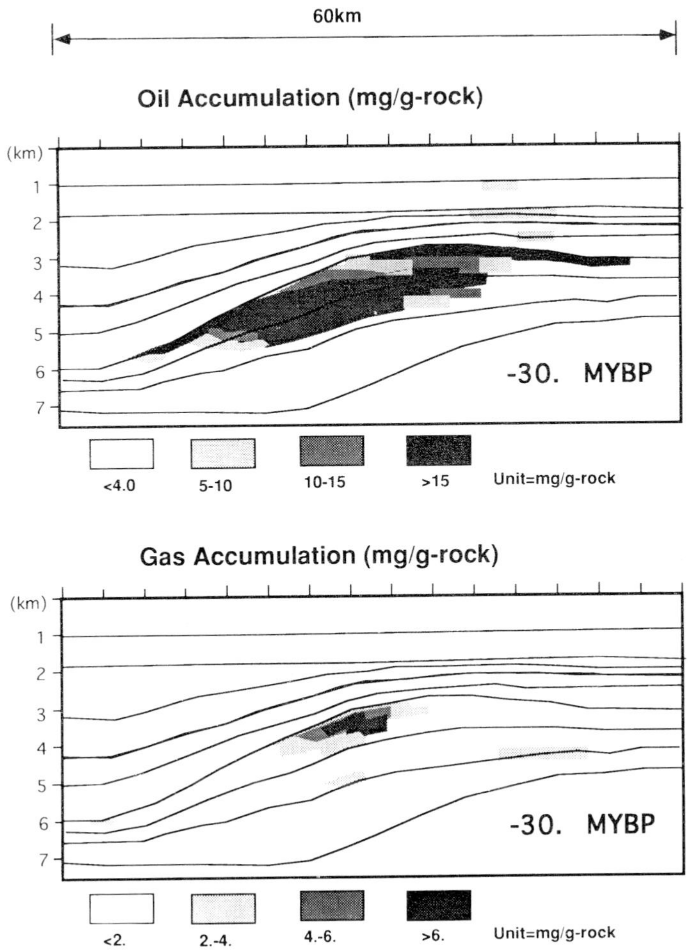


FIGURE 6.24 As for Fig. 6.19, but at 30 MYBP. Note the major oil and gas accumulations before the uplift.

1.0 as in set 1, Figs. 6.35 through 6.39 present the oil and gas accumulations under the same geological condition as set 1 with the gas solubility of 5 mg/g in oil. This setting makes oil more movable relative to set 1 results.

Comparing the figures of set 1 (e.g., Figs. 6.21, 6.22, 6.24, 6.27, and 6.29) with figures of this set at the same times, an obviously different pattern is shown of oil and gas accumulations in terms of amount and positions. At 55 MYBP a similar pattern of oil and gas accumulations occurs as for set 1 (see Figs. 6.35 and 6.21), because not enough time has elapsed for a migration difference in the early stage of oil and gas accumulations. How-

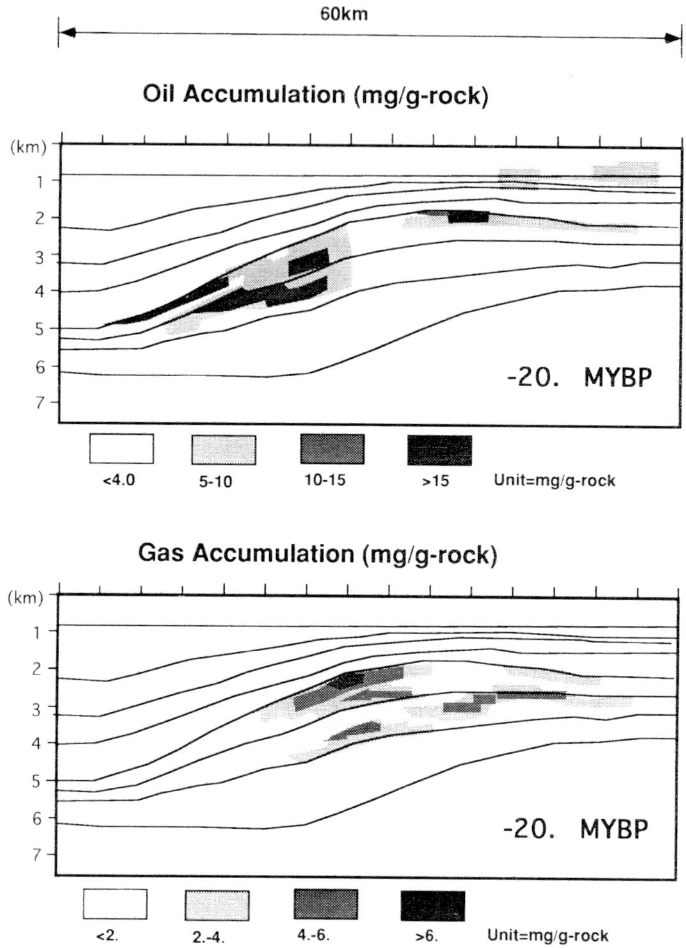


FIGURE 6.25 As for Fig. 6.19, but at 20 MYBP. Note the oil leakage from the sandstone reservoir.

ever, Figs. 6.36 and 6.37 at 45 and 30 MYBP, respectively, show much different oil and gas accumulations with less oil accumulated in the reservoir before uplift compared to set 1. Oil accumulation increases at 10 MYBP (Fig. 6.38) and considerable oil is lost due to leakage of the seal formation (Fig. 6.39) at the present day.

## D. DISCUSSION

This discussion of the examples indicates that the gas solubility in oil, and also the relative permeability indices, have significantly modified the

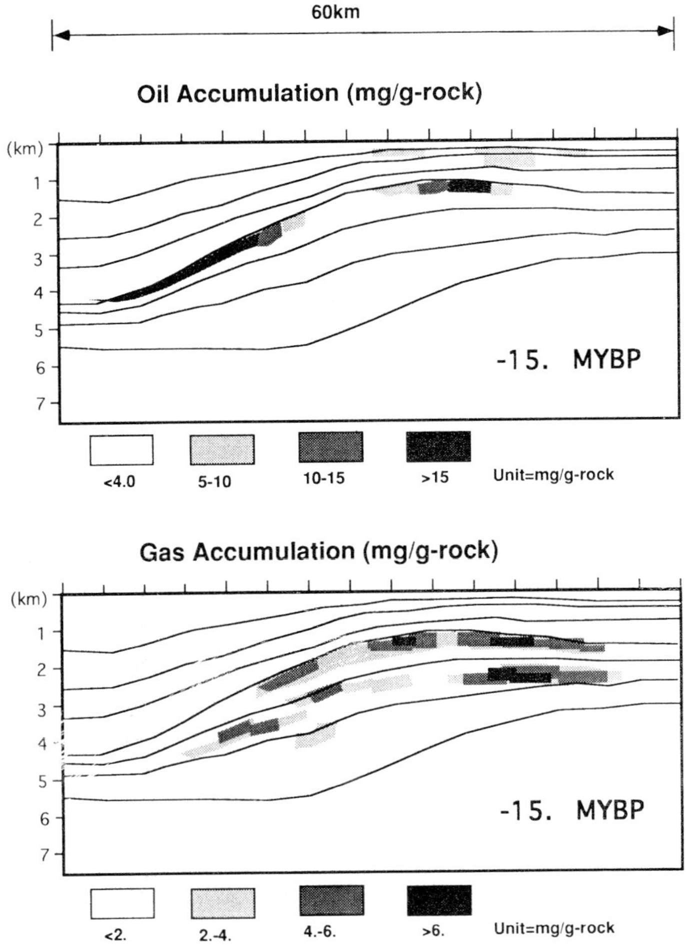


FIGURE 6.26 As for Fig. 6.19 but at 15 MYBP, showing oil and gas migration and accumulation.

process of hydrocarbon accumulation. For the same geological conditions, but different geochemical conditions with various gas and oil aqueous solubilities, very different hydrocarbon accumulation histories in the basin may occur.

Numerical modeling of sediment compaction and fluid flow provides a tool to study the processes of hydrocarbon generation, migration, and accumulation in evolving sedimentary basins. This study has shown the importance of solubilities and relative permeabilities in the processes of hydrocarbon accumulations in general, and of migration in particular. Migration of each phase in the three-phase system is dominantly controlled

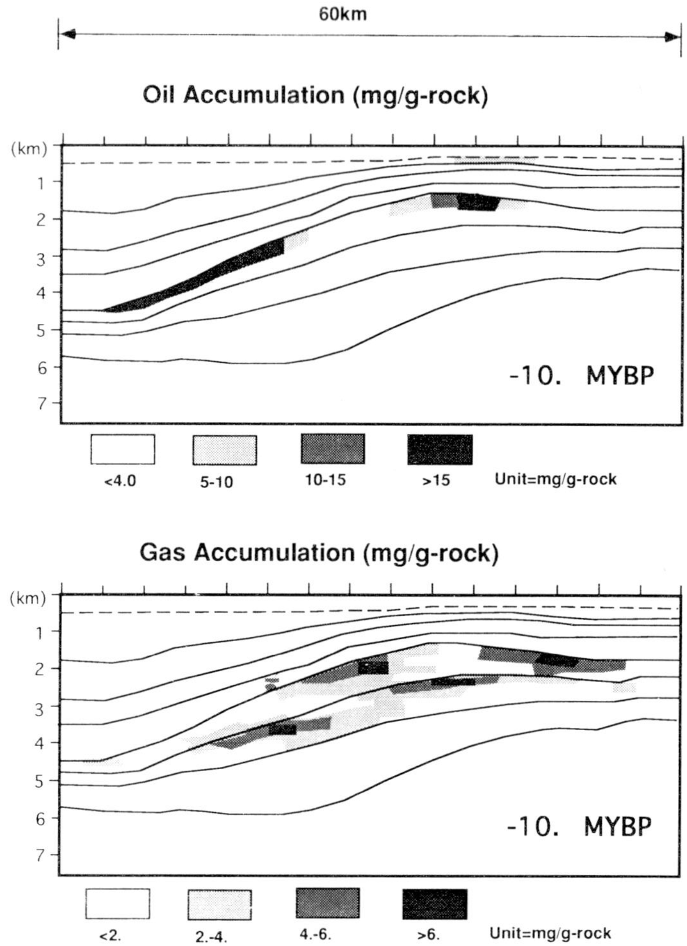


FIGURE 6.27 As for Fig. 6.19 but at 10 MYBP, showing oil and gas migration and accumulation.

by its relative permeability, related to the saturation of each phase in the pore space by a power law function with freely adjustable power index factors $a$, $b$, and $c$ [see Eqs. (6.9), (6.10), and (6.11)]. In this way individual free-phase migration can be accelerated or delayed by adjusting the power indices.

Saturation of a pore is related to the amounts of oil/gas in the free phase. The amounts of oil/gas that can be dissolved in solution are controlled by the solubility, dependent on temperature and pressure in the system, as

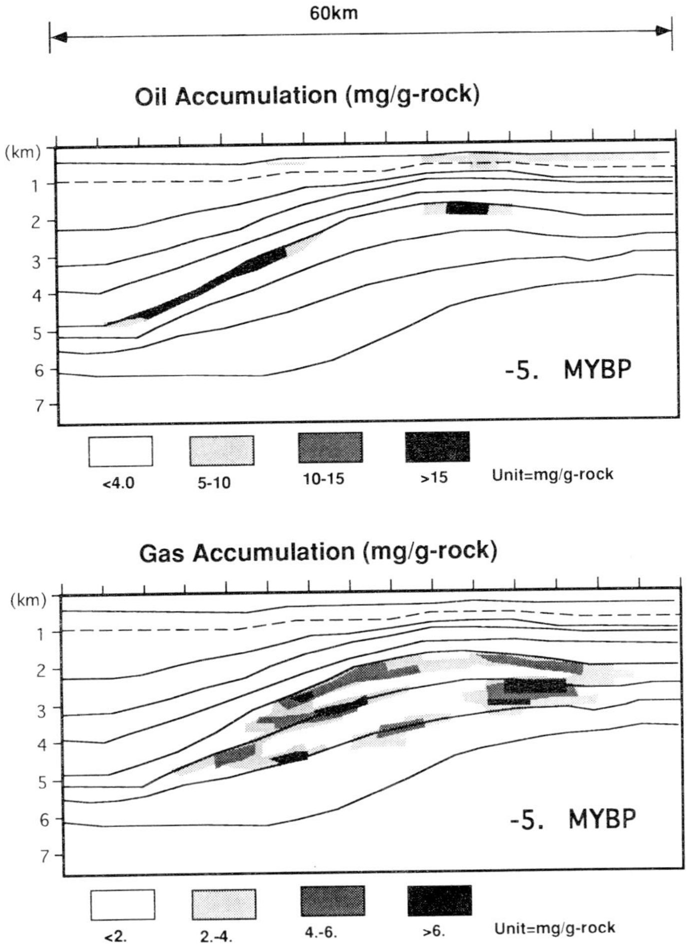


FIGURE 6.28 As for Fig. 6.19 but at 5 MYBP, showing oil and gas migration and accumulation.

well as by the availability of oil and gas. The relative permeability of water is related only to the movable water in the pore space and not to the irreducible water saturation.

The test results in this study indicate that oil and gas accumulation patterns vary substantially with solubility variations and with relative permeabilities. However, there is no single trend with solubility variation that encompasses hydrocarbon accumulation in reservoir rocks. It would seem that the specific geologic, geothermal, and geochemical conditions, as well

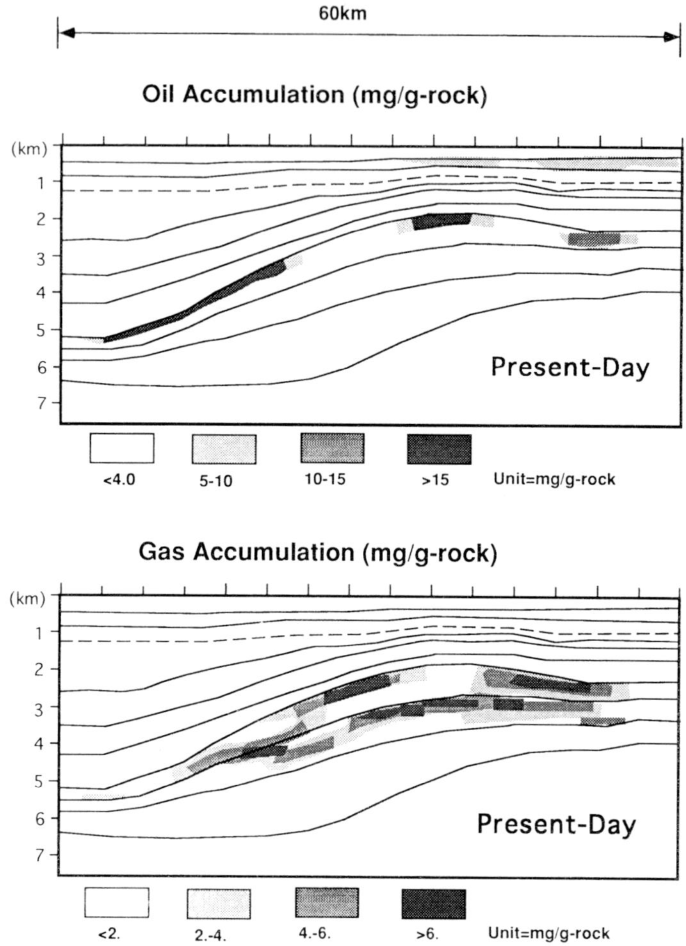


FIGURE 6.29 As for Fig. 6.19 but at present day, showing oil and gas migration and accumulation.

as capillary pressure, mixed-phase buoyancy, transport in solution, and free-phase transport, all of which are continually varying in the basin at different locations as the temperature and pressure are adjusting, are individually and collectively influencing the patterns of hydrocarbon accumulation.

The Hammerfest Basin study shows different hydrocarbon accumulation histories with various solubilities and relative permeabilities, illuminating the importance of oil and gas solubility effects. Some of the hydrocarbon accumulations are destroyed by Tertiary uplift and erosion, likely the cause of the majority of subcommercial exploration holes in the basin.

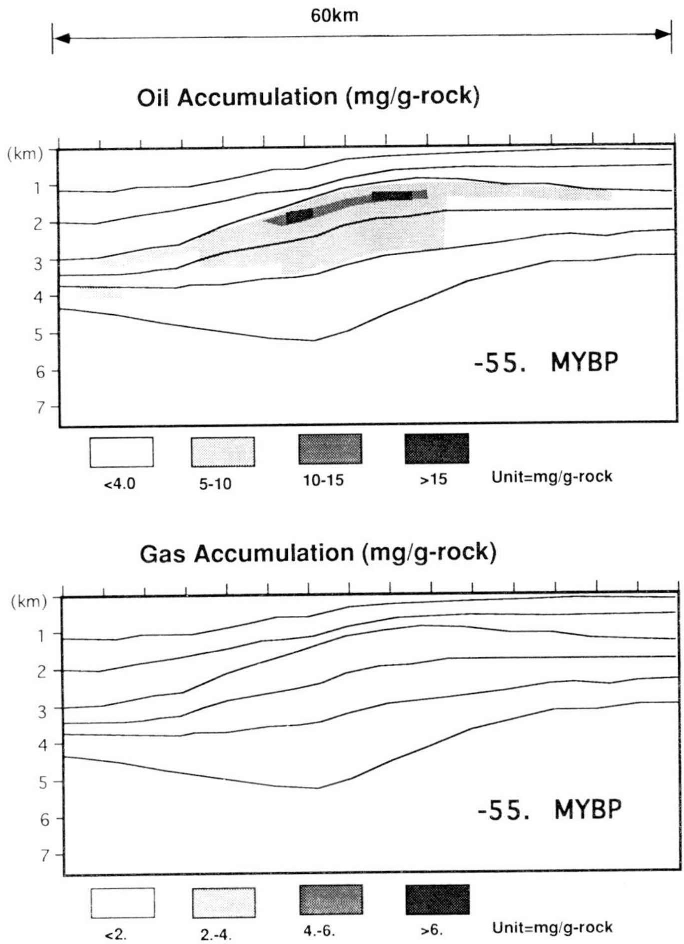


FIGURE 6.30 Oil and gas accumulation section at 55 MYBP for gas solubility set at 50 mg/g and relative permeability indices of 3.0 for water and oil and 1.0 for gas.

## APPENDIX A: REVIEW OF DEFINITIONS

Divide the source rock layer evenly into $n$ subdivisions (Fig. 6.2), then (Davies and Goldsmith, 1972): The arithmetic average TOC in the source rock layer is

$$\langle \mathrm{TOC} \rangle = \sum_{i=1}^{n} \mathrm{TOC}_i / n. \tag{A.1}$$

The standard deviation of the TOC distribution is

$$\sigma = \left[ \sum_{i=1}^{n} (\mathrm{TOC}_i - \langle \mathrm{TOC} \rangle)^2 / (n-1) \right]^{1/2} \tag{A.2}$$

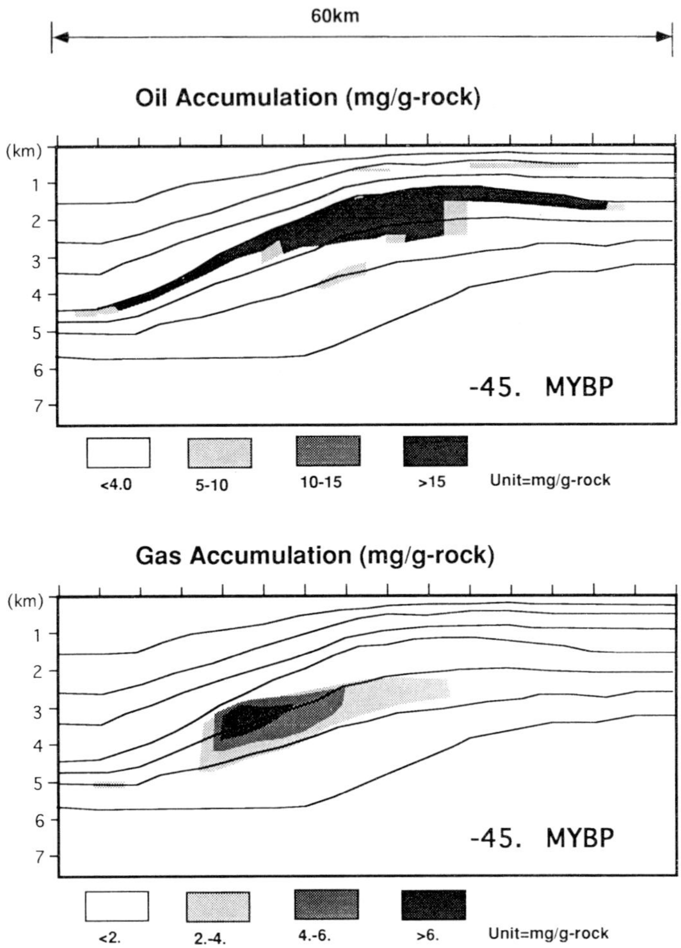


FIGURE 6.31 As for Fig. 6.30, but at 45 MYBP.

and the coefficient of skewness is defined as

$$b_1 = \sum_{i=1}^{n} W_i (X_i - \bar{X})^3 / c, \tag{A.3}$$

where

$$W_i = \frac{\mathrm{TOC}_i}{\sum_{i=1}^{n} \mathrm{TOC}_i} \tag{A.4}$$

and $\bar{X}$ is the center position of the source layer, $X_i = 1, \ldots, n$, number of positions within the layer from bottom to top, $c = \bar{X}^3$, to normalize skewness within $[-1, 1]$.

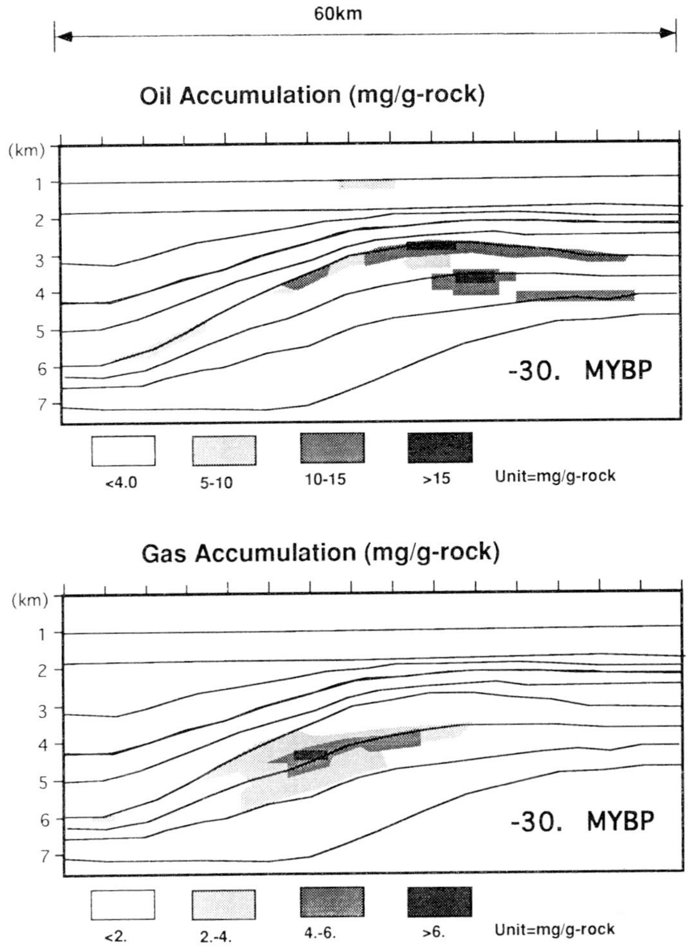


FIGURE 6.32 As for Fig. 6.30, but at 30 MYBP.

This skewness definition ensures a positive value for a distribution with more TOC weight in the upper part of the source formation and a negative value for a distribution with more TOC weight in the lower part. A value of zero skewness indicates a symmetric distribution.

## APPENDIX B: SOLUBILITIES IN A THREE-PHASE FLUID SYSTEM

### B.I. MIXING

Prior to hydrocarbon generation the sedimentary pore space is taken to be fully occupied by water. Thus for any unit volume (a "cell" in the two-

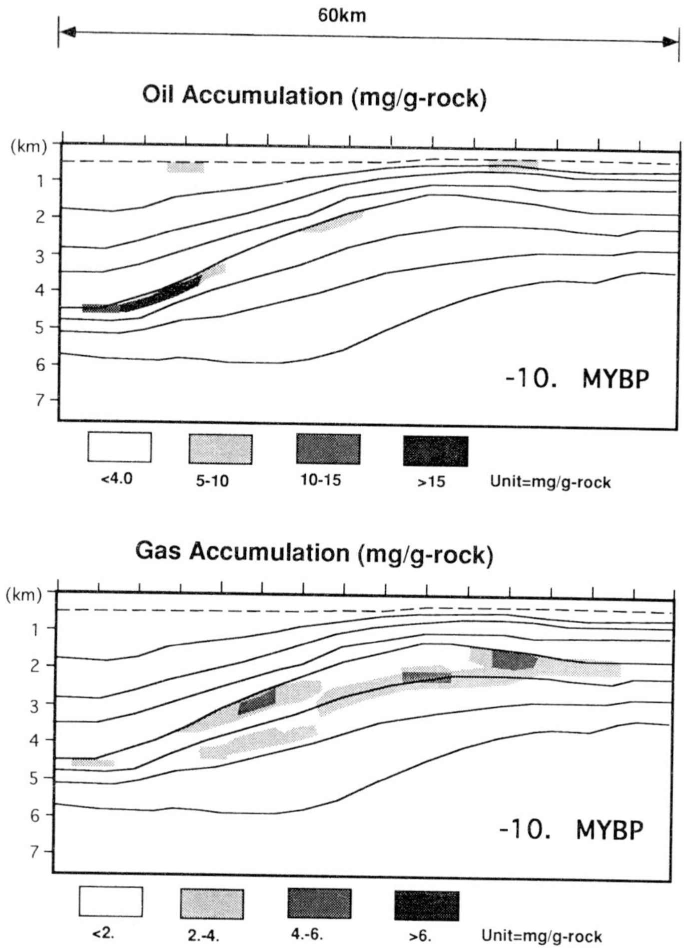


FIGURE 6.33 As for Fig. 6.30, but at 10 MYBP.

dimensional grid system of GEOPETII) of rock, we have

$$V_o = 0; \qquad V_g = 0; \qquad V_p = V_o + V_g + V_w, \tag{B.1}$$

where $V_p$ is the total void volume in the cell and $V_o$, $V_g$, and $V_w$ represent the void volume fractions taken by oil, gas, and water, respectively.

With the dynamic development of a sedimentary basin, the total masses of oil, gas, and water (in units of grams) in a cell at time $t$ are represented by $M_o$, $M_g$, and $M_w$, respectively. To consider the solubility mixing of oil, gas, and water in the three-phase system, the amounts of oil and gas in

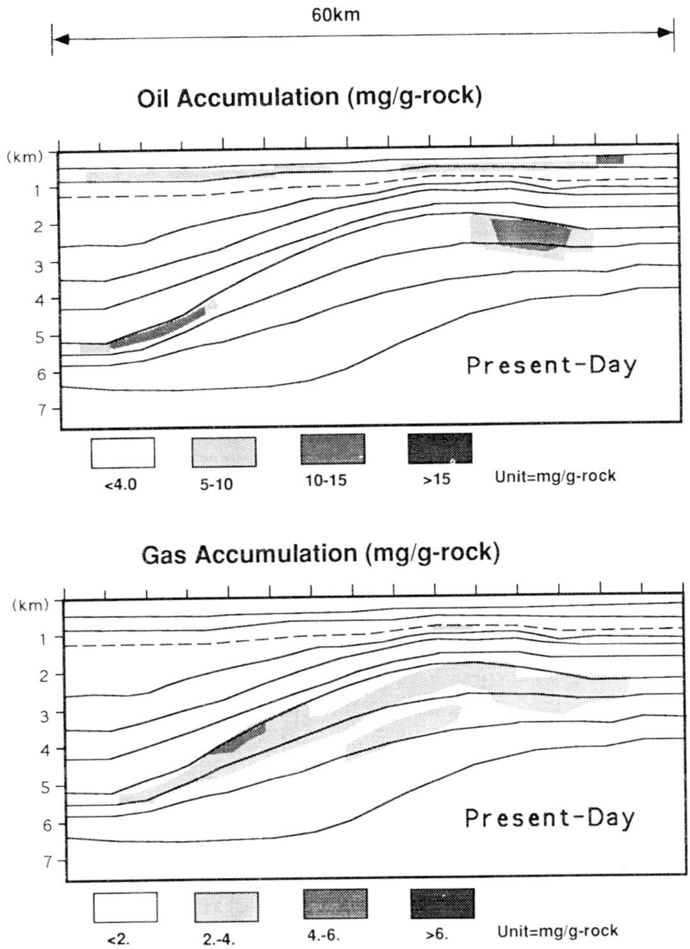


FIGURE 6.34 As for Fig. 6.30, but at present day.

water solution, as well as gas in oil solution, are given through

$$M_{\mathrm{wo}} = (M_{\mathrm{w}})S_{\mathrm{wo}} \times 10^{-3}, \qquad \text{oil in water;} \tag{B.2}$$

$$M_{\mathrm{wg}} = (M_{\mathrm{w}})S_{\mathrm{wg}} \times 10^{-3}, \qquad \text{gas in water; and} \tag{B.3}$$

$$M_{\mathrm{og}} = (M_{\mathrm{o}})S_{\mathrm{og}} \times 10^{-3}, \qquad \text{gas in oil} \tag{B.4}$$

provided $M_{\mathrm{wo}} \le M_{\mathrm{o}}$, $M_{\mathrm{wg}} + M_{\mathrm{og}} + M_{\mathrm{g}}$, where $S_{\mathrm{wo}}$, $S_{\mathrm{wg}}$, and $S_{\mathrm{og}}$ represent solubilities of oil in water, gas in water, and gas in oil, respectively, in units of mg/g, and all three can be temperature and pressure dependent.

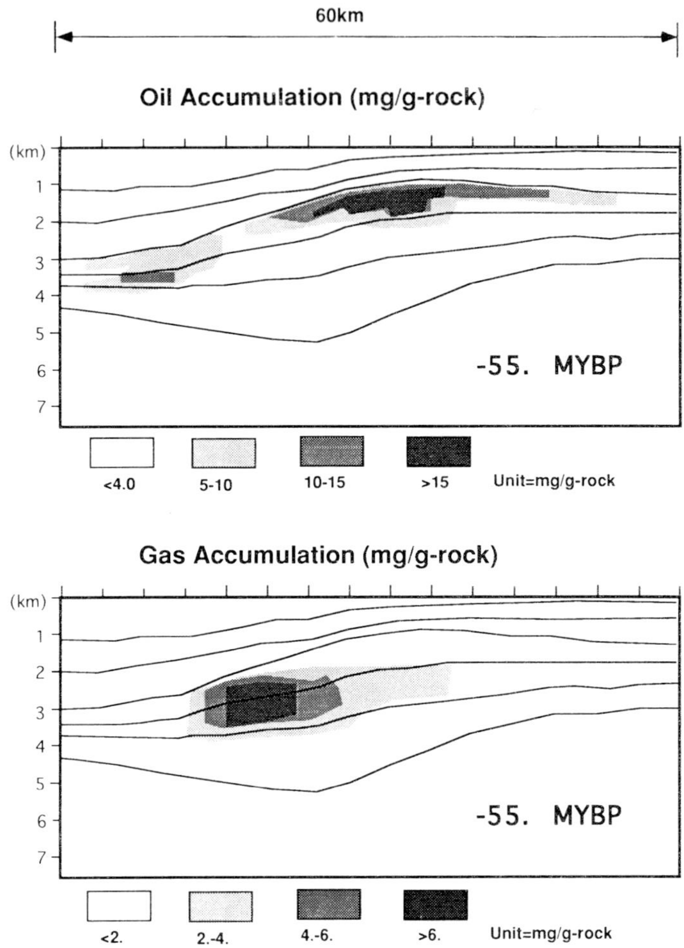


FIGURE 6.35 Oil and gas accumulation section at 55 MYBP for gas solubility set at 5 mg/g and relative permeability indices of 3.0 for water, oil, and gas.

So the amounts of gas, oil, and water in free-phase after mixing are given respectively as

$$M'_{\mathrm{g}} = \begin{cases} (M_{\mathrm{g}} - M_{\mathrm{wg}} - M_{\mathrm{og}}) & \text{in } M_{\mathrm{g}} > (M_{\mathrm{wg}} + M_{\mathrm{og}}) \\ 0 & \text{in } M_{\mathrm{g}} \leq (M_{\mathrm{wg}} + M_{\mathrm{og}}) \end{cases}, \tag{B.5}$$

$$M'_{\mathrm{o}} = \begin{cases} (M_{\mathrm{o}} - M_{\mathrm{wo}} + M_{\mathrm{og}}) & \text{in } M_{\mathrm{o}} > (M_{\mathrm{wo}} - M_{\mathrm{og}}) \\ 0 & \text{in } M_{\mathrm{o}} \leq (M_{\mathrm{wo}} - M_{\mathrm{og}}) \end{cases}, \tag{B.6}$$

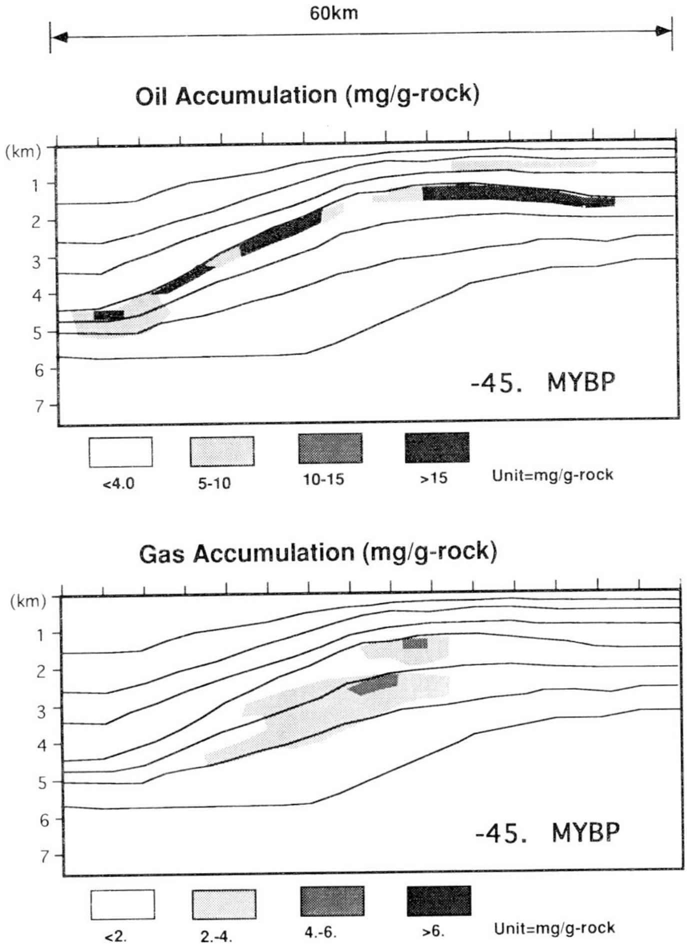


FIGURE 6.36 As for Fig. 6.35, but at 45 MYBP.

while

$$M'_{\mathrm{w}} = M_{\mathrm{w}} + M_{\mathrm{wo}} + M_{\mathrm{wg}}. \tag{B.7}$$

To calculate the fractional volumes of the pore space of volume $V_{\mathrm{p}}$, taken by each free phase, we have

$$V_{\mathrm{o}} = M'_{\mathrm{o}}/\rho_{\mathrm{o}}, \qquad V_{\mathrm{w}} = M'_{\mathrm{w}}/\rho_{\mathrm{w}} \tag{B.8}$$

and

$$V_{\mathrm{g}} = f_{\mathrm{vw}}(P, T, M'_{\mathrm{g}}), \tag{B.9}$$

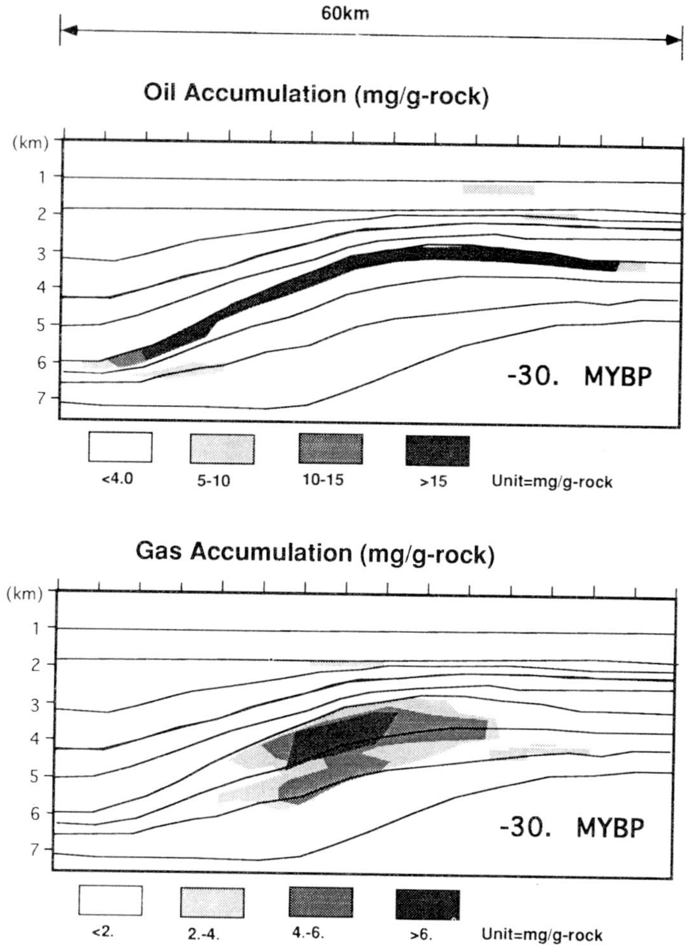


FIGURE 6.37 As for Fig. 6.35, but at 30 MYBP.

where $P$ is total pore pressure in Pa, $T$ is temperature in K, and $\rho_o$, $\rho_w$ are densities of oil and water, respectively. Here $f_{vw}$ represents the van der Waals' function for gas. The total pore pressure is adjusted so that $V_g/V_p = 1 - (V_w + V_o)/V_p \geq 0$, with saturations ($S_o$, $S_g$, and $S_w$ for oil, gas, and water, respectively) defined through $S_g = V_g/V_p$, $S_w = V_w/V_p$, $S_o = V_o/V_p$.

## B.II. MIGRATION

Once saturation of each phase in the pore space is determined, and the relative permeabilities of each flow phase assigned, or specified, as functions of the saturation, Darcy's law for three-phase migration can be applied, leading to the flow rates of each phase and therefore providing the amount

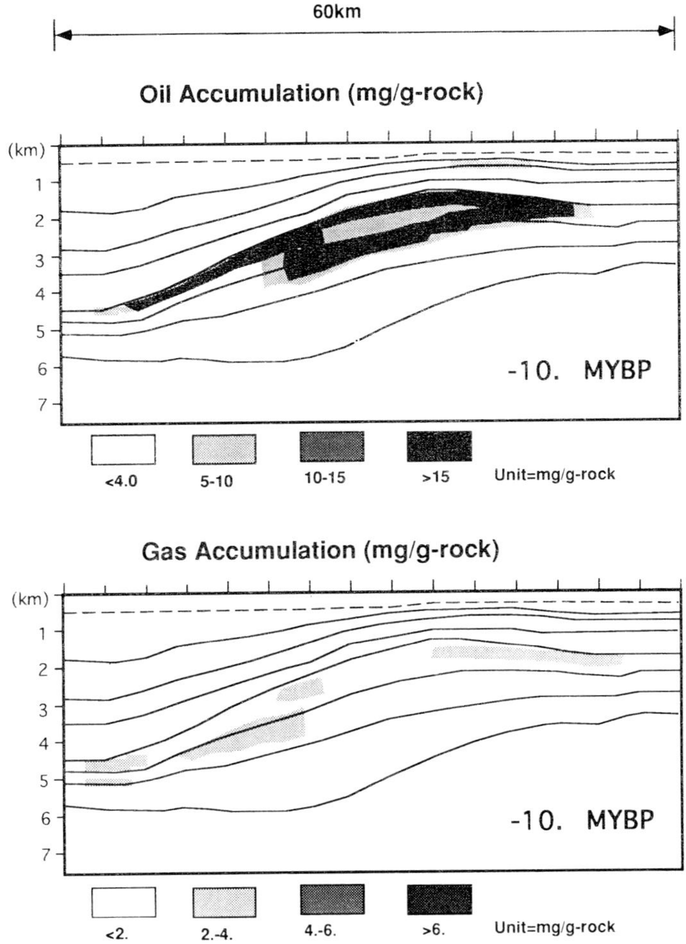


FIGURE 6.38 As for Fig. 6.35, but at 10 MYBP.

of each fluid migrated out of a pore volume within one time step ($\Delta t$) of the calculation. During the time step, because hydrocarbon generation continues, additional oil and gas masses are generated in the amounts, $\Delta M_o$ and $\Delta M_g$ for the grid cell of volume $V_p$. The total amounts of oil and gas at a time $(t + \Delta t)$ in the cell, before remixing, are

$$M_o = \frac{V'_o}{\rho_o} + M'_{wo} + \Delta M_o, \tag{B.10}$$

$$M_g = f_{vw}(P, T, M'_g) + M'_{wg} + M'_{og} + \Delta M_g, \tag{B.11}$$

$$M_w = V'_w/\rho_w. \tag{B.12}$$

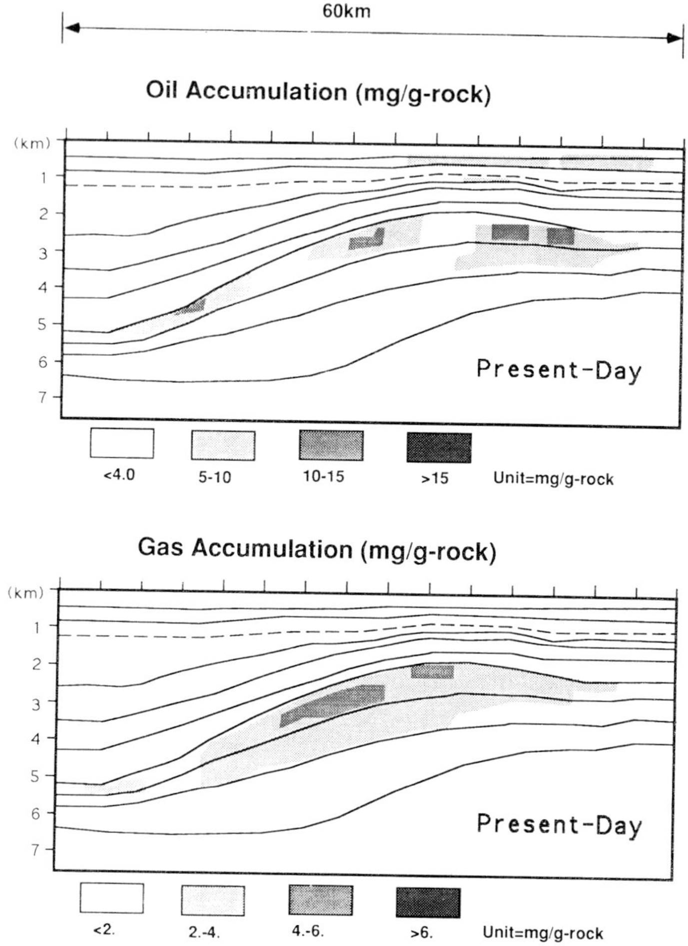


FIGURE 6.39 As for Fig. 6.35, but at present day.

The amounts of gas, oil, and water given by Eqs. (B.10), (B.11), and (B.12) are then mixed as per Eqs. (B.2), (B.3), and (B.4) and the process repeated through each subsequent time step until the present day is reached. The individual free-phase fluid components are tracked from cell to cell, as are the amounts in solution, and compaction of cells is also allowed for dynamically as the basinal system evolves with time. This appendix describes only the "mixing" procedure for a pore volume partially occupied by three fluid phases, including generation of oil and/or gas adding to the amounts available as time proceeds.

# 7

# FAULTS: OPEN OR SHUT?

## I. INTRODUCTION

The impact of faults in determining or controlling sedimentary basin fill, extensional and compressional tectonic regimes, structural evolution and stratigraphic juxtapositions, including roll-over structures and horst/graben features, fluid flow, sedimentary compaction, and present-day sites of potential hydrocarbon accumulations is well known (Hamblin, 1965; Crans *et al.*, 1980; Gibbs, 1983; White *et al.*, 1986; Nunns, 1991; Ocamb, 1961; Bally, 1983; Evamy *et al.*, 1978; James, 1984; Groshong, 1989; Rowan and Kligfield, 1989; Dula, 1991; Xiao and Suppe, 1988, 1989, 1992; Jackson and Galloway, 1984; Suppe, 1983, 1985; Cloos, 1968; Walsh and Watterson, 1988; Thomsen *et al.*, 1992; Kerr and White, 1996).

The geometrical relationship between faults and sediments has long been investigated quantitatively, but the corresponding dynamic impact of faults on fluid flow, compaction, thermal focusing, and hydrocarbon migration and accumulation has not yet been subjected to the same degree of thorough examination. The reason appears to be that the interactive dynamic development of subsurface fluids and faults requires that quantitative dynamic fluid flow/compaction codes be available that deal with at least three phases of fluids (water, oil, and gas) in terms of generation, migration, solubility, and in a minimum of two spatial dimensions at each instant of time; in addition, the ability to fault the sediments at some instant of time, or over a period of time, must also be included in such codes. The types of faults that need to be included are vertical, left and right lateral normal slant faults, left and right lateral reverse slant faults, arcuate faults, and listric growth faults. In addition, the faults have to be either hydraulically open or shut to fluid phase flow. There has been, and continues to be, a general and growing trend to include such faulting effects quantitatively in multidimensional dynamic fluid flow/compaction codes. For instance, in an investigation of the Pinedale Anticline "pop-up" structure, Wei and Lerche (1988) included a vertical fault which could be either open or shut to fluid flow; Liu *et al.* (1993) investigated the impact of multiple normal slant faults on the compaction and fluid flow in a two-dimensional basin; Bour *et al.* (1993) examined the interactive impact of two close faults in controlling the migration pathways taken by generated hydrocarbons; while Maubeuge and Lerche (1993) have shown how a dynamically evolving regional fault, driven by an underlying, uprising shale diapir, can be investigated quantitatively for its influence on fluid loss from syndepositional sediments.

Thus the inclusion of simple fault patterns in quantitative basin analysis models is burgeoning. Yet, many serious concerns remain that have to be addressed in a quantitative manner. For instance, the creation of a fault, with the concomitant vertical and/or lateral displacement of sediments from their prefault locations, generally requires a spatial volume accommodation (a lateral increase in the case of extensional faults, a vertical increase in the case of compressional faults). Such volume accommodations are long range in the sense that the total domain of the system under investigation must change to permit volume shifts. Thus either lateral boundaries or vertical boundaries must evolve during the time of a faulting event. Clearly, a problem of some concern is to evaluate quantitatively the importance of different boundary conditions in creating different geological scenarios that evolve with time and, simultaneously, satisfy present-day observations of faults and sedimentary beds.

In addition to such boundary effects influencing the composite system, there are also evolving internal changes in the sedimentary packages which are either caused or modified by the occurrence of a fault—such as the change in sedimentary location with time which, in turn, changes the local load, thereby causing a change in fluid flow so that the sedimentary thickness and sediment integrity (fracturing) can be changed with time.

Thus the evolution in time and space of liquid and gaseous hydrocarbons in terms of generation, migration, and accumulation is altered by the structural changes and variable stratigraphic juxtapositioning brought about by fault creation. In turn, the changes in fluid transport influence the volume of sediments so that space creation is also changed, leading to a shift in patterns of sediment roll-over along faults, and to changes in the position and curvature of faults with time and sedimentation. A simple example of such a change in fault curvature is presented in Xiao and Suppe (1989). Such internal changes with time in sedimentary beds also have roles to play in evaluating dynamic geological scenarios in relation to present-day observations.

When both boundary effects and dynamic sedimentary evolution occur together, the complexity and plethora of potential geologic scenarios increases. Some way must be found to assess the range of each parameter involved in describing a given scenario in relation to the observations made today, and of assessing the dominance of internal sedimentary effects and/or boundary effects. In such assessments, the honoring of present-day observations of multiple fault locations, sedimentary bed locations, and spatial thickness variations must be maintained. The purpose of the present chapter is to show that such a goal is achievable for multiple faults and multiple sedimentary beds. A paper by Kerr *et al.* (1993), in which deformation, compaction, and sedimentation are addressed in relation to fault behavior for 3-D normal faults, is of relevance to the present chapter in that it provides a start on the more complex problem of mltiple, overlapping, arbitrary faults, as is a recent paper by Lerche (1994).

One of the major concerns in quantitative basin analysis is to provide models that incorporate more geologic processes as sophistication increases so that measures of basinal development are available which more closely predict observations. To foster and promote such ends, the Basin Modeling Group at the University of South Carolina (USC) has constructed models that currently have reached their epitome in a two-dimensional fluid flow/compaction code called GEOPETII (Nakayama, 1987; Wei and Lerche, 1988; Yu, 1992). This particular code allows three independent phases (water, oil, gas) of fluid flow with solubility effects; incorporates erosion, spatially variable deposition, variable lithologies, cementation, and dissolution of formations, anisotropic permeabilities, and thermal conductivities of formations; allows for heat transfer by both conduction and convection with spatially and temporally variable heat flux; and permits hydrocarbon generation (through a multitude of possible kinetic models, if necessary), basement motion with time, and variations of paleo sea level and paleo surface temperature. The model incorporates capillary pressure, surface tension, buoyancy, thermal expansion of fluids, and overpressure as driving mechanisms for fluid motion. In addition the ability to handle up to 10 vertical faults (hydrodynamically open or shut to fluid flow) and to allow for fracture failure of rocks is included. The majority of these factors have been used in application studies to basins through the years and a summary of the types of processes included is contained in Lerche (1990a,b).

## II. NORMAL SLANT FAULT MODEL

### A. REPRESENTATION OF SLANT FAULT DEVELOPMENT

The purpose of this section of the chapter is to exhibit a new facet of the code: the ability to handle multiple slant faults (up to 10 at a time can be handled; here two faults are sufficient to illuminate the points). The faults may be open or shut at the user's discretion and may have arbitrary throws which develop at arbitrary times. The main restrictions in this section are as follows:

1. Only normal slant faults are handled; reverse faults are not included.
2. Faults may not overlap laterally.
3. Each fault is taken to form instantaneously at its creation time and, at that time, is rigorously a slant line fault. Later development may curve the line of faulting, but does not cause a further throw of the fault except for vertical compaction effects.
4. The fault may be open or shut, and reactivation of the fault does not occur.

Each sedimentary unit is taken to be composed of a set of subdivided layers. The subdivisions must move as well, not only within the layers offset by the fault, but also within the layer before the first layer offset by the fault and the layer after the last layer offset by the fault. For these two layers, the amount of the move is not the throw anymore but a displacement less than the throw, according to the position of the subdivision in the layer itself: The subdivisions close to the layers offset by the fault will have a displacement almost equal to the throw, and the subdivisions furthest from the fault, where the effects of the fault are not felt anymore, will have a displacement almost equal to zero. The main condition is that the subdivisions have to stay equally spaced within their own layer. Within that framework, when it is remembered that there is an instantaneous shift of lithologic units along the fault line, it is surprising to see some of the effects that can be coaxed out of a set of scenarios. In this chapter two such cases are exhibited to illuminate some of the possible situations that can be modeled.

The first case is for two slant faults, which are taken to be hydraulically open. The second case examines the behavior of graben development as two normal slant faults are created, with the graben "walls" (read: slant faults) shut to fluid flow.

### 1. Fluid Flow/Compaction

The fluid movements in the sediments follow Darcy's law:

$$\frac{\partial}{\partial t}(\rho e) = \frac{\partial}{\partial \xi}\left[\frac{K_z \rho}{\mu(1+e)} \frac{\partial P_{\text{ex}}}{\partial \xi}\right] + \frac{\partial}{\partial \text{x}}\left[\frac{K_\text{x} \rho}{\mu} \frac{\partial P_{\text{ex}}}{\partial x}\right], \tag{7.1}$$

where $P_{\text{ex}}$ is the excess fluid pressure, $\rho$ is the fluid density, $K_{x(z)}$ is the horizontal (vertical) permeability, $\mu$ is the fluid viscosity, $e$ is the void ratio, which is related to porosity by

$$e = \phi/(1 - \phi), \tag{7.2}$$

and $\xi$ is the fully compacted coordinate which avoids the distortion of the sediment framework (and the resulting numerical instabilities during the calculations) by creating a frame of reference in which the rock can be considered not to compact or move. Physical depth, $z$, is related to $\xi$ by:

$$\xi = \int_0^z \frac{dz'}{1 + e(z')} \tag{7.3a}$$

or

$$z = \int_0^{\xi} d\xi' \, (1 + e(\xi')), \tag{7.3b}$$

where $e(z')$ is the void ratio and is dependent on depth, lateral position, and time.

The excess pressure ($P_{ex}$) is the difference between the pore presure ($P_p$) and the hydrostatic pressure ($P_h$); the pore pressure itself is the difference between overburden pressure ($P_h$) and effective stress ($S$). Thus:

$$P_{ex} = (P_{ob} - S) - P_h. \tag{7.4}$$

The effective stress is defined in turn as

$$S = P_f - P_g + P_{scaling}, \tag{7.5}$$

where $P_f$ (frame pressure) is the part of the total pressure supported by the matrix, $P_g$ is the pressure created by hydrocarbon generation process, and $P_{scaling}$ is a scaling value determined by the requirement that $S = 0$ at the sediment–water interface. Then

$$S = (P_f - P_f^*) - (P_g - P_g^*), \tag{7.6}$$

where $P_f^*$ and $P_g^*$ are scaling values at the mudline for the respective pressures, (i.e., both frame pressure and fluid pressure are set to zero at mudline). Thus Eq. (7.4) becomes

$$P_{ex} = g \int_{\xi}^{\xi^*} (\rho_s - \rho_w) \, d\xi' - (P_f - P_f^*) + (P_g - P_g^*), \tag{7.7}$$

where $\rho_s(\rho_w)$ is the density of the sediment matrix (water), $g$ is the acceleration of gravity ($980 \text{ cm} \cdot \text{s}^{-2}$), and $\xi^*$ is the value of the compacted coordinate at the sediment mudline. Furthermore, empirical laws for different kinds of lithology are assumed (with the same notations as before):

$$e/e^* = (P_f/P_f^*)^A, \tag{7.8}$$

$$K_z/K^* = (e/e^*)^B, \qquad \text{for shale;} \tag{7.9}$$

$$e = e^* - C(P_f - P_f^*), \tag{7.10}$$

$$K_z/K^* = (e/e^*)^B, \qquad \text{for sand;} \tag{7.11}$$

where $e^*(K^*)$ is the depositional void ratio (depositional permeability) and $A$, $B$, and $C$ are lithology-dependent constants, which characterize the compaction behavior of the sediments. A high value of $C$ will characterize a slowly compacting sand, whereas a high value of $A$ will characterize a rapidly compacting shale. On the other hand, the higher the value of $B$, the more the rock lowers its permeability with increasing compaction (decreasing void ratio). Methods of calibration of the $A$, $B$, and $C$ values with observed data can be found in Cao (1987) and Zhao and Lerche (1993). Moreover, all the surface values (*) presented in Eqs. (7.8) to (7.11) have to be determined, as do the coefficients $A$ and $B$ (for both sand and shale lithologies) and $C$ in order to match the model calculations with observations.

## 2. A Basic Fault Model

A slant fault is treated as an instantaneous event that offsets the sedimentary sequence. The formation of the fault follows the deposition of the faulted sequence: One block is anchored, the other block moves either downward (normal fault) or upward (reverse fault). Here we treat normal slant faults only.

Input information (depth, lithology, etc.) of the formations is given for a number of equally spaced pseudo-wells, covering the whole section. Up to 10 slant faults can be included at once with each occurring at the same or different times. To date, the fault has to be implemented in the section according to two criteria: (1) Faults may not overlap at given pseudo-wells and (2) the faulting event has to be instantaneous (therefore growth faults are excluded). Of the 10 faults allowed in a section, a maximum of one per pseudo-well is permitted. The information needed for a fault is (1) the depth of both blocks, (2) the throw, (3) the first layer, and (4) the last layer offset by the fault. A fault can be considered as shut or open, depending on the strain field around the fault zone and the consequent hydraulic properties. For the open fault, users can input the value of permeability. However, whether the fault is open or shut relates solely to the fluid flow; lateral communication depends on the offset induced by the fault to the formations. For instance, if a high-permeability sandy formation overlaps a low-permeability shaly formation after faulting, the lateral fluid flow in the sand will largely decrease at the fault position, and then the only direction available to the fluid is to flow up or down the fault, if the fault is maintained open. Both open and shut faults are considered in this chapter because communication to the surface (or to less overpressured formations) can be influenced by the prescription of the hydraulic character of the fault.

Within the theoretical background described earlier, a graben case with two slant faults at both sides is taken to occur in a shale sequence as follows. Shale is taken to be deposited at a uniform rate in time over a 50-Ma interval, such that the present-day thickness is just over 5000 m laterally

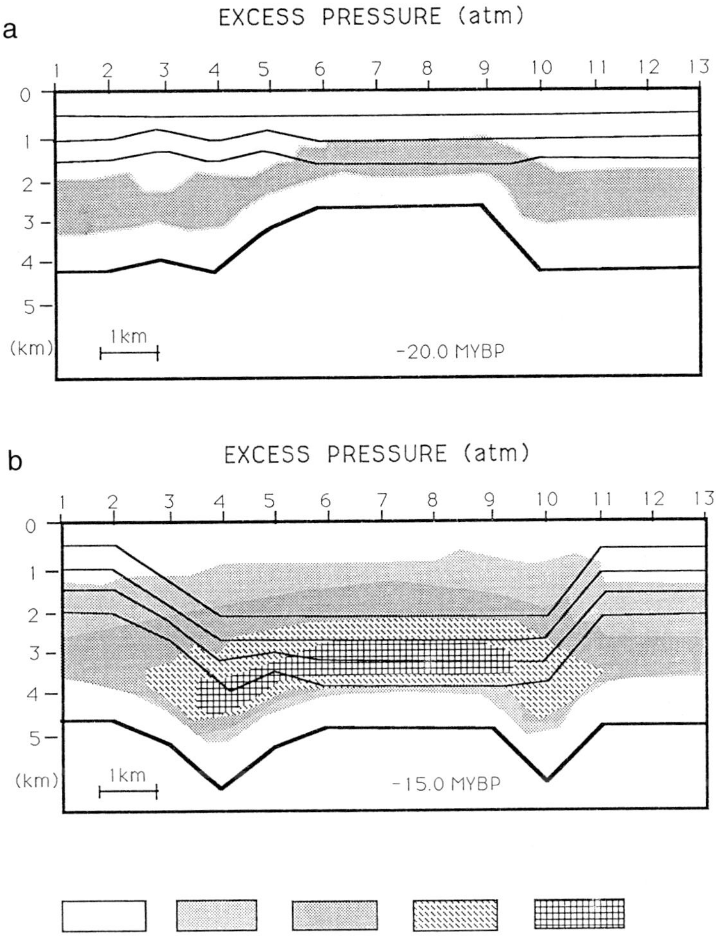


FIGURE 7.1 Graben development with two slant faults in shale sequence. Sections are displayed at 20, 15, 14.9, 10, and 5 MYBP and at present day. Excess pressure is superimposed, showing the "bleed off" of overpressure around the open slant faults. (*continues*)

far from the fault location. At 15 MYBP, two slant faults occurs instantaneously, offsetting formations presently between 1000 and 4500 m. The lateral section is taken to be 12 km long, and the left side fault offsets formations between pseudo-well locations 2 and 6, with 1-km spacing between pseudo-wells. The right side fault is between locations 9 and 11.

In the following discussion, the faults are taken to be open or shut hydraulically and the corresponding permeability values are to be assigned. A sketch of development of slant faults with superimposed excess pressure is shown in Fig. 7.1. Basement rock is taken to occur below 5200 m on the

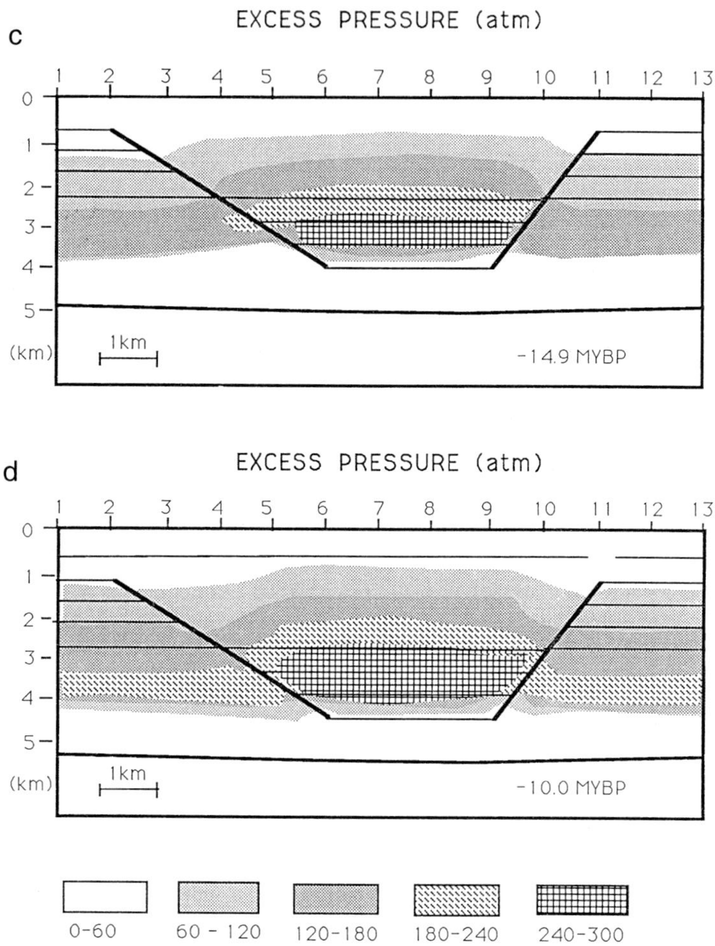


FIGURE 7.1 (*Continued*)

present-day section of Fig. 7.1, and is taken to be horizontal today. Response of the basement rock to applied load is taken to be purely isostatically compensated at each pseudo-well location. Thus, prior to the slant fault occurrence, when sediments "slide" down on the hanging wall block, the basement must have been subjected to a greater load at the locations on the downthrown block.

## B. FAULTED GRABEN FORMATION WITH "OPEN" FAULTS

The impacts of hydraulic open normal slant faults on fluid flow and compaction of sedimentary formations are investigated with a graben sce-

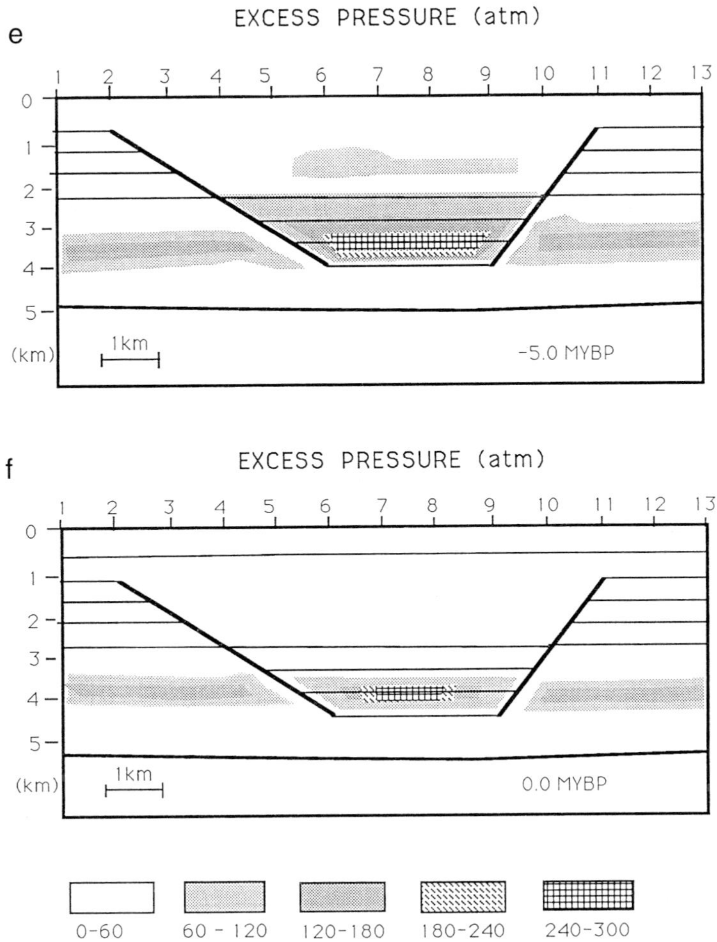


FIGURE 7.1 (*Continued*)

nario. Figure 7.2a depicts the development of permeability at the present-day across the section with the slant faults; Fig. 7.2b shows the permeability at location 3 (through the slant fault zone) with time and depth. Note that (1) high-permeability zones exist around the fault plane since the fault occurrence from 15 MYBP to the present day and (2) burial of the faulted formations continues as deposition proceeds. Because the faults could be dominant conduits for fluid flow, the development of overpressure across the section at different times is superimposed on Fig. 7.1. Prior to fault initiation, Figs. 7.1a and 7.1b show rapid development of excess pressure on the flanks of the horst, with a lesser development over the horst. However at, and after, fault initiation, overpressure is diminished on the nonmoving

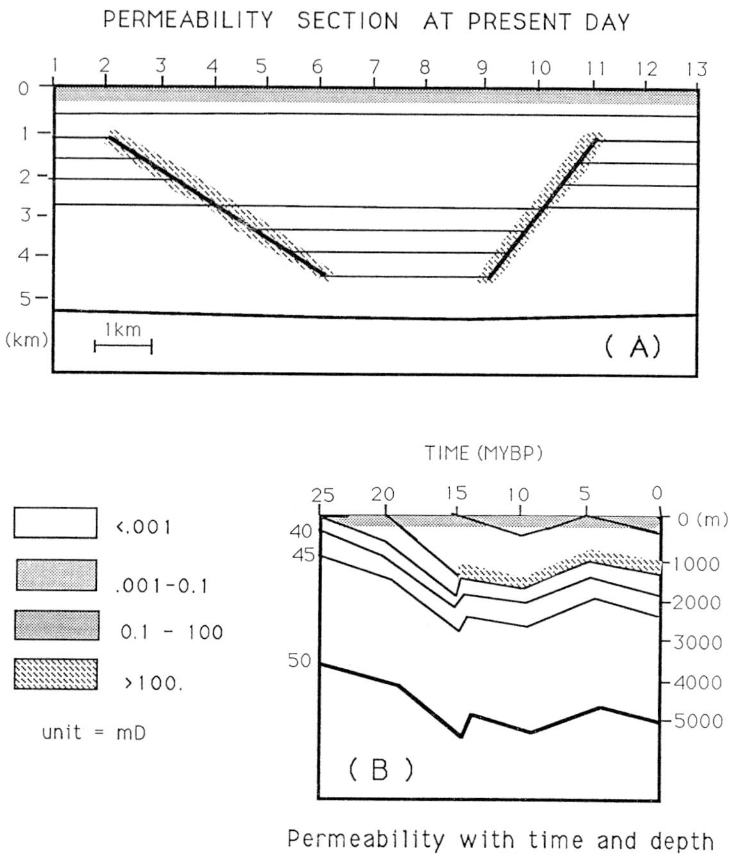


FIGURE 7.2 Permeability around open slant faults. (a) Cross-section at present day; (b) burial history at location 3 with depth and time.

blocks to either side of the graben, and is increased in the graben. The loss of fluid along the open fault planes after fault initiation then "bleeds off" overpressure both in the graben and in the nonmoving blocks so that, by present day as shown in Fig. 7.3, the overpressure decreases with closeness to the fault plane. The corresponding dominant effect on porosity with open faults across the section is to lower the porosity in the nonmoving block and to cause a diminution of porosity near the fault (Fig. 7.4) as fluid flows into the faults and escapes to shallower formations.

The two-dimensional porosity evolution indicates that the porosity decreases more rapidly with depth prior to faulting at locations outside the main fault zone, and, after faulting, a more nearly uniform lateral variation of porosity with time at each depth is observed due to the overpressure

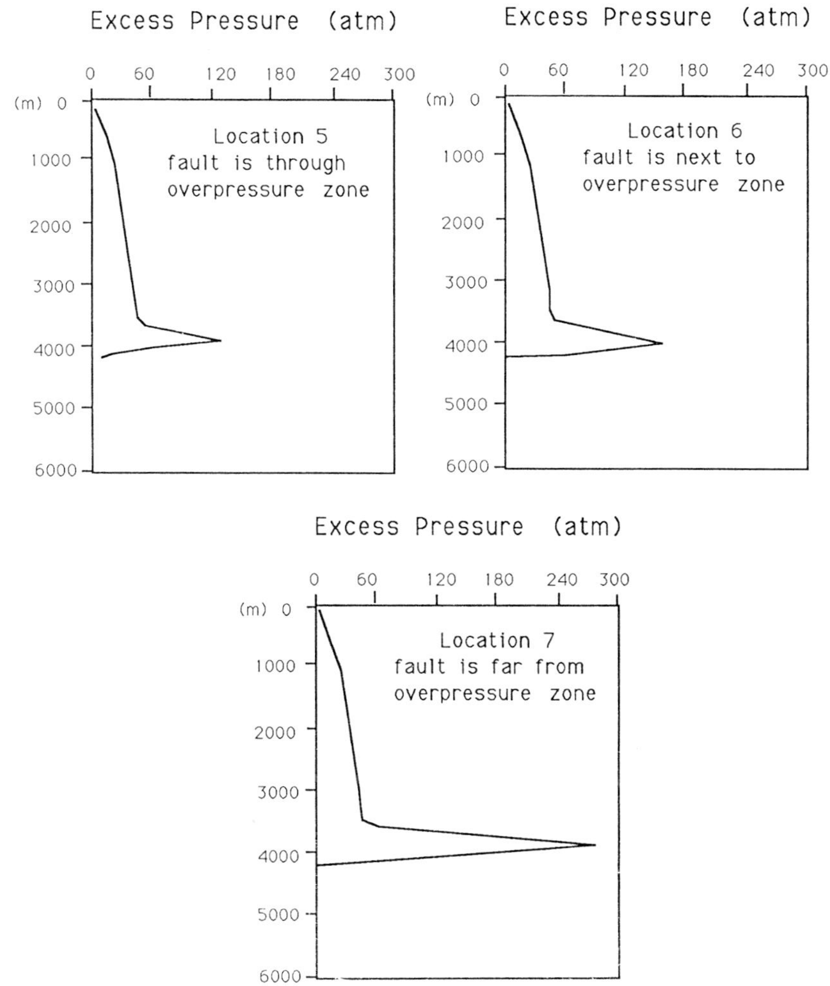


FIGURE 7.3 Excess pressure behaviors with depth for the open slant fault case at locations close to the fault plane showing that the closer the location is to the fault plane, the smaller the overpressure.

retention in the central graben sections and fluid flow release along the open faults constituting the graben walls.

Of interest from the point of view of hydrocarbon maturation is the temperature evolution with time in the presence of open faults. The corresponding evolution of temperature across the section is shown in Fig. 7.5 for the conditions of a constant surface temperature of 10°C and a constant heat flow from the basement of 1.5 HFU. Prior to faulting, the temperature for sediments of the same age is higher on the sides of the horst relative

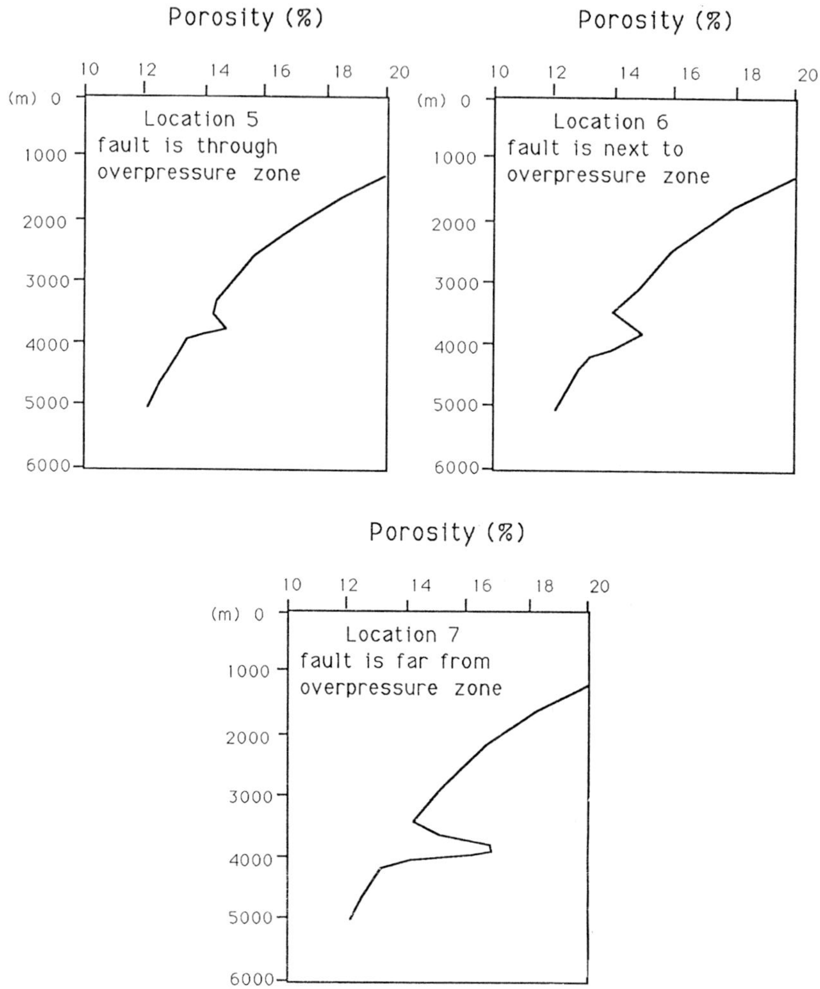


FIGURE 7.4 Porosity behaviors with depth for the open slant fault case, showing the diminution of porosity with approach to the fault.

to the center, but, after faulting, the dominance of the fault planes as fluid conduits and the lateral motion of the sediments both tend to covert the isotherms to a more horizontal pattern, with only a slight change in temperature laterally across the section. So sediments on the horst sides are not as overmature as they might have been without the fault, and sediments in the graben are not as undermature. Due to the narrow fault width relative to lateral and vertical variations of sediments, there is little impact on the overall spatial pattern of the temperature profile. The greatest effect would

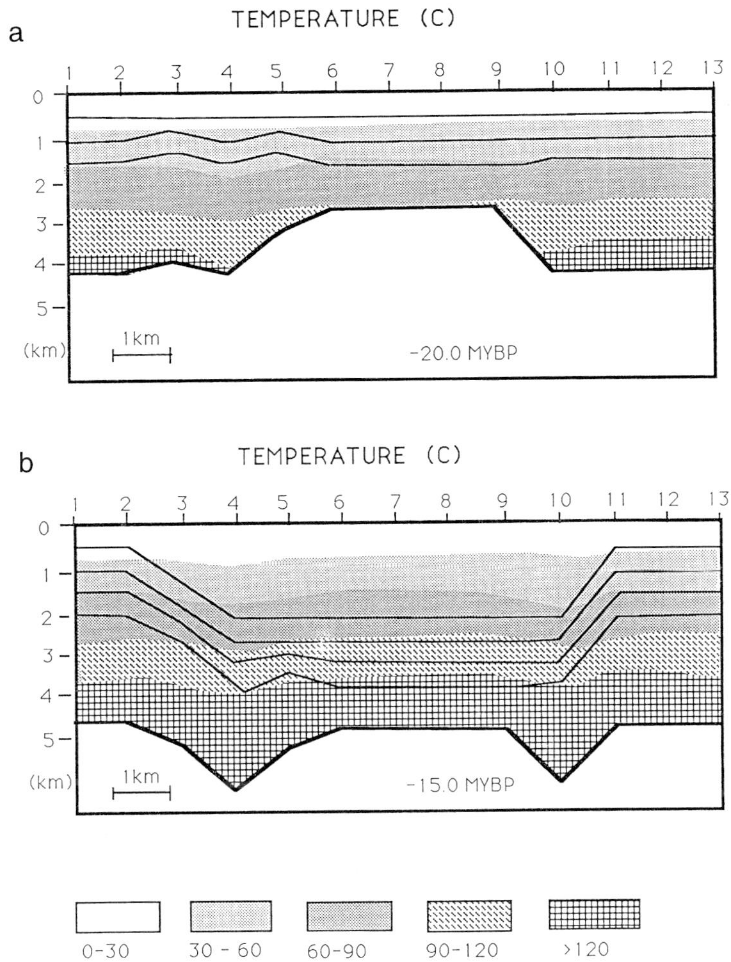


FIGURE 7.5 Temperature development with time for the open fault case, displayed at 20, 15, and 5 MYBP and at present day. (*continues*)

seem to be from the change in burial depth due to the motion of the sediments at fault creation time (15 MYBP).

## C. FAULTED GRABEN FORMATION WITH "CLOSED" FAULTS

To draw as close a correspondence as possible between the effects of hydraulically open and closed faults, precisely the same depositional situation, fault throw, and timing are considered in this section as in Section B. The sole difference is in the description of the slant fault hydraulic proper-

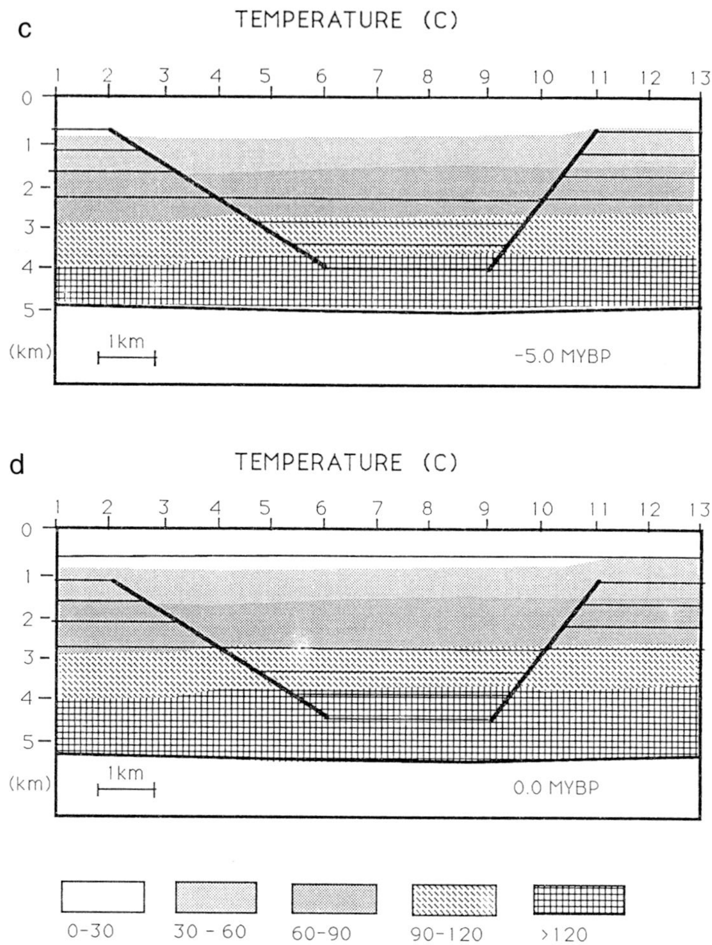


FIGURE 7.5 (*Continued*)

ties. The "open" fault of the previous section was treated by providing a high permeability along the fault; the case of the "closed" fault is taken to be represented by extending across the fault the lithologic permeability properties of formations on each side of the fault at a given depth. As differential compaction proceeds, these abutting permeability properties change, both because the lateral overlap of formations on each side of the fault changes with time as formations thin, and because intrinsic loss of fluids from juxtaposed lithologic units varies, thereby changing the internal permeability of each formation. The shift in differential load laterally as a consequence of the slant fault initiation also changes the excess fluid pressure that drives fluid flow. Thus both lateral and vertical fluid transport are

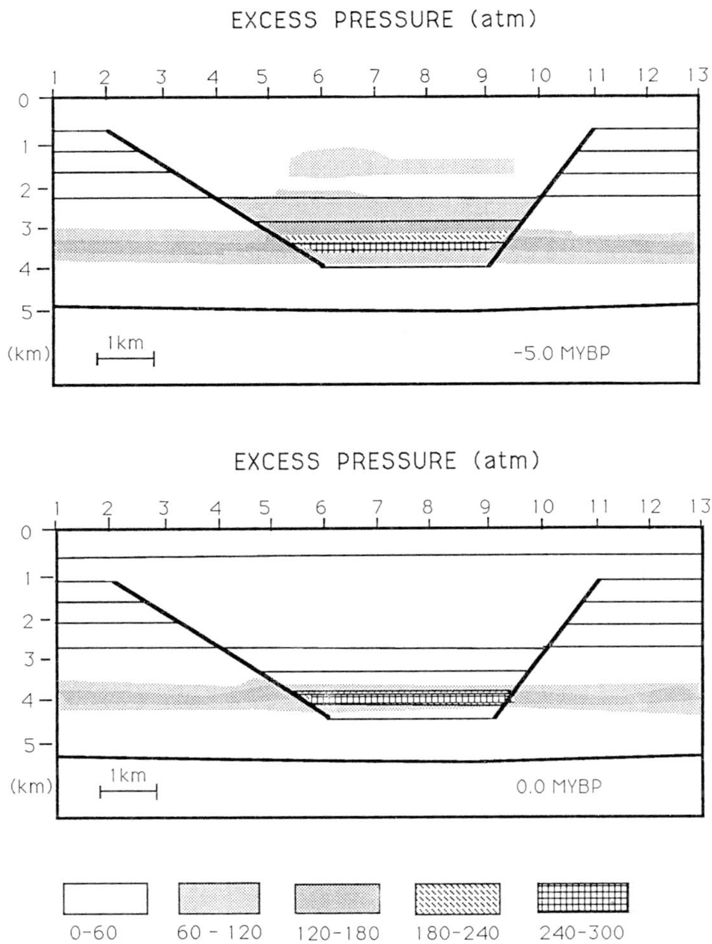


FIGURE 7.6 Excess pressure development with time for the closed fault case. No pressure bleed-off occurs of the magnitude of that in Fig. 7.1 for the case of open faults.

influenced directly and instantaneously by the slant fault (by sediment load shifting) and by behaviors continuing thereafter which are modified by creation of the fault.

Figure 7.6 shows the development of excess fluid pressure after the fault initiation time at 5 MYBP and at present day, whereas Fig. 7.7 shows the present-day excess pressure with depth at location 5 (through the fault plane) for open and closed faults. Note from comparison of Figs. 7.1, 7.6, and 7.7 that there is a difference in pressure evolution after fault development between the two cases of open and closed faults. The pressure decreases more quickly with time in the open fault case compared to the closed fault

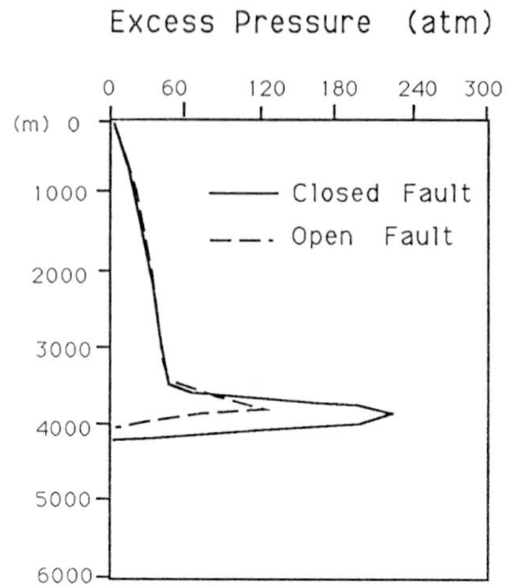


FIGURE 7.7 Comparison of present-day excess fluid pressure at location 5 for the open and closed slant fault cases showing the difference.

situation, and the domain around the faults where pressure is bled off is less extensive in the closed case compared to the open case. The overpressure is higher and lasts longer in the case of closed faults than in the open fault situation, and there is not as much of a variation caused by the graben development.

Because the numerical model uses a complete shale section to provide the maximum contrast between variability caused by the presence and properties of the slant faults, the permeability variation shows virtually no change from the case of unfaulted formations. As expected, the porosity development shows higher porosity values maintained for a longer period of time in the closed fault situation than in the open fault case (Fig. 7.8), with less of an impact of the fault on the lateral development of porosity.

The temperature development follows nearly the same dominance pattern, with high-temperature regions initially on the flanks of the section prior to graben formation and a more uniform variation of temperature across the section after graben formation. A slightly higher temperature around the fault plane might be expected due to the fluid retention as is shown in the closed fault case compared to the open fault case, which can lower rock thermal conductivity.

Thus it would seem that the dominant difference between an open and a closed slant fault is the change in excess pressure around the slant faults. In the numerical experiments reported here, the "bleed-off" of excess pressure is more rapid and more laterally extensive around the faults, and has a significant impact to a greater depth in the case of an open fault

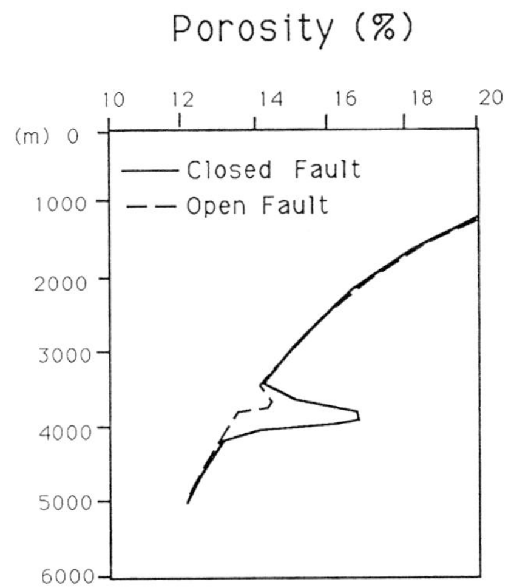


FIGURE 7.8 Comparison of present-day porosity at location 5 for the open and closed slant fault cases, showing the retention of porosity for the closed fault case.

compared to results for a closed fault. The dominant similarity between the open and closed slant fault cases would appear to be the motion of basement (under the isostatic assumption) as a consequence of lateral load shifting at fault initiation time. This effect is also coupled to a major change in temperature, with cooling in the nonmoving uplifted block and heating in the block subjected to the increased load at the time of faulting.

## D. DISCUSSION

Three dominant factors appear to be ruggedly stable characteristics arising from the numerical investigations of the effects of the graben-forming slant faults, whether they are open or shut. Perhaps one of the more obvious points is that while the excess pressure development in the local neighborhood of a slant fault is influenced by whether the fault is open or shut, there is only a fairly localized neighborhood around the fault where the difference between open or shut conditions can be felt. Thus, until a borehole is close to the fault (within about 0.5 km or so laterally), little observable dynamic consequences can be inferred about the state of "openness" or "closedness" of the fault. In this respect the "open" fault appears much as an isolated "sand" body in an encompassing shale matrix, such as that observed in the Paria Formation of the Orinoco Delta (Hunt, 1979).

A second point of some interest is that the lateral motion of sediments at the time of faulting implies that there is an uplift of sediments in the footwall block and a burial of sediments in the hanging block. Thus, prior to faulting, sediments on the flanks of a fault-initiated graben were deeper and so more thermally mature prior to fault development. Conversely, sediments which are downthrown were cooler and so less mature prior to faulting. The pattern of temperature variation is then tied closely to the structural development prior to, during, and after faulting. In turn, this paleothermal evolution has a role to play in influencing both hydrocarbon generation and subsequent migration pathways of hydrocarbons within the rapidly developing fault-initiated structures.

A third point of interest is the variation of porosity and permeability modified by fault development. In the case of sealed faults, porosity is generally maintained at a higher value for a longer period of time than in the open fault situation; the development of permeability shows the opposite pattern (at least in the spatial domain influenced by the open faults) in which a rapid loss of fluid along the fault planes leads to a "tightening" of the permeability in that local neighborhood.

Although the numerical experiments reported here are not exhaustive (no anisotropy of permeability included; sand layers abutting or crossing fault planes not included; variable throws of faults and major changes in timing of different faults not examined; no allowance for basement motion other than isostatic response; etc.), nevertheless the dominance of the effects reported here can be expected to prevail in all but the most exceptional geological circumstances. It is undoubtedly true that complications due to other factors, such as those mentioned, may modify the basic effects described in both local and regional senses, but it is difficult to see how the dominant patterns of behavior described could be eliminated, or reversed, by such extra factors.

It is this dominance of patterns of behavior that the numerical experiments described here have been designed to illuminate. One could, of course, argue that the effects described are due solely to two major causes: the requirement of normal slant faults which have terminations with depth today, and the presence of horizontal to subhorizontal formations below the faulted section. Such arguments, while formally correct, do not detract from the results presented. Indeed the presence of arbitrarily varying formation horizons below the faulted section can be included. Unless these sedimentary beds, which have to be deposited prior to fault initiation, are carefully arranged to produce horizontal beds prior to faulting *and* to be aware of the amount of throw of the faults which will develop later, then some form of structural high or low will prevail across the section in order that, when faulted, the beds take on their presently observed behavior. It seems unlikely that such a careful geological arrangement can be made under all conditions, suggesting the prevalence of fault-dominated structures pre- and/or postfaulting.

One could argue that the presence of normal slant faults which terminate at depth is an unlikely occurrence and that, in reality, all faults sole out along detachment surfaces but that we never have sufficient information to observe such effects. Perhaps so, but there seems to be no fundamental physical law forbidding absolutely normal slant faults. Clearly, extensional and compressional tectonic forces, together with three-dimensional motion considerations, can also provide mechanisms for slant fault initiation. But such problems are well beyond the scope of this section.

## III. APPLICATION TO AN OFFSHORE NIGERIAN BASIN

Here, using a 2-D fluid flow/compaction model [GEOPETII, constructed at the University of South Carolina (Nakayama, 1987)], we present an application of a case history for an offshore Nigerian basin.[1] Very high excess pressures, close to lithostatic pressure, have been measured. The basin was formed in a deltaic environment, where high sedimentation rate and very shaly facies produce high undercompaction and thus excess pressure development. This basin, governed by gravity tectonics, contains many faults which could have a major role in excess pressure development. The model should be able to reproduce not only the possible role of faults as pathways for fluid migration (provided the faults have a high permeability), but also the evolving structural geometry of the system which can influence directions of fluid-flow. The GEOPETII model is not a completely mechanistic model, in that it does not reproduce interactions between fluids and tectonics. In particular, it does not include the necessity to have high excess pressure in order to have basal decollement, which is usually described as the main mechanical feature of deltas (Crans *et al.* 1980; Mandl and Crans, 1981). However, as a fluid flow/compaction model, GEOPETII *does* reproduce the geometry of the basin and the possible pathways for fluids, such as faults.

The role of fault permeabilities in impacting fluid flow and overpressure development has already been described (Yu *et al.*, 1993), but layers offset by a normal slant fault had to be flat in that study. The improvement of the application presented here is to model slant faults including nonhorizontal sedimentary layers, and therefore to obtain a better 2-D modeling of fluid flow. Results are consistent with those previously obtained considering only flat layers (Yu *et al.*, 1993).

[1] For proprietary reasons, the name of the particular basin and the well names and locations are not given. Elf Aquitaine should be approached for such information. However, all data available are used so that the results of the study are appropriate without nomenclature attribution.

To analyze the African basin, both a 1-D fluid flow/compaction code and a 2-D code were used. The 1-D code allows parameters to be determined that govern the fluid flow equations by fitting observed data to predictions of the model with both an inverse method and a sensitivity analysis (Kai and Lerche, 1993). Data provided are fluid pressure measurements from two wells, lithologies, formations ages, and a seismic section between the two wells. After the most likely ranges of parameters were obtained with the 1-D model, the 2-D code GEOPETII was used, based on the same equations, but with parameters readjusted slightly to fit the data because of the influence of lateral fluid flow. The possible effects of faults, the lithologic fracturing coefficient, or other parameters on fluid flow can then be evaluated by seeking out those conditions which retain a good fit to the observed data.

## A. GEOLOGICAL SETTING

The offshore Nigerian basin has been formed in a deltaic environment where growth faults occurred. On the modeled section, faults do not have the classic curved shape of growth faults, but one can easily observe nonhorizontal beds and thickness variations due to syn-sedimentary deformation (Fig. 7.9a). These features will be of major importance in the model. Faults are modeled as instantaneous events happening during the geohistory of the basin. Faults in the GEOPETII code must be slant or vertical faults, and must have the same throw all along the fault. Moreover, no more than one fault can be input for each lateral location. In reality, faults are not instantaneous and, in general, are not slanted but more likely to be curved, and their throws vary with depth. As a consequence, the 2-D model does not honor completely the geologic complexity of nature but reproduces a simplified section (Fig. 7.9b).

To control the model, data from two wells (A and B in Fig. 7.9) are also provided. The data are lithologies, thicknesses of sedimentary layers, and fluid pressure measurements with depth (Fig. 7.10a). Lithologies are mainly sandy in the upper part of the basin and become more shaly with depth. Many sand intercalations are still present at intermediate depth, whereas from 2500 to 3700 meters depth lithologies are mainly shaly with few intercalations of sand. To reproduce the numerous intercalations of sand numerically is difficult; as a consequence, we simplified variations of lithology considering each formation as an homogeneous medium with a variable sand ratio (Fig. 7.10b).

The basin is young, not more than 10 million years old, and has a rapid rate of sedimentation, around 400 m/MY on average (Fig. 7.10b). This high rate of sedimentation, associated with deposition of shale having a low permeability, led to undercompaction of shale and overpressure generation. The top of the undercompaction is interpreted from well log measurements

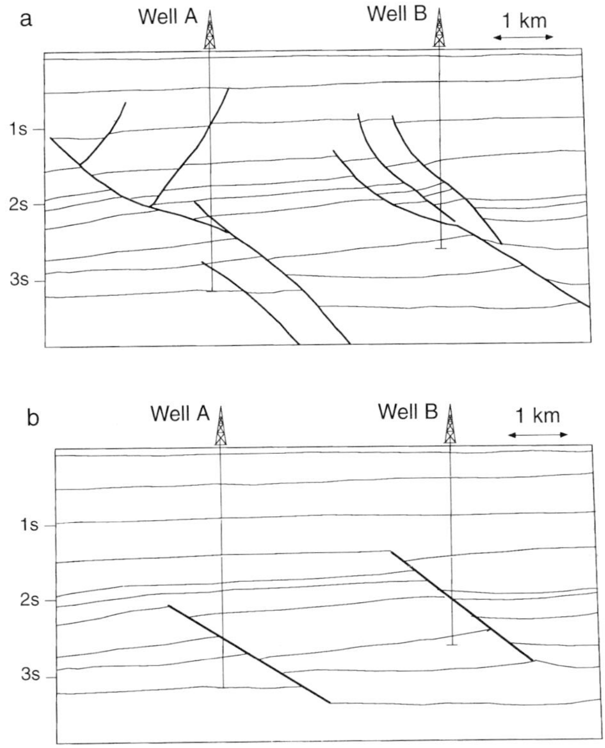


FIGURE 7.9 (a) Interpreted cross-section of the offshore Nigerian basin; (b) modeled cross-section of the basin.

to occur at a depth of 2440 m (Elf Aquitaine, private communication). From that depth to the bottom of well A, excess pressures increase until a maximum of 2.17 $g \cdot cm^{-3}$ equivalent mud weight (EMW) is reached. Abnormal pressures have been also measured on well B. But the maximum overpressure is less than 2.17 $g \cdot cm^{-3}$ EMW, probably because well B reached only 3087 m in depth.

The thermal gradient in the area is low, around 2–3°C/100 m (Elf Aquitaine, private communication), so that thermal expansion of water as a possible cause of such overpressure is unlikely. Moreover, under normal conditions the effect of mechanical loading is more important in generating excess pressure than aquathermal pressuring (Shi and Wang, 1986). As a consequence, we do not consider the possible effect of aquathermal pressuring in the numerical modeling.

To calibrate the model, a sensitivity analysis is used (Kai and Lerche, 1993; Cao and Lerche, 1990) which calculates the difference between the predictions of the model and observed data for each set of parameters.

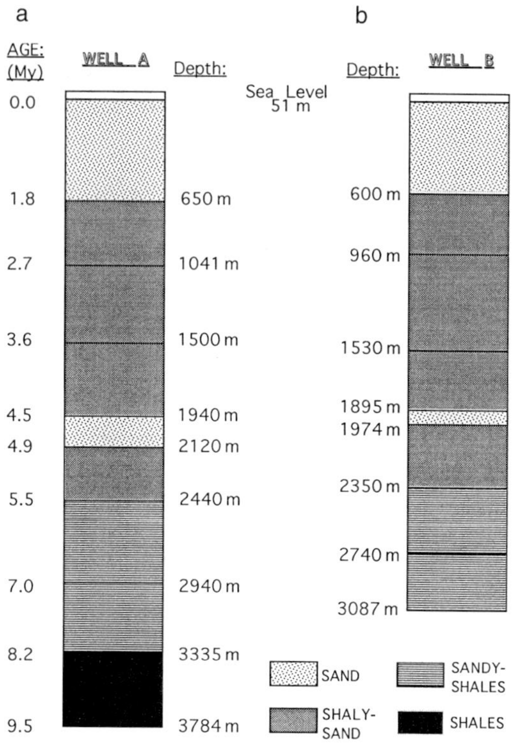


FIGURE 7.10 (a) Lithology and pressure data for well A; (b) lithology input for both wells.

Therefore a choice can be made of the best set of parameters that minimizes the mismatch between observations and predictions of the model. Because this procedure is very time consuming in two dimensions, it is applied only in one dimension. Good approximations of the best values of the parameters are then obtained which need to be only slightly recalibrated in two dimensions because of the influence of lateral fluid flow. To obtain "best values" of the parameters, one can vary one or two parameters at a time using a linear search procedure. In that case, the value of a parameter is varied within a fixed range by a fixed step size, both given by the user. An alternative procedure, which is not limited by the number of parameters to be varied, is a tomographic method which varies a set of parameters all together from first estimates in such a way that the differences between observations and predictions are always minimized (Lerche, 1991). These two methods are complementary and, in general, a linear search is used first to obtain a trend of variation, then a tomographic search is applied to determine if the minimum mismatch obtained corresponds to the best minimum or is only a local minimum.

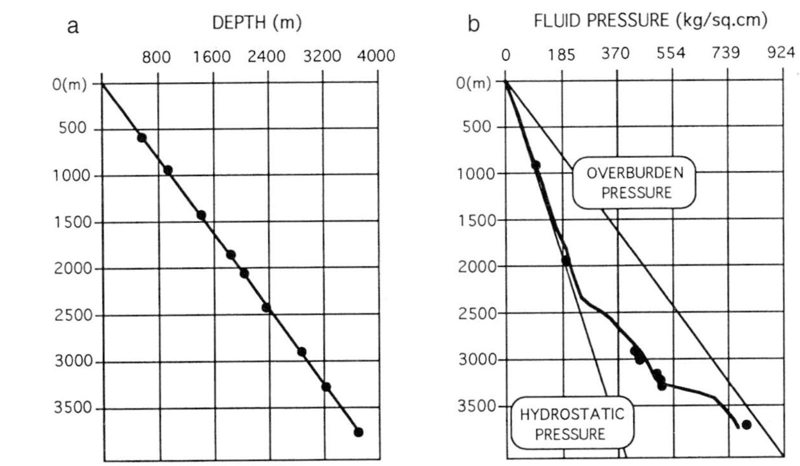


FIGURE 7.11 One-dimensional results after running a sensitivity analysis. (a) Comparison between input thicknesses (points) and modeled values (solid line); (b) fluid pressure profile with depth; points represent measurements.

## B. RESULTS

### 1. One-Dimensional Results

#### *a. Well A*

To obtain numerical results in least discord with observations, the model was calibrated using a sensitivity analysis incorporating fluid pressure measurements, thickness of each layer, and total depth of the well. The set of parameters so determined minimizes the mismatch between the input data and the predictions of the model. Predictions are in good agreement with observations (Fig. 7.11). Thicknesses measured (points) lie on the line of prediction (Fig. 7.11a). Similarly, the solid line of fluid pressure predictions from the model passes through the points of measurement (Fig. 7.11b). The top of excess pressure corresponds to that estimated from well logs. The upper part of the well is not overpressured because of the high permeability of sandy sediments. At the base of the well, the maximum of excess pressure is almost reached, so that excess pressure can be understood taking into account only undercompaction processes. However, although this first step results in a satisfactory model consistent with the data available, the second step is to deduce conditions required for such a model.

Two other rock characteristics derivable from the model (but not available from measurements) are permeability and porosity (Fig. 7.12). The predicted porosity decreases with depth (Fig. 7.12a), quickly at shallow depth and then more slowly at greater depth. Local increases of porosity can occur with depth according to lithology contrasts and to the locations of isolated sandy units within more shaly layers. Sandy units can be identi-

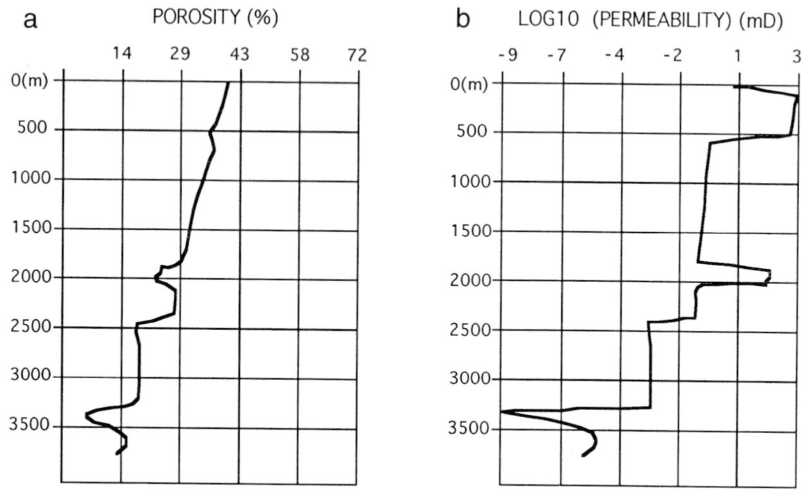


FIGURE 7.12 One-dimensional results: predicted porosity and permeability profiles with depth (there are no points of measurement).

fied easily from the permeability curve with depth (Fig. 7.12b). But at the base of the well, where major excess pressure occurs, porosities are low and consequently so is permeability. One can even observe that, just shallower than the depth at which major excess pressure occurs, porosities are as low as 4%. At greater depths, because of excess pressure, porosities increase with depth. At the boundary between a well-drained lithology and a low-permeability lithology, fluids flow so that compaction occurs and porosity decreases. Below such a boundary, fluids cannot escape so that porosity remains high. But are porosities too low? Moreover, are low permeabilities of $10^{-5}$ mD or even lower acceptable for shales (Turcotte and Schubert, 1982)? To answer such questions within a 1-D framework, a sensitivity analysis is performed.

Sensitivity analysis is useful not only to determine the best set of parameters according to the data available, but also to determine the sensitivity of each parameter. Figure 7.13 shows a plot of the variation of the mean square residual (MSR) for two parameters $A$ and $e^*$ governing compaction of shale. (The MSR is calculated from the difference between observations and predictions of the model.) The larger the value of $e^*$ (the surface depositional void ratio), the larger the porosity; the smaller the $A$ value, the faster the compaction of shale with depth. Usually $e^*$ is taken to vary from 0.41 to 3.0 (corresponding to a variation of depositional surface porosity from 30 to 75%), and $A$ varies from 1.0 to 6.0, corresponding to a variation from very fast compaction to extremely slow compaction with depth. In fact, according to the range of variation of each parameter, one can observe that $e^*$ is very sensitive because of the large variation of the

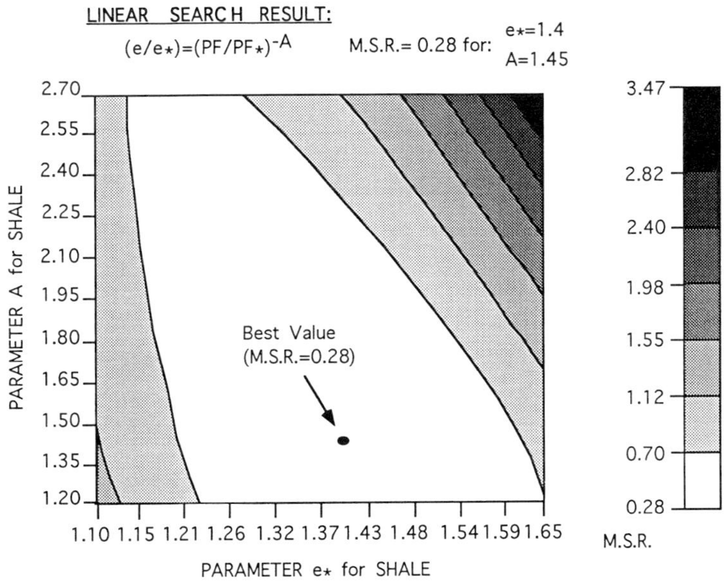


FIGURE 7.13 Linear search in two dimensions. The two parameters are $A$ and $e^*$ for shales. The gray scale corresponds to the values of the MSR, which represents the difference between observations and predictions of the model.

MSR, while $A$ is not a sensitive parameter (Fig. 7.13). Indeed, by increasing $A$ from 1.45 (the best model value) to 2.6 (Fig. 7.14), results for fluid pressure predictions with depth are quite similar (Fig. 7.15). By way of contrast, the porosity trend, especially toward the base of the well, is different, and permeability at depth has decreased by about 5 orders of magnitude. Then, even if the best model has been obtained and is well determined, because of the lack of particular types of data control (in this case permeability and porosity measurements) the model has large degrees of freedom for special parameters, which can cause different patterns of behavior with depth of unmeasured quantities. Accordingly, in the definition or use of a "best model" set of parameters, due care and attention must be paid to the range of uncertainty of each parameter.

### *b. Well B*

In the same way as for well A, a sensitivity analysis was run for the parameters that best fit the well B data on pressures and thicknesses. The parameters obtained differ slightly from those arising from well A (Table 7.1), but, considering the range of uncertainty of the parameters, differences are not significant enough to provide any conclusions about possible changes between the two wells in terms of parameter variations.

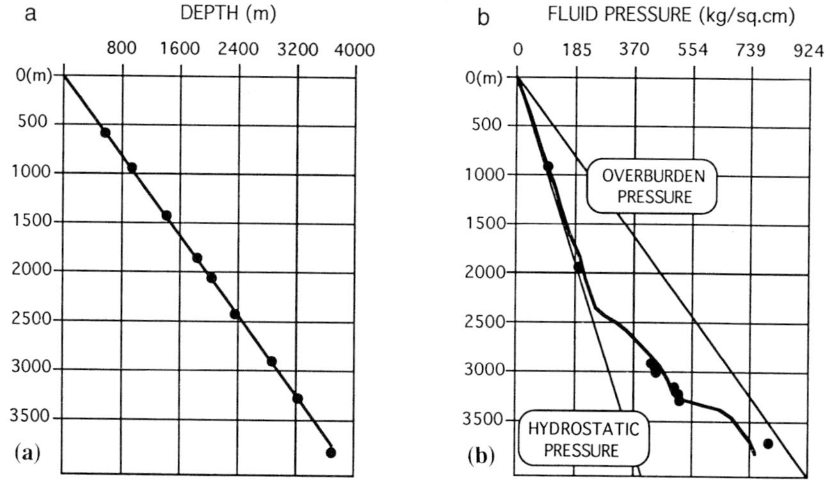


FIGURE 7.14 One-dimensional results with $A$ set to 2.6 rather than best value of 1.4. (a) Comparison between input thicknesses (points) and predicted values (solid line); (b) fluid pressure profile with depth; points represent measurements.

## 2. Two-Dimensional Results

### *a. Structure*

The first requirement is the ability of the model to reproduce slant faults. The GEOPETII model reproduces correctly the main characteristics of the basin at the present day (Fig. 7.16). The two normal slant faults are well

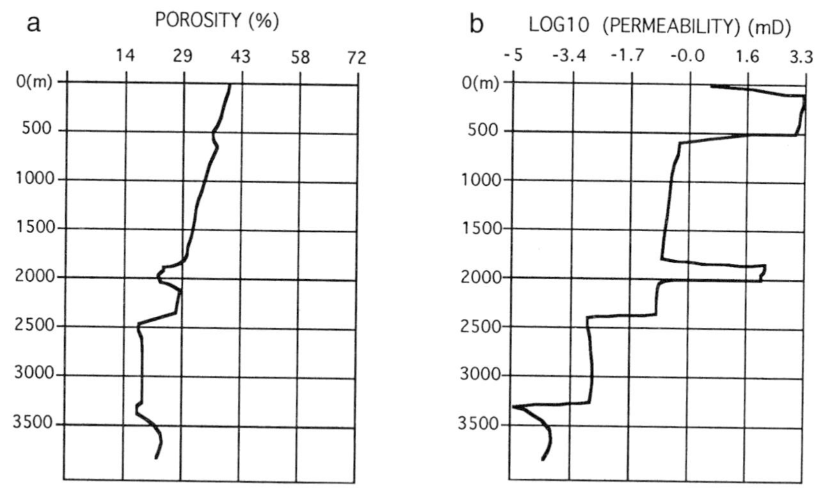


FIGURE 7.15 One-dimensional results with $A$ = 2.6 rather than the best value of 1.4 for porosity and permeability profiles (there are no points of measurement).

TABLE 7.1 Parameter Values Used in the 1-D Models

| | Shale | | | | | Sandy-shale | | | | |
|---|---|---|---|---|---|---|---|---|---|---|
| | *A* | *B* | *K** (mD) | *P** (atm) | *e** | *A* | *B* | *K** (mD) | *P** (atm) | *e** |
| Well A | 2.61 | 5.25 | 0.4 | 0.6 | 1.45 | 2.82 | 3.00 | 3.80 | 0.98 | 2.60 |
| Well B | 1.24 | 5.25 | 0.9 | 0.6 | 1.23 | 2.75 | 3.00 | 3.80 | 0.98 | 2.56 |

described in terms of position, angle, and throw. Flat layers at the bottom of the model are an assumption arising from ignorance of basement geometry. The main improvement relative to previous work (Yu *et al.*, 1993) is the possibility of having nonhorizontal layers offset by the faults, which allows a much better reproduction of the complete geometry of the system. This structural evolution is of primary importance in such a young basin where high excess pressures have developed, and where the geometry of low-permeability layers dominates in the control of fluid flow directions.

To obtain a good definition of a fault position, and of the layers offset by the fault, one of the conditions required is to have a fine array of grid points vertically as well as horizontally. A second condition, which is related to the first, is that there be no missing layer at any pseudo-well for nonhorizontal sedimentary layers. If there is a missing layer at one lateral location, the code is not able to assess the layer position because the sedimentary layers are not declared to be flat, as in the default situation. A consequence of the imposition of this second condition is that slant faults with nonhori-

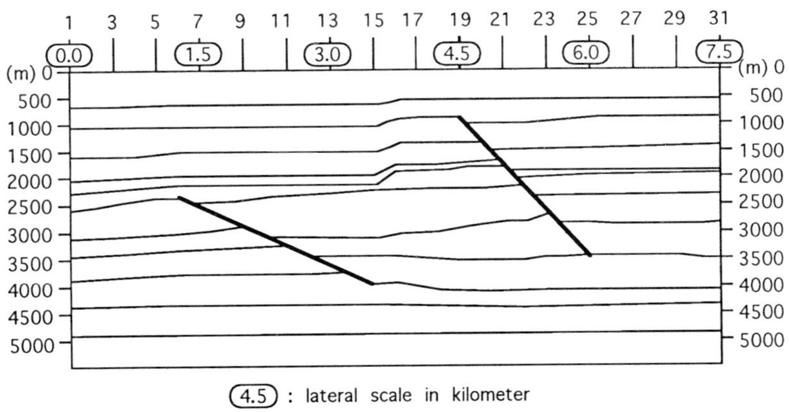


FIGURE 7.16 Two-dimensional section at the present day. Wells A and B are located at locations 8 and 23, respectively. Around the two faults, layers are not flat but respect the geometry of the beds.

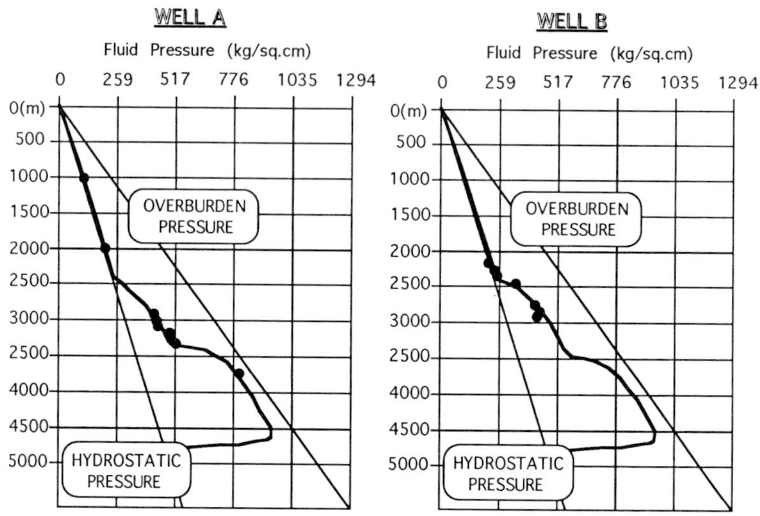


FIGURE 7.17 One-dimensional representation for wells A and B arising from projections of the 2-D section (faults are shut, with permeability corresponding to that of the neighboring sediments).

zontal offset layers cannot be modeled if the vertical fault throw is larger than the thinnest sedimentary layer offset by the fault. Indeed there would then be a "missing" position for one lateral location. At that location, the fluid flow distribution would be undefined.

### *b. Calibration of the Model*

Using the set of parameters provided from the 1-D models, good results were not obtained initially using the 2-D code GEOPETII because of the influence of lateral fluid flow. Parameters were then slightly readjusted to improve fits to data. In fact, because the sedimentary layers are mainly horizontal to subhorizontal, lateral fluid flow is not too important and fluid pressures were only slightly underestimated. To obtain the final best fit results (Fig. 7.17), we found it sufficient to vary only the coefficient $B$ from 3.0 to 3.5. This procedure shows how it can be useful to run data first on a 1-D model, using an inverse procedure to calibrate the model. (One computer run in 1-D takes 30 seconds while one computer run in 2-D takes 2 hours.) In addition the ability of the model to reproduce the data of both wells is clear. At well B (location 23 on the 2-D section of Fig. 7.9), the model predicts large excess pressure below 3500 m, but no data are available to confirm such a prediction. The "kickback" in pressure observed at 4750 m is a consequence of the boundary condition used to avoid any fluid flow through the basement. For that reason, a few layers were added at the base of the model so that the kickback in pressure occurs only in the deepest part of the basin and does not affect the modeled upper part. In the models,

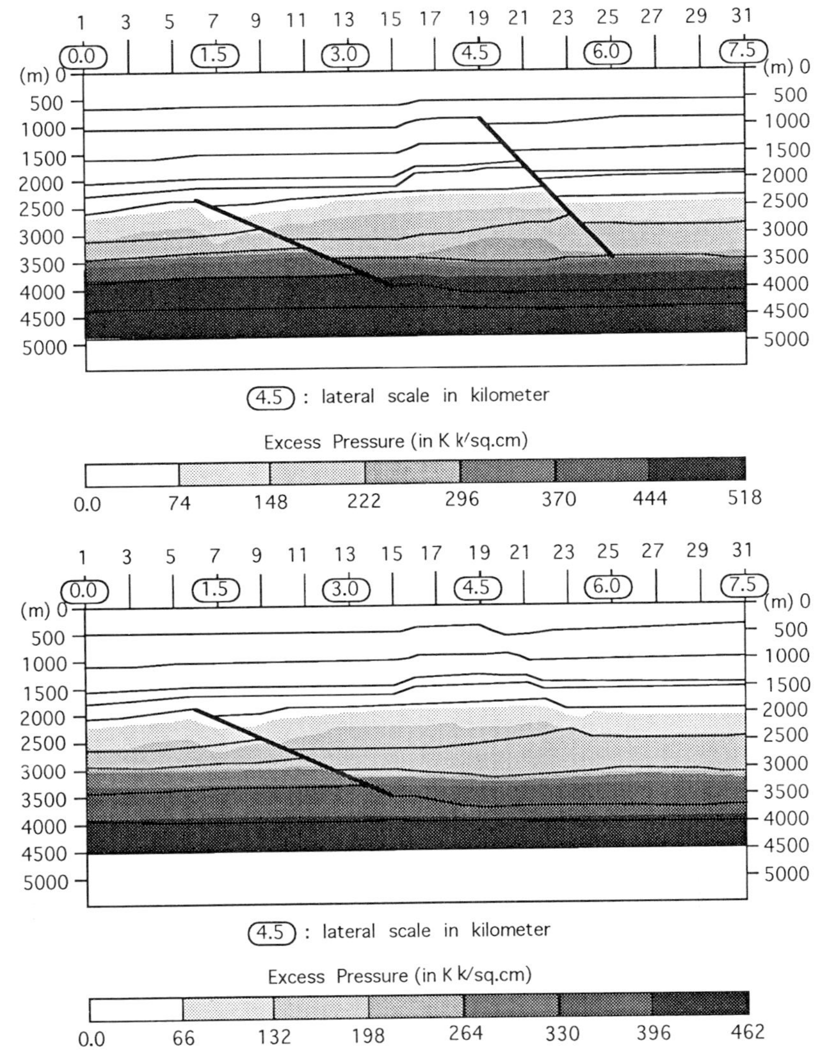


FIGURE 7.18 Evolution of excess pressure with time in relation to the development of faults (faults are shut, with permeability corresponding to that of the neighboring sediments).

faults have been assumed shut to fluid flow, with permeability equivalent to that of the surroundings sediments. Fluid flow is then only affected by the structural role of the fault and not with the fault as a possible high-permeability pathway for fluid migration.

### *c. Control, Geohistory, and Tests of the Model*

The evolution of the excess pressure distribution (Fig. 7.18) is consistent with previous synthetic results obtained using flat layers around faults (Yu

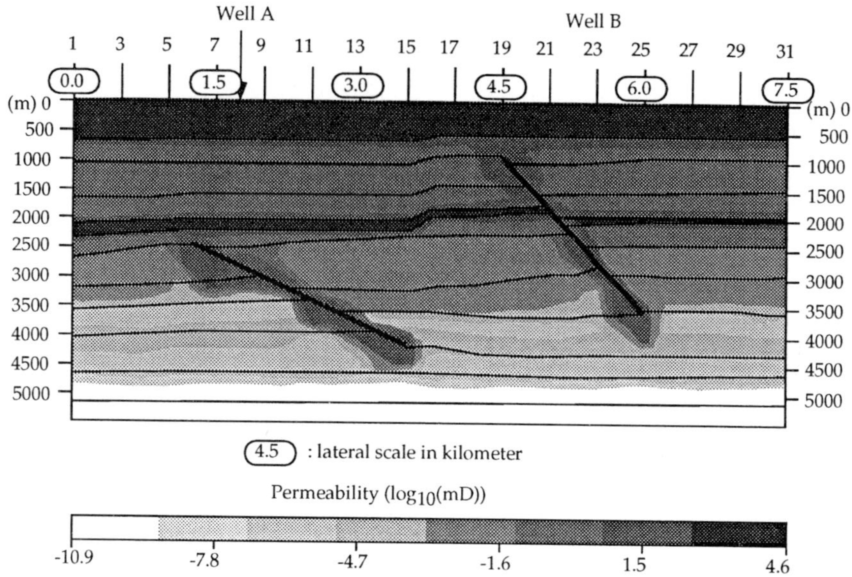


FIGURE 7.19 Two-dimensional permeability distribution at the present day: black = sandy layers; white = shaley layers.

*et al.*, 1993). When faults are assumed not to be open conduits to fluid flow, the effect on the excess pressure distribution with depth depends on the fault throw. If the angle of faulting is subperpendicular then there is an almost negligible effect of the lateral component of the throw, except for the footwall where little excess pressure is released, such as for the fault going through well A. In fact, on location 8 of Fig. 7.18 the fault makes a large overlap contact between shale and the upper sandy layer so that more shallow fluid can escape.

When faults are assumed open, with a permeability of 0.1 Darcy for example (Fig. 7.19), the excess pressure distribution is massively affected by such faults, especially in the footwall region (Fig. 7.20). In that case, fluid pressure measurements cannot be reproduced by the model (Fig. 7.21), so that it is unlikely that the faults have been open during most of the history of the basin.

The influence of other parameters, such as the fracture coefficient, on evolution of the composite system has also been investigated. The fracture coefficient is defined as the limiting ratio of fluid pressure versus load of sediment. At a fracture coefficient of unity, fluid pressure can be as large as the sedimentary load before fracturing occurs, while a fracture coefficient of 0.7 limits the maximum fluid pressure to 70% of the sedimentary load. When fracturing occurs, the rock permeability is increased to that of sand during a time sufficient to release enough excess pressure so that the ratio

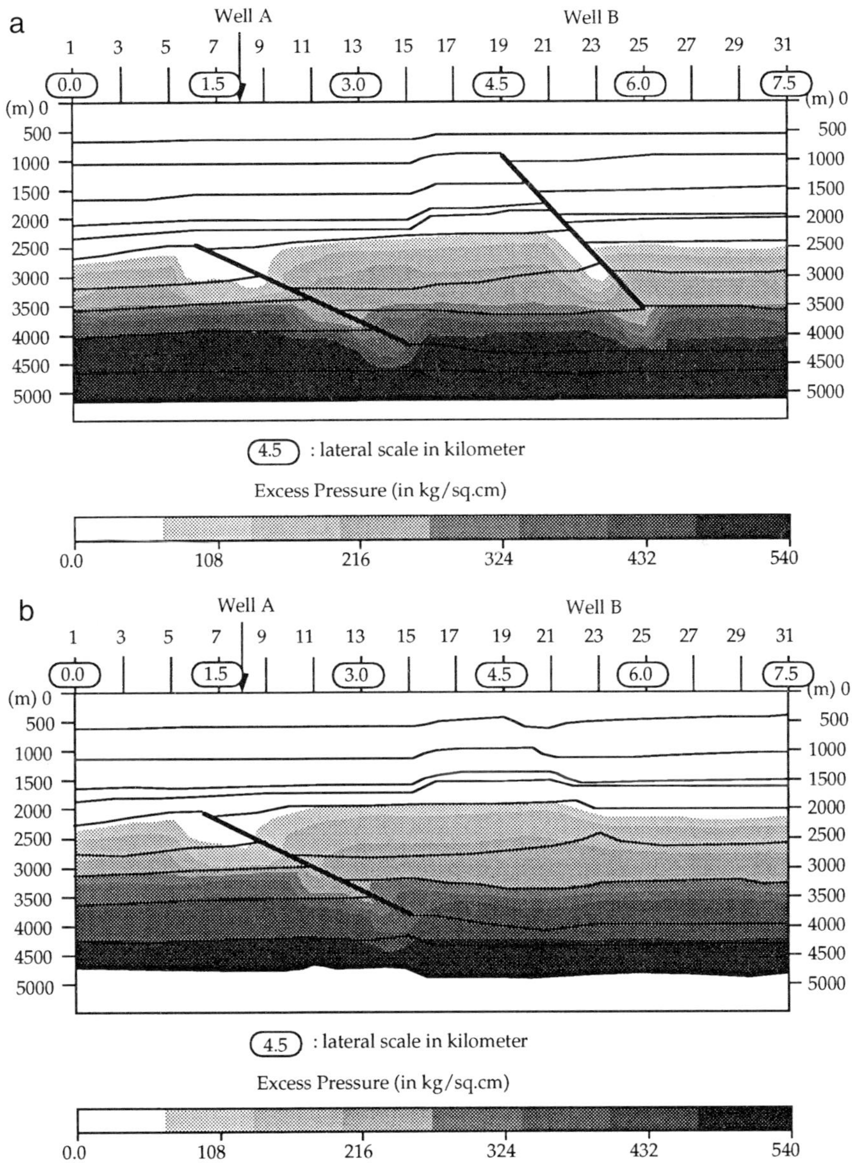


FIGURE 7.20 Evolution of excess pressure with time in relation to the development of faults (faults are open, with permeability set at 0.1 Darcy).

of fluid pressure versus sedimentary load decreases to a value equal to or less than the fracture coefficient.

To run the 1-D and 2-D models described, a fracture coefficient of 0.99 was used, so that fracturing was avoided; the aim of the exercise was to

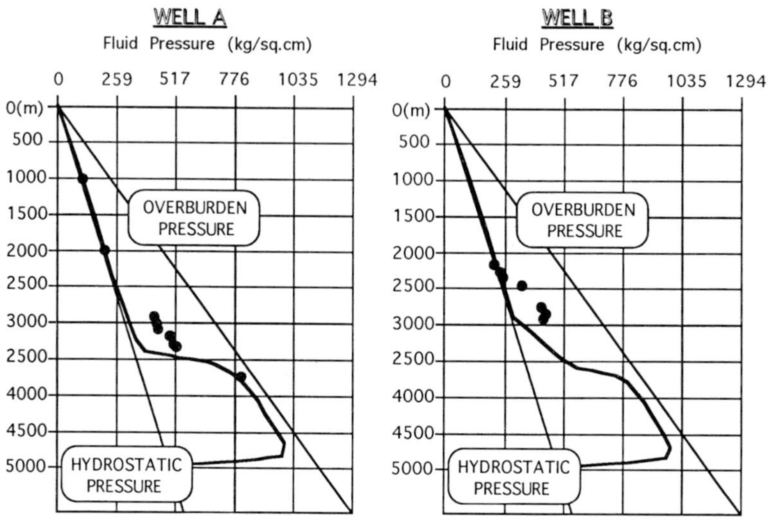


FIGURE 7.21 One-dimensional results for wells A and B obtained by projection from the 2-D section (faults are open, with permeability set at 0.1 Darcy).

find a set of parameters able to model the high excess pressures observed. When the fracture coefficient is set at a slightly lower value, 0.95 for example, the model slightly underestimates fluid pressure measurements because of fracture occurrence and escape of fluid; but, because of the uncertainty of the measurements, the model is still acceptable. However, for a lower fracture coefficient value of 0.8 (Fig. 7.22), the model does not reproduce the maximum of overpressure measured at well A. As a consequence the fracture coefficient should be higher than 0.9–0.95, in agreement with a previous proprietary study (Elf Aquitaine, private communication).

## C. DISCUSSION

The aim of this application was to model a situation of very high excess pressure occurring in a deltaic environment in offshore Nigeria. To achieve this goal, two problems have been solved. The first is the ability to model the observed high excess pressure at all, the second is the modeling of slant faults such that the geometry of the system is respected.

The modeling of high excess pressure, close to lithostatic pressure, is possible and has been accomplished by large undercompaction of shales. If faults are open throughout the geohistory of the basin, one cannot obtain such high excess pressures. If the fracture coefficient, defined as the critical ratio of fluid pressure versus sedimentary load, is lower than 0.9–0.95, the model cannot reproduce the data, even if a large uncertainty in data

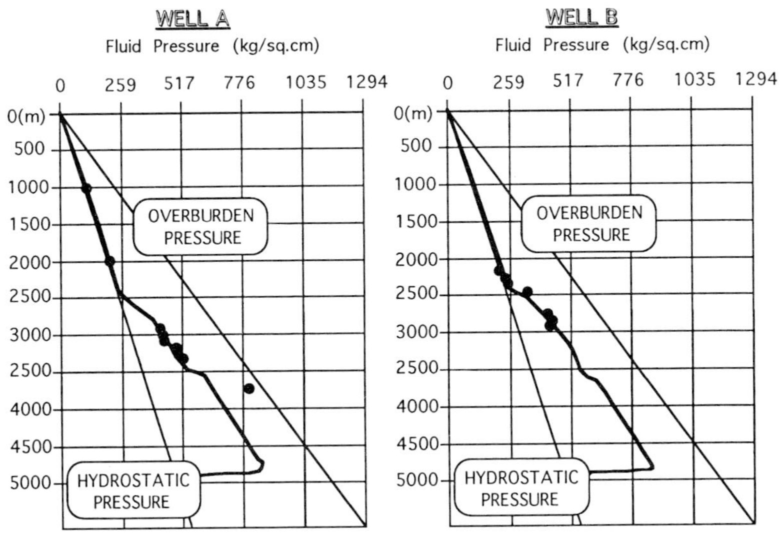


FIGURE 7.22 One-dimensional results for wells A and B obtained by projection from the 2-D section (the fracture coefficient is set to 0.8 instead of 0.99).

measurements is allowed for. Such a requirement for the fracturing coefficient to be very high suggests that not only the vertical stress but also the lateral horizontal stresses should be taken into account in the assessment of excess pressure (Grauls and Cassignol, 1993; Grauls and Kessler, 1992).

Concerning fault modeling, it has been shown that nonhorizontal layers which are offset by faults can be used. Requirements are that one invokes a fine grid of points and a fault throw less than the thickness of the smallest layer offset by the fault, so that no layer is missing at any lateral position along the fault. As a consequence, the basin geometry is then much better reproduced.

The main feature of this young (9.5-Ma) offshore Nigerian basin is rapid sedimentation of shales leading to very high overpressure development. The ratio of total fluid pressure to sedimentary load is estimated to be between about 0.9 and 0.95 at the bottom of well A, while the main tectonic activity appears to be gravitational, characterized by the development of growth faults. To model these conditions, where slant faults could have played an important role in the development of overpressure, a 2-D fluid flow/compaction model must reproduce the geometry of the basin according to the development of faults and in relation to their possible effects as pathways for fluid migration.

For this reason, the 2-D model is designed to respect, as much as possible, the geometry of the system. A significant improvement over previous uses of the 2-D fluid flow/compaction code is the ability to account for inclined layers around faults, instead of only horizontal layers. As a consequence, the basin geometry is much better reproduced. This improvement requires a fine grid of points and a fault throw less than the thickness of the smallest layer offset by the fault, so that no layer is "missing" at any position along the fault. The evolution of excess pressure or temperature with time, especially when faults occur, is consistent with results obtained on synthetic tests and using flat layers around faults. Finally, the model reproduces correctly the measured fluid pressure data. To obtain such overpressure, the fracturing coefficient must be higher than 0.9–0.95, as estimated by previous studies, and the faults cannot be always open.

## IV. ARCUATE AND GROWTH FAULT MODELS

### A. INTRODUCTION

Arcuate faults, which are often growth faults, occur commonly in extensional basins. The simulation of the influence of such faults in two-dimensional fluid flow/compaction models has long been a challenge because of the complexity of associated structures, compaction, fluid flow, and thermal effects modified by or triggered by the faulting. During the last few years, significant efforts have been made to incorporate various fault types into basin modeling codes in other than purely kinematic or graphical display ways. The ability to investigate the dynamic influence of single, precisely vertical faults (hydraulically open or closed to fluid-flow) was exhibited by Wei and Lerche (1988) for the case of a fault that occurred instantaneously; the situation of multiple vertical faults, occurring at different times, and with the same, or opposite, directions of throw was considered by Thomsen *et al.* (1992); the influence of multiple, instantaneous, planar, slant faults, either normal left lateral, normal right lateral, or both, occurring at different times and with different throws and open or shut to fluid flow, has been considered in a series of recent papers (Maubeuge and Lerche, 1993, 1994; Lerche *et al.,* 1993; Bour and Lerche, 1994; Bour *et al.,* 1993, 1994; Liu *et al.,* 1993). The main constraints in these attempts to model faulting influences on sedimentary basin evolution have been that (1) each fault occurs instantaneously in time (so that growth faults of any sort could not be included), (2) each fault had to be planar in character (so that arcuate faults could not be included), (3) faults could not "overlap" vertically (i.e., at a given pseudo-well location only one fault is permitted), and (4) reverse slant or reverse arcuate faults were not permitted.

However, fault throws could be arbitrary, as could the time of initiation of each fault, and the choice of having any particular fault hydraulically open or closed with respect to fluid flow was also at the user's discretion. Applications of the fault models available to particular basin studies have proven inordinately successful in delineating the broad sweep of conditions under which faults have a significant impact on basinal evolution (Wei and Lerche, 1988; Yu *et al.,* 1995; Bour *et al.,* 1993). Yet it is to be doubted that all basinal situations are encompassed by the simple expedient of slant normal fault models, whether the faults are hydraulically open or shut to fluid flow.

Indeed, there has been significant recognition during the last few years that the problems of arcuate faults (either relatively instantaneously, as in the cases of gravity slump faults and arcuate scoops, or in the form of continuously active growth faults) are of major importance. Perhaps the ongoing work of Xiao and Suppe (1992) serves to illuminate this point most vividly but, even there, the evolution of listric faults is treated as a geometric problem with either conservation of volume or with a prescribed variation of porosity with depth for all time—no allowance is made for fluid flow or overpressure development with time and/or hydraulic properties of faults, all of which influence the evolution of sediment compaction.

The purposes of this section of the chapter are to fill these lacunae to some extent. It is clear that the dynamical evolution of a faulted region can be extremely complex depending on the tectonic and depositional controls defining the temporal and spatial evolution of the region, and not all such situations can be investigated with currently available computer codes. Yet, as more processes are systematically added to the codes, presumably the ability to model numerically a greater number of situations, more closely representing geologically relevant conditions, can be investigated. It is from this perspective that the present section should be viewed.

Our aim here is to consider the combined evolution of two-dimensional arcuate faults and associated sediments in a self-consistent manner when the faults can be hydraulically open or shut to fluid flow, when the throw of the sedimentary beds can be large or small compared to sedimentary bed thicknesses, so that lateral hydraulic connectivity is either maintained (small throw) or realigned (large throw) depending on the lithologic facies of each sedimentary bed; for example, a sand bed sandwiched between two shale beds should show either a connected path for hydrodynamic flow across an "open" fault, or a nonconnected path if the throw is larger than the sand bed thickness. Thus the migration of hydrocarbons is influenced by both the hydraulic conductivity of the fault and by the sedimentary bed throw across the fault. In the case of a growth fault, the throw is progressive with time so that the connectivity of, say, a sand bed is progressively diminished as the throw increases. However, depending on the lithologic facies sequence, a downthrown young sand bed can reconnect hydraulically with a deeper, older, sand on the other side of the fault so that different hydrodynamic and hydrocarbon flows can then take place. Clearly, an appreciation of fluid flow, overpressuring development with time, and associated sedimentary compaction are influenced significantly by (1) whether the fault is hydraulically open or closed, (2) the throw of the fault in relation to bed thicknesses, and (3) the juxtapositional sequence of lithofacies.

Several synthetic case histories are examined in order to illuminate the types of patterns of development that can be anticipated to occur, depending on the "mix" of the three conditions mentioned. In addition, a field case history from south Louisiana is examined to provide an understanding of the likely conditions that prevailed in allowing development of the section to its present-day configuration.

## B. SYNTHETIC STUDIES

Four case studies are examined to illuminate the range of interactions possible. Although not exhaustive in total number, the case histories do provide a compendium of behaviors encompassing the patterns of development likely to occur in a wide variety of geological situations. This is the order of presentation of the case studies:

1. A hydraulically *closed* growth fault with a section composed of homogeneous sandy-shale
2. A hydraulically *open* growth fault with a section composed of homogeneous sandy-shale
3. A hydraulically *closed* growth fault with a section composed of interbedded sand and shale layers
4. A hydraulically *open* growth fault with a section composed of interbedded sand and shale layers

### 1. Closed Fault: Homogeneous Sandy-Shale

In this synthetic case a section 14 km in lateral extent, with 43 "pseudo-well" locations, has a total of 27,000 ft today of sedimentary material. The section consists of about 40% sand and 60% shale. A curved growth fault extends across the section, starting at about 3300 ft sub-mudline depth at pseudo-well 3, and soling out at about 24,000 ft at pseudo-well 40. Deposition is taken to be continuous across the section from about 70 MYBP through to the present day at approximately 400 ft/MY, but with deposition higher on the hanging wall side of the fault, of course. Fault initiation is taken to commence when approximately 3000 ft of sediments have been deposited at about 62.5 MYBP. The left-side sediments are taken to represent the footwall and, in these synthetic examples, are taken to be immobile (except for compaction) as the fault develops. The right-side sediments are taken to represent the hanging wall, and they both slide and roll over against the fault as growth faulting continues. The basement, today, is represented by a horizontal layer across the section at a depth of about 28,000 ft. Basement response to loading is taken to be isostatic so that, as sediment moves laterally in the section, the basement will uplift and depress in response to the instantaneous vertical load at each lateral position.

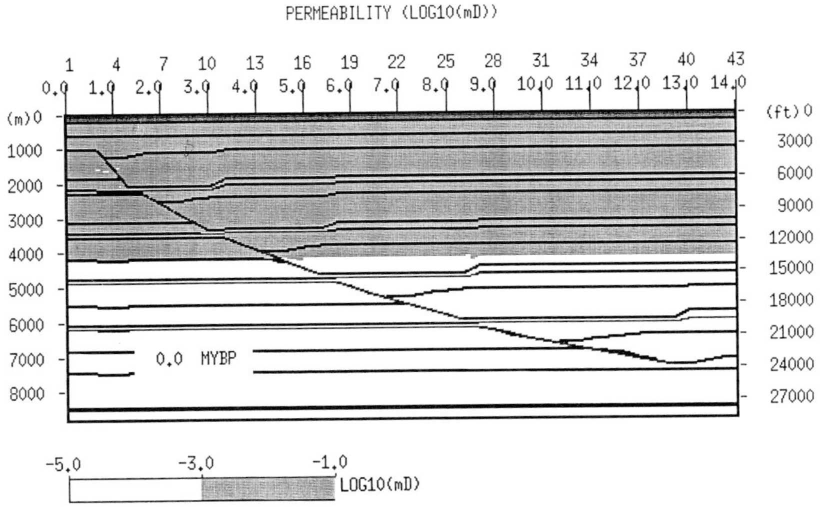


FIGURE 7.23 Present-day permeability profile of the homogeneous sandy-shale section for the case of a hydraulically closed fault. The section is 14 km in width, with 43 pseudo-wells.

A sketch of the present-day permeability across the section is shown in Fig. 7.23 where the permeability of sandy-shale is clearly decreased with depth, as are the steadily increasing throws of the hanging wall sediments along the fault with increasing depth and roll-over patterns of the formations against the fault. Because the principal purpose of this part of the chapter is to illustrate the development of excess pressure with time as the section evolves, there is no need to follow the complete evolution of the section from the moment of first shale deposition. Rather it is sufficient to examine the section only from the development of significant overpressure through to the present day. Because the section is sandy-shale dominant for about the first 35 MY of deposition, and because the rate of deposition is fairly low (~400 ft/MY), there is enough time for the sediments to compact at approximately equilibrium rates (i.e., fluid can leak through the sandy-shale rapidly compared to the rate at which the permeability of the sandy-shale is decreasing so that little overpressure develops). However, for times closer to the present day than 35 MYBP, the permeability has decreased enough that the ability of residual fluids to escape is much reduced, leading to overpressure development as time and overlying sediment deposition proceed. In addition, the progressive throw of the growth fault also starts to become significant, so that the lateral variation of preexisting and ongoing overburden of the hanging wall sediments increases, leading again to an impact on overpressure development.

To illustrate the development of overpressure, Figs. 7.24a–e show the progressive development of the section at 34, 16, 12, and 1 MYBP and

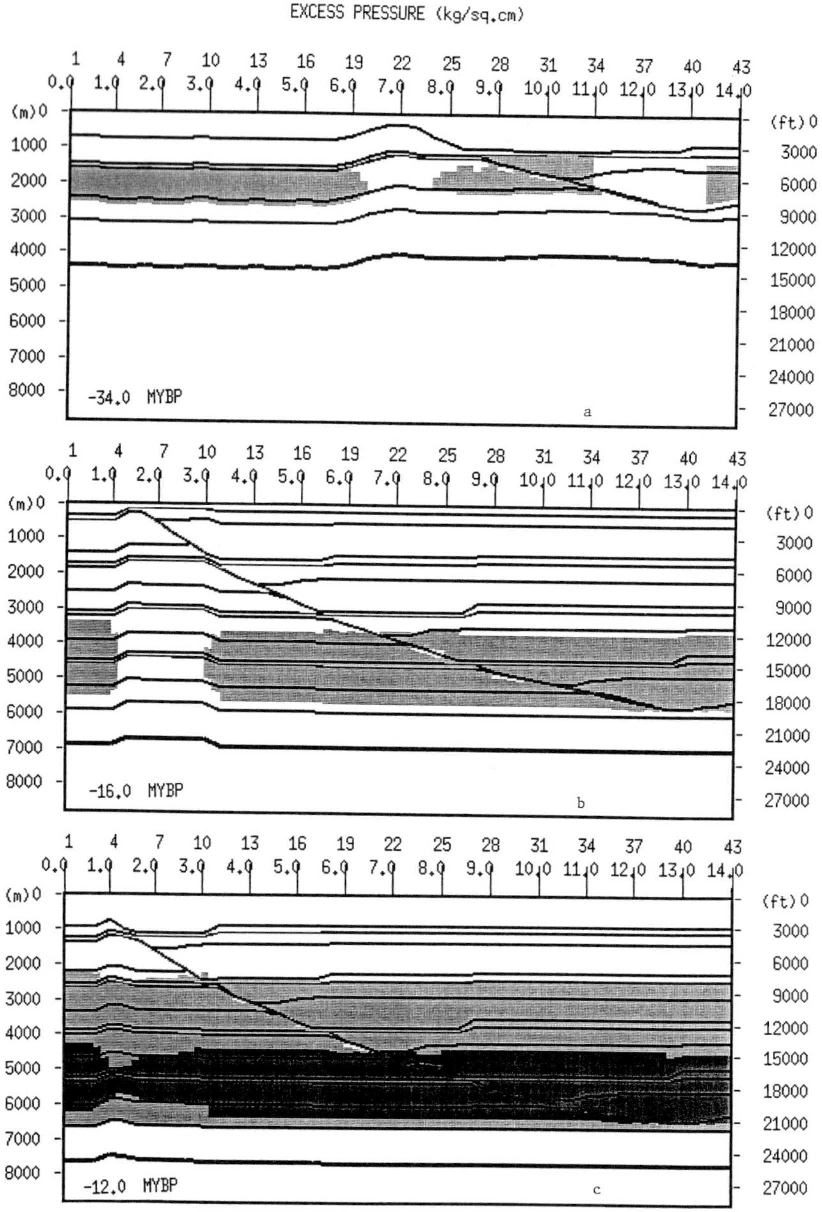


FIGURE 7.24 Excess pressure development with time in the case of a closed fault. (a) 34 MYBP; (b) 16 MYBP; (c) 12 MYBP; (d) 1 MYBP; (e) present day. The section is 14 km in width, with 43 pseudo-wells.

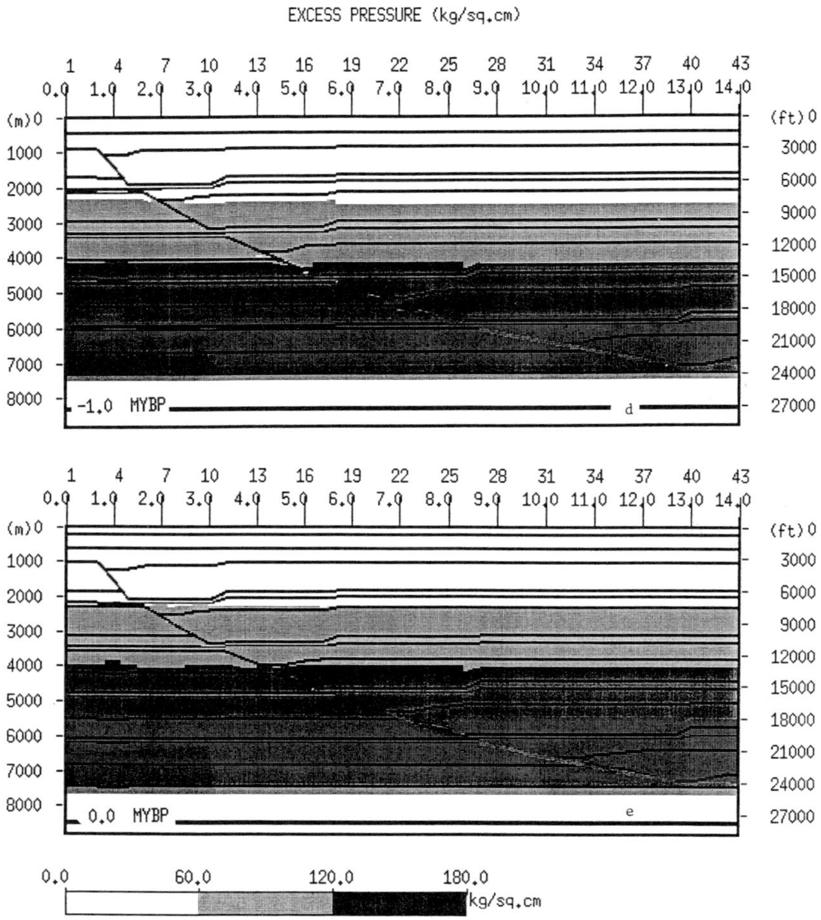


FIGURE 7.24 *(Continued)*

present day, respectively. Several factors are apparent by inspection of the figures: (1) The requirement of a horizontal basement today, together with the isostatic assumption of basement motion, provides a variable basement position across the section with time related directly to the motion of the hanging wall sediments along the growth fault. (2) The progressive "march" of the sub-mudline top of the fault to the left of the section is a direct consequence of the fact that the fault is a growth fault (one could equally have chosen a reference frame fixed to the top of the fault and then observed a progressive rightward march of the hanging wall sediments). (3) The curvature of the fault, particularly in the deeper parts of the section, changes progressively with time as the continually deposited sediments compact differentially. This phenomenon had been recorded earlier by Xiao and

Suppe (1992) in their geometric investigation of compaction-driven growth faults, and is due to the vertical thinning of deposited layers with time as compaction proceeds. (4) The development of overpressure with time across the section is dominated by two major effects: First, the permeability of the underlying formation must become small enough that fluid flow is inhibited (for the example reported here this effect starts to become significant around 34 MYBP); second, the displacement of hanging wall sediments along the growth fault must become large enough that significant alterations in vertical load cause a substantial lateral variation in overpressure. For the numerical example presented this effect becomes dominant about 16 MYBP. The fluid overpressure systematically increases from about 34 MYBP through to the present day when it reaches about 240 atm. at 24,000 ft, that is, a differential overpressure gradient of about 0.15 psi/ft, which, when added to the hydrostatic gradient of about 0.5 psi/ft, yields a total fluid pressure in the deeper formations of about 0.65 psi/ft. The development of such overpressure takes place in about 34 MY, corresponding to a systematic excess pressure build-up rate of about 100 psi/MY in the deeper units.

**2. Open Fault: Homogeneous Sandy-Shale**

Precisely the same numerical parameters and depositional events were run as for the previous case, the exception being the treatment of the fault. In the previous case the major role of the fault was to guide the motion of hanging wall sediments. The fault had no intrinsic permeability variation so that sedimentary facies abutting across the fault retained their own facies-derived permeability. In the present situation, as well as guiding the motion of the hanging wall sediments, the fault region is taken to be hydraulically open to fluid flow so that the region around the fault is assigned the permeability of an unconsolidated sand (~1 Darcy) at fault initiation time. As the deeper sections of the fault develop and sole out, the fault permeability is taken to decrease in the same manner as would a compacting sand.

A sketch of the present-day permeability with depth in the sandy-shale is given in Fig. 7.25 where the permeability along the open fault curve is also shown. In this case there is a significant difference in the development of overpressure with time compared to the previous case. The reason is that the highly permeable fault region provides a pathway for deep fluids to escape either to the surface or to the upper section. The competition here is between the rate at which fluids can escape from the sandy-shale into the fault region versus the rate at which permeability in the sandy-shale decreases, thus limiting further fluid loss. Indeed, the presence of an open fault exacerbates the fluid loss rate from the sandy-shale, thereby causing an even faster loss of permeability in those parts of the sandy-shale which are influenced by the fault, so that fluid flow into the fault is more quickly shut off.

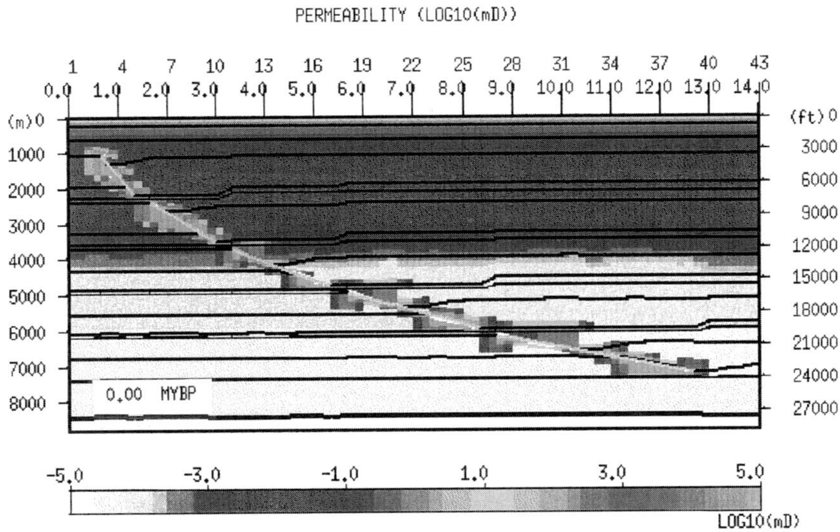


FIGURE 7.25 Present-day permeability profile of the homogeneous sandy-shale section for the case of an hydraulically open fault, showing the high permeability along the fault plane. The section is 14 km in width, with 43 pseudo-wells.

Figures 7.26a–e show the development of overpressure with time from about 20 MYBP through to the present day. While there is an increase in depth of isobars of excess pressure around the fault, the diminution does not typically extend laterally (or vertically) more than about 500 m around the fault position, representing the fact that beyond that distance the sandy-shale is sufficiently "tight" that little to no fluid flow can occur into the fault. This tightening of the sandy-shale occurs for both hanging wall and footwall sediments, thereby limiting the efficacy with which the fault can act as a conduit for fluid flow.

### 3. Closed Fault: Interbedded Shale and Sand Formations

With precisely the same section as in the previous two cases, here the major difference is in the manner of facies deposition. The prior cases considered a homogeneous sandy-shale deposition in the basin. In those cases there was no possibility of having reconnection of like-facies as sediments moved along the hanging wall of the growth fault. The present example interbeds sand and shale layers so that, depending on the evolution of bed compaction and on the progressively developing throw along the fault, like-facies sedimentary beds can connect, disconnect, and/or reconnect. Thus fluid flow can more easily occur across connected high-permeability sands, and so a completely different pattern of overpressure development is expected compared to the previous two cases. Figure 7.27

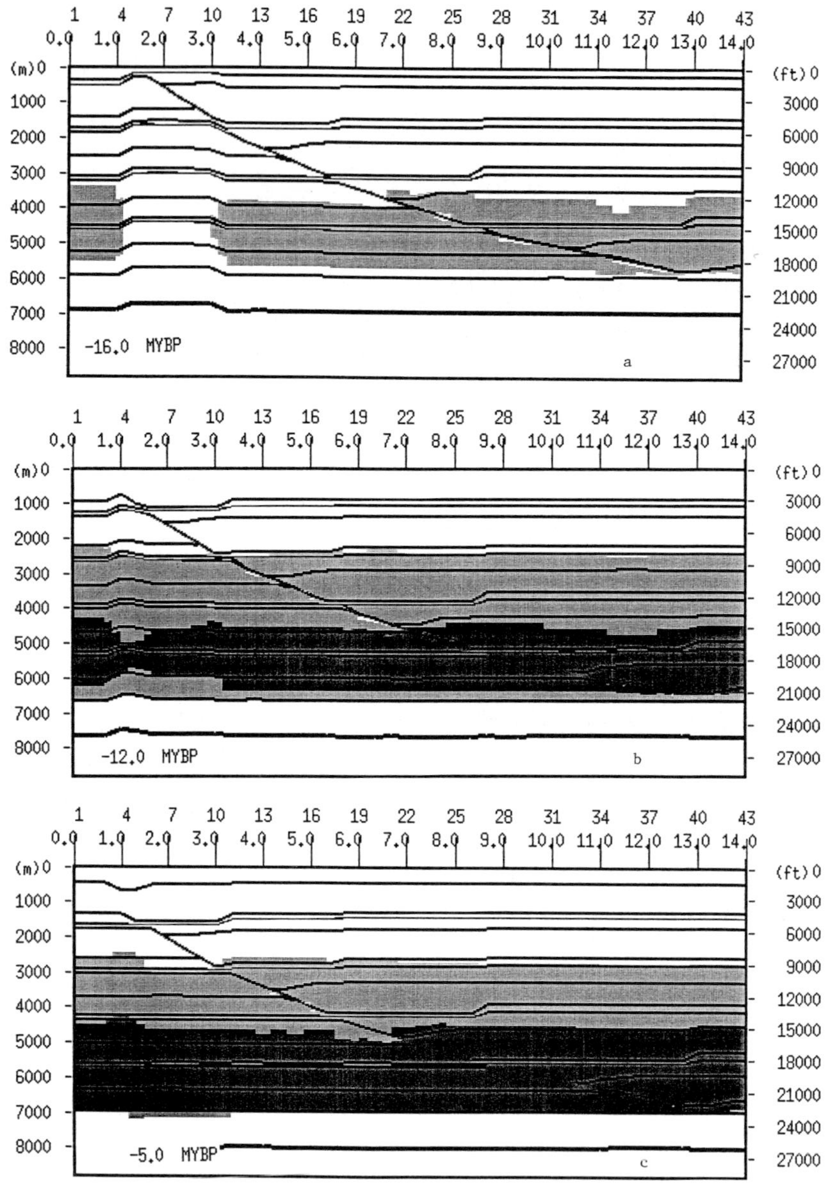


FIGURE 7.26 Excess pressure development with time in the case of an hydraulically open fault: (a) 16 MYBP; (b) 12 MYBP; (c) 5 MYBP; (d) 1 MYBP; (e) present day. Note the "bleed-off" of the excess pressure near the fault plane. The section is 14 km in width, with 43 pseudo-wells.

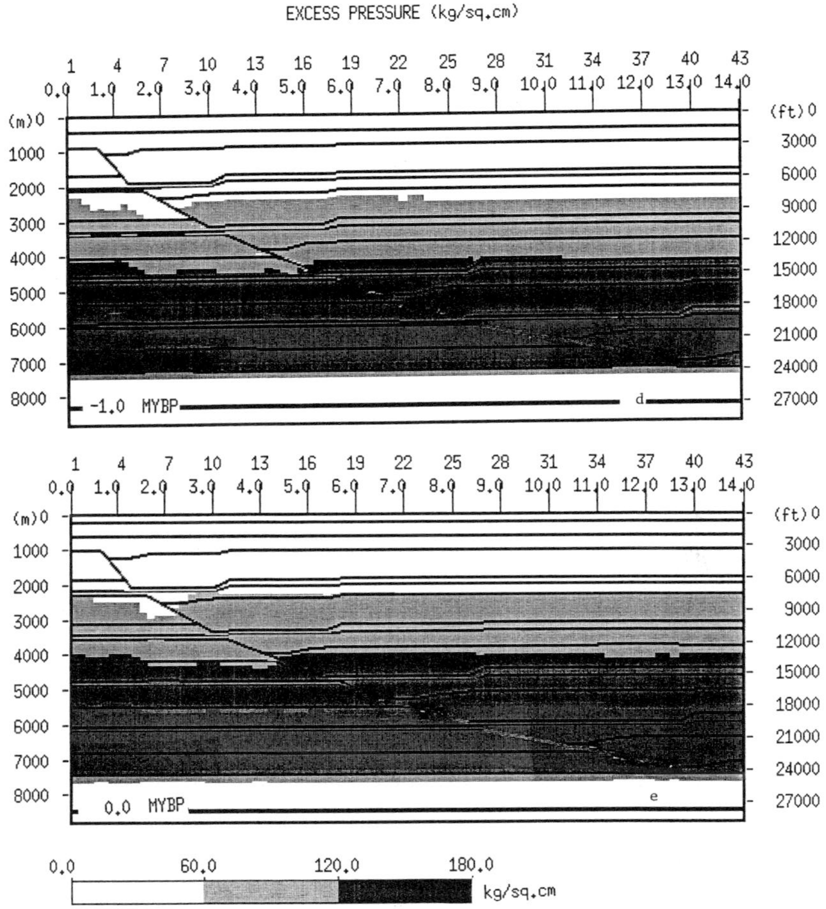


FIGURE 7.26 (*Continued*)

shows the present-day permeability variations across the section. The regions of shale (low permeability) and sand (high permeability) are obvious from the black/white contrast, as is the progressive nature of the throw along the growth fault, as well as the relative thinning of beds on the moving hanging wall side of the fault compared to the same beds on the stationary footwall side (which are subjected to a lesser overburden). Figures 7.28a–j provide the development of overpressure with time from 32 MYBP to the present day as the fault top steps progressively leftward and the basement moves under isostatic load to be in agreement with the requirement of a horizontal basement at the present day.

The major differences in this case compared to the two previous examples are (1) the more rapid increase with time of overpressure due to the more

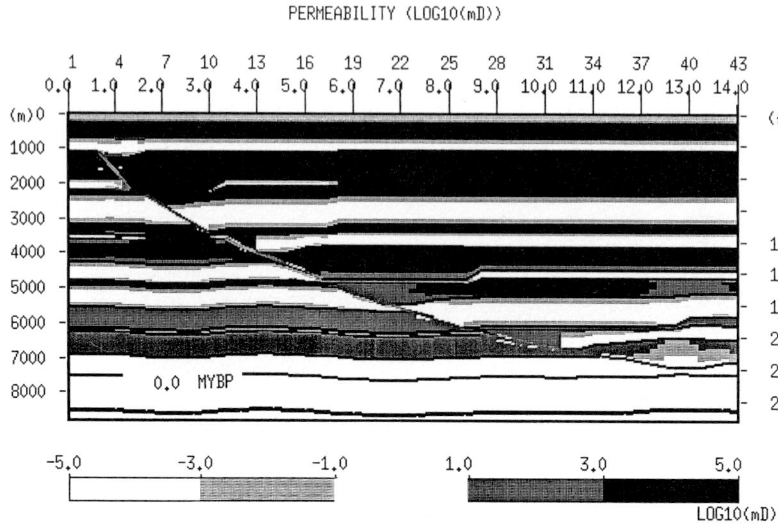


FIGURE 7.27 The present-day permeability variation in the interbedded sand–shale section for the case of a hydraulically closed fault. The section is 14 km in width, with 43 pseudo-wells.

interspersed shales providing progressive sealing capability throughout the whole section; development of overpressure now also occurs in the sands entrapped between shale layers; and (2) the lateral spread of massive overpressure across the highly permeability sands, which is retained in the isolated sand components when sand continuity is cut off due to progressive bed throw and also compaction.

The development of overpressure with time across the section is significantly different than in either of the two previous cases. For example, note the massive overpressure development (to nearly 300 atm) between about 4200 and 5000 m at the present day on the footwall side of the growth fault, and the corresponding development from 4200 to 6000 m on the hanging wall side. The implication here is that the interbedding of shales and sands provides more laterally extensive regions of high overpressure, and that such "compartments" are more prevalent throughout a section due to facies interbedding and progressive fault throw.

### 4. Open Fault: Interbedded Shale and Sand Formations

Precisely the same geometric configuration of deposition of interbedded shale and sand layers was used as in the previous case. The major difference is in the ascription of fault properties. Here the region around the fault is given sand permeability characteristics. The competition in this case, versus the previous example, is that fluids can now transit to the surface up the fault from the interbedded system, or to shallower sands from deeper sands.

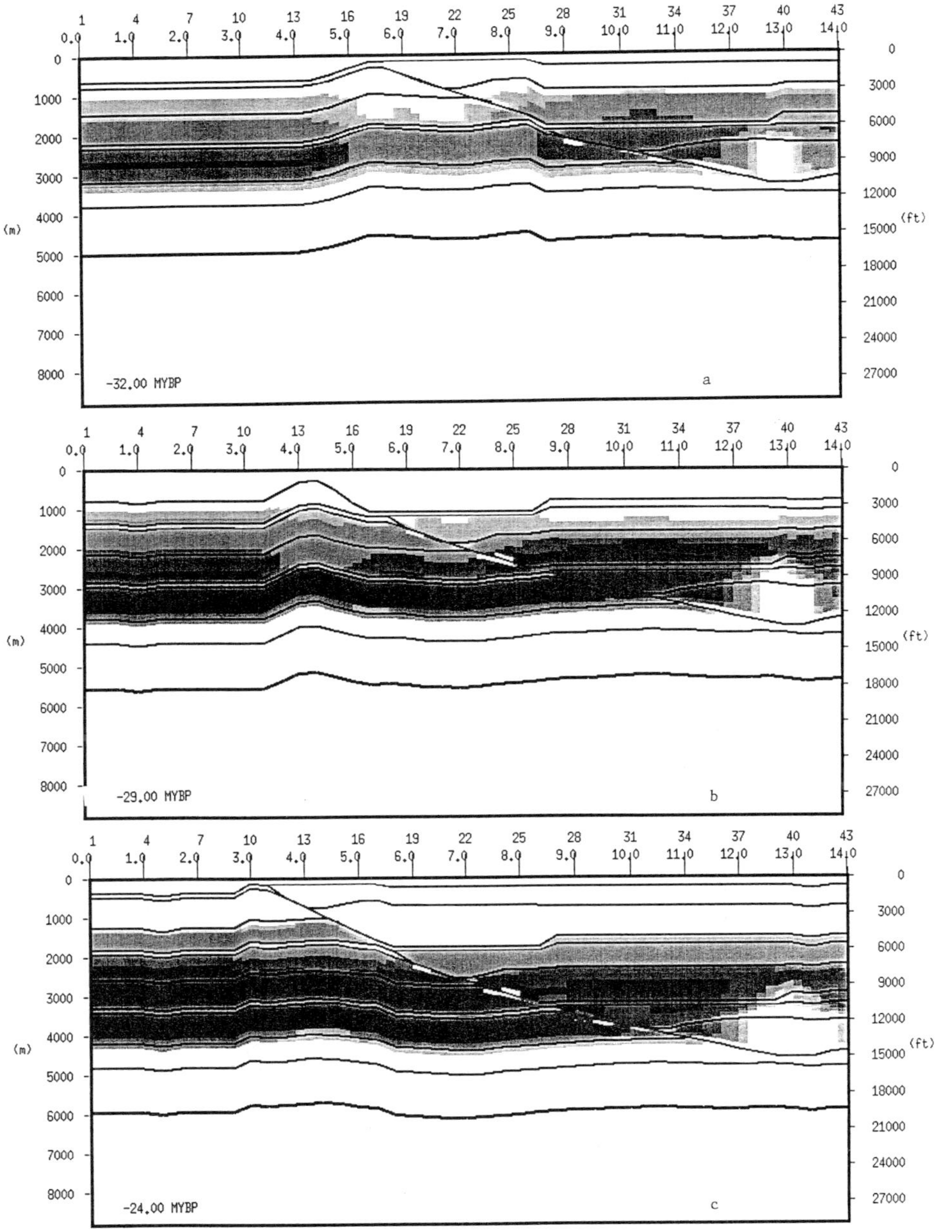


FIGURE 7.28 Excess pressure development with time in the case of the sand–shale interbedded section for an hydraulically closed fault. Parts (a) through (j) are at 32, 29, 24, 22, 20, 16, 12, 5, and 1 MYBP and present-day, respectively. The section is 14 km in width, with 43 pseudo-wells.

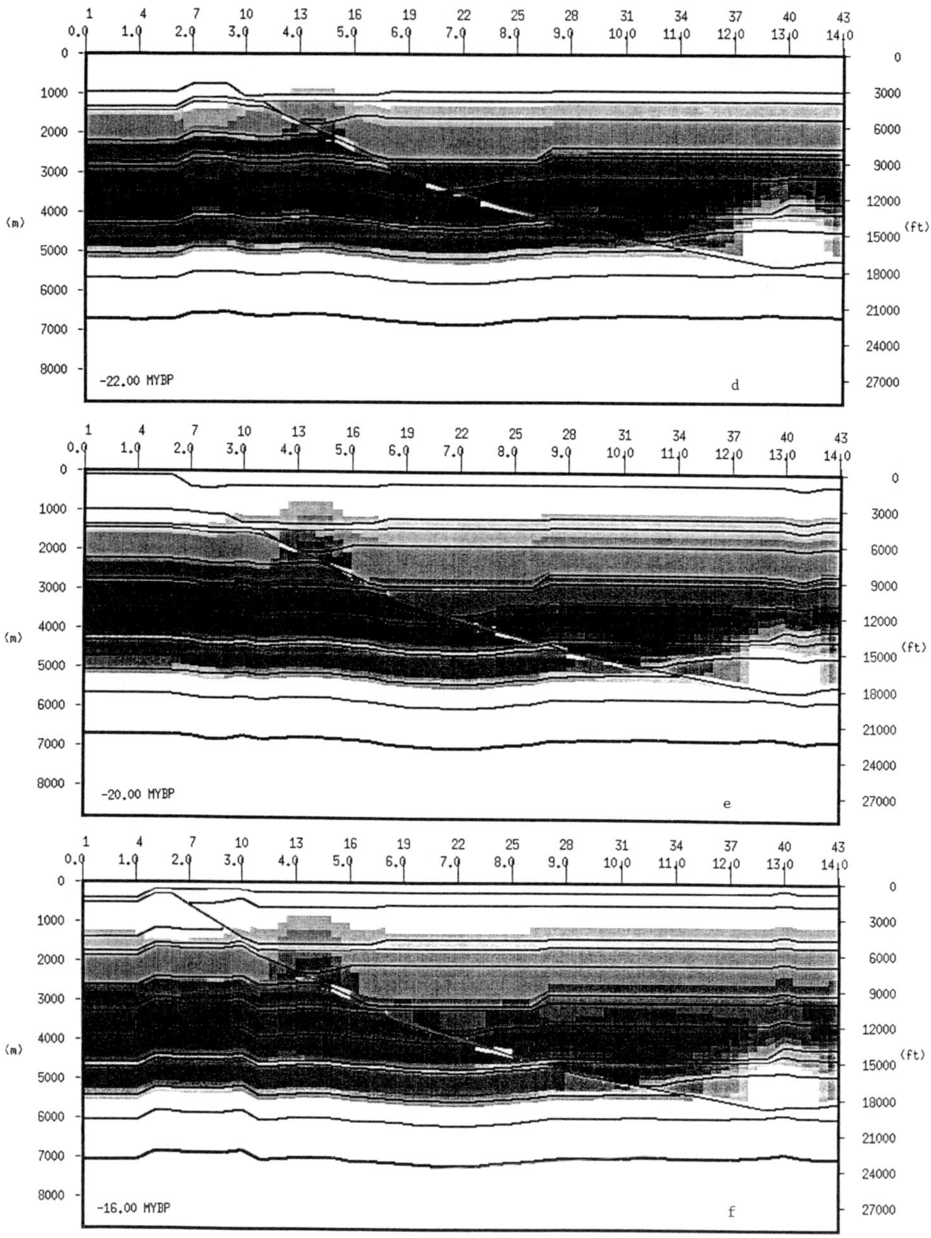


FIGURE 7.28 *(Continued)*

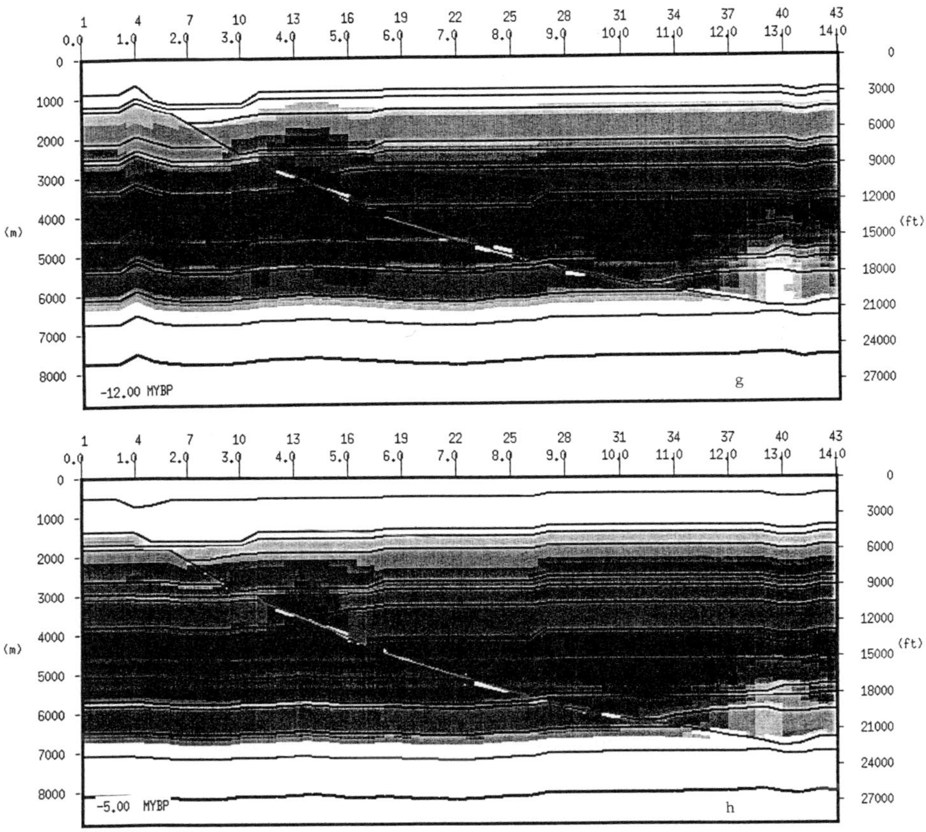


FIGURE 7.28 *(Continued)*

The permeability of the section at the present day is sketched in Fig. 7.29 where the dominance of the open fault is shown. The connectivity of high-permeability regions throughout the section is increased because an initially continuous sand body, when progressively thrown on the hanging wall side, now has permeability connectivity maintained through the open nature of the fault. For example Figs. 7.30a–k show the development of overpressure from 34 MYBP through to the present day. The dominant differences in this example compared to the previous case are (1) a much smaller overpressure development occurs; (2) the "compartments" of overpressure are both vertically and laterally smaller; (3) the region around the fault is always removing fluid, consequently the isobars of excess overpressure show significant "dimpling" around the fault to greater depths (representing flow); and (4) the influence of the open fault is laterally more pervasive in this interbedded situation than in the case of an homogeneous sandy-shale, primarily because the open fault continues to drain fluid from deep sands

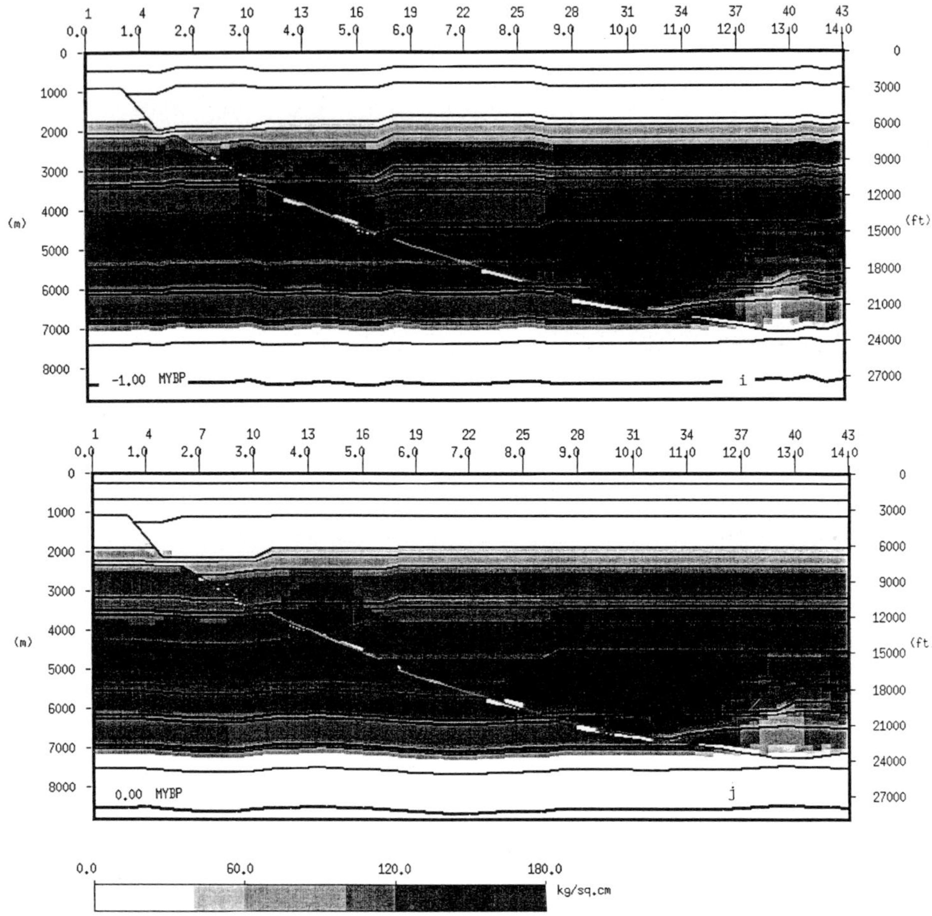


FIGURE 7.28 (*Continued*)

to both shallower sands and also to the surface. Continuous flow occurs from the deeper hanging wall sands to the shallower footwall sands, driven by the high overpressure of the deep sands. This flow continues until the shallower footwall sands, embedded between low-permeability shale layers, reach an overpressure comparable to that of the driving pressure along the fault; thereafter flow is to even shallower regions. Thus the compaction of interbedded layers, the spatial development of overpressure, and the flow of fluids all depend sensitively on (1) the interbedded nature of the system, (2) the progressive throw of the fault, and (3) the hydraulic connectivity of the fault.

These four synthetic cases have been chosen to illustrate the diversity of overpressure behaviors that can occur when a progressive growth fault

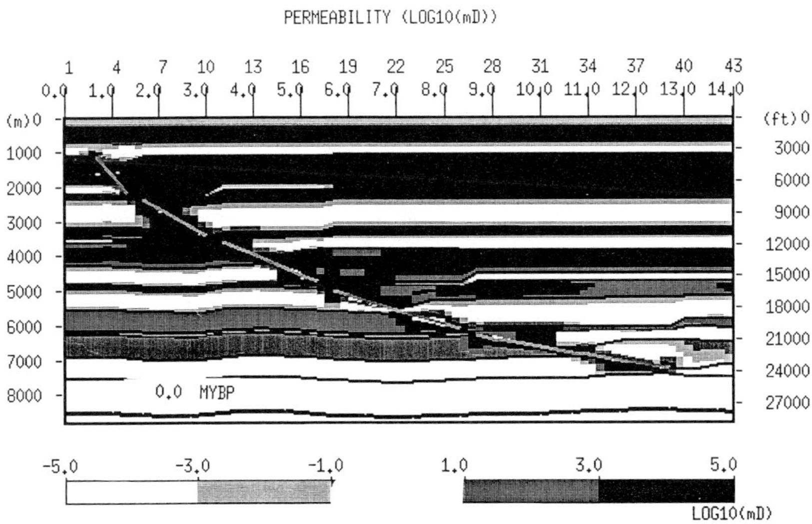


FIGURE 7.29 The present-day permeability variation in the interbedded sand–shale section for the case of a hydraulically open fault. The section is 14 km in width, with 43 pseudo-wells.

causes displacement of sedimentary beds. The overpressure development depends on the interbedded nature of the system, the throw of the beds, and the opennness or closedness of the growth fault.

## V. A FIELD EXAMPLE: SOUTH LOUISIANA ROLL-OVER STRUCTURE

Recently Xiao and Suppe (1992) have considered the geometric evolution of a growth fault associated with a southern Louisiana roll-over structure as depicted in their Figs. 20(a) and (b), with Fig. 20(a) showing the seismic section and Fig. 20(b) showing the depth-interpreted section. This figure is reproduced here as Fig. 7.31. Because Xiao and Suppe (1992) consider geometric evolution of the roll-over, they do not require information of sedimentary facies, and no information is recorded in their paper on facies development—possibly because no wells have been drilled through the section. To evaluate the fluid pressure evolution of the exhibited section, three extreme lithologies were used: (1) all sand, in which case compaction to equilibrium conditions occurs virtually instantaneously on a geological timescale because of the high sand permeability (~1 Darcy) and no overpressure ever develops; (2) all shale, in which case compaction occurs sufficiently slowly on the geologic timescale of the section that

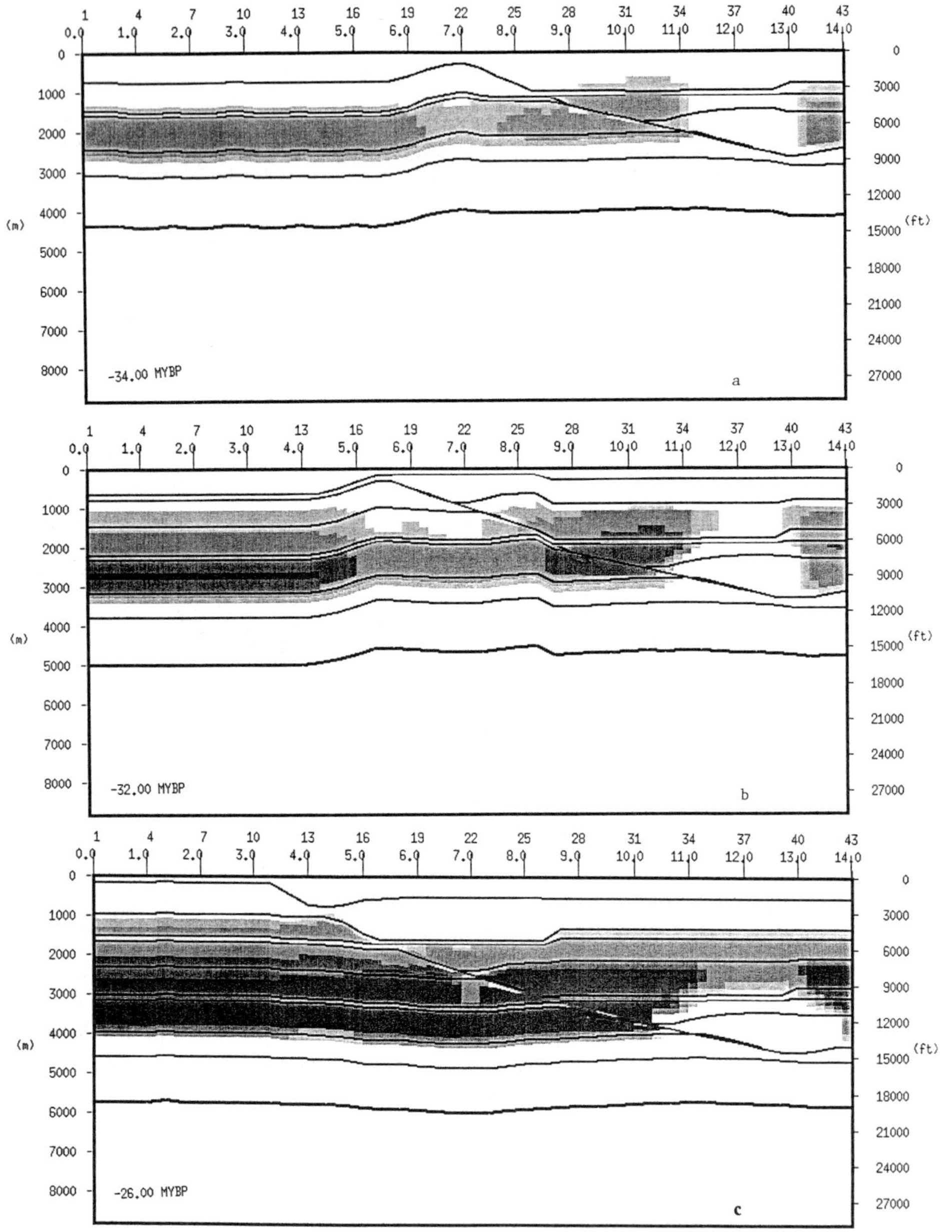


FIGURE 7.30 Excess pressure development with time in the case of the sand–shale interbedded section for an hydraulically open fault. Parts (a) through (k) are at 34, 32, 26, 24, 22, 20, 16, 12, 5, and 1 MYBP and present day, respectively. The section is 14 km in width, with 43 pseudo-wells.

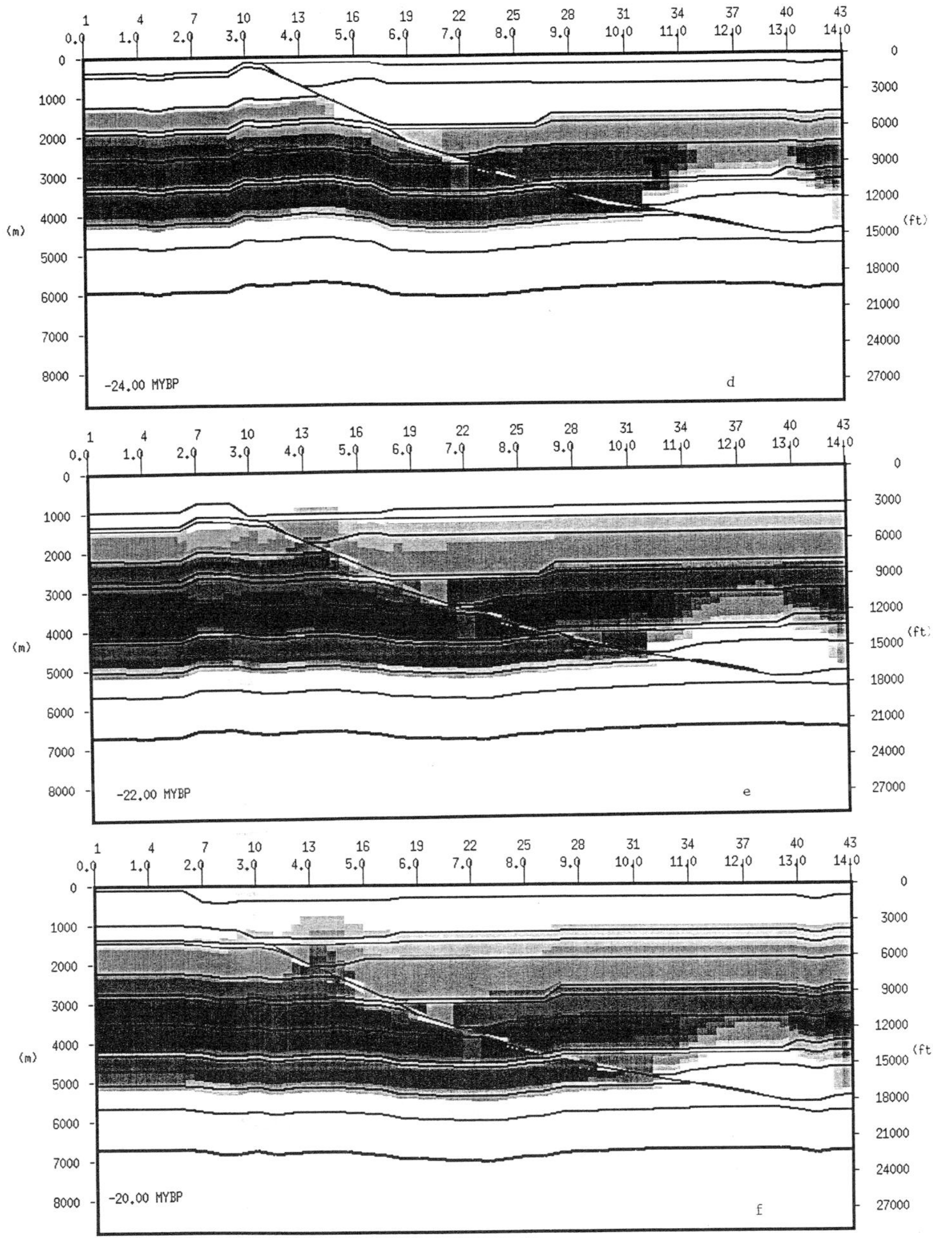


FIGURE 7.30 *(Continued)*

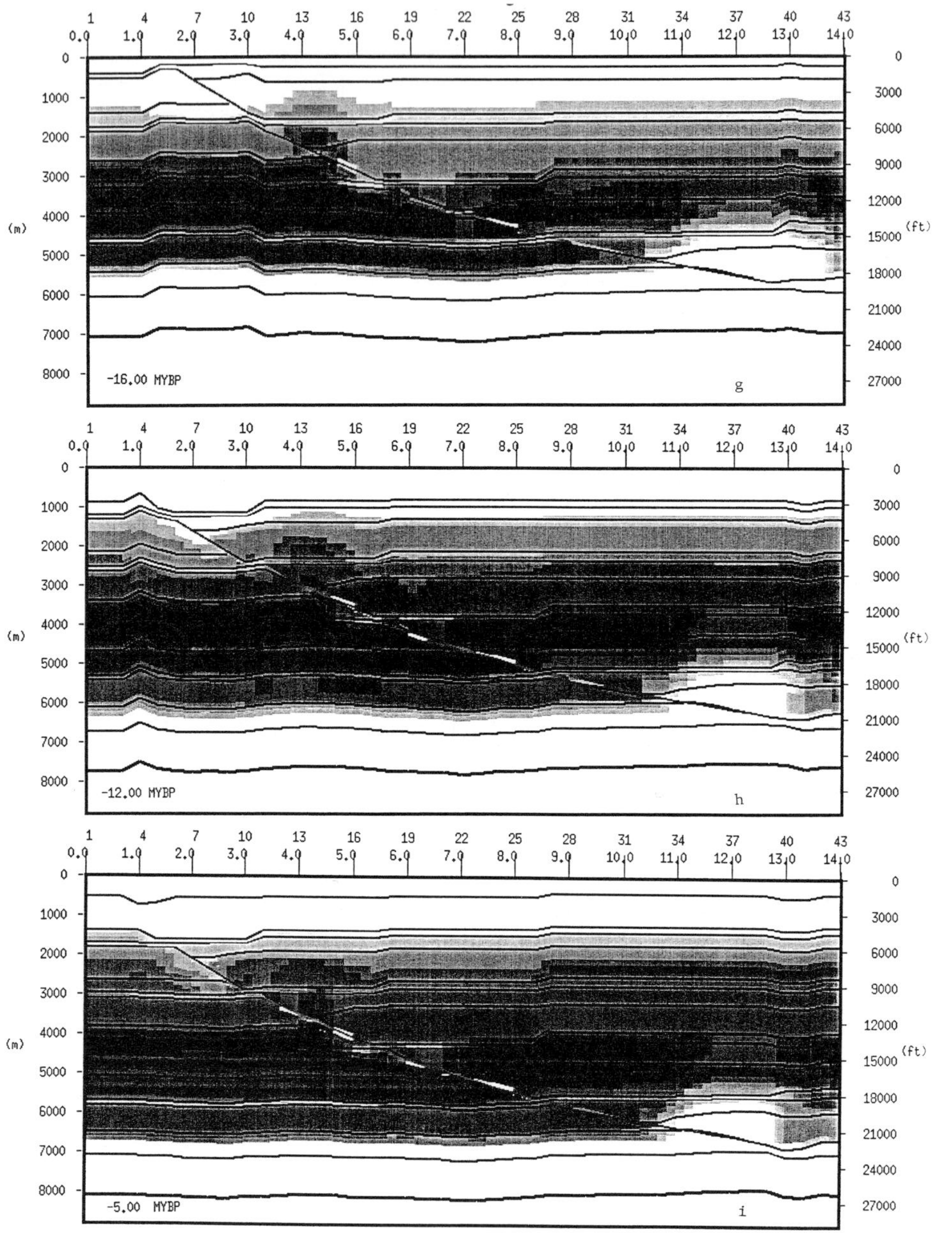


FIGURE 7.30 *(Continued)*

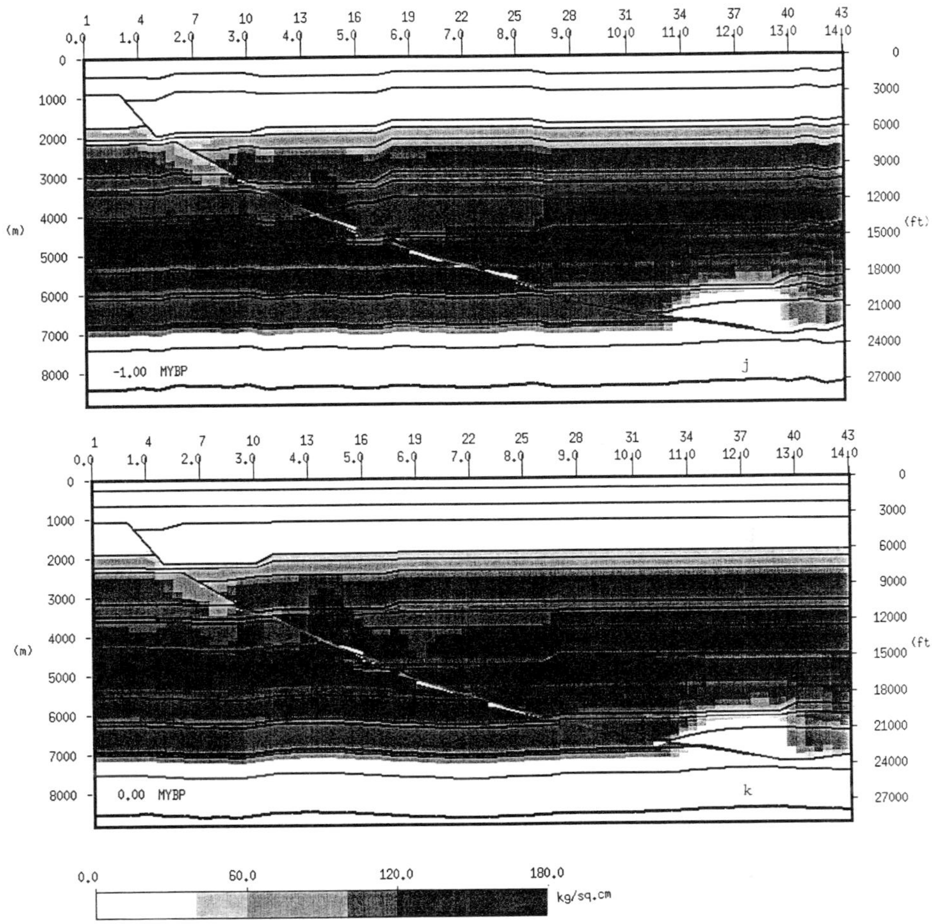


FIGURE 7.30 *(Continued)*

formations remain at or near to their depositional thicknesses for all time, and overpressure basically supports the total sediment load for all time since deposition. Geometrically this situation is most akin to that depicted by Xiao and Suppe (1992); and (3) a sandy-shale "mix" of lithologies on a spaital depositional scale finer than the spatial grid of the calculations, so that the system can be taken to be constructed of an homogeneous sand-shale lithology in which the permeability, $K_{ssh}$, is taken to be

$$K_{ssh} = K_s^S K_{sh}^{1-S},$$

where $K_s(K_{sh})$ is sand (shale) lithology, and S is the fraction of sand (0.5 in the present case). Because the permeabilities of both sand and shale are

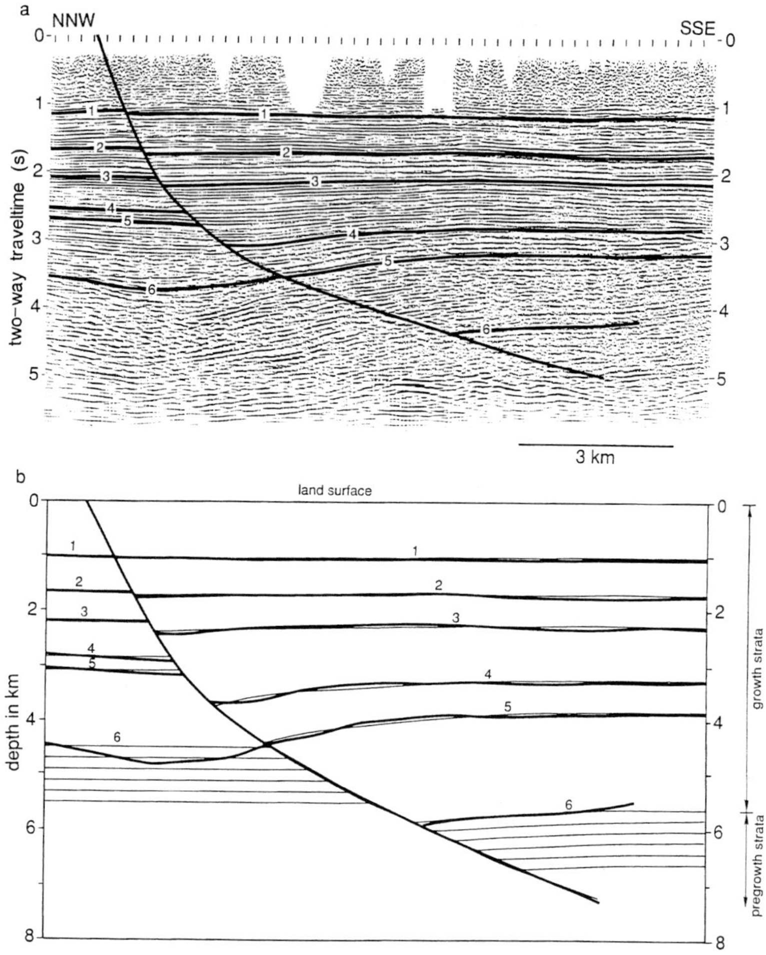


FIGURE 7.31 A reproduction of Fig. 20 from Xiao and Suppe (1992). (a) Seismic line showing an interpretation of a southern Louisiana rollover; (b) the depth conversion of the seismic interpretation in Fig. 20(a) is shown by thick lines. (H. Xiao and J. Suppe, © 1992, reprinted by permission of the American Association of Petroleum Geologists.)

taken to vary as power laws of the void ratio, $e$ [$\equiv \phi/(1 - \phi)$, where $\phi$ is the porosity] in the forms

$$K_{\mathrm{s}} = K_{\mathrm{so}}\,(e/e_*)^{b_{\mathrm{s}}},$$

$$K_{\mathrm{sh}} = K_{\mathrm{sho}}\,(e/e_*)^{b_{\mathrm{sh}}},$$

where $e_*$ is the depositional void ratio. It follows that $K_{\mathrm{ssh}}$ also varies as a power law with

$$K_{\mathrm{ssh}} = K_{\mathrm{ssho}}\,(e/e_*)^{a},$$

where

$$a = \mathrm{S}b_{\mathrm{s}} + (1 - \mathrm{S})\,b_{\mathrm{sh}}; \qquad K_{\mathrm{ssho}} = K_{\mathrm{so}}^{\mathrm{S}} K_{\mathrm{sho}}^{1-\mathrm{S}}.$$

Thus, at a 50/50 mix of sand and shale, the depositional permeability of a sandy-shale lithology is the geometric product of that for sand with that for shale.

In addition, Fig. 20(b) of Xiao and Suppe (1992) shows no vertical displacement of the bed labeled 1 or of shallower horizons so, for the purposes of computation, the fault top is terminated at bed 1. Missing on Fig. 20(b) of Xiao and Suppe (1992) is a horizon to represent the basement below the fault. We have added a horizontal bed at the 8500-m depth at the present day, which is taken to respond isostatically to the overlying vertical load at each location. Alternative options are possible, such as a nonhorizontal marker, but a horizontal "basement" provides a template marker from which all other options can be inferred. Also not included in Xiao and Suppe's (1992) presentation are the depositional ages of the horizons marked 1 through 6 on their Fig. 20b; in a geometric reconstruction such age dates do not play a significant role. However in the case of fluid flow through a compacting section, the ages of the formations *are* of significance because the time frames over which fluid pressure can build-up and/or be bled off *do* depend on the permeability variations. What we have done, then, to model pressure build-up in this south Louisiana section is to consider uniform deposition rates of 6000 m over a timescale of 35 MY, that is, 180 m/MY—typical of Gulf Coast sedimentation rates in south Louisiana. Development of overpressure in the section then depends on both the throw of sediments along the fault together with the "openness" or "closedness" of the fault to fluid flow. To examine the differences, both cases of fault properties were run.

The fault was taken to terminate at depth today at the two-way-travel time position marked on Fig. 7.31 [Fig. 20(a) of Xiao and Suppe (1992)] at about 5 s to the right of the figure. Below this depth two further horizon boundaries were added above basement to illustrate how continuity requirements of nonfaulted beds must behave below the faulted region. Because the calculations are done on a finite lateral grid of "pseudo-wells," instead of having a continuous roll-over of hanging wall sediments, a discrete set of step-down behaviors is present. The section used is 13 km long, with 40 pseudo-wells, so that the visual impression of the step-down effect is clearly seen. One could increase the number of pseudo-wells to better represent a larger number of smaller step-downs (with, of course, the same total roll-over) but, for the purposes of restricting computer run time, it is sufficient to treat here with the 40 pseudo-well case.

## A. OPEN FAULT CASE

Representation of the present-day permeability across the section is shown in Fig. 7.32, clearly indicating the presence of high permeability (~1 Darcy) along the fault. Also to be noted are regions of high permeability with the position above, but removed from, the fault; perhaps the most

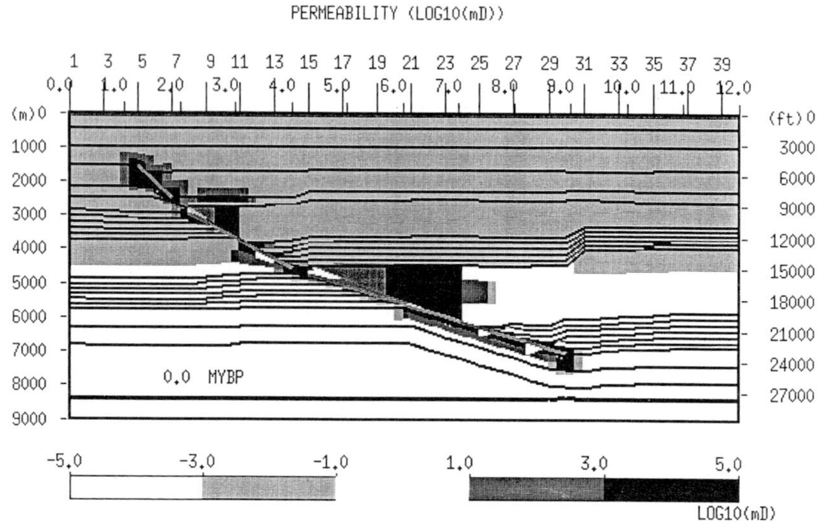


FIGURE 7.32 The present-day permeability profile with depth in the south Louisiana roll-over structure for a hydraulically open fault. The section is 12.5 km in width, with 40 pseudo-wells.

obvious of which is that occurring between pseudo-wells 20 and 24 at about a 4500-m depth, with a secondary region between pseudo-well locations 9 and 11 at about a 2500-m depth. These anomalies in the permeability of the section are caused by the fracture limit set in the code. Physically what is happening is that the open fault initially bleeds fluids from the neighboring sediments into the fault region; the sediments then compact, thereby lowering their permeability and so limiting the escape of further fluid. As deposition continues, the pore pressure in those sediments then increases rapidly until it reaches the point where the total fluid pressure exceeds a fraction (default is 0.85) of the total lithostatic overburden. At this point the sediments fracture and, during the fracture period, are assigned a sand permeability. Once the fluid flows sufficiently to decrease the pore pressure below the fracture limit "normal" sediment permeability is resumed. Thus, the situation is one of ongoing "pulses" of sediment fracturing and fluid release with resealing and refracturing occurring.

Development of the complete seciton with time has been tracked from 34 MYBP until the present day. However, significant overpressure occurs only in the last 10 MY due to the high permeability of the original sandy-shale lithology. Figures 7.33a–e show the excess pressure isobars with time as the system develops. Note the progressive "march" leftward of the top of the fault, the flattening of the deeper section of the fault, and the variation of sedimentary positions as a consequence of lateral sediment motion along the fault, together with the isostatic assumption for motion of the basement

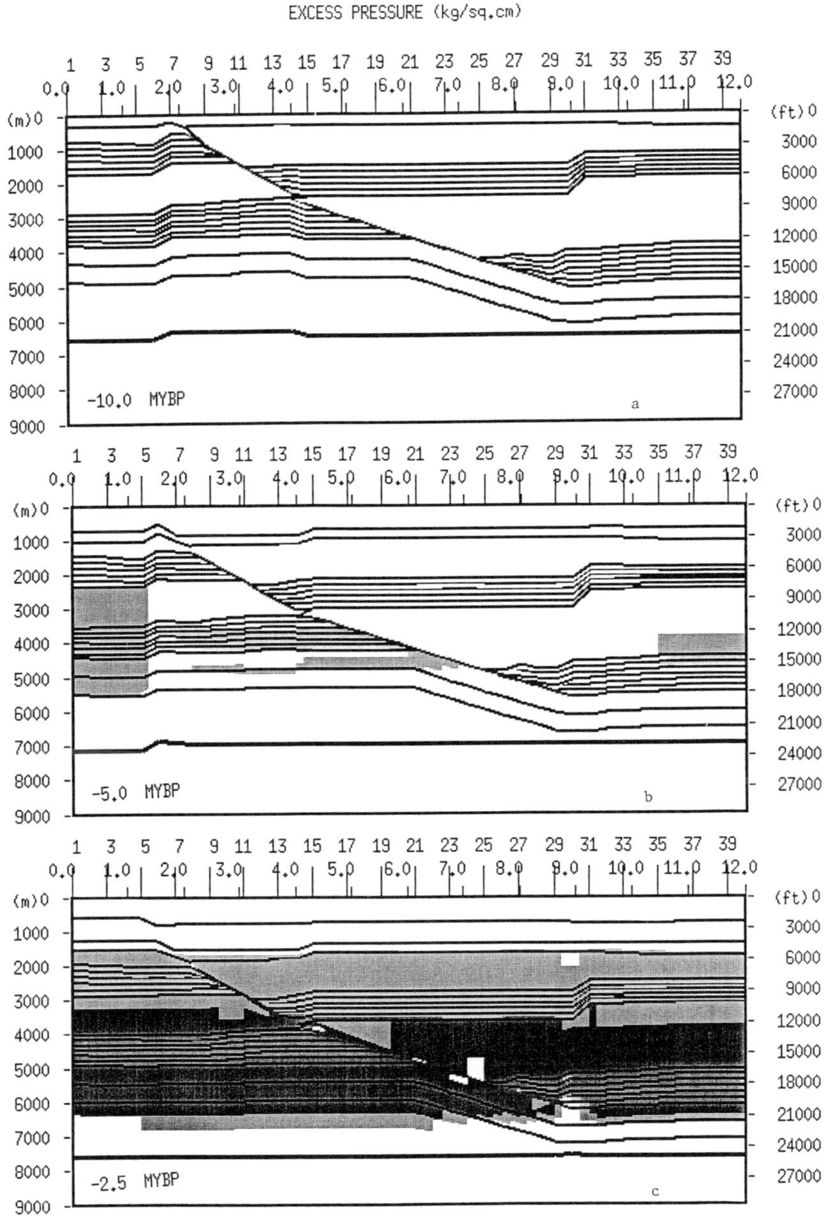


FIGURE 7.33 Excess pressure development with time in the south Louisiana roll-over section for a hydraulically open fault: (a) 10 MYBP; (b) 5 MYBP; (c) 2.5 MYBP; (d) 1 MYBP; (e) present day. The section is 12.5 km in width, with 40 pseudo-wells.

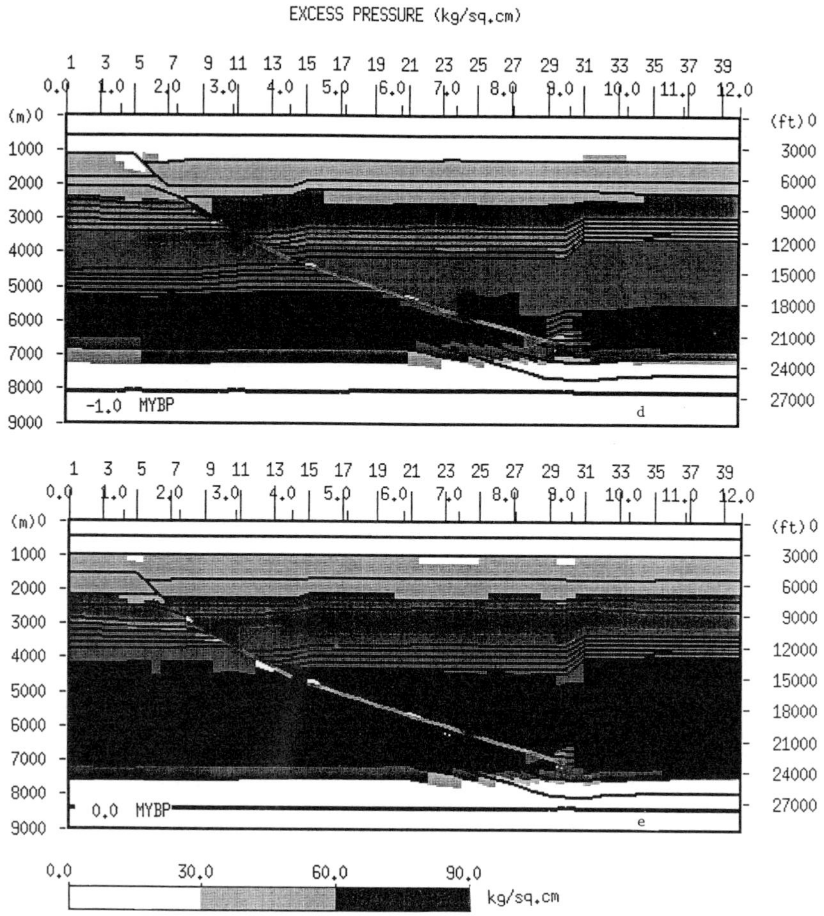


FIGURE 7.33 (*Continued*)

horizon. Major excess pressure does not develop until the last 5–2.5 MY—a direct consequence of the time at which the sandy-shale permeability is small enough to retain fluid. The variation of excess pressure with time shows the increased depth of isobars near the open fault, as well as the rapid increase in overpressure on the hanging wall side of the sediments in the last million years or so. In addition, comparison of the results here with those of the closed fault case (next subsection) shows an overall lesser development of overpressure throughout the hanging wall section. This effect arises because fluids can be bled off from the sandy-shale section into the fault for a considerable period of time after significant overpressure development but before the permeability has decreased so much that it effectively isolates the fluids. The overpressure today in the deep part of the section is about 150 atm—about half the value obtained with the synthetic example of sandy-shale. This lower excess pressure is due to the higher permeability of the sandy-shale section being retained to a greater depth (and so for a longer time) than in the synthetic case.

## B. CLOSED FAULT CASE

The present-day permeability pattern across the section shows considerable differences from the open fault case. Gone, naturally, is the high permeability region along the fault, but equally missing are the regions of high fracture permeability away from the fault, so prominent in the open fault case. The absence of the fracture regions is due to the slower, steadier loss of fluid through the sediments compared to the rapid loss in the open fault case. No longer do the sediments lose fluid quickly and so rapidly decrease their permeabilities; rather a gentler evolution of permeability occurs without the rapid lowering near the fault. As a consequence the fluid pressure does not build up as rapidly.

Indeed, in Figs. 7.34a–d the excess pressure build-up shows a thicker region of high pressure at depth compared to the open fault case, so that the overall excess pressure gradient is reduced. The general pattern of pressure development remains about the same; only with the decrease of permeability in the last 10 MY or so is there a slow enough fluid loss compared to deposition of overlying sediments to create overpressure. The basic overpressure pattern is close to the pattern obtained in the open fault case, except that the domains of overpressure, as represented by isobaric lines, are more nearly horizontal across the section and the physical thicknesses of domains between isobaric lines are larger than in the open fault case.

In either case the development of overpressure was fairly recent, extends to about 150 atm in magnitude, and is prevalent across the whole section from about 2000 m depth and greater. The effect of the openness or closedness of the fault on the present-day pressure section pattern is most noticeable only around the fault region and, even then, only to a minor extent. It would appear that the decrease of permeability is so rapid in this south Louisiana section that, once triggered, the overpressure development swamps any other factors attempting to represent significant regional differences in behavior.

## C. DISCUSSION

Both the synthetic case studies and the application to a south Louisiana section illuminate several major factors of growth fault development effects in quantitative basin analysis:

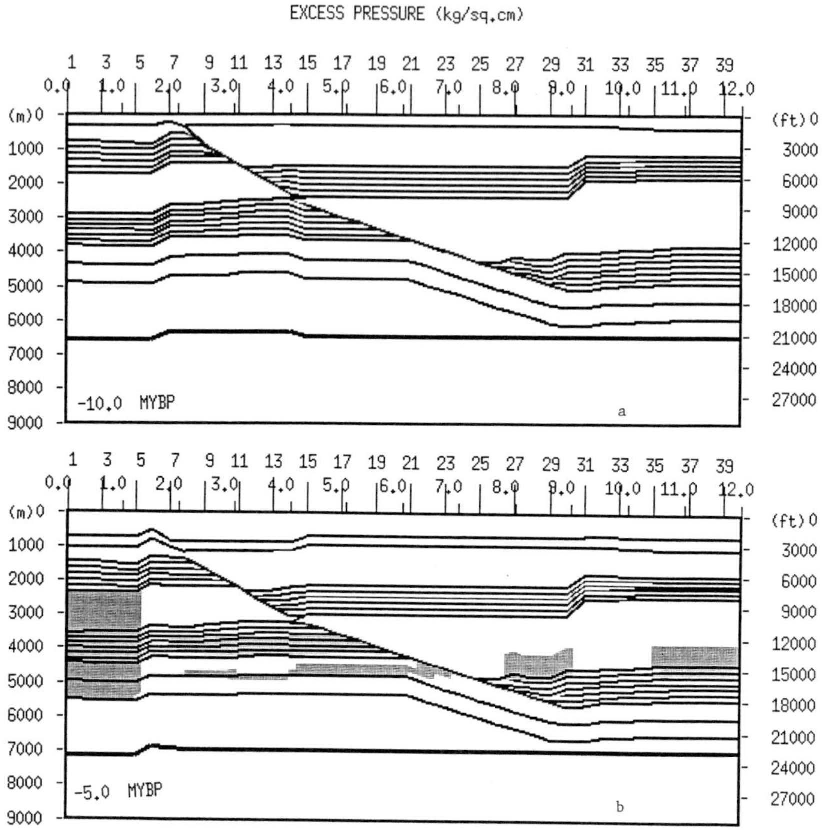


FIGURE 7.34 Excess pressure development with time in the south Louisiana roll-over section for a hydraulically closed fault: (a) 10 MYBP; (b) 5 MYBP; (c) 1 MYBP; (d) present day. The section is 12.5 km in width, with 40 pseudo-wells.

1. The curvature development with time of the fault is beholden to two major factors: (a) the evolution of porosity of surrounding sediments, which themselves are influenced by whether the fault is taken to be open or closed to hydrodynamic flow; and (b) the lateral motion of sediments along the growth fault, which is influenced by both the volume accommodation due to compaction, as well as by the prescribed or inferred motion of the underlying basement and the perceived shape of the basement today. In the examples a horizontal basement today is taken to respond isostatically to overburden load to illustrate both the shape and dynamic load effects.

2. The development of overpressure with time in a growth fault situation is dependent not only on the depositional rate across the section but, more dominantly, by the lithologic facies and their lateral and vertical juxtapositioning. Thus the effects on overpressure development of interbed-

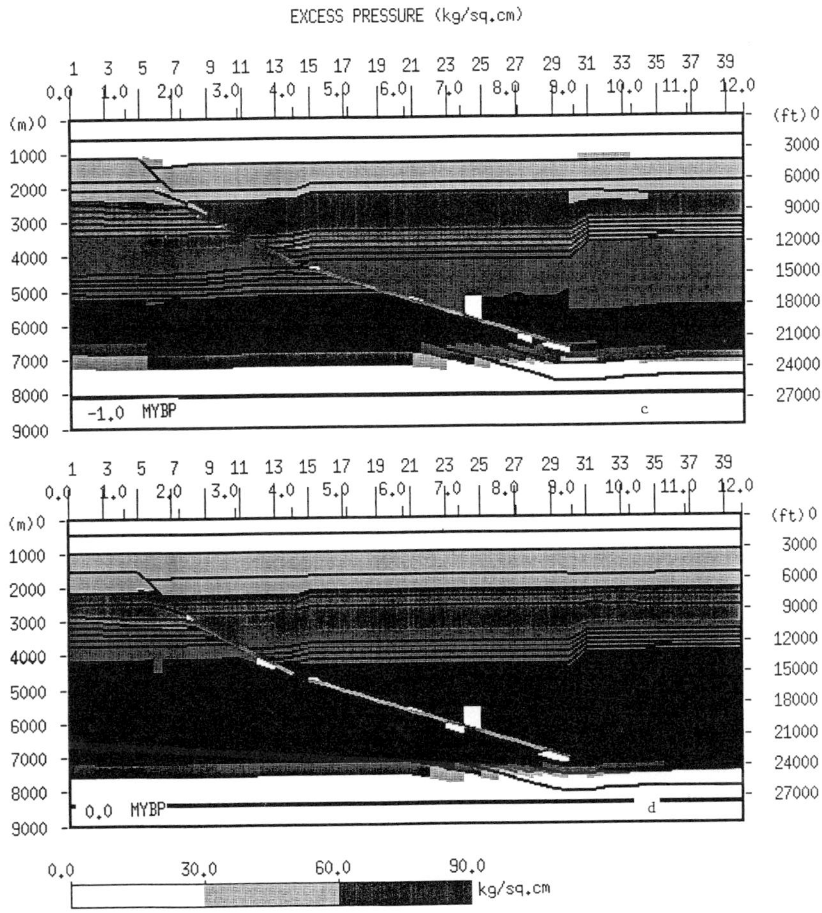


FIGURE 7.34 (*Continued*)

ded sand–shale formations are considerably different than in the case of thick sandy-shale deposition, irrespective of whether the fault is open or closed. The ability of an interbedded system to have sand formations connect, disconnect, and reconnect across the fault as evolution proceeds and the growth fault develops, with larger throws being progressively developed at depth, leads to a complex interplay of pressure variations across the section as fluids flow from one sand region to another or remain isolated in particular sands until later reconnection occurs.

3. The presence of an open fault can lead to a greater degree of fracturing in neighboring formations than occurs in the case of a closed fault. The reason is that fluids can flow quickly into the fault, thereby decreasing sedimentary porosity and permeability near the fault, which leads to a more

rapid build-up of overpressure in such regions followed by episodic fracturing.

4. The roll-over variation with time of sediments on the hanging wall side of the fault is controlled by (a) the compaction variation with time and so the lithofacies, (b) the depositional rate across the section, (c) the curvature variation of the fault as a consequence of compaction, and (d) the motion of the underlying basement and the shape evolution of a basement curve (horizontal in the examples and responding isostatically to applied load).

5. The geometrical structure and overpressure development with time are tied together through the permeability variations with porosity for each lithology. A rapidly compacting, high-permeability sand never develops overpressure on its own, but when sandwiched between very low permeability shale units can retain its thickness and be overpressured unless a conduit for fluid flow (e.g., an open fault) is available to drain the overpressure in the sand.

Although the calculations reported here for both the synthetic sections and the south Louisiana case history do not represent a complete encompassing of all effects that can be caused in sedimentary formations by growth fault development, they do nevertheless provide a compendium of some of the more obvious patterns of diverse development that can occur. It seems that the interbedded nature, the basement shape and motion, the lithofacies permeabilities, and the open or closed hydraulic nature of the fault all play roles of significance. In any specific situation all such factors must be included in order to evaluate quantitatively dominance of one or more factors in controlling the evolution of a system. It is this fact that the synthetic studies and the case history speak to most strongly.

# 8

# Models of Basinal Sediment Fill, Turbidite Flows, and Their Uncertainties

Sediment transport, deposition, and erosion are important geological processes in sedimentary basins. A better understanding of these processes, their relative timing, their interrelationships, and their magnitudes will help to reduce the risk in hydrocarbon exploration projects. Many process models have been developed during the last few years (Cross, 1990; Ross, 1995; Coakley and Pratson, 1996; Flemings and Carlson, 1996) to simulate sediment transport, deposition, and erosion in sedimentary basins. By and large most such sequence stratigraphic models tend to consider relatively smoothly continuous supplies of sedimentary material on fairly long timescales, and often do not include the ability to handle short-lived turbiditic sequences, which are episodic and caused by catastrophic collapse or failure on very short timescales.

This chapter examines some of the assumptions and parameters that go into constructing "forward" models for both sequence stratigraphic fill models and turbidite flow models. In the final analysis both types of models have to compare their model predictions against observations, so that the uncertainty in parameter values, intrinsic assumptions, and so on, must be addressed in relation to the uniqueness, resolution, sensitivity, and precision of output results.

## I. A SEQUENCE STRATIGRAPHIC SEDIMENT FILL MODEL

The classical inverse approach to interpreting the fill of a basin with sediment in terms of magnitude of eustatic changes, rates of change in tectonic movement, and sediment accumulation (Ferm, 1970; Vail *et al.*,

1977; Busch, 1983; Hallam, 1984) depends on assuming the size of at least two of these variables to determine the third. Such difficulties have led to development of forward (in time) simulations to handle eustasy, tectonics, and sedimentation, with the ultimate goal being to invert the forward models to determine sizes of the different variables which produced the observed sedimentary geometries or those synthetically generated within a given basin (Burton *et al.,* 1987; Johnson, 1996; Lessenger and Cross, 1996a,b; Tinker, 1996). Here, a system is outlined for reproducing the general geometries of both marine and alluvial clastic deposits. The sequential simulations of geologic processes are first described; then the modeling of the individual processes (including erosion, deposition, compaction, and tectonic movement); the simulation outputs; and finally the data and inputs required. The simulations are designed to generate two-dimensional (2-D) representations of time lines within sedimentary bodies, similar to those that can be observed on dip trending seismic lines. The geometries seen on seismic lines are the result of the interaction of eustatic changes, rates of sedimentation, and rates and directions of tectonic movement. The simulations vary parameters controlling these processes and plot the resulting geometries. When a favorable match is seen between a simulation plot and an interpreted seismic cross-section, statements can be made concerning the range of likely sediment accumulation rates, subsidence rates, and changes in eustasy that produced the geometry. Simulation models are based on the concept that gross features of the depositional geometry are controlled by the same macroprocesses that produce the stratigraphy of seismic sections. The aim, therefore, is to reproduce the sedimentary geometries, which are the results of such processes averaged over long periods of time.

## A. SEQUENTIAL SIMULATION OF GEOLOGIC PROCESSES

Sediment deposition over fixed time intervals is simulated by derivation from triangles of sediment of specified length and height. The length of the triangles is matched to the offshore sedimentary wedge found seaward of the shoreline, and triangles are a device to capture the geometry of this offshore wedge. The advantage of triangles is that the offshore wedge can be deposited arbitrarily without having to model all the physical controls. The geometrical configuration of time lines generated for each time interval is controlled by the following factors:

1. Configuration of the original depositional surface (Fig. 8.1a)
2. Variability of sediment (sand and/or shale) input (height of triangles) and the lateral distribution of sediment for each time interval (length of triangles, depositional angle of alluvial plain) (Figs. 8.1b and 8.2)

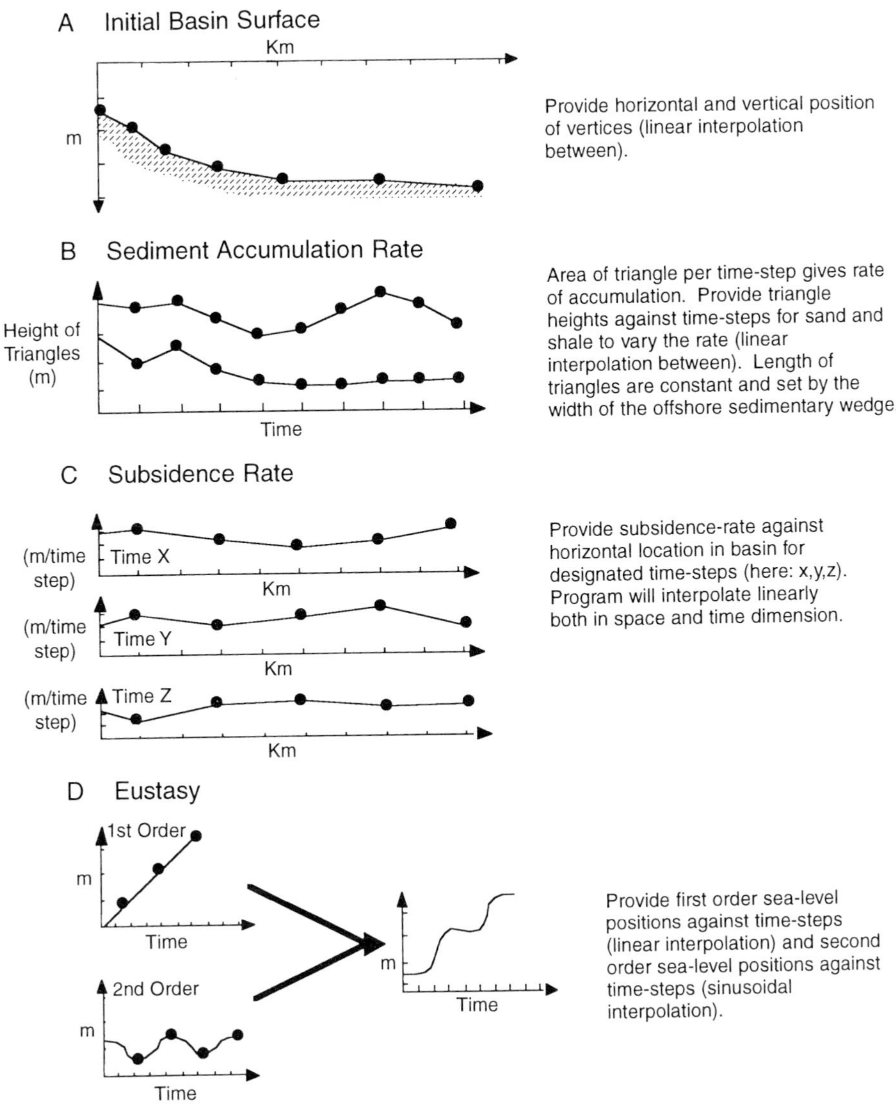


FIGURE 8.1 Main input parameters for controlling simulation geometries.

3. Subsidence (or uplift) behavior of each point for each time interval (including tectonic movement, subsidence in response to sediment and water loading, and compactional subsidence, Fig. 8.1c)
4. Variation of eustasy through time (Fig. 8.1d)

Sediments that are exposed above an "alluvial baseline" (defined by the depositional angle of the alluvial plain) at a given time are subject to

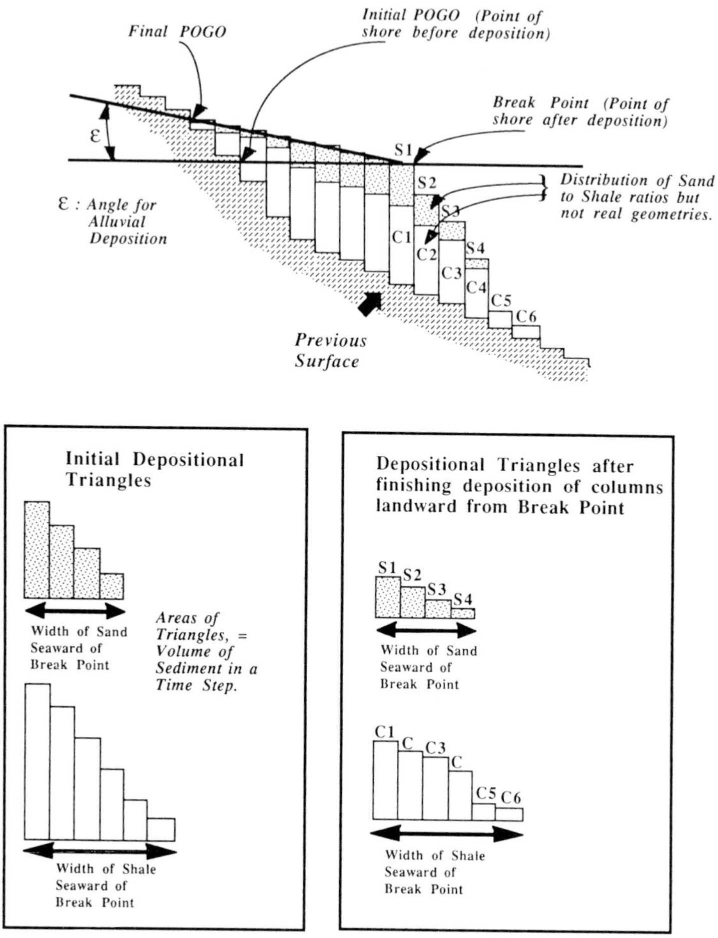


FIGURE 8.2 Distribution of sand and shale within a time step as defined by the program.

erosion. Moreover, if the inclination of the submarine slope is also above a critical value, sediments bypass this slope to the basin floor and are deposited there. For a given time, each depositional surface is defined by an array of points that corresponds to the vertical position of the surface between evenly spaced horizontal coordinates, which gives all sediment surfaces a staircase-like appearance (Fig. 8.2), which is not visible in the output because of the large number of horizontal points. The basic sequence of processes followed (Fig. 8.3) in each time step is:

1. Calculate sea level for the current time step.
2. Subside or uplift all surfaces according to the given tectonic movement for that time step.

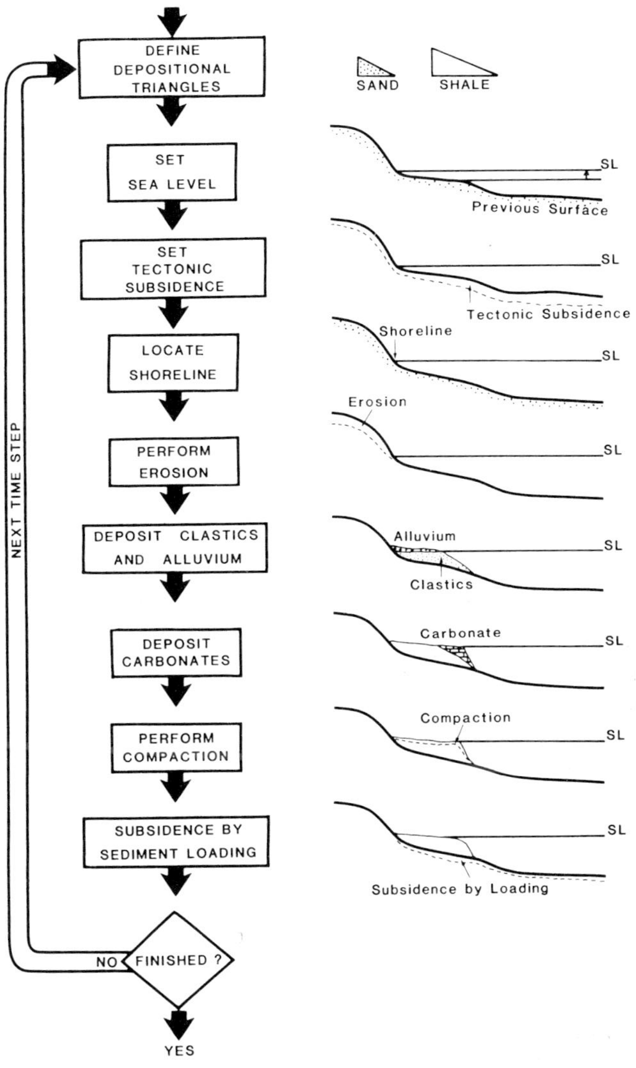


FIGURE 8.3 Flowchart of simulation sequence.

3. Locate the shoreline.
4. Erode sediments that are elevated above the "alluvial baseline" to a depth determined by an erosion formula.
5. Deposit sediments, including those to be deposited from the area of the triangles and those eroded in step 3. Submarine erosion is also possible including mass slumping.
6. Compact all sediment layers to equilibrium (note this strong assumption) according to the weight of the overburden.

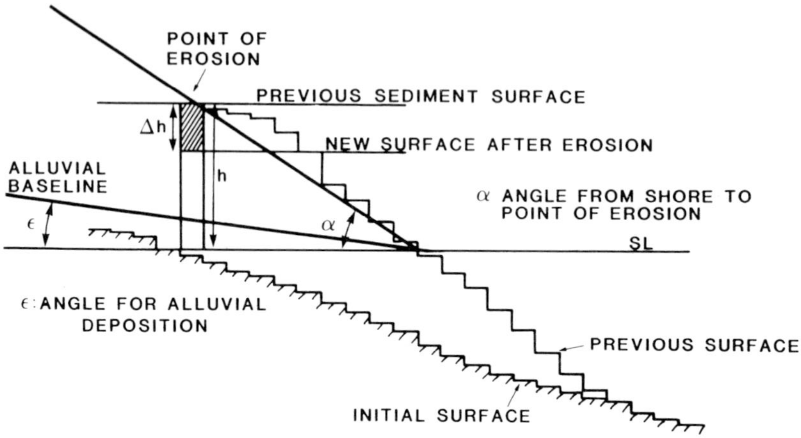


FIGURE 8.4 Sediment erosion model.

7. Subside all surfaces isostatically in response to sediment and water loading on the crust.

This sequence of subsidence, erosion, deposition, compaction, and loading is carried out for each time step. Deposition involves the creation of a new surface above the previous one. Both erosion and compaction involve reduction in the thickness of layers. Tectonic subsidence and loading cause surfaces to drop vertically, whereas erosion and unloading cause rises.

## B. INDIVIDUAL PROCESSES

### 1. Erosion

Erosion is simulated by stripping off sediments between the shoreline and the most landward location of deposition (Fig. 8.4). Erosion takes place only where depositional surfaces are elevated above the alluvial baseline (i.e., the surface connecting the shoreline with the "basement" on the landward side with a given angle). The erosion depth is determined by the erosion formula

$$\frac{\delta h}{\delta t} = \frac{h \sin\alpha}{T_{\max}}, \tag{8.1}$$

where $h$ is the height above sea level for each column, $t$ is time, $\alpha$ is the angle of the line connecting the shoreline with each column that is subject to erosion, and $T_{\max}$ is a constant (with the units of time) that can be varied by the user to match the observed geometries (e.g., from seismic sections). Thus, erosion depends both on the height above base level and the steepness of the topography to be eroded. However, erosion will not take place below the alluvial baseline, and only previously deposited sediments are available for erosion, that is, no erosion into the initial surface occurs. The layers of sediment available for erosion, sand (coarse), and shale (fine) are compacted increasingly with depth. Following the erosion of these variously compacted rocks to a depth $D$, the eroded sands and shales are uncompacted to their original porosity, increasing the thickness of any removed sediment to

$$d_2 = d_1(1 - \phi_1)/(1 - \phi_2), \tag{8.2}$$

where $d_1$ is the compacted thickness, and $\phi_1$ and $\phi_2$ are the compacted and depositional porosities, respectively (original porosities of sands and shales at deposition are 0.4 and 0.75 as defaults). The total eroded quantities of uncompacted sands and shales are added to the sand and shale already available for deposition (the two sedimentary triangles) in the alluvial time step.

**2. Deposition**

Following erosion and tectonic subsidence, deposition takes place. For a given time step, the area of sediment available for deposition is represented by sand and shale triangles (Figs. 8.2 and 8.5) with lengths that match the widths of the offshore sandstone and shale wedges. The model distributes sediments by adding sediments to each "stair" or column of the actual depositional surface (in a constrained manner as described later) with sand and shale being deposited together. As the model works its way along the columns depositing sediment, the areas of triangles are lessened by reducing their heights but preserving their lengths. Sediment supply can be varied through time by changing triangle heights for specified time steps. Sediments eroded (as discussed earlier) within a time step are added to the input triangles by increasing their heights. Sediments are deposited as described in the following paragraphs.

The model first locates the shoreline and then starts to deposit on the first column seaward (Fig. 8.5). Together, sand and shale are filled into the area between the depositional surface and sea level [Fig. 8.5(2A)] in a ratio determined by the height ratio of the two triangles. After having filled sediments to sea level on that particular column, the model starts to deposit sediments to the landward side of the column by filling the area between the depositional surface and the alluvial "baseline" [Fig. 8.5(2B)], which is defined by an inclined line that begins at the seaward edge or shoreline (after having filled the first column) and terminates at its intersection with the up-dip "basement." As with the marine deposition just described, sand and shale involved with the alluvial plain are removed from the triangles by reducing the triangle heights, and the ratio of these sediments deposited is determined by the height ratio of the two triangles.

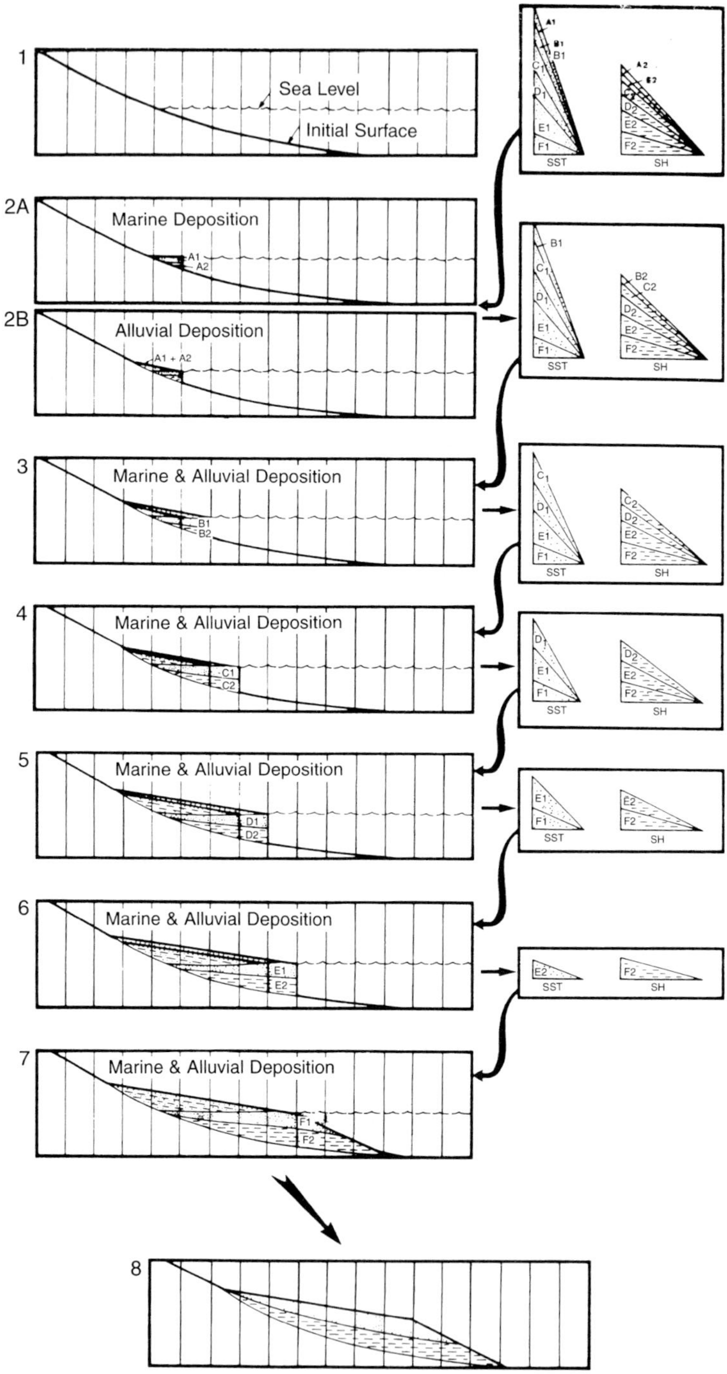


FIGURE 8.5 A sequence of deposition within a time step of the simulation (staircase-like appearance of sediment surface is not shown).

After having filled the alluvial segment corresponding to the first column seaward of the initial shoreline position (for that time step), the model shifts seaward one column, fills the column to sea level, and then fills a new alluvial segment [Fig. 8.5(3)], all the time reducing triangle heights according to the amounts deposited. These operations are repeated until the composite height of the two triangles matches, or is exceeded by, the depth of the sea over the depositional surface. When this point is reached, the remaining triangles are deposited seaward to their predetermined widths without changing their heights [Fig. 8.5(8) and Fig. 8.2 columns $S_1$ through $S_6$ (sand) and columns $C_1$ through $C_6$ (shale)].

In addition to the main elements of deposition outlined, additional constraints include a critical angle of marine deposition, which is given as an input parameter. Sediments are not deposited when the slope exceeds this critical value, but bypass to basinward positions with gentler gradients. An angle of submarine deposition, which is given as an input parameter, determines where the sediments bypass and are deposited. The sediments which bypass build up seaward of areas of nondeposition at a depositional slope that matches the angle of submarine deposition. Thus, the width of the offshore sediment wedge, although matching the length of the base of the depositional triangles, is also spread between areas of nondeposition, thus modeling offshore deep-water bypass slides and turbidites [Fig. 8.5(8)]. Moreover, the sediment surface is eroded to the critical angle if its inclination initially exceeded the critical angle (as in the case of subaerial erosion). The program will not erode into basement.

In addition to tracing the deposition of the ratio of sand to shale, carbonate deposition is also modeled (Fig. 8.3). In the simulation, carbonate buildup does not begin unless space exists between the clastic–seawater interface and the air–water interface. The rate at which this accommodation is filled is controlled by the user. In the simulation, the carbonate sedimentation rate is assumed to be a function of water depth. Water depth and sedimentation rate pairs are provided by the user, and the model interpolates between these pairs to determine the rate of accumulation corresponding to the water depth at a given point. This process takes place after the clastic deposition, but prior to the processes of compaction and subsidence. If the user inputs zero for the rate, the carbonate sedimentation is stopped.

**3. Compaction**

As with the decompaction of sediments performed in the erosion segment, compaction requires knowledge of the thickness of layers (sediment blocks) and their present porosities, that is, their present state of compaction. For each block, empirical expressions relating porosity and depth for sands and shales have been used. Shale porosity is taken to vary as ln $\phi \alpha z$, where $z$ is the depth burial below the sediment–water interface. Uncompacted porosity for shale is 75%, with a porosity at 3000 m of about 20%. For sandstones, porosity is taken as a linear function with depth. Uncompacted porosity for sandstone is 40% whereas porosity at 3000 m is about 27%. With depth $z$ given in meters, porosity for shale is defined here as

$$\phi = \phi_0 \exp(-cz), \tag{8.3a}$$

where $\phi_0 = 0.75$, and $c = 4.4 \times 10^{-4}\ \mathrm{m}^{-1}$. For sands

$$\phi = \phi_0 - az, \tag{8.3b}$$

where $\phi_0 = 0.4$ and $a = 4.33 \times 10^{-5}\ \mathrm{m}^{-1}$. In these empirical expressions, $z$ represents thickness of overlying sediment. If porosity is already less than the computed value for the depth, no compaction takes place, which can occur when erosion has reduced the depth of the sediment. When sediment is eroded, the remaining sediments do not compact. Thus, when weight is again added, further compaction does not occur until the new weight exceeds that previously eroded. Additionally, the simulated sedimentary sequence can be compacted further by considering the additional overburden of younger overlying sediments. In this case, the user must provide an estimate of the maximum burial of the sequence.

**4. Subsidence and/or Uplift**

Two independent types of subsidence (and/or uplift) are included:

1. Tectonic subsidence (or uplift)
2. Subsidence due to isostatic adjustment to sediment and water loading, as well as "uplift" if erosion takes place or sea level falls

The tectonic movement can be varied temporally as well as spatially by specifying rates against horizontal positions for designated intervals. The model interpolates coordinates linearly both in the time and in the space dimension (Fig. 8.1c).

Isostatic response to sediment loading is found by conserving the weight of water, sediment, and mantle to some fixed depth so that

$$\rho_w \Delta_w + \rho_s \Delta_s + \rho_m \Delta_m = 0, \tag{8.4}$$

where $\rho_w$, $\rho_s$, and $\rho_m$ are densities of water, fully compacted sediment, and mantle, and $\Delta_w$, $\Delta_s$, and $\Delta_m$ are the changes in water, sediment, and mantle thickness, respectively. The isostatic response is

$$\Delta_m = (\rho_w/\rho_m)\Delta_w - (\rho_s/\rho_m)\Delta_s. \tag{8.5}$$

Densities are taken to be $\rho_w = 1\ \mathrm{g \cdot cm^{-3}}$, $\rho_s \cong 2.7\ \mathrm{g \cdot cm^{-3}}$, $\rho_m = 3.4\ \mathrm{g \cdot cm^{-3}}$, and $\Delta_s$ is the newly added, uncompacted, sediment thickness.

## C. RESULTS OF THE SIMULATION

The purpose with this simulation was to incorporate both marine and alluvial clastic deposition at basin margins and to reproduce gross geome-

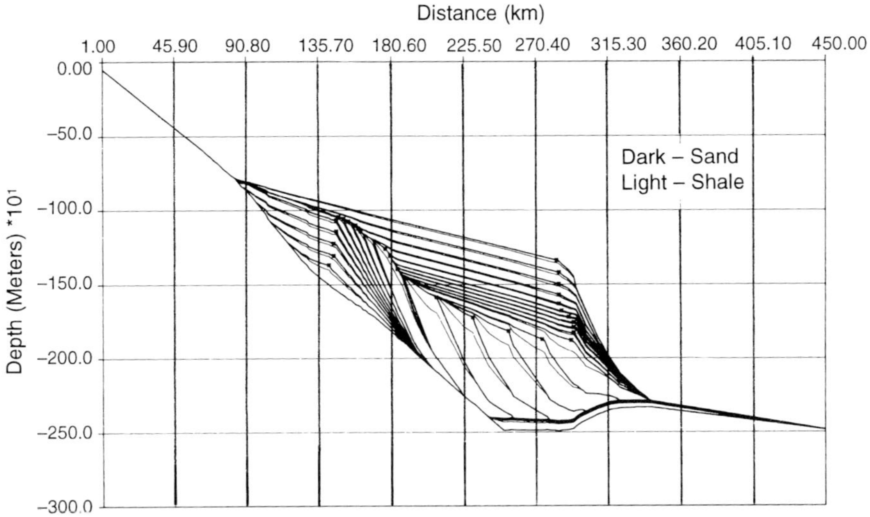


FIGURE 8.6 Simulation of gross sediment geometry in response to the type 1 unconformity of Vail *et al.* (1984). Geometries shown are of deposited surfaces at the end of any time step chosen by the user. The initial surface is plotted as a thick line. The subsequent packages of sediment deposited at each time step show the ratio of sands and shales deposited within a time step as alternating dark and light lines but is not intended to represent the exact lithological distribution. Thus, this ratio provides a basis for the compaction calculations but does not imply that shale is deposited before sand or vice versa. The horizontal scale is in kilometers and the vertical scale is in meters. Individual facies packages for each time step are not shown; instead the ratio of sandstone to shale in the sediment wedge of that time step is shown. Compaction and tectonic movement cause originally horizontal surfaces to rotate, whereas vertical exaggeration causes the dip of the individual sediment packages to be further exaggerated.

tries of these sediments (Fig. 8.6) but not individual facies units, such as channel sands, barrier island sands, lagoonal shales, point bars, or individual delta lobes. Thus, the simulation traces a general depositional history of the rock sequence which could aid in the interpretation of seismic data. The simulation highlights areas of little clastic input, potential zones for good source rocks, areas of massive mud accumulation (potential shale seals), and the general region in which potential stratigraphic traps (pinch-outs of areas prone to sandstone deposition within and/or beneath shale bodies) may occur.

The simulation can be used on examples from seismic and geologic sections. In initial stages of this model development, testing was done against some of the conceptual models generated by Vail *et al.* (1984). The results are instructive and expand our understanding of the fill of basin margins. At the same time, however, the true test of a simulation model is its comparison to real geologic sections.

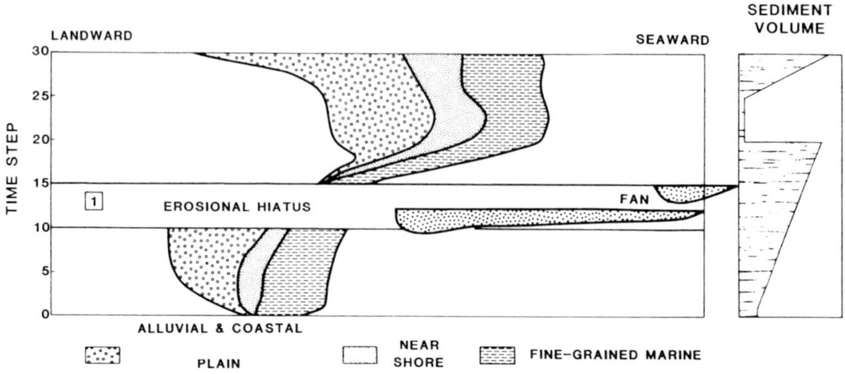


FIGURE 8.7 Simulation chronostratigraphic chart and rate of sediment input for simulation of type 1 unconformity of Vail *et al.* (1984).

Interim results (Figs. 8.6–8.17) trace the simulation of a type 1 unconformity (Figs. 8.6, 8.7, and 8.8). A type 1 unconformity occurs when sea level falls more rapidly than the rate of subsidence of the shelf edge. Consequently, sea level falls below the shelf edge and major erosion occurs on the shelf and the shelf edge. The model predicts that erosion coupled to fan development is prominent from the time of onset of an eustatic drop in sea level, whereas Vail *et al.* (1984, Fig. 9) show that erosion and fan generation began at a point of inflection on the eustatic curve, when the rate of fall is at its most rapid. This difference is probably because the morphology of the modeled shelf does not match that of Vail *et al.* (1984), and because the subsidence was less rapid at the onset of sea level fall than in the model used by Vail *et al.* (1984). By comparing both approaches, a better understanding of the interaction of variables responsible for the

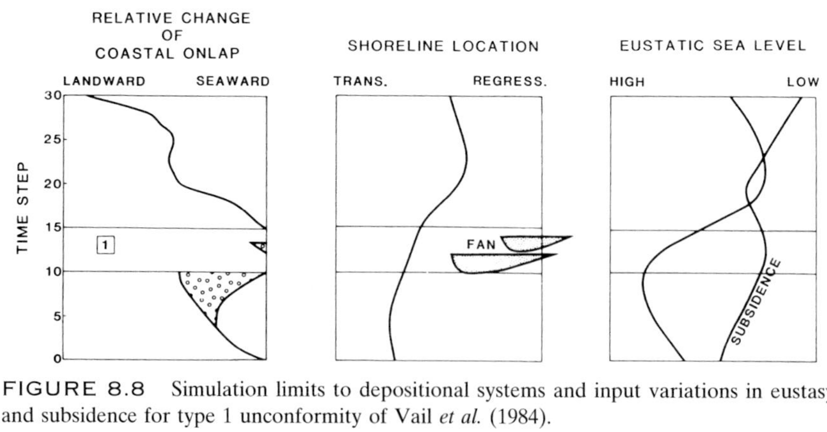


FIGURE 8.8 Simulation limits to depositional systems and input variations in eustasy and subsidence for type 1 unconformity of Vail *et al.* (1984).

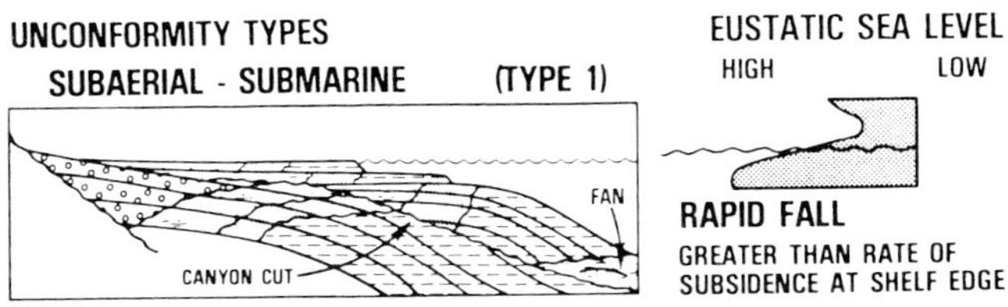


FIGURE 8.9 Diagram of type 1 unconformity of Vail *et al.* (1984).

observed geometries can be gained, and at the same time the concept is reinforced that a unique way does not always exist in which to produce a particular geometry.

A type 2 unconformity (Figs. 8.10, 8.11, and 8.12; Vail *et al.*, 1984, Fig. 13) is defined as one produced when the rate of eustatic fall is less than the rate of subsidence of the shelf edge so that sea level never falls below the shelf edge. Results here are similar to those of Vail *et al.* (1984), except the time at which the type 2 unconformity develops is when sea level is at its lowest, whereas Vail *et al.* (1984) show a type 2 unconformity developing at the time of greatest rate of eustatic fall. The Vail *et al.* (1984, Fig. 17) daigram, which encompasses several eustatic cycles, was also simulated (Figs. 8.14, 8.15, and 8.16). Much of the character is determined by a

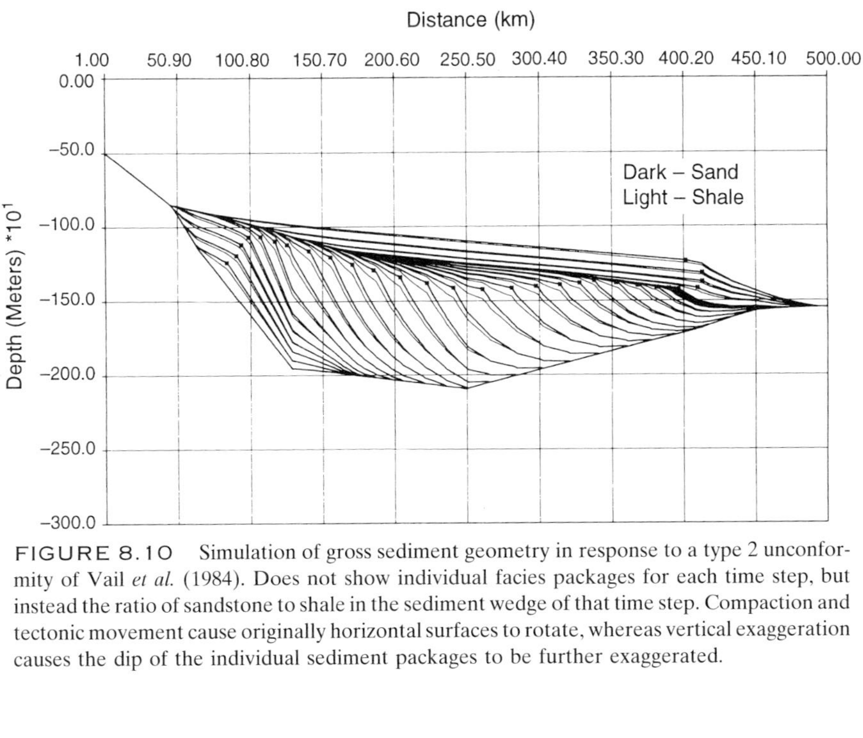


FIGURE 8.10 Simulation of gross sediment geometry in response to a type 2 unconformity of Vail *et al.* (1984). Does not show individual facies packages for each time step, but instead the ratio of sandstone to shale in the sediment wedge of that time step. Compaction and tectonic movement cause originally horizontal surfaces to rotate, whereas vertical exaggeration causes the dip of the individual sediment packages to be further exaggerated.

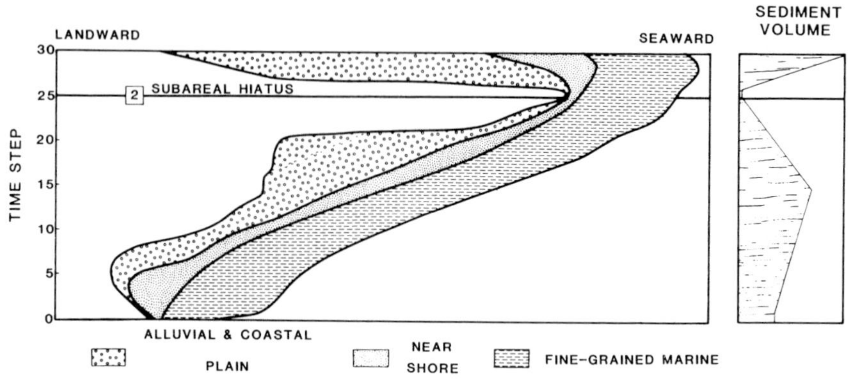


FIGURE 8.11 Simulation chronostratigraphic chart and rate of sediment input for simulation of type 2 unconformity of Vail *et al.* (1984).

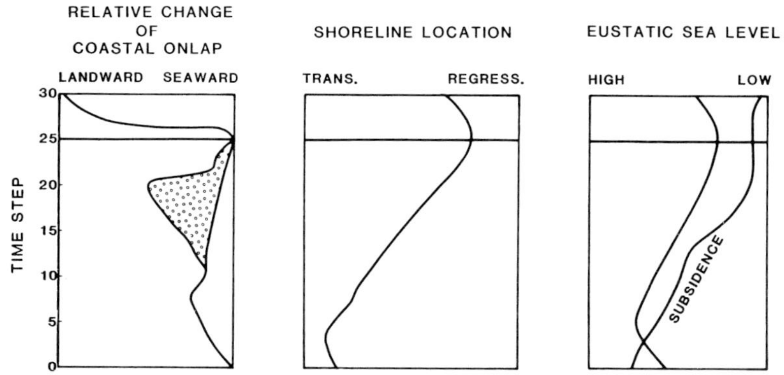


FIGURE 8.12 Simulation limits to depositional systems and input variation in eustasy at subsidence for type 2 unconformity.

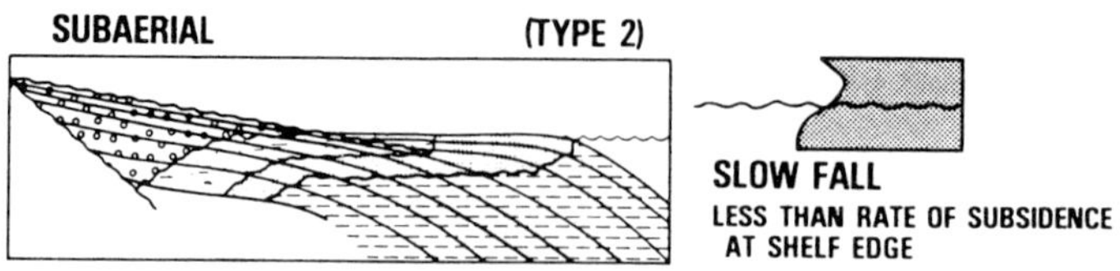


FIGURE 8.13 Diagram of type 2 unconformity of Vail *et al.* (1984).

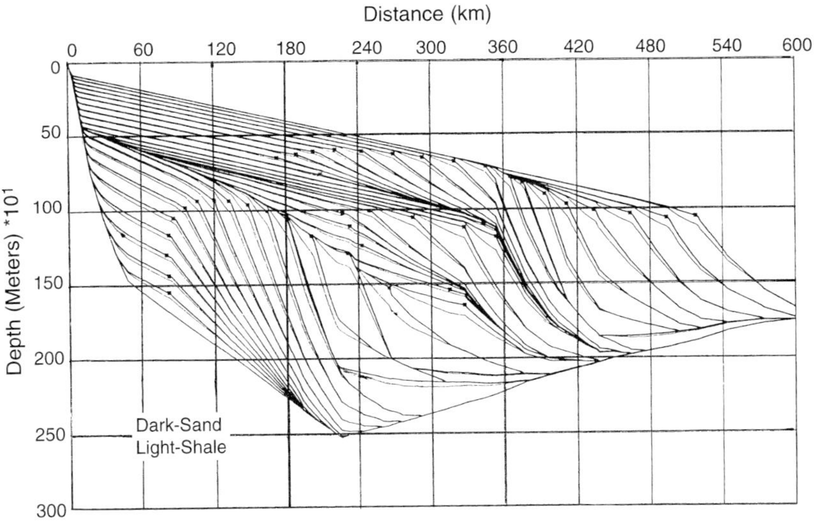


FIGURE 8.14 Simulation of gross sediment geometry built to match the Vail *et al.* (1984) diagram. The figure does not show individual facies packages for each time step, but instead the ratio of sandstone to shale in the sediment wedge of that time step. Compaction and tectonic movement cause originally horizontal surfaces to rotate whereas vertical exaggeration causes the dip of the individual sediment packages to be further exaggerated.

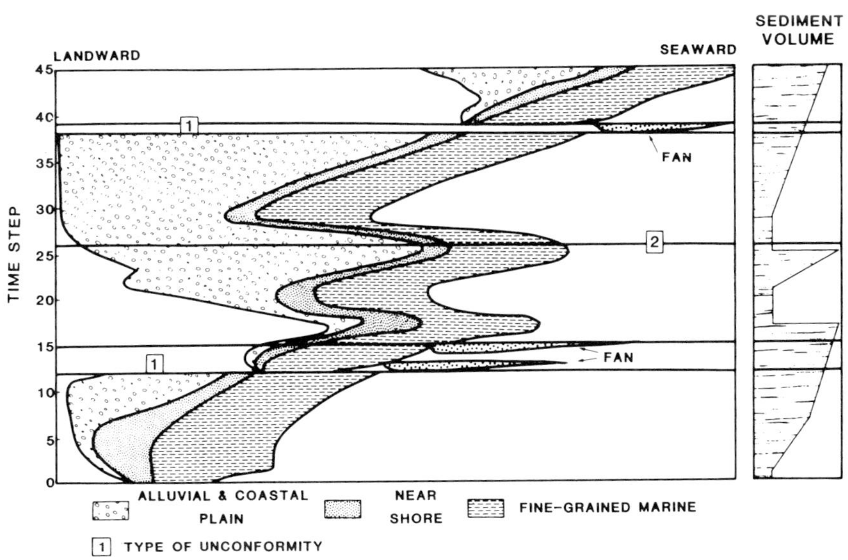


FIGURE 8.15 Simulation chronostratigraphic chart and rate of sediment input for diagram of Vail *et al.* (1984).

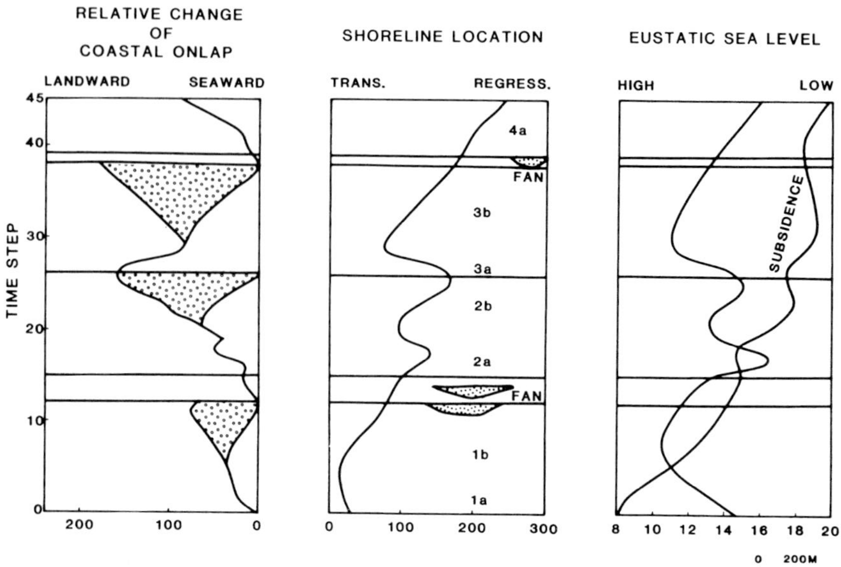


FIGURE 8.16 Simulation limits to depositional systems and input variations in eustasy and subsidence for diagram of Vail *et al.* (1984).

response to the sediment volume deposited within *each* sedimentary cycle or sequence. The time of maximum rate of eustatic sea level fall on Vail *et al.*'s (1984) diagram matches the development of submarine fans. However, submarine fans from the simulation begin their development a little higher on the eustatic cycle, when the rate of eustatic sea level fall was a little slower. Another difference between the Vail *et al.* (1984) diagram and the simulation is that the rate of subsidence is changed from a smooth curve to one with some greater frequency changes in rate.

## D. SUMMARY OF SIMULATION INPUTS

The following are descriptions of the simulation inputs. The sizes of inputs (Table 8.1) used to create the simulations of Figs. 8.14, 8.15, and 8.16 are given. The derivation of some of these inputs is from geologic data such as well logs and seismic cross-sections. To run the simulation, the following are required:

1. Horizontal range (Fig. 8.1a) in kilometers. Initial (left side) and final (right side) values used in output plots must be stated. The distance between each point, the horizontal resolution, is determined by the width of the simulated sequence and matches the length of the model, as defined by the starting and ending locations divided by the number of horizontal positions desired. Here 1 and 600 were used for the particular run exhibited.

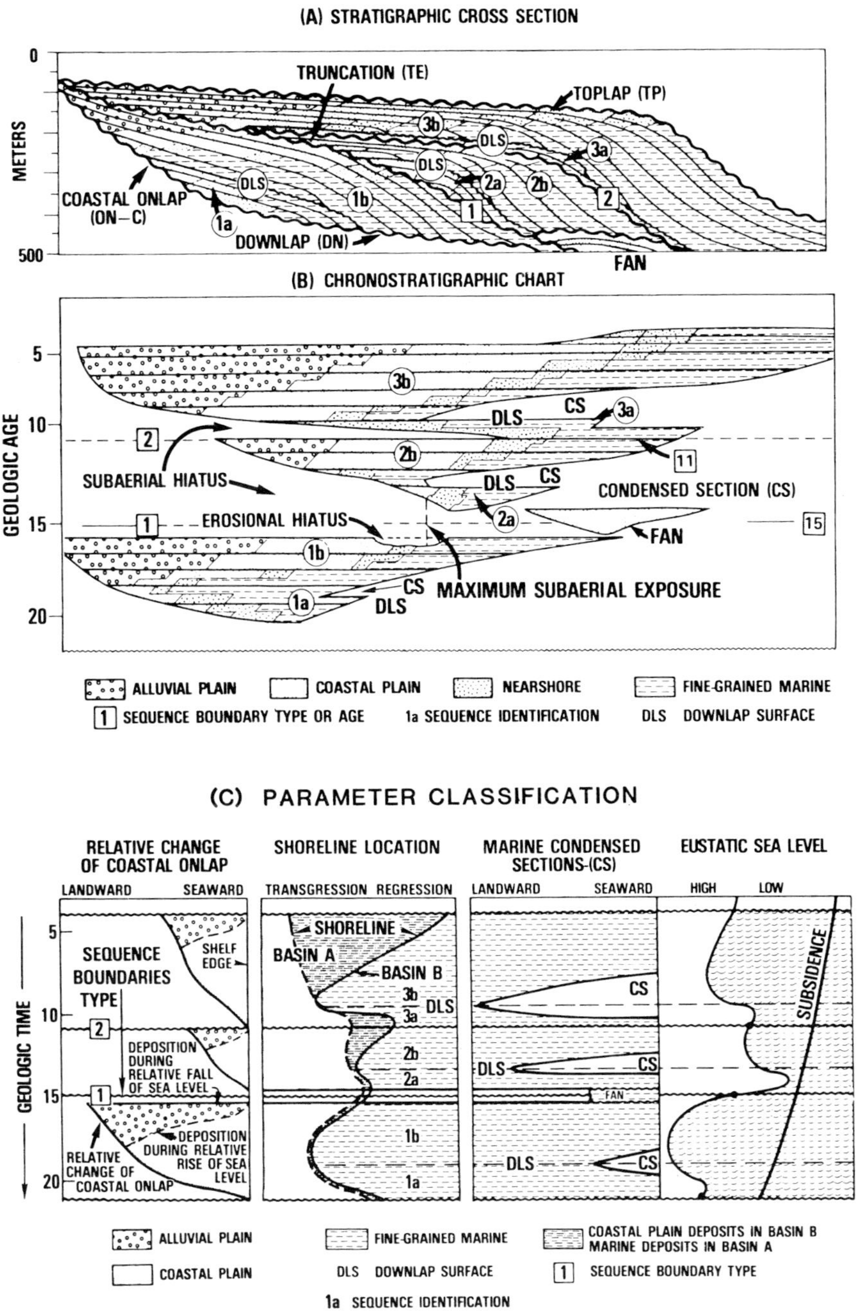


FIGURE 8.17 Diagram (Helland-Hansen *et al.,* 1988; after Vail *et al.,* 1984) showing (a) a stratigraphic cross-section with sequence boundaries, (b) a chronostratigraphic chart derived from the stratigraphic cross-section, and (c) a parameter classification for the stratigraphic sequences and their relationship to coastal onlap, relative sea level position, transgression/regression eustasy, marine condensed sequences, and subsidence.

TABLE 8.1 Input Information with Sample Data Set Used to Produce Simulation Outputs (Figs. 8.14–8.16)

| | |
|---|---|
| 1 | START HORIZONTAL POSITION (MI, KM) |
| | 1.00 |
| | END HORIZONTAL POSITION (MI, KM) |
| | 600. |
| | NUMBER OF HORIZONTAL POSITIONS (MAX = 300) |
| | 300. |
| | TIME STEP 3 FOR DETAILED PRINT OUT (0 = NO) |
| | 0 |
| | DIRECTION FOR GRAPH (0 = L − R, 1 = R − L) |
| | 0 |
| | # OF STEPS FOR GATHERING FORMATIONS |
| | 46 |
| | FLAG FOR UNIT (0 = ENGLISH, 1 = METRIC) |
| | 1 |
| 2 | START TIME (MY) |
| | 0.0 |
| | END TIME (MY) |
| | 20.0 |
| 3 | NUMBER OF TIME STEPS (MAX = 200) |
| | 45 |
| 4 | # VERTICES INITIAL SURFACE (MAX = 20) |
| | 7 |
| | SEDIMENTARY GEOMETRY PLOT (0 = NO; 1 = PLOT) |
| | 0 |
| | VAIL AGGRADATION PLOT (0 = NO; 1 = PLOT) |
| | 1 |
| | SEA LEVEL PLOT (0 = NO; 1 = PLOT) |
| | 1 |
| | SEDIMENTARY LIMITS PLOT (0 = NO; 1 = PLOT) |
| | 1 |
| | HORIZONTAL POSITION OF VERTICES (MI, KM) |
| | 1 20 30 40 50 225 450 |
| | VERTICAL POSITION OF VERTICES (FT, M) |
| | −50.0 −430. −590. −710. −800. −1450.0 −800. |
| 5 | # OF INPUT DEPOSITION TRIANGLE (MAX = 20) |
| | 9 |
| | LENGTH-SAND DEPOSITION TRIANGLE (MI, KM) |
| | 50.0 |
| | LENGTH-SHALE DEPOSITION TRIANGLE (MI, KM) |
| | 110.0 |
| | HEIGHT-SAND DEPOSITION TRIANGLE (FT, M) |
| | 60. 120. 180. 60. 60. 180. 60. 60. 120. |
| | HEIGHT-SHALE DEPOSITION TRIANGLE (FT, M) |
| | 150. 300. 400. 150. 150. 400. 150. 150. 400. |
| 6 | MAX ANGLE FOR STABLE DEPOSITION (DEG |
| | 1.30 (1°30′) |

(*continues*)

TABLE 8.1 *(continued)*

| | |
|---|---|
| 7 | ANGLE FOR ALLUVIAL DEPOSITION (DEG) |
| | 0.100 |
| 8 | ANGLE FOR SUBMARINE DEPOSITION (DEG) |
| | 0.05 (0°05′) |
| 9 | EROSION RATE CONSTANT (MY) |
| | 0.03 |
| 10 | # OF PAIRS OF DEPTH AND RATE OF CARBONATE SEDIMENTATION AT THIS DEPTH (MAX = 8 PAIRS) |
| | 3 |
| | MAX DEPTH FOR GRAPH (FT, M) |
| | 3500.0 |
| | SEDIMENTATION RATE FOR CARBONATE |
| | DEPTH (FT, M) RATE (FT, M/TIME STEP) |
| | 0.00 0.0 |
| | 5.00 0.0 |
| | 6.00 0.0 |
| 11 | # OF TIME STEPS FOR REGIONAL SUBSIDENCE (MAX = 10) |
| | 4 |
| | # OF LOCATIONS FOR REGIONAL SUBSIDENCE (MAX = 20) |
| | 2 |
| | HORIZONTAL POSITION (MI, KM) OF HINGE |
| | 0.0 |
| | RATE OF CHANGE OF SUBSIDENCE (FT/MI, M/KM) FROM HINGE POINT |
| | 0.0 |
| | LOCATIONS (MI, KM) |
| | 1 450 |
| | TIME STEP |
| | 1 |
| | REGIONAL SUBSIDENCE (FT, M/TIME STEP) |
| | 0.0 30.0 |
| | TIME STEP |
| | 23 |
| | REGIONAL SUBSIDENCE (FT, M/TIME STEP) |
| | 0.0 30.0 |
| | TIME STEP |
| | 24 |
| | REGIONAL SUBSIDENCE (FT, M/TIME STEP) |
| | 0.0 0.0 |
| | TIME STEP |
| | 45 |
| | REGIONAL SUBSIDENCE (FT, M/TIME STEP) |
| | 0.0 0.0 |
| 12 | ISOSTASY CONSTANT (MY) |
| | 0.0 |
| | MINIMUM DEPTH FOR SUBMARINE DEPOSITION (FT, MT) |
| | 4000.0 |
| | TIME STEPS |
| | 1 7 17 18 21 25 26 29 45 |

*(continues)*

TABLE 8.1 (*continued*)

| | |
|---|---|
| 13 | # OF INPUT PAIR FOR AVERAGE SEA LEVEL CHANGE |
| | 5 |
| | # OF INPUT PAIR FOR 2ND ORDER SEA LEVEL CHANGE (MAX = 30) |
| | 11 |
| | TIME STEPS FOR AVERAGE SEA LEVEL |
| | 1 15 20 32 45 |
| | AMPLITUDE FROM INITIAL SEA LEVEL (FT, M) |
| | −800.0 −700.0 −800.0 −800.0 −1000.0 |
| | TIME STEPS/QUARTER CYCLE |
| | 1 7 15 17 19 21 23 25 |
| | 27 29 45 |
| | AMPLITUDE OF SEA LEVEL (FT, M) |
| | 0.0 300.0 0.0 −400.0 0.0 100.0 0.0 −400.0 |
| | 0.0 300.0 0.0 |
| 14 | DENSITY OF SHALE MATRIX (G/CC) |
| | 2.70 |
| | DENSITY OF SAND MATRIX (G/CC) |
| | 2.65 |
| | DENSITY OF MANTLE (G/CC) |
| | 3.40 |
| 15 | MAX BURIAL FOR SEDIMENT (FT, M) |
| | 0.0 |
| | LENGTH FOR VERTICAL AXIS (INCH) |
| | 6.00 |
| | LENGTH FOR HORIZONTAL AXIS (INCH) |
| | 10.0 |
| | SHORE SYMBOL FLAG (0, 1, OR 2) |
| | 1 |
| 16 | TIME STEPS FOR GEOMETRY PLOTS |
| | NO PLOTS AFTER A ZERO |
| | 10 20 30 0 0 0 0 0 0 |

2. Range of time in millions of years. Beginning and ending values used in output usually will be determined from paleodata. Here 10 MY was used to match approximately the work of Vail *et al.* (1984).

3. Number of time steps. The duration of each step is determined by the length of the simulation (as defined by the beginning and end times) divided by the number of time steps. To simulate accurately the sediment geometry, which results in part from sea level variations, the time sampling must be finer than the time between maxima and minima of the sea level oscillations.

The length of time taken for sea level cycles may be estimated from paleontological dating. A variety of published timescales, including those for the Jurassic and Cretaceous by van Hinte (1976a,b), those for the Phanerozoic and Arehaen by Harland *et al.* (1982), and for North America

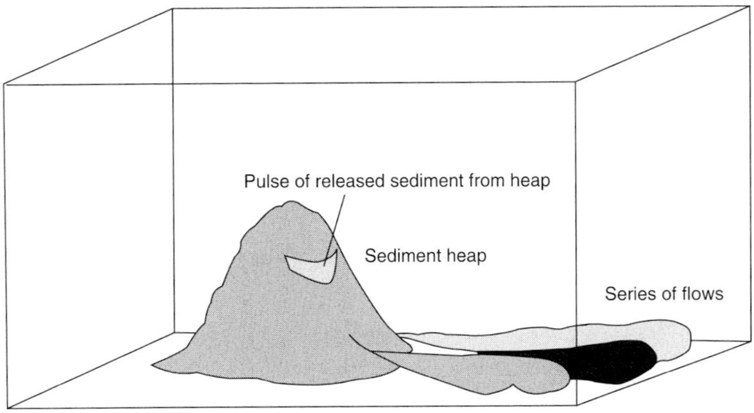


FIGURE 8.18 Sketch of the problem to be modeled.

by Palmer (1983), provide the relationship between geological stage names and an "absolute" chronometric timescale. An individual sea level cycle may last anywhere from a few thousand years to 12 MY, so the length of time steps will vary according to the length of the sea level cycles.

4. Initial surface (Fig. 8.1a) defined by a set of vertices between which the model linearly interpolates; vertical units are expressed in meters and horizontal units in kilometers which are determined either from seismic data, from which a time-to-depth conversion is made, or from a combination of seismic and well data. Paleobathymetric data may provide information

TABLE 8.2 Stress (Dimensionless) Values, $\theta$, for Different Lithologies

| Lithology type | Grain size (mm) | $\theta$ value | Lithology code description |
|---|---|---|---|
| Pebble | >4.000 | ~0.05 | L0 |
| Granule | 2.000–4.000 | ~1.040 | L1 |
| Very coarse sand | 1.000–2.000 | ~0.040 | L2 |
| Coarse sand | 0.500–1.000 | ~0.035 | L3 |
| Medium sand | 0.250–0.500 | ~0.030 | L4 |
| Fine sand | 0.125–0.250 | ~0.040 | L5 |
| Very fine sand | 0.062–0.125 | ~0.060 | L6 |
| Coarse silt | 0.020–0.062 | ~0.15 | L7 |
| Fine silt | 0.004–0.02 | ~0.500 | L8 |
| Clay | <0.004 | ~1.000 | L9 |

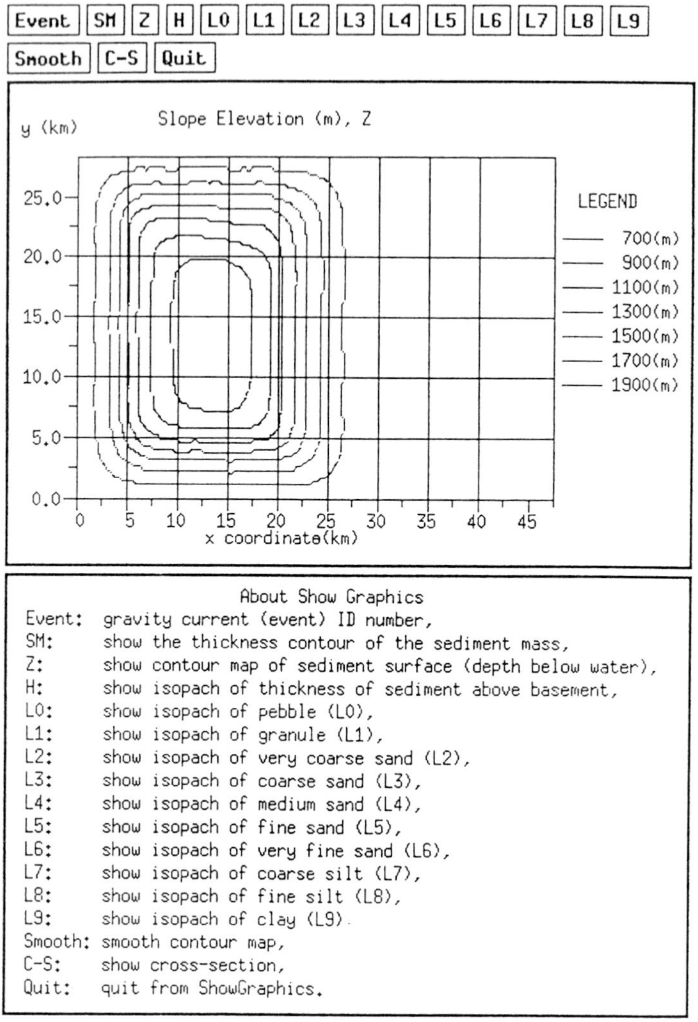


FIGURE 8.19 Contour map of the original topography for Test No. 1.

about the topography of the sediment interface in the marine environment. Clearly the present-day surface configuration only serves as a first estimate. In the examples shown, the number of vertices did not exceed 60. The seven vertices for which location and relief were picked are indicated (Table 8.1).

5. Length and height of sand and shale deposition triangles (Fig. 8.2). Height can be varied in time by specifying height against time (with linear interpolation between) (Fig. 8.1b). Lengths of triangles match average width of offshore sediment wedge. The sandstone wedge was chosen as 50 km

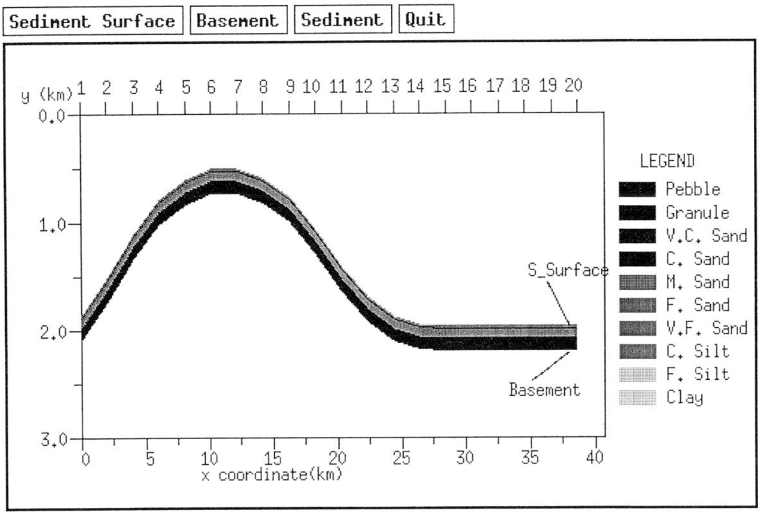


FIGURE 8.20 Cross-section profile of the original basin slope for Test No. 1.

and the shale wedge was chosen as 110 km. The quantity of sand and shale deposited within a particular time interval and the lateral distribution in the marine environment (not counting bypassed sediments) can be estimated in the following way:

a. Determine the amount of sediment in terms of cross-sectional area of the sediment body using at least two wells along the line of the simulation, in conjunction with a seismic cross-section.

b. Calculate the percentage of sand and the percentage of shale within this sediment volume. Using these percentages and the total area of the sediment, determine the area of sand ($A_s$) deposited and the area of shale ($A_c$) deposited per time step. The volumes of sand and shale should be uncompacted following this last calculation.

c. Set the lengths of triangles representing the lateral distributions of sand and shale within the time interval of the simulation. The shale triangle length ($L_c$) matches the distance from shore to the seaward limit of shale deposition and is less than the length of the total sediment body being simulated. The geometrical configuration of clinoforms from the "break-in-slope" to where they flatten in the offshore direction is probably a good estimate. The sand triangle length, which represents the distance from shore to the most seaward position of the "shoreline-attached" sand deposition, is more difficult to evaluate because reflectors match bedding planes rather than diachronous lithostratigraphic boundaries. Thus, the point where sand pinches out along a seismic reflector is difficult to estimate. If the seismic sequence contains many regressive–transgressive cycles coupled to a dense grid of well data, log correlations give some idea of the basinward extent

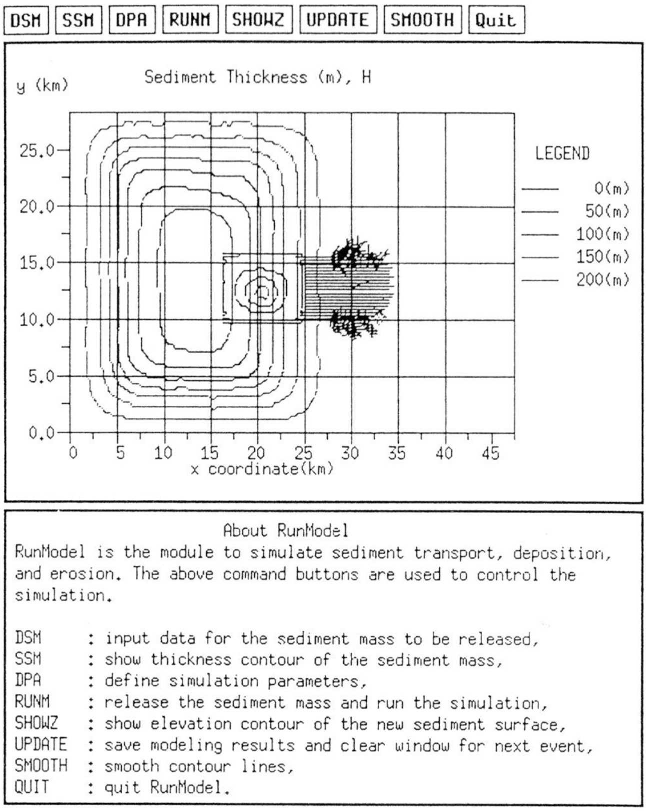


FIGURE 8.21 Trace pattern of the sediment current movement with contours of the released sediment mass for Test No. 1.

of sand bodies. If such information does not exist, sedimentological studies from cores and electric logs give some information about the mode of sand dispersal in the marine environment. When both areas of sand ($A_s$) and shale ($A_c$) and their triangle lengths ($L_s$ and $L_c$) are estimated, heights are given by

$$H_s = 2A_s/L_s \qquad \text{and} \qquad H_c = 2A_c/L_c. \tag{8.6}$$

6. Stable deposition angle, the sediment surface angle beyond which sediment is unable to come to rest and thus is not deposited, depends on many factors. These factors include deposition rate, grain size distribution, basinal energy, biological degradation, and so on, all of which may vary from basin to basin and also within basins. The inclination on clinoforms seen on seismic sections may provide a minimum value. In fact, compaction

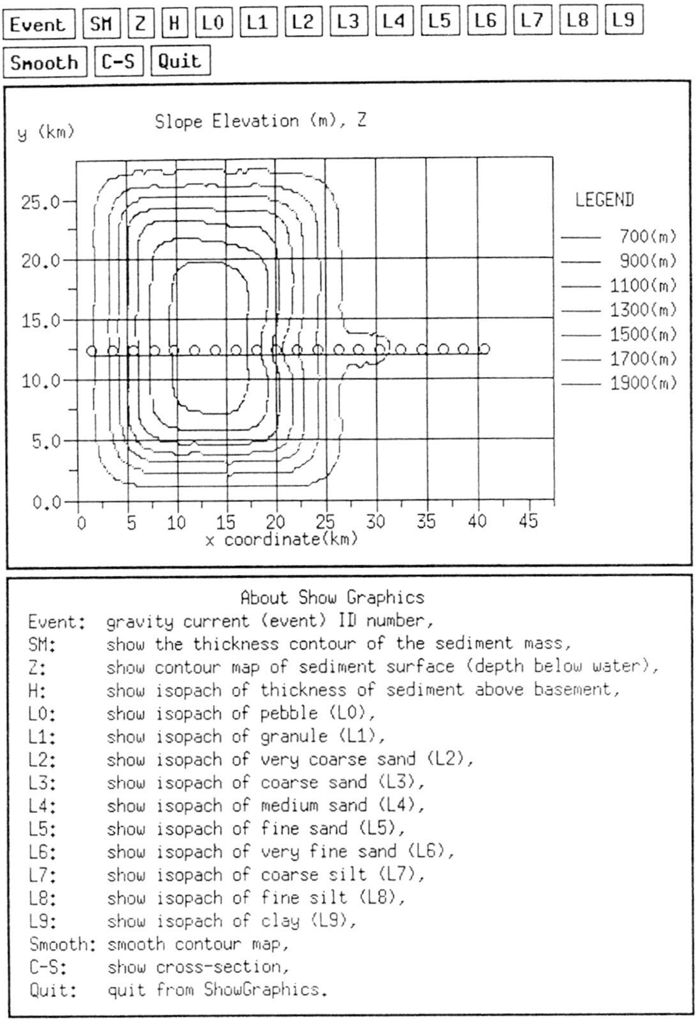


FIGURE 8.22 Contour map of the new topography with the location of a cross-section for Test No. 1.

has reduced the angle of the original depositional slope. Information on bypass angles is also available from recent depositional systems. In the models exhibited here an angle of $1^\circ\,30'$ was used, a little less than the $2^\circ$ of an average continental shelf.

7. Angle for alluvial deposition (Fig. 8.2). An approach for determining this angle is to estimate the angle of the reflectors landward of the break-in-slope (corrected for compaction) and from sedimentological studies,

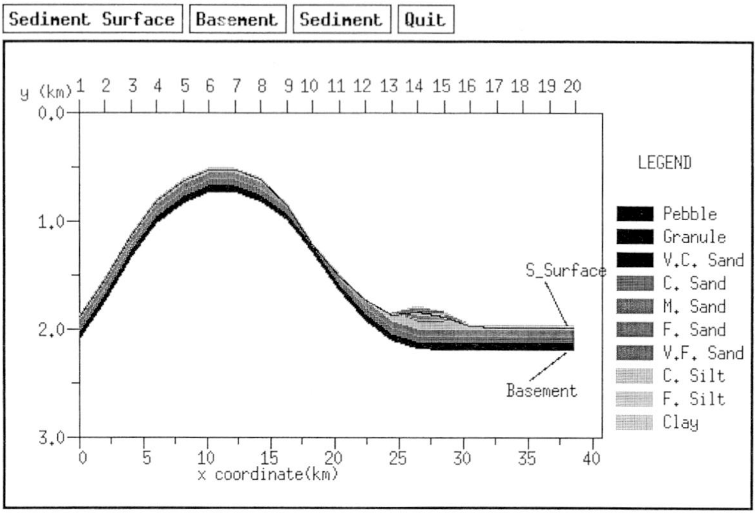


FIGURE 8.23 Cross-section profile (location drawn on Fig. 8.22) of the new basin slope for Test No. 1.

which use cores and logs, and to then make comparisons with modern systems. For the results exhibited here an angle of alluvial deposition of 0°10′ was chosen.

8. The angle for submarine deposition determines the angle at which sediments bypass or are deposited. Here, an arbitrary value of 0°5′ was chosen. Again, comparisons with modern systems and sedimentological studies of data from the sediment body to be simulated provide coarse constraints.

9. Erosion time constant ($T_{max}$). Evidence from modern coastal plains suggests that, unless the coastal plains are uplifted with respect to sea level, they are fairly stable.

10. Carbonate deposition (Fig. 8.3) is controlled by up to eight pairs of water depths with corresponding sedimentation rates for carbonate, chosen by the user, which can be varied to fit observed modern rates.

11. The tectonic movement/time step (Fig. 8.1c, Table 8.1) was varied through time and space in order to reproduce the Vail diagram. The rate of tectonic movement may be determined by analysis of geohistory curves (Guidish *et al.*, 1984). Alternatively, the geological literature can be a source of subsidence rates for the specific kind of basin being simulated.

12. Isostatic time constant. Literature studies to estimate this value for each particular basin are recommended. This parameter can be used to account for the relaxation time of the mantle to loading. The isostasy constant gives the time required for isostatic equilibrium to be achieved. Thus, a time equal to one time step indicates an instantaneous response.

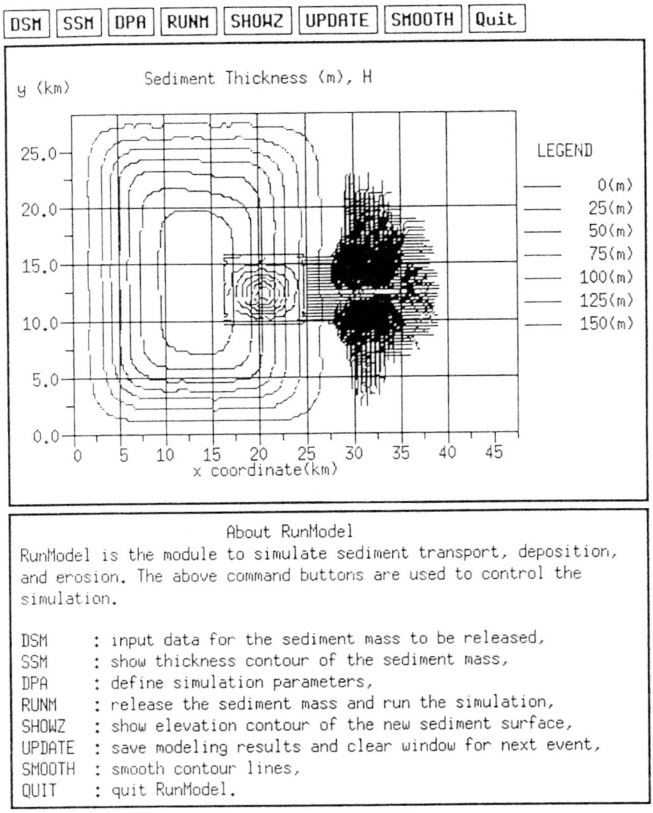


FIGURE 8.24 Tracer pattern of the sediment current movement with HHGO = 0.1 for Test No. 1.

13. Eustatic variation. Two orders of sea level changes are included in the simulation; both require operator-specified coordinates of height of sea level (against a fixed datum) versus time. Hopefully, the height of eustatic sea level variations can be derived, but currently eustatic sea level changes cannot be uniquely separated from the effects or rates of sediment accumulation and tectonic movement (Burton *et al.,* 1987). As a result, relative magnitudes can only be estimated. Initial use of Vail *et al.* (1977), Pittman (1978), or Vail and Hardenbol (1979) curves is suggested or, alternatively, curves based on the user's initial geologic interpretation of seismic data. Sea level curves can also be developed by varying an initial guess (along with other model inputs) and matching the simulation geometry of the true geologic geometry. Simulations interpolate linearly between first-order points and sinusoidally between second-order points as defined in the input data (Fig. 8.1d).

14. Matrix densities of shale and sand are both assumed to be 2.70 $g \cdot cm^{-3}$ and the density of mantle is taken as 3.4 $g \cdot cm^{-3}$.

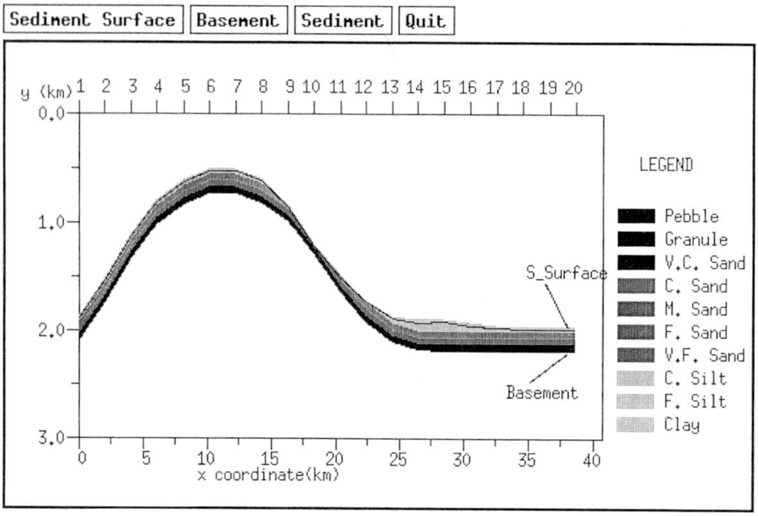


FIGURE 8.25 Cross-section profile (location drawn on Fig. 8.22) of the new slope with HHGO = 0.1 for Test No. 1.

15. Maximum burial sequence being simulated.
16. Time steps after which geometry plots are required.

## E. DISCUSSION

Results support the initial contention of Burton *et al.* (1987) that no unique single cause exists for sedimentary geometries that occur at basin margins; rather such geometries are the combined response to variations in sediment supply, subsidence, and sea level position. Despite limitations, simulations can be used to bracket the size of input variables that produce similar geometries. For instance, a particular sediment geometry might possibly be a product of pulsating up-and-down movements of the crust instead of variations in sea level. However, to reproduce the geometry, the simulation would have to use a tectonic behavior not commonly thought to occur in nature. An alternative cause for the sedimentary geometry could be varying eustatic sea level. Alternatively, of course, the geometries might be a product of pulsating sediment supply, with a rapid supply coinciding with an extremely rapid subsidence, which accommodates for the sediment load and slow rates of sedimentation matching slow rates of tectonic movement. However, although climatic reasons may be known for thinking that sediment input could vary in this way, strong coupling to tectonic behavior may be less likely. So, whereas a family of solutions may be large, probably only some are geologically "reasonable"—with a definition being required of "reasonable."

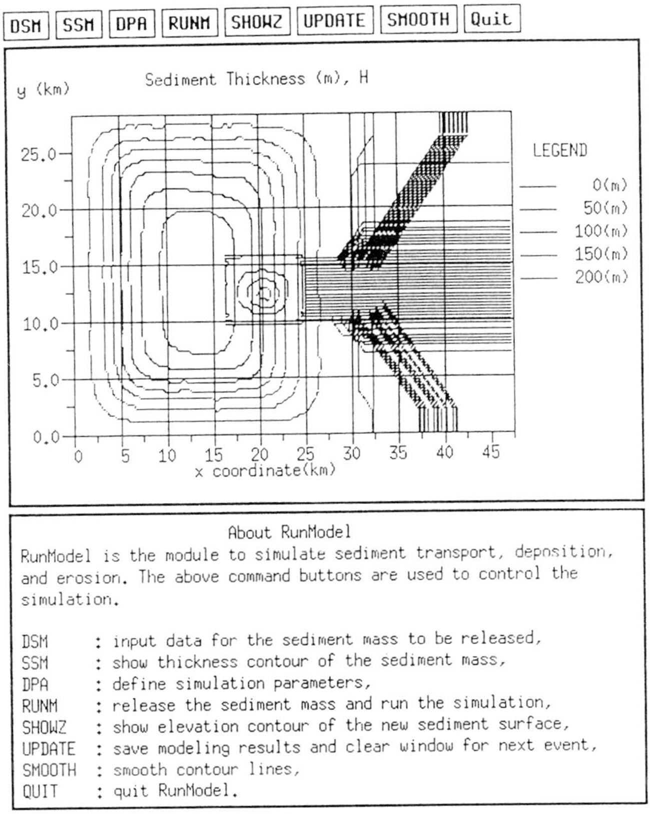


FIGURE 8.26 Tracer pattern of the sediment current movement with new lithology (L3 = 40%, L6 = 40%, and L9 = 20%) for Test No. 1.

Sediment geometries may possibly be a product of subsidence alone. For instance, sediment wedges could be expected to tend to verge on, and be close to, the hinge line. In contrast, in cases where sea level variations are responsible for the geometries, the sheetlike wedges that occur might be expected to spread over the whole basin marginal area. A series of sensitivity tests to determine the numerous combinations of parameters which generate substantially similar cross-sections (Fig. 8.14) is urgently needed. There seems to be little doubt that simulations give a general sense of the basinal response to sedimentation, sea level, and tectonism in the marine and alluvial sediments, although systematic tests of reliability and effectiveness of simulations as an aid to interpretation of real geologic examples are not readily available.

Because simulation geometries are so similar for a variety of inputs, no unique input parameters can be determined. However, simulations can be

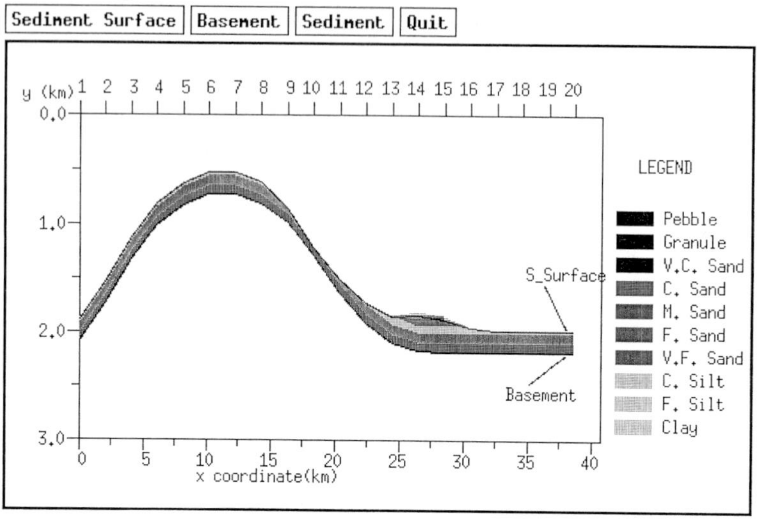


FIGURE 8.27 Cross-section profile (location drawn on Fig. 8.22) of the new basin slope with new lithology for Test No. 1.

used to produce a series of solutions which presumably bracket the correct behavior. Data that limit the solution family to a smaller set, discarding results that are physically impossible, can then be determined. At the same time, effects common to the many solutions are more readily identified.

## II. A TURBIDITE FLOW MODEL IN THREE DIMENSIONS

Sequence stratigraphic sediment fill models deal with more smoothly continuous supplies of sedimentary material than a "slump" deposition requires. The release of "quanta" of sediment may be triggered by earthquakes, by internal catastrophic failure due to fluid overpressuring, or by wetting a previously dry sediment locale (mudslides). The consequent development of turbidite sequences is but one instance where episodic and catastrophic "pulsing" of sediment models is required. This section describes a computer model that simulates the flow of sediment in three dimensions as a result of a sudden release of a quantum of sediment. The base outline of the problem to be modeled is sketched in Fig. 8.18. A quantum of sediment is released over a given volume centered at a location with high potential energy on a basin slope. The released sediment mixes with fluid (water, air, etc.), forming a gravity-driven current flowing downslope. The sediment will then be transported and deposited; slope sediments can also be eroded when the sediment current is strong enough.

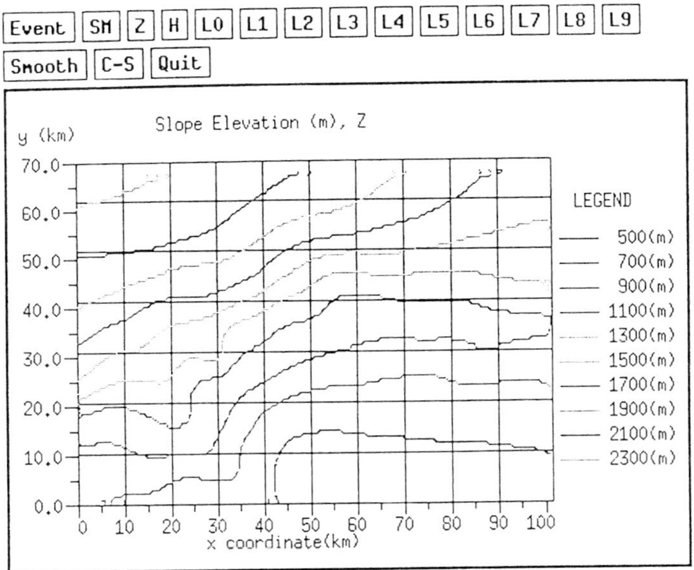


About Show Graphics

| | |
|---|---|
| Event: | gravity current (event) ID number, |
| SM: | show the thickness contour of the sediment mass, |
| Z: | show contour map of sediment surface (depth below water), |
| H: | show isopach of thickness of sediment above basement, |
| L0: | show isopach of pebble (L0), |
| L1: | show isopach of granule (L1), |
| L2: | show isopach of very coarse sand (L2), |
| L3: | show isopach of coarse sand (L3), |
| L4: | show isopach of medium sand (L4), |
| L5: | show isopach of fine sand (L5), |
| L6: | show isopach of very fine sand (L6), |
| L7: | show isopach of coarse silt (L7), |
| L8: | show isopach of fine silt (L8), |
| L9: | show isopach of clay (L9), |
| Smooth: | smooth contour map, |
| C-S: | show cross-section, |
| Quit: | quit from ShowGraphics. |

FIGURE 8.28 Contour map of the original topography for Test No. 2.

To simulate the transport, deposition, and erosion of sediment during such a sediment current, the following assumptions and conditions are used:

- Failure of the slope (release of sediment) is autochthonous (no external energy is provided) and instantaneous.
- Sediment current is constrained by the existing topography of the slope.
- Basin is filled with fluid (water, air, etc.), which is initially completely still (no eddy currents).

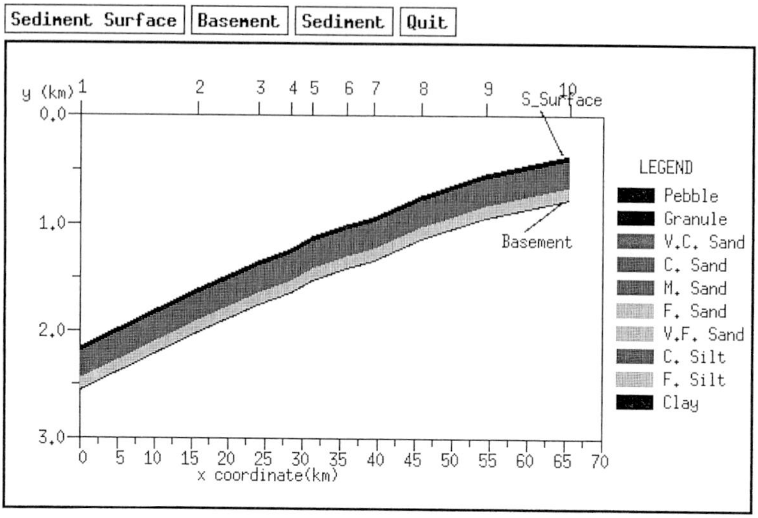


FIGURE 8.29 Cross-section profile A of the original basin slope for Test No. 2 (cross-section location drawn on Fig. 8.31).

- Sediment is not cemented and all grains are free to move independently.
- Slope sediments can have a different lithology than the quantum of released sediment, but with coarser grain sizes at the bottom and finer grain sizes at the top of the sediment column.
- Porosity for the slope sediments is unchanged during the simulation.
- Porosity for a given sediment current is unchanged during the flow, and different currents may have different porosity values.
- Only the thicknesses of the slope sediments and the height of the sediment currents are changed when deposition and erosion occur.
- Topography of the slope is updated after a given sediment current ends.
- All grain sizes have the same grain density.
- Series flows are allowed.

In the model, a discrete method, similar to Tetzlaff (1990), is used to simulate the transport, deposition, and erosion of sediments during a sediment flow in three dimensions. The advantages of such a discrete method are that less CPU time is required and there is more flexibility to control the simulation. The disadvantage is that the simulation is not explicitly time dependent, as in the deterministic methods of 3-D simulation of sediment flow (Bitzer and Pflug, 1990). As a consequence, not all potential energy in the sediment flow is necessarily converted fully before the sediment hits the boundary wall for a spatially small simulation. Care must be taken to

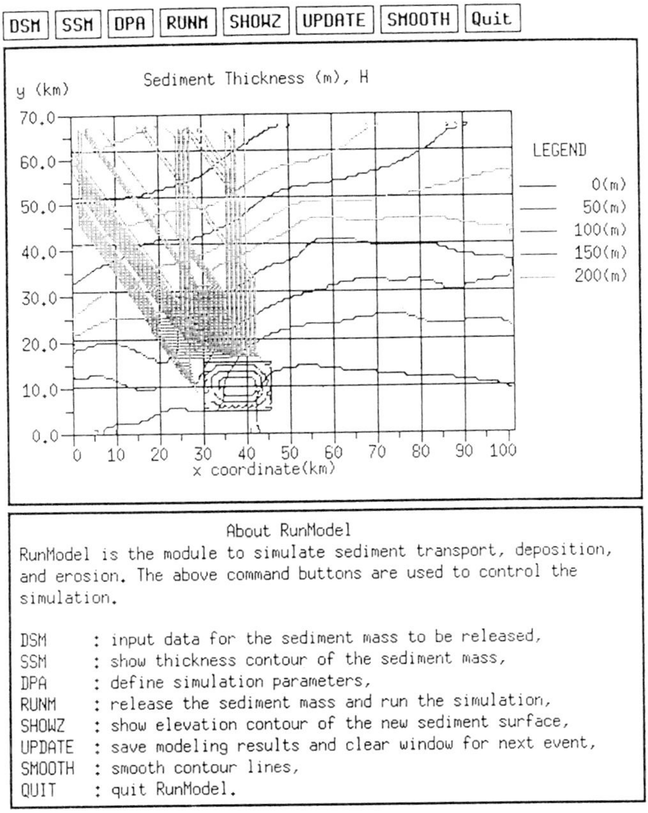


FIGURE 8.30 Trace pattern of the sediment current movement with contour of the released sediment mass A for Test No. 2.

ensure that the simulation volume is large enough so that boundary wall effects are far removed from the domain of interest.

## A. MATHEMATICAL FORMULATION

The model (called MOSED3D for MOdeling SEdiment Deposition in three dimensions) is based on the concept that a sediment current moves along a 3-D surface according to the balance between gravity (driving force) and friction (resistance force). Changes in the velocity of the current cause changes in the transport capacity, which, in large, controls the deposition and erosion of sediments. The rates of deposition and erosion also depend on both the basin slope and on the sediment type being carried by the current. The basic equations governing the current flow and also the transport, deposition, and erosion of sediment are taken from Allen (1985).

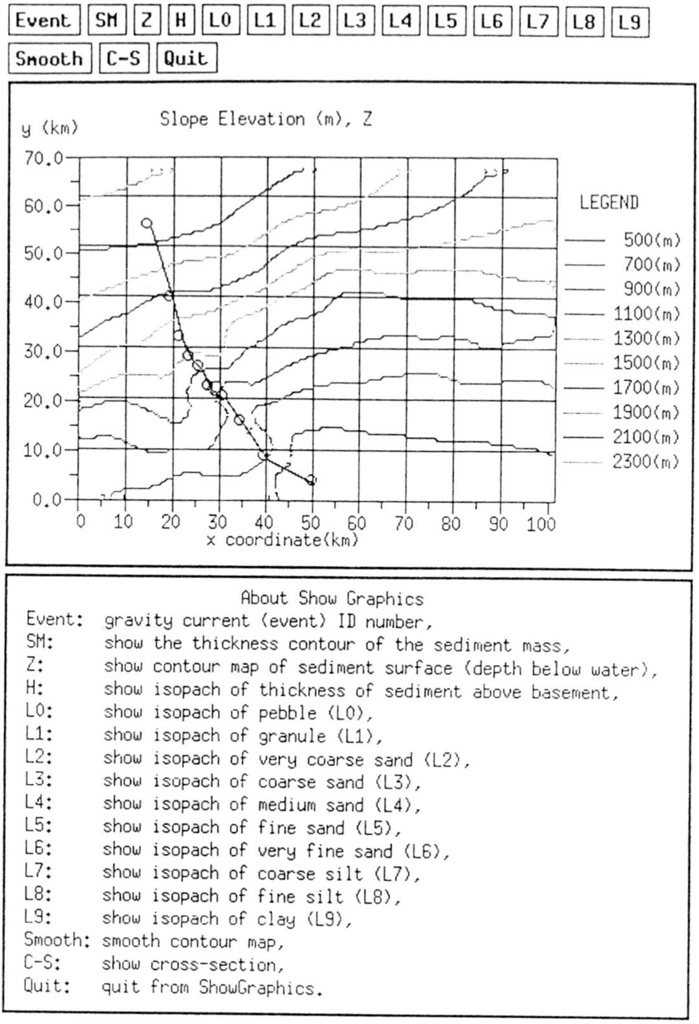


FIGURE 8.31 Contour map of the new topography with the location of cross-section A for Test No. 2.

Some modifications are made to the equations in order to fit the needs of MOSED3D.

### 1. Gravity Current Flow

Consider a flow element as a section of the flow of unit width and fixed streamwise length. The driving force per unit width is the downslope component of the immersed weight of the element. The driving force will

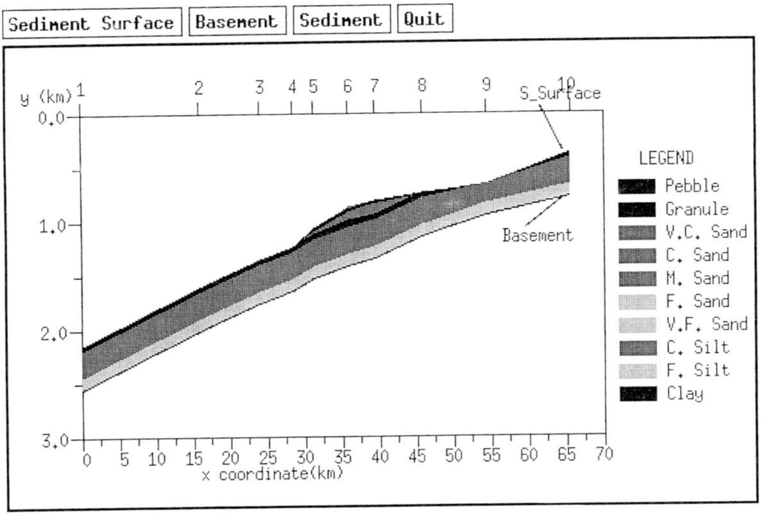


FIGURE 8.32 Cross-section profile A of the new basin slope for Test No. 2.

increase with the thickness and excess density of the current, and with the angle of the slope. The resisting force comes from friction between (1) the current and the bed and (2) the current and the medium. From Allen (1985), the mean flow velocity of the current is given as

$$V_a = \left[ \frac{8 \sin\beta}{(f_0 + f_1)} \frac{(\rho_2 - \rho_1)}{\rho_2} gh \right]^{1/2} \qquad \mathrm{ms}^{-1}, \tag{8.7}$$

where $V_a$ is the mean flow velocity of the current, $h$ the thickness of the flow element, $\rho_1$ and $\rho_2$ the densities of the ambient medium (fluid) and the current respectively, $g$ the acceleration due to gravity, $\beta$ the slope angle, and $f_0$ and $f_1$ the Darcy–Weisbach friction coefficients for the bed and the medium, respectively. Values of $f_0$ usually vary from ~0.006 to ~0.06 for rivers, and values of $f_1$ are usually less than 0.01 (Middleton, 1966).

**2. Sediment Transport**

The theoretical sediment load which a current can carry is (Allen, 1985)

$$M_t = M_b + M_s = \frac{J_b}{V_b} + \frac{J_s}{V_s} \qquad \mathrm{kg \cdot m^{-2}}, \tag{8.8}$$

where $M_t$ is the total theoretical sediment load; $M_b$ and $M_s$ are the theoretical bed load and suspended load, respectively; $J_b$ and $J_s$ are the mass transport rates for bed load and suspended load, respectively; and $V_b$ and

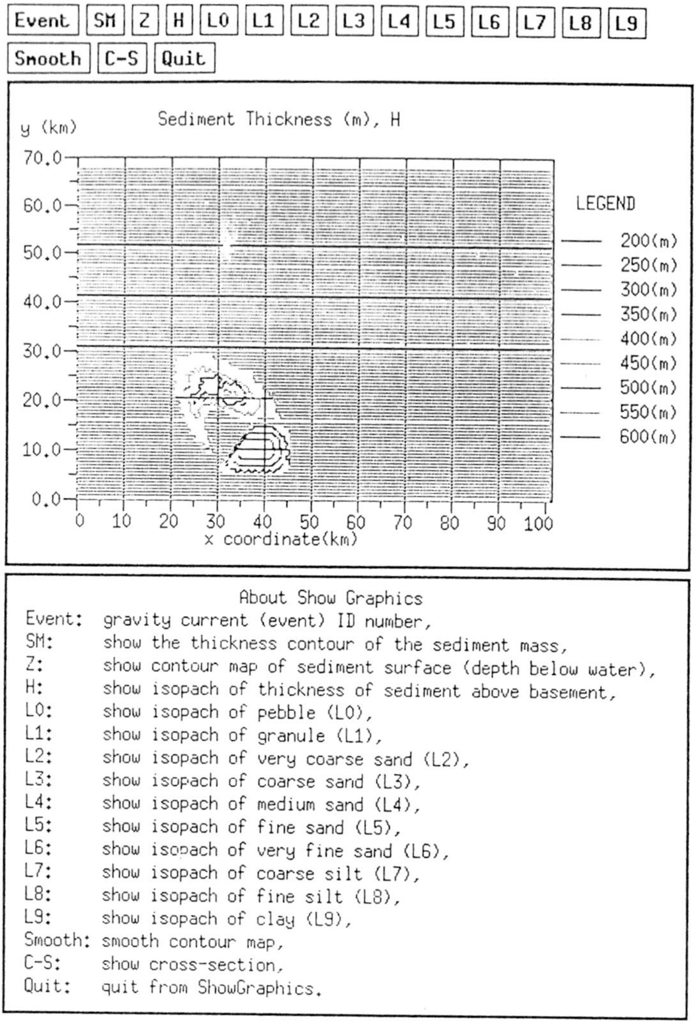


FIGURE 8.33 Isopach map of the new sediment for Test No. 2.

$V_s$ are the respective transport velocities (usually $V_b \approx 0.2\ V_a$ and $V_s \approx V_a$). Equation (8.8) shows that the total transport load consists of two components, bed load and suspended load. According to Bagnold's (1966) semiempirical formulas, we have

$$J_t = J_b + J_s = \frac{\rho}{(\rho - \rho_1)g}\left(0.148 + 0.01\,\frac{V_a}{8}\right) \times \frac{f_0}{V_f}\rho_2(V_a)^3 \qquad [\mathrm{kg \cdot m^{-1} \cdot s^{-1}}] \tag{8.9}$$

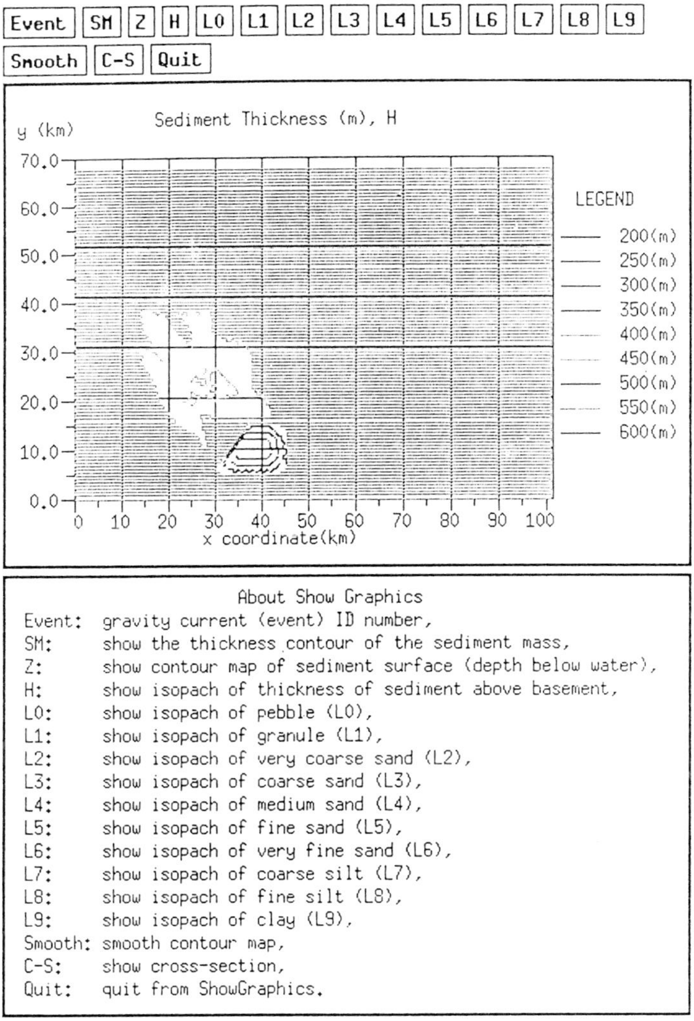


FIGURE 8.34 Isopach map of the new sediment with HHGO = 0.1 for Test No. 2.

and

$$J_s/J_b = 0.068\frac{V_a}{V_f}, \tag{8.10}$$

where $J_t$ is the total-load transport rate, $\rho_1$ the density of the ambient medium, $\rho$ the grain density, and $V_f$ the terminal fall velocity of the sediment particles in the current. The terminal fall velocity $V_f$ is defined as (Allen, 1985)

$$V_f = \frac{1}{18}\frac{(\rho - \rho_1)}{\eta}gD^2 \qquad [\mathrm{ms^{-1}}], \tag{8.11}$$

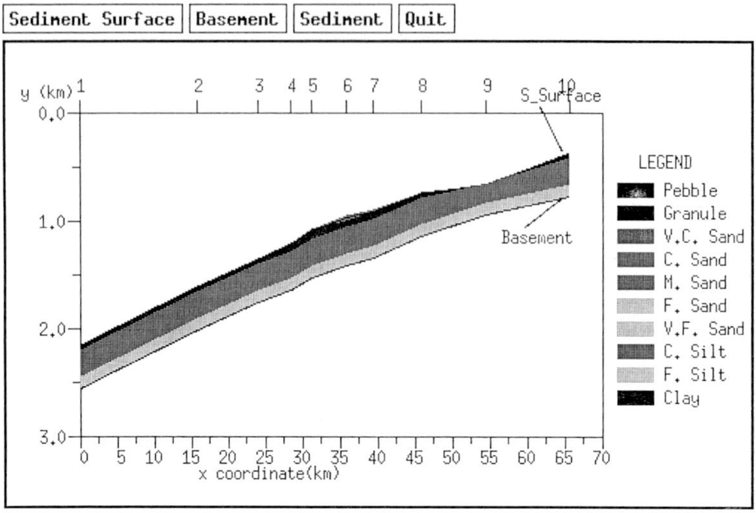


FIGURE 8.35 Cross-section profile A of the new basin slope with HHGO = 0.1 for Test No. 2.

where $\rho$ and $\rho_1$ are the density of grain and fluid, $\eta$ the fluid viscosity, and $D$ the grain diameter. The power factor of 3 in Eq. (8.9) is set as a user-defined constant (CON_T) in MOSED3D so that users can better control the transport capacity calculation.

### 3. Sediment Deposition

Whether deposition will actually occur, with consequent losses of excess density and possibly thickness, depends on the balance between the sediment load actually present in the gravity current and the sediment load that can be theoretically supported by the forces due to the motion of the current. The sediment can be deposited only if it is present in excess of the theoretical maximum transportable load, $M_t$. Hence no sediment deposition will be expected if the actual load at all times is equal to or less than the theoretical value (Allen, 1985). The actual load $M_a$ can be defined as

$$M_a = \rho_2 h \qquad [\mathrm{kg \cdot m^{-2}}], \tag{8.12}$$

where $\rho_2$ is the density of the current and $h$ the height or thickness of the current.

If $M_a > M_t$, sediment deposition occurs and the amount of sediment deposited in mass will be $M_a - M_t$. The coarser grain size material will deposit first and the finer grain size sediments later. In MOSED3D the

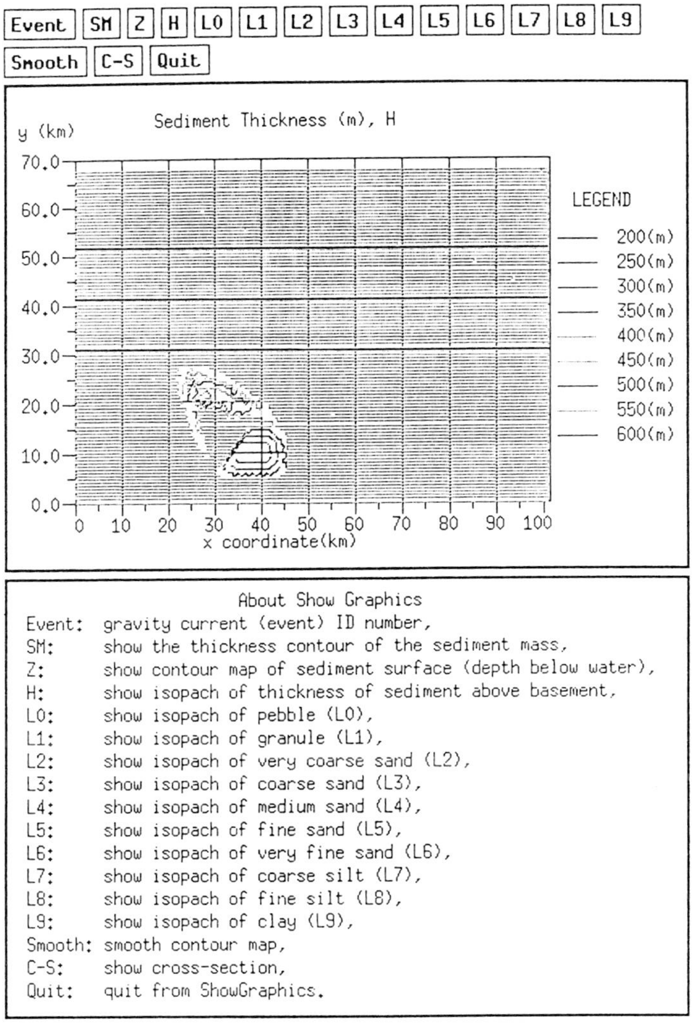


FIGURE 8.36 Isopach map of the new sediment with HHGO = 0.9 for Test No. 2.

deposited amount of mass is converted to thickness and a fractional constant (HHGO), ranging from 0.0 to 1.0, is introduced to control the actual thickness to be deposited, with HHGO = 1.0 being the total thickness.

### 4. Erosion

Two criteria are used to examine whether erosion will occur for the slope sediments. The first criterion is that the total theoretical load must be greater than the actual load, that is, $M_t > M_a$, and the second criterion is that the shear stress ($\tau$) of the current on the slope sediments must be

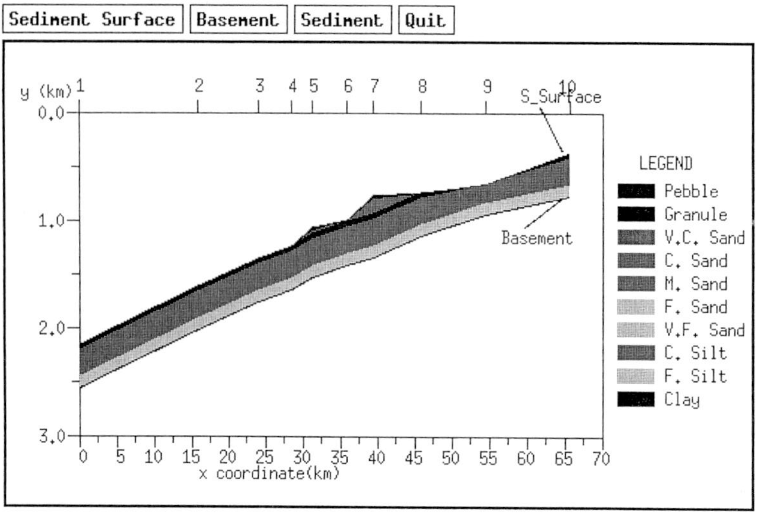


FIGURE 8.37 Cross-section profile A of the new basin slope with HHGO = 0.9 for Test No. 2.

greater than the critical shear stress ($\tau_c$) for the sediments to be eroded, that is, $\tau > \tau_c$. The shear stress $\tau$ is defined as (Allen, 1985)

$$\tau = \frac{f_0}{8}\rho_2(V_a)^2 \qquad [\mathrm{N \cdot m^{-2}}]. \tag{8.13}$$

In MOSED3D, the power factor of 2 in Eq. (8.13) is set as a user-defined constant (CON_E) for more control on the erosion capacity. The critical shear stress $\tau_c$ is defined as

$$\tau_c = \theta(\rho - \rho_1)gD \qquad [\mathrm{N \cdot m^{-2}}], \tag{8.14}$$

where $\theta$ is the dimensionless threshold shear stress, $\rho$ and $\rho_1$ the density of grain particles and fluid, respectively, and $D$ the grain diameter (Allen, 1985). As in the case of deposition, the eroded amount is converted from mass to thickness with the factor HHGO. The dimensionless threshold shear stress, $\theta$, has different values for different grain sizes. In MOSED3D, 10 grain sizes are allowed and the corresponding $\theta$ values are taken from Allen (1985, Fig. 44, p. 58) as shown in Table 8.2.

## B. TEST CASES AND SENSITIVITY

Two test cases are given here to illustrate the use of the MOSED3D program and to demonstrate the sensitivity of the program to some modeling parameters.

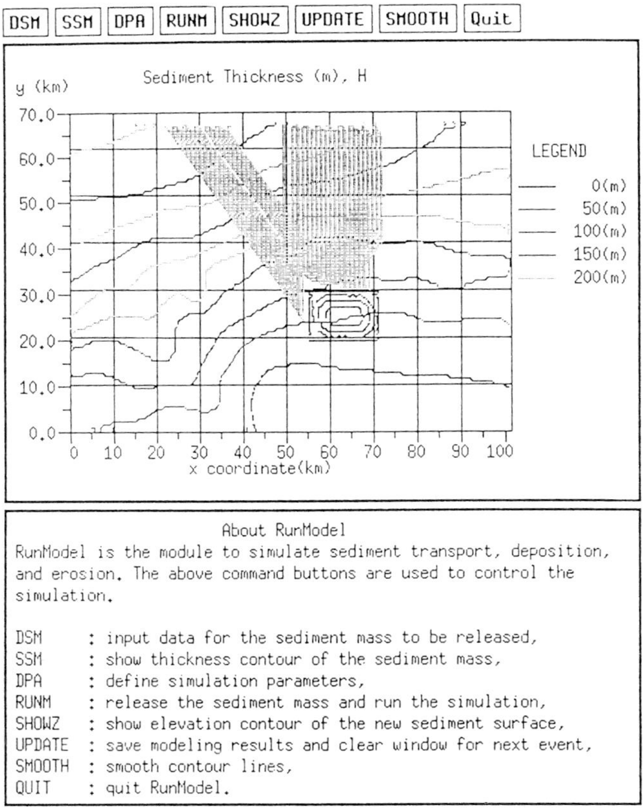


FIGURE 8.38 Tracer path of the sediment current movement with contours of the released sediment B for Test No. 2.

## 1. Test No. 1

Figure 8.19 show the topographic surface of Test No. 1 with a structural dome on the left, filtering out to the right. A sediment thickness of 200 m is used with 25% very coarse sand (L2), 25% coarse sand (L3), 25% fine sand (L5), and 25% coarse silt (L7). Figure 8.20 shows the cross-section with no controlling points from left to right of the dome (see Fig. 8.22 for the location of the cross-section). The sediment mass to be released is located on the right slope of the dome (Fig. 8.21). Default values of 12 modeling parameters are used for the first run. These default values are as follows:

PSB = 70% (porosity of the basin slope sediment)
PGC = 70% (porosity of the sediment current)
DG = 2650 $kg \cdot m^{-3}$ (density of the sediment matrix)
DF = 1025 $kg \cdot m^{-3}$ (density of the fluid)

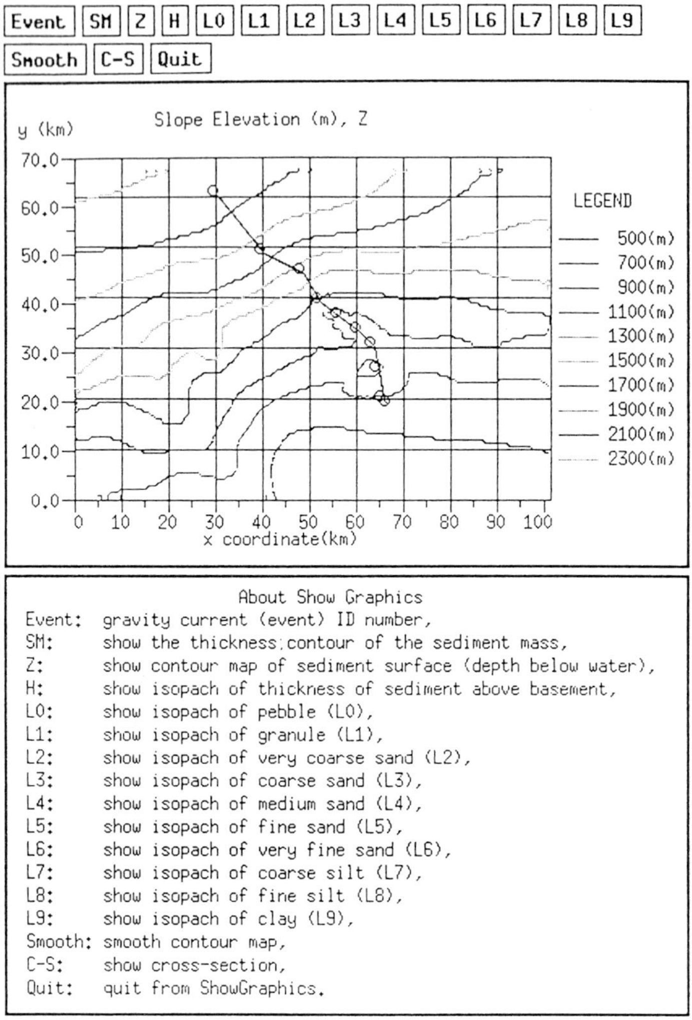


FIGURE 8.39 Contour map of the new topography with the location of the cross-section B for Test No. 2.

ISMV = 0 (using average height of sediment mass)
VF = 0.0009 N · s/m$^2$ (viscosity of the fluid)
DWF0 = 0.008 (Darcy–Weisbach friction coefficient for bed)
DWF1 = 0.01 (Darch–Weisbach friction coefficient for fluid)
CON_T = 2.50 (constant in calculation of transport capacity)
CON_E = 1.20 (constant in calculation of shear stress for erosion)
HHGO = 0.50 (erosion/deposition coefficient)

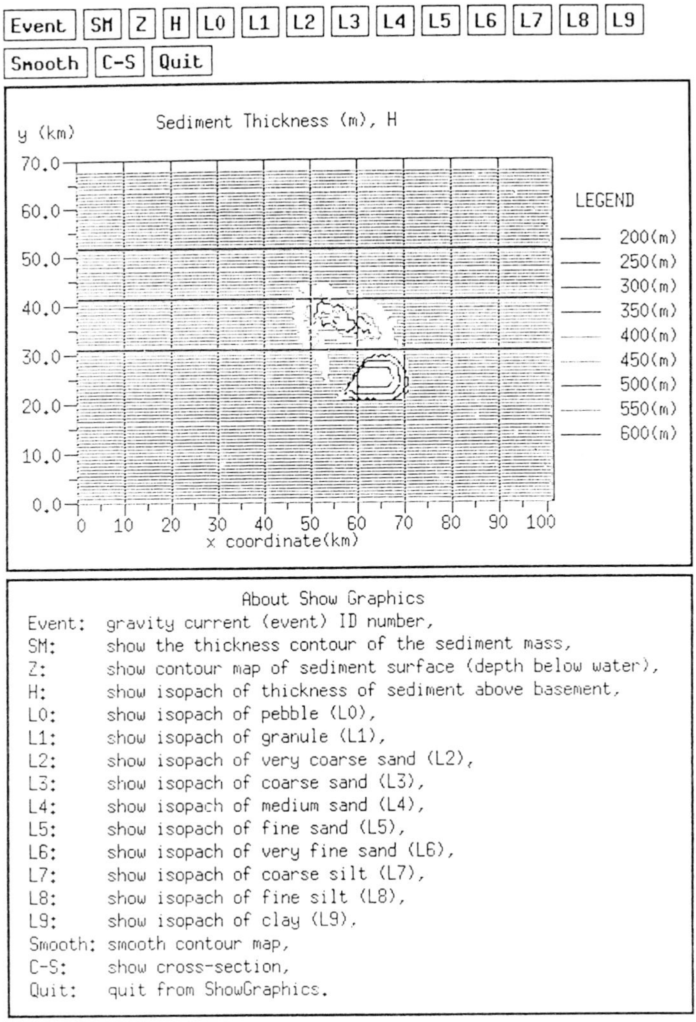


FIGURE 8.40 Isopach map of new sediment with HHGO = 0.5 (default value) for Test No. 2.

ELEFT = 0.70 [remaining energy due to energy loss during deposition (erosion)]

Figure 8.21 shows the track of the sediment current movement with contours of the released sediment mass. Figure 8.22 shows the contour maps of the new topography after the end of the first run with the location of the cross-section lines, indicating a sediment loss at the location of the

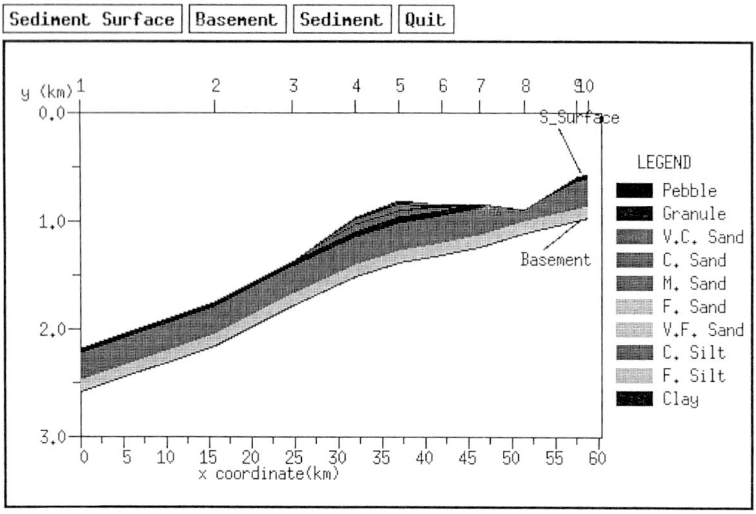


FIGURE 8.41 Cross-section profile B of the new basin slope with HHGO = 0.5 for Test No. 2.

release and sediment deposition down the slope. Figure 8.23 shows the cross-section of the new basin slope; the sediment is released, a concave-in surface shape (locations 8–12) is left in the basin slope with most of the coarse sand, fine sand, and coarse silt moving out. A pile of sediment is deposited at locations 13–15 with interbedded coarse sand, fine sand, and coarse silt superposed on the original basin slope.

Next, change the default value of a particular modeling parameter and keep other parameters unchanged to evaluate the sensitivity of the modeling results to the change. The parameter values changed in turn are PSB = 40%, PGC = 90%, DG = 2400, DF = 1500, ISMV = 1, VF = 0.009, DWF0 = 0.06, CON_T = 2.0, 3.0 and 4.0, CON_E = 1.5 and 2.0, and ELEFT = 0.1. All these changes cause some effects in (1) the tracer pattern of the sediment current movement, (2) contours of the new topography, and (3) lithology distribution in the cross-sections, but all these effects are too small to observe clearly with the graphical resolution available. Thus one may conclude that the changes in the parameters are not very sensitive for the test case. However, the modeling results of the test case are sensitive to changes in parameter HHGO. Figure 8.24 show the tracer pattern of the sediment current movement with HHGO = 0.1. Because less sediment can be deposited (HHGO = 0.1 is smaller than the default value of 0.5), more sediment is spread out in a wider area. Figure 8.25 show the cross-section profile of the basin slope with a pile of thin sediment deposited at locations 13–15.

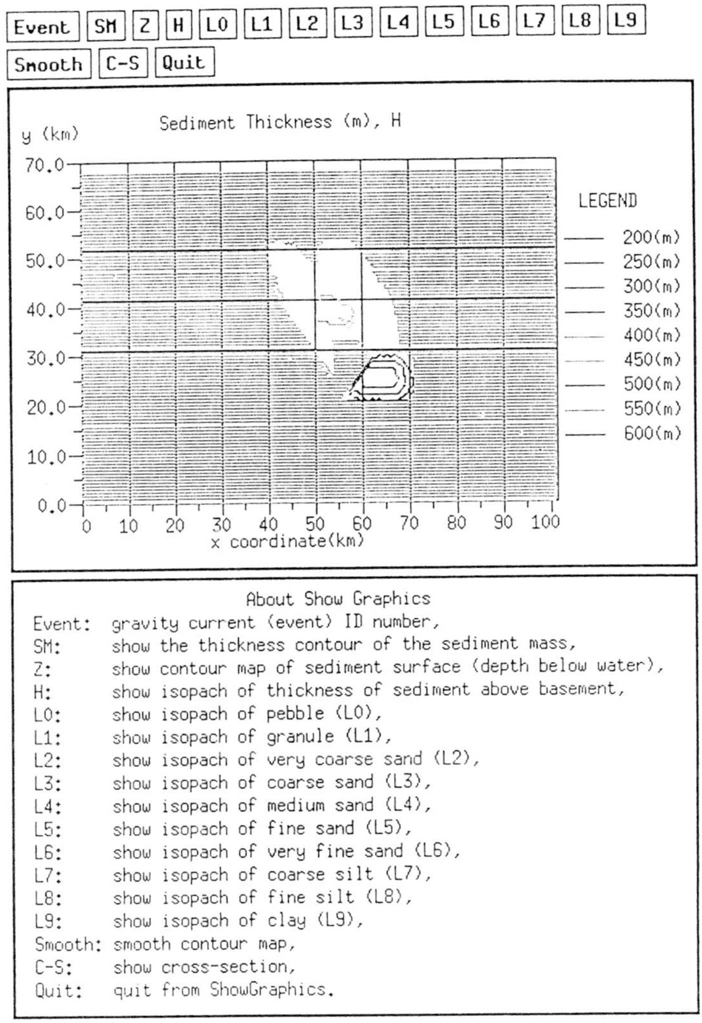


FIGURE 8.42 Isopach map of the new sediment with HHGO = 0.1 for Test No. 2.

To see the effects of lithology type, we changed the lithology type of the slope sediments to 40% coarse sand (L3), 40% very fine sand (L6), and 20% clay (L9), while keeping the modeling parameters at the default values. Figure 8.26 shows the tracer pattern of the sediment movement, with clay sediment moving out of the field of view of the total simulation volume. Figure 8.27 shows the cross-section profile of the new basin slope, indicating that only very fine sand is deposited at locations 13–15, and no clay deposition occurs at all.

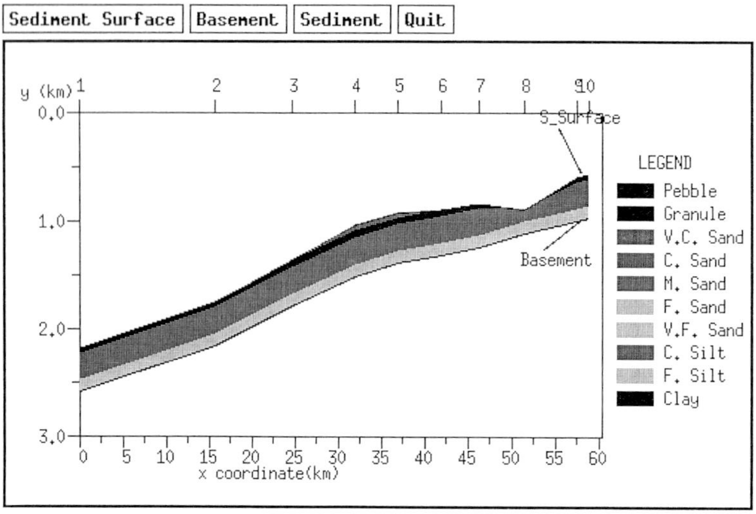


FIGURE 8.43 Cross-section profile B of the new basin slope with HHGO = 0.1 for Test No. 2.

## 2. Test No. 2

Figure 8.28 show the original topography of Test No. 2 (provided by BP/Statoil). The slope sediment has a thickness of 400 m (Fig. 8.29), with 28% fine sand (L5), 60% coarse silt (L7), and 12% clay (L9). The slope sediment porosity is set at 54%, and the sediment current porosity at 80%. First release a sediment mass at location A ($x \approx 40$ km, $y \approx 10$ km) with other parameters at their default values. Figure 8.30 shows the tracer pattern of the sediment current movement with the contours of the released sediment mass. Some sediment is moved out of the field of view (mostly clay). Figure 8.31 is the contour of the new topography after the release, with the control points of a cross-section also exhibited. Figure 8.32 shows the cross-section profile of the new basin slope with fine sand (from locations 8–10) deposited on the top of clay at locations 5–8. Figure 8.33 is the isopach map of the sediment thickness after the release.

Next change the value of HHGO to 0.1 and 0.9, as opposed to the default value of 0.5. Figures 8.34 and 8.35 show, respectively, the isopach map of the new sediment and cross-section profile of the slope for HHGO = 0.1, whereas Figs. 8.36 and 8.37 show, respectively, the isopach map of the new sediment and cross-section profile of the slope for HHGO = 0.9. The parameter HHGO sensitively controls how much sediment will be deposited or eroded. With HHGO = 0.1, less sediment is deposited at each movement of the sediment current from one grid to another grid. As a result, sediment spreads out in a large area with a thin sediment thickness. On the other hand, with HHGO = 0.9, more sediment is deposited at each grid-to-grid movement of the sediment current, and thicker sediment will be deposited in a smaller area.

Another sediment mass with the same geometry as the first mass is then released at location B ($x$ = 60 km, $y$ = 20 km), keeping all parameters the same as above. Figure 8.38 shows the tracer pattern of the current movement, and Fig. 8.39 shows the contour map of the new topography with the control points of a cross-section. Figures 8.40 and 8.41 are, respectively, the isopach of the sediment and cross-section of the slope for HHGO = 0.5 (default value). Figures 8.42 and 8.43 present, respectively, the isopach map and cross-section for HHGO = 0.1, whereas Figs. 8.44 and 8.45 give, respectively, the isopach map and cross-section profile for HHGO = 0.9.

## C. DISCUSSION

MOSED3D is a computer program that simulates in three dimensions the transport, deposition, and erosion of sediments during flow of a sediment current formed as an instantaneous release of sediment. Two kinds of input data are required before the MOSED3D code can be run: (1) input data for the original basin geometry, defined as the topography of sediment thickness and lithology composition of the basin slope, and (2) input data for the sediment mass to be released, defined as a matrix of sediment thicknesses.

Twelve modeling parameters are used in the program to define and control the simulation. These parameters are preset with default values, but the values can be changed during the simulation. The parameters can be separated into two groups: one for the physical properties of the sediment, sediment current, and fluid; and the other for the physical laws conserving the transport, deposition, and erosion of sediment. In general, the model results are not very sensitive to the first group of parameters, but are sensitive to the second group of parameters, in particular to CON_T (power constant in calculation of transport capacity) and HHGO (deposition and erosion coefficient)—at least these are the sensitive variables for the test cases exhibited here.

The lithology types have significant effects on the simulation, with coarser grain sizes producing more stable results and finer grain sizes giving unstable results, in the sense that transport of fine-grained material completely out of the simulation volume can occur more readily so that the simulation volume must be chosen large enough to minimize this edge effect. For example, a clay-dominated current has relatively greater transport power to carry the clay sediments further and, as a consequence, little sediment will be deposited.

Test cases presented here have shown the basic function of the MOSED3D model, including a series of sediment currents. Also, the model

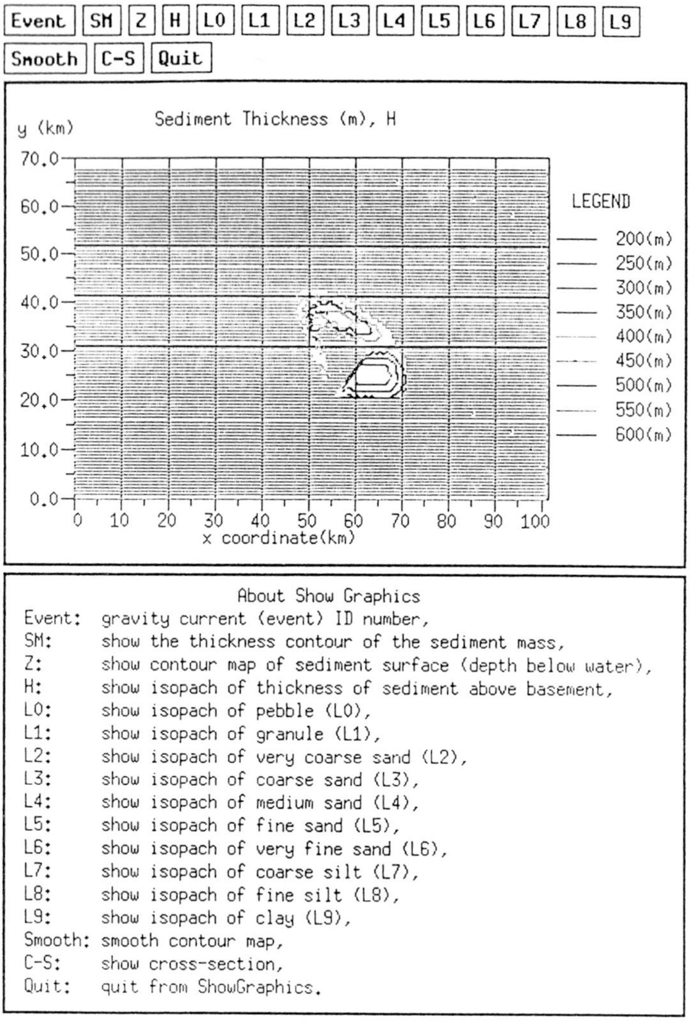


FIGURE 8.44 Isopach map of the new sediment with HHGO = 0.9 for Test No. 2.

can be improved in both scientific calculation and computer programming to provide more accurate and flexible simulations. For example:

1. The constants CON_T and CON_E in the calculations of transport capacity and erosion shear stress, respectively, should be made lithology-type dependent. In the current version of MOSED3D, these constants have the same value for all grain sizes.
2. The equation used to calculate the theoretical transport capacity [Eq. (8.9)] is sensitive to the terminal fall velocity. In the case of

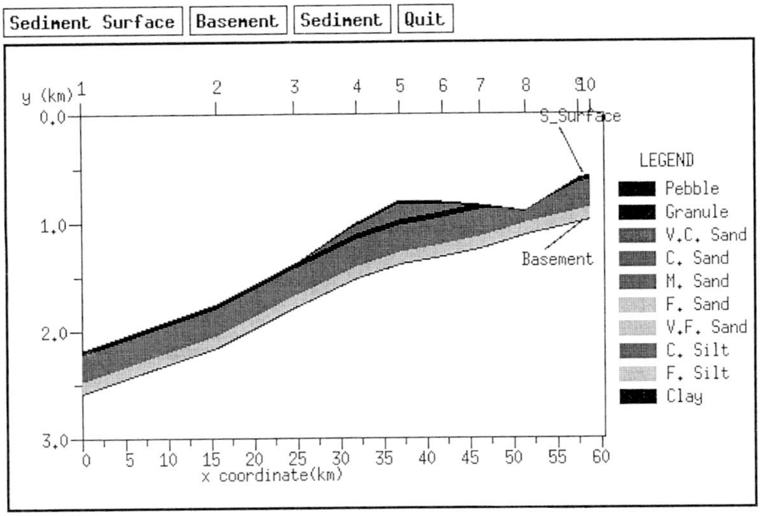


FIGURE 8.45 Cross-section profile B of the new basin slope with HHGO = 0.9 for Test No. 2.

clay particles, the calculated transport capacity is high because the full value for clay terminal fall velocity is extremely small (of the order of $10^{-7}$ units), compared to the values for coarse grain sizes (or order $10^{-2}$–$10^{-1}$ units). Presumably other equations for calculation of the theoretical transport capacity should also be used as alternative prescriptions.

3. The lithology distribution (vertically) of the original basin slope sediments should be a user input. The current version automatically puts the coarser grain sizes at the bottom and the finer grain sizes at the top of the basin slope sediments.
4. More case studies should be done in order to test the flexibility, sensitivity, and accuracy of the model in relation to observed sediment flow behaviors.

PART

# II

# Scientific Probability, Risk, and Strategy

Part I of this volume presented methods used to analyze the paleo-uncertainty of basin analysis components. The basic technique is clear: (1) Construct a model, or models, of the relevant physical, chemical, or geologic processes, being aware of the intrinsic assumptions of the model; (2) compare the model output results with observations, then systematically adjust parameters, boundary and initial conditions, and intrinsic assumptions within the model until a minimum mismatch of predicted and observed behaviors is obtained; and (3) evaluate quantitatively the accuracy, precision, resolution, sensitivity, and, if possible, uniqueness of the model results. In this way an appreciation is gained of the likely paleo-evolution, subject to the constraints of the data quality, quantity, and sampling frequency used as control criteria.

As is clear from Part I, it is possible to have more than one model satisfy a given set of data (e.g., flexural basement, hydrocarbon kinetics, fission tracks). It is also possible for some parameters in a model to be extremely insensitive to the observational control criteria, and so to have a wide range of indeterminate uncertainty (e.g., unconformity onset times, fracture limits), whereas other parameters may be extremely well determined (low variance) and so provide better control of the paleo-evolution.

In cases of more than one model satisfying a given set of data, extraneous criteria, not contained within the models, must be used in attempts to narrow down the uncertainty of which model (or models) to use.

In cases of sensitivity and resolution of parameters within a particular model, the question is this: When is the uncertainty on parameters of significance in influencing particular outputs? (For instance, in the hydrocarbon kinetic model illustration, at least three radically different distributions satisfied the data set of S2 measurements to the same degree of mismatch, but the generation of hydrocarbons with time from each of the three is virtually indistinguishable.)

Within this encompassing framework of models, parameters, and control by mismatch against observed data, the one underpinning assumption so far is that the data are themselves not uncertain.

Uncertainty in data used as control criteria will then add to the uncertainty of models, parameters, and output results; this point is addressed in Part II.

In addition, the combined uncertainty of models, parameters, and data means that, in general, paleo-evolution of *any* quantity must be uncertain. Ths task is to provide a method for handling quantitatively the uncertainty so that probable results, and ranges of the probability, can be given. In this way one can come full circle and both understand and use the ranges of results for hydrocarbon assessments presented in Chapter 1. As a by-product of such a procedure, we are then able to assess where effort needs to be put to improve resolution. An illustration of this assessment method is also given in Part II.

The tie-in of hydrocarbon charge estimates, and their uncertainties, from such basin analysis considerations to economic exploration aspects is then readily given. The point here is that in making economic worth assessments, uncertainties due to hydrocarbon selling price, rig costs, inflation, and so on, are routinely applied to plays and prospects, so that measures of risk and value are available. These techniques from the economic arena are transportable almost wholesale to the scientific arena, so that an integrated exploration assessment can be given with dependence on *both* basin analysis and economic factors. Determination can then be made of the relative importance of individual uncertainties from scientific and economic in contributing to the decision to risk money on a particular prospect or play type. This point is addressed in Part II.

Finally, from a scientific point of view there is a question of strategy of technical operation. (There are also questions of economic strategy, but they would take us too far afield in this already long tome.) The point is that various numerical codes are available with which we can perform quantitative basin analysis calculations. Some way must be provided to determine when, and in what order, application of increasingly sophisticated tools should be used so that an efficient strategy is available for obtaining trustworthy results. This problem is also addressed in this part.

# 9

# THE UNCERTAINTY OF AGE

## I. INTRODUCTION

One of the more important factors in quantitative basin analysis models is the use of an accurate timescale for formation horizons. The inferred elapsed duration of events can have significant impacts on processes related to geological, physical, geochemical, and hydrologic control. For example, sedimentation rates depend on the timescale used, and the buildup of fluid overpressure is strongly related to the rate at which shale deposition occurs. A deposition rate of shale greater than about 30 to 60 m/MY is usually indicative of overpressure, so changing a timescale will decrease (increase) the deposition rate, thereby leading to a different overpressure development with time. Because the flow rates of water, oil, and gas depend on both the permeability and on the excess pressure gradient, and because the permeability depends on the void ratio, which, in turn, depends on the escape of fluid from the basin, the migration of subsurface fluids is also influenced by the timescale used. In addition, under prescribed quantitative rules for basement motion with time, which are dependent on the applied sedimentary load to that time, the subsidence of a basin is also dependent on the timescale chosen, so that the inferred evolution of basin geometry and structures is influenced by the timescale.

Because kinetic models of hydrocarbon production depend on both the evolution of paleotemperature with subsurface depth and on the elapsed time, both the sediment deposition and later compaction, together with total hydrocarbon generation, are beholden to the timescale invoked. In turn, a shift in the timing of hydrocarbon generation then influences the timing and directions of secondary migration of hydrocarbons, so that accumulation sites and accumulation amounts may differ depending on the timescale used in the calculations.

Although other factors and processes in quantitative basin modeling calculations are also influenced by the timescale chosen, the listing given here is sufficient to make the point that it is appropriate to undertake an investigation of the influence of timescales on basin modeling calculations. Two points are considered in this chapter: (1) choosing a timescale and (2) the influence of different timescales on quantitative basin modeling calculations.

## II. CHOICES OF TIMESCALES

Over the years, starting at least as early as Holmes (1937) and continuing to the present day, various timescales have been proposed for age-dating sedimentary horizons (Holmes, 1937, 1947, 1959; Kulp, 1961; Lambert, 1971; Van Eysinga, 1975; Armstrong, 1978; Walters, 1980; Hambrey and Harland, 1981; Haq *et al.,* 1987; Harland *et al.,* 1982, 1989; GSA, 1991; Gradstein *et al.,* 1994). The timescales differ between each other and also, occasionally, internally. Lerche and Thomsen (1994) investigated the distribution of ages for a variety of sedimentary horizons ranging from late Tertiary to as far back as the age distribution for the Cambrian–Precambrian boundary using 10 timescales. What was found is that the age distribution of each horizon was very closely approximated by a log-normal distribution. However, in that previous investigation timescales developed prior to World War II were included together with those based on radiometric methods, procedures not developed until the last two to three decades. Thus the log-normal nature uncovered for the age distribution of individual sedimentary horizons could have been influenced by the choices of different methods for producing ages. We did remark in that earlier work that the age distributions were even more closely log-normal in character if the pre–World War II timescales were omitted.

The purpose of this section is twofold: First, we use four of the five most recent timescales to obtain a statistical age distribution for a variety of horizons, which all turn out to be remarkably close to log-normally distributed. From the cumulative age distribution plot for each horizon we then estimate the mean age and uncertainty in the mean based on the four timescales.

Second, we then add in the latest timescale of Gradstein *et al.* (1994) to the age distribution plot for each horizon. As will be shown, that latest timescale sits squarely on the log-normal curve obtained from the previous four timescales and is amazingly close to the expected age value for each horizon calculated solely from the previous four timescales. The four basic timescales used in the first part of this section are those of Harland *et al.* (1982), Haq *et al.* (1987), Harland *et al.* (1989), and the GSA timescale (1991); the fifth timescale used in the second part of this section is that of Gradstein *et al.* (1994).

From each of the first four timescales, ages were taken for nine horizons labeled as Pliocene, Mid Pliocene, Cretaceous (Maastrichtian), Turonian, Cenomanian, Aptian, Jurassic, Callovian, and Triassic. These ages were then plotted on cumulative log-probability axes as shown in Figs. 9.1a to 9.4a, respectively. Note that straight lines pass through all the points for all the data, suggesting that the cumulative distribution of age dates for each horizon is well approximated by a log-normal distribution, as had already been remarked earlier was the case (Lerche and Thomsen, 1994).

For a precisely log-normal distribution the mean value occurs at about a cumulative probability of 68% (Feller, 1957). Reading off the age values at the 68% cumulative probability point, $P(68)$, for each of Figs. 9.1a through 9.4a, we obtain the values listed in column 2 of Table 9.1. Figures 9.1b through 9.4b add in the corresponding age dates from the Gradstein *et al.* (1994) timescale without changing the distribution. The added points sit directly on the log-normal probability curve derived from the previous four age scales. Indeed, in Table 9.1 (column 3) we have presented the Gradstein *et al.* (1994) age dates next to the $P(68)$ values obtained from the prior four timescales. There is no difference to within intrinsic error on the age determinations from each timescale.

Two conclusions can be drawn:

1. Once the different timescales have been examined, they are very accurately described by a cumulative log-normal age distribution. An accurate, reproducible, and quantifiable measure of uncertainty in age dating can then be given at the 90% confidence level by writing the age, $A$, of a horizon as

$$A = A(68)^{+[A(90)-A(68)]}_{-[A(68)-A(10)]}, \tag{9.1}$$

where $A(68)$ is the expected age at $P(68)$, $A(90)$ is the age at $P(90)$ [it is 90% certain the true age is *less* than $A(90)$], and $A(10)$ is the age at $P(10)$ [it is 90% certain the true age is *greater* than $A(10)$].

2. Modern timescales would appear to have reached some sort of limit in refining the "absolute" timescale, because the newest published timescale (Gradstein *et al.*, 1994) lies squarely within the distribution from the four previous "best" published timescales (Harland *et al.*, 1982, 1989; Haq *et al.*, 1987; GSA, 1991).

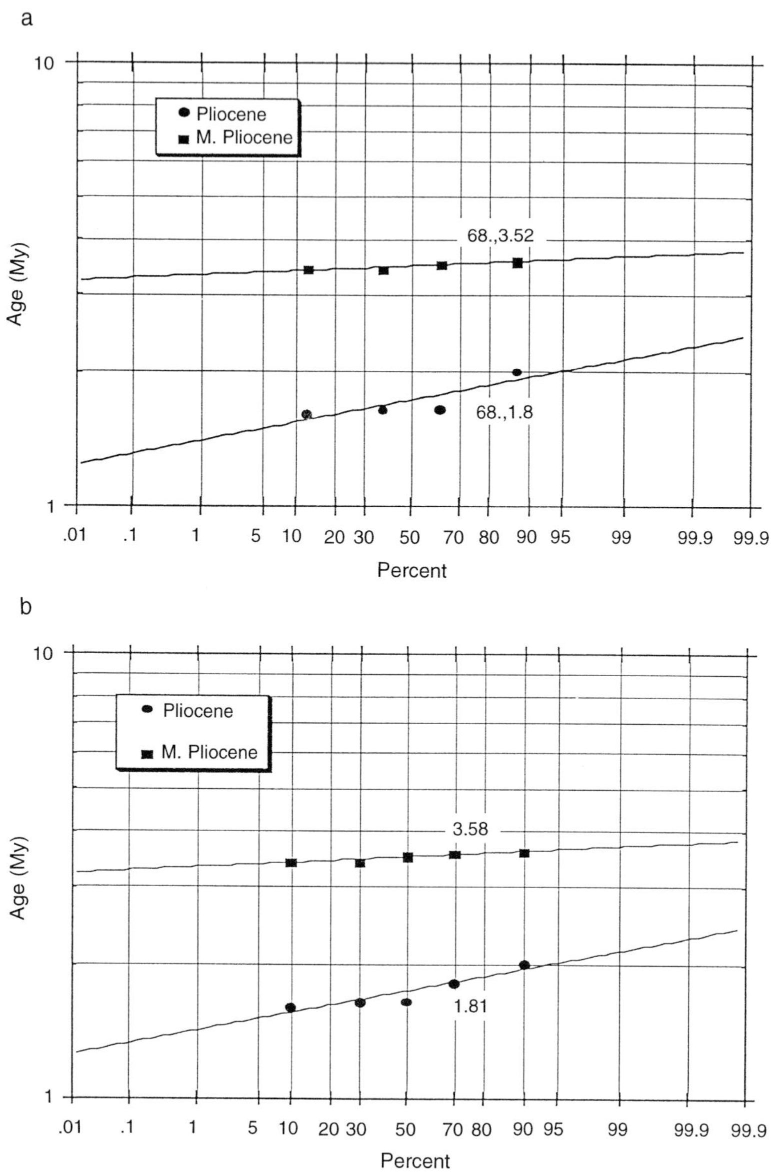


FIGURE 9.1 Cumulative probability plots for the Pliocene and Mid-Pliocene age distributions: (a) for the four timescales prior to 1994 used in text; (b) with the age dates superposed from the Gradstein *et al.* (1994) timescale.

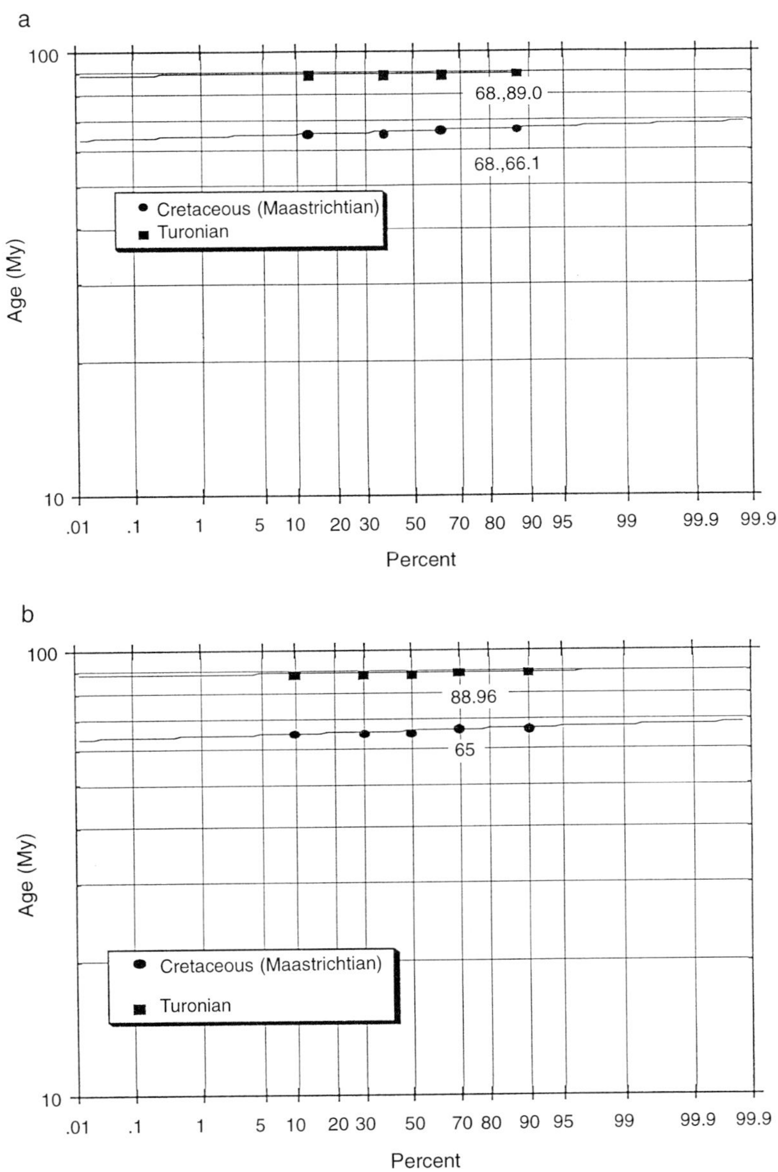


FIGURE 9.2 As for Fig. 9.1 but for the Cretaceous (Maastrichtian) and Turonian horizons: (a) for the four timescales prior to 1994 used in text; (b) with the age dates superposed from the Gradstein *et al.* (1994) timescale.

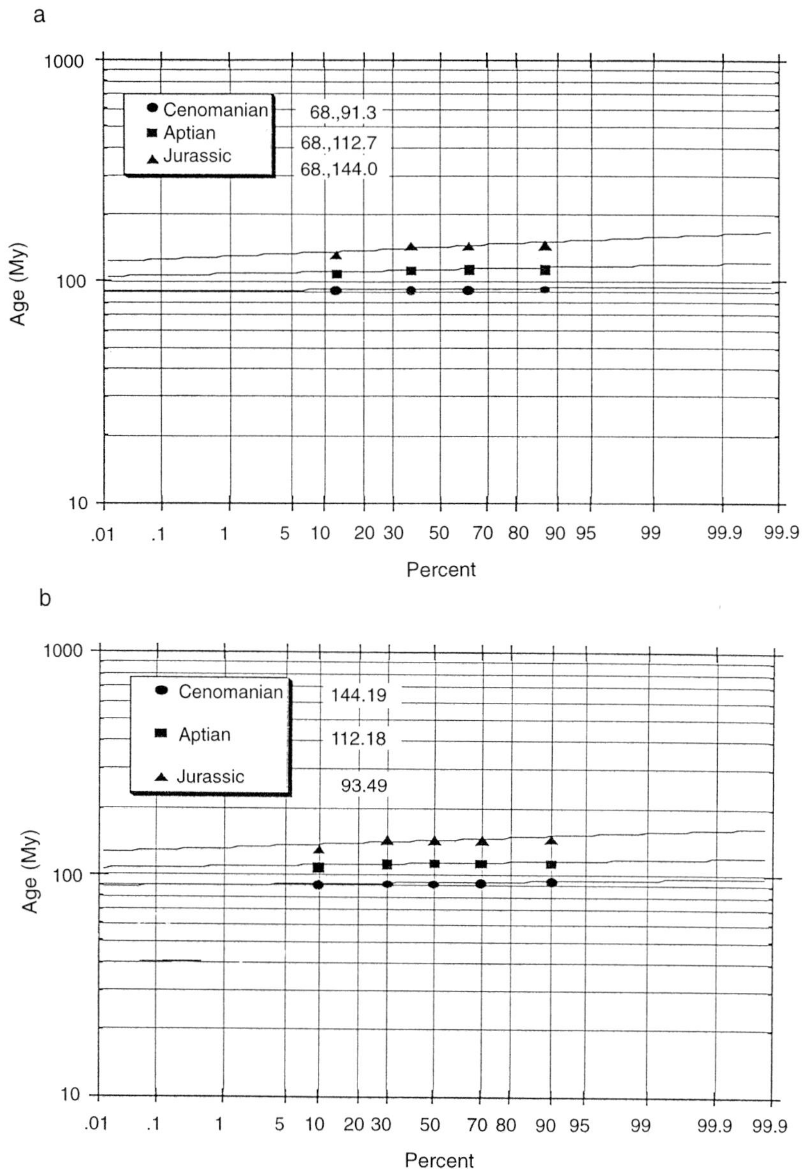


FIGURE 9.3 As for Fig. 9.1 but for the Cenomanian, Aptian, and Jurassic horizons: (a) for the four timescales prior to 1994 used in text; (b) with the age dates superposed from the Gradstein *et al.* (1994) timescale.

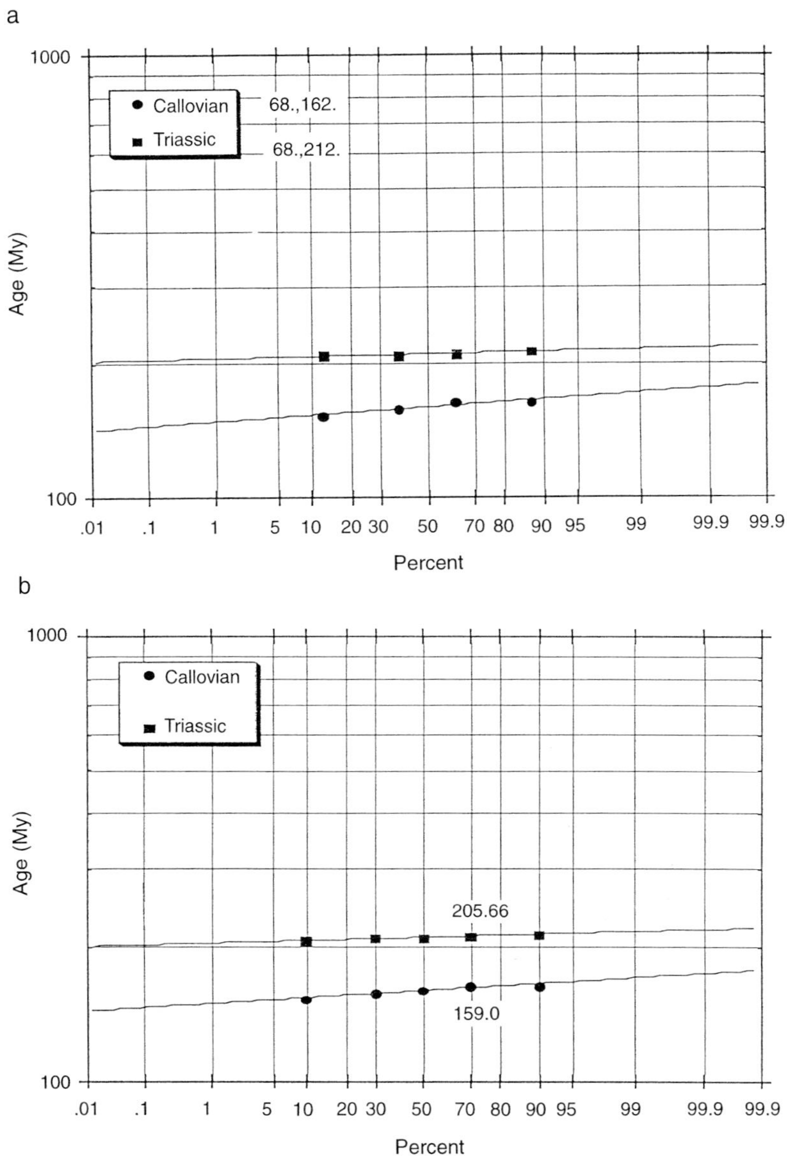


FIGURE 9.4 As for Fig. 9.1 but for the Callovian and Triassic horizons: (a) for the four timescales prior to 1994 used in text; (b) with the age dates superposed from the Gradstein *et al.* (1994) timescale.

TABLE 9.1 Ages for Selected Horizons

| Horizon | $P(68)$ Age (MY) | Gradstein *et al.* (1994) |
|---|---|---|
| Pliocene | 1.8 | 1.81 |
| M. Pliocene | 3.52 | 3.58 |
| Cretaceous (Maastrichtian) | 66.1 | 65.0 |
| Turonian | 89.0 | 88.9 |
| Cenomanian | 91.3 | 93.5 |
| Aptian | 112.7 | 112.2 |
| Jurassic | 144.0 | 144.2 |
| Callovian | 162.0 | 159.0 |
| Triassic | 212.0 | 205.7 |

## III. INFLUENCES OF TIMESCALES ON BASIN MODELING RESULTS

To illustrate any significant variations in evolution of basin modeling quantities as a direct consequence of different choices of timescales, in this section we consider the one-dimensional burial history of a North Sea well dominated by shale and sandy-shale deposition throughout most of its evolution. Five calculations were run with identical lithology inputs, surface depositional porosities, permeability–porosity, and effective stress–porosity equations of state. The only difference between the calculations was the timescale used, encompassing the five most recent timescales of the previous section. Excess pressure development and cumulative oil generation are two of the quantities more sensitive to variations in timescales, so we focus our illustration solely on them. The code used to examine the one-dimensional evolution of fluid overpressuring and cumulative oil generation with time is the 1D-DTIT code (dynamic and thermal indicator tomography), developed at the University of South Carolina.

### A. OVERPRESSURE DEVELOPMENT

For the first illustration, only the dynamic section of the code was activated in order to concentrate on excess pressure development due solely to undercompaction rather than any component produced by hydrocarbon generation. In addition, the critical fracturing coefficient for release of overpressure was set so that overpressure had to equal lithologic overburden before fracturing could occur. In this way overpressure is developed and retained in the sedimentary column as long as possible.

The stratigraphic column of the North Sea well consists of a basal thin (200 m) Triassic sand overlain by approximately 4000 m of pre-Turonian shale and sandy-shale formations, with about three-quarters shale and one-quarter sandy-shale, and a preponderance of the sandy-shale deposited in early Jurassic time. In Late Jurassic to Turonian time the deposition is effectively all shale with a thickness today of about 1000 m. From Turonian time to the present day, a further thickness of about 3000 m was deposited with about 80% being sandy-shale, and with thin shales (of order 200 m thick) deposited at both early Tertiary and Oligocene-Miocene times. The dominant sealing unit of the system is a 1000-m-thick shale formation laid down almost continuously and uniformly during the Late Jurassic to Turonian period. One anticipates that this unit will mark the onset time of significant overpressure in deeper lying formations, with a continued increase of excess pressure development through to the present day.

Figures 9.5a–e exhibit the development of excess pressure (in units of $kg \cdot cm^{-2} \cong 1.03$ atm) with burial for the five timescales of Harland *et al.* (1982) (Fig. 9.5a), Haq *et al.* (1987) (Fig. 9.5b), Harland *et al.* (1989) (Fig. 9.5c), GSA (1991) (Fig. 9.5d), and Gradstein *et al.* (1994) (Fig. 9.5e). Shown across the top of each figure are the age values for formation boundaries together with depositional lithologies. The color bar scale gives excess pressure in $kg \cdot cm^{-2}$.

Inspection of Figs. 9.5a–e shows that there is virtually no difference to be seen in the onset time or depth of overpressure, in the magnitude of maximum overpressure, in the positioning with respect to time and depth of contours of overpressure, or in the evolutionary thickness variations of formations with time and depth. The only significant difference we observe is a maximum time shift of about 6 to 7 MY for the onset of overpressure at about 155 MYBP, and this difference is negligible in relation to the later development of overpressure throughout the sedimentary column.

The conclusion that can be drawn from this example, which was chosen to maximize potential differences in excess pressure development (one of the more sensitive quantities in basin modeling), is that to within the uncertainty of current modern timescales, the choice of one timescale over another does not significantly influence the results from basin modeling.

## B. CUMULATIVE OIL GENERATION

In this illustrative example, the thermal history module of the 1D-DTIT code was enabled together with a hydrocarbon generation kinetic module to convert TOC to oil and/or gas. A constant heat flux for all time of 1 HFU (42 $mW/m^2$) and a constant surface temperature of 10°C were used in the computer model.

Precisely the same five age scales were used as in the overpressure history investigation with the same North Sea well. Figures 9.6a–e show the cumulative oil generation (in mg/g TOC) through the burial history. The color bar scale provides measures of contours for cumulative oil generation. Inspection of Figs. 9.6a–e shows virtually no differences in the total

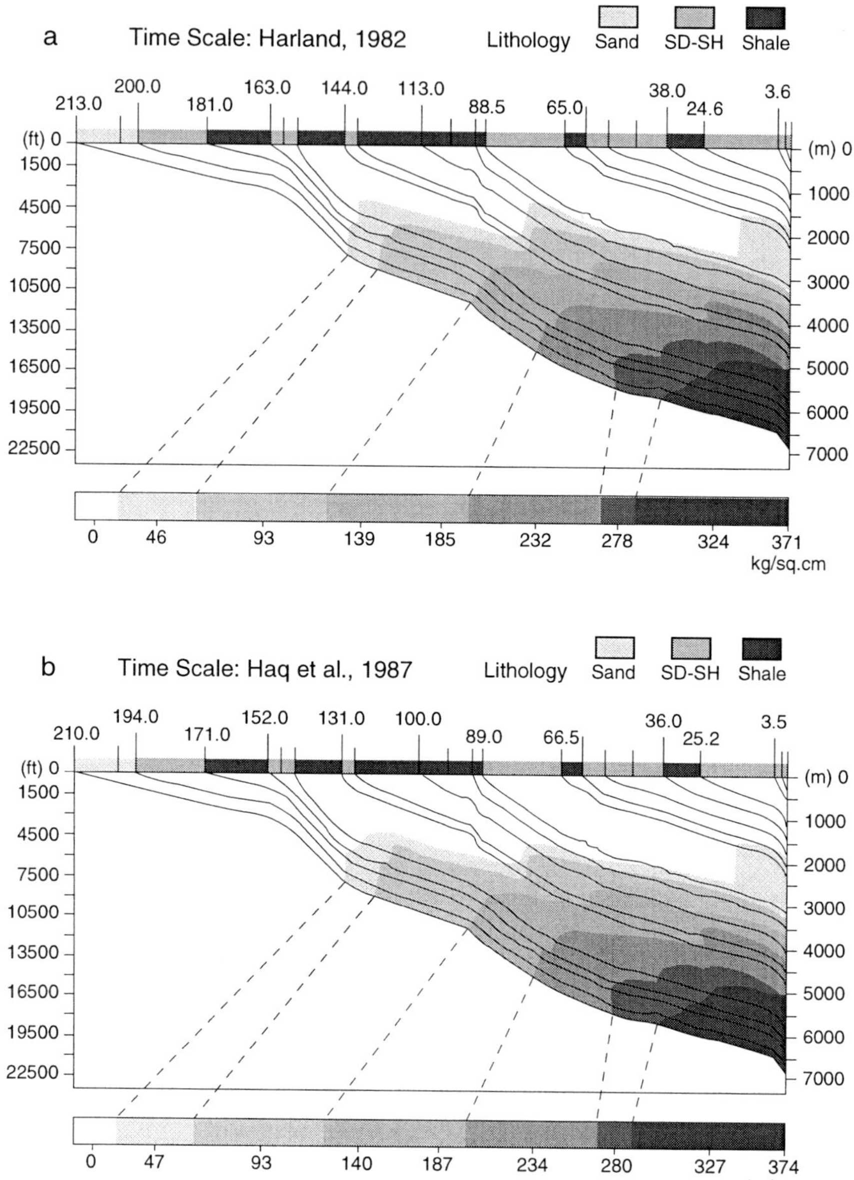


FIGURE 9.5 Excess fluid pressure ($kg \cdot cm^{-2}$) development with time and burial depth for a North Sea well with different timescale choices: (a) Harland *et al.* (1982); (b) Haq *et al.* (1987); (c) Harland *et al.* (1989); (d) GSA (1991); (e) Gradstein *et al.* (1994).

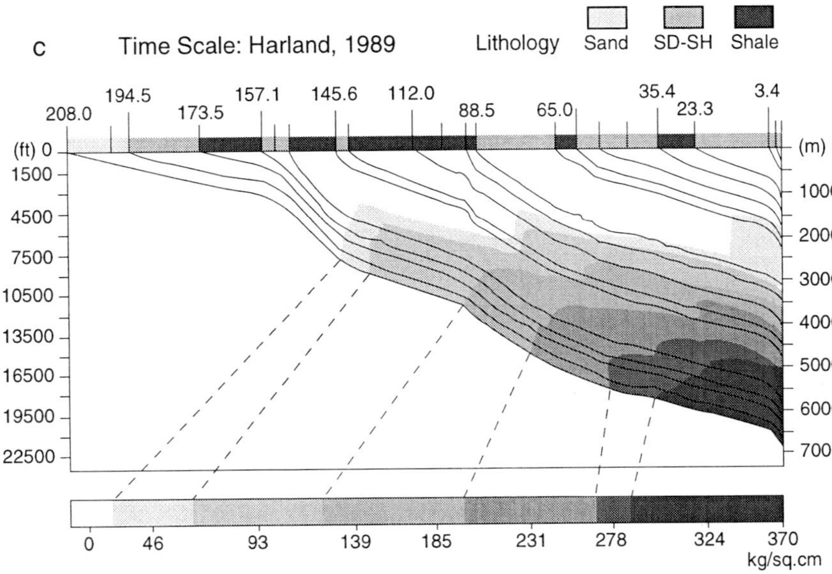


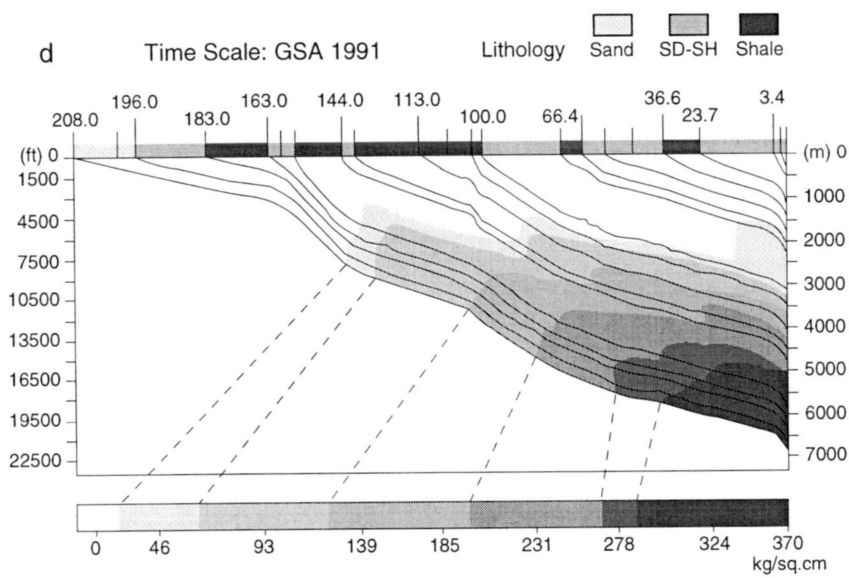


FIGURE 9.5 *(Continued)*

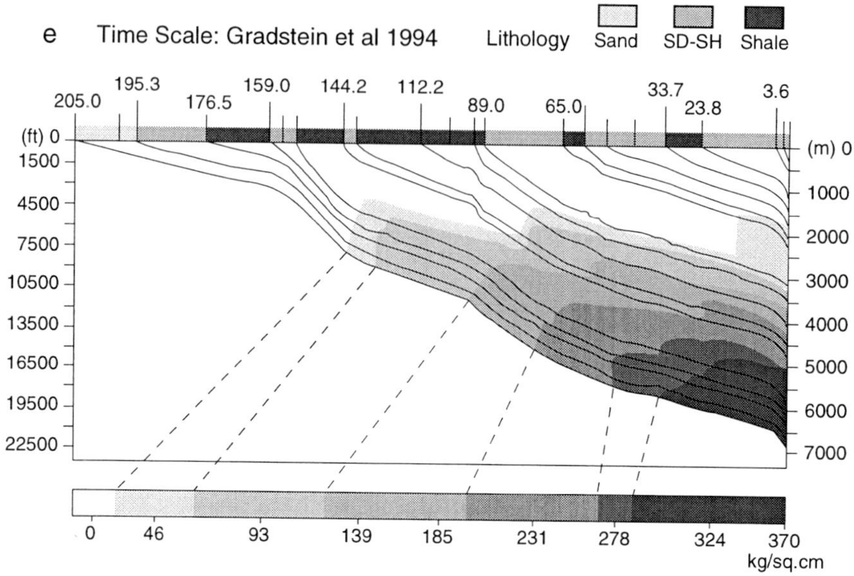


FIGURE 9.5 (*Continued*)

amounts, timing, depth ranges, or variations of the cumulative oil generation. There is a nominal numerical dynamic maximum range from 219 to 235 mg/g TOC, corresponding to a $\pm 3\%$ variation around the average maximum, which is negligible at the accuracy level of the calculation. Thus, even for the sensitive quantity of cumulative oil generation there is no difference, to within resolution, regardless of which timescale is chosen.

## IV. DISCUSSION

Three major conclusions follow from the work reported here:

1. Once different timescales have been evaluated within the framework of a cumulative probability plot, it is then simple to identify both the expected age, $P(68)$, of a horizon together with the 90% confidence interval [$P(10)$ and $P(90)$] around the mean expected age. For the five most recently published timescales, the $P(10)$ and $P(90)$ values are already within the resolution claimed by the individual timescale crafters.

2. It would appear that some sort of limit has been reached in attempts to refine timescales given that (a) the latest published timescale of Gradstein *et al.* (1994) lies squarely on the log-normal probability distribution of age scales formed from the four previous recent timescales, and (b) the age values for horizons from the Gradstein *et al.* (1994) timescale are remarkably

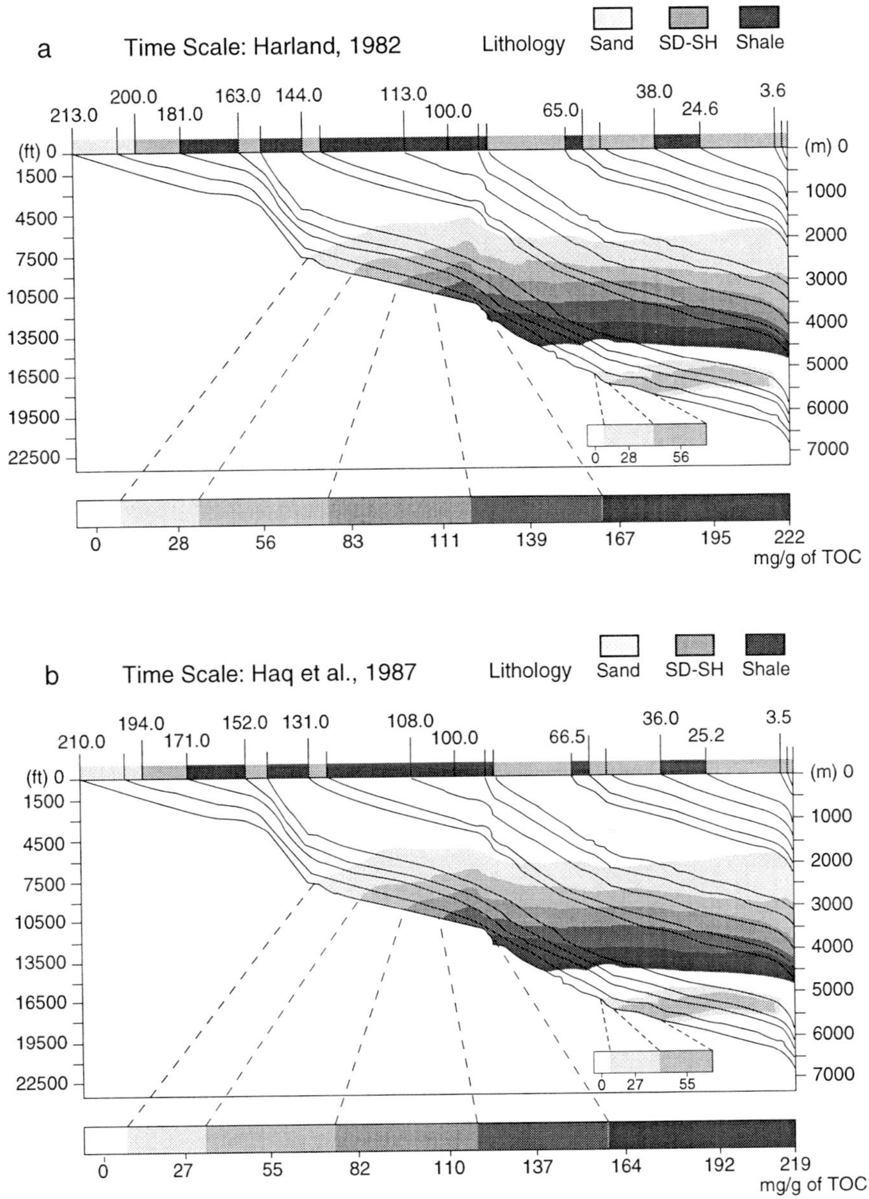


FIGURE 9.6 Cumulative oil generation (mg/g TOC) development with time and burial depth for a North Sea well with different timescale choices: (a) Harland *et al.* (1982); (b) Haq *et al.* (1987); (c) Harland *et al.* (1989); (d) GSA (1991); (e) Gradstein *et al.* (1994).

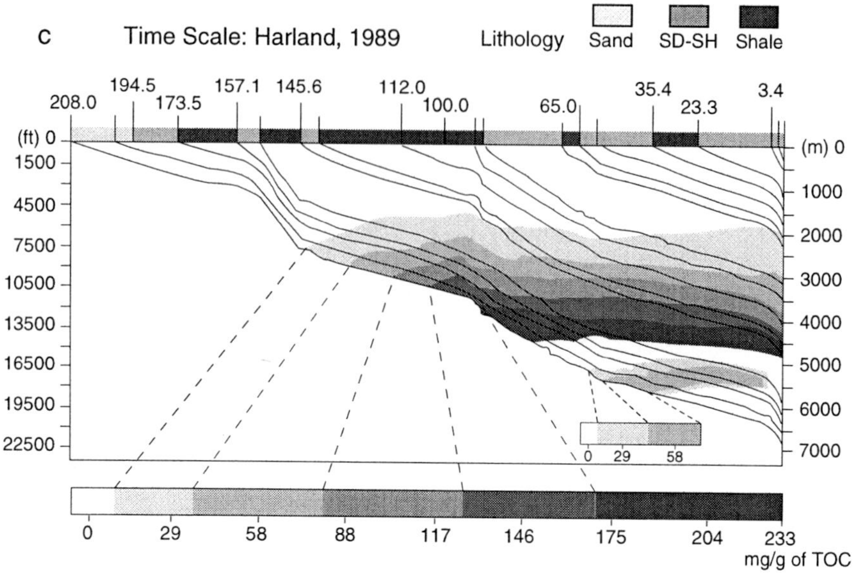


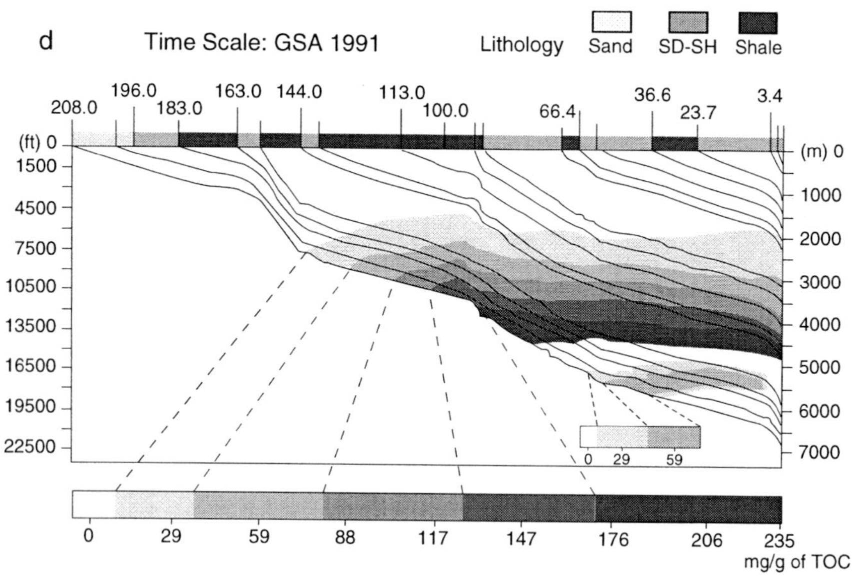


FIGURE 9.6 (*Continued*)

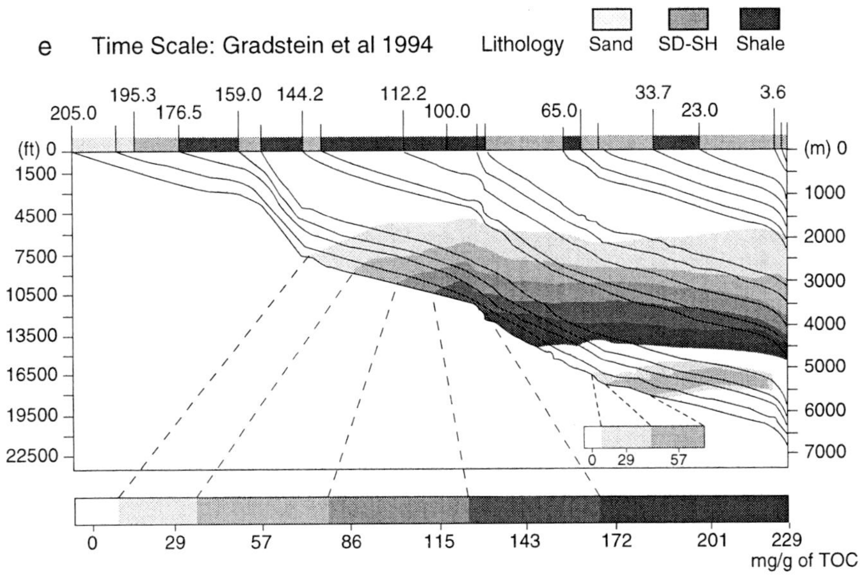


FIGURE 9.6 *(Continued)*

close to the $P(68)$ values that can be deduced directly from the prior four timescales.

3. It would seem that, to within the uncertainty ranges, the choice of timescale has little impact on results from basin modeling computations, as exemplified by analysis of excess pressure development and of cumulative oil generation with burial history.

# 10

# THE UNCERTAINTY OF INFORMATION

As noted in earlier chapters, comparison of model results with observations is the dominant theme in attempting to constrain model behaviors. However, the problem at issue is the reliability of the observations against which model predictions are to be compared and contrasted. Elsewhere (Lerche, 1990a) an investigation has been given of the reliability of thermal indicator data. Here we concentrate on information derived directly from downhole logs and inferentially from seismic data. The point to be made by illustration is about the methods used to assess reliability because porosity, water saturation, total organic carbon content (TOC), excess pressure, formation thicknesses, and so on are all used as constraints in the evaluation of paleo-evolution of a basin. To illustrate the main problems, we concentrate here on determinations of porosity, water saturation, and TOC content from well logs; and on the estimation of erosion from both well log and seismic information. (In Chapter 4 we already evaluated the uncertainty of erosion estimates from thermal indicator data.)

## I. LOG-DERIVED QUANTITIES

### A. INTRODUCTION

The accuracy with which one can determine quantities of physical interest, such as porosity, water saturation, total organic carbon content, etc., from individual or multiple downhole logs rests on the accuracy with which we can determine parameters entering the equations connecting log measurements with the physical quantities being sought, as well as on the accuracy of the measurements themselves. This remark is appropriate regardless of whether the log is resistivity, bulk density, neutron die-away, SP, or any other type.

For example, in a bulk density log, the measured bulk density, $\rho_b$, is usually connected to the porosity, $\phi$, through (Hearst and Nelson, 1985; Schlumberger, 1987)

$$\phi = (\rho_M - \rho_b)/(\rho_M - \rho_w), \tag{10.1}$$

where $\rho_M$ is the solid rock matrix density, and $\rho_w$ the water density. But the matrix density for different sedimentary lithologies varies. Values used typically range from $\rho_M = 2.65$ g/cm$^{-3}$ to $\rho_M = 2.85$ g/cm$^{-3}$ depending on the rock lithology supposedly present at the depth where the bulk density is measured. That rock lithology itself is usually a composite mixture of rock types so that there is always some uncertainty about the value to use for $\rho_M$. Likewise, the water density at depth depends on several factors. For instance, the chloride content of water in boreholes is highly variable. If the chloride content (in ppm) is taken to be reflected in the water density then $\rho_w$ can vary from 1.00 g/cm$^3$ (fresh water) to an extreme of 1.3 g/cm$^3$ for maximum brine saturation. Typical marine water is usually taken at $\rho_w = 1.03$ g/cm$^{-3}$ If the pores contain gas as well as water, then the equivalent water density in the pores will be $\rho_w S_w + \rho_g(1 - S_w)$ where $S_w$ is the fractional water saturation and $\rho_g$ is the gas density. Thus the water density itself can be highly variable.

Apart from uncertainties in the intrinsic parameters connecting well log measurements to quantities of physical interest, there is also uncertainty involved in the actual measured quantity itself. Thus in the case of the bulk density it is unlikely (Hearst and Nelson, 1985) that the bulk density is more accurate than about 5% so that a recorded value of, say, $\rho_b = 2.2$ g/cm$^{-3}$ could, in reality, lie in the range $2.1 \leq \rho_b \leq 2.3$ g/cm$^3$ at the 5% accuracy level. Thus, as an example, with $\rho_M$ as large as possible (2.85 g/cm$^3$), $\rho_b$ as low as possible (2.1 g/cm$^3$) for a nominal recorded value of 2.2 g/cm$^3$, and $\rho_w$ as low as possible (1.0 g/cm$^3$), it follows that the porosity estimate would be $\phi = 0.40$. If, however, $\rho_M$ is taken at its low value (2.65 g/cm$^3$), $\rho_b$ at its high value (2.3 g/cm$^3$), and $\rho_w$ at its high value (1.3 g/cm$^3$), then the estimate of porosity would be $\phi = 0.26$.

The point is that there will inevitably be a range of possible values of porosity from the well log. The importance of providing an estimate of the dynamic range of uncertainty of a physical attribute cannot be underscored too strongly, because knowledge of physical values is often the basis on which viable economic interest is assessed. An equal concern about error assessment of fractional water saturation, $S_w$, has been expressed by Wahlstrom *et al.* (1967), Moore and Kaufmann (1981), Hearst and Nelson (1985), Schlumberger (1987), Carpentier *et al.* (1989), and Passey *et al.* (1990) based on resistivity log and logs related to porosity. As noted in Hearst and Nelson (1985), Moore and Kaufmann (1981) "have discussed the problems in getting a good determination of $R_f$ and $R_w$. Both the resistivity determination itself and the extrapolation of a room temperature value to a formation temperature value are subject to error. They compared fluid resistivity measurements and found that a 25.7 percent difference existed in the $R_w$ value determined by two service companies. This error translates into a 15.5 percent error in $S_w$."

The point, then, is to provide a procedure for assessing the likely range of uncertainty on a physical quantity determined from a well log so that an appreciation is obtained of the appropriateness of using the physical values in endeavors related to geology and geological fluids. Wahlstrom *et al.* (1967) have discussed one method for evaluating errors based on numerical procedures involving Monte Carlo simulations. Water saturation $S_w$ is related to porosity $\phi$, resistivity $R_f$, water resistivity $R_w$, and formation factor $F$ through

$$S_w = (aR_w/R_f)^{1/n}\phi^{-m/n}, \tag{10.2}$$

where Archie's (1942) law has been used to relate porosity to resistivity through

$$R_f = R_w\phi^{-m}F \tag{10.3}$$

with the index $m$ of order 2, and where fractional water saturation $S_w$ is related to the formation factor through

$$F = aS_w^{-n} \tag{10.4}$$

with an index $n$ also taken to be of order 2. Equations (10.3) and (10.4) imply Eq. (10.2). The porosity was then determined from a bulk density log via Eq. (10.1). Wahlstrom *et al.* (1967) basically let each parameter have a range of possible values, for example, $n$ was chosen to vary between 1.8 and 2.2. Then a series of random Monte Carlo computer runs was performed with the value of each parameter being randomly assigned within its allowed range. The corresponding value of $S_w$ was then recorded. From the total series of runs, it was then possible for Wahlstrom *et al.* (1967) to record the cumulative probability of obtaining a value less than a specified $S_w$. The corresponding cumulative probability plot [modified from Wahlstrom *et al.* (1967) by being redrawn on log-probability paper] is given in Fig. 10.1. Note that a straight-line fit, as drawn, provides the information that there is a 90% chance that the fractional value of $S_w$ is greater than 0.29 and also a 90% chance that $S_w$ is less than 0.42, with a 68% probability of the value being less than 0.37, that is, one would assign

$$S_w = 0.37^{+0.05}_{-0.08}\% \tag{10.5}$$

at the 90% confidence level, based on the ranges assigned to each of the parameter values. Note from Fig. 10.1 that the behavior of $S_w$ with cumulative probability follows a log-normal distribution.

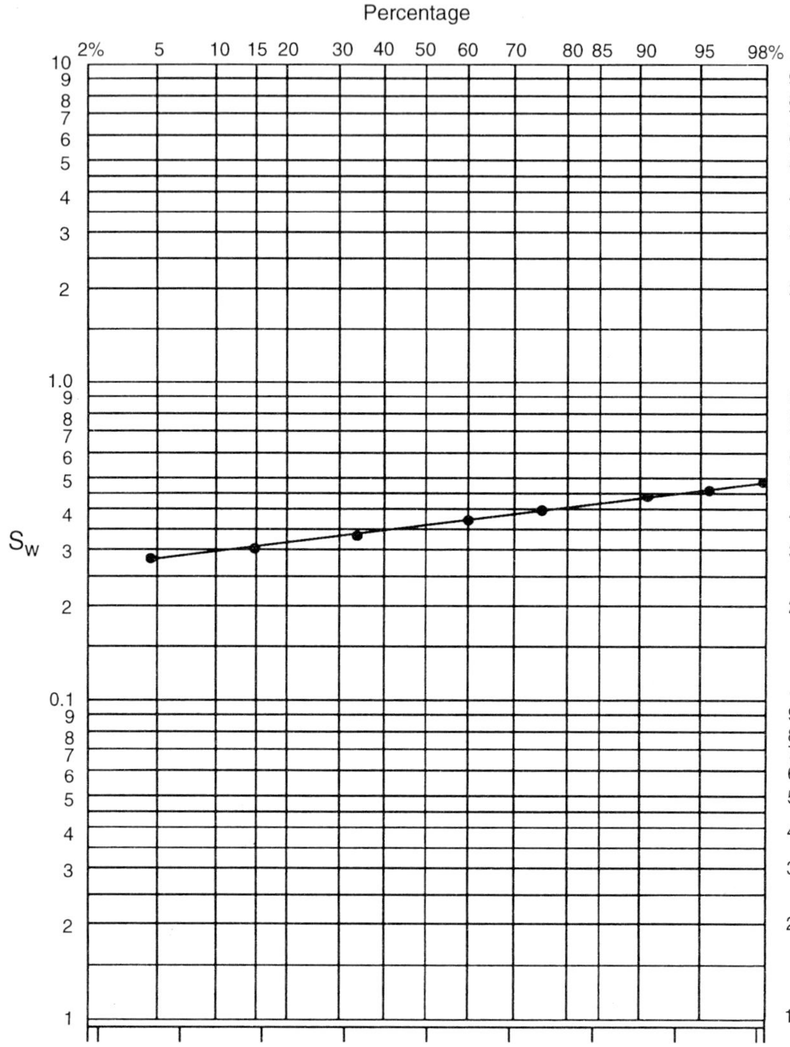


FIGURE 10.1 Cumulative probability plot of water saturation, drawn on probability paper, for the results reported by Wahlstrom *et al.* (1967).

While the Monte Carlo procedure advocated by Wahlstrom *et al.* (1967) is an effective method for assessing error, it suffers from the drawback that for each measured pair of resistivity, $R_F$, and bulk density values, $\rho_b$, at a given depth, a series of Monte Carlo runs must be made to assess error range on a physically required parameter. Such a numerical investigation can be extremely time consuming and may be part of the reason that such error determination is not yet routinely applied to well log derived properties. In its favor, we can remark that the Monte Carlo procedure is completely general and can be applied to any log response equation once the range and underlying probability distributions for each parameter *and* for the measurement uncertainties have been provided.

The question we address here is whether it is possible to produce a probabilistic assessment of error and uncertainty on quantities of physical interest *without* having to perform the tedious, time-consuming, Monte Carlo calculations for each measurement value with assigned ranges for intrinsic parameters and an assigned range of measurement uncertainty. Clearly, if such a procedure can be developed, then it becomes a relatively simple matter to assign error ranges to quantities of relevance derived from logs. The purpose of this chapter is to demonstrate that such a procedure is an achievable goal. The relevant mathematical manipulations used to provide a quantitative framework for assessing probabilistic measures of uncertainty were provided in Chapter 1. Armed with this quantitative methodology we provide two applications of the method, one to porosity determined from a bulk density log, and the second to water saturation determined from a resistivity log.

The general framework can be applied to any log, or any suite of logs, once the response equations between physical quantities and log measurements are available, together with range assessments for the intrinsic parameters connecting measured values to physical parameters. We should, however, point out that errors in derived parameters can also result from systematic effects such as washouts, thin beds, tool standoff, heavy mudcakes, mud additive changes, etc. Such systematic effects cannot easily be evaluated with the statistical treatment provided in the body of the chapter without major changes to allow for systematic "bias." Such allowances can be made but would take us far from the main point of this chapter, which is to exhibit the basic technique as cleanly as possible incorporating solely statistical error at this stage.

The balance to be struck between statistical and systematic effects in any given situation is one that is better evaluated with more complete well log information than we use here as illustrations of the procedure.

## 1. Porosity from Bulk Density

The response equation connecting porosity, $\phi$, and bulk density measurements, $\rho_b$, is Eq. (10.1), which we rewrite here as

$$\ln\phi = \ln(\rho_M - \rho_b) - \ln(\rho_M - \rho_w). \tag{10.6}$$

The matrix density is taken to vary in $2165 \leq \rho_M \leq 2.85$ with $\rho_{min} = 2.65$, $\rho_{max} = 2.85$, and a most likely value of $\rho_M = 2.75$ g/cm$^3$. The water density we take to vary in $1.0 \leq \rho_w \leq 1.05$ with $\rho_{w,min} = 1.0$, $\rho_{w,max} = 1.05$, and a most likely value of $\rho_{w,prob} = 1.03$ (corresponding to marine). We consider the bulk density measurements to have a fractional error $f$ such that

$\rho_{b,min} = (1 - f)\rho_{b*}$, $\rho_{b,max} = (1 + f)\rho_{b*}$, with $\rho_{b*}$ being the recorded value, which we take to be the most likely value. Then

$$\phi = \rho_M/(\rho_M - \rho_w) - \rho_b/(\rho_M - \rho_w). \tag{10.7}$$

Following the procedure given in Chapter 1 it follows that

$$E_1(\phi) = E_1(\rho_M/(\rho_M - \rho_w)) - E_1(\rho_b/(\rho_M - \rho_w)). \tag{10.8}$$

Now it also then follows by the same procedure that

$$E_1(\rho_M/(\rho_M - \rho_w)) \cong E_1(\rho_M)E_1(1/(\rho_M - \rho_w)) \tag{10.9a}$$

and

$$E_1(\rho_b/(\rho_M - \rho_w)) = E_1(\rho_b)E_1(1/(\rho_M - \rho_w)) \tag{10.9b}$$

so that

$$E_1(\phi) = E_1(1/(\rho_M - \rho_w))[E_1(\rho_M) - E_1(\rho_b)]. \tag{10.9c}$$

With $2.65 \leq \rho_M \leq 2.85$ and $1.0 \leq \rho_w \leq 1.05$ then $1.60 \leq \rho_M - \rho_w \leq 1.85$ so that $1/1.85 \leq 1/(\rho_M - \rho_w) \leq 1/1.60$, with a most likely value of about $1/1.72$ [$\equiv 1/(2.75 - 1.03)$].

Then $E_1(1/(\rho_M - \rho_w) = 0.58$ so that

$$E_1(\phi) = 0.58\,[2.75 - \rho_{b*}]. \tag{10.10}$$

It also follows that

$$E_2(\phi) \cong E_2(\rho_M - \rho_b)E_2(1/(\rho_M - \rho_w)). \tag{10.11}$$

Then

$$E_2(1/(\rho_M - \rho_w)) \cong 0.38, \tag{10.12a}$$

while

$$E_1(1/(\rho_M - \rho_w))^2 \cong 0.34. \tag{10.12b}$$

Equally

$$\begin{aligned} E_2(\rho_M - \rho_b) &= E_2(\rho_M) - 2E_1(\rho_M)E_1(\rho_b) + E_2(\rho_b^2) \\ &\cong 7.74 - 5.5\rho_{b*} + \frac{1}{6}(7 + f^2)\rho_{b*}^2. \end{aligned} \tag{10.13}$$

Then

$$\begin{aligned} \mu^2 &= \ln\{E_2(\phi)/E_1(\phi)^2\} \\ &= \ln\left\{0.38\frac{(7.74 - 5.5\rho_{b*} + \rho_{b*}^2(7 + f^2)/6}{0.34(2.75 - \rho_{b*})^2}\right\}. \end{aligned} \tag{10.14}$$

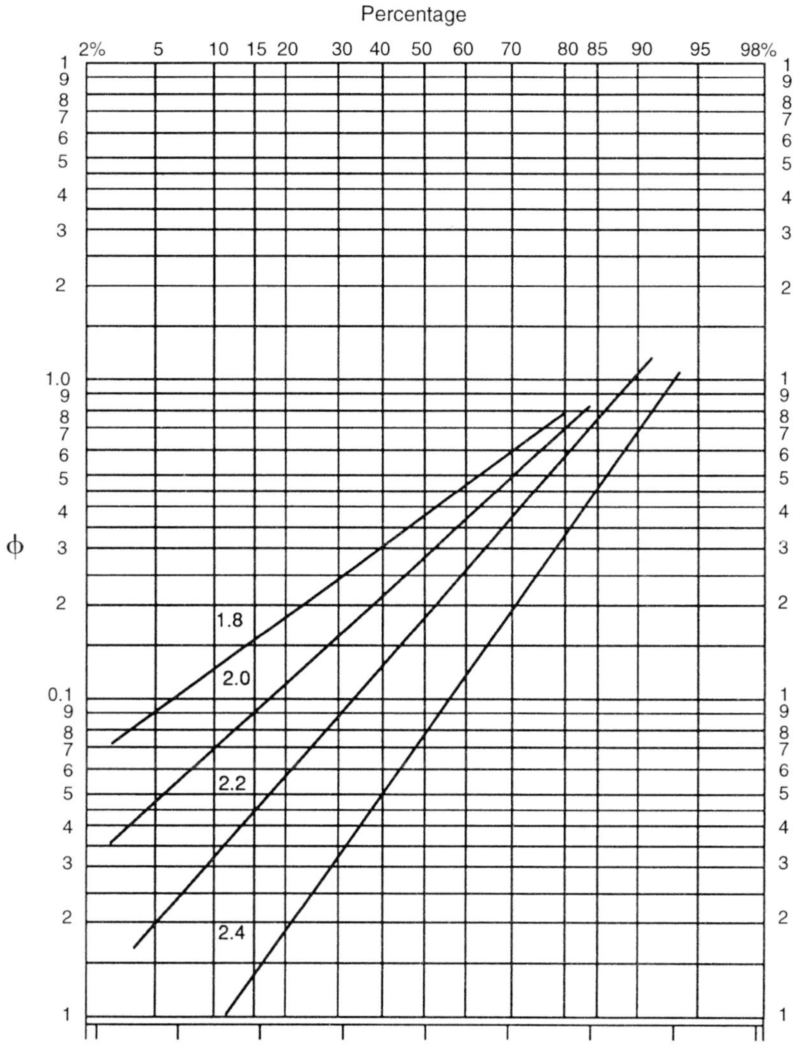


FIGURE 10.2 Cumulative probability plot for porosity for different values of bulk density ($\rho_b$ = 1.8, 2.0, 2.2, and 2.4 g/cm$^3$) using the uncertainty ranges of intrinsic parameters specified in the text.

The 68% cumulative probability point for porosity then occurs on about

$$\phi_{68} = E_1(\phi)\left\{0.38\frac{0.34(2.75-\rho_{b*})^2}{(7.74-5.5\rho_{b*}+\rho_{b*}^2(7+f^2)/6)}\right\}^{3/2}. \qquad (10.15)$$

The lines on Fig. 10.2 show the cumulative probability of porosity for various nominal measurements of $\rho_{b*}$ assuming $f = 0.1$, that is, 10% fractional error

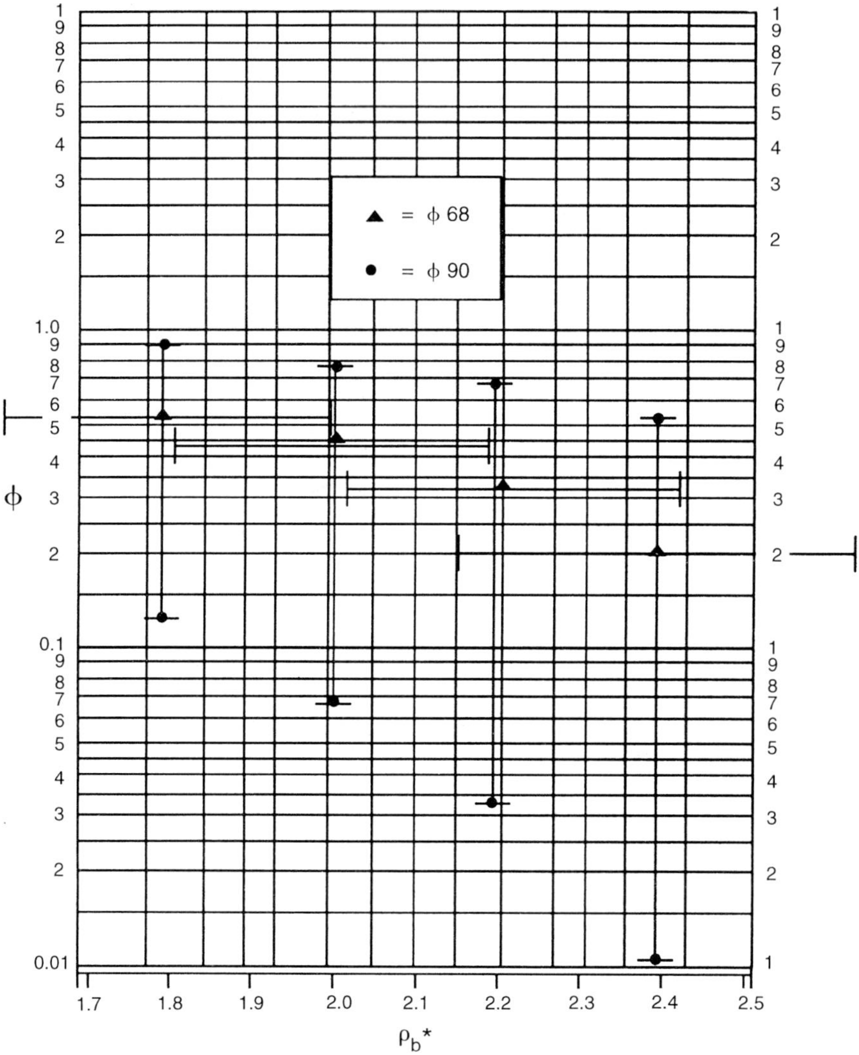


FIGURE 10.3 Cross-plot of estimated porosity versus bulk density, including the 90% likelihood values obtained from the cumulative probability plot of Fig. 10.4. The 10% error assessment on a bulk density measurement value is also plotted as an error assessment.

on the measurements of bulk density. Reading off from the lines of Fig. 10.2 the 10% and 90% cumulative probability values then enable a plot to be made of expected (68% cumulative probability) porosity values, together with the 90% confidence intervals, as a function of the nominal recorded value of $\rho_{b*}$. Such a plot is given in Fig. 10.3 for the parameter ranges assessed. Note that the relative accuracy of determining porosity *decreases* as the mean (68% point) porosity decreases and that, for the illustrative parameter ranges chosen, there is still an extremely broad uncertainty on the porosity value—perhaps such an uncertainty accounts for some of the scatter and differences recorded in porosity with depth plots as measured by different groups or with different devices or with different assumptions on the intrinsic parameter values used to convert a logging response to porosity.

It can, of course, be argued that the parameter ranges chosen in this illustrative example are too broad and that in more practical case histories a much narrower range is appropriate for each and every parameter. Perhaps so, but the point of the illustration was to demonstrate how, with a given choice of ranges, it was possible to construct analytically an estimate of the relevant likely range of determinable physical values for porosity. For each specified choice of parameter ranges, it is a simple matter to perform the preceding analysis very quickly and so obtain the appropriate error assessment. Indeed, a pocket calculator program quickly performs the required calculations as in the manner given earlier.

**2. Water Saturation from Resistivity**

As recorded in Eq. (10.2), fractional water saturation is calculated from

$$\ln S_w = 1/n\{-m \ln \phi + \ln a + \ln R_w - \ln R_f\}, \tag{10.16}$$

where $R_f$ is the measured resistivity (usually in $\Omega \cdot \text{m}$), $a$ is the formation factor at 100% water saturation, $R_w$ is the water resistivity, and $\phi$ is the porosity estimated or inferred from another log. The parameters $m$ and $n$ are usually set to precisely 2 (Schlumberger, 1987; Carpentier *et al.*, 1989; Passey *et al.*, 1990) despite unequivocal evidence that a range of values for $m$ and $n$ is necessary centered on 2 but varying typically from 1.8 to 2.2 (see, e.g., many references to experimental determinations in Hearst and Nelson, 1985). For illustrative purposes take $m$ to vary in $1.8 \leq m \leq 2.2$ with a most likely value of 2; take $n$ to vary in the same range as $m$, also with a most likely value of 2; take the formation porosity to vary in $0.1 \leq \phi \leq 0.3$ with a most likely value of 0.2; take the water resistivity to vary in $0.8 \leq R_w \leq 2.4\ \Omega \cdot \text{m}$ with a most likely value of $1\ \Omega \cdot \text{m}$; and take the formation factor $a$ to vary in $0.5 \leq a \leq 0.7$ with a most likely value of 0.6. These parameter values, gleaned from Hearst and Nelson (1985), are chosen because they are not atypical of ranges used in individual well logging situations. We also assume that the measured resistivity, $R_f$, has a fixed fractional logarithmic uncertainty $f$, that is, $(1 - f) \ln R_{f*} \leq \ln R_f \leq (1 + f) \ln R_{f*}$ where $R_{f*}$ is the nominal measured value. Then from Eq. (10.16), we have

$$E_1(\ln S_w) = E_1(1/n)\{-E_1(m)E_1(\ln \phi) + E_1(\ln a) + E_1(\ln R_w) - E_1(\ln R_f)\} \tag{10.17}$$

and

$$\begin{aligned} E_2(\ln S_w) = E_2(1/n)\{&-E_2(m)E_2(\ln \phi) + 2E_1(m)E_1(\ln \phi) \\ &\times E_1(\ln a) + E_1(\ln R_w) - E_1(\ln R_f) \\ &+ E_2(\ln a) + 2E_1(\ln a)(E_1 \ln R_w - E_1 \ln R_f) \\ &+ E_2(\ln R_w)E_2(\ln R_f) - 2E_1(\ln R_w)E_1(\ln R_f)\}. \end{aligned} \tag{10.18}$$

For the ranges given, we have

$$\begin{array}{lll} E_1(1/n) = 0.5; & E_1(m) = 2; & E_1(\ln \phi) = -1.703; \\ E_1(\ln a) = -0.44; & E_1(\ln R_w) = 0.217; & E_1(\ln R_f) = \ln R_{f*}; \\ E_2(1/n) = 0.25; & E_1(m) = 4.01; & E_1(\ln \phi) = 2.951; \\ E_2(\ln a) = 0.162; & E_2(\ln R_w) = 0.104; & E_2(\ln R_f) = (7 + f^2)(\ln R_{f*})^2/6; \end{array}$$

so that

$$E_1(\ln S_w) = 1.59 - \frac{1}{2}\ln R_f, \tag{10.19a}$$

$$E_2(\ln S_w) = 3.36 + 1.91 \ln R_f + \frac{(7 + f^2)}{24}(\ln R_f)^2. \tag{10.19b}$$

Then the 68% cumulative probability point occurs on about

$$\ln(1/S_w) = \frac{1}{2}\ln R_f - 1.59, \tag{10.20}$$

while the 16% cumulative probability point occurs on

$$\ln(1/S_w) = \left(\frac{1}{2}\ln R_f - 1.59\right)\left|1.59 - \frac{1}{2}\ln R_f\right|^3 \Big/ \left(3.36 + 1.91 \ln R_f + \frac{(7 + f^2)}{24} \times (\ln R_f)^2\right)^{3/2}. \tag{10.21}$$

For 10% fractional error on log resistivity measurements then $f = 0.1$.

A plot of cumulative probability of water saturation is given in Fig. 10.4 for specific values of resistivity. Note that the 90% likelihood points for water saturation, together with the average (68% probability) point for water saturation, then yield the curve shown in Fig. 10.5. Note from Fig. 10.5 that the fractional saturation has a greater chance of being found in a range that systematically decreases as the measured resistivity increases. Perhaps the salient point is that the 90% likelihood of obtaining as range of water saturation decreases with increasing resistivity, as recorded in Fig. 10.4. The variation in resistivity (and therefore in water saturation) can be assessed rapidly analytically once a set of ranges is provided for each of the parameters that enters into the conversion of log derived measurement values of resistivity to porosity, and the variation of the measurements is also easily included.

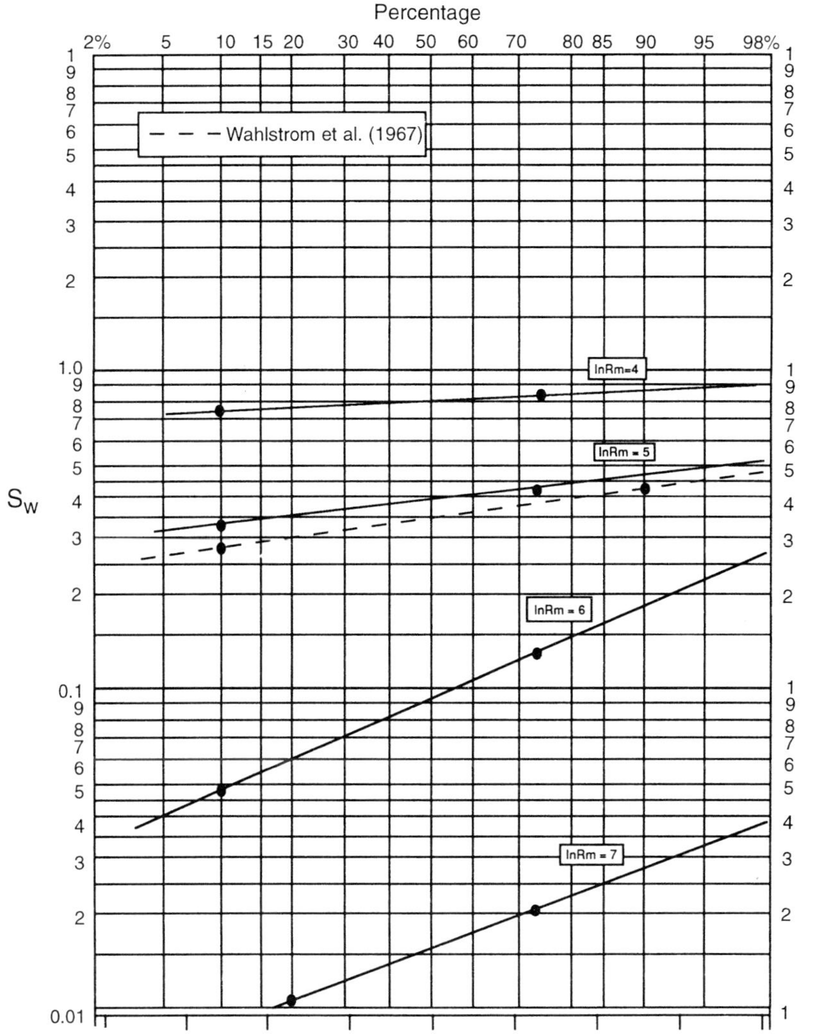


FIGURE 10.4 Cumulative probability plot for water saturation for different values of resistivity ($\ln R_f = 4, 5, 6, 7\ \Omega \cdot m$) using the uncertainty ranges of intrinsic parameters specified in the text. The superposed dashed line is the cumulative probability for water saturation obtained by Walstrom *et al.* (1967) using different parameter values and ranges.

Although the specific ranges of parameters chosen for illustrative purposes vary for each particular geologic setting and usesr-preference values for parameters, the point is that the log-normal distribution behavior, recorded by Wahlstrom *et al.* (1967) using time-consuming Monte Carlo methods, can also be obtained from analytic procedures as outlined here

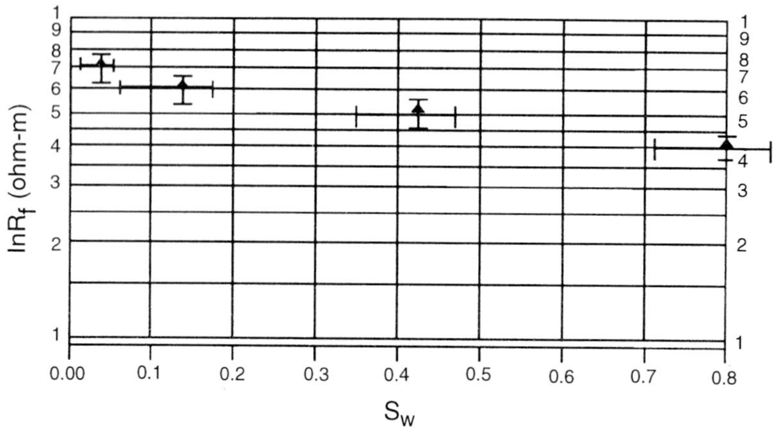


FIGURE 10.5 Cross-plot of estimated water saturation, including the 80% likelihood values, obtained from the cumulative probability plot of Fig. 10.6. The 10% error assessment of measured resistivity values is also plotted as an error assessment.

for each set of measured values of $R_f$ and $\phi$, thereby avoiding the tedious numerical methods of random variations.

### 3. Total Organic Carbon Content from Sonic and Resistivity Logs

There has long been an interest in assessing the total organic carbon (TOC) content of potential source rocks in sedimentary basins in order to better define the risk factor associated with drilling for hydrocarbon accumulations (see, e.g., the many references in Hearst and Nelson, 1985). Recently Carpentier *et al.* (1989) and Passey *et al.* (1990) have examined the possibility of estimating the *in situ* amount of organic matter content of sediments directly from wireline logs, in particular from sonic and resistivity logs. The purpose of this section is several-fold: First we provide an overview of the responses of different logs to physical properties of the sediments related to hydrocarbons and then illustrate how to extract quantities of physical interest from the logs. Second, we provide a procedure for assessing uncertainty in the quantities derived from the logs and also to provide a method for assessing the best parameter values to use in converting log responses to quantities of physical interest. Third, we examine the procedure suggested by Carpentier *et al.* (1989) and that by Passey *et al.* (1990) in light of the general framework of the first two points. As we shall see, the ability to honor the relevant uncertainty criteria, and thus provide a reasonably accurate measure of TOC, is not as simple as it might appear to be at first glance.

#### *a. Log Responses: An Overview*

As noted in the Schlumberger volume, *Log Interpretation Principles/Applications* (1987), by Hearst and Nelson (1985) and by Stocks and Lawrence (1990), many logs have some response to the presence of organic matter. As Hearst and Nelson (1985), among others, point out carbon-to-oxygen logs (C/O) and direct geochemical logs are usually the mainstay for TOC estimates. But when such specialist logs are not available, other more customarily available logs are often used to infer TOC indirectly. Four logs taken in concert appear to provide some dominant measure of organic matter: logs generally labeled gamma ray, resistivity, acoustic, neutron, and density devices. Two problems that need to be addressed are, first, the veracity of the basic assumptions that convert response measurements to physical quantities and, second, the estimates needed of basic quantitative parameters that enter the underlying equations connecting log response to a physical variable. There are also the problems of intrinsic measurement error and reproducibility, as well as of the finite sampling interval of log responses with borehole depth. These last two problems have been addressed in fairly extensive detail by White (1984), Hearst and Nelson (1985), Williams *et al.* (1988), and Cressie (1991), and we refer the reader to those articles for an appreciation of the statistical treatments involved.

Our concern in this section is more with the intrinsic assumptions made and explicit parameter values used to obtain quantitative assessments of physically relevant quantities from log response characteristics. In view of the current concentration on resistivity and sonic logs in estimating TOC (fractional total organic carbon content) for a formation, or with borehole depth as illuminated by the works of Carpentier *et al.* (1989) and Passey *et al.* (1990), we concentrate here on these two logs. We point out, however, that because the combination of sonic and resistivity logs reacts only indirectly to TOC the combination should be used only if more direct measures (such as from C/O logs) are not available.

*Resistivity Log Response.* Details of tool design and resistivity determination are provided in the Schlumberger (1987) manual as well as in Hearst and Nelson (1985). For our purpose we take it that resistivity, $R$, has been measured at a borehole depth $z$. The connection between resistivity and formation porosity, $\phi$, is usually made through Archie's (1942) empirical law

$$R = R_w F \phi^{-m}, \tag{10.22}$$

where $F$ is the formation factor, $R_w$ is water resistivity, and $m$ is a lithology-dependent exponent, conventionally set at about $m = 2$. For later convenience in manipulation we leave $m$ as a free parameter. It follows that Eq. (10.22) can be written

$$\ln R/R_w = -m \ln \phi + \ln F. \tag{10.23}$$

The intrinsic assumption here is that the porosity is water filled. The problem then is to express the water porosity in terms of some other suitable measurable log value, and to then relate the log values to TOC content.

The implicit assumption made is that there is no direct TOC response in Archie's law or any of its variants. Because both Carpentier *et al.* (1989) and Passey *et al.* (1990) concentrate on the sonic log, we follow their lead in order to provide a conventional framework for interpretation of results.

*Sonic Log.* As with the resistivity log, details of tool design and conversion of measurements to a transit time per foot, $\Delta t$, are available in Schlumberger (1987) and Hearst and Nelson (1985). For our purposes it suffices to note that Carpentier *et al.* (1989) and Passey *et al.* (1990) both use a transit time written in terms of a Wyllie *et al.* (1956) formula in the form

$$\Delta t = \Delta t_{\mathrm{m}} V_{\mathrm{m}} + \sum_{i=1}^{N} \Delta t_i V_i + \Delta t_{\mathrm{OM}} V_{\mathrm{OM}} + \Delta t_{\mathrm{w}} V_{\mathrm{w}}, \tag{10.24}$$

where $\Delta t_{\mathrm{m}}$ is the intrinsic transit time across 100% matrix material, $\Delta t_i$ ($i = 1, \ldots, N$) is the intrinsic transit time across lithology $i$ (shale, sand, clay, etc.), $\Delta t_{\mathrm{OM}}$ is the intrinsic transit time across 100% organic material, and $\Delta t_{\mathrm{w}}$ is that across water. Following the preceding references, the matrix response of acoustic propagation to TOC is taken to be virtually independent of TOC. The positive volume fractions $V_{\mathrm{m}}$, $V_i$, $V_{\mathrm{OM}}$, and $V_{\mathrm{w}}$ refer to a given unit being measured with

$$V_{\mathrm{w}} + V_{\mathrm{m}} + V_{\mathrm{OM}} + \sum_{i=1}^{N} V_i = 1, \tag{10.25}$$

so that Eq. (10.24) can be rewritten

$$\Delta t = \Delta t_{\mathrm{m}} + \sum_{i=1}^{N} (\Delta t_i - \Delta t_{\mathrm{m}}) V_i + (\Delta t_{\mathrm{OM}} - \Delta t_{\mathrm{m}}) V_{\mathrm{OM}} + (\Delta t_{\mathrm{w}} - \Delta t_{\mathrm{m}}) V_{\mathrm{w}}. \tag{10.26}$$

For porosity, which is 100% water filled, it follows that $V_{\mathrm{w}} = \phi$. Under this condition Eq. (10.26) can be rearranged to yield

$$\phi = [\Delta t_{\mathrm{w}} - \Delta t_{\mathrm{m}}]^{-1} \left\{ \Delta t - \Delta t_{\mathrm{m}} - \sum_{i=1}^{N} (\Delta t_i - \Delta t_{\mathrm{m}}) V_i - (\Delta t_{\mathrm{OM}} - \Delta t_{\mathrm{m}}) V_{\mathrm{OM}} \right\}. \tag{10.27}$$

*Combining Resistivity and Transit Time Responses.* Equation (10.27) for porosity can be substituted into Eq. (10.23) for log resistivity to yield

$$\ln R/R_{\mathrm{w}} = m \ln \left\{ \Delta t - \Delta t_{\mathrm{m}} - \sum_{i=1}^{N} (\Delta t_i - \Delta t_{\mathrm{m}}) V_i - (\Delta t_{\mathrm{OM}} - \Delta t_{\mathrm{m}}) V_{\mathrm{OM}} \right\} + \ln F + m \ln [\Delta t_{\mathrm{w}} - \Delta t_{\mathrm{m}}]. \tag{10.28}$$

Consider now a given range of $\Delta t$ values in $\Delta t_{\max} > \Delta t > \Delta t_{\min}$. Then, with $0 \leq u \leq 1$ and

$$\Delta t = \Delta t_{\max} - (\Delta t_{\max} - \Delta t_{\min})u, \tag{10.29}$$

the right-hand side of Eq. (10.28) can be rearranged to read

$$\begin{aligned}\ln R/R_{\mathrm{w}} = -m \ln \bigg\{ \Delta t_{\max} - \Delta t_{\mathrm{m}} - \sum_{i=1}^{N} (\Delta t_i - \Delta t_{\mathrm{m}}) V_i - u(\Delta t_{\max} - \Delta t_{\min}) \\ - V_{\mathrm{OM}}(\Delta t_{\mathrm{OM}} - \Delta t_{\mathrm{m}}) \bigg\} + \ln F + m \ln[\Delta t_{\mathrm{w}} - \Delta t_{\mathrm{m}}].\end{aligned} \tag{10.30}$$

For brevity, set

$$\Delta t_* = \Delta t_{\max} - \Delta t_{\mathrm{m}} - \sum_{i=1}^{N} (\Delta t_i - \Delta t_{\mathrm{m}}) V_i \qquad \text{with } \Delta t_* > 0. \tag{10.31}$$

Choose the range $|\Delta t_{\max} - \Delta t_{\min}| \ll \Delta t_*$ and also assume $|V_{\mathrm{OM}}(\Delta t_{\mathrm{OM}} - \Delta t_{\mathrm{m}})| \ll \Delta t_*$.

Then expand the first logarithmic factor on the right-hand side of Eq. (10.30) to obtain

$$\begin{aligned}\ln R/R_{\mathrm{w}} \cong m \ln \left[ \frac{\Delta t_{\mathrm{w}} - \Delta t_{\mathrm{m}}}{\Delta t_*} \right] + \ln F \\ + \frac{m}{\Delta t_*} [(\Delta t_{\max} - \Delta t_{\mathrm{m}})u + V_{\mathrm{OM}}(\Delta t_{\mathrm{OM}} - \Delta t_{\mathrm{m}})].\end{aligned} \tag{10.32}$$

Equation (10.32) has the form

$$\begin{aligned}\ln R/R_{\mathrm{w}} = \ln F + m \ln \left( \frac{\Delta t_{\mathrm{w}} - \Delta t_{\mathrm{m}}}{\Delta t_*} \right) + m \frac{\Delta t_{\max}}{\Delta t_*} \\ - m \frac{\Delta t}{\Delta t_*} + m \frac{(\Delta t_{\mathrm{OM}} - \Delta t_{\mathrm{m}})}{\Delta t_*} \frac{\mathrm{TOC}}{\rho_{\mathrm{OM}}},\end{aligned} \tag{10.33}$$

where we have replaced the volume fraction of organic matter $V_{\mathrm{OM}}$ by the weight fraction, TOC, divided by the organic matter density, $\rho_{\mathrm{OM}}$.

Now rewrite Eq. (10.33) as follows. Set the constant values $R_{\mathrm{b}}$ and $\Delta t_{\mathrm{b}}$ through the relation

$$\ln R_{\mathrm{b}} + m \frac{\Delta t_{\mathrm{b}}}{\Delta t_*} = \ln R_{\mathrm{w}} + \ln F + m \ln \left( \frac{\Delta t_{\mathrm{w}} - \Delta t_{\mathrm{m}}}{\Delta t_*} \right) + m \frac{\Delta t_{\max}}{\Delta t_*}. \tag{10.34}$$

Then Eq. (10.33) takes the form

$$\ln(R/R_{\mathrm{b}}) = \frac{-m}{\Delta t_*} (\Delta t - \Delta t_{\mathrm{b}}) + m \frac{(\Delta t_{\mathrm{OM}} - \Delta t_{\mathrm{m}})}{\Delta t_*} \frac{\mathrm{TOC}}{\rho_{\mathrm{OM}}}. \tag{10.35}$$

Rearranging Eq. (10.35) we then have

$$\mathrm{TOC} = \rho_{\mathrm{OM}} \times \frac{\Delta t_*}{(\Delta t_{\mathrm{OM}} - \Delta t_{\mathrm{m}})} \left[ \ln R/R_{\mathrm{b}} + \frac{m}{\Delta t_*}(\Delta t - \Delta t_{\mathrm{b}}) \right]. \quad (10.36)$$

Equation (10.36) provides a general relation relating organic matter content to resistivity and transit time measurements under specific conditions. Several of these conditions follow:

1. $|\Delta t_{\mathrm{max}} - \Delta t_{\mathrm{min}}| \ll \Delta t_*$. (10.37a)
2. $|\mathrm{TOC}\ (\Delta t_{\mathrm{OM}} - \Delta t_{\mathrm{m}})| \ll \rho_{\mathrm{OM}} \Delta t_*$. (10.37b)
3. $\Delta t_* > 0$; that is,

$$\Delta t_* = \Delta t_{\mathrm{max}} - \Delta t_{\mathrm{m}} - \sum_{i=1}^{N} V_i(\Delta t_i - \Delta t_{\mathrm{m}}) > 0. \quad (10.37c)$$

4. Parameter values can be assigned for

$$m, \Delta t_{\mathrm{max}}, \Delta t_{\mathrm{min}}, \Delta t_{\mathrm{OM}}, \rho_{\mathrm{OM}}, V_i, \Delta t_i, \Delta t_{\mathrm{m}}, R_{\mathrm{b}}, \Delta t_{\mathrm{b}}. \quad (10.37d)$$

5. The value of TOC [from Eq. (10.36)] is positive, that is,

$$\ln R/R_{\mathrm{b}} + \frac{m}{\Delta t_*}(\Delta t - \Delta t_{\mathrm{b}}) \gtrless 0 \qquad \text{according to } \Delta t_{\mathrm{OM}} \gtrless \Delta t_{\mathrm{m}}. \quad (10.37e)$$

6. Extraneous factors such as cave-ins, etc., do not make invalid the assumptions of the veracity of Archie's law for resistivity and of Wyllie's law for transit time.
7. Maximum and minimum velocity values must not change at each and every depth.

*Parameter Assessments.* The problem that needs solving is to obtain values for $R_{\mathrm{b}}$, $m/\Delta t_*$, $\Delta t_{\mathrm{b}}$, $\rho_{\mathrm{OM}}$, and $\Delta t_{\mathrm{OM}} - \Delta t_{\mathrm{m}}$, for use in Eq. (10.36). The measurements of resistivity and transit time would then automatically prescribe a value for TOC. The constraints of Eqs. (10.37) must be in place in such an assessment.

There seems to be a wide range of basic petrophysical parameter values from which one can obtain the quantities needed. Inspection of formula (10.36) shows that the parameters split into two groups: the three values $R_{\mathrm{b}}$, $\Delta t_{\mathrm{b}}$, $m/\Delta t_*$, which are independent of organic matter and are solely either user determined ($R_{\mathrm{b}}$ and $\Delta t_{\mathrm{b}}$) or are related to intrinsic petrophysical characteristics ($m/\Delta t_*$); and $\rho_{\mathrm{OM}}$, $\Delta t_{\mathrm{OM}} - \Delta t_{\mathrm{m}}$, which depend on organic matter properties. While $m$ is a directly measurable parameter for each lithology, $\Delta t_*$ is a composite [through Eq. (10.31)] of rock property transit times and the user value chosen for $\Delta t_{\mathrm{max}}$. Thus, uncertainties in intrinsic rock transit time values, and in the user choice of $\Delta t_{\mathrm{max}}$, propagate through to uncertainties on $\Delta t_*$.

TABLE 10.1 Parameter Values

| Lithology | $\Delta t_m$ ($\mu$s/m) | $\Delta t_f$ ($\mu$s/m) | $m$ |
|---|---|---|---|
| Sand | 182 (P) | 620 (P) | 2 (P,C) |
| Shale | 216 (P) | 620 (P) | 2 (P,C) |
| Limestone | 156 (P) | 620 (P) | 2 (P,C) |
| Dolomite | 143 (P) | 620 (P) | 2 (P,C) |

| | |
|---|---|
| $\Delta t_{\rm OM} = 1100\ \mu$s/m (C1) | $\Delta t_{\rm OM} - \Delta t_m = 324\ \mu$s/m (C2) |
| $\Delta t_{\rm max} = 400\ \mu$s/m (P) | $= 924\ \mu$s/m (C1) |
| $\Delta t_{\rm min} = 210 \pm 14\ \mu$s/m (P) | $\Delta t_{\rm OM} = 415\ \mu$s/m (C2) |
| $\Delta t_* = 300\ \mu$s/m (P) | $\Delta t_{\rm OM} = 1350\ \mu$s/m (C3) |
| | $\Delta t_{\rm OM} = 1100\ \mu$s/m (C4) |

*Key:* P = Passey *et al.* (1990); C = Carpentier *et al.* (1989).

To provide some idea of the parameter values that have been ascribed by different workers, in Table 10.1 we provide a synthesis of rock property matrix transit times used by four different references (Hearst and Nelson, 1985; Schlumberger, 1987; Carpentier *et al.,* 1989; Passey *et al.,* 1990). Equally, an uncertainty in $\Delta t_{\rm OM} - \Delta t_{\rm m}$ can be assessed from the four examples given in Carpentier *et al.* (1989) who record $\Delta t_{\rm OM}$ (example 1) = 1100 $\mu$s/m; $\Delta t_{\rm OM}$ (eample 2) = 415 $\mu$s/m; $\Delta t_{\rm OM}$ (example 3) = 1350 $\mu$s/m; $\Delta t_{\rm OM}$ (example 4) = 1100 $\mu$s/m. Again, the uncertainty in Archie's exponent $m$, is variable. Both Carpentier *et al.* (1989) and Passey *et al.* (1990) arbitrarily set $m = 2$ (with no quoted uncertainty), whereas Hearst and Nelson (1985) argue that $m$ varies from around 1.45 to around 2.2 depending on lithology and even for the same lithology.

For $\Delta t_*$, Passey *et al.* (1990) use the value 300 $\mu$s/m with no assigned error, together with $\Delta t_{\rm max} \cong 400\ \mu$s/m, and $\Delta t_{\rm min} \cong 210\ \mu$s/m again with no assigned error. There is also the question of how the values are to be chosen for $R_{\rm b}$ and $\Delta t_{\rm b}$. Passey *et al.* (1990) argue that a "baseline" set of values is to be chosen where the sonic and resistivity logs match in shape identically (presumably such regions of the logs are also to be identified as those having zero TOC). Presumably a more systematic procedure is to determine the *smallest* value of the combination $\ln R + m\ \Delta t/\Delta t_*$ and to declare that this value corresponds to zero TOC. All other values of $\ln R + m\ \Delta t/\Delta t_*$ must then be in excess of this minimum. Alternatively, an arbitrary value can be specified for $\ln R_{\rm b} + m\ \Delta t_{\rm b}/\Delta t_*$. While some measurements of $\ln R + (m/\Delta t_*)\ \Delta t$ will then fall below the minimum arbitrarily set value, as long as they correspond to a low TOC value, it is possible to default to zero by ignoring TOC below a threshold. Passey *et al.* (1990) follow this arbitrary value prescription. Presumably, a better procedure for calibration is as follows. Consider a set of wells in which TOC, ln $R$, and $\Delta t$ have been directly measured at a variety of depths in each well. Then from Eq. (10.35) we have

$$\ln R = -a\,\Delta t + b\mathrm{TOC} - c, \tag{10.38}$$

where

$$a = \frac{m}{\Delta t_*}, \qquad b = \frac{m(\Delta t_{\mathrm{OM}} - \Delta t_{\mathrm{m}})}{\Delta t_* \rho_{\mathrm{OM}}}, \qquad c = \ln R_{\mathrm{b}} - m\,\Delta t_{\mathrm{b}}/\Delta t_*.$$

Consider a single well where a set of measurements of ln $R_i$, $\Delta t_i$, and TOC have been made at a set of depths $z_i (i = 1, \ldots, N)$. Then a least-squares fit to Eq. (10.38) yields

$$cN - b\sum \mathrm{TOC}_i - a\sum \Delta t = \sum \ln R_i, \tag{10.39a}$$
$$c\sum \mathrm{TOC}_i - b\sum (\mathrm{TOC}_i)^2 - a\sum \Delta t_i \mathrm{TOC}_i = \sum \mathrm{TOC}_i \ln R_i, \tag{10.39b}$$
$$c\sum \Delta t_i - b\sum \mathrm{TOC}_i \Delta t_i - a\sum \Delta t_i^2 = \sum \Delta t_i \ln R_i, \tag{10.39c}$$

so that the values of $a$, $b$, and $c$ can be determined by solution of the three simultaneous equations (10.39). Then, for each well in turn of the suite of $M$ wells for which measurements of TOC, ln $R$, and $\Delta t$ are available, the same procedure is followed. The result is clearly a suite of values for $a_j$, $b_j$, and $c_j$ $(j = 1, 2, \ldots, M)$. This suite of values for $a$, $b$, and $c$ can then be used to obtain estimates of resolution for each from

$$\langle Q \rangle = \frac{1}{M}\sum_{j=1}^{M} Q_j, \tag{10.40a}$$

$$\delta Q = \left\{ \frac{1}{(M-1)} \sum_{j=1}^{M} [Q_j - \langle Q \rangle]^2 \right\}^{1/2}, \tag{10.40b}$$

where $Q$ is any of $a$, $b$, or $c$, with the most likely values given by

$$Q = \langle Q \rangle \pm \delta Q. \tag{10.40c}$$

This procedure provides a measure of the uncertainty on each of the parameter values from a controlled suite of wells.

It is then a relatively simple matter to assess the determination of TOC from exploration wells in which *only* ln $R$ and $\Delta t$ have been measured. This procedure operates as follows.

*Estimates of TOC.* From Eq. (10.38) we have

$$\mathrm{TOC} = (\ln R + a\,\Delta t + c)/b, \tag{10.41}$$

with each of $a$, $c$, and $b$ having an estimated value together with a range of uncertainty determined either from the calibration procedure outlined in the previous subsection or from user-defined values and ranges for $a$, $b$, and $c$. The problem then is to take measurements of ln $R$ and $\Delta t$, together with the mean values and allocated uncertainties of the parameters $a$, $b$, and $c$, and thus provide an assessment of TOC, including uncertainties in the estimates of the most likely TOC value.

An appropriate procedure for evaluating the likelihood of correctness of TOC estimates is based on cumulative probability measures. The general sense of argument operates as follows. To make full use of the powerful techniques of probability measures, knowledge would be required of the intrinsic probability of obtaining an event. This knowledge is, by and large, based on data to do with scientific conditions. But the data available to work with are very limited at the exploration assessment stage, usually imprecise or derived by analogy, or dependent on conditions about which little knowledge (as opposed to surmise) is available. For these reasons, approximations and assumptions are introduced in attempts to obtain relatively robust estimates of data-related quantities from which some form of assessment can be made.

### *b. Estimates of Parameters Needed to Determine TOC*

Estimates of the parameter values $a$, $b$, and $c$ are provided in the various references given earlier. For example, Passey *et al.* (1990) provide $1/b = 10^{2.297}$, $a = 0.06$ m/$\mu$s, $c \equiv -a\ \Delta t_b - \ln R_b$ with $\Delta t_b = 300$ $\mu$s/m, and $R_b = 1\ \Omega \cdot$m for their Fig. 1 example, while in their Fig. 11 example, $R_b = 2.4\ \Omega \cdot$m above 2285 m, and $R_b = 1\ \Omega \cdot$m deeper than 2285 m. However, their Fig. 15 uses $R_b = 12\ \Omega \cdot$m, while Fig. 16 uses $R_b = 2\ \Omega \cdot$m, and their Fig. 17 has $R_b = 4\ \Omega \cdot$m with all other parameters held at the values given earlier. From Carpentier *et al.* (1989) it is possible to derive estimates for the values of $a$, $b$, and $c$ used from their figures and text, although explicit values are not given. We find several values depending on which of the four cases is being discussed (see Table 10.2).

Thus, note the following cases:

| | | | |
|---|---|---|---|
| C1 | $a = 0.67$ m/$\mu$s, | $b = 0.910^{-2.3}$, | $c = -22$; |
| C2 | $a = 0.6$ m/$\mu$s, | $b = 0.28 \times 10^{-2.3}$, | $c = -20$; |
| C3 | $a = 0.5$ m/$\mu$s, | $b = 1.2 \times 10^{-2.3}$, | $c = -22$; |
| C4 | $a = 0.06$ m/$\mu$s, | $b = 10^{-2.3}$, | $c = -20$; |

with $\ln R_b = 0$, that is, $R_b = 1\ \Omega \cdot$m. Values of $m$ vary from 1.45 to 2.2 (Hearst and Nelson, 1985; Schlumberger, 1987) so that the values of $a$ and $b$ will vary proportionately, that is, 25% variation of both $a$ and $b$ around the corresponding values for $m = 2$ (as given by Passey et al., 1990, Carpentier *et al.*, 1989). Thus, instead of the common range for both the Carpentier and Passey values of $0.05 \leq a \leq 0.067$ m/$\mu$s with a declared most likely value of 0.06, the 25% range will then produce

$$a_{\min} \approx 0.045\ \text{m}/\mu\text{s}; \qquad a_{\text{probable}} \approx 0.06\ \text{m}/\mu\text{s}; \qquad a_{\max} \approx 0.075\ \text{m}/\mu\text{s};$$

TABLE 10.2 Values of Logging Parameters from Shales, Coals, and Evaporites[a]

| | $V_{ma}$ (ft/s) | $\Delta t_{ma}$ ($\mu$s/ft) | $\Delta t_{ma}$ ($\mu$s/ft) (commonly used) |
|---|---|---|---|
| Sandstones | 18,000–19,500 | 55.5–51.0 | 55.5 or 51.0 |
| Limestones | 21,000–23,000 | 47.6–43.5 | 47.5 |
| Dolomites | 23,000 | 43.5 | 43.5 |
| Anhydrite | 20,000 | 50.0 | 50.0 |
| Salt | 15,000 | 66.7 | 67.0 |
| Casing (iron) (from Schlumberger, 1987) | 17,500 | 57.0 | 57.0 |

| Material | $\gamma$-ray (API) | Density (g/cm$^3$) | Sonic ($\mu$s/ft) | Neutron $I_H$ (LS) | Resistivity ($\Omega \cdot$m) |
|---|---|---|---|---|---|
| Shale | 80–150 | 2.4–2.6 | 70–150 | 0.3–0.4 | 2–10 |
| Shale | | | 60–180 | | 1–100 |
| Chlorite | 1 | 2.6–2.96 | | 0.34–0.51 | |
| Kaolinite | 6 | 2.42–2.61 | | 0.35 | |
| Illite | 70 | 2.5–2.61 | | 0.12–0.3 | |
| Smectite | 3 | 2.1–2.7 | | 0.13–0.4 | |
| Clay mixture 1 | 21 | | | 0.2–0.3 | |
| Clay mixture 2 | 85 | | | 0.1–0.2 | |
| Clay mixture 3 | 35 | | | 0.1–0.2 | |
| Anthracite | 10–30 | 1.5–1.7 | 100–110 | 0.20–0.35 | 2–8 |
| Anthracite | | 1.4–1.8 | | | |
| Anthracite | 0 | 1.47 | | 0.37 | |
| Bituminous | 20–45 | 1.2–1.4 | 115–127 | 0.35–0.45 | 50–200 |
| Bituminous | | 1.3–1.5 | 100–150 | | |
| Bituminous | 0 | 1.24 | | 0.5–0.6 | |
| Subbituminous | 20 | 1.4 | 128 | 0.62 | |
| Lignite | 0–25 | 1.05–1.25 | 100–110 | 0.2–0.35 | 2–10,000 |
| Lignite | | 0.7–1.5 | 140–170 | 0.47–0.52 | |
| Halite | 0–10 | 2.0–2.2 | 40–67 | −0.02–+0.02 | 500–1000 |
| Trona | 0 | 2.11 | 65 | 0.25–0.4 | 0 |
| Anhydrite | 0–10 | 2.8–3.0 | 40–55 | −0.01–+0.03 | 500–1000 |
| Gypsum | 0–10 | 2.2–2.35 | 40–55 | 0.3–0.6 | 500–1000 |
| Sylvite | 500 | 1.98 | 74 | −0.03–0 | |
| Carnallite | 200 | 1.61 | 78 | 0.41–0.65 | |
| Polyhalite | 180 | 2.78 | 58 | 0.15–0.25 | |
| Langbeinite | 275 | 2.83 | 52 | −0.02–0 | |

(*continues*)

TABLE 10.2 *(continued)*

| Material | γ-ray (API) | Density (g/cm³) | Sonic (μs/ft) | Neutron $I_H$ (LS) | Resistivity (Ω·m) |
|---|---|---|---|---|---|
| Kainite | 225 | 2.12 | | 0.40–0.60 | |
| Quartz | 0 | 2.65 | 50–55 | −0.04 | |
| Dolomite | 0 | 2.88 | 43.5 | 0.04 | |
| Limestone | 0 | 2.71 | 47.5 | 0 | |

From Hearst and Nelson, 1985.

[a] Typical values of parameters for sedimentary rocks also are shown. It is assumed that the neutron log is calibrated for a limestone matrix.

while in place of the range $0.28 \leq b \times 0^{2.3} \leq 1.2$, a 25% uncertainty on each value will produce

$$b_{\min} \times 10^{2.3} \approx 0.21, \qquad b_{\text{probable}} \times 10^{2.3} \approx 1, \qquad b_{\max} \times 10^{2.3} \approx 1.6.$$

Equally the value of $c$ will vary due to variations in $m$ around $m = 2$. For $\ln R_b = 0$ we obtain

$$c_{\min} \cong -27, \qquad c_{\text{probable}} \cong -20, c_{\max} \cong -15;$$

while taking $m = 2$ and using the variaton in $R_b$ of 1 to 12 $\Omega \cdot \text{m}$ as presented by Passey *et al.* (1990), we have:

$$c_{\min} \cong -20 - \ln_e 12, \qquad c_{\text{probable}} \cong -20 - \ln_e 4, \qquad c_{\max} \cong -20.$$

The minimum and maximum values of $c$ are dominated more by the variations in $m$ than by the variations in the baseline resistivity. Accordingly we take

$$\begin{aligned} &c_{\min} = -27, \qquad c_{\text{probable}} \cong -20, \qquad c_{\max} \cong -15; \\ &b_{\min} \times 10^{2.3} \approx 0.21, b_{\text{probable}} \times 10^{2.3} \cong 1, b_{\max} \times 10^{2.3} \cong 1.6; \\ &a_{\min} \cong 4.5 \times 10^{-2}, a_{\text{probable}} \cong 6 \times 10^{-2}, a_{\max} \cong 7.5 \times 10^{-2} \text{m}/\mu\text{s}. \end{aligned}$$

Using these values, we obtain the statistical parameter estimates given in Table 10.3. It then follows from Table 10.3 that

$$E_1(\text{TOC}) = 10^{-2.3}[2.2 \ln R + 18(\Delta t/100) - 23] \qquad (10.42a)$$

and

$$\sigma^2(\text{TOC}) = 10^{-4.6}[0.9\,(\ln R)^2 + 94.5 \times (\Delta t/100)^2 + 42], \qquad (10.42b)$$

where $\Delta t$ is in units of $\mu$s/m, $R$ in $\Omega \cdot \text{m}$, and logs are taken to base $e$.

As has been noted earlier the transit time minimum and maximum values are taken to the range of 200 to 400 $\mu$s/m, while the resistivity varies typically from 1 to $10^3$ $\Omega \cdot \text{m}$; that is, $7 \geq \ln R \geq 0$. It then follows that $\sigma^2$

TABLE 10.3 Parameter Ranges for TOC Evaluation

| Parameter | Minimum | Probable | Maximum | $E_1$ | $\sigma^2$ |
|---|---|---|---|---|---|
| $a$ | $4.5 \times 10^{-2}$ | $6 \times 10^{-2}$ | $7.5 \times 10^{-2}$ | $6 \times 10^{-2}$ | $0.3 \times 10^{-4}$ |
| $b$ | $0.21 \times 10^{2.3}$ | $1 \times 10^{2.3}$ | $1.6 \times 10^{-2.3}$ | $0.94 \times 10^{2.3}$ | — |
| $1/b$ | $0.7 \times 10^{-2.3}$ | $1 \times 10^{-2.3}$ | $5 \times 10^{-2.3}$ | $2.2 \times 10^{-2.3}$ | $0.9 \times 10^{-4.6}$ |
| $-c$ | 15 | 20 | 27 | 21 | — |
| $a/b$ | $3.2 \times 10^{-4.3}$ | $7 \times 10^{-4.3}$ | $37.5 \times 10^{-4.3}$ | $18 \times 10^{-4.3}$ | $94.5 \times 10^{-8.6}$ |
| $-c/b$ | $10 \times 10^{-2.3}$ | $20 \times 10^{-2.3}$ | $38 \times 10^{-2.3}$ | $23 \times 10^{-2.3}$ | $42 \times 10^{-4.6}$ |

(TOC) is dominated by the second uncertainty terms in Eq. (10.42b). Indeed, at a $\Delta t$ of 400 $\mu$s/m, the second term is 1500, while the first term is 360. Thus $10^{4.6}\ \sigma^2$ (TOC) varies between about 1600 and 400 as ln $R$ and $\Delta t$ cover their maximum and minimum ranges. It is then an adequate approximation for the parameter ranges given to write

$$\sigma^2(\text{TOC}) \cong 10^{-4.6}\, 10^2(\Delta t/100)^2.$$

Then

$$\sigma^2(\text{TOC})/E_1(\text{TOC})^2 \cong \frac{10^2(\Delta t/100)^2}{[2.2 \ln R + 18(\Delta t/100) - 23]^2}.$$

For $R \geq 1\ \Omega\cdot$m, and $\Delta t \geq 200\ \mu$s/m it follows that

$$\sigma^2/E_1^2 \leq (10/13)^2 \approx 0.6.$$

Then

$$\mu^2 = \ln[1 + \sigma^2/E_1^2] \approx \sigma^2/E_1^2$$

and $\mu \approx \sigma/|E_1| \leq 0.8$.

A cumulative probability plot of TOC can then be given because the cumulative probability, $P$, takes on the values:

$$\begin{aligned} P &= 68\% \text{ on } x_1 \equiv E_1(\text{TOC}), \\ P &= 16\% \text{ on } x_1 \cong E_1(\text{TOC}) \exp(-\mu - \mu^2/2), \\ P &= 84\% \text{ on } x_1 \cong E_1(\text{TOC}) \exp(\mu - \mu^2/2). \end{aligned}$$

The pragmatic procedure is then clear:

1. First determine, or specify, the minimum, most likely, and maximum values of $a$, $b$, and $c$.
2. Then construct $E_1(\text{TOC})$ and $\sigma^2(\text{TOC})$ for each measurement pair of $\ln_e R$ and $\Delta t$.
3. On an ordinate that is logarithmic in TOC, plot on probability paper the corresponding 84%, 68%, and 16% probability points as ab-

scissas. The straight-line connection between the values can be extrapolated yielding estimates of the cumulative probability and the associated TOC values.

For example, suppose that measured values are $R = 100\ \Omega\cdot\text{m}$, and $\Delta t = 300\ \mu\text{s/m}$. Then we have

$$E_1(\text{TOC}) = 17.6 \times 10^{-2} \qquad \text{that is, } 17.6\%\ \text{TOC},$$

and

$$\sigma^2(\text{TOC}) = 239 \times 10^{-4},$$

so that

$$\sigma^2(\text{TOC})/E_1^2 = \frac{239}{310} \cong 0.77; \qquad \text{and } \mu^2 = \ln[1 + \sigma^2/E_1^2] \cong 0.65.$$

Then $P = 68\%$ occurs on TOC $\cong$ 17.6%, while $P = 16\%$ occurs on TOC $\cong$ 7%. A straight-line plot, as given by the curve labeled A in Fig. 10.6, then enables the assessment to be made that there is a 90% chance that the TOC is greater than 6% and a 90% chance that the TOC is less than 24% with a mean value of 17.6%. Thus at the 90% confidence interval (for the parameter values $a$, $b$, and $c$ and their ranges of uncertainty used to illustrate the point) one can write a TOC value for $R = 100\ \Omega\cdot\text{m}$ and $\Delta t = 300\ \mu/\text{m}$ as TOC $= 17.6^{+6.2}_{-11.6}\%$. If a resistivity of $R = 10\ \Omega\cdot\text{m}$ and a transit time of $\Delta t = 150\ \mu\text{s/m}$ are recorded, then $E_1(\text{TOC}) = 10^{-2} \times 4.5$, that is, 4.5% TOC; and $\sigma^2(\text{TOC}) = 10^{-4} \times 68$; so that

$$\sigma^2(\text{TOC})/E_1^2 = \frac{68}{20} \cong 3.4;$$

and $\mu^2 = \ln\{1 + \sigma^2/E_1^2\} = 1.5$. Then $P = 68\%$ occurs on TOC = 4.5% and $P = 16\%$ occurs on TOC = 2.2%.

A straight-line plot, as given by the curve labeled B in Fig. 10.6, then provides the assessment that there is a 90% chance that the TOC is greater than 1% and a 90% chance the TOC is less than 6%. Thus the estimate of TOC at the 90% confidence interval would be TOC $= 4.5^{+1.5}_{-3.5}\%$.

The point to be made is that each pair of measurements for resistivity and sonic transit time can be used, in conjunction with the assessed uncertainty in the scaling constants relating $\ln R$ and $\Delta t$ to TOC, in order to provide a probabilistic assessment of TOC content.

### c. Discussion

The ability to assess TOC content from log responses, particularly resistivity and sonic transit time, has had a long history of successes and failures (Hearst and Nelson, 1985). Part of the problem appears to have been the specification of very precise values for the intrinsic parameters that enter

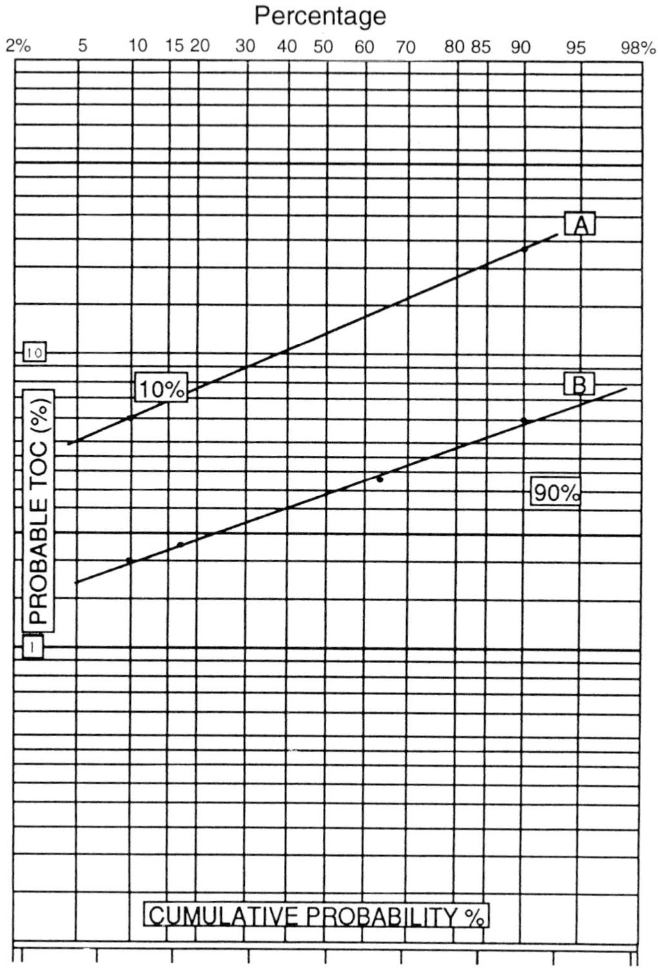


FIGURE 10.6 Two curves representing different TOC estimates as given in the text. Curve A represents $R = 100\ \Omega \cdot \text{m}$, $\Delta t = 300\ \mu\text{s/m}$; curve B represents $R = 10\ \Omega \cdot \text{m}$, $\Delta t = 150\ \mu\text{s/m}$. The 10% and 90% cumulative probability values are marked.

the relevant formulas. But such specification has differed among different groups over the years and from different basins or indeed even from the same borehole over different depth ranges (Passey *et al.*, 1990). The problem is that one must somehow specify a range of intrinsic rock transit times, a power for Archie's law, and scaling values for baselines for resistivity and sonic transit time when it is known that such values are site specific.

What is often known, as attested to by the many investigations undertaken (see, e.g., references in Hearst and Helson, 1985; Schlumberger, 1987;

Carpentier *et al.*, 1989; Passey *et al.*, 1990), is the likely range of a parameter or suite of parameters which enter the relevant formulas.

Instead of insisting, therefore, that particular values of parameters be selected from the encompassing ranges and then attempting to provide a unique (but imprecise and poorly resolved) value for the TOC content [a procedure that is in general going to fail for data in boreholes not part of a calibration set (Stocks and Lawrence, 1990)], it is preferable to use the range of uncertainty of each of the parameters to provide a probability measure at some given confidence interval (90% in the numerical illustrations given) to bracket the likely value of TOC.

### B. DISCUSSION AND CONCLUSION

The analytic probability procedure given here for obtaining estimates of uncertainty of any particular log-derived quantity, in terms of assessments of ranges of uncertainty of intrinsic parameters and of measurement error in factors entering the response equations, is quite general and will apply to any log or suite of logs.

The advantage of the analytic procedure provided is that cumulative probability measures of uncertainty can be obtained very rapidly from rough estimates of intrinsic parameter uncertainty and, in addition, can be used to assess quickly sensitivity of a derived physical parameter to changes in the assumptions concerning the range of particular intrinsic parameters in the relevant response equation, or indeed to the error in measurement. In this way it can quickly be made apparent which factor or factors need to be determined to a better degree, and the appropriate resolution of those factors can also be evaluated in order to keep a derived quantity of physical meaning within particular bounds.

## II. EROSION AND UPLIFT ESTIMATES

### A. INTRODUCTION

Determining the amount and timing of post-Cretaceous uplift and erosion in the Barents Sea is one of the more important factors limiting our ability to improve on estimates for likely hydrocarbon accumulations. The relative timing of hydrocarbon generation, migration, and accumulation in relation to structural and stratigraphic dynamic evolution, and the question of whether hydrocarbon reservoir sealing conditions were breached after uplift and erosion, are of considerable concern to economic assessments of oil potential in the Barents Sea [see, e.g., many papers and references in *Arctic Geology and Petroleum Potential,* Vorren *et al.* (1993)].

Many attempts have been made, using different techniques, to provide some measures of the erosion/uplift amount. Each method or technique has its own strengths and weaknesses related to intrinsic assumptions of the method; intrinsic assumptions of the method; intrinsic parameters required in order to determine a quantity being sought; and the quality, quantity, and sampling frequency of data used as constraint information. In addition, any output from a method has to be evaluated in terms of precision, resolution, uniqueness, and sensitivity. It can, and does, happen that a particular method is per se highly regarded as an accurate device for assessing erosion; but it can also happen that the data available from which to extract an estimate of erosion are sufficiently sparse, or of poor quality, or sampled so nonuniformly that the ability of the method to determine erosion is compromised by the vagaries of the data. In such a case it is not the method which is at fault, but rather the sharpness of resolution attainable with a particular set of data.

Clearly one has to be aware of such effects because in the case of an accurate method but poor data, significant improvements can be obtained by collection of better data, while in the case of a coarse method even good, high-quality data will not improve the ability of the method to resolve any better than its intrinsic limitations.

In the case of the Barents Sea, methods used to assess erosion fall into three broad categories: (1) measures of thermal maturity; (2) measures of dynamic burial history and rock properties; (3) measures based on modeling the geometric patterns of observed development from seismic, gravity, and downhole data.

The purpose of this section is to consider some of the limitations of procedures and methods within the second two categories in order to provide an appreciation of the causes on the resolution attainable with each method. Limitations on thermal indicator methods have already been addressed in Chapter 4.

## B. DYNAMIC ROCK PROPERTY METHODS

Two of the most popular dynamic rock property methods (but not necessarily the most precise) for assessing the amount of erosion are based on either (1) stacking velocities from seismic or, equivalently, from sonic log and/or density log readings or (2) drilling exponent methods. We consider each in turn.

### 1. Stacking Velocity, Sonic Log, and Density Log Methods

The essence of each of these three methods is exemplified by an analysis of density log procedures as follows. A connection is inferred between measured density, $\rho_i$, at depth $z_i$ and porosity $\phi_i$ at the same depth, usually (Hearst and Nelson, 1985) assumed to be in the form

$$\rho_i = \rho_M(1 - \phi_i) + \rho_w \phi_i, \tag{10.43}$$

where $\rho_M$ is the rock matrix density and $\rho_w$ the density of the phase filling the pore space (usually taken to be water). If $\rho_M$ and $\rho_w$ are specified then a porosity inferred from the measured density can be written

$$\phi_i = (\rho_M - \rho_i)/(\rho_M - \rho_w). \tag{10.44}$$

Hence, by plotting the inferred porosity values with depth, one can obtain some curve which best fits the observed measurements. For instance, if it is further *assumed* that the porosity decreases exponentially from the present-day mudline with depth in the form

$$\phi = \phi_* \exp(-cz) \tag{10.45}$$

then least-squares best estimates of $\phi_*$ and $c$ can be obtained by fitting the form of Eq. (10.45) to the inferred values of $\phi_i$. However, if there has been an erosional event and no subsequent deposition, then the value $\phi_*$ will not match the surface depositional porosity $\phi_s$ taken from other regions that have not undergone erosion. The thickness amount of eroded material, $H$, is then inferred from

$$\phi_s = \phi_* \exp(-cH), \tag{10.46a}$$

which yields

$$H = c^{-1} \ln(\phi_s/\phi_*). \tag{10.46b}$$

Several intrinsic assumptions permeate this method: (1) The model connects downhole density measurements and porosity; (2) knowledge must be provided of the intrinsic "constants" $\rho_M$ and $\rho_w$; (3) an assumption must be made on the functional form to fit to the inferred porosity measurements; (4) the constants in the functional form must be determined and *so must their uncertainties*; and (5) an assumption must be made of the depositional surface porosity and that the inferred functional form for the porosity remains appropriate for the eroded material, and does not change with time. In addition, the uncertainty on the measured density log values must be given if one is to assess reliability of any and/or all of the parameters and assumptions entering determination of the erosion amount.

Similar behaviors and assumptions are made for sonic log transit times and stacking velocities. Usually, the sonic log transit times, $t_i$, are related to matrix transit times, $t_M$, and water transit time, $t_w$, through the Wyllie *et al.* (1956) formula

$$t_i = t_M(1 - \phi_i) + t_w\phi_i, \tag{10.47}$$

yielding a sonic log derived porosity of

$$\phi_i = (t_M - t_i)/(t_M - t_w) \tag{10.48}$$

so that the logic applied to this sonic log behavior to determine erosion amounts is then identical to that applied to the density log along with similar caveats on assumptions. In the case of stacking velocity analysis the argument is virtually identical. Indeed Løseth *et al.* (1993) have stated:

The method used here is based on the relationships between velocity, porosity and depth. Interval velocities are related to depth through a compaction consistent velocity (CCV) function. A statistical procedure optimizes the CCV-function to the (depth, velocity) data set and the adjusted input parameters are then used to compute the amount of missing section.

The exponential porosity versus depth function used for backstripping by Sclater and Christie (1980) is combined with the Wyllie equation into the CCV function:

$$v = [V_{ma}^{-1} + \phi_0(V_w^{-1} - V_{ma}^{-1})\exp(-cz)]^{-1}, \tag{10.49}$$

where $v$ = bulk velocity (m/s); $V_{ma}$ = matrix velocity, $V_w$ = water velocity; $z$ = depth (m); $\phi_0$ = optimized porosity at the surface (after erosion); and $c$ = compaction constant ($m^{-1}$). Once the parameters $\phi_0$ and $c$ have been determined, the amount of missing section ($H$) is computed from the "surface porosity" before compaction ($\phi_1$):

$$H = c^{-1}\ln(\phi_1/\phi_0). \tag{10.50}$$

We use an optimized procedure which automatically changes the input variables i.e., $V_{ma}$, $\phi_0$ and $c$, until an error function reaches a minimum.

Unfortunately, Løseth *et al.* (1993) do not consider uncertainties in the analysis due to resolution in the manner described earlier in Chapter 4 for thermal indicators, nor in density, sonic, and stacking velocity analyses. They do consider different depth ranges from which to infer erosion (corresponding to different sets of values, *without* associated uncertainties on each, from which to write down a set of erosion estimates). In this way, Løseth *et al.* (1993) obtain a *minimum* uncertainty of erosion of 1750 ± 500 m. The true uncertainty could, of course, be significantly larger if the remaining points on resolution were to be included.

Consider in more detail the shale compaction method, for which a sonic log and estimated uplift have been published by Skagen (1992) for a well in the Barents Sea. Indeed, Skagen (1992) has written:

Both the sonic log and the density log readings follow a compaction trend. Most published reference curves are based on the porosity of shale or shaly sands versus depth. Magara (1976) suggests a relation between porosity and the sonic transit time for the Western Canada Basin. This proved to be quite good also for the Barents Sea area compared with the results from other methods. Sonic log readings do not give the correct porosities of shales. Therefore, the equation presented by Wyllie *et al.* (1956) for porosities from the sonic log is limited to sandstones with porosities within a range of 25 to 30% (Raymer *et al.*, 1980). Porosities outside this range have to be corrected by a compaction factor $c$. This factor is known for shale and is variable with depth. A method has been published for calculating porosities in shale from the sonic log, but this method uses the density of the shales from the density log in the calculations (Raymer *et al.*, 1980). Critical for both the sonic and the density log readings is the reference curve of shale porosities versus depth. The best reference curve so far has been presented by Sclater and Christie (1980), based on Cretaceous shales in the North Sea. Checking these parameters in a Barents Sea well where only a small amount of uplift has taken place indicates good agreement with their proposal. A valid reference curve should be developed for each area if possible.

In Fig. 10.7 we reproduce thet two panels of Skagen's Fig. 2 from which uplift estimates can be made; due consideration should, of course, be given to the intrinsic uncertainties. Skagen (1992) has remarked that the mean value of the histogram of Fig. 22b (of Skagen, 1992) is most commonly used and he further notes that "The method is regarded by Statoil to be more accurate than ±200 m." To evaluate the accuracy of the method, we have taken the frequency histogram of uplift generated by Skagen (his Fig. 2a) and integrated across the area under the curve, thereby generating a cumulative probability plot of erosion amount, as shown in Fig. 10.8. Note from Fig. 10.8 that the cumulative probability curve of Skagen's (1992) estimated frequency histogram is remarkably close to a log-normal probability curve. The mean value of a true log-normal probability distribution is well known (Feller, 1957) to occur at a cumulative probability of about 68%, which, from Fig. 10.8, is at 1400 m. Measures of uncertainty are usually given at the 90% and 10% cumulative probability points which, from Fig. 23 (of Skagen, 1992), are at 1700 and 150 m, respectively. Thus, on the basis of the case history analyzed by Skagen (1992), his data indicate that it is 90% certain that the eroded amount is *less* than 1700 m, 90% certain that the eroded amount is *greater* than 150 m, with a two-thirds chance (68%) that the mean estimated eroded amount is less than 1400 m.

The cumulative probability result is customarily written as

$$H = 1400^{+300}_{-1250}\ \text{m}. \tag{10.51}$$

reflecting the uncertainty band. It is certainly *not* the situation that Skagen's (1992) case history is accurate to better than ±200 m for determining erosion amounts. This comment is made without including the "extra" uncertainties due to assumptions made, in much the same sense as the assumptions in density log interpretations or stacking velocity methods. Such inclusions would add to the uncertainty, suggesting that Skagen's (1992) estimate of ±200 m uncertainty is too conservative by a large factor, as also found with his own example.

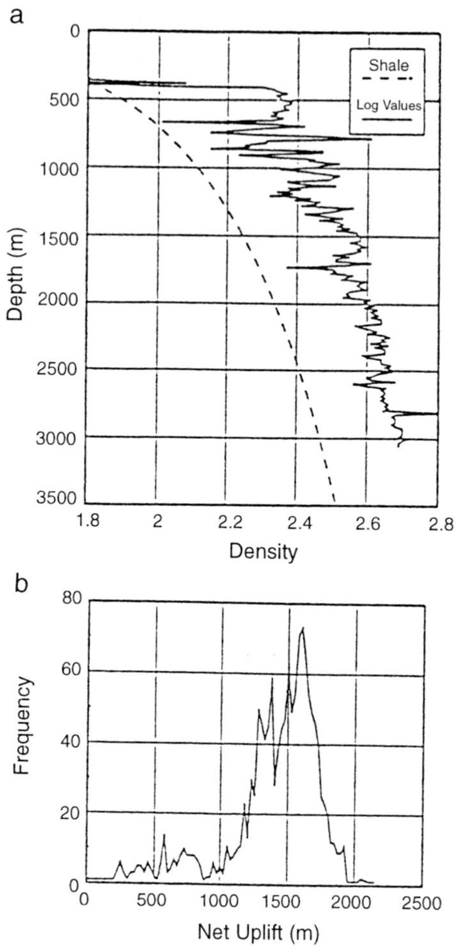


FIGURE 10.7 Direct reproduction of Fig. 2 of Skagen (1992). The dashed line in (a) is the shale density curve as suggested by Sclater and Christie (1980), whereas the solid line in (a) is the observed log response. Part (b) of the figure shows the distribution of all the calculations of uplift obtained by Skagen (1992) using the matrix method. (Reprinted from "Methodology applied to uplift and erosion," by J. I. Skagen from *Norsk Geologisk Tidsskrift*, 1992, Vol. 72, pp. 307–311, by permission of Scandinavian University Press.)

## 2. Drilling Exponent Methods

Because the lithification of sedimentary formations is dependent on the applied stress (Hearst and Nelson, 1985), as well as on a large number of other factors, the ability to use the penetration rate of a drill bit to assess the porosity of a formation has long been used as a rough estimate of overpressure (either high porosity or unconsolidated formation) development in a drill hole. Skagen (1992) has suggested that drilling exponent parameters can be used to determine erosional uplift amounts. Skagen (1992) has written:

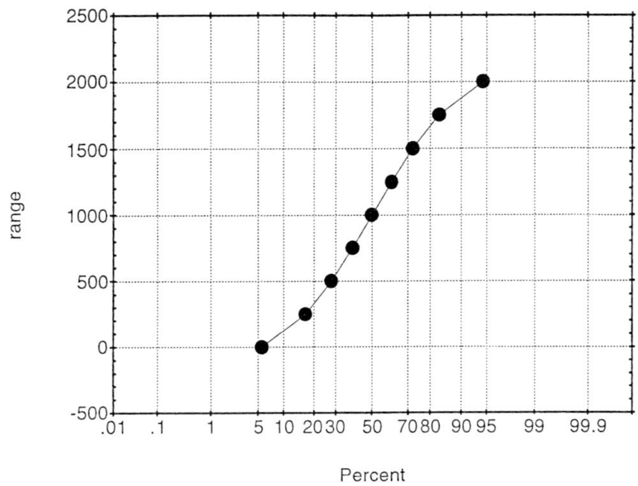


FIGURE 10.8 Cumulative probability plot of the histogram of uplift values given in Fig. 10.7b. The ordinate, labeled "range," is the uplift amount in meters.

The principles of the method are as follows: The rate of penetration (ROP) is a function of compaction versus depth. So are the porosities of shales, the densities of shales and thereby also the overburden pressure gradient which is calculated by integrating the densities.

The drilling exponent (dc-exponent) is a logarithmic function of the ROP. Several mechanical drilling parameters like weight on bit, revolutions per minute, bit diameter and mud weight are needed to calculate the dc-exponent. These parameters should be kept constant while drilling. The dc-exponent is used to calculate the pore pressure for a well and the pore pressure is a function of the overburden gradient. The dc-exponent can be expressed as a function of the pore pressure, the overburden pressure and depth. The method developed by Eaton (1975) and the equivalent depth method, also called the matrix stress method (Hottman and Johnson, 1965), are commonly used for the calculation of the pore pressure from analysis of the dc-exponent. Both methods are based on the same analytical technique. By combining the two methods a simple expression of the dc-exponent can be achieved as a function of the pore pressure, the overburden pressure and depth. This expression describes the increase of the dc-exponent with increase in depth if the pore pressure is normal, the mechanical drilling parameters are kept constant and a reference overburden gradient exists. By picking an initial dc-exponent from the actual drilling in normally pressured shales, any new dc-exponents on a linear scale shows that the dc-exponents follow a straight line. By extrapolating this line to the depth axis the net uplift is achieved directly from the plot. This is similar to the technique applied on vitrinite reflectance values.

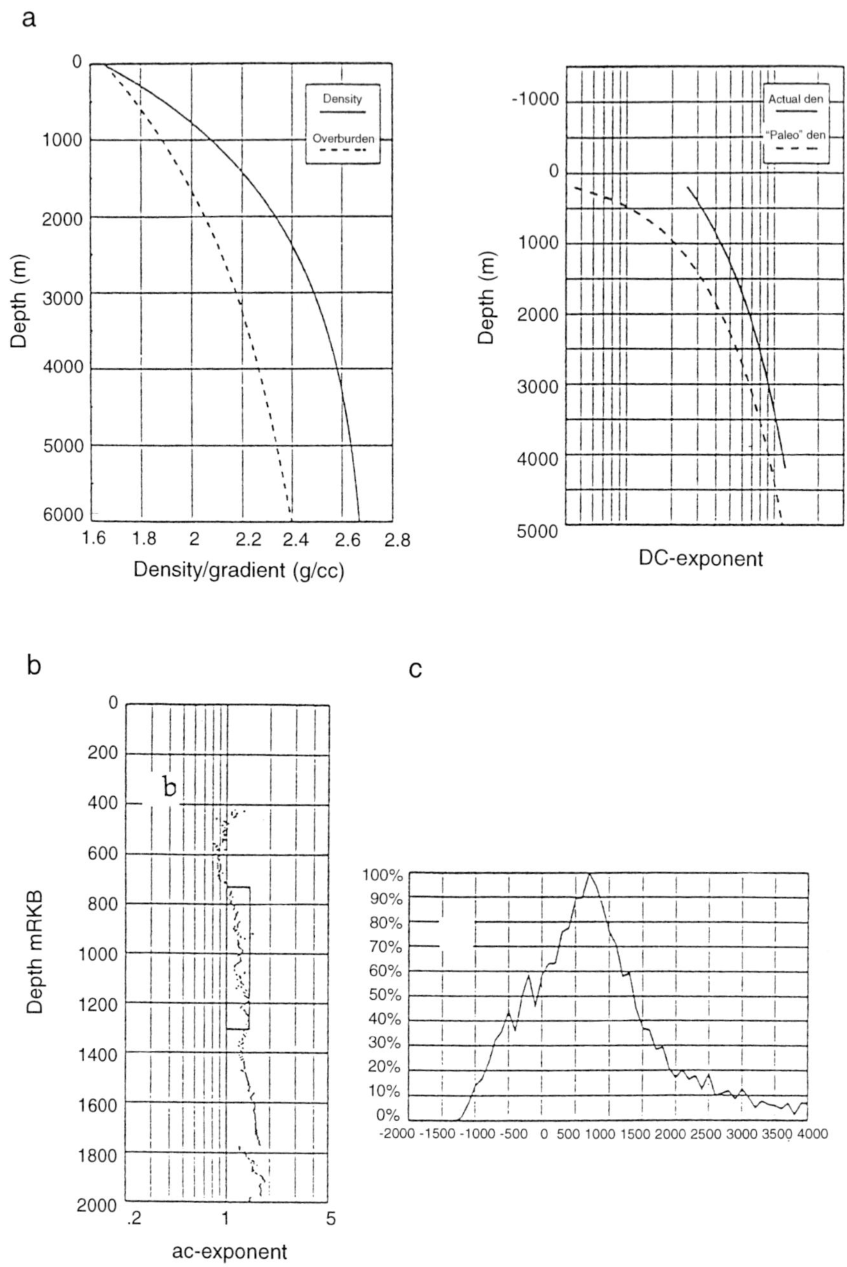


FIGURE 10.9 All three panels are taken directly from Skagen (1992). Panel (a) shows "compaction curves based on the suggestion by Sclater and Christie (1980) of shale porosities versus depth. Shale density and overburden pressure with depth." Panel (b) shows "Principles for uplift calculations from dc-exponent on logarithmic scale. The dashed line is the dc-exponent for normally pressured shales (dcn) where no uplift and erosion have occurred. The

Skagen goes on to note that

The method is in our experience thought to be more accurate than ±200 m. This method requires a 'paleo' overburden pressure gradient. By integrating shale densities calculated from the shale porosities suggested by Sclater & Christie (1980), an overburden pressure gradient is achieved which can be applied to this method. Figure 3 shows the shale densities from this method together with the resulting overburden pressure gradient. The dc-exponent is normally plotted on a semilogarithmic scale versus depth. Plotting the calculated 'paleo' dc-exponents and the actual dc-exponents observed in a well gives two curved lines (left graph, Fig. 4). Plotting the same dc-exponents on a linear grid gives two almost straight lines (right graph, Fig. 4). In a zone of interest, the uplift can either be calculated from a linear regression of the dc-exponents within the interval or by a matrix operation. The matrix operation will give rise to numerous sets of calculations within an interval of normally pressured shales and with constant mechanical drilling parameters. All pairs of dc-exponents within that interval might be solved as a straight line, except for the cases where the slope is negative, i.e., where the initial dc-exponent is greater than the other dc-exponents. By doing that, thousands of calculations are performed, which will help to achieve a statistically more reliable result for the net uplift. It is recommended that 1 m data are obtained from the mud logging company.

Only calculations of the net uplift greater or equal to zero are valid. The method is dependent on the lithology, and calculations of the net uplift should only be carried out in homogeneous shales. The mechanical parameters have to be constant within the calculation zone, and the normal pore pressure analysis technique has to be followed. This kind of analysis requires a software package to ease the amount of data computation. Figure 5 is an example from a well in which this method has been applied. Calculations were done in the boxed interval, where 4899 calculations were valid. The histogram at the right of Fig. 5 shows the intensity of the calculations where the most frequent range always indicates 100% intensity.

Skagen's (1992) Figs. 3, 4, and 5 are reproduced here as Figs. 10.9a, b, and c, respectively. Note from the net uplift histogram of Fig. 10.9c the broadly spread range of possible values of uplift. Skagen (1992) quotes a mean value of 834 m for the estimated uplift, but without presenting any

---

solid line is the dc-exponent for normally pressured shales where an uplift and erosion have occurred." Panel (c) shows "example of calculation of the net uplift in the Tertiary by use of dc-exponent from a Barents Sea well. Linear regression of values within the boxed interval in this example gives a correlation coefficient of 0.775 and a net uplift of 618 m. A matrix operation gives 4899 calculations with a mean value of the net uplift of 834 m. The histogram shows the distribution of all calculated results." (Reprinted from "Methodology applied to uplift and erosion," by J. I. Skagen from *Norsk Geologisk Tidsskrift,* 1992, Vol. 72, pp. 307–311, by permission of Scandinavian University Press.)

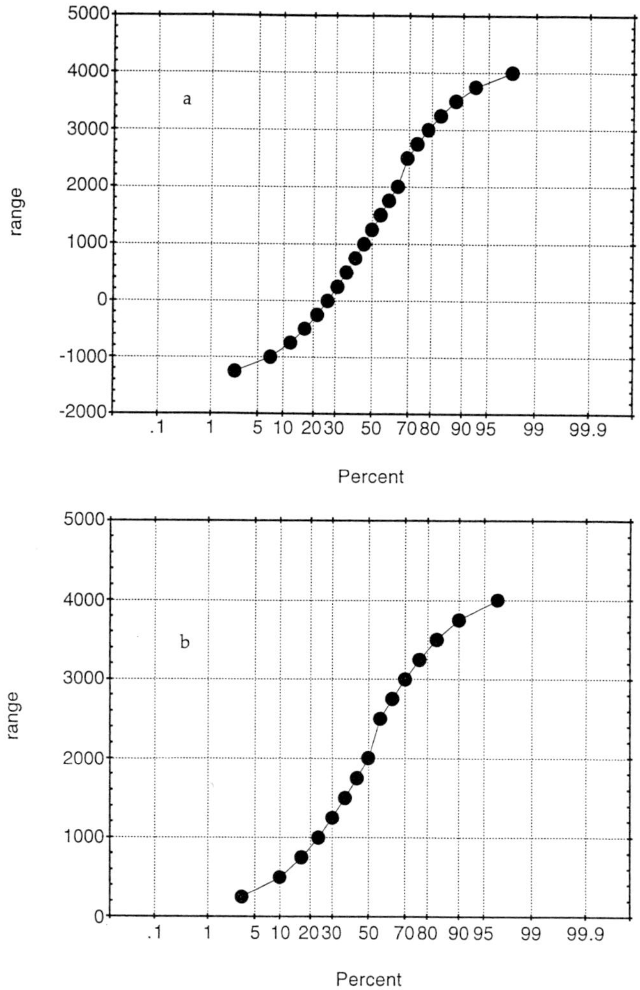


FIGURE 10.10 The cumulative probability plot of the histogram of uplift values recorded in Fig. 24c of Skagen (1992) is shown (a) when all values (including negative uplift) are included; (b) when only the positive values are included. In both cases the distribution is almost exactly log normal in character. The ordinate, labeled "range," is the uplift value in meters.

estimate of the uncertainty on this value. We have taken his frequency distribution of uplift estimates and, normalizing to unit area, have integrated across the distribution in two ways: (1) with *all* estimates of erosion/uplift, including the negative values; and (2) ignoring the negative values of uplift and integrating over psoitive values only. The corresponding cumulative probability estimates of erosion amount are given in Figs. 10.10a and b, respectively. Both curves are remarkably close to log-normal probability distributions. Including all data [which is in violation of the theme expressed by Skagen (1992) in terms of usage of the dc-drilling exponent method] yields an uplift amount of

$$H = 2600^{+1000}_{-3400}\ \text{m}, \tag{10.52}$$

suggesting very little resolution capability. If, however, Skagen's (1992) dictum is followed by removing all the negative estimates of erosion, then Fig. 10.10b provides the estimated uplift amount of

$$H = 3000^{+800}_{-2500}\ \text{m} \tag{10.53}$$

at the 90% confidence interval. Thus, in this example the confidence interval is much broader than the $\pm 200$ m reported by Skagen (1992) and, furthermore, neither is the expected value at the value quoted by Skagen (who seems to be quoting the mode value), nor is the accuracy of the method at the $\pm 200$-m mark, rather the accuracy is around $\pm 1$ to 2 km, comparable to both thermal indicator methods and to the shale compaction and stacking velocity methods. It seems difficult to improve on these estimates using Skagen's own data. And including uncertainties in the assumptions would make it even more difficult to improve the resolution. It is certain that the accuracy of the method is much worse than $\pm 200$ m.

The upshot of these investigations suggests that dynamic rock property methods to assess erosion amounts in the Barents Sea are likely no more accurate, at best, than $\pm 500$ m and, for the cases exhibited, are worse than $\pm 1$ to 2 km, making such methods no better than results derived from thermal indicator methods, and possibly considerably worse when the uncertainties in assumptions and parameter values are allowed for. The accuracy is well outside the $\pm 200$ m that Skagen (1992) considers relevant, by his own data.

Overall, the uncertainties for all methods so far discussed are generously encompassed by the conservative range of $\pm 1$ km.

## C. DYNAMIC GEOMETRY MODEL METHODS

Perhaps because of frustrations with attempts to bracket to better than about $\pm 1$ km the amount of erosion determined by downhole thermal indicator methods and/or dynamic rock property methods, ingenious attempts have been made to use observed geometries of sedimentary beds in relatoin to salt cap motion, or in terms of quantification of glacial erosion amounts, to estimate erosion amounts. Of course, in a similar vein to the uncertainties attached to downhole methods, there are also uncertainties in estimates from geometry models, which depend on the model assumptions, model parameters, and quality and quantity of data used to constrain the geometry models.

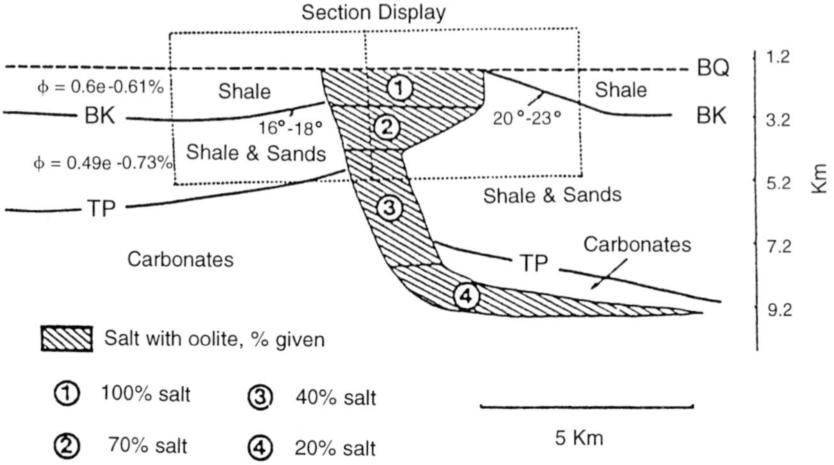


FIGURE 10.11 Interpreted salt dome in the Barents Sea, Norway (model 1 with half-side cap). BQ = base of Quaternary; BK = base of Cretaceous; TP = top of Permian.

Here we look at two such models and examine their strengths and weaknesses.

**1. Salt Diapir Mushroom Caps and Erosion Estimates**

In the Nordkapp Basin of the Barents Sea, copious salt diapirs are present which, at present day, reach almost to the sediment surface. Seismic interpretations of some of the diapirs have been given. The section displayed in Fig. 10.11 shows a Quaternary sediment thickness of about 800 m lying unconformably on an upturned Cretaceous section with little or no intervening Tertiary.

***a. Data Description***

Figure 10.11 shows the interpreted cross-section of a salt dome with a half-sided mushroom cap in the Barents Sea, Norway. The total thickness of the basal salt is not clear at this stage. Oolites were mixed in the salt dome while the upper parts of the dome have cleaner salt. The lithologies are labeled in the section. The Triassic plus Jurassic sediment thickness on the right side of the salt dome is 2000 to 3000 m thicker than on the left side. The dip angle of the base Cretaceous on the right side is 20 to 23° and on the left is 16 to 18°. There are no Tertiary sediments in the section. An important feature of the salt dome is that it developed a cap on the right side but not on the left side.

From this section some questions arise: When did the salt move upward? Why are the sediment thicknesses different on the two sides of the salt dome? Why is the dip angle of the base Cretaceous on the two sides different? Finally, can we reproduce such a structure by using a simulation model?

### b. *Timing of Salt Uplift*

We first provide an estimate of the minimum thickness of overburden for the basal salt layer to gain enough buoyant force to uplift the salt. From Fig. 10.11 we took empirical values for the surface porosity and depth scale of porosity reduction in the Barents Sea basin and typical salt, matrix, and fluid densities were used.

Three major properties of salt cause it to play a dominant role in sedimentary basin development (see, e.g., Nettleton, 1955; Biot and Ode, 1965; Berner *et al.,* 1972; Bishop, 1978; Lerche and O'Brien, 1987a,b):

1. On a geologic timescale, salt flows as a nearly incompressible fluid under applied stress, thereby both distorting sedimentary patterns and influencing further basin evolution.
2. Salt has a density of approximately 2.0 to 2.2 g/cm$^3$, which is intermediate between sedimentary densities at deposition (1.6 to 1.9 g/cm$^3$) and the densities of fully compacted sediments (2.6 to 2.8 g/cm$^3$). The density of salt varies little during burial under an increasing overburden weight. Thus at some point in a basin's evolution salt will become buoyant and will attempt to rise up through the overlying formations.
3. Salt has a thermal conductivity approximately three times greater than that of "typical" sedimentary formations and so salt bodies act as conduits for heat transport from depth. In the vicinity of salt, local thermal effects caused or modified by this conductivity contrast will influence chemical precipitation and dissolution, hydrocarbon maturity, and fluid flow.

The effects of subsurface salt bodies are often reflected in changes in sesdimentary patterns such as uplift, thinning and deformation of surrounding beds, development of faults in the overburden, fracturing of formations, rim syncline development, and so forth. In recent years, computer simulation techniques have been employed to demonstrate the physical processes of a subsurface salt body and its effects on the surrounding sediments (Cao *et al.,* 1989), so as to provide better geologic interpretations. As mentioned previously, salt has a density of approximately 2.0 to 2.2 g/cm$^3$, which is intermediate between sediment densities at deposition (1.6 to 1.9 g/cm$^3$) and those of fully compacted sediments (2.6 to 2.8 g/cm$^3$). On a large scale, porosity of clastic sediments is dominated by compaction effects, with density progressively increasing as porosity decreases in response to the increasing weight of overburden. The porosity at the critical point where sediment density equals that of salt is given by

$$\phi_c = (\rho_m - \rho_s)/(\rho_m - \rho_f), \tag{10.54}$$

TABLE 10.4 Results of Minimum Thickness Estimation for Salt Motion

| Lithology | Surface porosity, $\phi_0$ (%) | Porosity depth scale (m) | Salt density (g/cm$^3$) | Rock matrix density (g/cm$^3$) | Fluid density (g/cm$^3$) | Thickness estimated (m) |
|---|---|---|---|---|---|---|
| Shale | 60 | 2000 | 2.2 | 2.7 | 1.0 | 900 |
| Shale + sand | 49 | 1400 | 2.2 | 2.7 | 1.0 | 1700 |
| Shale + sand | 49 | 1400 | 2.4 | 2.7 | 1.0 | 2400 |

where $\rho_m$, $\rho_f$, and $\rho_s$ represent the densities of the rock matrix, the fluid saturating the rock, and salt, respectively. For typical values of these parameters ($\rho_m = 2.65$ to $2.72$ g/cm$^3$, $\rho_s = 2.2$ g/cm$^3$, $\rho_f = 1.0$ g/cm$^3$) the sediments must compact from a surface porosity of $\phi_0$ to a critical porosity, $\phi_c$, of less than 27% to 30%, before a density inversion develops. If we use an empirical function, $\phi(z) = \phi_0 \exp(-z/a)$, to describe the porosity decrease with increasing depth of burial, the critical porosity corresponds to an overburden thickness $H_{crit}$, given by

$$H_{crit} = a \ln(\phi_0/\phi_c), \tag{10.55}$$

which gives a minimum depth of sediment cover over salt before buoyancy prevails. Overburden thickness $H_{crit}$ varies in different basins, with typical values in the range of about 900 to 3000 m.

The onset of buoyant salt motion in a given basin can be estimated with the above argument to be about $H_{crit}/R$ where $R$ is the sediment deposition rate. Other factors such as overpressure and lithification strength of sediments delay the salt's rise. The results (Table 10.4) show that the minimum thicknesses of overburden before salt motion is 900 m for shale and 1700 m for mixed shale and sand. From the section of Fig. 10.11, a 1700-m thickness corresponds to Early or Mid Triassic periods for initiation of salt motion. Thus the salt moved after the deposition of carbonates and after the deposition of Triassic and Jurassic started. The salt did not reach the depositional surface during the Triassic or Jurassic, which means it did not reach the top of the Triassic or Jurassic formations until after Cretaceous deposition started.

The carbonate layer on the immediate top of the salt layer could delay the initiation of salt motion. Another fact that could delay the salt rise is the mixing of oolite in the salt diapir, which probably increased the "dirty" salt density. If we take 2.4 g/cm$^3$ as the mixed diapir density rather than the 2.2 g/cm$^3$ of clean salt, then the estimated minimum thickness of sedimentary cover before salt can rise would be 2400 m for shale and sands. The thicker deposit of Triassic and Jurassic shales and sands on the right side of the salt diapir can be accounted for as follows. Let the salt begin to move during Early to Mid Triassic and Jurassic (say, 180 MaBP to fix ideas) and, at that time, sediment deposition was also occurring. The rising salt dome and the loss of salt volume out of the mother layer would result in an increased volume accommodation. The unbalanced supply induces different volume accommodation rates on the two sides of the salt dome. This action would be an analog of the effects of faulting with the salt dome serving as a fault plane. Consequently, the depositional faulting associated with salt motion would result in the right side of the salt receiving a thicker volume of Triassic and Jurassic shale and sand deposits. This feature also provides evidence that the salt motion started while the Triassic and Jurassic sediments were being deposited.

The sedimentary deposition rates during the Triassic and Jurassic are about 90 m/Ma on the right side and 40 m/Ma on the left side of the salt. The *average* rates for the Triassic, Jurassic, and Cretaceous (ignoring the Tertiary, which is eroded) are 50 m/Ma on the right and 27 m/Ma on the left side of the salt. The salt rate of uplift can be estimated by calculation of the salt volume moved when a roughly steady-state salt flux is assumed. For the mushroom shape (estimated by calculating a larger cylinder sitting on a smaller cylinder) we obtain the rough estimate:

$$V_{\text{salt}} R_{\text{stock}}^2 t = R_{\text{stock}}^2 H_{\text{stock}} + R_{\text{cap}}^2 H_{\text{cap}}, \tag{10.56}$$

where $V_{\text{salt}}$ is the salt rate of uplift; $R_{\text{stock}}$ and $R_{\text{cap}}$ are the radii of salt stock and salt cap, respectively; $H_{\text{stock}}$ and $H_{\text{cap}}$ are the heights of salt stocks and salt cap, respectively; and $t$ is time measured from the beginning of the salt motion to its end.

For the Barents Sea salt dome, estimated salt motion started during the Scythion, about 240 MaBP. The rest of the values needed in Eq. (10.56) can be measured from Fig. 10.11. The calculated vertical salt speed is then about 150 ± 50 m/Ma. We note here that this estimate does not account for salt eroded at the surface, so the real salt velocity could be a little higher.

### c. Deformation of Formations

Consider now the structural deformation of formations surrounding a developing salt dome. For simplicity, neglect the lateral strength of the formations. Structural deformation can then be modeled as the response of a perfect fluid being penetrated by an uprising salt dome. The detailed model and calculations are given in Lerche and O'Brien (1987a).

To determine the influence of a rising salt dome on sedimentary structure, we map the path followed by an element of sediment during burial. Consider a horizontal sedimentary formation located sufficiently far above the salt dome that it is initially uninfluenced by the motion of the salt. The location and shape of this horizon may be determined at any later time by tracing the path followed by each element of the horizon. This path tracing can be achieved numerically because if we know the location of an element of

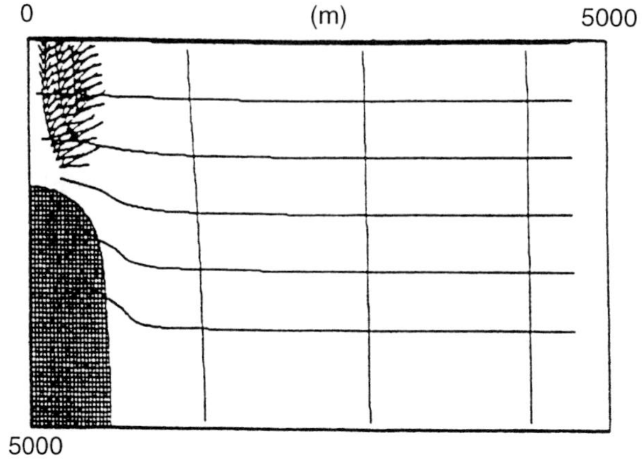


FIGURE 10.12 Simulated sediment structure for the left side of the salt dome in Fig. 10.11. The dip angle of BK is 10 to 15°, which is consistent with the observations on Fig. 10.11. The directions of principal stress, and of primary and secondary fracturing, are shown over the top of the diapir.

sedimentary material at time $t$ we can determine both the radial and vertical components of velocity, allowing us to determine the location of that element at time $(t + dt)$. By repeating this process we can map the path followed by a sedimentary element during burial.

### *d. Modeling Results*

Because the shape of the salt dome in the Barents Sea, Norway, is a half mushroom cap, with different sediment thickness on the two sides, we simulated the two sides separately as follows. Figure 10.12 gives the sediment structure on the left side of the dome without the salt cap. The dip angle of the base of the Cretaceous is about 10 to 15°, consistent with the observations of Fig. 10.11 on the salt section. Figure 10.13 shows the left side behavior during the vertical salt motion stage before development of the salt cap.

Figure 10.14 shows the result on the right side of the salt dome with the mushroom cap developed. The salt scales in this picture are comparable with Fig. 10.13. Note that the dip of the base Cretaceous is about 13 to 15°, which is *less* than on the salt section (Fig. 10.14). But in the simulation, the present day *observed* thickness of salt cap was used, which does not include the eroded thickness of the salt cap. Therefore, add thickness to the salt cap to see how much salt is needed to obtain a Cretaceous bed dip angle consistent with observations.

Figure 10.15 shows the result of adding a 1000-m thickness to the cap. The total thickness (prior to erosion) is 3500 m, relative to 2500 m for the observed thickness today. The dip angle of the base Cretaceous is now 15

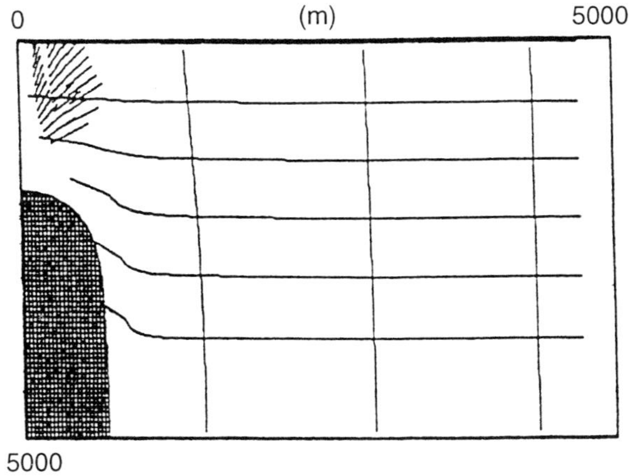


FIGURE 10.13 Simulated sediment structure for the left side of the salt dome in Fig. 10.11 before development of the horizontal salt motion.

to 23°, which is in good accord with the observations on Fig. 10.11. Simulations have been run in which more thickness was added to simulate the observed Cretaceous dip angle; 1000 to 1500 m best reproduces the observations, although up to 2000 m cannot be excluded as an extreme. Thus there is at least 1000 m of salt cap and sediments eroded during the Tertiary. To summarize the results:

1. During the Early Triassic, the salt started to rise in the Barents Sea when the critical overburden thickness range of 1700 to 2400 m

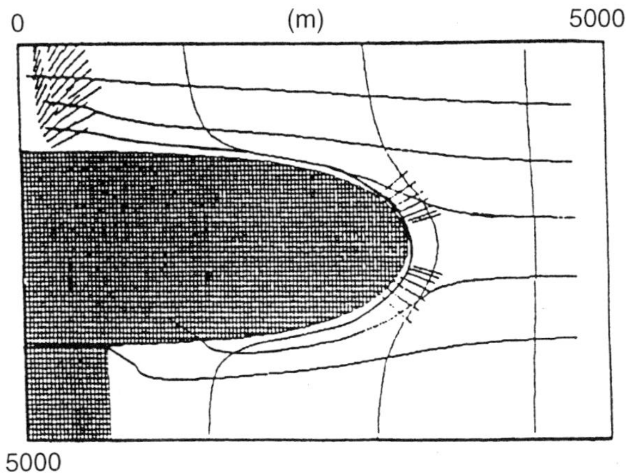


FIGURE 10.14 Simulated sediment structure with the mushroom cap on salt for the right side. The measured dip angle for BK is 13 to 15°. Note the primary fracturing directions indicated around the "nose" of the salt mushroom cap.

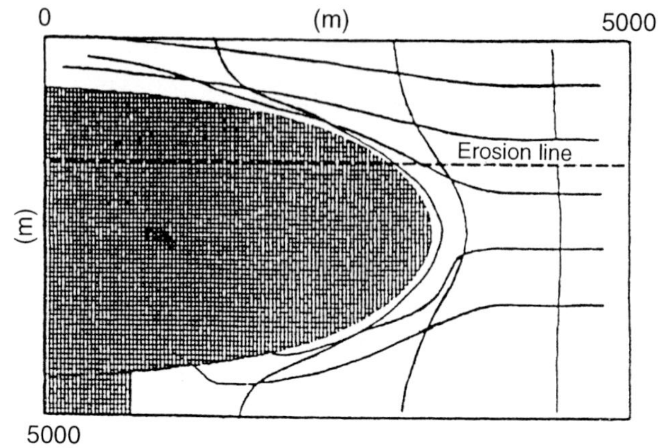


FIGURE 10.15 Simulated sediment structure for the right side with a 1000-m cap thickness eroded. The BK dip angle is 15 to 23°, which is consistent with the observations on Fig. 10.11.

was reached. The salt diapir associated faulting happened while the salt dome was moving upward at about 150 ± 50 m/Ma, which results in different sedimentary rates on the two flanks.

2. The eroded thickness of the salt cap is estimated to be at least 1000 m (but probably not more than 1500 to 2000 m) by modeling of sediment dip angles in the Cretaceous sequence.

Thus, the geometrical model development is likely uncertain by around ±500 m on the amount of erosion in the Nordkapp Basin section of the Barents Sea. The range of uncertainty is comparable to that reported in the thermal indicator section, and is likely as good, or better, than the estimates made using dynamic parameters.

## 2. Glacial Erosion Estimates

Regional glacial erosion involves mantle compensation. On the assumption that the compensation is purely isostatic, in the local Airy isostatic manner, Oerlemans (1984), and later Løseth *et al.* (1993), estimated erosion thickness, $E$, by replacing the present-day column of seawater of depth $w$, and a column of air of height $h$ (if the pre-eroded section is assumed to be that of an elevated land area) by a column of sedimentary rocks. Isostatic compensation then demands

$$E\left\{\rho_{\mathrm{M}} - \int_0^E \rho(z')\,dz'/E\right\} = h(\rho_{\mathrm{M}} - \rho_{\mathrm{a}}) + w(\rho_{\mathrm{M}} - \rho_{\mathrm{w}}) + q\left(\rho_{\mathrm{M}} - \int_0^q \rho(z')\,dz/q\right), \tag{10.57}$$

where $q$ is the sedimentary column thickness on top of the present-day erosional surface, $\rho_M$ is mantle density, $\rho_a$ is air density, and $\rho_w$ is water density. The sedimentary density variation with sub-mudline depth, $\rho(z)$, must be known, or specified, both pre-erosion for the eroded sediments and posterosion for the later deposited sediments in order that the average sediment densities,

$$\rho_E \equiv E^{-1} \int_0^E \rho(z')\, dz' \qquad \text{and} \qquad \rho_q = q^{-1} \int_0^q \rho(z')\, dz'$$

can be computed. Thus the porosity with depth variation must be known for both the eroded and the later deposited sediments, implying that a lithology variation in the eroded sediments can make a significant difference in the estimated eroded thickness, even under the restricted framework of an Airy isostatic calculation. Both Oerlemans (1984) and Løseth *et al.* (1993) ignore porosity variations and replace $\rho_E$ and $\rho_q$ with the constant values of 2.2 g/cm$^3$, respectively, which, for the same porosity with depth behavior for the eroded sediments and the later deposited sediments, automatically constrains $E > q$. In addition, the estimates of eroded thickness require local Airy isostatic compensation, that is, the balance of uplift and erosion, per Eq. (10.57), refers specifically to the time *after* completion of erosion and rebound. Continuing uplift would not permit the Airy isostatic assumption to be in force. Further, one must surely examine a basin undergoing uplift/erosion for the possibility of responses to flexural, viscoelastic, and/or tectonic forces, which could change completely the assumptions (and determinations!) of eroded amounts of sedimentary material (Turcotte and Schubert, 1982).

In the Barents Sea, current water depth is around 200 to 400 m, while the Quaternary sediments overlying the erosional truncation surface vary from about 100 m to 1 km in thickness. If the elevated land section is ignored then, from Eq. (10.57), we have

$$E \geq [w(\rho_M - \rho_w) + q(\rho_M - \rho_q)]/(\rho_M - \rho_E). \tag{10.58}$$

The smallest estimate of the right-hand side of inequality (10.58) occurs when $w$ and $q$ are smallest, $\rho_q$ is largest, and $\rho_M - \rho_E$ is largest. For $w \cong$ 200 m, $q \approx 100$ m, $\rho_q \approx 2.65$ to 2.85 g/cm$^3$ (fully compacted density for most sedimentary lithologies) and $\rho_e \cong 1.6$ to 1.9 g/cm$^3$ (the surface depositional range of density for most sedimentary lithologies), it follows that, with a mantle density of $\rho_M \approx 3.3$ g/cm$^3$, then

$$E_{\text{smallest}} \geq 330 \text{ m}. \tag{10.59a}$$

The largest estimate of the right-hand side of inequality (10.58) occurs when $w$ and $q$ are largest, $\rho_q$ is smallest, and $\rho_E$ is largest. With $w = 400$ m, $q = 1000$ m, $\rho_q = 1.6$ g/cm$^3$, and $\rho_E = 2.85$ g/cm$^3$, then

$$E_{\text{largest}} \geq 5800 \text{ m}. \tag{10.59b}$$

Thus, the *minimum* estimate of erosion (ignoring completely the elevated land area possibility) under Airy isostatic rebound is difficult to pin down in its extreme end member cases to better than lying between 0.3 to 5.8 km, for an average of about 3.2 ± 2.6 km. Also, when models of glacial erosion and rebound other than Airy isostasy are considered, presumably estimates of erosion and the estimated uncertainty on the eroded thickness for each such model need to be addressed.

## D. DISCUSSION

One of the major factors influencing the Barents Sea Basin is the erosion of most of the Tertiary section. From many perspectives it is extremely important to be able to assess the amount of erosion and to provide some measure of resolution on the uncertainty of the estimated eroded thickness.

Three factors confound the ability to evaluate erosional amounts. First, some model or models are required of processes that are determined by, or modified by, an erosional event. The models may reflect any sort of activity (chemical, physical, geometric, geological) in which erosion plays a role. In general, the models are dependent on intrinsic assumptions and there may be, and often is, more than one model behavior suggested in the literature for the characteristic activity response. The confounding factor here is that the functional form of behavior for the model has to be prescribed, and many functional forms are available. The determination and resolution of the amount of the erosion (and the time of erosion if possible) must be analyzed with respect to variations in intrinsic model asumptions.

Second, within the framework of particular models, each such model has its own intrinsic parameters describing response characteristics (doubling temperatures, reaction rates, surficial depositional porosity, and so on). Such parameters are often not only variable from location to location in each and every sedimentary basin but have to be determined (or provided) by the data to hand and the uncertainty on the intrinsic parameter values used has to be allowed for.

Third, the ability to determine, or at least bracket, the size of an erosional event is limited by the quality, quantity, and sampling frequency of the available data being used as control constraints to extract an estimate of erosion amount from a process model.

What we have done in this section is examine some of the different types of methods that are often used to assess erosion amounts. Three dominant factors stand out:

1. In terms of thermal indicators (see Chapter 4), which do contain some information on erosion events, the different intrinsic model assumptions, the physicochemical parameters and their uncertainties, together with the need to use the thermal indicator to extract not only the erosion amount

and/or time, but also the paleothermal history, make it difficult to extract a well-constrained estimate of erosion even when copious, high-quality and well-sampled data are present. The bulk of the constraints from the data apparently are needed to determine other parameters, with low sensitivity left over for erosion resolution determination.

2. In terms of dynamic rock property indicators, it appears that both the intrinsic model assumptions and the uncertainty in parameter values associated with proposed connections of porosity to depth completely swamp the ability of such models to be at all accurate in their estimates of erosion amounts. Indeed, Løseth *et al.* (1993) made precisely this point when they remarked "The compaction parameter ($c$) of Sclater and Christie (1980) is difficult to determine from the available data base due to a lack of well data and a fairly thin interval which is studied. Therefore the analysis has been done with a predefined set of compaction parameters." The same problem occurs in sonic log and drilling exponent methods, as is also clear from Skagen's (1992) work.

3. In terms of geometric and geological process models of behavior (of which the salt mushroom case and the glacial isostatic rebound case are typical illustrations), the problems here have to do with the intrinsic assumptions of the models, the parameter values that have to be assigned, and the coarser scale of data than is afforded by downhole measurements.

In all events, it would seem difficult to provide an unequivocal estimate of the erosion amount in the Barents Sea that is *less* uncertain for any one method than at least 500 to 1000 m, and that is often much more poorly resolved. Perhaps attempts to use multiple indicators of a common erosion event, in a simultaneous manner, would be more successful because, in such a case, all indicators are required to produce the same amount of erosion. Such multiple indicator methods used simultaneously have been of considerable help in thermal indicator work (Lerche, 1990a) and in dynamic indicator tomography problems (Lerche, 1991).

# 11

# ASSIGNING RISK TO SCIENTIFIC RESULTS

## I. RISK AND PROBABILITY IN RESOURCE ASSESSMENT

### A. INTRODUCTION

Geological processes related to petroleum generation, migration, and accumulation are complicated and no model can simulate these processes exactly. A comprehensive examination is required of how the results of the model are influenced by changes in the assumptions and parameters of the model and by errors in the data, thereby making the model a more effective tool in basin analysis (Tetzlaff, 1996; Burnett, 1996). Integrated models consist of three parts: a geohistory model, a thermal history model, and a hydrocarbon generation model.

1. *Geohistory model.* Through a simulation of the fluid flow movement in sediments caused by the compaction of the sediments, the geohistory model reconstructs the burial history, basement subsidence, vertical and horizontal fluid flow, and the changes of porosity, permeability, pressure, and fluid flow rate with both time and depth across a basin. Also, the evolution of cementation, dissolution, and fracturing, caused by abnormally high pore pressure, are simulated in terms of the change in formation permeability. The input data required to run a geohistory model are the depth and age of each formation base at a suite of locations in the basin and the lithology and paleowater depth of each formation.
2. *Thermal history model.* Based on the burial history created in a geohistory model, the thermal history model reconstructs the thermal history by comparing predicted thermal indicator values (such as vitrinite reflectance) to measurements down a borehole, and by adjusting the paleoheat flux to

minimize discordances. The outputs are a heat flow change with time, a temperature change with time and depth, and vitrinite reflectance changes with time and depth. The input data required to run the thermal history model are the temperature at the sediment surface, bottom hole temperature (or thermal gradient/present-day heat flow), and some thermal indicator measurements with depth.

3. *Hydrocarbon generation model.* The hydrocarbon generation model is based on the kinetics of kerogen degradation, and the general scheme of evolution uses mathematical models to simulate hydrocarbon generation. Based on the burial history and temperature history produced in the model, the generation model gives the absolute amount of hydrocarbons generated (per gram kerogen) with time and depth. The input data required to run the generation model are the content of different kerogen types of each formation.

Results that need to be examined in the analysis include formation thickness, formation porosity, formation permeability, formation pressure, heat flow with time, temperature with time and depth, vitrinite reflectance with time and depth, "oil window" in terms of time and depth, and amount of hydrocarbons generated.

Three groups of variables influence the results:

1. *Input data.* Very commonly there are errors in the measurements of the geological and geochemical data and, in addition, the data are nonuniformly sampled with depth at different locations as well as being of different total number at each location. The variables that need to be examined in the input data are depth and age of each formation base, lithology and paleowater depth of each formation, unconformity time and eroded thickness, temperature at the sediment surface, bottom hole temperature, heat flow at present day, thermal gradient, vitrinite reflectance, kerogen type, and content of each formation.

2. *Equation parameters.* Most of the parameters/constants in the equations used in the models are based on empirical data (for example, a depositional porosity of shale, 0.62, is often used as a default in geohistory models). The following parameters need to be examined for the sensitivity of output results to variations in the input parameters: depositional porosity, permeability and frame pressure, viscosity of the fluid in the sediments, parameters in void ratio/frame pressure functions and in the void ratio/permeability functions for each lithology (see later), critical temperature and doubling temperature in the vitrinite reflectance equation, thermal conductivity of each lithology, activation energies and frequency factors in the generation model, etc.

3. *Intrinsic assumptions.* Most models contain intrinsic assumptions such as the 1-D nature versus a 2-D problem, such as the replacement of complete isostatic movement of the basement by partial flexural compensation, etc.

Examination of these assumptions would make for a very long chapter, and on this ground are not included, although we recognize the pressing need for such a development (Cao and Lerche, 1989). The results of models have to be evaluated for four factors: (1) the sensitivity of the model results to changes in input assumptions, parameters or data, that is, how much model results vary with input changes; (2) the resolution of the model results, that is, whether the mean output value of a quantity is well resolved (low variance) or is poorly resolved (high variance); (3) the precision of output results (e.g., there is little point in quoting output temperature values with depth to decimal level values if the input controls are uncertain by much larger factors); and (4) the uniqueness of the model results (e.g., it can happen that two very different models of paleoheat flux variations with time are consistent with observed thermal indicator data at the present day. In that case the model results are not unique). Thus, on the path to achieving an evaluation of hydrocarbon accumulation site determinations, the four major concerns on model results are (1) resolution, (2) sensitivity, (3) uniqueness, and (4) precision. These four factors are controlled to a greater or lesser extent by the three overriding inputs of (1) intrinsic assumptions, (2) parameter values, and (3) quality, quantity, and frequency distribution of data.

A detailed analysis of the evaluation of basins taking into account the three input factors and the four output concerns has been given elsewhere (Lerche, 1990a,b) for dynamic, thermal, and hydrocarbon assessments. It cannot be stated too strongly that every case history investigated should be evaluated in relation to the points made here. However, the appropriateness of any quantitative basin analysis result in relation to the evolution of a basin can be no better than the quality and quantity of the modeled behaviors. At any given stage of knowledge of a sedimentary basin, there are always the questions of determining what can be garnered without equivocation, what can be achieved with a degree of equivocation that is acceptable, what cannot be evaluated without further data, and what measures of "acceptable equivocation" can be provided. The purpose of this section is to provide a risk analysis assessment for quantitative basin analysis studies when uncertainty is allowed for.

The development of quantitative basin analysis can be separated into three stages (Fig. 11.1): the first stage of *model development* concentrates on the construction and use of mathematical and numerical models of basinal evolution; the second stage is *uncertainty and sensitivity,* in which the main aim is to examine the stability of model results related to the uncertainties and ranges of variations in assumptions, parameters, and input data; the third stage, and the one we are dominantly concerned with here, is *risk analysis,* in which model results are assigned risk factors dependent on the uncertainties and sensitivities of the models used, as in stage two.

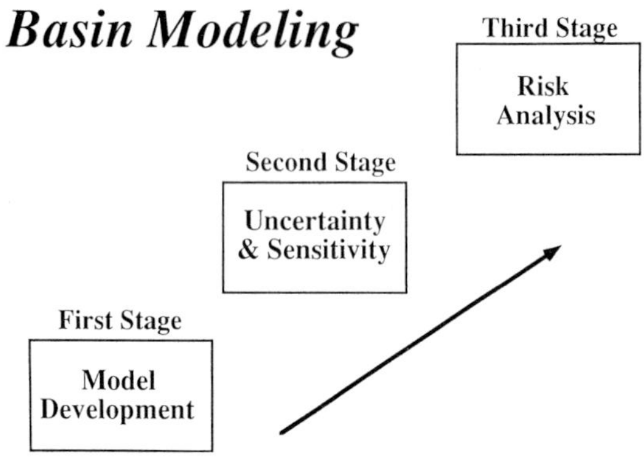


FIGURE 11.1 Development of basin modeling.

The purpose, then, is to provide quantitative probability measures to outcomes of model calculations in order to have objective, reproducible estimates of the likelihood of uncertainty. From such measures strategic exploration decisions can be made, tied also to the probability of finding economic hydrocarbon accumulations (Lerche, 1992a).

According to Yukler and Kokesh (1984), mathematical models are used to simulate complex processes with one or more variables and thus they are essential in the assessment of hydrocarbon resources. Mathematical models are applied either as statistical models or as deterministic models to reconstruct and predict geological processes assuming that these processes are deterministic. The statistical models are mainly used in estimation of hydrocarbon resources because they cannot directly analyze the dynamic processes of hydrocarbon generation, migration, and accumulation. A Monte Carlo simulation is usually employed in the statistical models to construct various probability curves, with the most commonly used assessment methods being geological analogy (Weeks, 1952; Conybeare, 1965; Bally, 1975; Pitcher, 1976; Warren, 1979), delphi (Miller *et al.,* 1975), areal and volumetric yields (Stoian, 1965; Wahlstrom *et al.,* 1967; Smith, 1968; Jones, 1975; Newendorp, 1975; Roadifer, 1975), geochemical yields (Conybeare, 1965; McDowell, 1975), and field distributions (Atwater, 1956; Roy *et al.,* 1975). The need for such measures is illustrated by the following specific example.

The Chukchi Sea Basin, Alaska, was (prior to 1989) a true frontier basin in that only seismic information was available from five regional lines; no wells were then available. Nevertheless, an estimate of likely hydrocarbon proneness of the basin was required in order to decide on future drilling and leasing commitments.

The "background" geology of the basin, in relation to paleogeography and continental plate motion, can be estimated so that a rough idea of basin setting, paleotemperature (warm/cold), likely tectonic events influencing the basinal development, and likely lithologies are at hand. With this background information, it is then possible to identify, from the seismic sections, tectonic sequences (compression/extension/rotated blocks, etc.) and to suggest, from sequence stratigraphy, the possible chronostratigraphy of sedimentary units as a function of sea level variations. From the structural and stratigraphic identifications, and from estimates of sedimentary load at each location, some inference can then be made of basement motion (isostatic/flexural/viscoelastic) and of fault development with time. Seismic velocities, tied to a Wyllie-type (1963) model relating velocity and porosity, can often be used to assess a rough porosity with depth behavior, as can gravity (i.e., density) modeling across the basin. At this minimal data stage, the determination of paleoheat flux, or even present-day heat flux, is a problem because there are only a few indications to guide and control. Some form of model is required. At least four possibilities are available:

1. Arbitrarily assign a basement heat flux with time according to idiosyncratic preference. This procedure is recommended only as a last resort.
2. Tie the paleoheat flux model at each location to a basement motion (McKenzie-type rift; Royden intrusive model, etc.).
3. Tie the motion of the section as a whole, with sediment loading, to the depth to detachment under a model of regional load compensation assume this detachment is always at a fixed brittle/ductile temperature (450° to 550°C is conventional).
4. Use gravity/aeromagnetics or seismics to figure an estimate of the depth today to the brittle/ductile transition. Estimate the temperature at the transition from analogy. Construct a present-day temperature gradient estimate and then allow for an uncertainty range (e.g., $2.5 \pm 0.5$°C/100 m); assume this range of gradients covers all possibilities.

For preference it is appropriate to try to do the last three of the above four, thus running several cases. In this way some idea of the range of output results can be achieved even with minimal data.

Such considerations have been given in detail elsewhere (Cao *et al.*, 1993). For illustration purposes, we concentrate here on the paleoheat flux. As noted in Cao *et al.* (1993), there is uncertainty in the paleoheat flux because there were no wells at all in the Chukchi Sea Basin prior to 1989. Three possible extreme heat flux patterns were proposed based on geological and geothermal data, with the geothermal data coming from the nearest wells, which are several hundred kilometers from the center of the basin. Figure 11.2 shows the three paleoheat flux patterns used (labeled Increase, Constant, and Decrease) based on both individual well studies (shown by dashed lines) and on geothermal models of rifting: Increase

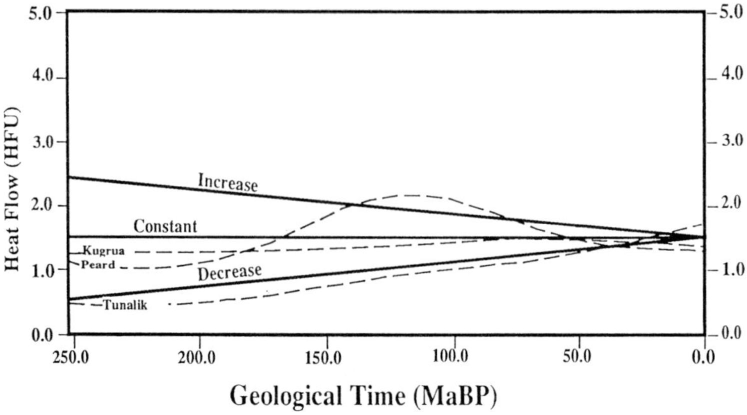


FIGURE 11.2 Three possible heat flow patterns in the Chukchi Sea Basin, Alaska.

(Decrease) implies higher (lower) heat flux in the past. (The extreme values were obtained by varying uplift amounts and uplift timing, together with parameters describing equations of state, for the wells. The point is that, regardless of how determined, the three extreme estimates of heat flux are used here to illustrate application of the probability procedure).

The three paleoheat flux patterns give three considerably different temperature histories for sediments in the basin, as shown in Table 11.1 for the Pebble Shale Unit (considered to be a likely source rock for hydrocarbons based on extrapolation from the National Petroleum Reserve, Alaska, where the Pebble Shale is known from drilling to be a good source rock). The differences in paleotemperature are sufficient to suggest significant

TABLE 11.1 Paleotemperature at the Bottom of the Pebble Shale Unit in the Chukchi Sea Basin, Alaska

| | Heat flow pattern | | |
|---|---|---|---|
| Depth (m) / Time (MaBP) | Decreasing heat flow with time (lower in the past) | Constant heat flow with time | Increasing heat flow with time (higher in the past) |
| 4,800 / 75 | 179 | 224 | 268 |
| 5,000 / 55 | 190 | 223 | 255 |
| 5,900 / 23 | 229 | 244 | 259 |

Temperatures are in degrees centigrade.

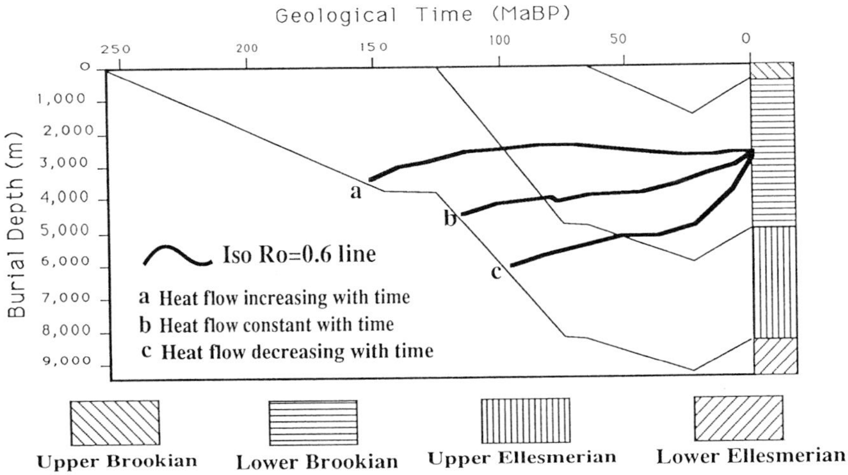


FIGURE 11.3 Burial history at a location of the Chukchi Sea Basin with superimposed vitrinite isoreflectance lines of 0.6% for each of the three paleoheat flux cases.

uncertainties in the amount and timing of thermal maturity, and also in the "oil window" depth and time in relation to the development of likely reservoir structures.

For example, Fig. 11.3 shows a burial history at a location in the Chukchi Sea Basin with superimposed vitrinite isoreflectance lines of 0.6% for each of the three paleoheat flux cases. The reflectance value of 0.6 is often considered to be indicative of entry into the oil window for sediment maturity (Hunt, 1979; Tissot and Welte, 1984). Inspection of Fig. 11.3 shows that there is about a 50 Ma spread in time in the $R_0 = 0.6$ onset estimates, and a spread of more than 3 km in depth for onset of the hydrocarbon oil window. The large ranges of uncertainty are direct consequences of the paleoheat flux uncertainty. A serious concern arises about providing some probabilistic measure of correctness of these behaviors in order to lessen the risk of estimating hydrocarbon charge to reservoirs.

## B. CUMULATIVE PROBABILITY AND ERROR ASSESSMENT

In principle, Monte Carlo procedures should be applied to the study of risk assessment and most probable outcomes of basin analysis models (Smith and Buckee, 1985). The probability distribution of each model output result should be evaluated by considering all possible values of each assumption, each parameter, and each datum point, within determined ranges. Then, an intrinsic probability distribution of finding each such input control, assigned within its given range (e.g., uniform, Gaussian, triangular,

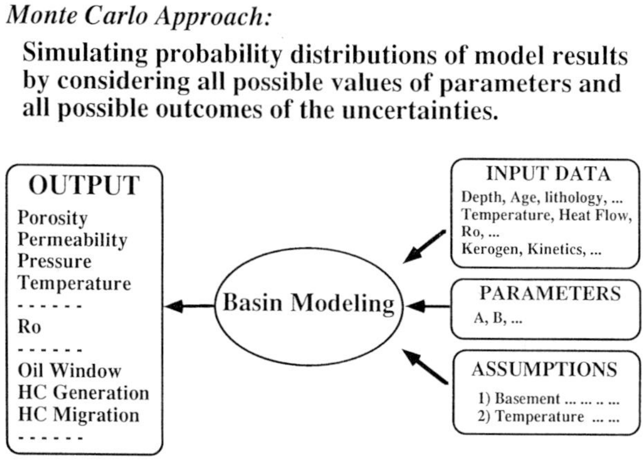


FIGURE 11.4 Monte Carlo approach for risk analysis in basin modeling.

log-normal), should be computed. Using a multi–Monte Carlo approach to sample values for each and every input control a large number of times, a basin modeling calculation can then be run through to produce a set of outputs. The probability distributions of the respective outputs can then be recorded and used in risk assessment (Fig. 11.4). Although the Monte Carlo procedure is completely general and can be applied to any basin analysis model, once the range and underlying probability distributions for each parameter (and for the measurement errors in data) have been provided, the problem is that such a numerical investigation can be extremely computer intensive, perhaps one of the reasons that such methods are not routinely applied to the results derived from basin analysis models.

It would, therefore, be of help if alternative procedures could be used, which would provide the same probability outcome, but which would not demand such massive computer time investment as do the Monte Carlo methods. This chapter demonstrates how to use cumulative probability techniques to achieve such a goal.

Appendices A and B give not only the general mathematics showing how to develop a cumulative probability methodology for handling the uncertainty, but also how to use the procedure practically. This relegation of the quantitative mathematical procedures to the appendices is done deliberately so as not to interrupt the flow of the main thrust of the chapter, which is the use of quantitative measures to handle the uncertainty in basin modeling computations. This focus in the text is aimed at actual applications in order to illustrate, by direct example, the ability of the cumulative probability method to generate measures of probable behavior (Fig. 11.5) for particular attributes of interest.

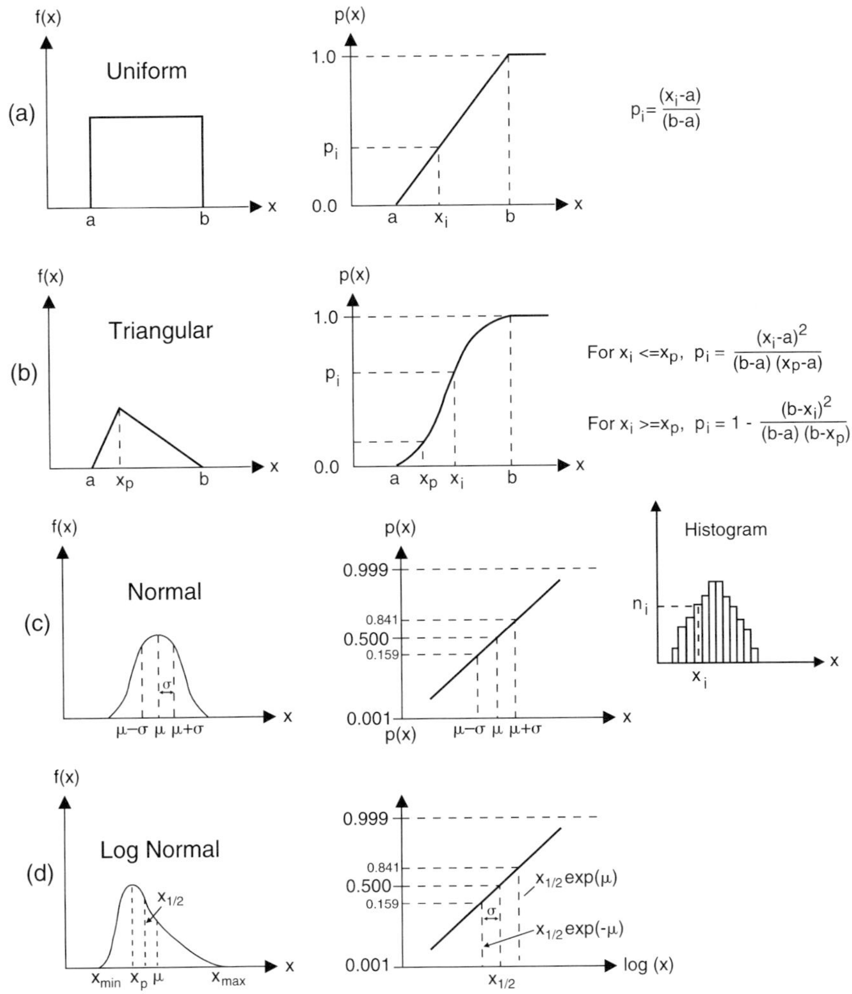


FIGURE 11.5 Four common probability distributions and their corresponding cumulative probability distributions (a) uniform; (b) triangular; (c) normal; (d) log-normal.

## C. APPLICATION TO THE NAVARIN BASIN COST NO. 1 WELL

A detailed quantitative geological investigation of the dynamic, thermal, and hydrocarbon histories of the COST No. 1 well has been given elsewhere (Cao and Lerche, 1989) and does not need to be repeated here. The location of the COST No.1 well is shown in Fig. 11.6a, and the corresponding geological column is shown in Fig. 11.6b. For the purposes of illustrating how the cumulative probability and risk assessment procedure operates,

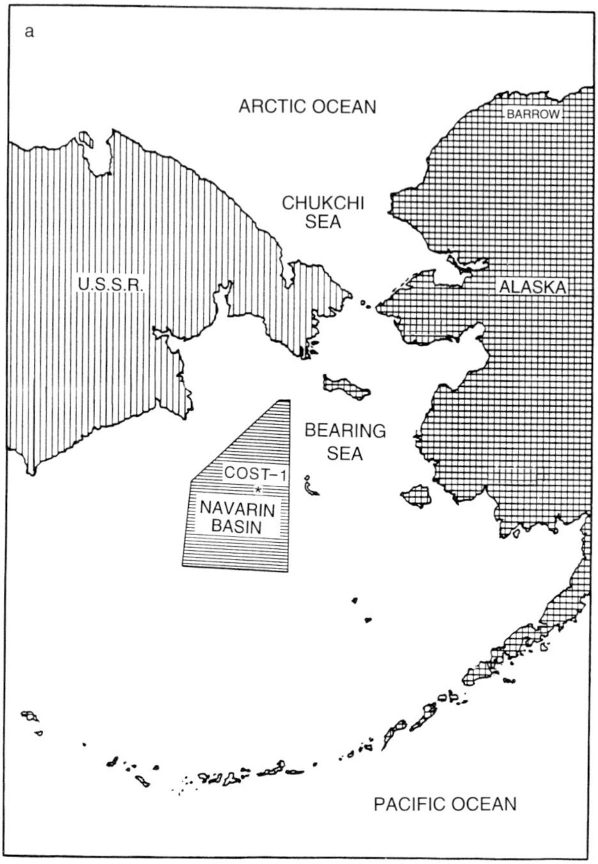


FIGURE 11.6 (a) Location of COST No. 1 Well in the Navarin Basin, Bering Sea, Alaska. (b) Corresponding geological column for the well (modified from Turner *et al.*, 1984).

we use the one-dimensional fluid flow/compaction code, 1D-DTIT, to investigate dynamic and thermal maturity evolution with time in the well. The code allows deposition of each lithologic unit with time, ascribes a relation between permeability and porosity (void ratio) to each lithology, as well as a frame pressure/void ratio equation of state for each lithology, and then allows the sediments to compact according to fluid loss rates controlled by Darcy's law in a compacting system. Sediment overburden is allowed for as is the ability of fluid to escape from formations with time, depending on both the permeability and fluid pressure drive. Empirically (Dutta, 1989) it would appear that each lithology has a permeability/void ratio equation of state of the form:

$$k = k_*(e/e_*)^{\mathrm{B}},$$

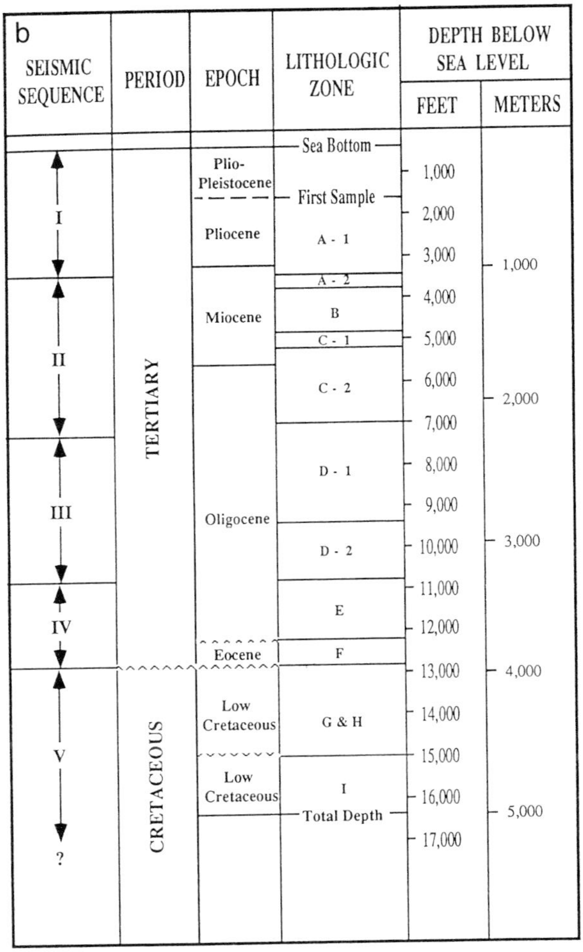


FIGURE 11.6 (*Continued*)

where $k_*(e_*)$ is the depositional permeability (void ratio) of the lithology; $k(e)$ are the values after burial, and B is a parameter in the rough range of about 3 ± 3 for most lithologies. [Porosity, $\phi$, is related to void ratio, $e$, by $\phi = e/(1 + e)$.] Equally, the empirical relation between frame pressure (effective stress), $p_f$, and void ratio, $e$, for each lithology is of the general form:

$$p_f = p_{f*}(e/e_*)^{-1/A},$$

where $p_{f*}$ is a scaling value, and A is a constant, also in the rough range 3 ± 3. Each lithology (sand, shale, sandy-shale, carbonate, salt, etc.) has

TABLE 11.2 Uncertainties in the Nine Parameters Used in the Model, Presented as Minimum, Maximum, and Most Likely Values

| Parameter value | Minimum value | Maximum value | Most likely value |
|---|---|---|---|
| Heat flow $\beta$ (1/Ma) | −0.015 | 0.450 | 0.015 |
| Shale A | 0.60 | 5.00 | 1.77 |
| Shale B | 1.10 | 5.10 | 2.48 |
| Shale $k^*$ (md) | 70.0 | 220.0 | 160.0 |
| Sandy shale A | 1.20 | 10.00 | 4.97 |
| Sandy shale B | 0.80 | 4.70 | 2.50 |
| Sandy shale $k^*$ (md) | 10.0 | 530.0 | 300.0 |
| Sandy shale $p^*$ (atm) | 0.21 | 3.00 | 1.70 |
| Sandy shale $e^*$ | 0.16 | 2.03 | 1.07 |

different values and ranges for each of the parameters $k_*$, $e_*$, $p_{f*}$, A, and B. The evolution of excess fluid pressure, porosity, and other dynamic quantities of interest is then influenced by choices made for the parameter values. The difficulty, as always, is that we do not know the values of these parameters with precision; so any dynamic predictions of, say, the buildup of excess fluid pressure with time will entail an uncertainty. The task of the cumulative probability technique is to address that uncertainty in some objective, reproducible, quantitative manner.

To illustrate the method, consider the two lithologies, shale and sandy-shale, which constitute the lithologic zones of the Navarin Basin COST No. 1 well from mudline to near base Oligocene (at 4000 m). Start with the empirical ranges thought to represent approximately the values of parameters considered geologically acceptable based on whatever criteria are considered controlling factors on the ranges. For instance, observations and extrapolations of present-day porosity with depth trends from many wells suggest that it is very unlikely that the surface void ratio for a sandy-shale lithology lies outside the broad range of $0.16 \leq e^* \leq 2.03$ with a prevalent value of 1.07. In Table 11.2 we present the minimum, maximum, and most likely values of nine parameters that influence both the dynamical (eight parameters) and thermal (eight dynamic plus the rate of change of heat flux with time) evolutions. The thermal parameter, $\beta$, is determined from the heat flow as

$$Q(t) = Q_0 \exp(\beta t),$$

where $Q_0$ is present-day inferred heat flux, $t$ is geological time (with $t = 0$ being the present day), and $\beta$ (units of $\text{Ma}^{-1}$) measures whether paleoheat flux was higher ($\beta > 0$), constant ($\beta = 0$), or lower ($\beta < 0$) in the past.

The range of $\beta$ can be assessed either from a model behavior (e.g., a rifting model) or from uncertainty in $\beta$ when attempting to perform a thermal inversion with control against present-day observed thermal indicators with depth. With the range and most likely value of each parameter provided, the burial and thermal maturity histories of the COST No. 1 Well can be run in the manner outlined in the previous section. Two simple pictures can be drawn: first one can plot present-day predicted behavior with depth for a quantity of interest and superpose on the plot both any observed or inferred values of the quantity as well as the cumulative probability likelihood values. We do so here for porosity, fluid pressure, and vitrinite reflectance. Second, one can plot the burial history with time (which varies for each computer run depending on the parameter values chosen for each run) and superpose the development with time of each quantity of interest (porosity, fluid pressure, and vitrinite reflectance) together with the cumulative probability. In addition to these two simple pictures, one can also plot the present-day cumulative probability of a particular iso-value being reached at a given depth and, with time, the corresponding cumulative probability. From the perspective of when fluids were most likely in motion in the subsurface and when thermal maturity was sufficient to generate hydrocarbons, the cumulative probability plots with time are extremely useful.

### 1. Porosity and Probability

Following the prescription given in the last section, and using the ranges of parameter values of Table 11.2, Fig. 11.7 plots the predicted probability of porosity with depth at the present day. In addition, the nominal measured porosity values are represented by filled circles. No attempt has been made to allow for uncertainty on the measured porosity values, although the scatter in the data raises a cautionary flag that some error of measurement (or inference) is likely present, depending on how the porosity was measured. Nevertheless, to focus on the essence of the cumulative probability procedure, the error in measurement is not discussed here (see, however, Chapter 10 and Lerche, 1993).

Four curves are drawn on Fig. 11.7, representing cumulative probability iso-values of 30%, 50%, 70%, and 90% respectively. For instance at a present-day depth of 3000 m there is a 70% chance that the porosity will be less than about 17% but only a 30% chance the porosity will be less than 8%. A differential probability can be obtained by subtraction; thus the probability that the porosity lies between 28% to 8% at a 3000-m depth is 60% (90% cumulative probability of less than 28%; 30% cumulative probability less than 8%). Clearly, as the ranges allowed for different parameters are chosen to vary in different ways so, too, will the cumulative probability values.

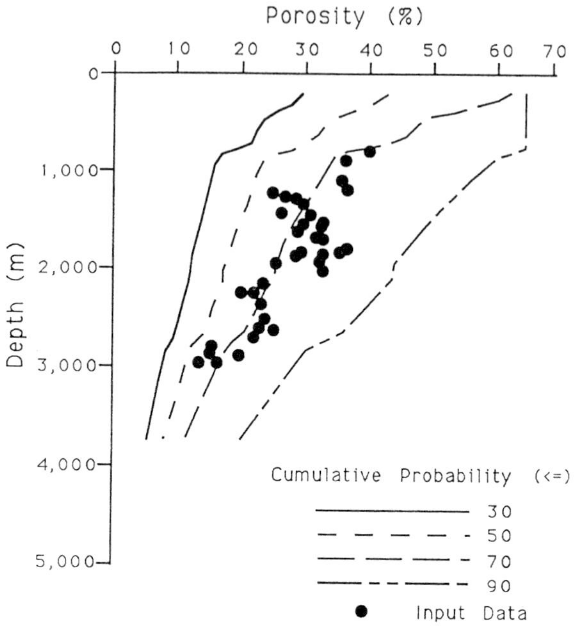


FIGURE 11.7 Predicted cumulative probability of porosity with depth at present day for Navarin COST No. 1 Well.

An alternate way to view the uncertainty in output is to plot cumulative probability with depth for *fixed* values of the porosity. Thus, Fig. 11.8 gives the likelihoods of obtaining a fixed porosity. For example, at 3000 m the cumulative probability that the porosity is greater than 10% is 45%, while the cumulative probabilty that the porosity is less than 20% is 80%. Again, a differential statement can be constructed by subtraction: The probability that the porosity lies between 10% to 20% at 3000 m depth is about 35% (80% cumulative probability of less than 20% porosity; 45% cumulative probability of less than 10% porosity). In one view (Fig. 11.7) we plot the variation of porosity with depth for constant cumulative probability values, while in the other view (Fig. 11.8) we plot the variation of cumulative probability with depth for constant porosity values. Both plots allow information to be assessed quickly as to the degree of uncertainty of information at the present day in different formats.

Of equal or greater importance than assessing the values of present-day porosity outputs is the development of porosity with time, which influences migration of fluids, development of excess pressure, and thermal gradients. Again two different types of plots show different aspects of the uncertainty. For example Figs. 11.9a–c show the burial history versus porosity plots at different constant cumulative probability values. Thus on Fig. 11.9a, for

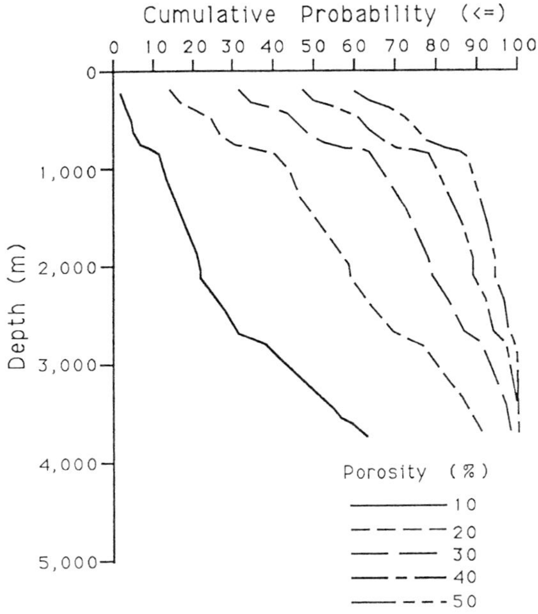


FIGURE 11.8 Porosity versus depth with different cumulative probability curves for Navarin COST No. 1 Well.

example, there is only a 30% chance that the porosity values are less than or equal to the values shown on the figure; while if the cumulative probability is set to 70%, as in Fig. 11.9b, then there is a 70% chance that the porosity values are less than the values given; while at the extreme case of a 90% value for the cumulative probability, the porosity is 90% likely to be less than the values recorded on the burial history curve of Fig. 11.9c. For instance at 20 MYBP it is 90% certain that the porosity at 2000 m depth is less than about 40% to 50%, it is 70% certain that the porosity is less than 20% to 30%, and it is only 30% certain that the porosity is less than 10% to 20%. Thus an evaluation of likely porosity evolution with time for each formation, or with burial history depth, can be obtained.

By flipping the argument around we can ask: For a fixed porosity value, what is the likely cumulative probability evolution with burial history? Such a situation is sketched in Figs. 11.10a–c for porosities of 10%, 20%, and 30%, respectively. Note that at 20 MYBP at a depth of 2000 m, the probability is 30% that the porosity is less than 10%, the probability is 40% that the porosity is less than 20%, and the probability is 90% that the porosity is less than 30%. Thus it is unlikely (3 chances out of 10) that a porosity less than 10% occurred in the formation that was at 2000 m depth at 20 MYBP,

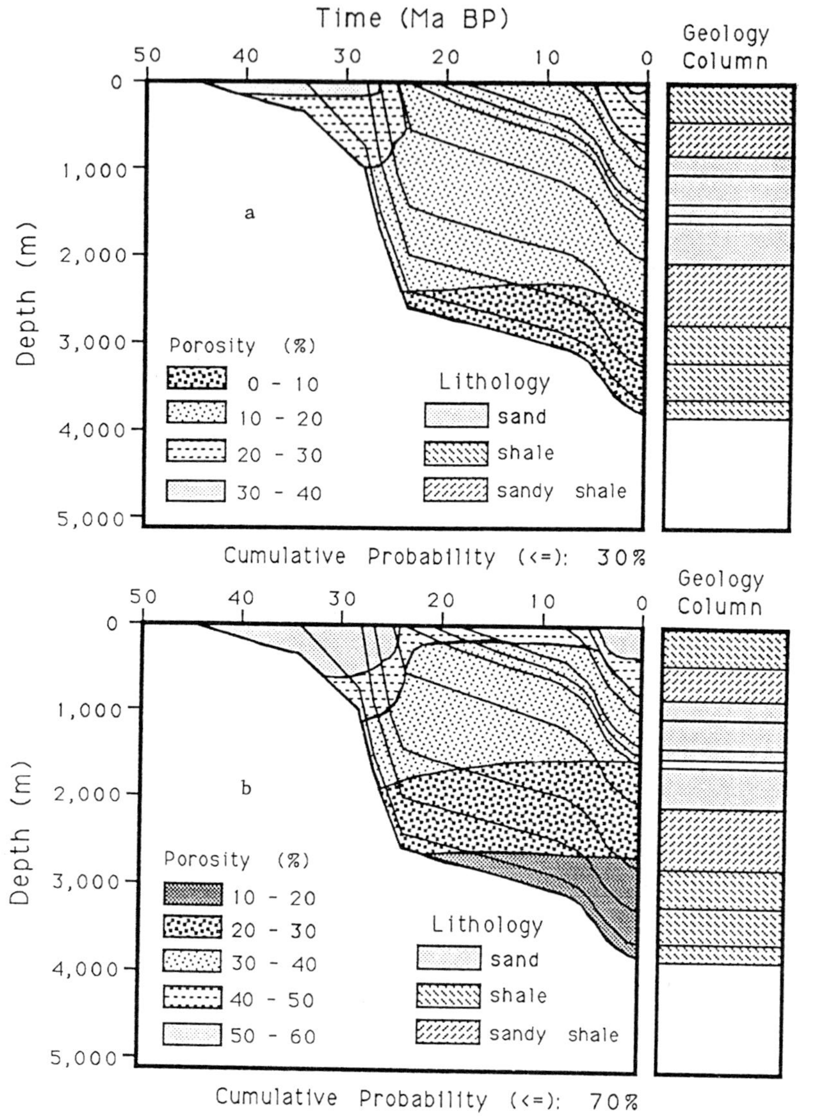


FIGURE 11.9 Burial history versus porosity at different cumulative probability values. (a) 30% cumulative probability; (b) 70% cumulative probability; (c) 90% cumulative probability.

but it is also unlikely (1 chance in 10) that the porosity was greater than 30% at that time and depth.

In short an estimate can be made rather quickly of the expected evolution of porosity and the likely range of variation of porosity based on the range of variation of the nine input parameters of Table 11.2.

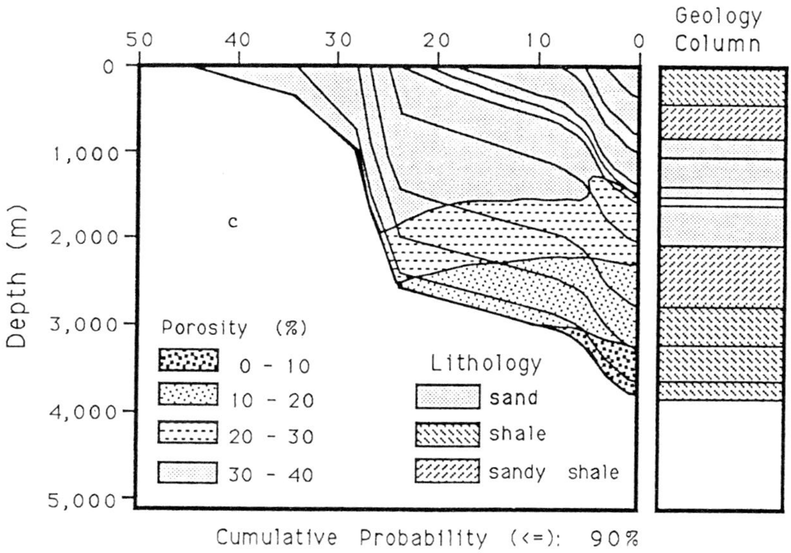


FIGURE 11.9 *(Continued)*

**2. Fluid Pressure and Probability**

Because the fluid flow/compaction code allows fluids to escape according to Darcy's law, and because the difference between overburden weight and frame pressure is supported by a fluid pressure in excess of hydrostatic, it becomes a relatively simple matter to calculate the influence of varying ranges of parameters on the likely total fluid pressure (excess fluid pressure plus hydrostatic pressure) with depth at the present day. For the parameter ranges given in Table 11.2, Fig. 11.11 presents observed fluid pressure measurements with depth (again without ascribing any error or uncertainty to the measurements), together with cumulative probability curves. One can observe, for instance, that there is only a 30% predicted chance of a fluid pressure less than 150 kg/cm$^2$ at 3000 m, a 70% chance of less than 300 kg/cm$^2$ and a 90% chance the fluid pressure is less than 600 kg/cm$^2$.

Curves of present-day predicted behavior can be projected in a different manner, as in Fig. 11.12, where cumulative probability of fluid pressure with depth is plotted for selected values. For instance, at 2000 m depth, Fig. 11.12 indicates a less than 10% chance of obtaining a fluid pressure of less than 50 kg/cm$^2$, about a 40% chance of a fluid pressure of less than 100 kg/cm$^2$, a 65% chance of fluid pressure less than 150 kg/cm$^2$, and a nearly 80% chance of a fluid pressure less than 200 kg/cm$^2$. Viewed from the perspective of Fig. 11.12, this way of presenting results is important for drilling operations because a guide can be given as to the likely pressures expected to be encountered and the chance of encounter.

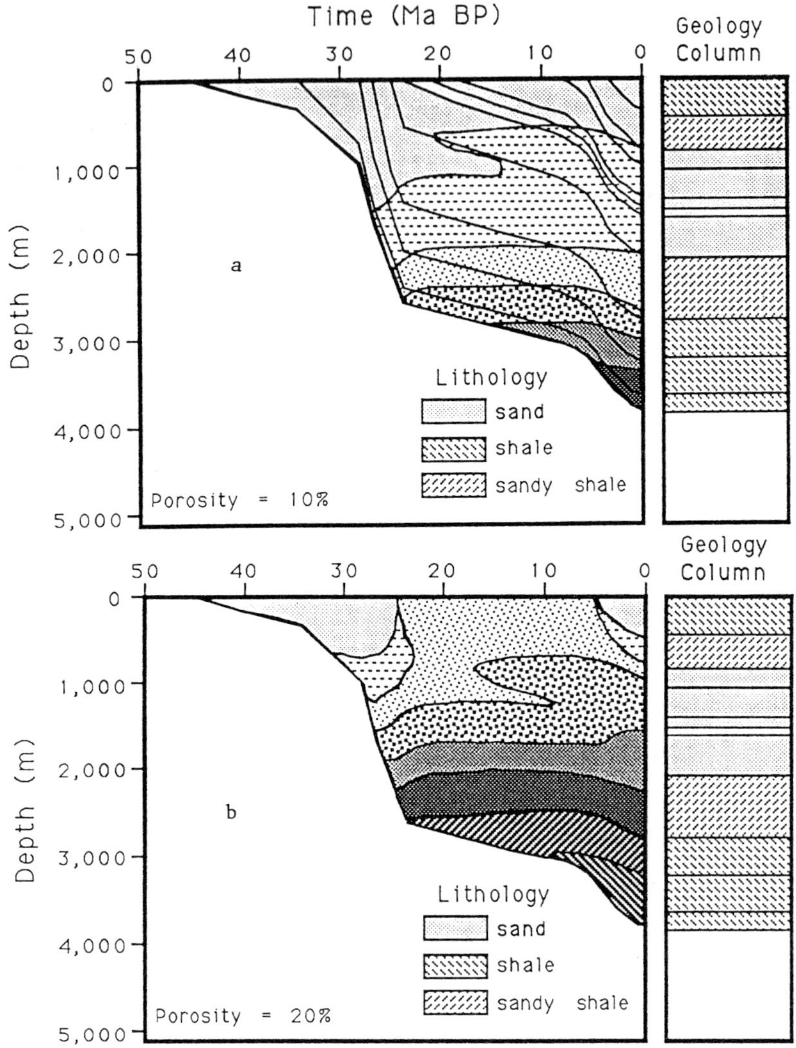


FIGURE 11.10 Burial history versus cumulative probability for porosity. (a) 10% porosity; (b) 20% porosity; (c) 30% porosity.

The buildup of overpressure with time is also of significance in that such buildup indicates the development of a sealing capability in the system for likely hydrocarbon retention, the preservation of porosity, and a thermal gradient increase due to lower thermal conductivity than would occur under nonoverpressured sediment compaction—of interest in developing earlier genesis of hydrocarbons than might otherwise have been thought to occur. Figures 11.13a–c show the burial history with cumulative probability curves

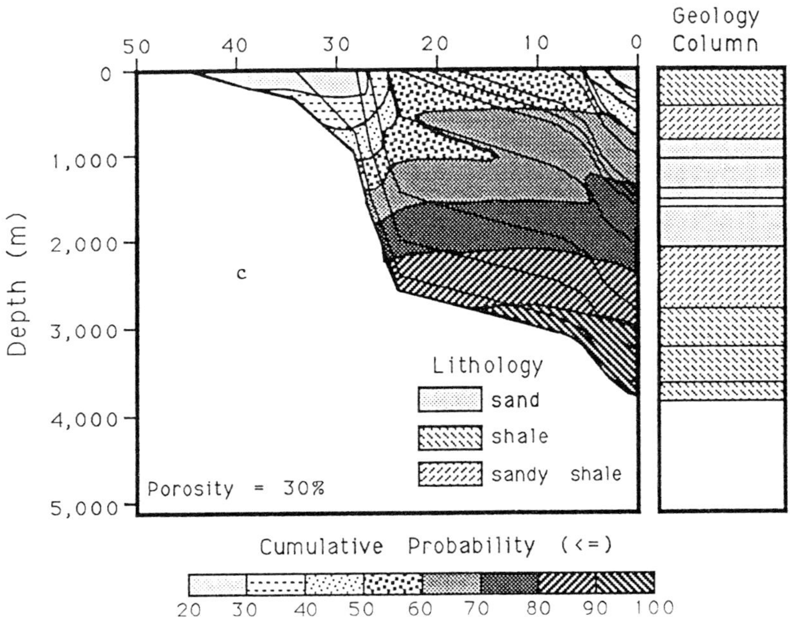


FIGURE 11.10 *(Continued)*

superposed. Thus, in Fig. 11.13a, for a fixed value of 50 kg/cm$^2$ fluid pressure, the cumulative probability curves indicate that the chance of a regime deeper than 1000 m having less than 50 kg/cm$^2$ fluid pressure is less than 40% throughout the whole burial history. Figure 11.13b, drawn for a fluid pressure of 100 kg/cm$^2$, indicates that the probability is less than 40% throughout the whole burial history that the fluid pressure is less than 100 kg/cm$^2$ at depths in excess of 2000 m (60% chance the pressure is greater than 100 kg/cm$^2$). Figure 11.13c (drawn for 200 kg/cm$^2$ fluid pressure) indicates that it is 80% likely that the fluid pressure is less than 200 kg/cm$^2$ during the whole burial history at all depths shallower than about 3000 m.

The other projection of information is given in Figs. 11.14a–c, which show fluid pressure development with burial history for different, fixed, cumulative probability values. Thus, Fig. 11.14a shows that there is only a 30% chance that fluid pressure development will attain values *less* than those given, at any time and depth, on Fig. 11.14a; Fig. 11.14b shows that there is a 70% chance of obtaining values *less* than those drawn; while Fig. 11.14c provides a 90% chance that fluid pressure values will be *less* than those given. Thus one can assess the likelihood of the amount and timing of fluid pressure development in terms of probabilistic ranges determined by the intrinsic assigned uncertainties in the input parameters of Table 11.2.

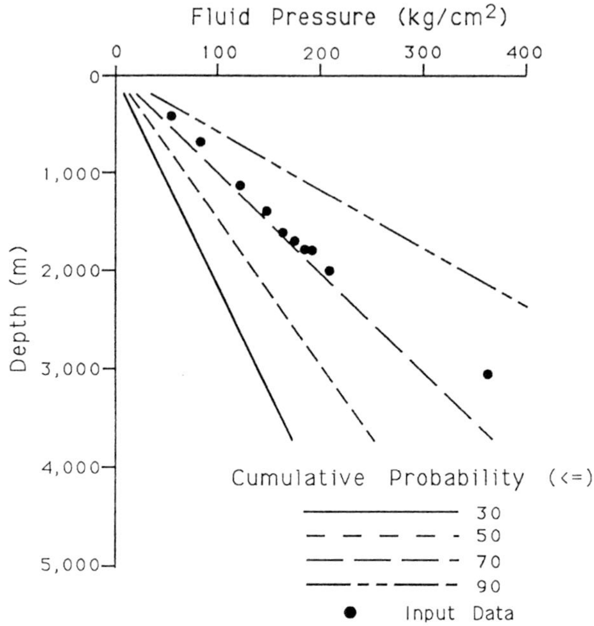


FIGURE 11.11 Fluid pressure versus depth with cumulative probability curves for Navarin COST No. 1 Well.

### 3. Thermal Maturity and Probability

Vitrinite reflectance provides an indirect measure of thermal maturity of hydrocarbon proneness in a basin. The reflectance is dependent not only on parameters influencing the dynamic behavior of the system, but also on both the present-day heat flux and the paleoheat flux. Indeed measurements of vitrinite reflectance with depth have often been used in an inverse sense to determine, or at least bracket, the paleoheat flux (Lerche, 1991). In Table 11.2 we provide a range of paleoheat flux values considered likely to bracket the extremes of the true paleoheat flux. The predicted evolution of vitrinite reflectance depends on both the paleoheat flux and on parameters controlling paleotemperature due to dynamic conditions (thermal conductivity is tied to porosity; formation thicknesses and depths are tied to permeability; etc.). In Fig. 11.15 we provide present-day measured values of vitrinite reflectance with depth. (Once again, the data are taken "as is, where is" and no attempt has been made to include error or uncertainty measures of the data per se.) Superposed on Fig. 11.15 are the corresponding cumulative probability curves based on the ranges of the nine parameters of Table 11.2.

For instance, if one takes a value of reflectance of 0.6% as marking the onset of the oil window, then Fig. 11.15 indicates there is 90% chance that

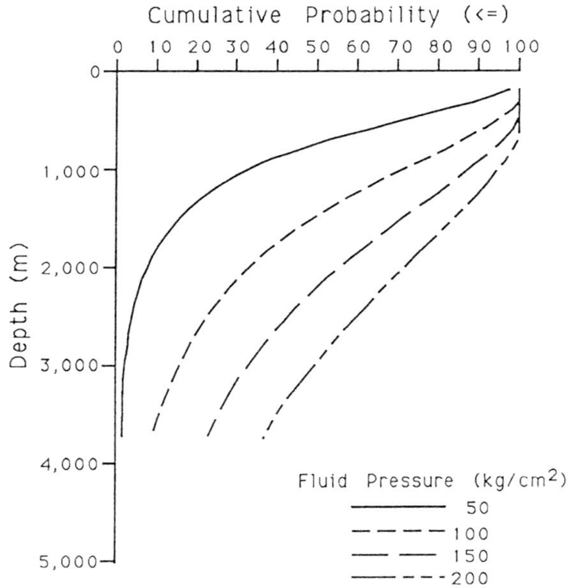


FIGURE 11.12 Predicted cumulative probability with depth at fixed fluid pressure values at present day for Navarin COST No. 1 Well.

this value of reflectance is currently at a depth greater than 1200 m, 70% chance of the value being at a depth greater than about 2500 m, 50% chance of being at a depth greater than 3500 m, and only 30% chance of occurring at a depth in excess of 3800 m. Equally, if a reflectance value of 1.2% is taken as marking the onset to the gas window, then Fig. 11.15 indicates that there is 90% chance of the value occurring at a depth greater than 2700 m, and 70% chance of being at a depth greater than 3700 m. The burial history goes no deeper, today, than 3800 m so it is difficult to establish any further probability statements for greater depths. But the point is made. For instance, it is 70% likely that onset of the oil window (as measured by a reflectance of 0.6%) occurs at a depth today *greater* than 2500 m but *shallower* than 3900 m; it is 70% likely that the gas window today occurs only at depths in excess of 3800 m. Again another projection throws light on the information extractable.

Figure 11.16 shows plots of cumulative probability with depth for various values of vitrinite reflectance. From Fig. 11.16, we see, for instance, that at 2000 m depth there is 55% chance of a reflectance less than 0.4, 80% chance of a reflectance less than 0.6, 90% chance of a reflectance less than 0.8; while at 3000 m depth, there is only 40% chance of a reflectance less than 0.4, 70% chance of a reflectance less than 0.6, and 85% chance of a reflectance less than 0.8. Thus the onset of the oil window can be ranked

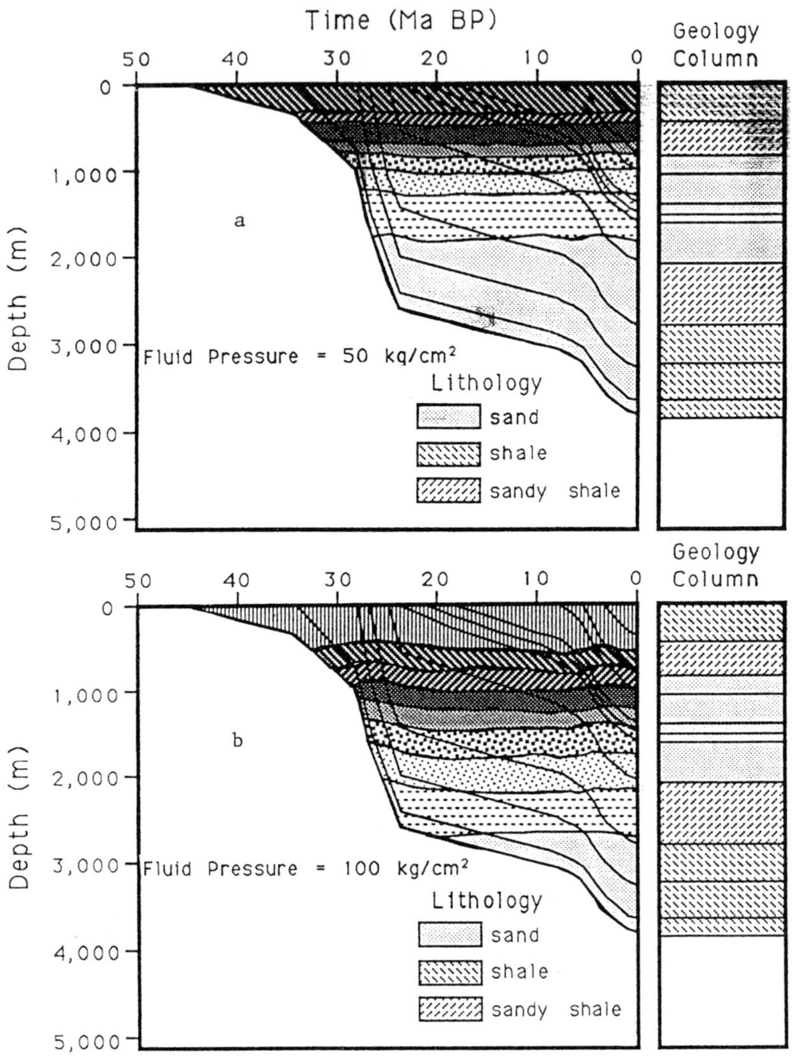


FIGURE 11.13 Burial history with cumulative probability curves for different fluid pressure values. (a) 50 kg/cm$^2$; (b) 100 kg/cm$^2$; (c) 200 kg/cm$^2$.

either by the choice of a fixed vitrinite reflectance value (and then an assessment provided of the probable depth range at which this value occurs), or one can choose a fixed depth and then assess the probability that the vitrinite reflectance range at that depth will encompass the oil window onset today.

Apart from present-day estimates of depths to oil and gas windows based on vitrinite reflectance, there are also the questions of the timing of the onset

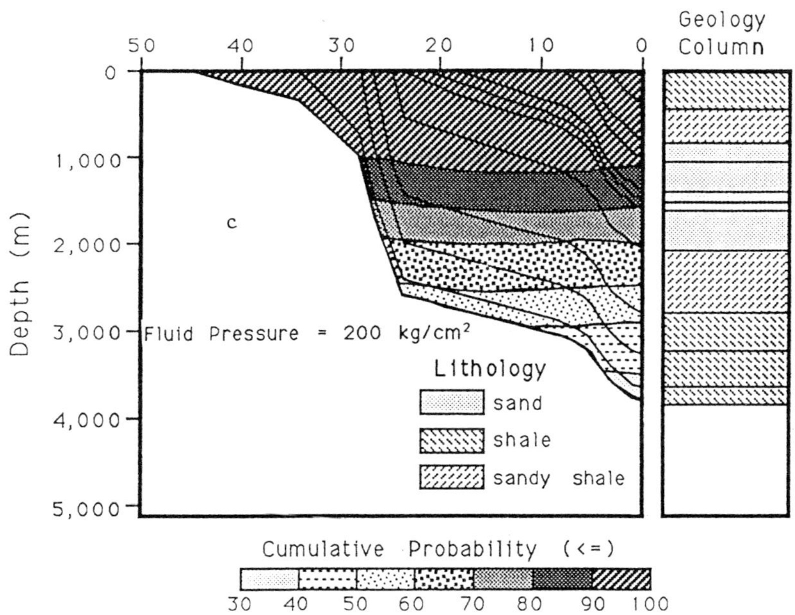


FIGURE 11.13 (*Continued*)

of maturity throughout the burial history of the sediments. Accordingly, in Figs. 11.17a–c the cumulative probability behaviors are plotted as a function of burial history for different fixed values of vitrinite reflectance. For example, Fig. 11.17a (drawn for a vitrinite reflectance of 0.6%) shows that most of the sediments are likely immature (reflectance < 0.6%) throughout most of the burial history. There is a 90% chance that reflectance will be less than 0.6% for all formations except below a line commencing at about 2500 m at 20 MYBP and rising to 1500 m today.

Similarly in Fig. 11.17b, drawn for a reflectance of 1.2%, it can be observed that almost all formations have a 90% chance that, throughout the whole burial history, they never reached 1.2% reflectance, with the exception of the deepest formation, currently at 3800 m, which had a 20% chance of exceeding a reflectance of 1.2% in the last 5 million years. Figure 11.17c shows that there is almost a 100% chance that no formation ever reached a vitrinite reflectance of 1.6% at any time in the burial history.

The plots of vitrinite reflectance with burial history for *fixed* cumulative probability values are recorded in Figs. 11.18a–c. Figure 11.18a, drawn for a probability of 30%, indicates the depth–time range values. There is only a 30% chance that the reflectance values will be small than those shown; Fig. 11.18b provides the 70% chance of obtaining a reflectance value smaller than that shown, while Fig. 11.18c provides the 90% chance burial history–reflectance behavior. Again one sees the shift in pattern of behavior as the

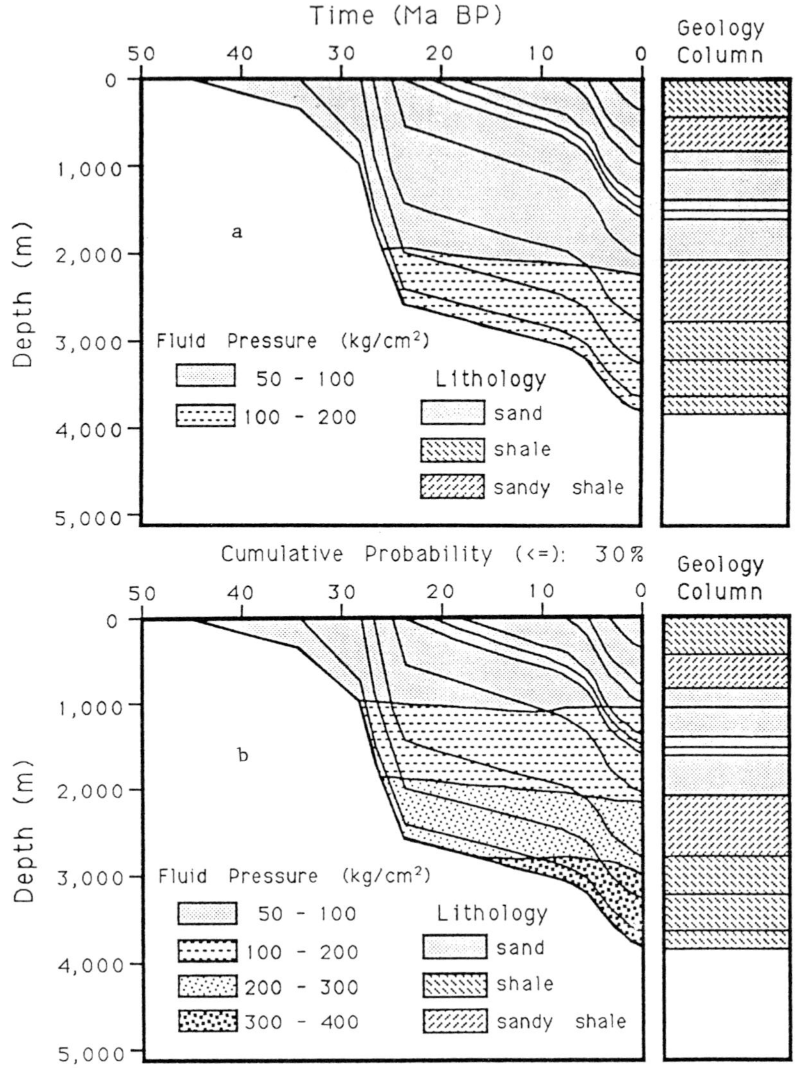


FIGURE 11.14 Burial history with fluid pressure for different cumulative probability values. (a) 30%; (b) 70%; (c) 90%.

probability is increased. For example, reading from Fig. 11.18c there is a 90% likelihood that the vitrinite reflectance at 2000 m at 10 MYBP is greater than 0.5% but less than 1%.

The points to note about the Navarin Basin COST No. 1 Well example are that the uncertainties in dynamic parameters, and the degree of uncertainty of the paleoheat flux, can both be used to assess the likely dynamic and

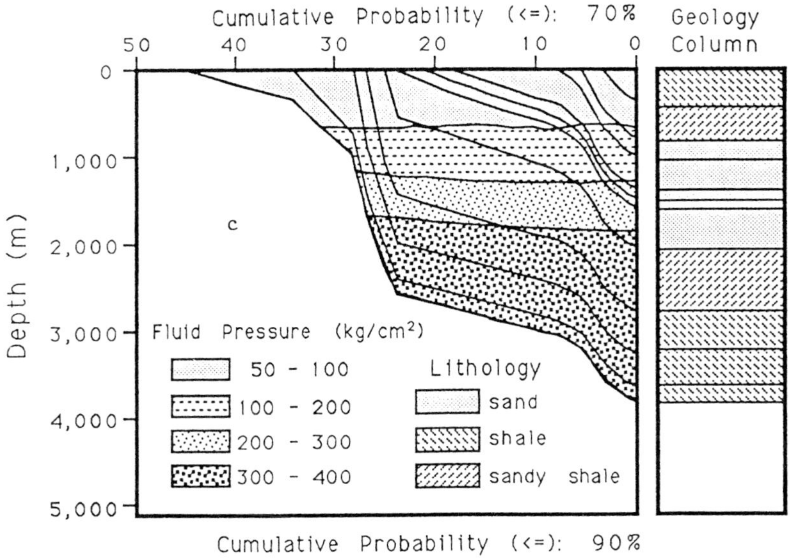


FIGURE 11.14 *(Continued)*

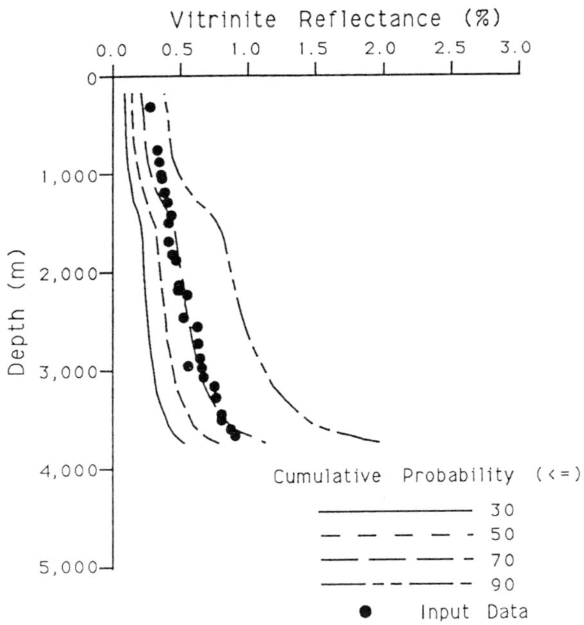


FIGURE 11.15 Vitrinite reflectance versus depth with cumulative probability values for Navarin COST No. 1 Well.

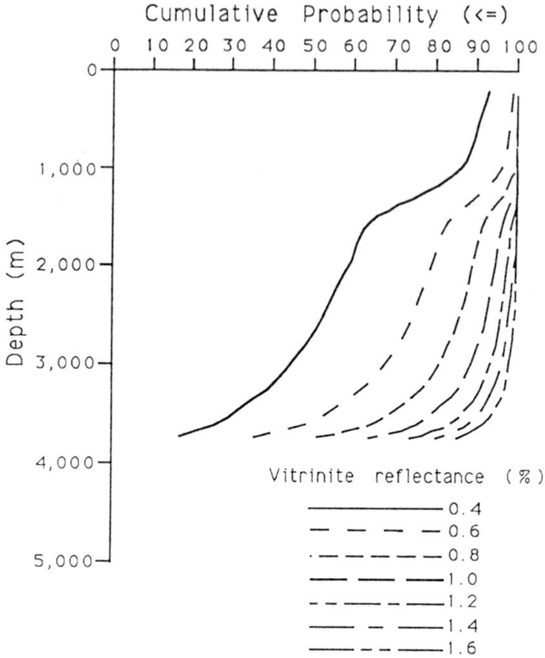


FIGURE 11.16 Predicted cumulative probability of vitrinite reflectance with depth for Navarin COST No. 1 Well.

thermal maturation histories with a quantitative degree of error assignment directly tied to the uncertainty of each and every input parameter. The risk of, say, reaching a particular overpressure or of not reaching a particular vitrinite reflectance value can then be assessed both for the present day (where some control data are available to constrain the degrees of uncertainty) and in the past.

The control of the probabilistic values in this example is facilitated to some extent by the use of the measured data as though those data (1) contained no error, (2) were of sufficient number to control the system uncertainties, and (3) were sampled sufficiently densely with depth to provide accurate control of the system evolution throughout the total burial history.

When such a situation does not occur, then errors and uncertainties due to sparse, poorly sampled, and uncertain measurements can also provide a degree of uncertainty on the likely values to be expected for outputs of basin models. We illustrate this point with a second example.

## D. APPLICATION TO KUGRUA NO. 1 WELL, ALASKA

The Kugrua No. 1 Well is located in the western section of the National Petroleum Reserve, Alaska (NPRA), just to the south of Peard Bay. The

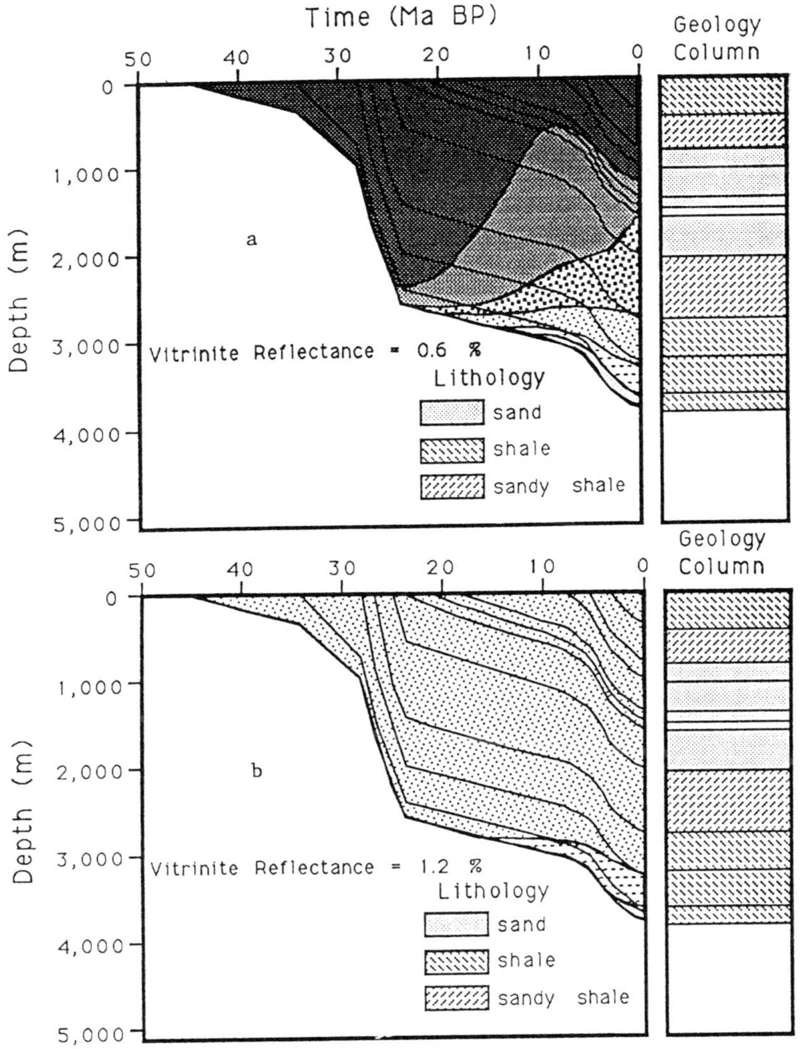


FIGURE 11.17 Burial history with cumulative probability for different vitrinite reflectance values. (a) $R_0$ = 0.6%; (b) $R_0$ = 1.2%; (c) $R_0$ = 1.6%.

well was drilled to a total depth of 4127 m (12,580 ft). The present-day oil window in the well, based on vitrinite reflectance, encompasses all formations from the Sadlerochit Group through the Torok Formation as depicted in Figure 11.19. The major problem here is that only six vitrinite reflectance measurements are available for the well, so that it is difficult to provide any reasonable mesure of paleoheat flux based on thermal inversion methods with such sparse data. Fortunately three other wells, Tunalik, Peard, and S. Meade, share the general regional area, and all three also have

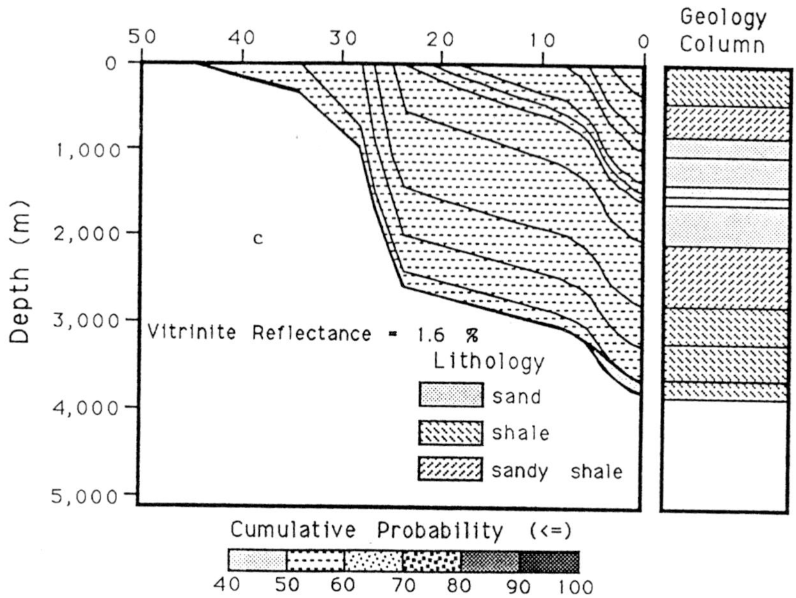


FIGURE 11.17 *(Continued)*

vitrinite reflectance values that can be used with thermal inversion procedures to obtain paleoheat flux estimates for each well. While each well has only 6 to 10 measurement values, one would naively anticipate that a more or less average consistent paleoheat flux behavior would result from each well. Figure 11.20 shows the predicted 'best fit'' vitrinite reflectance for each well, while Fig. 11.21 shows the best paleoheat flux for each well. The problem, as can be seen from Fig. 11.21, is that the individual paleoheat fluxes are fairly different, suggesting that the reflectance data from any single well are not sufficient to have a high degree of confidence in the quality of the "best" paleoheat fluxes. In turn, this uncertainty leads to an uncertainty in the thermal maturation onset for any one, and all, of the wells in the area. To assess the degree of risk and uncertainty, three heat flow patterns with time were used to encompass in a broad sense the possible paleoheat flux values based on regional geologic and geothermal information. These three patterns are drawn in Fig. 11.21 and labeled "Decrease," "Constant," and "Increase."

All other dynamic and thermal parameters were then held constant and only the paleoheat flux was allowed to vary with time. The Decrease pattern was taken as representing the most likely minimum heat flux at each instant of time, the Constant pattern as representing the likely value, and the Increase pattern as representing the most likely maximum value. It would appear that this range of values generously encompasses the

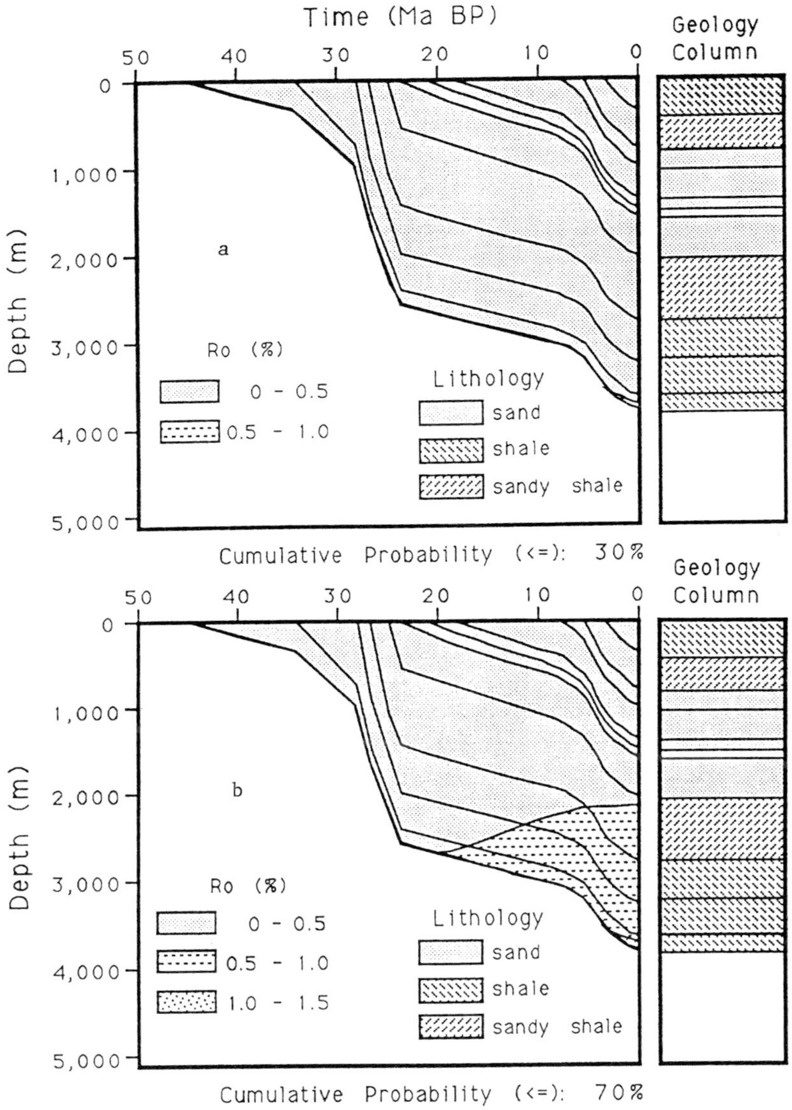


FIGURE 11.18 Burial history with vitrinite reflectance for different cumulative probability values. (a) 30%; (b) 70%; (c) 90%.

best estimate made for the Kugrua No. 1 Well. We then followed the procedure laid out in the preceding section. Figure 11.22 presents the observed vitrinite reflectance measurements with depth with superposed cumulative probability curves based on the three paleoheat flux estimates described earlier.

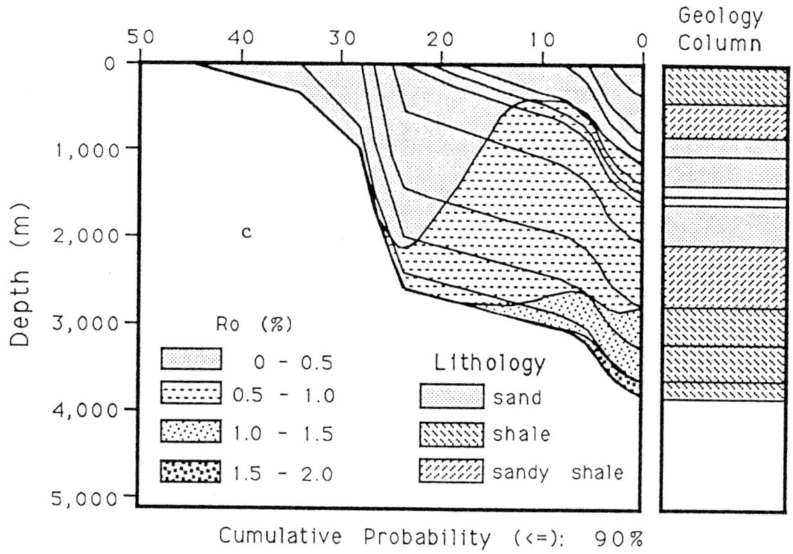


FIGURE 11.18 (*Continued*)

Inspection of Fig. 11.22 suggests that the two vitrinite reflectance measurements at shallow depth (less than 500 m) may need to be reevaluated because the cumulative probability for which they are correct is less than about 50% for each. Likewise, the deepest three vitrinite measurements also fall at about the 60% cumulative probability, suggesting that they, too, may be slightly low. The expected value occurs at a cumulative probability of about 68% so that the combined information would indicate that the paleoheat flux in the Kugrua No. 1 well needs to be higher than the value most consistent with the sparse downhole vitrinite measurements. There is about a 40% overall chance that the total thermal maturity is underestimated.

To provide a different presentation of information, cumulative probability values with depth are given in Fig. 11.23 for different fixed values of vitrinite reflectance. Note from Fig. 11.23 that there is an 80% chance that the surface reflectance value is less than 0.4%, suggesting that there is a 20% chance of a surface reflectance value greater than 0.4%—likely due to late Cretaceous and Tertiary erosion. Also note that at 3000 m depth there is a 70% chance that the reflectance should be less than about 1.4%, and a 40% chance that a value *greater* than 1.2% should be present. Observations (Fig. 11.22) indicate reflectance values of between about 1.1% to 1.7% at 3000 m, suggesting that the high values (1.7%) are probably too high and, equally, that the low values (1.1%) are likely to low—or that the intrinsic uncertainty is about ±0.3% in any one measurement—or both, of course.

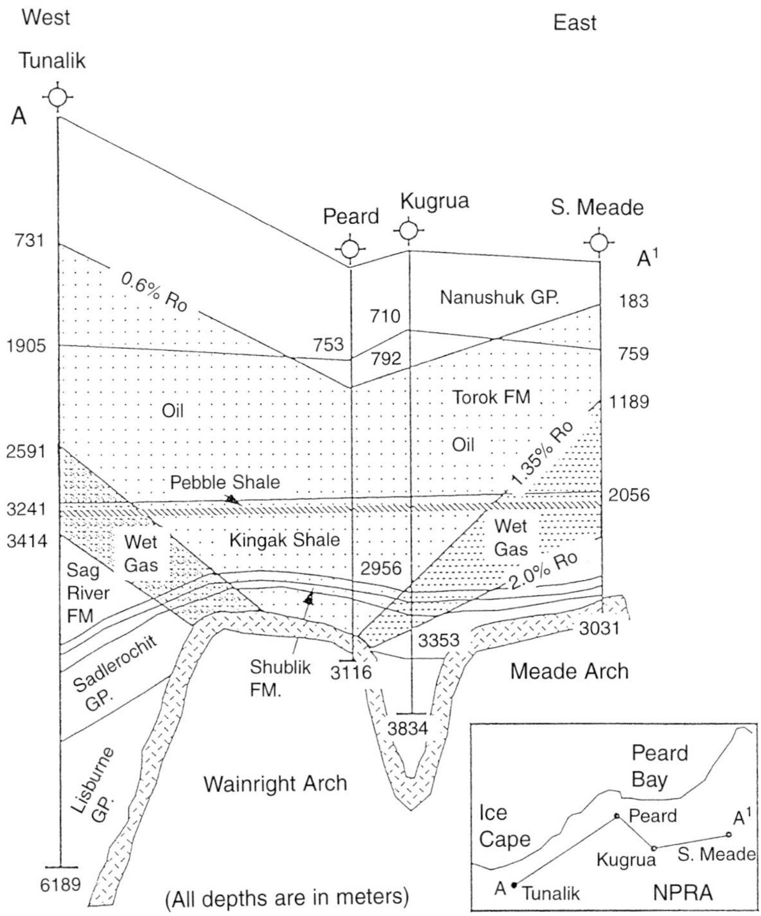


FIGURE 11.19 Present-day oil window in the northwest of the National Petroleum Reserve, Alaska, based on the vitrinite reflectance data from five wells, including Kugrua No. 1 (modified from Magoon and Bird, 1986).

To complete the investigation, we have also looked at the probability of obtaining particular values of vitrinite reflectance in relation to the burial history of the sediments. Thus, Figs. 11.24a–c show the cumulative probability behaviors with time for obtaining vitrinite reflectance values of less than 0.6% (Fig. 11.24a), 1.2% (Fig. 11.24b), and 2.0% (Fig. 11.24c), while Figs. 11.25a–c show the conjugate variation of vitrinite reflectance behaviors with depth and time at fixed estimates for the cumulative probability of 30%, 70%, and 90%. One notices from Fig. 11.25c that it is very likely (90%) that at no time or any depth was there any value of reflectance in excess of 2.5%, indeed it was most likely (70% chance) that reflectance

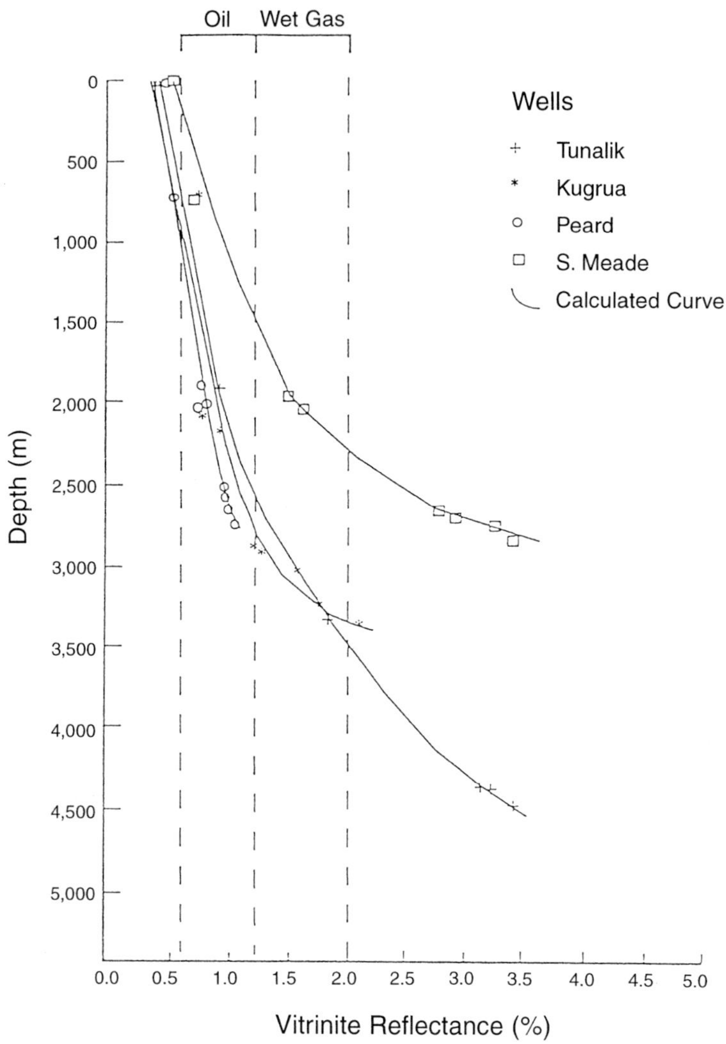


FIGURE 11.20 Predicted "best fit" vitrinite reflectance for the four wells in the northwestern NPRA.

was always less than 2.0%. Thus, although the vitrinite reflectance data from the Kugrua No. 1 Well are themselves too sparse to provide a good and precise measure of paleoheat flux, data from neighboring wells can be used to assess probable ranges of uncertainty of the paleoheat flux for the Kugrua No. 1 well. These ranges of paleoheat flux, in turn, can be used to assess likely regimes of timing of thermal maturity either at a given level of probability or at a given value of vitrinite reflectance.

The results from this example are complementary to those from the

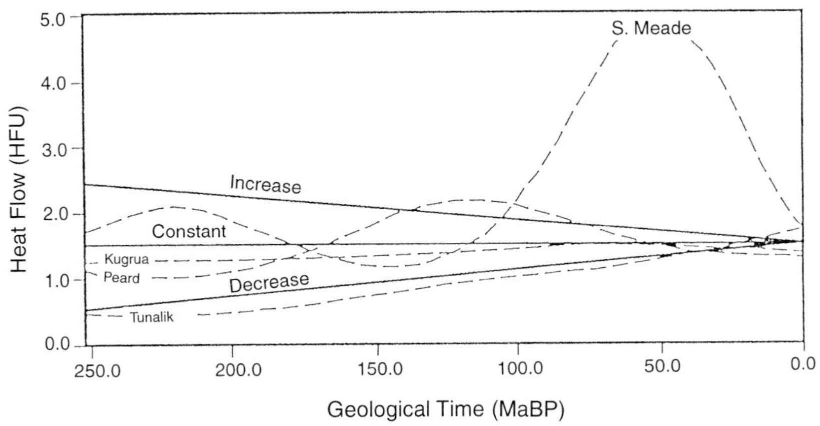


FIGURE 11.21 Modeled "best" paleoheat flux for the four wells in the northwestern NPRA.

Navarin Basin COST No. 1 Well example, in the sense that only with further well information can an appreciation be obtained of the likelihood of the correctness (or lack of correctness) of any particular choice for thermal maturity.

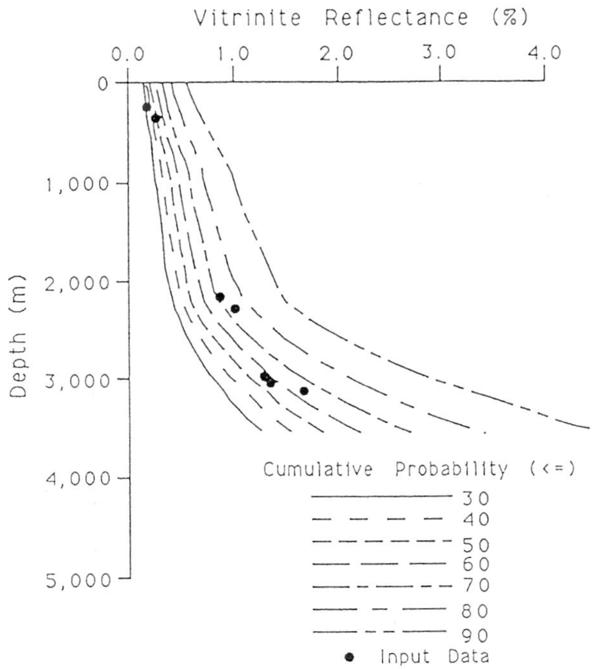


FIGURE 11.22 Vitrinite reflectance versus depth with different cumulative probability values for Kugrua No. 1 Well.

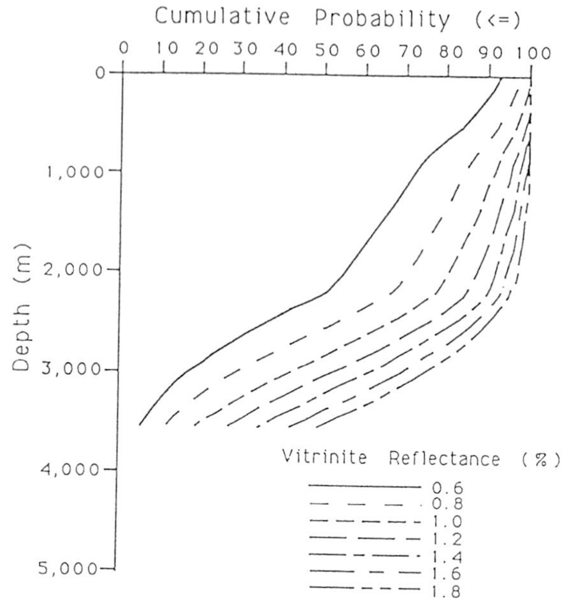


FIGURE 11.23 Cumulative probability of vitrinite reflectance with depth for Kugrua No. 1 Well.

Perhaps the salient point from both examples is an understanding of how probability procedures can be used to assess risk for hydrocarbon proneness in the face of uncertainty of parameter values, or in the face of uncertainty in geological factors caused by a lack of good quality, numerous, well-sampled data. It would seem that the two examples presented here illustrate both of these points rather well.

## II. RISK AND PROBABILITY IN RESERVE ASSESSMENT

### A. INTRODUCTION

As noted already, one of the standard reasons for performing quantitative basin modeling calculations is to provide measures of (1) hydrocarbon generation, (2) expulsion of generated hydrocarbons into domains where (3) migration can then occur so that one can assess the fractional loss of hydrocarbons during migration, and (4) the total area of source rock draining into a particular trap. In this way, a quantitative assessment can be made of the likely amount of hydrocarbons to be found in a trap (Burnett, 1996; Aurell and Zyczkowski, 1996). The difficulty is that each basin model

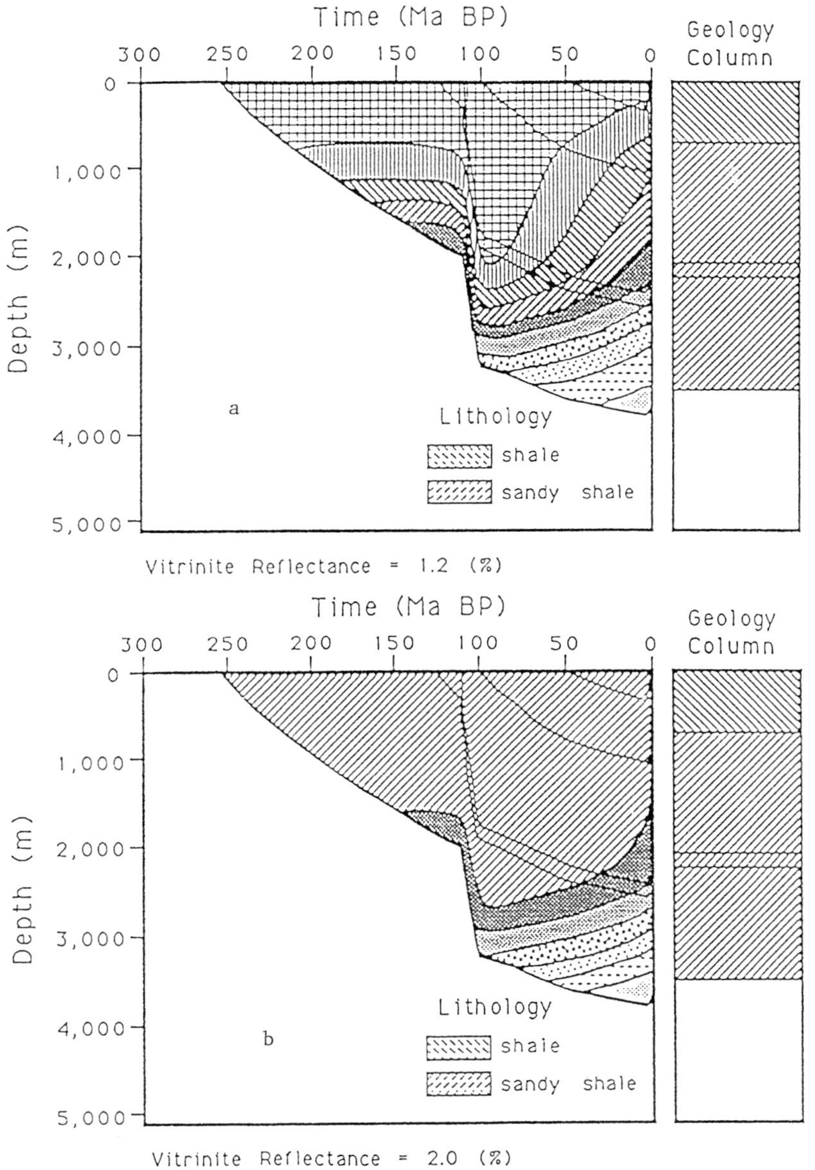


FIGURE 11.24 Burial history with cumulative probability for different vitrinite reflectance values. (a) $R_0$ = 0.6%; (b) $R_0$ = 1.2%, (c) $R_0$ = 2.0%.

code produces results that are limited by the intrinsic assumptions, by the numerical resolution, by the initial values chosen, by boundary conditions, by the intrinsic parameters used with the model, and by the quality, quantity, and distribution frequency of data used as control information.

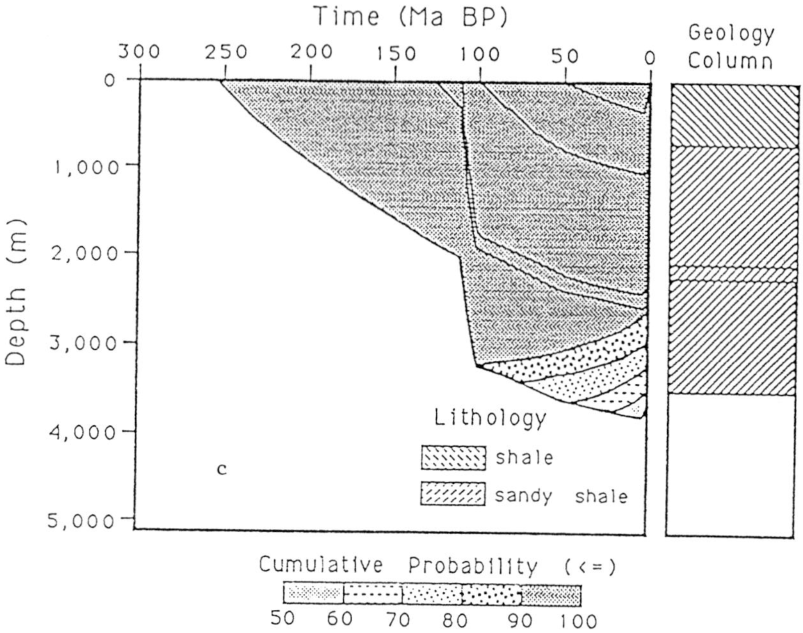


FIGURE 11.24 *(Continued)*

Thus, every result from every model has some degree of nonuniqueness, is neither accurate nor precise, and is both imperfectly resolved as well as being sensitive (to different degrees) to the assumptions, parameters, and data involved in producing the model results. As a consequence there are ranges of uncertainty for all the parameters that are used to assess hydrocarbon charge to a trap, so that the charge estimate is, itself, also uncertain.

The volumetric modeling is usually taken to be described (Thomsen, 1993) as follows:

1. For each unit area of source rock, of thickness $H$, containing a weight percent TOC of organic material, with a potential generation efficiency of $P_g$, a porosity $\phi$, with source rock density $\rho_{SR}$, and with generated hydrocarbons at source having a density $\rho_{HC}$, the generated volume $G$ of hydrocarbons is

$$G\ [\mathrm{m^3/km^2}] = \phi H(\rho_{SR}/\rho_{HC})\ \mathrm{TOC} \times P_g. \tag{11.1}$$

2. If the source rock has a fractional expulsion threshold, $E_{TH}$, for release of hydrocarbons to the secondary migration system, then the volume of hydrocarbons expelled, $V_e$, per unit area of the source rock, is given by

$$V_e\ [\mathrm{m^3/km^2}] = G - \phi H E_{TH} > 0. \tag{11.2}$$

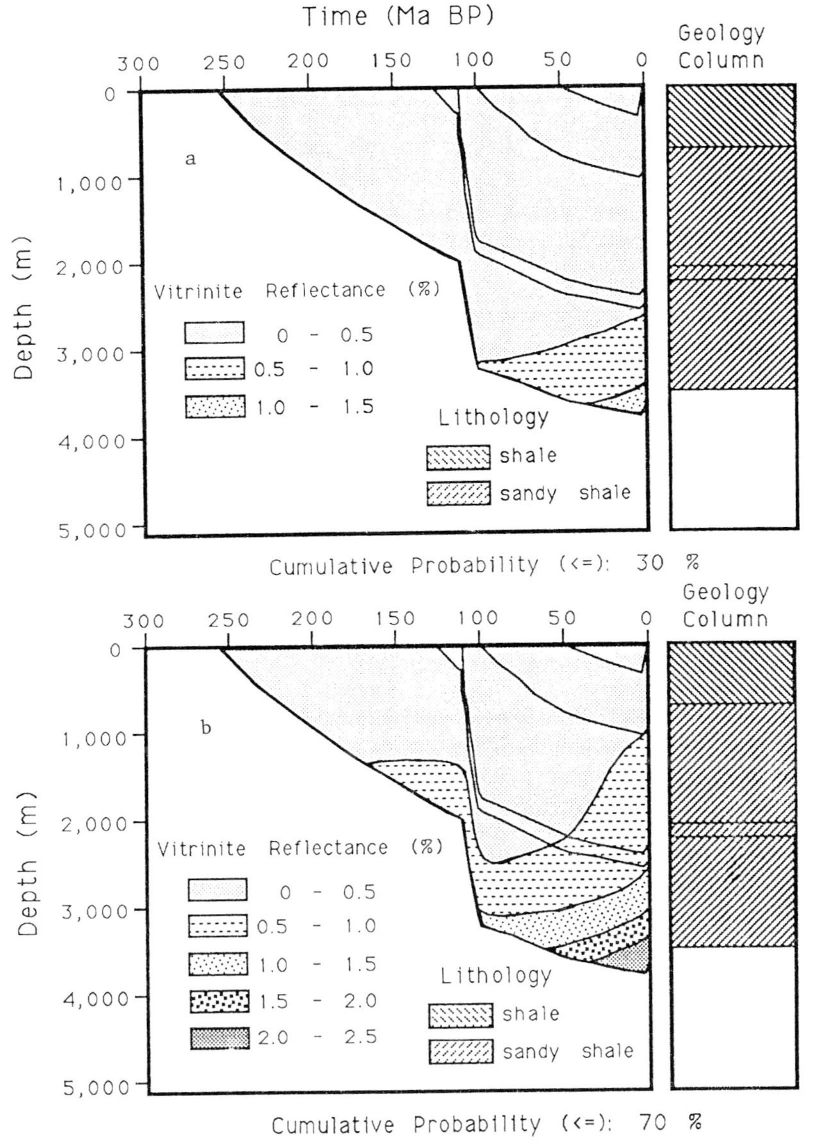


FIGURE 11.25 Burial history with vitrinite reflectance for different cumulative probability values. (a) 30%; (b) 70%; (c) 90%.

The residual saturation, $S_R$, of the source rock is just

$$S_R\ [\mathrm{m^3/km^2}] = \phi H E_{TH}. \tag{11.3}$$

3. Of the expelled volume, $V_e$, a fraction ML is lost during migration to a particular trap, so per unit area of the source rock the hydrocarbon

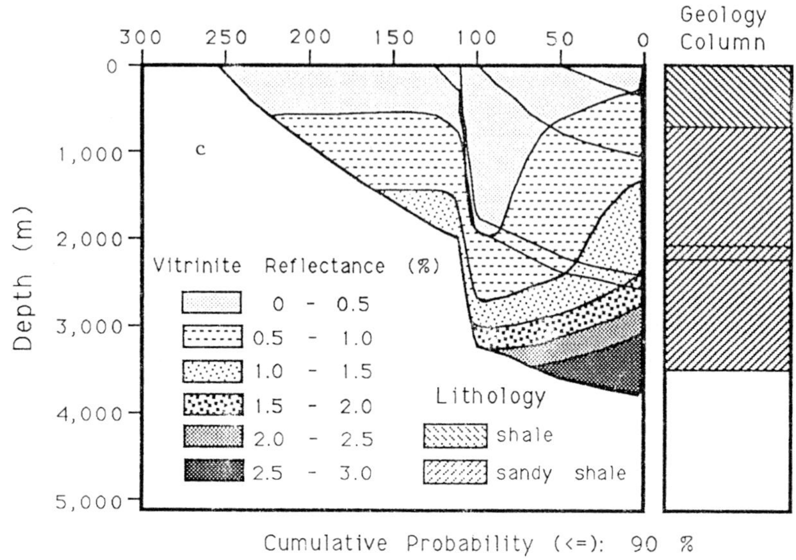


FIGURE 11.25 (*Continued*)

volume, $V_T$, charging the trap is

$$V_T\ [\mathrm{m}^3/\mathrm{km}^2] = V_e\ (1 - \mathrm{ML}). \tag{11.4}$$

4. The total area, $A$, of the source rock draining to a particular trap then allows the total volumetric hydrocarbon charge to be written

$$\text{Charge}\ [\mathrm{m}^3] = V_T A. \tag{11.5}$$

The function of basin models, which involve dynamic, thermal, and hydrocarbon kinetic components, is to provide assessments of the values of the parameters $\phi$,$H$, $\rho_{SR}$, $\rho_{HC}$, TOC, $P_g$, $E_{TH}$, ML, and $A$, so that the total charge can be estimated. Decisions on risk, rank, economic worth, and so on can then be made, based on the charges assessed for available traps. Irwin *et al.* (1993) and Hermanrud *et al.* (1989) have applied such types of charge estimates to the Egersund Basin.

Two of the major problems that arise are as follows:

1. How do we use the degrees of uncertainty to evaluate scientific risk and uncertainty on reserve estimates?
2. How do we determine which of the factors (generation, expulsion, migration, drainage area) is causing the largest uncertainty in hydrocarbon charge reserve estimates, so that one then has some idea of where to expend resources to improve knowledge? In addition, we must also have some idea of when it is pointless to continue the

exercise of knowledge improvement relative to the anticipated better delineation of hydrocarbon reserves from any such knowledge advance.

The purpose of this section is to illustrate how the uncertainty ranges of parameters from basin modeling calculations are used to address these two problems simultaneously within the framework of the volumetric estimates of Eqs. (11.1) through (11.5). This section is *not* concerned with the technical details of basin modeling calculations per se but rather with the ability to convert the output results, and their ranges of uncertainty, into volumetric estimates together with their ranges of uncertainty. However, the basin analysis codes used must, of course, be sufficiently sophisticated to include modules describing hydrocarbon generation, expulsion mechanisms, secondary migration dynamics, and drainage, or else there is a lack of available quantitative descriptions from which to deduce parameters needed for the volumetric estimates.

The illustration chosen here to illuminate the procedure is a North Sea case study with the basin modeling code GEOPETII (developed at USC). The history of the project in relation to uncertainties uncovered is also included.

The method used to assess uncertainty is that of cumulative probability of occurrence. So as not to break the train of the discussion, the mathematical details of the cumulative probability method are presented in Appendix B.

## B. A NORWEGIAN SEA ILLUSTRATION

In an area of the Norwegian Sea, in the general region of Haltenbanken to Traenabanken, seismic observations together with well information suggested that a Jurassic source rock, about 100 m thick, with a TOC content of about 8%, and with a current porosity of about 5% had generated oil. The generated fraction of the TOC seemed to be around 200 mg HC/g TOC based on pyrolysis studies, while the source rock density was around 2400 kg/m$^3$, and oil shows yielded oil densities of around 800 kg/m$^3$. The total source rock area, which drained toward the trap under investigation, was initially estimated at around 160 km$^2$ based on seismic information.

As part of the exercise, those involved were initially asked to assess minimal, most likely, and maximal values for each of the variables—source rock thickness, TOC, generated fraction of TOC, source rock porosity, source rock density, hydrocarbon (oil) density, and drainage area—based on the observational data available.

Basin analysis code GEOPETII was then run to provide estimates of expulsion threshold, secondary migration loss, generation potential, porosity, and drainage area, and so forth. Computer runs were done under a variety of conditions, for example, low to high fracture criteria, different kinetic models for hydrocarbon production, different paleoheat flux estimates, different mechanisms for expulsion (solubility limits, capillarity, fracturing), and different ranges of parameter values within the framework of particular mechanisms. Thus, variable values of surface tension were used in capillarity-controlled primary expulsion, and so on. The upshot of all these computer machinations, taken in concert with the observational data constraints, was to provide initial minimum, likely,and maximum estimated values for the nine parameters needed in Eqs. (11.1) through (11.5). These values are reported in Table 11.3 as case 1. Based on the values given, the cumulative probability method of Appendix B was used to generate cumulative probability plots of total generated volume of hydrocarbons (Fig. 11.26a) together with the total volume expelled and volume migrated (Fig. 11.26b), which is the probable range of volume of hydrocarbons supplied to the trap. The range of uncertainty in the hydrocarbon charge (volume migrated) to the trap is large in case 1, as shown on Table 11.4 by the $P(16)$, $P(68)$, and $P(84)$ cumulative probability values of reserve estimates. The parameter $\mu^2$ in Table 11.4 characterizes the slope of the log-normal cumulative probability plot (see Appendix B); a relatively small value of $\mu^2$ implies little error in the mean $[P(68)]$ estimate, whereas a correspondingly large value of $\mu^2$ implies a significantly larger error in the $P(68)$ estimate.

Using the cumulative probability method given in Appendix B, one can construct the contributions of each of the nine parameters used in Eqs. (11.1) through (11.5) to the total uncertainty factor, $\mu^2$. Thus a plot of *relative* contribution (%) of each parameter to reserve uncertainty can be given, as shown in Fig. 11.26c. Note that Fig. 11.26c implies that case 1 has a dominance of secondary migration loss as the main contributor to the uncertainty in reserve estimates, which is likely because the maximum potential estimated loss is 100% from Table 11.3, case 1. Note from Fig. 11.26c that drainage area provides a relatively small contribution to the total uncertainty of reserve estimates compared to secondary migration loss uncertainty.

However, prior to being able to convince the decision makers that more effort should be put on narrowing down the uncertainty in secondary migration loss (particularly the remarkable initial estimate of 100% maximum loss!), it had already been decided that effort would be placed on improving the drainage area uncertainty. After some reprocessing of seismic information, paleotopographical revisions, and allied information, the range of uncertainty in the drainage area was considered to have been narrowed sufficiently that it was worthwhile to reassess the hydrocarbon charge. This investigation is shown in Table 11.3 as case 2. However, note from Table 11.4 that there is little difference in $\mu^2$ compared to case 1, nor much

TABLE 11.3 Parameter Ranges Used in the Nine Case Histories Detailed in Text

| | | Source rock thickness (m) | TOC (%) | Potential generation (mg/g TOC) | Source rock porosity (%) | Source rock density (kg/m$^3$) | Hydrocarbon density (kg/m$^3$) | Expulsion threshold (%) | Secondary migration loss (%) | Drainage area (km$^2$) |
|---|---|---|---|---|---|---|---|---|---|---|
| Case 1 | Maximum | 150 | 12 | 250 | 7 | 2700 | 900 | 40 | 100 | 180 |
| | Most likely | 110 | 8 | 190 | 5 | 2400 | 800 | 25 | 30 | 160 |
| | Minimum | 40 | 4 | 160 | 3 | 2200 | 600 | 15 | 10 | 100 |
| Case 2 | Maximum | 150 | 12 | 250 | 7 | 2700 | 900 | 40 | 100 | 180 |
| | Most likely | 110 | 8 | 190 | 5 | 2400 | 800 | 25 | 30 | 160 |
| | Minimum | 40 | 4 | 160 | 3 | 2200 | 600 | 15 | 15 | 150 |
| Case 3 | Maximum | 150 | 12 | 250 | 7 | 2700 | 900 | 40 | 50 | 180 |
| | Most likely | 110 | 8 | 190 | 5 | 2400 | 800 | 25 | 30 | 160 |
| | Minimum | 40 | 4 | 160 | 3 | 2200 | 600 | 15 | 20 | 150 |
| Case 4 | Maximum | 150 | 12 | 250 | 7 | 2700 | 900 | 30 | 40 | 180 |
| | Most likely | 110 | 8 | 190 | 5 | 2400 | 800 | 25 | 30 | 160 |
| | Minimum | 40 | 4 | 160 | 3 | 2200 | 600 | 20 | 20 | 150 |
| Case 5 | Maximum | 150 | 12 | 250 | 7 | 2700 | 900 | 30 | 40 | 180 |
| | Most likely | 110 | 8 | 190 | 5 | 2400 | 800 | 25 | 30 | 160 |
| | Minimum | 40 | 4 | 160 | 3 | 2200 | 750 | 20 | 20 | 150 |
| Case 6 | Maximum | 150 | 12 | 250 | 7 | 2500 | 900 | 30 | 40 | 180 |
| | Most likely | 110 | 8 | 190 | 5 | 2400 | 800 | 25 | 30 | 160 |
| | Minimum | 40 | 4 | 160 | 3 | 2200 | 750 | 20 | 20 | 150 |
| Case 7 | Maximum | 150 | 12 | 250 | 7 | 2500 | 900 | 30 | 40 | 180 |
| | Most likely | 110 | 8 | 190 | 5 | 2400 | 800 | 25 | 30 | 160 |
| | Minimum | 40 | 4 | 160 | 4 | 2200 | 750 | 20 | 20 | 150 |
| Case 8 | Maximum | 150 | 10 | 270 | 7 | 2500 | 900 | 30 | 40 | 180 |
| | Most likely | 110 | 8 | 190 | 5 | 2400 | 800 | 25 | 30 | 160 |
| | Minimum | 40 | 8 | 160 | 4 | 2200 | 750 | 20 | 20 | 150 |
| Case 9 | Maximum | 120 | 10 | 220 | 6 | 2500 | 900 | 30 | 40 | 180 |
| | Most likely | 110 | 8 | 190 | 5 | 2400 | 800 | 25 | 30 | 160 |
| | Minimum | 100 | 6 | 180 | 4 | 2200 | 750 | 20 | 20 | 150 |

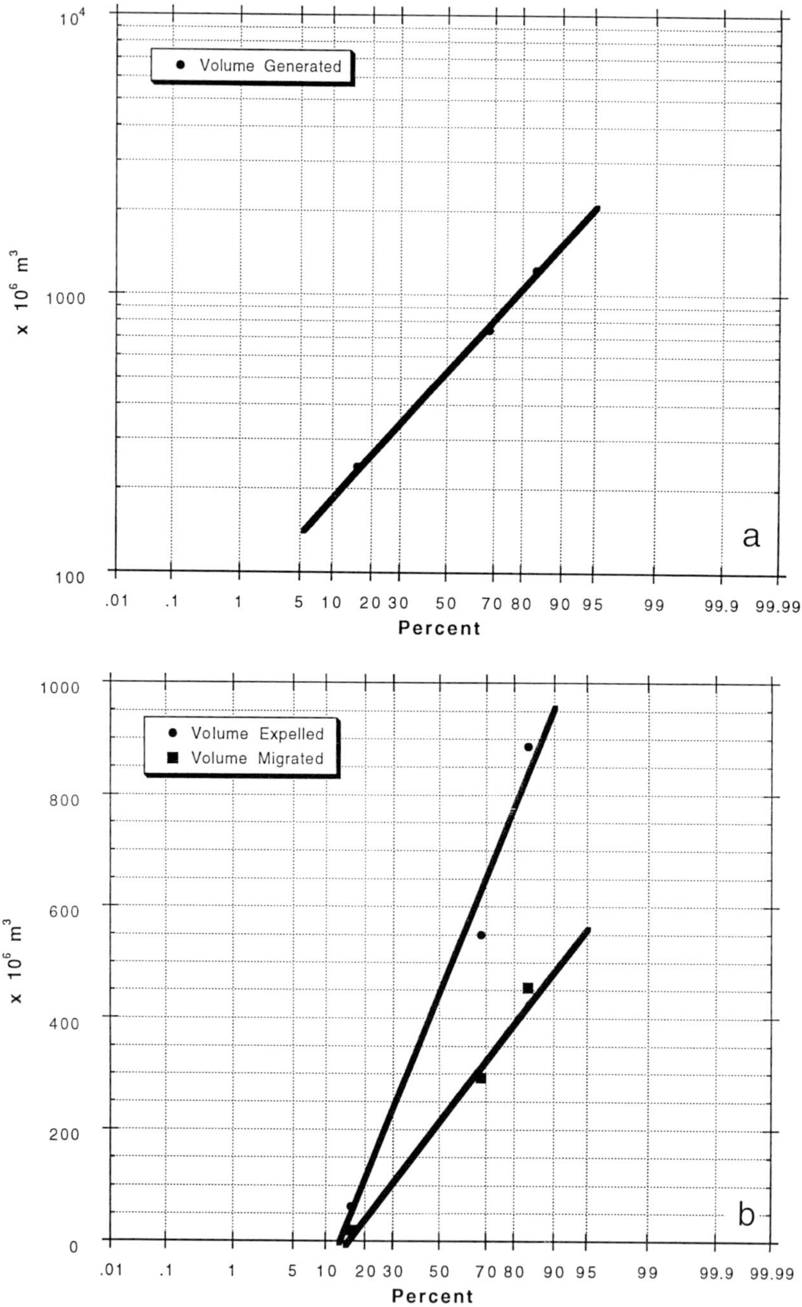


FIGURE 11.26 Cumulative probability of plots of (a) generated volume of hydrocarbons; (b) volume expelled and volume migrated for the parameter ranges in Table 11.3, case 1; (c) relative contributions (%) to the reserve uncertainty from the nine parameters used in Eqs. (11.1) through (11.5).

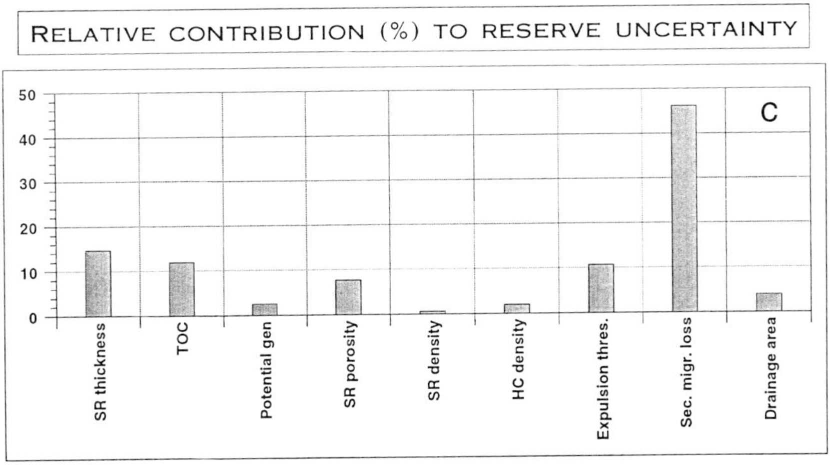


FIGURE 11.26 *(Continued)*

difference in the $P(68)$, $P(84)$, or $P(16)$ estimates of total reserves. This point is also reinforced by the cumulative probability plots of generated volume (Fig. 11.27a), expelled volume (Fig. 11.27b), and volume migrated (Fig. 11.27c), which are remarkably close to the corresponding figures (Figs. 11.26a and 11.26b, respectively) for case 1. Indeed, a plot of relative contribution (%) to the reserve uncertainty for this case (Fig. 11.27c) points out that the narrowing of uncertainty in the drainage area just makes the uncertainty in secondary migration loss that much more of an important factor to constrain better. In particular, the 100% loss as an upper limit had to be reduced.

TABLE 11.4 Estimates of Cumulative Probability of Reserves for the Different Cases Reported

| Case 1 | $\mu^2$ | $P(68)$ ($\times 10^6 m^3$) | $P(84)$ ($\times 10^6 m^3$) | $P(16)$ ($\times 10^6 m^3$) |
|---|---|---|---|---|
| 1 | 0.342 | 293 | 455 | 20 |
| 2 | 0.310 | 316 | 500 | 27 |
| 3 | 0.207 | 408 | 661 | 51 |
| 4 | 0.161 | 438 | 717 | 68 |
| 5 | 0.156 | 402 | 660 | 66 |
| 6 | 0.155 | 387 | 636 | 65 |
| 7 | 0.142 | 377 | 621 | 68 |
| 8 | 0.116 | 395 | 652 | 81 |
| 9 | 0.049 | 416 | 667 | 174 |

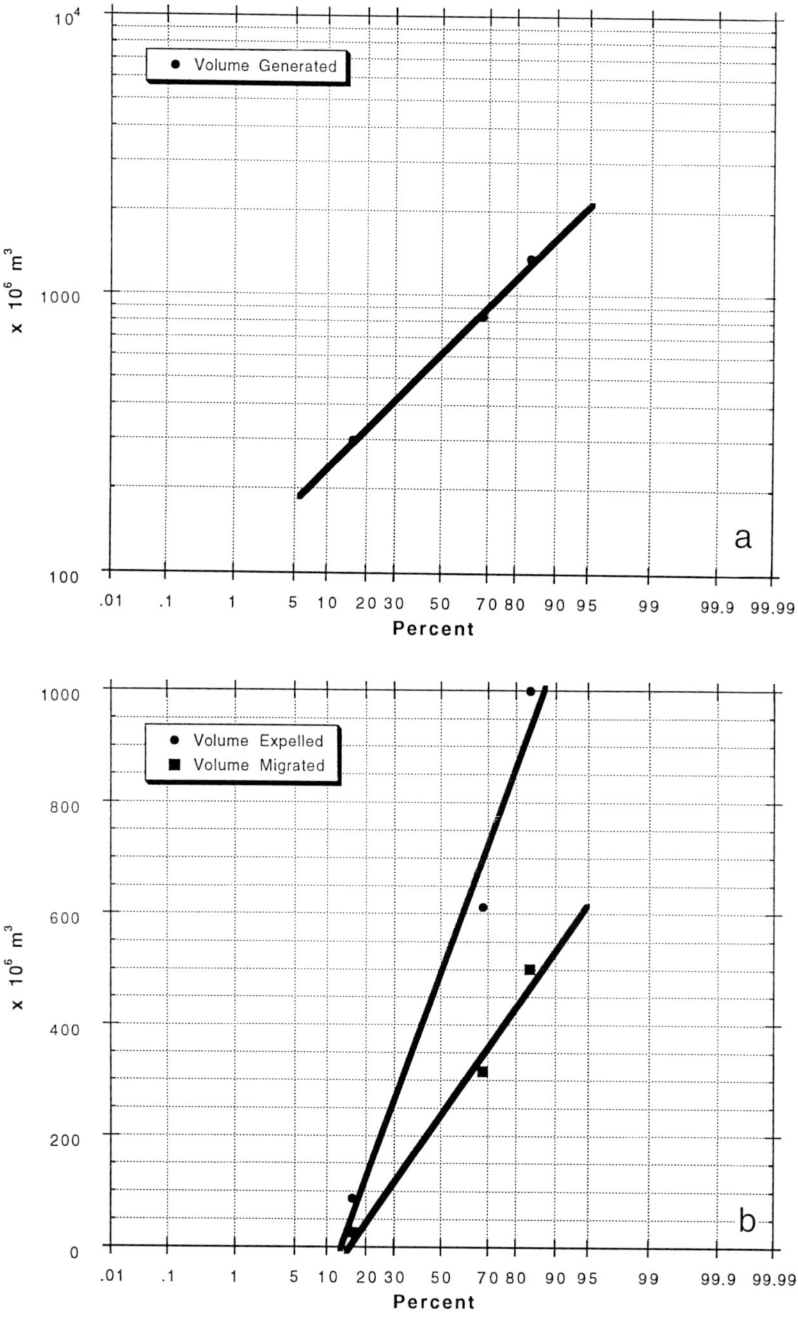


FIGURE 11.27 As for Fig. 11.26 but for the parameter ranges of Table 11.3, case 2.

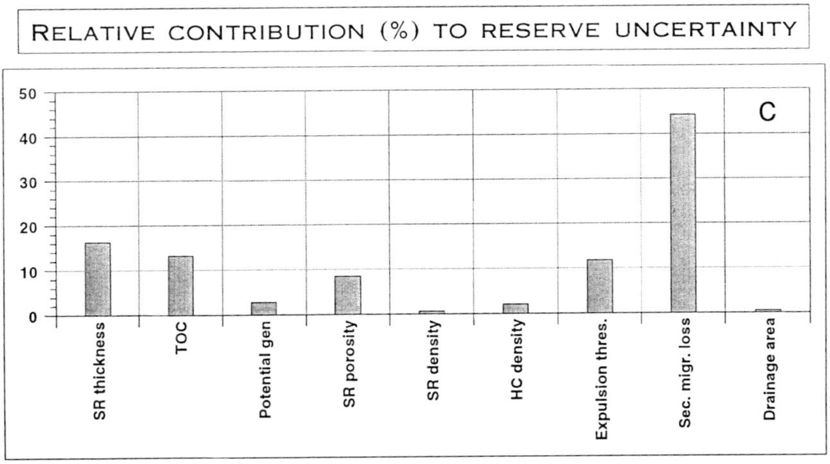


FIGURE 11.27 (*Continued*)

At this stage it was recommended that core samples from the wells be examined for oil staining along the aquifer path(s) and for shifts in GGMS signatures with respect to distance between the expected reservoir and the basinal kitchen. Any signature indicating passage of hydrocarbons could then be used, in conjunction with the basin modeling code computations, to limit fractional secondary migration losses. The implementation of such considerations between the geochemical group and the basin modeling group led to a reduction in the estimated maximum secondary loss to 50% (rather than 100%) and a slight increase in the minimum estimate from 15% to 20% as recorded in Table 11.3, case 3. While these improved estimates are, perhaps, somewhat subjective, they do fit within the available knowledge base. Correspondingly, the hydrocarbon volume migrated (which is the volume available at the trap) has a different cumulative probability behavior, as recorded in Fig. 11.28a and also in Table 11.3, case 3, with a smaller degree of uncertainty than for case 2. The relative uncertainty of the secondary migration losses then influences the reserve uncertainty estimate to a lesser extent than uncertainty of source rock thickness, TOC content, and expulsion threshold (as shown in Fig. 11.28b).

In efforts to narrow further the degree of uncertainty, we then focused on assessing better the expulsion threshold limits. A detailed literature survey indicated that while the most likely value (25%) for expulsion threshold was in accord with many other prior estimates, the minimum and maximum estimates were too parsimonious and too generous, respectively, at least for estimates from other similar North Sea case histories which had been produced. Accordingly, the expulsion threshold range was then set to the 20% and 30% extremes that were considered appropriate for the

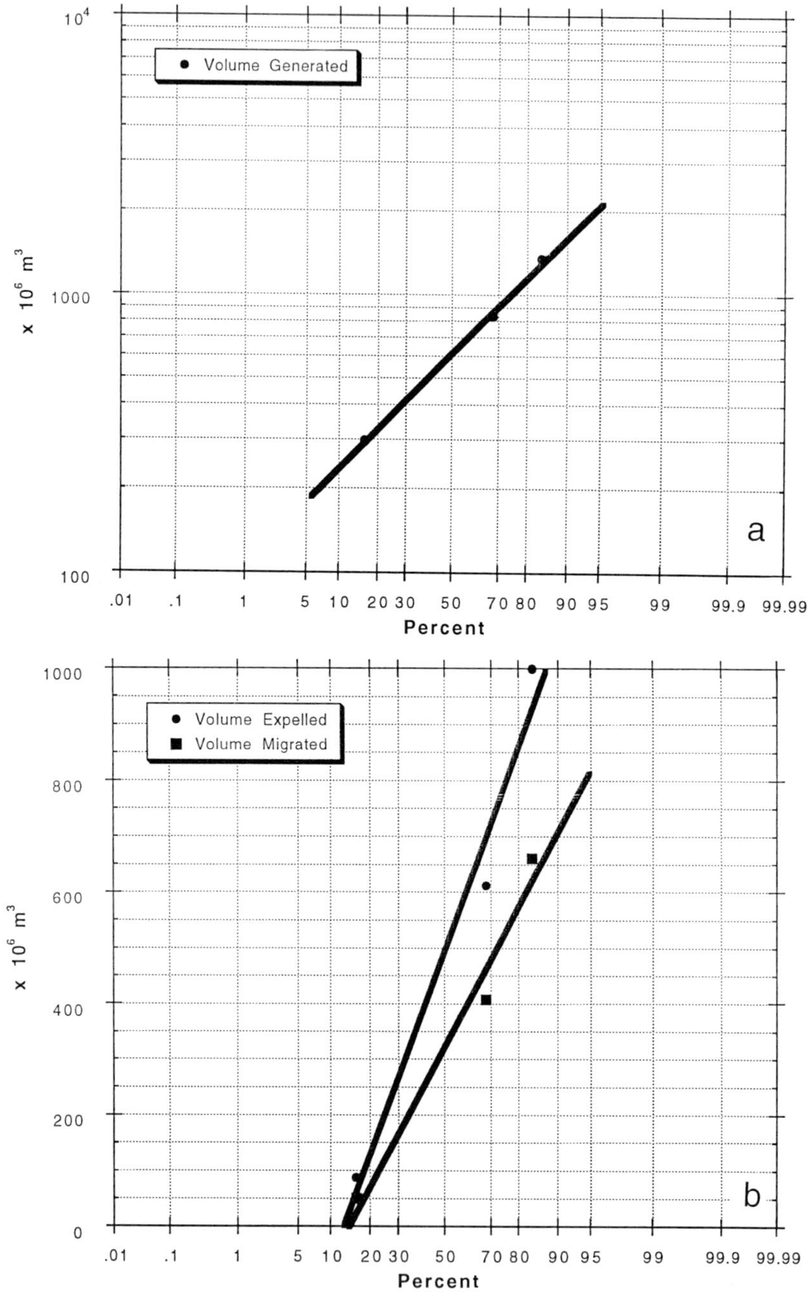


FIGURE 11.28 As for Fig. 11.26, all parts, but for the parameter ranges of Table 11.3, case 3.

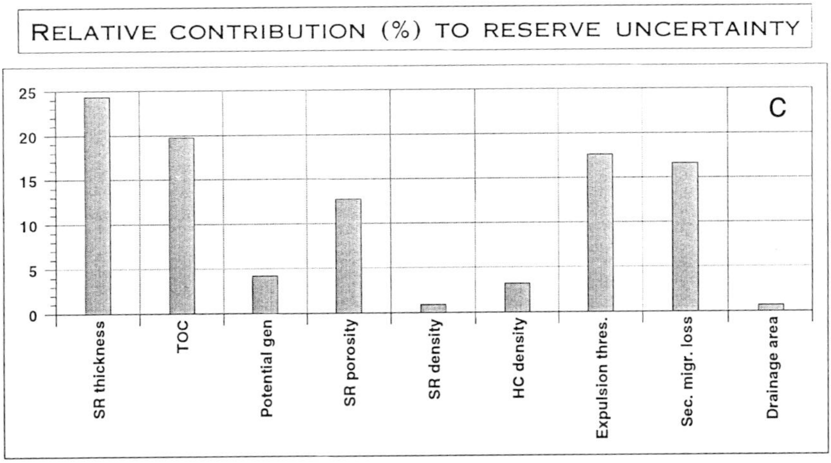


FIGURE 11.28 (*Continued*)

North Sea region. The results of this investigation (case 4) are recorded in Fig. 11.29a for the relative contribution to reserve uncertainty, showing the dominance now of source rock parameters rather than expulsion, migration, or drainage area considerations. The volume generated remains the same as in case 2, but the volume expelled and the volume trapped probabilities are both modified (Fig. 11.29b) in line with the lesser uncertainty taken for the expulsion threshold. As shown in Table 11.4, case 4, the uncertainty on the expected $P(68)$ volume of hydrocarbons in the trap is now reduced considerably.

At this stage, attention focused more on the generation parameters. The uncertainty of the hydrocarbon density at production, which had been set to between 0.6 g/cm$^3$ (representative of condensate) and 0.9 g/cm$^3$ (representative of heavy oil) was reinvestigated in terms of the GEOPETII model results with different heat flux and hydrocarbon kinetic models for kerogen breakdown. The computer runs indicated that while gas was produced along with oil, the relative amount of gas to oil was quite small, so that a condensate density of 0.6 g/cm$^3$ was quite probably too low a value. After some discussion with the geochemists on the project, a compromise position of 0.75 g/cm$^3$ was taken for the lowest likely hydrocarbon density at generation. While it had been pointed out to the geochemical group that the value of the hydrocarbon density was not a major player in contributing to the reserve uncertainty based on the prior four cases, nevertheless it was requested that this case be run to prove the point. Table 11.3, case 5, illustrates the values used, while Fig. 11.30a reinforces the lack of any sensitivity of reserve uncertainty to the hydrocarbon density relative to uncertainties on source rock thickness, TOC content, and source rock poros-

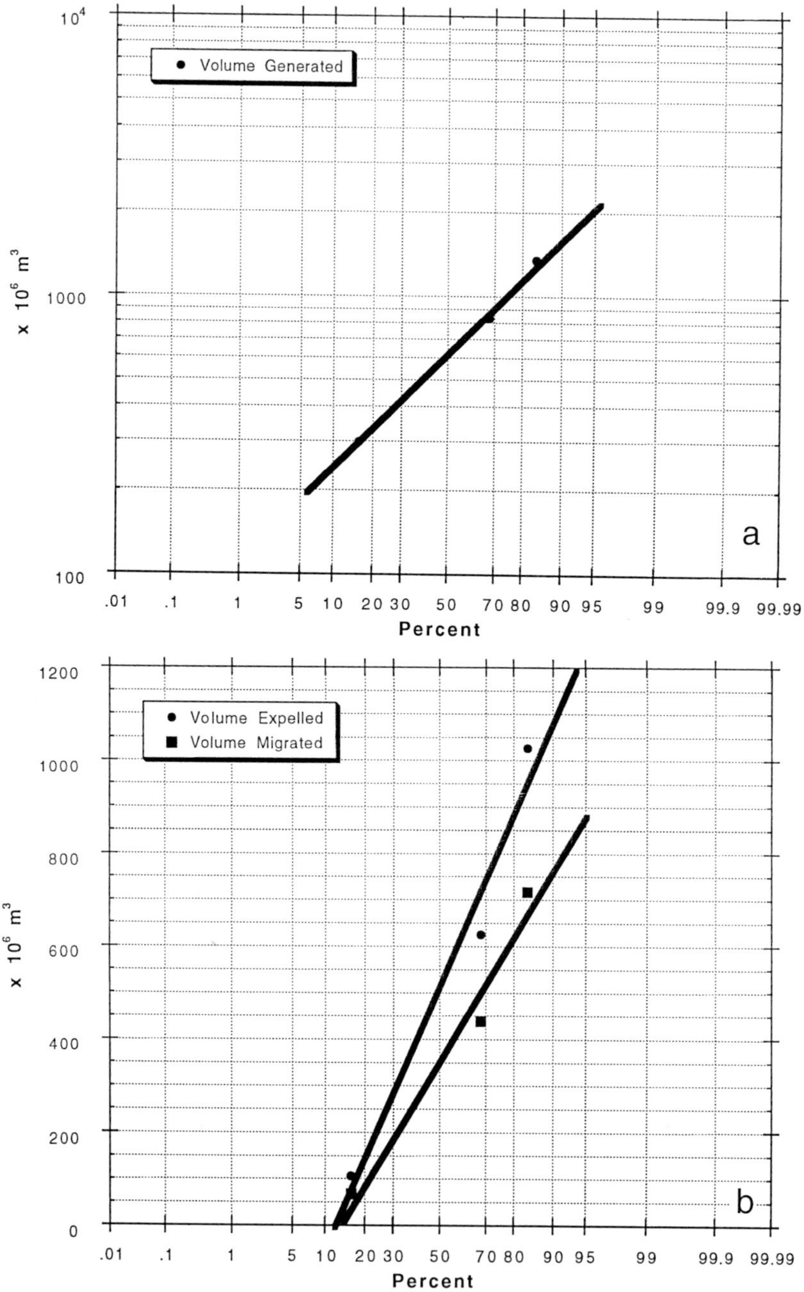


FIGURE 11.29 As for Fig. 11.26, all parts, but for the parameter ranges of Table 11.3, case 4.

RELATIVE CONTRIBUTION (%) TO RESERVE UNCERTAINTY

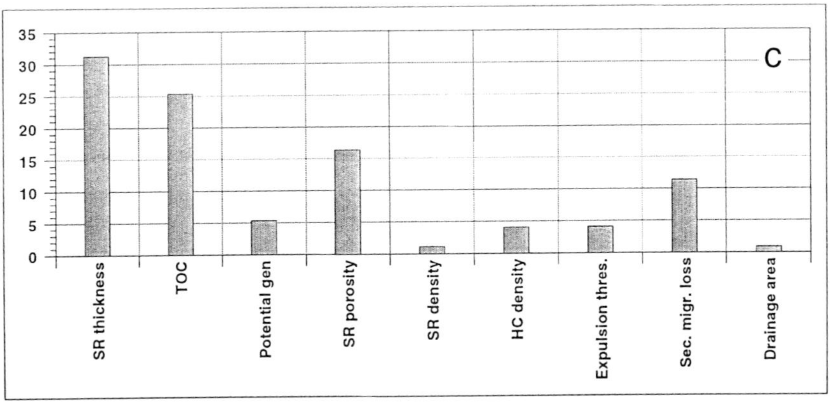


FIGURE 11.29 *(Continued)*

ity. The same point is made differently in Table 11.4, case 5, where little to zero improvement is obtained in the range of cumulative probability values for volume migrated—as is also hammered home by the cumulative probability plots of volume generated and volume expelled and migrated given in Figs. 11.30a–c, respectively. A similar pattern of behavior of lack of resolution and sensitivity was also obtained when source rock density was varied (which never had a significant role to play in any of the prior estimates of reserve uncertainty). Accordingly this case is not shown graphically here and is recorded only through values in Table 11.3, case 6, and Table 11.4, case 6.

The three dominant factors causing significant uncertainty in the reserve calculations were source rock porosity, TOC, and source rock thickness. Because source rock porosity was easier to remeasure than source rock thickness or TOC, attention was then concentrated on improving that value. Because the source rock porosity was low (typically 5%), a shift in the minimum estimate from 3% to 4% had a significant influence on the volume generated and on the volume expelled and so, indirectly, on the volume migrated to the trap. As shown for the parameters in Table 11.3, case 7, narrowing the minimum porosity estimate to 4% from 3% made the influence of uncertainty in TOC and source rock thickness even more paramount (Fig. 11.31a) in contributing to reserve uncertainty. The uncertainties around $P(68)$ are narrowed some (see Table 11.4, case 7), but not significantly compared to improvements arising from improved estimates of the expulsion, migration, and drainage parameter ranges, as is also seen in Figs. 11.31a–c for cumulative probability plots of volumes generated, expelled, and migrated, respectively.

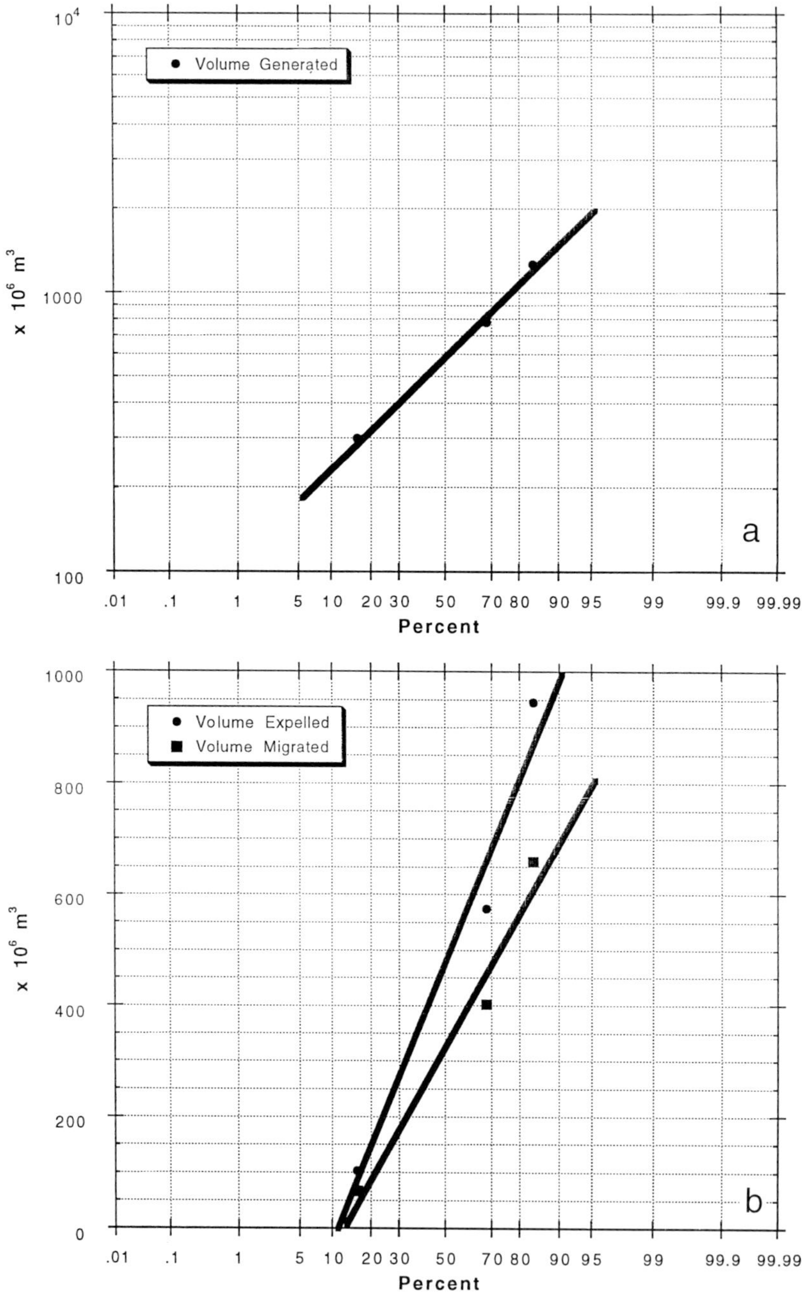


FIGURE 11.30 As for Fig. 11.26, all parts, but for the parameter ranges of Table 11.3, case 5.

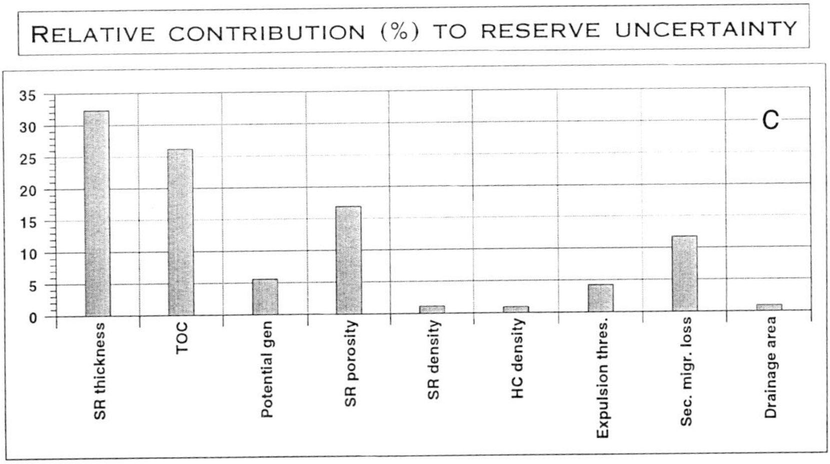


FIGURE 11.30 *(Continued)*

At this stage it became clear that some serious effort had to be undertaken to narrow the TOC range of uncertainty and also the uncertainty in source rock thickness estimates. Two factors dictated that the TOC would be the next easiest parameter to constrain better. First, direct measurements of TOC content were available from the same source rock in different parts of the region; these values were then used as analogs from which was constructed a better representation histogram (not shown for proprietary reasons) of the frequency of occurrence of TOC content. Second, direct measurements (and duplicate measurements) were done on source rock samples, so that the confidence of TOC values, obtained from different groups with different equipment and laboratory protocols, was improved.

The upshot of these measurements was that the TOC uncertainty was narrowed to 6% (minimum) and 10% (maximum) from the prior values of 4% and 12%, respectively. Table 11.3, case 8, records the values that were then used to assess reserve uncertainty. As shown in Table 11.4, case 8, the range of uncertainty on volume migrated was then reduced significantly. Figure 11.32a shows that as a result of improving the TOC estimated range, dominance of relative contributions to the reserve uncertainty was then controlled by uncertainties in the source rock thickness; while Figs. 11.32a–c record the shift in the estimated cumulative probability ranges for generation, expelled, and migrated hydrocarbons, indicating narrowing of the range of uncertainty.

At this stage the variations in the estimated source rock thickness played a major role in reserve uncertainty. Reprocessing of seismic data, together with a better seismic interpretation in relation to knowledge of paleodeposi-

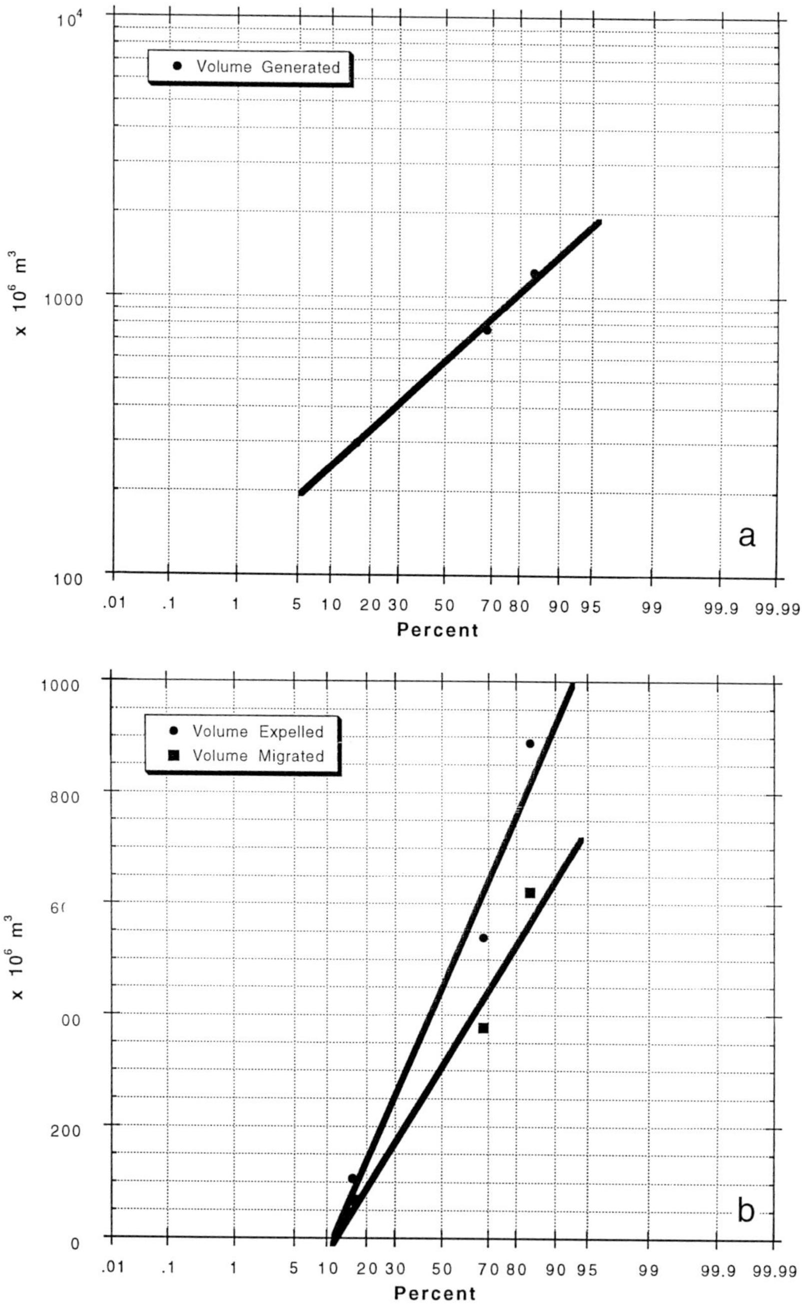


FIGURE 11.31 As for Fig. 11.26, all parts, but for the parameter ranges of Table 11.3, case 7.

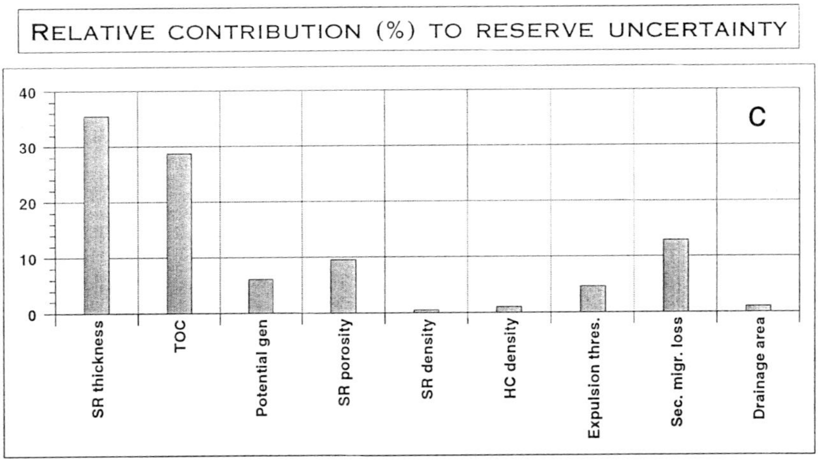


FIGURE 11.31 (*Continued*)

tion in the Norwegian Sea, was then used to constrain the thickness estimates to a higher degree. This case took the most time to evaluate (and also cost the most, which is why it was delayed until absolutely necessary) due to the seismic reprocessing and reinterpretation. While some improvement in the maximum source rock thickness took place (from 150 to 120 m), by far the biggest change was in the minimum thickness estimate (from 40 to 100 m). At the same time, it was requested that, due to time and budget constraints, (1) the project be completed, (2) all final estimates of ranges of all parameters be used, and (3) a final estimated range of uncertainty on reserves be made so that the information could be used in terms of economic ranking and risking strategies. These mandates then resulted in a final set of calculations being done using the parameters given in Table 11.3, case 9. Figure 11.33a shows the resulting relative contributions to reserve uncertainty, indicating that secondary migration would need to be reevaluated in any further attempt to improve matters. The final estimated probabilities for generated, expelled, and migrated hydrocarbons are given in Figs. 11.33a–c, respectively, while the final estimated ranges of uncertainty are given in Table 11.4, case 9.

Note that while the *relative* contributions to reserve uncertainty now vary between dominance from secondary migration and, to a lesser extent, TOC content, the *absolute* uncertainty [as determined by the $P(90)$ and $P(10)$ values around the expected $P(68)$ point of Table 11.4] has stabilized—indicating that it is difficult to improve matters much more without significant, expensive effort. The cost in terms of time and money to provide such improvements was considered to be inappropriate relative to commitments of personnel to other projects, time limitations to submit information

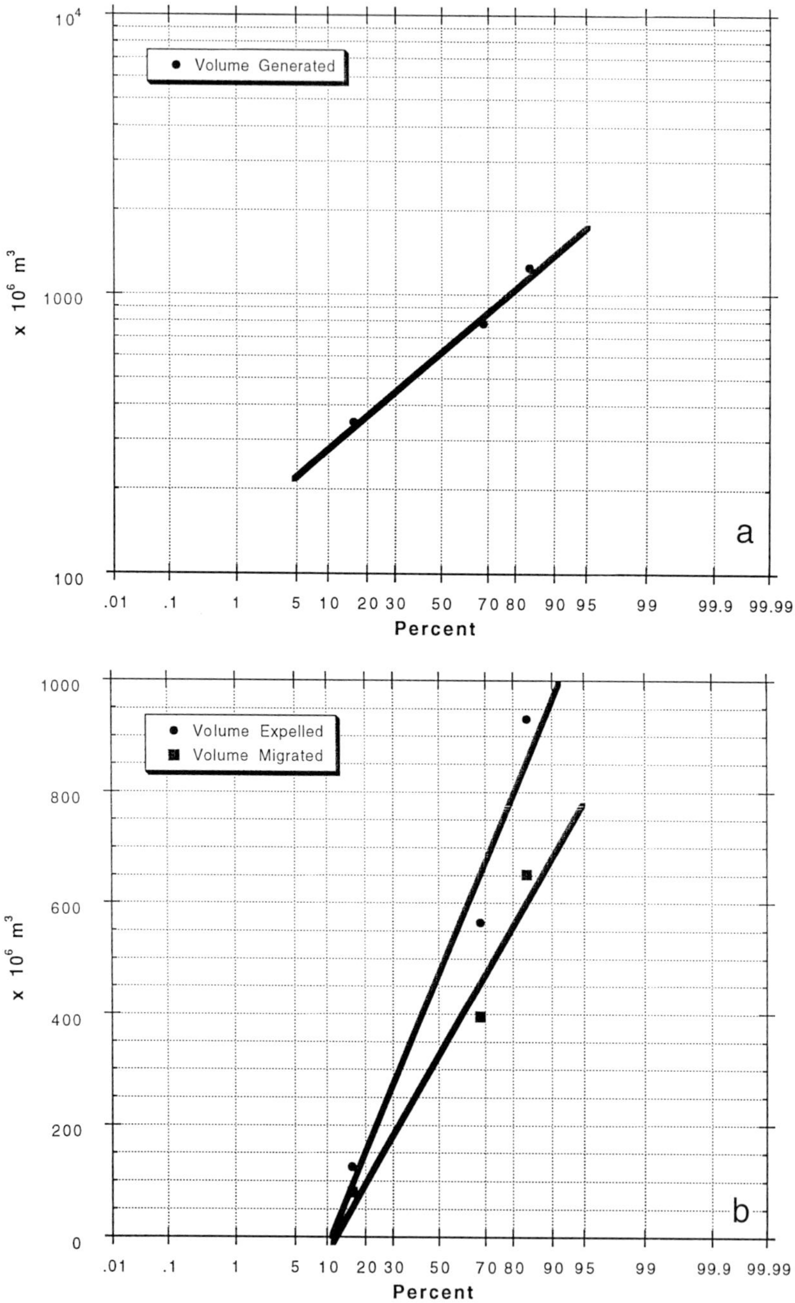


FIGURE 11.32 As for Fig. 11.26, all parts, but for the parameter ranges of Table 11.3, case 8.

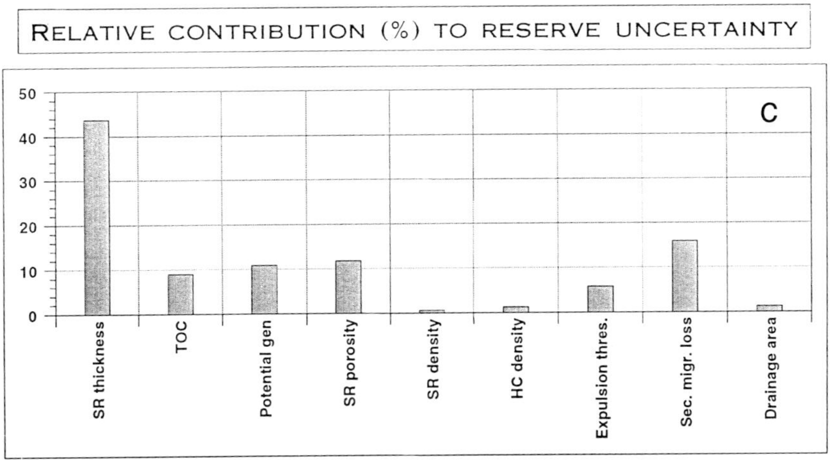


FIGURE 11.32 *(Continued)*

to the appropriate governmental authorities, and the need to make economic assessments [in the sense that if the $P(10)$ reserve estimate were to be economically worthwhile then all higher reserve assessments would be even more worthwhile, so that a minimum ranking and priority could already be given].

Note from Table 11.4 that the dominant factor in all cases was the progressive increase in hydrocarbon charge estimate at the $P(10)$ level, while the $P(68)$ and $P(84)$ values do not shift as much overall. As a consequence the project was deemed to have fulfilled its objectives.

## C. DISCUSSION

The purpose of this work has been to illustrate how to use the relative contributions of scientific factors influencing uncertainty of reserve assessments in order to determine where to concentrate efforts to improve the range of uncertainty.

The integrated use of modeling methods, geochemistry, geology, and seismic processing can all impact the determination of reserves. The project described here shows how and when the individual subdisciplines have to be addressed to improve matters. Perhaps one of the more interesting facets of such assessments is to have objective, reproducible methods expressed in simple form graphically to demonstrate to the personnel in the various subgroups *why* the improvements are needed and the level of increased resolution that needs to be obtained. At the same time the exigencies of reality must be kept firmly in mind in terms of work cost, personnel, and

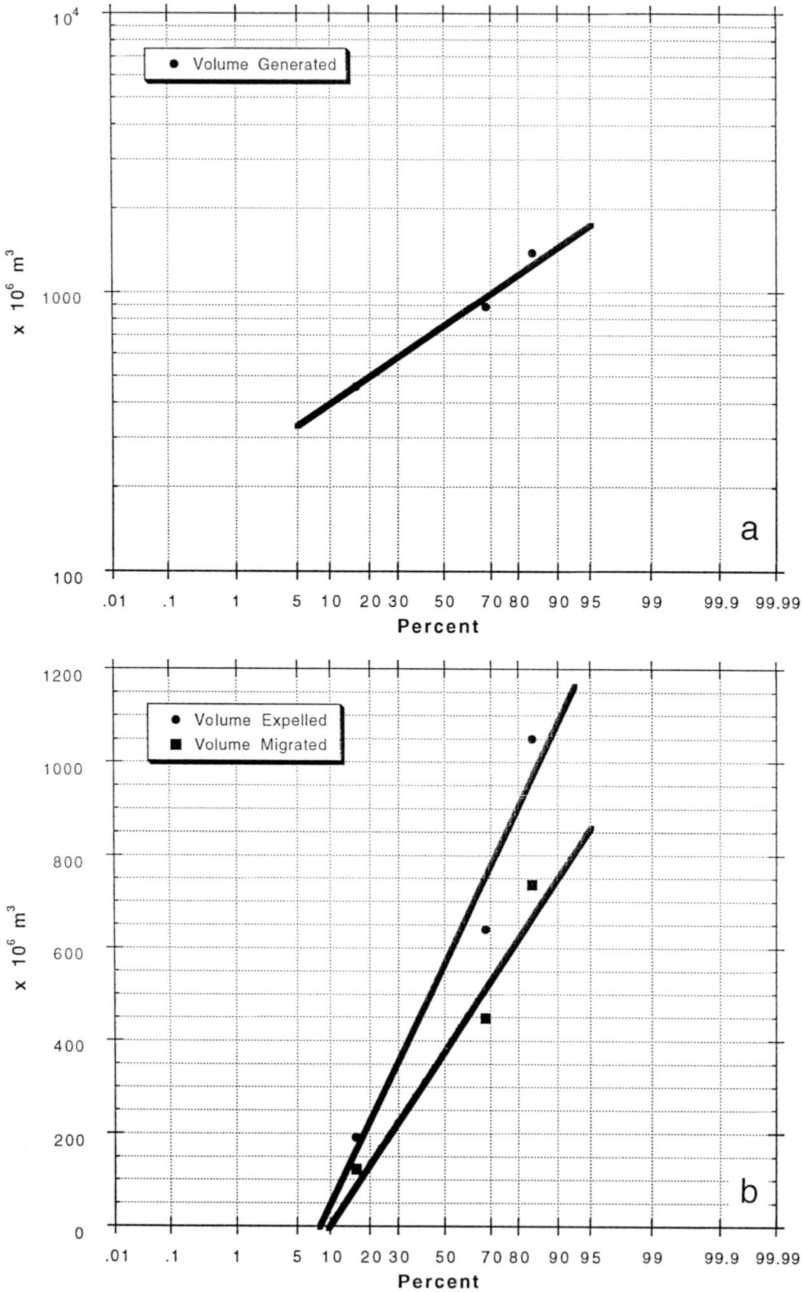


FIGURE 11.33 As for Fig. 11.26, all parts, but for the parameter ranges of Table 11.3, case 9.

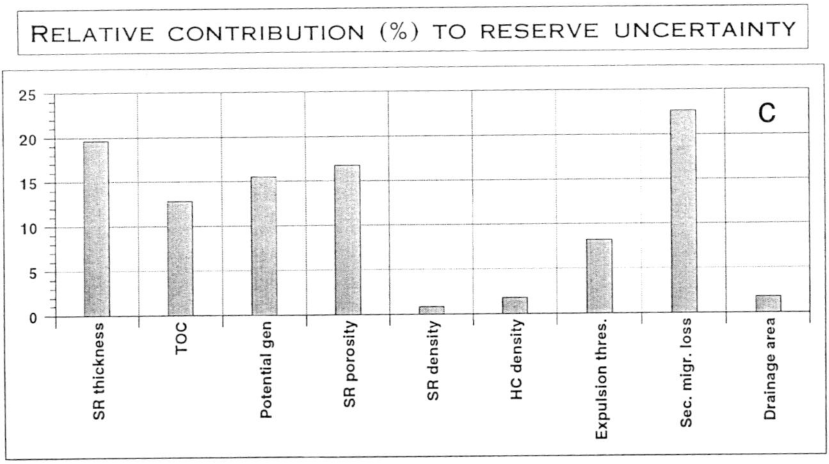


FIGURE 11.33 (*Continued*)

reporting deadlines. In addition, the need for composite groups of scientists to interact positively almost mandates some form of objective measures for the manner in which individual factors influence reserve estimates; the alternative is subjective measures, with the attendant inferences for selective bias by individuals in what they believe needs improvement rather than what can be demonstrated needs to be improved.

For the example investigated, it is clear that TOC, source rock thickness, and migration efficiency dominate all other factors in controlling the uncertainty of reserve estimates. It does not matter how many basin model calculations are done or under what conditions; the dominance of the three major factors can be neither gainsaid nor mitigated.

Not only are the methods presented here of value scientifically in terms of assessing reserve estimates and their reliability, but the social interaction needed to operate such a project also proved to be a very valuable learning experience.

## III. DRILLING RISK FOR SUBSALT PLAYS

### A. INTRODUCTION

The morphology of salt bodies in the subsurface of the Gulf of Mexico is highly variable, but the presence of massive salt bodies in the form of sheets lying horizontal to subhorizontal is well documented (West, 1989). Such salt sheets can impact the potential for the occurrence of hydrocarbon accumulations in the formations underlying the salt sheets (1) by providing

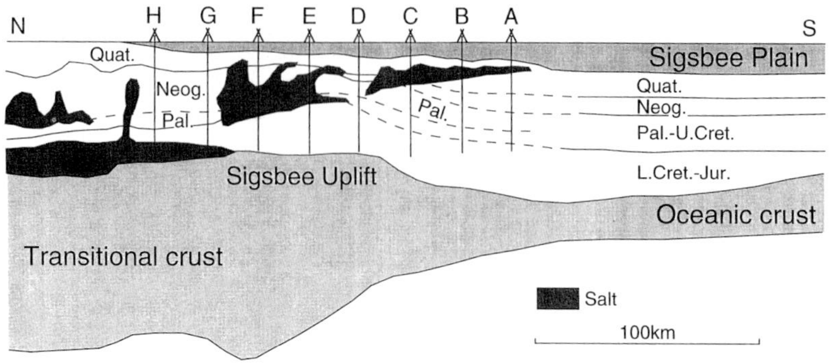


FIGURE 11.34 A scaled section along the Gulf of Mexico from north to south. The locations of pseudo-wells are marked, together with interpreted horizons and salt mass locations.

an impermeable sealing cover; (2) by lateral motion through the sediments, thereby promoting fracturing of the surrounding sediments and so enhancing fracture permeability pathways for hydrocarbon migration; (3) by creating a thermal regime that is cooler under the salt sheet than at the same subsurface depth far from the salt, thereby delaying the conversion of oil to gas; and (4) by dissolving in subsalt waters, thereby promoting a hypersaline environment that is not only anoxic, but in which hydrocarbons are less soluble, thus creating volume space for hydrocarbon trapping. Although these general consequences of the basic rules of physics would seem to be difficult to avoid, nevertheless, the precise and specific magnitudes, and lateral and vertical scales, of salt-induced or modified effects would seem to be of crucial importance to estimate better the likelihood of hydrocarbon occurrences under salt sheets.

The purposes of this section are twofold. First, from available seismic data from the Gulf of Mexico an idealized section is constructed (Fig. 11.34), which shows that salt sheets have been influential in disturbing underlying sedimentary formations and in changing the depositional patterns of surrounding formations; thus, an understanding can be achieved of some of the physical conditions under salt sheets. As demonstrated in the next subsection, some general, pervasive patterns are emerging. Second, sophisticated computer codes are available that allow for deposition, fluid pressure development, and thermal development including salt insertion. Given the control information and inferences from the seismic data, the extent of salt sheet effects (for instance, excess fluid pressure development) can be investigated. As is shown in the next subsection, some general considerations then follow that are consistent with inferences from wells (Yu *et al.*, 1993).

TABLE 11.5 Summary Table Listing Salt Depths and Thicknesses, Mud Weights, and Log Values for Wells Drilled Through Salt Sheets

| Well name | Depth of top salt (ft) | Salt thickness at well (ft) | Well TD (ft) | MW above salt (lb/gal) | MW below salt (lb/gal) | Density profile (gm/cm$^3$) | Sonic log transit times below salt (ms/ft) |
|---|---|---|---|---|---|---|---|
| West Cameron Block 505#2 | 14,100 | 1,580 | 18,500 | 12 | 18 | 2.45/2.05/2.25 | 140 |
| Garden Banks Block 171#1 | 8,500 | 1,096 | 10,597 | 12 | 14 | 2.15/2.05/2.25 | 140 |
| Vermillion Block 356#1 | 8,504 | 2,098 | 17,000 | 13 | 18 | No data | 140 |
| South Marsh Island Block 200#1 | 8,810 | 990 | 13,500 | 15.6 | 17.2 | 2.15–2.25/ 2.05/ 2.15–2.25 | 130 |

## B. PHYSICAL CONDITIONS IN FORMATIONS UNDERLYING SALT SHEETS

We consider four Gulf of Mexico wells: West Cameron 505#2; Garden Banks 171#1; Vermilion 356#1, and South Marsh Island 200#1 penetrated salt sheets of thicknesses ranging from 1000 to 2000 ft (300 to 600 m), at sub-mudline depths to salt top of 8000 to 14,000 ft (2300 to 4700 m). Table 11.5 provides a summary of (1) the mud weights (pounds per gallon) above and below salt, (2) the density profiles, and (3) the sonic transit time values below salt. The salt thicknesses at the well bores are also given, as well as depths to top salt and TD of each well. The log values quoted represent measurements over scales of 300 to 500 ft (100 to 170 m), a scale often considered appropriate for salt and fluid motion and allied physical phenomena. The sonic transit time values and the log density values are both extremely low for the formations under the salt sheets compared to regional values at the same depths. Figure 11.35a plots the inverse of the sonic transit times for the wells, together with the regional behavior as exemplified by data from Southern High Island, while Fig. 11.35b shows the corresponding density values, again with reference to High Island data as a convenient regional reference. The several regional trend curves drawn on Figs. 11.35a and b are characterized by their total fluid pressure regimes in the Gulf of Mexico: normal (hydrostatic) pressure, moderate overpressure (less than about 0.6 psi/ft), high overpressure (less than about 0.8 psi/ft), and ultrahigh overpressure (greater than 0.8 psi/ft).

Mud weight data (Table 11.5) also indicate penetration into high over-

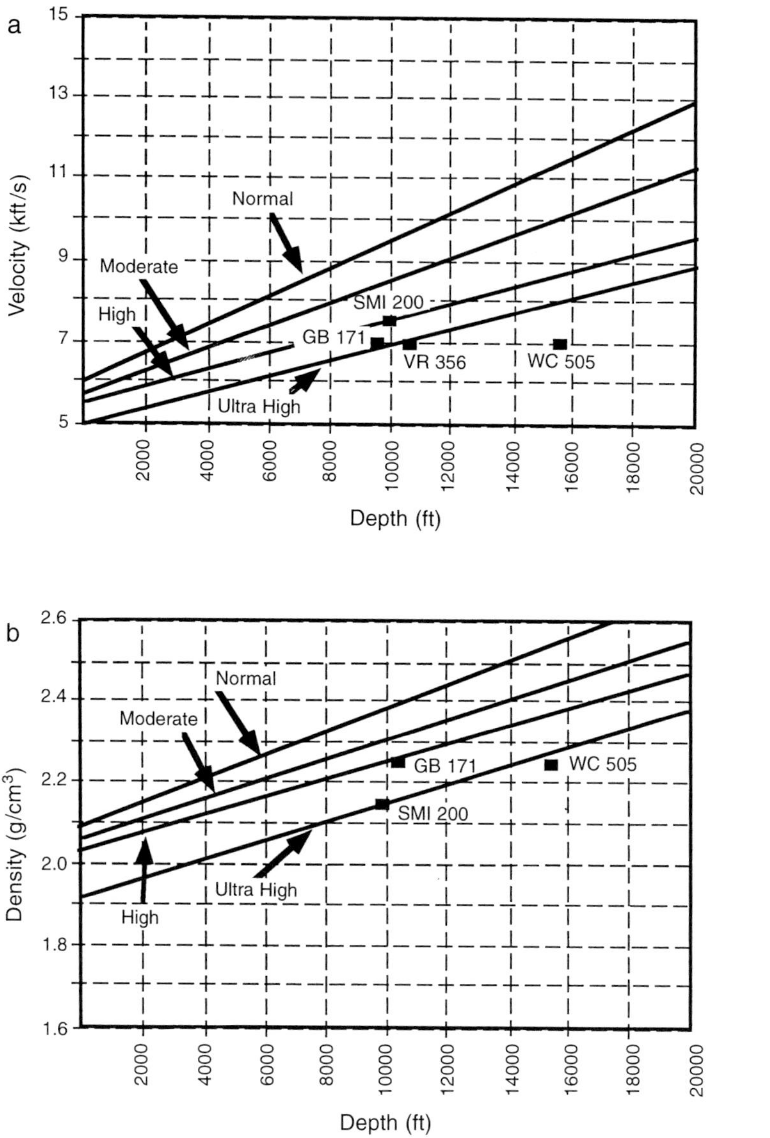


FIGURE 11.35 (a) Sonic transit times with depth; (b) density variations with depth for West Cameron 505#2; Garden Banks 171#1; Vermilion 356#1; and South Marsh Island 200#1.

pressure regimes: mud weights of 17 to 18 lb/gal were used below the salt in three of the wells, while 14 lb/gal mud was used in Garden Banks 171#1. A combination of the pressure indicators given, as shown in Table 11.6, indicates that the subsalt formations in West Cameron 505#2, Vermilion

TABLE 11.6 Summary of Pressure Indicators for Formations Immediately Underlying Salt Sheets Penetrated by Wells

| Well name | Pressure regime from mud weight | Pressure regime from sonic log | Pressure regime from density log |
|---|---|---|---|
| West Cameron Block 505#2 | Ultrahigh | Ultrahigh | Ultrahigh |
| Garden Banks Block 171#1 | Moderate | Ultrahigh | High |
| Vermilion Block 356#1 | Ultrahigh | Ultrahigh | No data |
| South Marsh Island Block 200#1 | Ultrahigh | Ultrahigh | Ultrahigh |

356#1, and South Marsh Island 200#1 all lie in the ultrahigh overpressure regime. The subsalt section in Garden Banks 171#1 is also overpressured but to a lesser extent than the other three wells; this lower pressure may be linked to proximity of the well to the edge of the salt sheet, which is only 4000 ft (1300 m) away toward the southwest. In this case the salt sheet may only partially block the flow of fluids, thereby leading to a less well-developed overpressure condition at the well location in relation to the other three wells, which are considerable distances (several miles) from the edges of their respective salt sheets. The data from these four wells, each associated with a different salt sheet, strongly indicate that the presence of a thick, laterally extensive, salt sheet can easily result in extreme overpressure development in subsalt formations, impeding subsequent compaction and so aiding overpressure growth. Such conditions provide significant mechanical and dynamic drilling risks for subsalt exploration.

## C. DYNAMIC MODELS OF PHYSICAL PROPERTIES IN SUBSALT FORMATIONS

The majority of the Louann salt wedge was deposited on continental crust. The rates of salt migration from deposition sites to present locations are computed (see Worrall and Snelson, 1989, their Fig. 15) based on these assumptions: All salt deposition was during Callovian (157 to 152 Ma, stratigraphy from Haq *et al.*, 1987, simplified to 150 Ma); all salt movement from potential deposition sites to present locations has been in a generally straight line; and rates calculated are maximal (greatest distances are used). Rates are derived from constant movement over the 150 Ma since deposition and, also over those times, are characterized by low sea level, presumably a time of greater deposition and salt extrusion. Rates for salt migration averaged over the time since deposition would be 0.3 cm/year; rates of movement during lowstands solely would be 0.6 cm/year with the motion due south. As depicted in Fig. 5 of Lowrie *et al.* (1990), the original sediment deposition site was on the most basinward continental crust, and then migrated some 500 km southward during the next 150 My, approximately from Callovian time, upper Middle Jurassic. Even if much of the present Louann salt wedge had traveled from hundreds of kilometers farther inland, the average salt migration rate would have still been less than a centimeter per year. We suppose that the salt movement was intermittent rather than constant. It must be emphasized that this calculation pertains to movements by the entire Louann salt wedge from the time of deposition to the present.

West (1989) has persuasively argued that salt served as the outer wall of the Upper Miocene-Pliocene (UM-P) (10 to 2.5 MYBP) depocenter near the present shelf-break. Presumably, the salt had begun migrating southward over continental crust and was "buoyed up" over the Sigsbee uplift, as suggested by Jackson *et al.* (1988, their Fig. 65). The position of the UM-P Sigsbee escarpment can then be inferred. Possible migration rates of the UM-P Sigsbee escarpment to its present position yield the following: for constant movement, 160 km/2.5 MY $\cong$ 6 cm/year; for movement during lowstand only, 160 km/1 MY $\cong$ 16 cm/year. These rates give an impression of a very dynamic continental margin, particularly when it is remembered that the calculation is for an average migration rate for the entire Louann salt wedge from the Upper Miocene–Pliocene to the present. This rapid motion has immediate consequences for sediment fracturing and thermal maturity (Lerche and Lowrie, 1992) and for subsalt overpressure development as investigated here.

The Gulf of Mexico section displayed in Fig. 11.34 can be used to infer extremely rapid motion of the salt sheet currently intersecting locations A, B, and C, as exhibited by the thickness changes in Neogene sediments from location D through to location A, and by the distortions (relative to horizontal) of the sedimentary beds. Given the ages of the formations, and the spacing between locations A through D, horizontal speeds for the salt sheet of about 5 to 15 cm/year are implied. Thus, the most southerly "nose" of the salt intersected location A at about 0.1 MYBP and location D at about 0.9 MYBP, while the most northerly "tail" of the salt left location D at about 0.1 MYBP. A sketch of the motion of the salt sheet with time is given in Fig. 11.36, indicating the southerly passage of salt past locations D, C, B, and A with time. Because the location spacing is about 20 km between consecutive wells, analysis of the effects of the rapid salt insertion and withdrawal at each location can be modeled most simply by treating the fluid flow, compaction, and overpressure development at each location as though the system behavior is one dimensional, and then by contour display on a two-dimensional section of the relevant quantities of interest at different instances of times. There is not enough time for lateral flow of fluid to be significant in influencing the results.

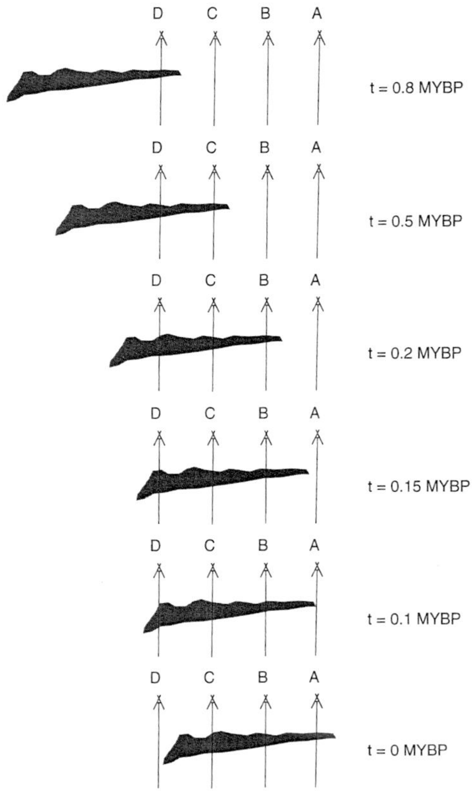


FIGURE 11.36 Sketch of the interpreted motion with time of the southerly salt mass currently intersecting locations C, B, and A. Speed of lateral salt motion is assessed from West (1989) and Lowrie (1994) to be between 5 to 15 cm/year.

## 1. Individual Pseudo-Well Behaviors

Because salt insertion occurs at all pseudo-well locations close to the present day, burial histories are shown only from about 4 MYBP through to the present day. The numerical results are generated by the 1-D DTIT code developed at the University of South Carolina. For the studies reported here only the dynamic indicator section of the code was enabled because the only information available to act as control criteria on modeled evolution is just formation thickness at present day. The lithology for each pseudo-well was taken from general background literature of the Gulf of Mexico, as compiled by Lowrie (1994), and is dominantly shale or sandy-shale throughout the last 4 Ma.

### *a. Pseudo-Well D*

Figure 11.37a shows the evolution of excess pressure on a burial history diagram. The lithology variation is given across the top of the figure with

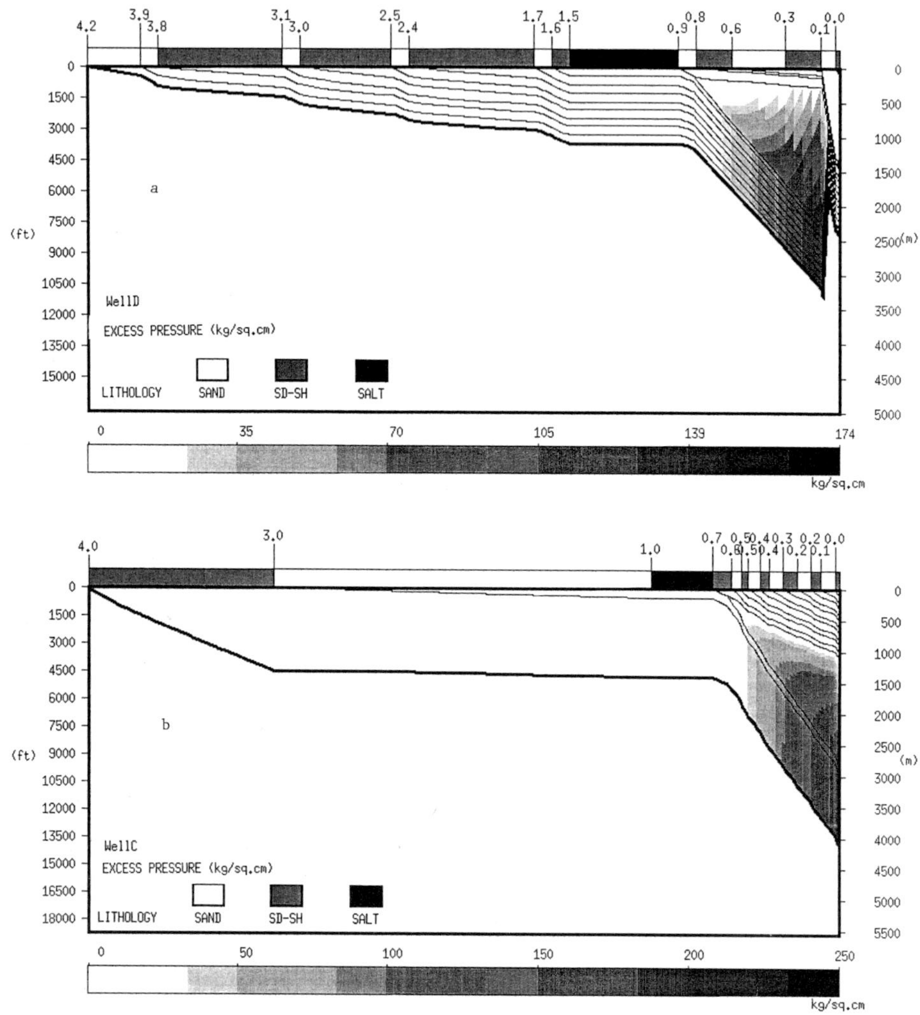


FIGURE 11.37 Burial history and overpressure evolution with time at different pseudo-well locations as salt is inserted at each location: (a) at well D; (b) at well C; (c) at well B; (d) at well A.

a legend inserted. Note the insertion of the salt "nose" at 0.9 MYBP and the withdrawal of the thick salt section at 0.1 MYBP, followed by deposition of a further 2.5 km of sandy material, with a thin Holocene cover of sandy-shale. The evolution of excess pressure shows a significant buildup with time after salt insertion, reaching a peak value of about 174 kg/cm$^2$ ($\approx$170 atm) just below the salt at maximum salt thickness time ($\cong$0.1 MYBP). As the salt wedge thickens with time at location D, the overpressure buildup

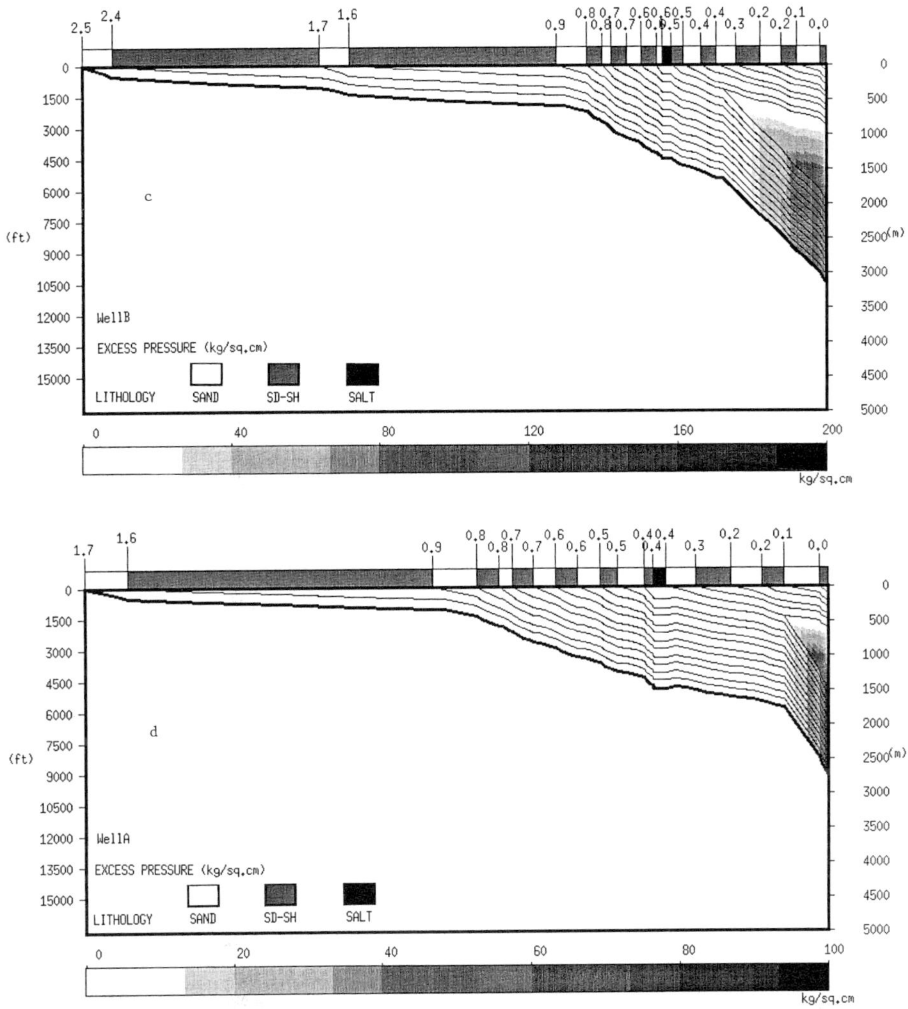


FIGURE 11.37 *(Continued)*

is in approximate proportion to the salt thickness, corresponding to a buildup rate of some 210 atm/MY. The rapid withdrawal of the thick salt section immediately removes overburden from the underlying formations and, more importantly, removes the tight permeability seal of the salt. As a consequence overpressure drops rapidly, as indicated on Fig. 11.37a.

Also of interest are the porosity and permeability variations with depth as the salt is inserted, as shown in Figs. 11.38a and 11.39a at 0.1 MYBP, the time of maximum salt thickness. The low porosity and low permeability zones representing the salt are obvious by inspection, while the interbedded

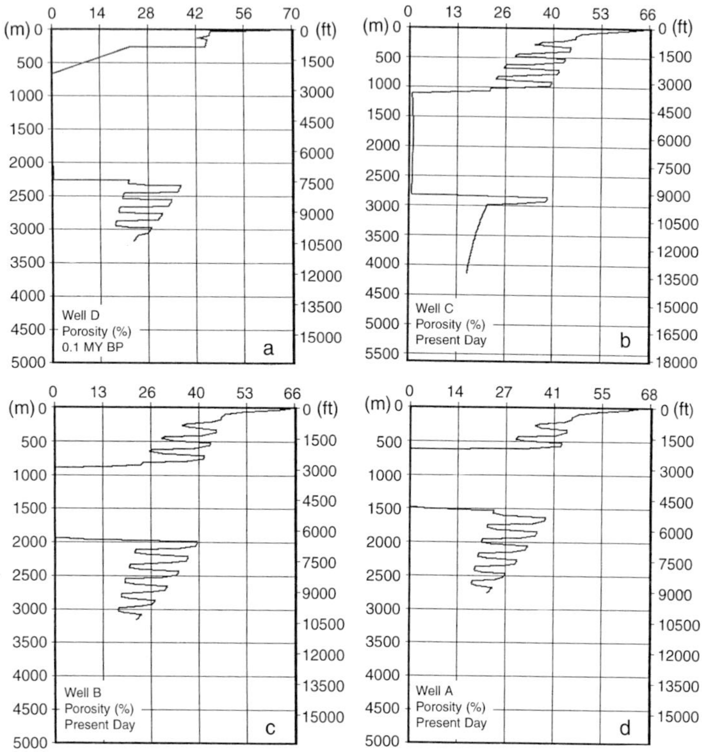


FIGURE 11.38 Porosity variation with depth at present day at well sites C, B, and A or, in the case of pseudo-well D, at 0.1 MYBP, the latest time salt was present. Note the preservation of porosity below the salt: (a) at well D; (b) at well C; (c) at well B; (d) at well A.

sand/sandy-shale variations immediately below the salt are also clear. The dynamic indicator control of formation thickness is represented in Fig. 11.40a, showing that observed and predicted thicknesses of formations today are in excellent agreement, providing corroborative support for the consistency of the model procedures.

### b. Pseudo-Well C

In the case of pseudo-well C, lying about 20 km further southward than pseudo-well D, the regional geologic information (Lowrie, 1994) would suggest that the last 4 MY of deposition were dominated by an early phase (4 to 3 MYBP) of sandy-shale, followed by about a 2 MY period of mainly sand. Insertion of the salt nose occurred around 0.6 MYBP, with salt continuing to increase in thickness through to the present day where it is 1.5 km in width, with the salt top at about 1 km sub-mudline depth. The sediment deposition on top of the salt is taken to be intercalated sand and sandy-

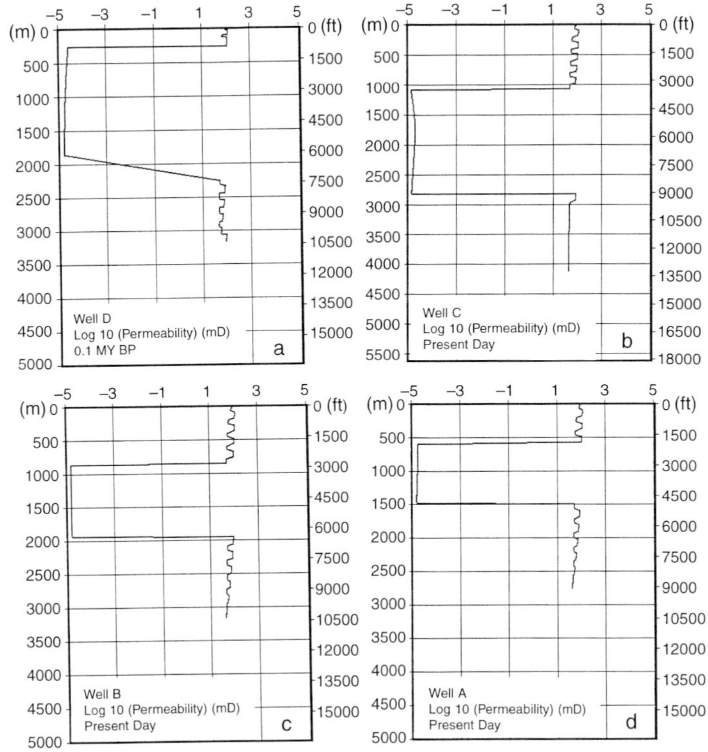


FIGURE 11.39 As for Fig. 11.38 but for permeability variations with depth, indicating the preservation of high permeability in formations underlying the salt: (a) at well D; (b) at well C; (c) at well B; (d) at well A.

shale layers, representative of the deepening water and increased topographic slope at pseudo-well C compared to pseudo-well D. Noted from fig. 11.37b the massive buildup of overpressure below the salt, starting at about 0.5 MYBP and reaching approximately 250 atm today, for a buildup rate of some 500 atm/MY—about twice the rate occurring at pseudo-well D. Again, the salt insertion is sufficiently rapid that the buildup of overpressure is essentially directly proportional to the salt thickness at any instant of time after salt insertion, corresponding to about 170 atm/km salt. Figures 11.38b and 11.39b show the variations of porosity and permeability with depth at the present day, clearly exhibiting the presence of salt, and also show the preservation of porosity in the sand underlying the salt because of the sealing nature (low permeability) of the salt to fluid flow. Thus, once the salt sheet is breached, there is a high breakout subsalt fluid pressure but also well-preserved sand porosity and permeability for extraction of any entrapped hydrocarbons. Again, the dynamic indicator

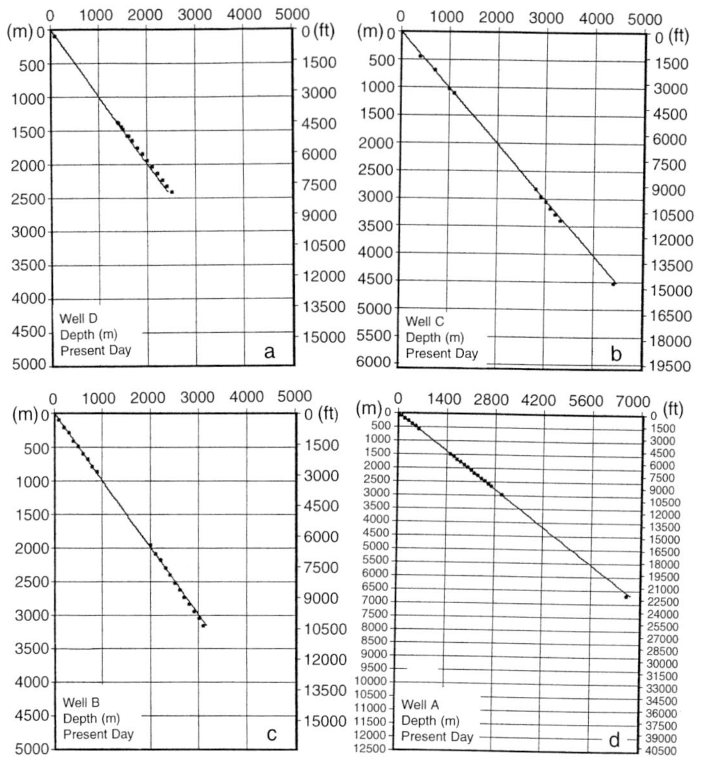


FIGURE 11.40 The variation of model predicted versus observed formation thicknesses at each pseudo-well, which act as a control criterion on the variation of dynamic parameters in the model. A mismatch that is perfect would yield a 45° straight line: (a) at well D; (b) at well C; (c) at well B: (d) at well A.

behavior of matching observed and measured formation thicknesses is shown in Fig. 11.40b, indicating that a consistent pattern was obtained by the dynamic code.

### *c. Pseudo-Well B*

At this pseudo-well site, located about 40 km southward of pseudo-well D, salt nose insertion did not take place until about 0.35 MYBP and salt has reached a current thickness of only just over 1 km at present day. The salt top is also shallowing, being only about 0.9 km below the mudline, but water depth has increased as shown in Fig. 11.36. The dominant deposition in the interval 2.5 to 0.9 MYBP consists mainly of sandy-shale with a few thin sand bodies; thereafter the intercalation of thin sands and sandy-shales is more prevalent both prior to, and after, salt insertion and continuing through to the present day. In this case, as shown in Fig. 11.37c, the overpressure below salt reaches a maximum of about 200 atm corresponding to a buildup rate of about 600 atm/MY, and is again roughly proportional to inserted salt thickness, corresponding to about 220 atm/km salt. The porosity and permeability variations are shown in Figs. 11.38c and 11.39c, respectively. They show not only the salt effect, but also the preservation of porosity and permeability in the intercalated sands below the salt sheet. Again, a good match is obtained of predicted and observed formation thicknesses today (Fig. 11.40c).

#### *d. Pseudo-Well A*

The southernmost pseudo-well did not experience salt insertion until about 0.1 MYBP, with a current salt thickness of about 0.9 km and a shallow salt top at the 0.6-km sub-mudline. The dominant lithology here consists of sandy-shale material from about 1.6 to 0.9 MYBP, followed by intercalated sands and sandy-shales through to the present day. The burial history diagram of Fig. 11.37d shows an excess pressure buildup only to about 100 atm over a period of 0.1 MY, corresponding to 1000 atm/MY; or, in terms of salt thickness, about 110 atm/km salt. Figures 11.38d and 11.39d, respectively, show the preservation of porosity and permeability in the intercalated sands below the salt sheet, while Fig. 11.40d shows the match of predicted and observed formation thicknesses. The clear implications from these calculations are as follows:

1. Rapid salt insertion effectively seals the formations underlying the salt, thus preventing the further escape of fluids.
2. As a consequence of item 1, both porosity and permeability are retained at high values in underlying sedimentary units, thus permitting good quality reservoir sands to be preserved.
3. Also as a consequence of item 1, the buildup of overpressure in the subsalt formations is remarkably fast, corresponding to around 250 to 1000 atm/MY or about 100 to 300 atm/km salt thickness. The consequences for drill penetration through salt are then to expect an increase in excess pressure at breakout of some 1400 to 4000 psi/km salt.
4. Equally as effective as the rapid buildup of overpressure is the rapid decrease once salt withdraws from a location, as at pseudo-well D. The rapid removal of the salt overburden almost immediately reduces the overpressure directly proportional to overburden loss. This effect can be likened to a hydraulic jack supporting a car: As long as the car is in place the hydraulic pressure in the jack is proportional to car weight; once the car is removed, the hydraulic pressure rapidly returns to its hydrostatic value.

### 2. Sectional Behaviors

Patterns of pressure behavior across the section of Fig. 11.34 can be drawn at different times as the salt progressively transits into and through

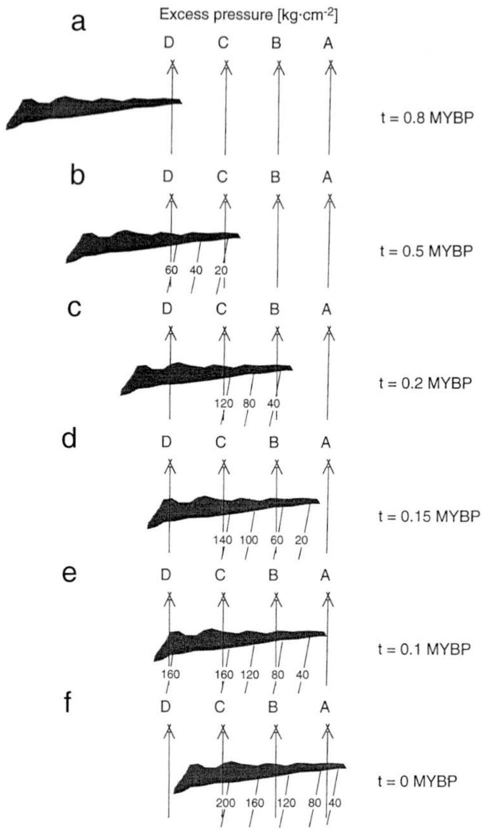


FIGURE 11.41 Overpressure values directly below the salt wedge at different times, indicating the evolution of overpressure laterally across the section of Fig. 11.34 as the salt both moves across the thickens at the pseudo-well positions: (a) 0.8 MYBP; (b) 0.5 MYBP; (c) 0.2 MYBP; (d) 0.15 MYBP; (e) 0.1 MYBP: (f) present day.

the pseudo-well locations. Figure 11.41 presents a series of "snapshots" of overpressure just below the salt as the salt encroaches on the various well locations. Figure 11.41a shows overpressure at 0.8 MYBP when the salt nose had just been inserted at pseudo-well D; Fig. 11.41b shows the corresponding behavior at 0.5 MYBP when the salt nose has crossed pseudo-well C location but has not yet reached pseudo-well B. Note that both wells D and C now signal the presence of massive overpressure. In Fig. 11.41c we present the excess pressure pattern at 0.2 MYBP, when the salt has spread across well sites D, C, and B but not A, indicating a considerable increase in pressure development at the most northerly two wells. Figure 11.41d presents the pressure profile at 0.15 MYBP when the thick tail end of the salt has just exited the position of well D, thus reducing overpressure, and the nose of the salt has just crossed over the location of well A, starting the rise of overpressure. In this case the excess pressure is confined to the central two wells (C and B) with virtually no overpressure at wells D and A. At the present day, Fig. 11.41f, overpressure continues to increase in wells C and B, and has already reached about 100 atm in well A.

Thus, even though the salt top is shallowing as the salt steps laterally southward, the impermeability of the salt and the rapid increase in salt wedge thickness passing each location, when taken together, are sufficient to induce massive overpressure buildup across the section as the salt progressively moves. The overpressure is dominantly confined to the lateral extent domain of the salt (much as a shadow created by a cloud tracks with the cloud), but the importance of assessing the dynamic range of overpressure development as the salt moves is evident.

## D. DISCUSSION

Several major points result from the study presented here:

1. Perhaps the dominant point of the numerical experiments is that they indicate the development of significant overpressure under the salt sheet in the last 0.8 MYBP through to present day. The exact degree of overpressure and its precise buildup rate are, of course, controlled by local (to the salt) variations of lithology (intercalated sands) and dip angle of the salt, and by salt speed, but the buildup of overpressure always occurs for a laterally extensive salt sheet.
2. The buildup of overpressure slows the compactional loss of porosity in formations under the salt relative to the regional behavior far from the salt sheet.
3. The presence of isolated sands does not stop overpressure buildup. The sands themselves are overpressured as are the more impermeable shale formations.
4. The presence of a dipping salt layer lowers the overpressure on the up-dip underside of the salt both because of shallower burial of the salt at the up-dip end and because fluids can flow more easily away from the salt at shallower depths (more permeable formations). Dipping salt is also known to produce a significant asymmetric thermal anomaly on the up-dip end of the salt, provided the dip is in excess of about 5 to 10° (Lerche and Petersen, 1995);
5. An overpressure increase (relative to regional) of about 100 to 300 atm occurs under and in the vicinity of the salt.

Although the results of the numerical experiments reported here are not exhaustive, in that they do not cover all geological scenarios that could be envisioned, nevertheless the experiments provide a good appreciation of the magnitude, lateral and vertical extends, and timing of ruggedly stable physical effects initiated or modified by the presence of salt sheets. It would seem that the behaviors of ultrahigh overpressure and low density (observed in four subsalt sheet wells) are phenomena that fit comfortably within the framework of the numerical models given. The dominant physical characteristics observed are captured by the models. More detailed modeling with increased downhole and seismic information could help to refine the statistical sharpness of the estimates of buildup time and magnitude of excess pressure. In this way a more relevant assessment could be made of the risk of hydrocarbon proneness of each salt sheet, as well as of the risk of drilling, so that a balanced assessment of potential gains and potential problems and costs could be developed.

## E. STOCHASTIC BEHAVIORS

For particular and specified parameter values for moving salt sheets, calculations have been given here and elsewhere (O'Brien and Lerche, 1994) that illuminate the excess fluid pressure values to be expected. However, three factors confound the use of the particular calculations in pragmatic strategies for assessing drilling risk. First, the lithology under a salt sheet is imperfectly known at best, so that the permeability of formations underlying a salt sheet is unclear. Because of the many orders of magnitude variation in permeability between sands and shales (which are the dominant lithologies in the Gulf of Mexico), the overpressure beneath a salt mass could then vary considerably depending on lithology variations. Second, it is often difficult to be precise about the salt thickness ahead of drilling due to the problem of determining the base of salt from seismic information. The thicker the moving salt, the greater the load produced by the salt and the greater the underlying fluid pressure needs to be in order to support the load. Thus the uncertainty in salt thickness also has a role to play in determining the degree of overpressure in a subsalt formation. Third, the timing of salt emplacement at a given location in a basin is also not too well determined; a high-speed salt mass presumably is emplaced much earlier geologically than a slower salt mass, thereby trapping fluids (including hydrocarbons) for a longer period of time. Thus a more rapid, and earlier, emplacement of salt is expected to be responsible for a higher degree of overpressure. Fourth, the sub-mudline depth of the salt top also influences the overpressure of subsalt formations. For a given salt sheet thickness, a shallow salt top means that the salt sheet has thinner sediments of higher porosity above it than a more deeply buried salt sheet. Thus the weight of the sediments overlying the salt is less in the case of a shallow salt top. Because the weight of such sediments is transferred through the salt to be balanced by overpressure in the subsalt formations, a shallower buried salt sheet will contribute less subsalt overpressure, all other factors being equal.

For these and other reasons the problem of addressing the uncertainty of subsalt overpressure values becomes relevant. Now the influence of uncertainties in the preceding factors on the subsalt overpressure is by no means equal due to the nonlinearity of response of compaction and fluid flow in sediments to both geological conditions, and their variations (1) to parameters in equations of state controlling the intrinsic evolution of sediments under applied loads and (2) to the basic nonlinear dynamic response of the geologic system. Thus, we need to determine what the relative importance (RI) is of uncertainties in each factor [salt thickness, subsalt lithology, salt speed (or emplacement time), salt top depth] in contributing to the estimates of both subsalt overpressure and its uncertainty. RI provides information on *which* of the factors causes the greatest uncertainty on overpressure determination. RI also then permits an evaluation to be made that *if* one needs to limit the uncertainty on overpressure determination further, then it is clear which of the factors one has to concentrate on to narrow the range of uncertainty. Note that the RI does not determine that one *does* have to narrow the range of uncertainty of the input factors; it may be that the output uncertainty on expected subsalt overpressure is considered small in relation to other anticipated drilling uncertainties.

In any event, what is clear is that some evaluation is necessary of the relative importance of the factors in relation to subsalt overpressure buildup, so that one has available the information needed to make a decision regarding likely drill conditions at breakout.

The purposes of this section are to illustrate how the uncertainties in salt thickness, burial depth of salt top, subsalt lithology, and emplacement speed of salt contribute to the relative uncertainty in subsalt overpressure, and how to organize the resulting information in a manner that addresses the concerns of uncertainty and mean values.

### 1. Model Behaviors

The generalized present-day cross-section of the Gulf of Mexico, displayed in Fig. 11.34, was used as the basis for input information to quantitative modeling. In this section we are concerned only with the influence of the salt sheet at location C of Fig. 11.34. West (1989) and Lowrie (1994) have persuasively argued that the salt sheet displayed on the cross-section has moved to its currently observed location at a rate of between 1 and 20 cm/year. At location C, the emplacement of the salt, which is interpreted to be currently situated in the midst of Quaternary sediments, is treated as a wedge of inserted salt, currently between 1700 and 2300 m thick. The top-of-salt depth below mudline is not clearly delineated on the section of Fig. 11.34, but original seismic information leads to the suggestion of a salt top 300 to 500 m below the mudline. Equally, the lithology under the salt at location C is not well known; accordingly we have run computations with the underlying formations treated as 100% sand, 50% sand and 50% shale, and 100% shale, respectively.

**2. Burial History and Excess Pressure**

The procedure for evaluating model behaviors is as follows. A one-dimensional fluid flow/compaction code (1D-DTIT, developed by the Basin Modeling Group at the University of South Carolina) was used to describe the deposition, fluid flow, and overpressure developments at location C of the cross-section shown in Fig. 11.34. A detailed description of the DTIT code is available elsewhere (Kai and Lerche, 1993). However, a précis of the information needed and procedures used for the salt insertion problem is now provided. The ages, present-day thicknesses of each formation (including salt), and water depth are provided to the code. Prior to salt insertion, each formation is then sequentially deposited and continuous evolution of porosity, permeability, fluid flow, and overpressure development computed for prescribed lithologies (see later) for each of the formations.

At salt insertion time (determined by taking the "nose" of the salt sheet in Fig. 11.34, which is to the right of location A, leftward at a constant speed until the nose intersects location C) the salt is considered to develop as an ever increasing thickness unit inserted at a precise depth. (For instance, if a lateral salt speed of 10 cm/year is taken, with the present-day salt top 500 m below the present-day mudline, then for a salt nose 10 km to the right of location A, the salt would have been emplaced at location A at 0.1 MYBP. At that time, if the sediment deposition rate above the salt is 100 m/MY, then 10 m of the present overburden above salt would need to be removed. Thus the emplacement depth of the salt would be 490 m, that is, 500 − 10 m). In fact, the calculations are a little more complex because porosity evolution with time and depth of each formation, including the influence of salt, needs to be included in the calculation of salt insertion depth. Fortunately, however, the DTIT code was developed to handle precisely the determination and resolution of parameters influencing paleo-evolution at a single well location using present-day data as constraint information.

The code uses a tomographic procedure (hence the acronym DTIT—dynamic and thermal indicator tomography) to evaluate the best parameter values most consistent with present-day information on formation thicknesses and depths. In addition, for each prescribed lithology, the DTIT code evaluates the parameter values in equations of state connecting both permeability to porosity and effective stress to porosity so that the differences between present-day calculated and measured formation thicknesses, porosity variations with depth, fluid overpressure with depth, and permeabilities with depth result in minimum discord.

In the present situation only formation thicknesses were available so that, once parameter values were determined by the DTIT code, output

TABLE 11.7 Parameter Ranges Used in Uncertainty Calculations

| | Minimum | Most likely | Maximum |
|---|---|---|---|
| Case 1: Present-day depth of burial = 300 m | | | |
| Lateral speed (cm/yr) | 8 | 10 | 12 |
| Thickness (m) | 1800 | 2000 | 2200 |
| Lithology (% sand) | 0 | 50 | 100 |
| Case 2: Present-day depth of burial = 500 m | | | |
| Lateral speed (cm/yr) | 5 | 10 | 15 |
| Thickness (m) | 1700 | 2000 | 2300 |
| Lithology (% sand) | 0 | 50 | 100 |

results with geological time provide an evolution at location C, including a predicted fluid overpressure variation with depth at the present day, the quantity of primary concern in this section.

Two cases were examined: case 1 in which the present-day salt top is 300 m below the mudline and case 2 in which the present-day salt top is 500 m below the mudline. For each case, the salt thickness was varied, the lateral insertion speed of the salt was varied, and the lithology underlying the salt was varied from 100% sand through a 50% sand/50% shale mix, to 100% shale. Table 11.7 provides the ranges of values used for each of the two cases. The DTIT code was then used for each set of parameter values for each case.

Figures 11.42a and b show, as an example, the predicted excess fluid pressure (in units of kg/cm$^2$ $\cong$ 1.03 atm) with depth for a salt insertion speed of 10 cm/year, with the current salt top at 300 m sub-mudline, for a 50% sand/50% shale lithology under the salt and with the salt thickness set at 2000 m (Fig. 11.42a) and 2200 m (Fig. 11.42b), respectively. Figure 11.42c shows the difference (higher subsalt excess pressure) from Fig. 11.42a when the lithology is changed to 100% sand and all other parameters values are held fixed. For each case similar figures can be drawn, each providing an excess fluid pressure value at breakout just below the salt sheet at present day.

### 3. Probability, Uncertainty, and Relative Importance

Because the exact salt thickness, salt speed, and underlying lithology are not known, it follows that the estimated excess fluid pressure at sub salt breakout is uncertain. Two factors are needed: (1) the probability of encountering an excess pressure less than (greater than) a prescribed value

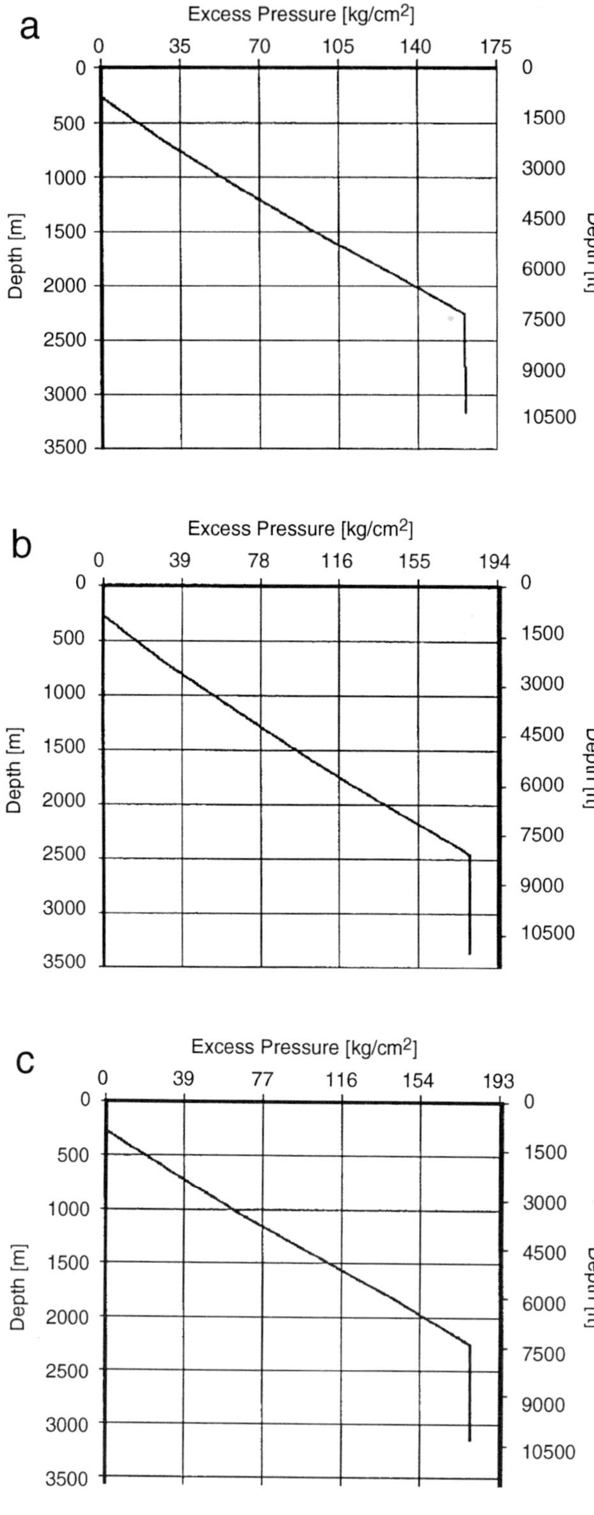

a
Excess Pressure [kg/cm2]
0 35 70 105 140 175
Depth [m]
0 500 1000 1500 2000 2500 3000 3500
Depth [ft]
0 1500 3000 4500 6000 7500 9000 10500
b
Excess Pressure [kg/cm2]
0 39 78 116 155 194
Depth [m]
0 500 1000 1500 2000 2500 3000 3500
Depth [ft]
0 1500 3000 4500 6000 7500 9000 10500
c
Excess Pressure [kg/cm2]
0 39 77 116 154 193
Depth [m]
0 500 1000 1500 2000 2500 3000 3500
Depth [ft]
0 1500 3000 4500 6000 7500 9000 10500

at breakout and (2) the degree of uncertainty of the estimate made, together with information on which parameter variation (salt thickness, salt speed, underlying lithology) is providing the greatest fraction of the uncertainty in excess pressure variability, that is, the relative importance of variations in each contributory factor to excess pressure uncertainty. In this way one can assess which factor first needs to have its range of uncertainty narrowed if a better job of limiting excess pressure uncertainty is needed. This RI then enables one to concentrate on the significant parameters, without having to expend effort on improving parameter values that have a lesser effect on uncertainties in excess pressure. Consider each component in turn.

### *a. Cumulative Probability and Uncertainty*

Suppose that, for each case, the individual parameters (salt thickness, salt speed, underlying lithology) are held at their most likely values of Table 11.7, with the exception of each one of the parameters in turn, which is allowed to vary from its minimum value through to its maximum value. Each such parameter value then produces an excess pressure value at subsalt breakout. For each parameter variation, the excess pressure values are then ordered from smallest to largest, and cumulative probability plots of obtaining an excess pressure less than or equal to a prescribed value are obtained, as shown in Figs. 11.43a–c for each case and with each of the three parameters varying. For instance, in Fig. 11.43a, case 1—Excess Pressure (Thickness) means present-day excess pressure at subsalt breakout due to variations in salt thickness (as prescribed in Table 11.7) when salt speed and lithology are held at 10 cm/year and 50/50 sand/shale, respectively. The values $x = 134, p = 10; x = 160, p = 50; x = 187, p = 90$ in Fig. 11.43a mean an excess pressure of 134 atm is found at 10% cumulative probability chance; 160 atm at 50% chance; and 187 atm at a cumulative probability of 90%, respectively. For example, it is 90% likely that for the situation being described the breakout excess pressure will be less than 187 atm.

For each parameter in turn, similar plots can be drawn as shown in Figs. 11.43b and c, so that some idea of the variation in excess pressure caused by each individual factor is available. A summary of these results for each of the two cases is provided in Table 11.8 with, in addtion, an estimate of the volatility of the excess pressure uncertainty. The volatility, $v$, is defined as

$$v\ [\%] = 100[P(90) - P(10)]/P(50) \qquad (11.6)$$

---

FIGURE 11.42 Excess pressure (in kg/cm$^2$ $\cong$ 1.03 atm) variations with present-day depth at location C of Fig. 11.34 for a salt insertion speed of 10 cm/year and present-day salt top at 300 m sub-mudline: (a) salt thickness of 2000 m; (b) salt thickness of 2200 m; (c) as for case (a) but the underlying formation lithology is changed to 100% sand from the 50% sand/50% shale lithology used in (a) and (b).

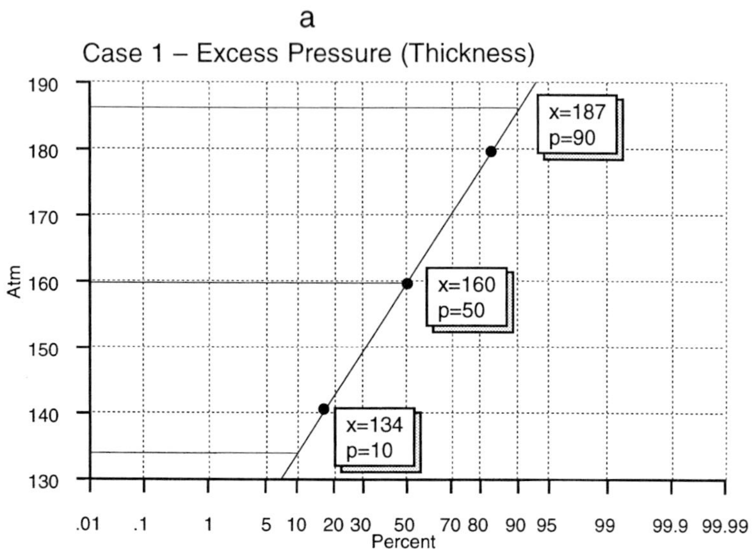


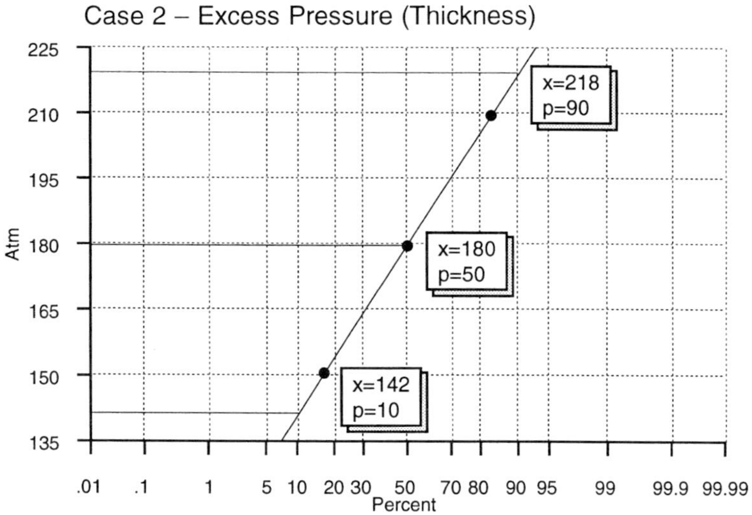


FIGURE 11.43 Cumulative probability excess pressure plots as each parameter is varied individually for the two cases of salt top at 300 and 500 m sub-mudline depth, respectively: (a) thickness; (b) lateral speed; (c) subsalt lithology.

and provides a measure of the spread in uncertainty of excess pressure around the 50% chance value caused by variations in each parameter. A high volatility ($v > 100\%$) implies a significant uncertainty, whereas a low volatility ($v \ll 100\%$) implies a relatively well-constrained value. For

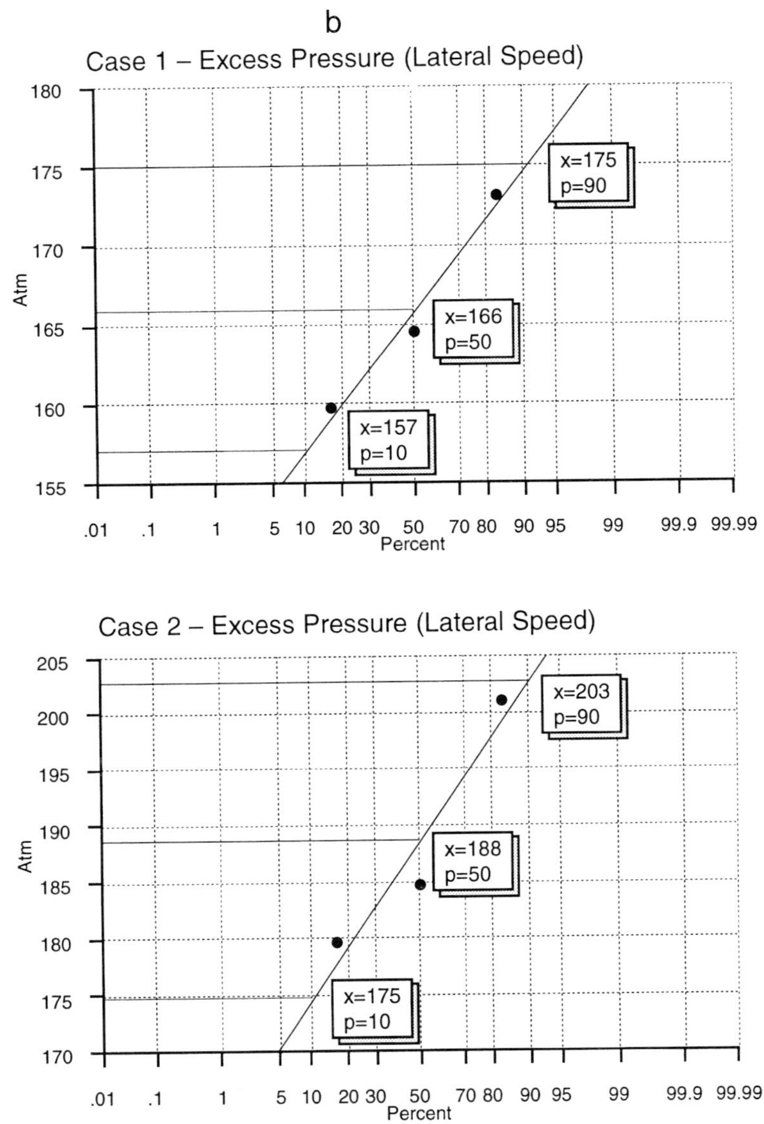


FIGURE 11.43 *(Continued)*

instance, note from Table 11.8 that in case 1 (salt top at 300 m today) the uncertainty in salt thickness causes about three times the uncertainty in volatility compared to either salt speed or subsalt lithology uncertainties. Note that in both cases 1 and 2, the volatilities are less than around 40%, and even as low as 7% (case 2, Lithology), so that there can be substantial

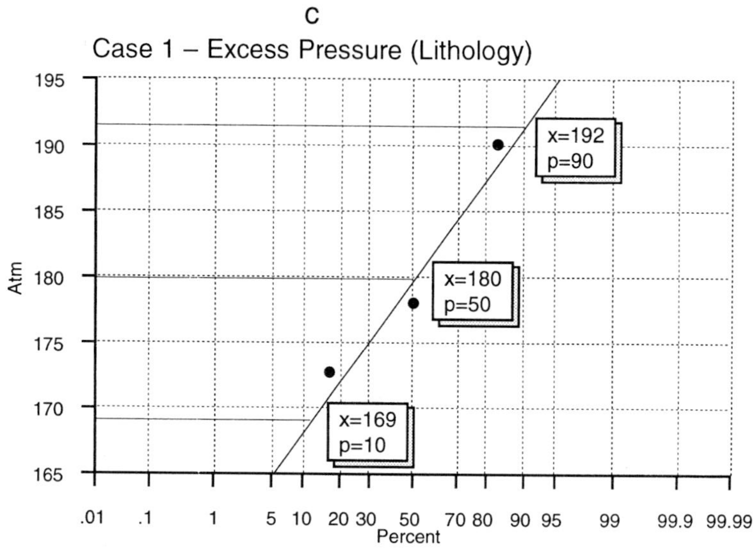


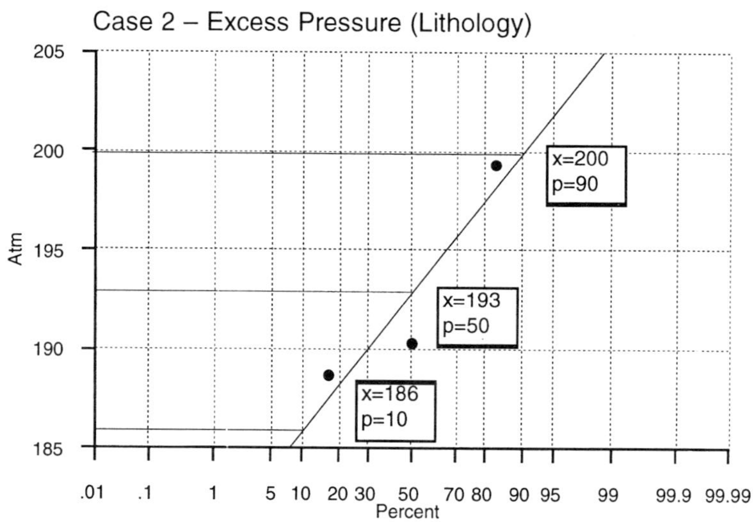


FIGURE 11.43 *(Continued)*

uncertainties in excess pressure at breakout. All of the parameters can be varied simultaneously to provide a cumulative probabilistic assessment for all uncertainties taken together. At each subsurface depth (and not just at salt breakout depth) one can then provide two different perspectives of excess pressure development as shown in Figs. 11.44a and b. A cumulative

TABLE 11.8 Subsalt Fluid Pressures and Volatilities

| Category | Value (atm) at $P$ (10) | Value (atm) at $P$ (90) | Value (atm) at $P$ (50) | Volatility (%) |
|---|---|---|---|---|
| *Case 1* | | | | |
| Lateral speed | 157 | 175 | 166 | 11 |
| Thickness | 134 | 187 | 160 | 33 |
| Lithology | 169 | 192 | 180 | 13 |
| *Case 2* | | | | |
| Lateral speed | 175 | 203 | 188 | 15 |
| Thickness | 142 | 218 | 180 | 42 |
| Lithology | 186 | 200 | 193 | 7 |

probability versus depth behavior, as shown in Fig. 11.44a for different isobaric excess pressure values, gives the cumulative probability of obtaining an excess pressure less than or equal to the posted values. Thus, at a 2000-m depth there is a 90% chance the excess pressure will be less than 100 atm and only about a 5% chance the excess pressure will be less than 50 atm; whereas at a 3500-m depth, there is only about a 5% chance the excess pressure will be less than 100 atm, and about a 95% chance of obtaining an excess pressure of less than 150 atm.

Equally, one can plot the converse information as shown in Fig. 11.44b where, at different depths, one plots the excess pressure for fixed iso-probability values. Thus, there is a 90% chance that at 4500 m the excess pressure will be less than 220 atm but only a 10% chance of less than 130 atm, as shown in Fig. 11.44b. Alternatively, a 130-atm isobar has less than a 10% chance of *not* being found at 4500 atm or greater, and a 90% chance

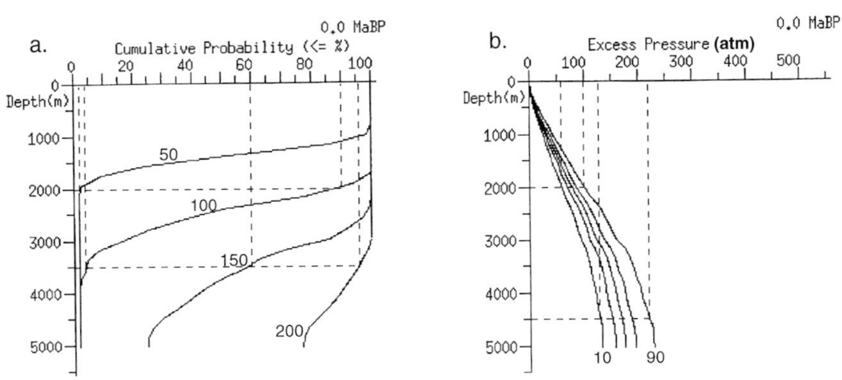


FIGURE 11.44 (a) Cumulative probability versus depth plot of isobaric excess pressure contours; (b) excess pressure versus depth plot of constant cumulative probability values.

of *not* being found shallower than 2000 m. Thus, both the probable depth range of achieving given excess pressure values and the probability of achieving a specified range of excess pressure at a given depth are determined.

### *b. Relative Importance*

Because each parameter of salt thickness, lateral speed, and subsalt lithology is varied independently, each causes a different variation of excess pressure with depth. Of concern is to determine which of the factors is providing the largest uncertainty in the excess pressure, so that one has an appreciation of which factor(s) to focus on first in attempts to narrow the range of uncertainty in excess pressure. To answer this question, proceed as follows. As we saw in Fig. 11.42, as each factor varies, it produces an uncertainty in the excess pressure, and for the cumulative probability the distribution of excess pressure values recorded is close to a log-normal distribution (represented by the straight lines on Fig. 11.42). For a precisely log-normal distribution, with $E_2$ the second moment of the excess pressure, $p$, distribution, and $E_1$ the first moment (mean value), the scale value

$$\mu^2 = \ln\{E_2(p)/E_1(p)^2\} \tag{11.7}$$

describes the spread in uncertainty around the mean $E_1(p)$, because the variance, $\sigma^2$, is given by

$$\sigma^2 = E_2 - E_1^2 \equiv E_1^2(\exp(\mu^2) - 1). \tag{11.8}$$

The variation of each individual parameter then provides a contribution $\mu_i$ to the total $\mu$ for excess pressure. The relative importance, RI(%), of each factor in contributing to the uncertainty in excess pressure is defined by

$$\mathrm{RI}_i\,[\%] = 100\,\mu_i/\sum_{j=1}^{N} \mu_j, \tag{11.9}$$

where $N$ is the number of parameters being varied (three in the present situation, for each of cases 1 and 2, respectively). Plotted on Fig. 11.45 are the RI values for the two cases where the salt top today is 300 and 500 m, respectively, below the sediment surface. In both cases note that it is the uncertainty in salt thickness that contributes between 58% and 65% of the uncertainty in excess fluid pressure directly beneath the salt with the lateral speed of salt and lithology uncertainty below salt contributing between 18% and 23%, and 23% and 11%, respectively.

Overall, then, if one wishes to narrow the range of uncertainty on excess fluid pressure at salt breakout, primary effort should be placed on attempting to cut the estimated uncertainty of ± (200–300) m on present-day salt thickness, with a secondary, but lesser, effort on salt speed delineation and lithology improvement. The dominance of the salt thickness uncer-

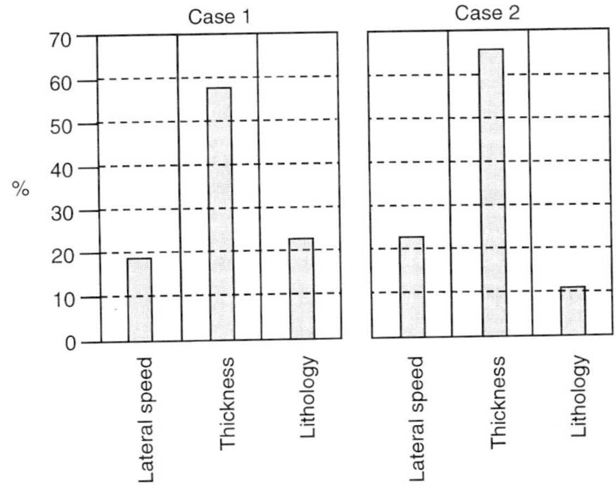


FIGURE 11.45 Relative importance (%) of the uncertainty in salt thickness, salt speed, and subsalt lithology in contribution to excess pressure uncertainty: (a) case 1, salt top at 300 m sub-mudline; (b) case 2, salt top at 500 m sub-mudline.

tainty is about a factor of 3 more important to resolve than the uncertainties in either salt speed or lithology.

## F. DISCUSSION AND CONCLUSIONS

The existence of laterally extensive, rapidly moving, thick salt sheets of low permeability in the Gulf of Mexico causes a change in the paths available for fluid flow (and so salt sheets provide improved hydrocarbon seals) and an excess fluid pressure to be expected at breakout into subsalt sediments, both of which are of considerable concern to drillers.

The problem addressed here is to estimate the excess pressure and the range of uncertainty likely to be expected, and to determine which factors are dominant in causing the uncertainty. To determine *a* value for excess pressure is easy; one merely runs some basin modeling code that permits fluid flow, with a set of parameters prechosen for salt insertion. The difficulty is that the particular value of excess pressure obtained may be sensitively dependent on the precise details of the model and on the precise values used for salt description. Other parameters may give very different excess pressure values. Because one often does not know with a high degree of precision from observations the parameters entering the calculations, it is important to vary the parameters (in this case salt thickness, salt speed, and lithology underlying salt) in ranges around likely mean values to obtain an appreciation of the uniqueness, resolution, sensitivity, and precision of calculated excess pressure values to such changes. In this way one obtains not only a mean excess pressure estimate but also determines which of the parameters is causing the greatest impact on uncertainty in the estimated excess pressure. It then becomes clear which factors to concentrate on in order to narrow their ranges of uncertainty to the point at which they have less dominance on the excess pressure uncertainty. Thus, effort is not expended on attempting to narrow the uncertainty of factors that have less important roles to play in contributing to the excess pressure. Equally, the ability to provide some measure of the stability of the estimated mean value of the excess pressure is needed, and is provided here by cumulative probability plots of expected present-day excess pressure with depth. Thus, one obtains precisely the factors needed.

In the present situation it is clear that the $\pm(200–300)$ m uncertainty on a salt thickness of around 2000 m is causing more than half of the uncertainty in excess pressure values, with salt speed and subsalt lithology uncertainties providing roughly equal contributions of about $25 \pm 10\%$ each to the total uncertainty in excess pressure. The variations in excess pressure estimates at breakout are about $\pm 30$ atm on a mean value of about 170 atm. Thus, if one can accept a high breakout value of 200 atm while drilling, then there is no need to attempt to improve the uncertainty on salt thickness prior to drilling, which is responsible for about 65% to 75% of the $\pm 30$ atm. On the other hand, if one cannot accept that level of uncertainty on excess pressure, then a concerted effort should be made first to improve salt thickness estimates rather than concentrating on salt speed or subsalt lithology uncertainties.

## APPENDIX A: PROBABILISTIC PROCEDURES

### A.I. PRODUCTS OF PARAMETERS

It is well known that multiple independent random variables $X$, $Y$, $Z$, ..., from *log-normal* distributions combine in generic product form $X^aY^bZ^c\ldots$ to give a distribution for the product that is also precisely log-normally distributed with mean value:

$$E_1(X^aY^bZ^c\ldots) = E_1(X^a)E_1(Y^b)E_1(Z^c)\ldots, \tag{A.1a}$$

with second moment

$$E_2(X^aY^bZ^c\ldots) = E_2(X^a)E_2(Y^b)E_2(Z^c)\ldots \tag{A.1b}$$

and scale factor $m$ given through

$$\mu^2 = \ln[E_2(X^aY^bZ^c\ldots)/E_1(X^aY^bZ^c\ldots)^2]. \tag{A.1c}$$

Empirically it appears that $N$ independent random variables from any frequency distributions (not necessarily log-normally distributed) tend to combine to produce a product $P_N$ ($\equiv X_1^a X_2^b X_3^c \ldots X_N^d$) that is approximately log-normally distributed with

$$E_1(P_N) = E_1(X_1^a)E_1(X_2^b)\ldots, \tag{A.2a}$$

and scale parameter

$$\mu^2 \cong \ln[E_2(P_N)/E_1(P_N)^2]. \tag{A.2b}$$

## A.II. NUMERICAL PROCEDURES

Empirically, the determination of $\mu^2$, which controls the slope of the cumulative log probability curve at each coordinate, proceeds as follows.

Consider, first, that a vector set of parameters, $\mathbf{p}$, is to be used in a model. The output of a specific positive quantity, $R$, from the model is then available at spatial coordinates $\mathbf{x}$, at time $t$, and is dependent on the specific values, $\mathbf{p}$, used for the parameters, that is, $R = R(\mathbf{p}; \mathbf{x}, t)$. For brevity, throughout this development, the dependence of $R$ on $\mathbf{x}$ and $t$ is not written explicitly, although one must keep in mind that an output quantity is being calculated at each $\mathbf{x}$ and $t$ of the model computation.

Then we write

$$\mathbf{p} = E_1(\mathbf{p}) + \delta\mathbf{p} \tag{A.3}$$

with $E_1(\delta\mathbf{p}) \equiv 0$, where $E_1(\mathbf{p})$ is the average value of the vector $\mathbf{p}$. With given minimum $\mathbf{p}_{\min}$, most likely $\mathbf{p}_{\text{likely}}$, and maximum $p_{\max}$, values for $\mathbf{p}$, the Simpson triangular rule yields

$$E_1(\mathbf{p}) \simeq \tfrac{1}{3}(\mathbf{p}_{\min} + \mathbf{p}_{\text{likely}} + \mathbf{p}_{\max}). \tag{A.4}$$

Taylor series expansion of $R$ yields

$$\begin{aligned} R(\mathbf{p}) = R(E_1(\mathbf{p}) + \delta\mathbf{p}) \simeq R(E_1(\mathbf{p})) &+ \sum_i \delta p_i \left(\frac{\partial R(\mathbf{p})}{\partial p_i}\right)_* \\ &+ \frac{1}{2}\sum_i \delta p_i\, \delta p_j (\partial^2 R(\mathbf{p})/\partial p_i \partial p_j)_* + \ldots, \end{aligned} \tag{A.5}$$

where $(\,)_*$ means evaluate the bracketed quantity on $\mathbf{p} = E_1(\mathbf{p})$. It follows that

$$E_1(R(\mathbf{p})) = R(E_1(\mathbf{p})) + \frac{1}{2}\sum_i \left(\frac{\partial^2 R}{\partial p_i^2}\right)_* E_1(\delta p_i^2) + \ldots. \tag{A.6}$$

It also follows that

$$\begin{aligned} E_2(R(\mathbf{p})) &\equiv E_1(R(\mathbf{p})^2) \\ &\simeq E_1(R(E_1(\mathbf{p})))^2 + \sum_i E_1(\delta p_i^2)(\partial R/\partial p_i)_*^2 + \ldots. \end{aligned} \tag{A.7}$$

Then

$$\mu^2 \equiv \ln(E_2(R(\mathbf{p}))/E_1(R(\mathbf{p}))^2) \simeq \ln\left\{1 + \sum_i E_1(\delta p_i^2)(\partial \ln R(\mathbf{p})/\partial p_i)_*^2 + \ldots\right\}. \tag{A.8}$$

Now

$$\left(\frac{\partial R}{\partial p_i}\right)_* \simeq [R(\mathbf{p}_{\max}) - R(\mathbf{p}_{\min})]/(p_{\max,i} - p_{\min,i}), \tag{A.9}$$

while

$$E_1(\delta p_i^2) = E_2(p_i) - E_1(p_i)^2. \tag{A.10}$$

With

$$\sin^2\theta_i = [p_{\text{likely},i} - p_{\min,i}]/(p_{\max,i} - p_{\min,i}), \tag{A.11}$$

it follows from the Simpson triangular rule applied to $\delta\mathbf{p}$ that

$$E_1(\delta p_i^2) = \tfrac{1}{18}(p_{\max,i} - p_{\min,i})^2(\sin^4\theta_i + \cos^2\theta_i). \tag{A.12}$$

Hence

$$\mu^2 \simeq \ln\left\{1 + \frac{1}{2}\sum_i (\sin^4\theta_i + \cos^2\theta_i) \times [R(\mathbf{p}_{\max}) - R(\mathbf{p}_{\min})]^2 \times [R(E_1(\mathbf{p})) + R(\mathbf{p}_{\max}) + R(\mathbf{p}_{\min})]^{-2}\right\}, \tag{A.13}$$

where

$$E_1(p) \simeq \tfrac{1}{3}(\mathbf{p}_{\max} + \mathbf{p}_{\text{likely}} + \mathbf{p}_{\min}). \tag{A.14}$$

Thus the two quantities, $E_1(R)$ and $\mu^2$, which control the cumulative log probability distribution of an output quantity, $R$, are evaluated at each vector spatial coordinate, $\mathbf{x}$ and for each time, $t$, in terms of computer runs with the parameter vector $\mathbf{p}$ set to its maximum, minimum, and average values, respectively.

Accordingly, several strategies are clear for determining parameters that sensitively control an output. For $N$ parameters one can (1) run all at once; (2) run each independently; or (3) run groups of parameters. Cumulative probability outputs can then be evaluated for each case, so that sensitive dependencies on a particular parameter and its range of uncertainty, or to a group of parameters, can be evaluated quickly.

In terms of minimizing the computer run time for basin analysis models, the mathematical procedure developed earlier has the following pragmatic operational format:

1. For the impact of each of $N$ input parameters on the behavior of output results, the basin analysis model is run $N$ times for the $N$ uncertainty ranges of the $N$ parameters. This procedure produces a set of output values of interest.
2. The cumulative probability for each output value is then determined from the $N$ runs of the model.
3. The probability uncertainty range of particular output behaviors can then be assessed, and a risk factor assigned to the likelihood of, say, porosity being in the reservoir window, or of excess pressure being above a given value, etc.

## APPENDIX B: APPROXIMATE UNCERTAINTY CONTRIBUTIONS

Dealing in large parts with measurements of physical properties, interpreted quantities, and equations simulating geological processes of interest, we are strongly limited by the accuracy with which any of these quantities can possibly be obtained. As a consequence, we have to accept ranges of uncertainty on all parameters entering the equations connecting measurements with the physical quantities being sought. Thus the accuracy with which one can determine physical quantities, such as volume of hydrocarbons generated over a given period of time, depends on the accuracy of the parameters entering the equations. The goal of this appendix is to provide a procedure for assessing the likely range of uncertainty on a physical quantity of interest, determined from available data and observations through process simulation, so that the appropriateness of using the quantity is sharpened. Also, if the modeled quantity is used in an economic evaluation, assigning risk to the quantity then becomes a more straightforward task.

One extremely valuable procedure used to assess the probability of a particular result being correct, based on ranges of uncertainty of input parameters, is to define ranges of possible values for each parameter and to then perform a series of random Monte Carlo computer runs. Although the Monte Carlo method is an effective procedure for assessing errors on simple systems, the computer time needed for assessing errors on geological modeling results prohibits its routine application, particularly if large numbers of basin analysis calculations are done, each of which can take several computer hours.

Fortunately, other means are also available for assessing uncertainty and risk that are much less demanding on computer time. The method to be used here has been described in detail elsewhere (Lerche, 1992a), but a brief review seems appropriate. First, we need to be concerned with the cumulative frequency distribution $P(y > x)$ and therefore with the probability that a modeled quantity $x$ is either greater than or less than a given value $y$, where $p(x)$ is the differential probability of obtaining $x$ in the range $x$ to $x + dx$. Then

$$P(y > x) = \int_y^\infty p(x)\,dx, \tag{B.1}$$

$$P(y < x) = \int_0^y p(x)\,dx \equiv \int_0^\infty p(x)\,dx - \int_y^\infty p(x)\,dx \equiv 1 - P(y > x). \tag{B.2}$$

Second, we have to make an assumption of how the differential probability is distributed. In general, three types of frequency distributions are appropriate for the types of problems associated with exploration assessments: binomial, normal, and log-normal. The binomial distribution is useful when we suspect the probability of an outcome is either one of two possibilities. The normal distribution is particularly useful when concerned with small errors, and the log-normal distribution occurs physically in many situations in nature and is appropriate when there is a constraint on a variable (e.g., the volume generated must be positive; porosity cannot be negative; drainage area cannot be negative). In the case of basin modeling results, a log-normally distributed frequency therefore seems appropriate, in which case only a few numbers are needed in order to be able to apply a cumulative probability method: the mean value of $x$, the mode value, and the variance. The mean value of $x$, $E_1(x)$, is given by

$$E_1(x) = x_{1/2}\exp(\mu^2/2), \tag{B.3}$$

the mode value by

$$x_m = x_{1/2}\exp(-\mu), \tag{B.4}$$

and the variance in $x$ by

$$\sigma^2 = E_1(x)^2[\exp(\mu^2) - 1], \tag{B.5}$$

where $x_{1/2}$ is the median value (Fig. B1).

It is often quite difficult to obtain enough information to determine the precise shape of the frequency distribution of a particular parameter or variable. Often it is considered an achievement to make an estimate of a likely minimum, $x_{min}$, a likely maximum, $x_{max}$, and a likely most probable value, $x_p$, for a parameter. Using Simpson's triangular rule (Fig. B2) estimates of relevant mean and variance are then

$$E_1(x) \cong \tfrac{1}{3}(x_{min} + x_p + x_{max}) \tag{B.6}$$

and

$$\sigma^2 \cong \tfrac{1}{2}E_1(x)^2 - \tfrac{1}{6}[x_{min}x_{max} + x_p(x_{min} + x_{max})] > 0 \tag{B.7}$$

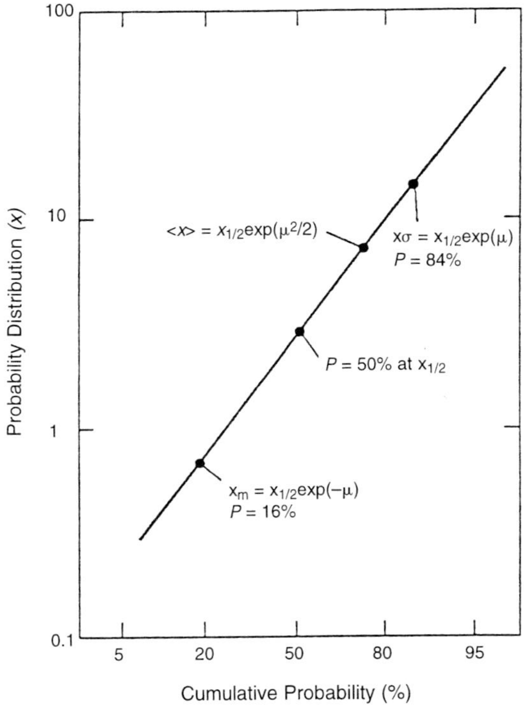


FIGURE B1 Cumulative log-normal probability distribution for an arbitrary variable showing the relative positions of the mean, median (50% probability point), mode, and the 84% probability point.

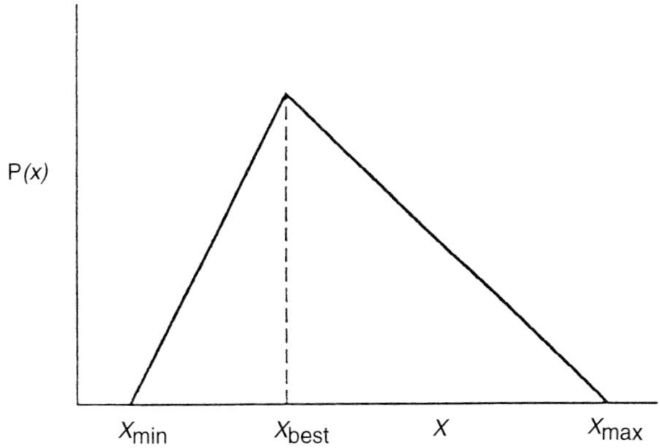


FIGURE B2 Sketch of triangular probability distribution used with Simpson's rule for estimating mean and variance from three values.

so that

$$E_2(x) = E_1(x)^2 + \sigma^2. \tag{B.8}$$

Estimates of $\mu$, $x_{1/2}$, and $x_m$ are then given by

$$\mu = [\ln(E_2(x)/E_1(x)^2)]^{1/2}, \tag{B.9}$$

$$x_{1/2} = E_1(x)\exp(-\mu^2/2) \equiv E_1(x)/(E_2/E_1^2)^{1/2}, \tag{B.10}$$

$$x_m = E_1(x)\exp(-\mu^2/2 - \mu). \tag{B.11}$$

Thus, the cumulative probability, $P$, takes on the values:

$$P = 0.5 \text{ on } E_1(x)\exp(-\mu^2/2), \tag{B.12}$$

$$P \cong 0.68 \text{ on } E_1(x), \tag{B.13}$$

$$P = 0.84 \text{ on } E_1(x)\exp(\mu - \mu^2/2), \tag{B.14}$$

$$P = 0.16 \text{ on } E_1(x)\exp(-\mu^2/2 - \mu). \tag{B.15}$$

Because parameters describing volume generated, expelled, and migrated enter formulas (11.1) through (11.5) as products, the relative contributions to reserve uncertainty can be written in terms of $\mu^2$, with

$$\mu^2 = \sum_{i=1}^{N} \mu_i^2, \tag{B.16}$$

where $N$ is the number of independent variables in Eqs. (11.1) through (11.5), and $\mu_i^2$ is the value of $\ln\{1 + \sigma_i^2/E_i(x)^2\}$ for each independent variable. Here $\mu^2$ measures the variation of the output probability distribution around a centered value, and so is a measure of the uncertainty of the centered value, that is, of the "variance" in the output.

The relative contribution (RC) (%) made by the $i$th variable to reserve uncertainty is then measured by

$$\mathrm{RC}_i = 100\,\mu_i^2/\mu^2,$$

which is the quantitative parameter used in text to provide the relative contributions shown in the figures.

# 12

# STRATEGIES FOR MODEL USE

## I. BASIN ANALYSIS CODES AND DATA AVAILABILITY

### A. INTRODUCTION

From an oil industry perspective the subject of quantitative basin analysis has three major components: (1) determination of the spatial and temporal dynamic evolution of the sediments in relation to evolving basinal structure, and with respect to the spatial variations of lithologic and diagenetic fabrics in connection with ongoing stratigraphic and structural evolution; (2) determination of the thermal history of sedimentary units in relation to their burial histories; and (3) determination of the hydrocarbon generation, migration, and accumulation histories of all hydrocarbons generated in the basin, so that the relative timing of hydrocarbon evolution can be assessed in relation to the development of traps and their seals, and in relation to the continued integrity of sealed traps after hydrocarbon emplacement. In this way a better appreciation can be obtained of potential hydrocarbon sites in a basin.

On the path to achieving such an evaluation of hydrocarbon site determination, four major concerns have to be considered in relation to the output from any basin analysis code. The concerns are (1) resolution, (2) sensitivity, (3) uniqueness, and (4) precision. These four factors are controlled to a greater or lesser extent by three overriding inputs: (1) intrinsic assumptions; (2) parameter values; and (3) quality, quantity, and sample spacing of data.

A detailed analysis of the evaluation of basins taking into account the three input factors and the four output concerns has been given elsewhere (Lerche, 1990a,b) for dynamic, thermal, and hydrocarbon assessments. We cannot state too strongly that *every* case history investigated should be evaluated for its rugged stability (or lack thereof) in relation to the points just made.

However, the appropriateness of any quantitative basin analysis result in relation to the evolution of a basin can be no better than the quality and quantity of the data that serve to provide control criteria on the modeled behaviors. At any given stage of knowledge of a sedimentary basin, there are always the questions of determining what can be garnered without equivocation, what can be achieved with a degree of equivocation that is acceptable, and what cannot be evaluated without further data. This section provides strategies for quantitaive basin analysis studies when minimal, moderate, and maximal data are available. However, a caveat is in order. Although we have found the general strategies outlined here to be of use as overall patterns in analyzing basins, there are enough exceptions for any individual basin that care should be taken when applying the strategy templates as though they were infallibly correct in every case. With this caveat, in mind we consider each case in turn.

## B. MINIMAL DATA AVAILABLE

We consider "minimal" data to be the availability of seismic sections, perhaps some gravity or aeromagnetic studies and/or outcrop information, but certainly no well information. The first study usually done is to set up the background geology of the basin in relation to paleogeography and continental plate motion so that a rough idea of basin setting, paleotemperature (warm/cold), likely tectonic events influencing the basinal development, and likely lithologies are at hand. With this background information, it is then possible to identify from the seismic, paragenetic sequences in the basinal setting (compression, extension, rotated blocks, etc.) and from sequence stratigraphy the possible chronostratigraphy of units in relation to some estimate of sea level variations. From the structural and stratigraphic identifications, and from estimates of sedimentary load at each location, some inference can often be made of basement motion (isostatic, flexural, viscoelastic) and of fault development with time. Seismic velocities, tied to a Wyllie-type model relating velocity and porosity, can often be used to assess a rough porosity with depth behavior, as can gravity (i.e., density) modeling across the basin.

At this minimal data stage, the determination of paleoheat flux, or even present-day heat flux, is a problem because there are only a few indications to guide and control. Some form of model is required. At least four possibilities are available: (1) Arbitrarily assign a basement heat flux with time according to idiosyncratic preference. This procedure is recommended only as a last resort. (2) Tie the basement motion at each location to a paleoheat flux model (McKenzie-type rift; Royden intrusive model, etc.). (3) Tie the motion of the section *as a whole,* with sediment loading, to the depth to detachment motion under a model of regional load compensation. Assume this detachment is always at a fixed brittle/ductile temperature (450° to 550°C is conventional). (4) Use gravity/aeromagnetics and/or seismic data to estimate the current depth to the brittle/ductile transition. Estimate the temperature at the transition from analogy. Construct a present-day temperature gradient estimate and then allow for an uncertainty range (e.g., 2.5 ± 0.5°C/100 m); assume this range of gradients covers all possibilities. It is appropriate to try to do the last three of the preceding four possibilities, thus running several cases. In this way some idea of the range of output results can be achieved even with minimal data.

Although the total organic carbon (TOC) content of each formation is an unknown, some rough idea of expected rich and lean behaviors can be achieved from depositional environment estimates. One of the remaining uncertainties at this stage is the hydrocarbon kinetics. Some form of kinetic behavior must be prescribed to assess hydrocarbon maturity. Three options are commonly available under these minimal data conditions: (1) Use the expected depths to the 100° and 150°C isotherms (as approximated from a heat flux model) as rough indications of the onset of oil and gas, respectively. (2) Use estimates of TTI = 15 and TTI = 75 to mark oil and gas onsets, respectively. (3) Put in one or preferably more hydrocarbon kinetic models. Determine ranges of depths and times for hydrocarbon generation from these three different procedures and estimate range of uncertainty.

As this stage of data knowledge it is rarely worthwhile to do a 2-D analysis (including fluid flow) because the data, by and large, are not available to control the outputs. It is preferable, and more reliable, to do a set of 1-D calculations at pseudo-well sites across the basin and iso-contour appropriate outputs. Usually if a 2-D run is done after this procedure, it is most often the case (but not always) that the 2-D run confirms the iso-contouring within the resolution of the assumptions, parameters, and data used. One should run at least one 2-D section just to be sure that the 1-D iso-contouring results are appropriate because, once in a while, a major difference will result, well outside the realm of uncertainties, indicating that a more detailed 2-D run is warranted. This difference is a rare event; it usually occurs when a major sheet sand or continuous flow path is held open to the surface for a long period of time.

A summary of the basic strategy when minimal data are available is given in Table 12.1.

## C. MODERATE DATA AVAILABLE

We consider "moderate" data to be the minimal data statements of Section I.B plus a few scattered wells in the basin. For exploration wells, the information available is often considerably less than for, say, a scientific

TABLE 12.1 Minimal Data Available

| | |
|---|---|
| Dynamic aspects | Evaluate basinal setting and paleogeography.<br>Estimate basement motion.<br>Determine fault timing and evolution.<br>Estimate lithology.<br>Use 1-D dynamics and iso-contours.<br>Check if 2-D improves matters. |
| Thermal aspects | Estimate paleoheat flux from several model connections to basement motion.<br>Use 1-D dynamics to get paleotemperature for each model.<br>Check to see if answers are affected to within resolution by changes in faults open/shut, estimated lithologies varied, fluid flow fracturing.<br>Use a 2-D model to see if heat flux is focused by structural, stratigraphic variations, fluid flow direction. |
| Hydrocarbon aspects | Estimate likely source rocks and TOC from (1) sequence stratigraphy, (2) paleogeography, and (3) depositional environment.<br>Vary hydrocarbon generation models to estimate uncertainty.<br>Use 1-D code and iso-contour generation, migration, and accumulation.<br>Use 2-D code to assess lateral migration and accumulation differences from 1-D code.<br>Iso-contour results. |
| *Warning:* | This minimal data picture usually provides only a regional picture on a broad lateral scale. Occasionally it is not too bad on a fairway lateral scale, but is usually fairly poor on a prospect-size lateral scale. |

or COST well. In addition, most wells are drilled on structural highs so their representation as indicators of general conditions in the basin is often less than optimal. Nevertheless, well information is of great benefit for a variety of reasons. First, formation ages, thicknesses, lithologies, and porosities are routinely available, thereby providing control of the more crudely estimated values from seismic responses. Seismic ties to wells can then be used to better constrain the system. Second, most often downhole temperature and thermal indicator data (vitrinite reflectance, apatite fission tracks, etc.) are available. These data provide a means to better assess present-day heat flux and paleoheat flux, at least at the well sites, thus providing some measure of improvement over background modeling. Two options for the remainder of the basin are available: (1) If enough wells are present, contour present-day and paleoheat flux to provide regional statements. (2) If not enough wells are present to make the contouring a worthwhile idea, then use the well information on heat flux to modify parameters of the background heat flux model to be in conformity with the scattered observations.

Often some data on TOC and S2 pyrolysis are available together with GCMS data. In addition, excess fluid pressure data are often available with depth. The organic data can be used to assess residual kerogen availability and to evaluate the likely fraction of kerogen that already produced hydrocarbons. The pressure data provide an indication of the sealing nature of the sediments and, if enough pressure data are available, can be contoured to suggest regional flow directions both vertically and horizontally.

The combined use of scattered well data together with the minimal data of Section I.B usually allow a systematic improvement to be achieved in understanding of the basin. One of the strategies that stands a fair chance of success is as follows.

First, run each well with a 1-D fluid flow/compaction code, which also incorporates thermal indicator procedures for assessing paleoheat flux. When predicted present-day thicknesses, porosities, fluid pressure, downhole temperature, and thermal indicator values are in optimal accord (minimum degree of mismatch) with observed values then some consistency has been achieved (Mudford, 1996). Then assess whether the hydrocarbon kinetic predictions are in accord with the TOC, Rock-Eval pyrolysis, and GCMS data for each well. Contour the behaviors across the basin based on 1-D wells, in-filling where necessary with pseudo-wells determined from seismic data tied to well data.

Second, run 2-D sections using well data and seismic pseudo-wells to determine if, within resolution, any significant changes are needed in dynamic, thermal, or hydrocarbon sections of the strategy. Unless major faults are crossed in 2-D, or massively continuous sands exist, the 1-D contouring often tends to be fairly close to the 2-D results because isolated "pods" of sand tend not to be continuous on a lateral scale larger than around 10 to 20 miles. Thus, as long as the 2-D section is quite long compared to the sand continuity scale, stringing together contours based on 1-D wells tends to be a fairly accurate approximation. This procedure will certainly fail if the lateral scale size being investigated is comparable to sand bed scales, horst scales, salt diapir scales, or if it crosses major, open faults. In those cases it is relatively clear that a 2-D effort is normally necessary, that is, the medium data case generally works best on a play or fairway scale.

The basement motion is, in general, a 2-D problem depending on flexure, rifting, etc. It is often the case that a local isostatic balance under the applied distributed sedimentary load is a relatively good approximation, but nonisostatic basement motion should certainly be used on a lateral scale greater than prospect size, and should be tested on shorter scales for departure of basement motion from the isostatic assumption. In a rough sense the medium data case allows information to be gathered on the tens of miles scale laterally but often is not accurate enough to be a very good approximation on the local scale of a prospect. An outline of the general strategy for the medium data case, remembering that copious exceptions exist, is given in Table 12.2.

TABLE 12.2 Moderate Data Available

| | |
|---|---|
| Dynamic aspects | Use minimal data strategy *but* control layer age, thicknesses, porosity, pressure, lithology by tying seismic to well data.<br>Use 1-D code at real wells and pseudo-wells.<br>Iso-contour results.<br>Check if 2-D improves matters. |
| Thermal aspects | Use downhole temperatures to get present-day heat flux at well sites. Contour in 2-D.<br>Use thermal indicators to get paleoheat flux at each site.<br>Contour across region of interest. Check to see if this procedure improves matters from the minimal data strategy of Table 12.1.<br>Use 1-D code at real and pseudo-wells.<br>Iso-contour results.<br>Check if 2-D improves thermal history. |
| Hydrocarbon aspects | Use TOC from wells then contour to get regional idea of source rock variation.<br>Use Rock-Eval, S2 peak, GCMS to get an idea of the amount of kerogen burnt. Contour in region of interest.<br>Vary hydrocarbon kinetics to be consistent with pyrolysis and residual kerogen analyses. Use uncertainties in parameters of kinetics to bracket range.<br>Use 1-D code and iso-contour results.<br>Use 2-D code to assess generation, migration, and accumulation sites. |
| *Warning:* | This moderate data picture usually provides a fairly good regional picture and a fair-to-middling fairway picture. It usually is poor on a prospect-size scale, but can be relatively good once in a while depending on complexity of basin. |

## D. MAXIMAL DATA AVAILABLE

We consider "maximal" data to be the moderate data statements of Section I.C plus many wells, fairly densely spaced (less than about 10 miles spacing between the majority of wells), in which information on hydrocarbon reservoirs is probably available (e.g., height of oil/gas column; sands in which hydrocarbons exist). In this case a variety of strategies are available depending on motivation and philosophy. Three end-member cases are indicative of strategies to employ:

1. Analysis required on a prospect scale
2. Analysis required on a fairway scale
3. Analysis required on a regional basin scale

To illustrate strategies consider each case in turn.

### 1. Analysis Required on a Prospect Scale

In attempting to delineate the size of a hydrocarbon accumulation using quantitative basin analysis methods, with copious wells across the field and good "infill" from reservoir control of wells and seismic information, a relatively successful strategy is as follows.

First, determine the extent to which 1-D simulations can be successfully contoured to permit prediction in both interpolation and extrapolation senses at well sites not used in the contouring. This procedure provides some idea of the extent of trustworthiness of 1-D simulations. If the contouring provides an accurate assessment of well conditions, to within resolution, then there is no need to use more sophisticated modeling. However, while such contouring procedures are often acceptable for interpolating to "infill" well locations, they tend to become poorer the further one extrapolates outside the control database. In general, it would seem that 2-D fluid flow/compaction modeling provides a better representation of the data field and more accurate contour diagrams. Here the idea is to use 2-D models in several lateral directions and "fence-diagram" those results with minimal discord one to another. The resolution of such procedures in relation to intrinsic assumptions, parameter input values, and data quality and quantity can be tested in several ways:

1. Use half the wells (chosen at random) and determine the extent to which the remainder can be predicted correctly.
2. Use the wells in one-half of the region and determine the extent to which the other half of the region is predicted correctly.
3. Use all the wells, but miss out various components of the data available (e.g., ignore porosity in half the wells, ignore thermal indicators in half the wells, etc.) and determine to what extent the "missing" data are predicted correctly.

These strategies normally provide a very good indication of (1) systematic biases in code predictions, (2) the lateral and vertical spatial scales over which code predictions are trustworthy in relation to observed data, and (3) the extent to which extrapolations can be made with known degrees of equivocation.

**2. Analysis Required on a Fairway Scale**

On the larger lateral scale (several tens of miles) the appropriate strategy depends on the degree of spatial resolution required. If only a coarse picture is required, then it is appropriate to default to the moderate data case where only a few wells are used and to determine the extent to which predictions match the observations of the many "unused" wells in the area under investigation. Inconsistency of predictions with observations in many of the "unused" wells can often be taken to imply that either (1) model parameters need to be improved or (2) the wells used are atypical of the majority (e.g., on top of salt diapirs). In either event a more thorough investigation is then clearly to be encouraged. If a fine-scale picture is needed, then all of the real well information, together with pseudo-wells based on the seismic data tied to real well data, should in general be used, with due regard taken of anomalously sited wells and/or anomalous data in wells.

### 3. Analysis Required on a Regional Basin Scale

On the large scale of a basin, the local, high-density information is usually not as useful as one might initially suppose, largely because fluid flow of either connate waters or hydrocarbons is rarely across a major fraction of a basin, being more restricted to the several tens of mile scale. A strategy here is to fall back to the moderate data case, run 1-D wells, contour the results, and attempt to sort out regional contour patterns from local "bulls-eye" types of features, which usually represent prospect level variations and have less to do with basinal behavior as a whole, that is, compartmentalization tends to be a more prevalent response.

Some factors are often basin wide, such as the basement evolution, basinal fill, and paleoheat flux. But once these factors are identified, the remainder tend to be scaled down in size relative to the basinal character. A good strategy is to try 1-D well characterization first and then see if 2-D computations make that much difference. Table 12.3 provides an outline of such strategies.

## E. INTEGRATED BASIN MODELING AND SEISMIC DATA

The outcome of any basin analysis model, whether on a regional, fairway, or prospect level, is the prediction of likely locations and expected amounts of hydrocarbons accumulated in both gas and oil forms. But the predictions made can sometimes be checked out for appropriateness, prior to drilling, as follows:

1. If, on seismic, a diagenetic flatspot crossing stratigraphic boundaries is found at the predicted hydrocarbon location, the flatspot could be an indicator of gas/oil, gas/water, or oil/water contacts.
2. If, on seismic, a "leakage" chimney of degraded seismic data is seen over a structural or stratigraphic trap predicted to be a hydrocarbon-filled reservoir, then the chimney could be an indicator of gas leakage from the reservoir.
3. If, on seismic, a "bright spot" reflection is seen, then a surmise that gas has produced a strong impedance mismatch might be appropriate.
4. If, on seismic, a low-velocity zone is found at the predicted hydrocarbon location, such a zone could mark the replacement of pore water by gas, again acting as a hydrocarbon indicator. Other seismic attributes can also act as hydrocarbon indicators (Lerche, 1987).

TABLE 12.3 Maximal Data Available

| | |
|---|---|
| Dynamic aspects | Use moderate data strategy of Table 12.2 *but* control the local (prospect-size) behavior tightly with multiple-well information. Tie observed grain size distribution, diagenesis, faults and fractures more closely to downhole information from well logs, excess pressure variations.<br>Determine if prospect, fairway, or regional evolution picture is being addressed. If fairway or regional, identify well behavior typifying prospect (anomalous wells, e.g., on salt diapirs, should be avoided), and use moderate data strategy.<br>If prospect size and evolution are needed, use predictive interpolation and extrapolation in prospect with majority of data test accuracy by halving data field.<br>Use 1-D in regional or fairway picture and check if 2-D improves picture.<br>Use 2-D in prospect-size picture, and then see if 1-D would have been equally acceptable. |
| Thermal aspects | Use multiple thermal indicators, fluid inclusions, paragenetic diagenesis sequences, geochemical information, etc., to determine paleoheat flux. Contour across prospect and/or basin.<br>Use multiple downhole temperatures to get present-day heat flux; contour across prospect and/or basin. Check if there is significant improvement on the heat flux modeled from basement motion, thereby improving tectonic motion—tie basement motion better to isostacy, flexure, etc.<br>Use 2-D or 1-D model as appropriate. Check resolution and effects of heat focusing by fluid flow, structural, stratigraphic, and fluid composition with time.<br>Iso-contour and check for consistency with "unused" wells or "unused" data in wells. |
| Hydrocarbon aspects | Use geochemical data, oil shows, hydrocarbon column thicknesses, GCMS, regional maps of hydrocarbons in prospect. Tie hydrocarbon kinetics to hydrocarbon production and residual kerogen to "new" data from all wells and to data from moderate data strategy. Determine uncertainty on kinetic parameters for each source rock.<br>Use reservoir information to evaluate resolution of kinetics with respect of generation, migration directions, and known accumulations.<br>Contour results with respect to prospect.<br>Check using predictions at "infill" wells and with "unused" well data.<br>Use 2-D code for migration, structure, and stratigraphy development. |
| *Warning:* | This maximal data picture is usually fairly good within a single prospect both in terms of interpolation and extrapolation. It usually improves the fairway picture by a significant degree, but usually does not provide too much improvement to the regional picture—although copious exceptions occur to all warnings for Tables 12.1, 12.2, and 12.3 depending on the complexity of basinal evolution. |

TABLE 12.4 General Strategy for Basin Analysis Code Usage

1. Determine basement motion (isostatic/flexural/etc.) for basin: (a) tectonic; (b) applied sediment load effect with time.
2. Determine sediment evolution: (a) 1-D; (b) 2-D.
3. Determine paleotemperature evolution from models or thermal indicators. Apply to thermal maturity of sediments and fluid flow transport.
4. Determine hydrocarbon kinetics from laboratory experiments and/or from downhole data. Apply to source rocks and evaluate generation in relation to structural and stratigraphic development. Do same for migration and accumulation.
5. Check variations caused by (a) input assumptions; (b) input parameters; (c) data quality, quantity, and sampling frequency (both vertically and laterally). Make assessment of resolution, precision, uniqueness, and sensitivity of outputs.
6. Check predicted behaviors with seismic attributes to evaluate whether common or contradictory stories are being presented.
7. Improve analysis with "postmortem" downhole data or newly acquired seismic data.

But the salient point from each and every such indicator on seismic is that the *common* identification of potential hydrocarbon accumulation sites from basin analysis models and from seismic data are doubly reinforcing factors, lowering the risk that either seismic or basin modeling on their own could be providing a false signal of hydrocarbon occurrence.

## F. SUMMARY AND CONCLUSIONS

### 1. Summary

Three factors control the output from any given basin analysis model: input assumptions; input parameters; and quality, quantity, and sample spacing of control data. Four factors have to be addressed in output results: precision, resolution, uniqueness, and sensitivity. Within that paramount framework, patterns of strategy have been addressed for performing quantitative basin analysis calculations with respect to dynamic, thermal, and hydrocarbon behaviors when minimal, moderate, and maximal data are available. These general strategies are summarized in Tables 12.1, 12.2, and 12.3. A rough idea of the sequential order in which basin analysis codes, addressing different factors of basin evolution, should be used is given in Table 12.4—although many exceptions exist to both the general strategies of Tables 12.1, 12.2, and 12.3 and to the usual sequential order of analysis of Table 12.4.

Perhaps the main point to be made is that serious attention must be paid to the uncertainties attendant on any model. Without due regard being paid to such uncertainties, it is all too easy to provide predictive behaviors that are precise, because they arise from a computer result, but which are inaccurate because of limitations on the model, the data, or the interpolations/extrapolations.

## 2. Conclusions

Apart from the general strategies outlined, two sorts of basic types of conclusions are possible depending on corporate philosophy and mandates. In a broad sense these recommendations can be classified as applications or research.

### *a. Applications*

From a corporate perspective of enhancing capital value and shareholders' dividends the sine qua non of any scientific device is whether it helps to find hydrocarbon accumulations. From this perspective, an appropriate strategy for an oil company is to obtain relevant basin analysis codes from research institutes and commercial vendors. Any such codes should first be evaluated prior to their application to real situations, particularly in frontier basin senses where the results are not known. One way to perform such an evaluation is to test the codes in areas where the answers *are* known to see if the predictions match the observations: a "blind test" mode of operation. Once this phase of development is accomplished, application to real situations can proceed with knowledge of the degree of systematic and statistical uncertainties that will be predicted, based on the blind test applications done under the minimal, moderate, and maximal data availability scenarios.

### *b. Research*

Research into various aspects of basin modeling comes in a couple of guises. Direct applications often suggest new lines of investigation to be carried out based on the applications. There is also an intrinsic interest in understanding the basic interaction of the geologic components that constitute the sedimentary basins of the world. By and large, it would seem that the development of quantitative basin analysis codes is best being carried out at various not-for-profit research organizations, institutes, and universities. It makes sense for oil companies to support financially those organizations which are developing new procedures and methods.

It also makes sense for oil companies to influence the directions of such research through proprietary projects for the organizations, through release of data, through interactive development of ideas, personnel, and evaluations of procedures and methods. However, at the same time, each oil company should have a dedicated group in-house whose function is to do research and to develop its own ideas—particularly in areas where the company wishes to maintain a technological edge over its competitors—so that both the basic ideas and their technical execution are kept out of the public arena. This aspect has two advantages: competitors do not know what devices and procedures a company is using in evaluations, and competitors are unlikely to recognize for a goodly while the direction in which scientific research is heading.

### c. Commentary on Practical Illustrations

The number of case histories that have been analyzed, both in this book and by various groups using different aspects of the general strategies considered here, is large. Inclusion of all such case histories would make for a very long book indeed, and one that would defeat the point of bringing together a set of strategic rules to be followed in a general sense. Yet the importance of case histories, treated in different ways by different groups, should not be understated, because it is the combined results of many such case histories that enables rules of strategy to be formulated at all.

Accordingly, we now provide a few basic references to case histories so that the interested reader can decide the relative interleaving of such case histories to the general strategy scheme spelled out in the body of the chapter. Perusal of the very many detailed references quoted in the reference list at the end of the book will extend the list of case histories almost indefinitely. Volumes of some interest are *Quantitative Dynamic Stratigraphy* (Cross, 1989), *Thermal Modeling in Sedimentary Basins* (Burrus, 1986), *Migration of Hydrocarbons in Sedimentary Basins* (Doligez, 1986), *Thermal History of Sedimentary Basins* (Naeser and McCulloh, 1989), *New Perspectives in Basin Analysis* (Kleinspehn and Paola, 1988), and *Basin Modeling: Advances and Applications* (Dore, 1992). The case histories reported in these books, together with those reported here and in professional journals during the last few years, form a practical basis of case histories from which the general strategy rules set out earlier were developed.

## II. SCIENTIFIC RISK AND RELATIVE IMPORTANCE

### A. INTRODUCTION

The implementation of basin analysis codes to provide measures of likely hydrocarbon amounts and their locations requires a fairly intensive and sustained effort by an oil exploration company. Clearly, some form of prioritization of effort in different basins, different parts of a particular basin, or different play concepts needs to be addressed prior to spending an inordinate amount of effort on one area at the expense of other areas. Strategies for prioritization are as many and as varied as the individual corporations involved in hydrocarbon exploration, but the choices are often guided by estimates of likely anticipated reserves based on prior finds, by economic considerations, or by estimates based on simple rough assessments of reservoir volume. Indeed, in Chapter 1 we say how the relative importance of different categories for gas could be prioritized using assessments by different groups of the likely amounts of gas still remaining.

TABLE 12.5 Estimated Discoverable Reserves, Norwegian Shelf[a]

| | Oil | | | Gas | | |
|---|---|---|---|---|---|---|
| | $P(5)$ | $P(68)$ | $P(95)$ | $P(5)$ | $P(68)$ | $P(95)$ |
| North Sea | 475 | 750 | 1100 | 660 | 960 | 1320 |
| Norwegian Sea | 110 | 320 | 685 | 75 | 520 | 1550 |
| Barents Sea | 30 | 190 | 615 | 380 | 930 | 2000 |
| Norwegian Shelf | 750 | 1260 | 1980 | 1310 | 2410 | 4020 |

[a] From NPD, 1993.

In this section we return therefore to scientific risk and relative importance in order to illustrate how the competing requirements of both oil and gas can be evaluated for rank importance. The easiest way to make the point is with an illustration. We have chosen the Norwegian Continental Shelf, where copious data exist, to provide a precise illustration. The concern then is to address estimates made of future oil and gas reserves for the Norwegian Continental Shelf. Current proven reserves of oil and gas are about 5.6 btoe (with 1.1 btoe produced, leaving a residual of 4.5 btoe with approximately 40% oil and 60% gas). With a mean annual oil production of 80 mtoe/year, and gas production of 25 mtoe/year, the estimated lifetime is about 22 years for oil and 100 years for gas. Clearly some assessments have to be provided of potential oil and gas reserves, their likelihood of being found, and the types of conditions under which oil and gas can be found so that decisions can be made for investment in and development of such reserves. In addition, the assessments have to be evaluated for their probability of being correct.

## B. BASIC DATA AND PROBABILISTIC INTERPRETATION

During the last few years the NPD (Norwegian Petroleum Directorate) has organized reserve estimates in different categories, using different methods, and with different qualities and quantities of data. The NPD (1993) has summarized the estimates made and that summary is given here as Table 12.5, taken directly from the NPD report (p. 20). Note from Table 12.5 that all reported potential resource estimates have been adjusted to the end of 1992.

A perusal of the numerical values for expected oil and gas resources in Table 12.5 for the three areas—North Sea, Norwegian Sea, and Barents

Sea—is quite interesting. Just as for the U.S. potential gas picture discussed in Chapter 1, again we have that each individual grouping gives *precision* statements of three and four figures on the assessments of potential oil and gas reserves; yet, again, we doubt that such precision is significant because the range of values reported for each of the three areas spans a factor of more than 3 and ranging as high as 20. Thus, once more, the *accuracy* of each assessment can be called into question; the same difficulty arises when individual areas are examined in Table 12.5. For example, the entries under North Sea (oil) vary from 475 to 1100 mtoe, a range of a factor of 2.3. At a fixed current production of, say, 40 mtoe/year, this category alone would yield a lifetime estimate to depletion of between 12 to 28 years, a fairly broad range.

As for the U.S. gas situation, the difficulty lies in figuring out how to plan for the future when faced with such broad ranges of uncertainty. To provide some measure of the accuracy of potential oil and gas resources, here we use cumulative probability methods, which have proven to be extremely powerful devices in the exploration and economic theathers (see, e.g., Chapter 1 and also the many references in Lerche, 1992a,b). The cumulative probability provides an estimate of the chance of *not* exceeding a particular magnitude of a quantity under consideration and is often taken to be describable by a log-normal behavior (Warren, 1983a).

Except at the extreme ends of the range of the cumulative probability plot, the log-normal distribution is a remarkably good fit to most data. This empirical point has been noted many times previously in other arenas of exploration and economic assessment (Warren, 1983a), which is why the log-normal probability method (on a cumulative probability basis) is used most often to categorize uncertainty. Therefore, we use it here. The important points about a log-normal cumulative distribution are (1) that the expected value, $E(x)$, of a quantity $x$ occurs close to 68% [i.e., there is a two-thirds chance of the actual value being less than or equal to $E(x)$]; and (2) that the 5% to 10% and 90% to 95% cumulative probability values are stable measures of the range within which an exactly log-normal distribution satisfies as best as it can a set of histogram values (Feller, 1957). Accordingly the 5% to 10% and 90% to 95% cumulative probability points are often used to provide a measure of the "volatility" of an estimate of a quantity $x$. The NPD (1993) report takes the 5% and 95% probability values as measures of range uncertainty.

For instance, in Table 12.5, North Sea (oil) category, the 68% [$P(68)$] cumulative probability value is 750 mtoe, the 5% [$P(5)$] cumulative probability value is 475 mtoe, while the 95% [$P(95)$] cumulative probability value is 1100 mtoe. Thus, there is a two-thirds chance of a value of *less* than 750 mtoe, a 95% chance of a value *less* than 1100 mtoe, and a 95% chance of a value *greater* than 475 mtoe. The estimate is then often written as

$750^{+350}_{-275}$ mtoe to show specifically the range at the 95% level of confidence. A volatility measure, $v$, often written (Warren, 1978) as

$$v = [P(95) - P(5)]/P(68)$$

provides an estimate of how the uncertainty of the $P(68)$ value. In the case of Table 12.5, the North Sea (oil) category, we have $v = (1100 - 475)/750 = 625/750 \approx 83\%$. Thus there is an 83% volatility to the two-thirds chance of having less than 750 mtoe. Once again, risk-weighted measures of potential reserves of each category and relative importance factors can be assigned based on the information available. For $N$ categories being investigated, each with its expected value $E(x_i)$, corresponding to $P(68)$, and each with its volatility, $v_i$, a risk-weighted measure of total potential reserves, $R$, can be made from all categories as[1]

$$R_i = E(x_i)(1 + v_i^2)^{-1/2}$$

with

$$R = \sum_{i=1}^{N} R_i,$$

which weights the contribution of each category to the total according to its relative volatility. The risk-weighted relative importance, RI($i$) (in %) of the $i$th category in relation to the other categories is given by

$$\mathrm{RI}(i) = 100(R_i) \Big/ \left( \sum_{i=1}^{N} R_i \right).$$

This relative importance factor measures the likelihood of a particular category contributing potential resources compared to all categories. As such RI is a useful measure of where most exploration effort should be put to enhance the likelihood of reserve appreciation being realized, with allowance made for the speculative nature of individual categories. Such a device is very useful in the decision-making process of where to commit capital investment and resource exploration.

## C. RISK-WEIGHTED POTENTIAL RESERVES: NORWEGIAN CONTINENTAL SHELF

Each individual category (North Sea, Norwegian Sea, and Barents Sea) for oil and gas, as given in Table 12.5, enables one to construct Table 12.6 showing the volatility of each category. Inspection of Table 12.6 shows that the North Sea (gas) category has the lowest volatility, which is not surprising given the amount of data from which the NPD could select to make the estimate. It is surprising, however, that the volatility is as high as nearly

[1] As in Chapter 1, different risk-weighted measures are available.

TABLE 12.6 Relative Volatilities, $[P(95) - P(5)/P(68)]$

| | Oil | Gas |
|---|---|---|
| North Sea | 0.83 | 0.7 |
| Norwegian Sea | 1.80 | 2.8 |
| Barents Sea | 3.1 | 1.73 |
| Norwegian Shelf | 0.98 | 1.13 |

69%, and it is even more surprising that the two remaining gas categories have between about 170% (Barents Sea) to 280% (Norwegian Sea) volatilities, suggesting that they are extremely poorly determined relative to the North Sea (gas) category—a factor of between about 2 to 4 more uncertainty based on relative volatilities. It would seem that assessments of the categories are less than accurate.

Similar arguments can be put forward for the oil resource estimates given in Table 12.5. The volatilities are 83% for North Sea (oil); 180% for Norwegian Sea (oil); and 310% for Barents Sea (oil)—again illustrating rather high uncertainties in estimated resources. Table 12.7 gives the risk-weighted assessments of reserves and the relative importance for the three categories. Note that, even without including the volatilities of each subcategory that make up the total resource estimates of the Norwegian Shelf, there is a high volatility of 98% on an unweighted estimate of total oil reserves as obtained from the last line of entries of Table 12.5; and a volatility of 113% on total gas reserves. Risk-weighted estimates of potential oil and gas reserves in the sum of categories can also be made and yield

$$R(\text{oil}) = 800^{+460}_{-325}\ \text{mtoe},$$

$$R(\text{gas}) = 1420^{+700}_{-630}\ \text{mtoe}.$$

TABLE 12.7 Risk-Weighted Reserves (RWR) and Relative Importance (RI) (%)

| | Oil | | Gas | |
|---|---|---|---|---|
| | RWR (mtoe) | RI (%) | RWR (mtoe) | RI (%) |
| North Sea | 580 | 73 | 790 | 56 |
| Norwegian Sea | 160 | 20 | 175 | 12 |
| Barents Sea | 58 | 7 | 465 | 32 |
| Total, Norwegian Shelf | 798 | — | 1420 | — |

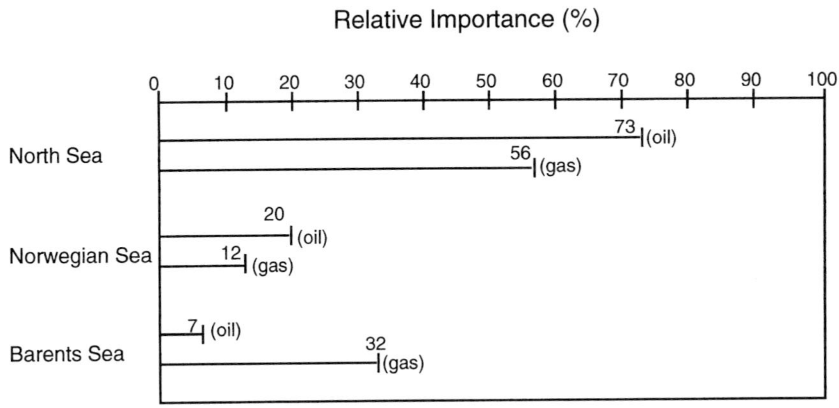


FIGURE 12.1 Risk-weighted relative importance of each category in terms of likely *discoverable* reserves. Relative importance lines are drawn using the volatility measures of Table 12.6.

These risk-weighted estimates argue for a two-thirds chance of *less* than about 800 mtoe (oil) and 1400 mtoe (gas) likely to be found, with a 95% chance of *less* than 1260 mtoe (oil) and 2100 mtoe (gas) likely to be found, and a 95% chance of *greater* than 475 mtoe (oil) and 800 mtoe (gas) likely to be found.

Thus, instead of just adding all of the categories together as though each category estimate has the same level of uncertainty as all others, if one allows for the relative worth of each estimated category, then a more representative measure can be given of what fraction is *likely* to be found of the total estimated potential reserves. It would seem that total ranges of 320 to 1300 mtoe (oil) and 450 to 2100 mtoe (gas) with a likelihood of less than about 800 mtoe (oil) and 1400 mtoe (gas) more than generously covers the statistics of Table 12.5. The suggestion, then, is that not all categories are equally attractive based on the individual sets of assessments made.

## D. RELATIVE IMPORTANCE OF BARENTS SEA, NORTH SEA, AND NORWEGIAN SEA

As well as estimating the total potential oil and gas reserves likely to be found, it is of equal importance to know which categories carry the most likely chance of being those containing discoverable reserves. Such assessments provide information on where to concentrate resources and capital to maximize the probability of actually finding the reserves. Figure 12.1 provides the risk-weighted relative importance (RI) for oil and gas, respectively, for the North Sea, Norwegian Sea, and Barents Sea categories,

TABLE 12.8 Rank Values for Different Categories

| | Oil | | Gas | |
|---|---|---|---|---|
| Category | Ranking order | Rank value (%) | Ranking order | Rank value (%) |
| North Sea | 1 | 73 | 1 | 56 |
| Norwegian Sea | 2 | 20 | 3 | 12 |
| Barents Sea | 3 | 7 | 2 | 32 |

using the data of Table 12.5 and the volatility measures given in Table 12.6. Each RI is evaluated using values at $P(68)$, $P(5)$, and $P(95)$, so that one has an appreciation of the stability and rank of each category's relative importance. Inspection of Fig. 12.1 strongly suggests that, at any level, the relative importance of finding likely oil and gas reserves is dominated by the North Sea category with the Norwegian Sea category a distant second for oil, and the Barents Sea a distant second for gas.

A ranking directly proportional to risk-weighted relative importance can then be made, as shown in Table 12.8. The likelihood is 2.5 times as great of finding oil reserves in the North Sea as it is in the Norwegian Sea plus Barents Sea combined, with the Norwegian Sea likely to be three times as great an oil producer compared to the Barents Sea. However, the likelihood is only 1.1 as great of finding gas reserves in the North Sea as it is in the Norwegian plus Barents Sea combined, while the Barents Sea is about 2.5 times as likely to have gas reserves compared to the Norwegian Sea.

## E. DISCUSSION

The purpose of this section has been to show how the very large ranges of uncertainty in individual assessments of categories of potential oil and gas resources can be organized, allowing both for the vagaries of individual assessments and for the speculative nature of some of the categories.

Probabilistic methods provide an eminently satisfactory vehicle for enabling both the absolute uncertainty within a category, and relative ranking between categories, to be assessed (1) for the relative importance of each category of potential oil and gas reserves; (2) for providing a risk-weighted likelihood of actually finding the potential reserves; and (3) for determining which categories are likely to yield the potential reserves. In this way a more accurate appreciation can be obtained of how, where, and when to commit capital investment and exploration effort.

For the assessments reported in several categories by the NPD, it would seem that there is about a 73% likelihood that the majority of the discoverable oil reserves will be found in the North Sea with about a 20% probability they will be found in the Norwegian Sea; about half of the discoverable gas reserves are likely to be found in the North Sea with about one-third likely to be found in the Barents Sea.

The likely risk-weighted amounts of oil and gas that are discoverable would appear to lie in the 95% probability ranges of 320 to 1300 mtoe (oil) and 450 to 2100 mtoe (gas), with a most likely discoverable amount of less than about 800 mtoe (oil) and 1400 mtoe (gas).

## III. INTEGRATED SCIENTIFIC AND ECONOMIC RISK

On the one hand, we now have available quite sophisticated quantitative models that estimate likely hydrocarbon reserves, with calculable degrees of uncertainty; on the other hand quite sophisticated quantitative models are available that estimate the likely economic worth of a potential hydrocarbon reserve, also with calculable degrees of uncertainty.

The integration of both of these disciplines provides an encompassing exploration strategy, which is clearly the direction in which the subject must head in order to provide maximum efficiency of hydrocarbon discovery and of economic returns. There is little point to spending inordinate efforts on scientific detailing of a likely hydrocarbon prospect if the economic assessment is bleak, nor is there much point in performing a cursory economic evaluation if the scientific likelihood of a major find is high. Clearly, prioritization of effort at the exploration stage is called for. Procedures for assessing such priorities were provided in Chapter 11 where estimates of hydrocarbon charge and uncertainty were evaluated. Thus the uncertainties of both parameters that enter quantitative economic worth assessments, as well as those entering quantitative basin analysis assessments, are now firmly tied together. In this way the assignment of risk, ranking, and relative importance of improving the scientific or the economic uncertainties are addressed on the same footing. We can then provide a strategy for prioritization:

1. Make a rough estimate of likely hydrocarbon charge and uncertainties; evaluate the relative importance of contributions to the uncertainty.
2. Make a rough estimate of likely economic worth and uncertainties; evaluate the relative importance of contributions to the uncertainty.
3. Make prioritization lists of the roughly estimated, risk-weighted, likely worth and relative importance factors for each prospect/basin/play type.

4. Use the prioritization list to do a more quantitative assessment of the economic and scientific factors, thereby improving the rough approximations of both components.
5. Determine whether improvements in the factors providing the largest contributions to uncertainties in the worth are needed, or whether the worth uncertainty is already tightly enough constrained that there is little point in expending the effort (time, money, personnel, data acquisition, better model construction, etc.) to improve matters.
6. Repeat steps 3, 4, and 5 at ever increasing levels of computational sophistication until it can be proven quantitatively that no further resolution on worth assessment is possible.

A final prioritized list is then available for corporate decision makers, with relative importance quantified properly from both the scientific and economic arenas.

# References


Airy, G. B. (1855). On the computation of the effect of the attraction of mountain-masses as disturbing the apparent astronomical latitude of station of geodetic surveys. *Philos. Trans. R. Soc. London* **145,** 101–104.

Aki, K., and Lee, W. H. K. (1976). Determination of three-dimensional anomalies under a seismic array using first P arrival times from local earthquakes: A homogeneous initial model. *J. Geophys. Res.* **81,** 4381–4399.

Allen, J. R. L. (1985). *Principles of Physical Sedimentology.* Allen & Unwin, London.

Andrews-Speed, C. P., Oxburgh, E. R., and Cooper, B. A. (1984). Temperature and depth-dependent heat flow in West North Sea. *AAPG Bull.* **68,** 1764–1781.

Armagnac, C. (1985). The determination of paleoheat flow in the National Petroleum Reserve, Alaska through the inversion of vitrinite reflectance. M.S. Thesis, University of South Carolina, Columbia.

Armagnac, C., Kendall, C., Kuo, C., Lerche, I., and Pantano, J. (1988). Determination of paleoheat flux from vitrinite reflectance data and from sterane and hopane isomer data. *J. Geochem. Explor.* **30,** 1–28.

Armagnac, C., Bucci, J., Kendall, C. G. St. C., and Lerche, I. (1989). Estimating the thickness of sediment removed at an unconformity using vitrinite reflectance data. In *Thermal History of Sedimentary Basins* (N. Naeser and T. H. McCulloh, eds.), Chapter 13, pp. 217–238. Springer-Verlag, Berlin.

Armstrong, R. L. (1978). Pre-Cenozoic Phanerozoic time scale—computer file of critical dates and consequences of new and in-progress decay-constant revisions. *Stud. Geol.* (*Tulsa, OK*) **6,** 73–91.

Athy, L. F. (1930). Density, porosity and compaction of sedimentary rocks. *Am. Assoc. Pet. Geol. Bull.* **14,** 1–24.

Atwater, G. I. (1956). Future of Louisiana offshore oil province. *Am. Assoc. Pet. Geol. Bull.* **40,** 2624–2634.

Aurell, E., and Zyczkowski, K. (1996). Option pricing and partial hedging: Theory of Polish options. *J. Political Econ.* (submitted for publication).

Bagnold, R. A. (1966). An approach to the sediment transport problem from general physics. *Geol. Surv. Prof. Pap.* (*U.S.*) **422-I.**

Bally, A. W. (1975). A geodynamic scenario for hydrocarbon occurrences. *Proc.—World Petr. Congr.* **9**(2), 33–44.

Bally, A. W., ed. (1983). *Seismic Expression of Structural Styles,* Stud. Geol. 15, Vol. 2, variously paginated. Am. Assoc. Pet. Geol., Tulsa, OK.

Battersby, D. G. (1976). Cooper Basin gas and oil fields. *Monogr. Ser.—Australas. Inst. Min. Metall.* **7,** 321–368.

Beaumont, C. (1981). Foreland basins. *Geophys. J. R. Astron. Soc.* **65,** 291–329.

Berner, H., Ramberg, H., and Stephansson, O. (1972). Diapirism in theory and experiment. *Tectonophysics* **15,** 197–218.

Bethke, C. M. (1985). A numerical model of compaction-driven groundwater flow and heat transfer and its application to the paleohydrology of intracratonic sedimentary basins. *J. Geophys. Res.* **90,** 6817–6829.

Bezdek, J. C. (1981). *Pattern Recognition with Fuzzy Objective Function Algorithms.* Plenum, New York.

Bezdek, J. C., Ehrlich, R., and Full, W. E. (1984). The fuzzy c-means clustering algorithm. *Comput. Geosci.* **10,** 2–12.

Bharucha-Reid, A. T. (1968). *Probabilistic Methods in Applied Mathematics.* Academic Press, New York.

Biot, M. A., and Ode, H. (1965). Theory of gravity instability with variable overburden and compaction. *Geophysics* **30,** 213–227.

Bishop, R. S. (1978). Mechanism for emplacement of piercement diapirs. *AAPG Bull.* **62,** 1561–1583.

Bitzer, K., and Pflug, R. (1990). DEPO3D: A three-dimensional model for simulating clastic sedimentation and isostatic compensation in sedimentary basins. In *Quantitative Dynamic Stratigraphy* (T. A. Cross, ed.) pp 335–348. Prentice-Hall, Englewood Cliffs, NJ.

Bomford, G. (1971). *Geodesy,* 2nd ed., pp. 441–443. Oxford University Press, London.

Bornholdt, S., and Westphal, H. (1996). Optimization using genetic algorithms. *Numer. Exp. Stratigr.,* Lawrence, KS, *1996,* p. 61.

Bostick, H., and Freeman, L. (1984). Tests of vitrinite reflectance and paleotemperature models at the Multiwell Experiment Site, Piceance Creek Basin, Colorado. *Geol. Surv. Open-File Rep. (U.S.)* **84-754.**

Bour, O., and Lerche, I. (1994). Numerical modeling of abnormal fluid pressure in the Navarin Basin, Bering Sea. *Mar. Pet. Geol.* **11,** 491–500.

Bour, O., Grauls, D., and Lerche, I. (1993). Modeling of high overpressure in an offshore Nigerian Basin: Slant fault models with non-horizontal beds. *J. Af. Earth Sci.* **17,** 307–321.

Bour, O., Grauls, D., and Lerche, I. (1995). Quantitative models of very high fluid pressure: Possible lateral stresses and their impacts on sedimentary basin evolution. *Terra Nova* **7,** 68–79.

Bower, T. H. (1951). Mud flow occurrence in Trinidad, BWI. *Am. Assoc. Pet. Geol. Bull.* **35,** 908–912.

Braun, R. L., and Burnham, A. K. (1987a). Analysis of chemical reaction kinetics using a distribution of activation energies and simpler models. *Energy Fuels* **1,** 153–161.

Braun, R. L., and Burnham, A. K. (1987b). General kinetic model of oil shale pyrolysis. *Energy Fuels* **1,** 162–173.

Braun, R. L., and Burnham, A. K. (1988). *Petroleum Geochemistry Project Quarterly Report October-December 1987,* Doc. UCID-21303. Lawrence Livermore Laboratories, Livermore, CA.

Brooks, J., and Welte, D. (1984). *Advances in Petroleum Geochemistry,* Vol. 1. Academic Press, London.

Burke, W. H., and von Rosenberg, D. U. (1985). A graphical method for studying models of oil generation. *J. Pet. Geol.* **8,** 221–228.

Burnham, A. K. (1989). Private communication.

Burnham, A. K., Braun, R. L., Gregg, H. R., and Samoun, A. M. (1987). Comparison of methods for measuring kerogen pyrolysis rates and fitting kinetic parameters. *Energy Fuels* **1,** 452–458.

Burnett, T. L. (1996). Petroleum exploration risk reduction using new geoscience technology. *Energy Explor. Exploit.* (in press).

Burrus, J., ed. (1986). *Thermal Modeling in Sedimentary Basins.* Editions Technip, Paris.

Burton, R., Kendall, C. G. St. C., and Lerche, I. (1987). Out of our depth: On the impossibility of fathoming eustatic sea level from the stratigraphic record. *Earth Sci. Rev.* **24,** 237–277.

Busch, R. M. (1983). Sea level correlation of punctuated aggradational cycles (PAC's) of the Manlius Formation, Central New York. *Northeast. Geol.* **5,** 82–91.

Caldwell, R. H., and Heather, D. I. (1995). Why our reserves definitions don't work anymore. *Soc. Pet. Eng. Hydrocarbon Econ. Eval. Symp.,* Dallas, TX, *1995,* SPE No. 30041, pp. 85–94.

Cao, S. (1985). A quantitative dynamic model for basin analysis and its application to the Northern North Sea Basin. M.S. Thesis, University of South Carolina, Columbia.

Cao, S. (1987). Sensitivity analysis of 1-D dynamical model for basin analysis. Ph.D. Dissertation, University of South Carolina, Columbia.

Cao, S., and Lerche, I. (1989a). Geohistory, thermal history and hydrocarbon generation history of Navarin Basin COST No. 1 well, Bering Sea, Alaska. *J. Pet. Geol.* **12,** 325–352.

Cao, S., and Lerche, I. (1989b). One-dimensional modeling of episodic fracturing: A sensitivity study. *Terra Nova* **1,** 177–181.

Cao, S., and Lerche, I. (1990). Basin modeling: Applications of sensitivity analysis. *J. Pet. Sci. Eng.* **21,** 523–542.

Cao, S., and Lerche, I. (1994). Thermal history reconstruction by inversion of S2 pyrolysis data. *Appl. Geochem.* **9,** 93–117.

Cao, S., Glezen, W. H., and Lerche, I. (1986). Fluid flow, hydrocarbon generation and migration: A quantitative model of dynamical evolution in sedimentary basins. *Proc.—Annu. Offshore Technol. Conf.,* Houston, TX, Pap. OTC 5182, pp. 267–276.

Cao, S., Lerche, I., and Hermanrud, C. (1988). Formation temperature estimation by inversion of borehole measurements. *Geophysics* **53,** 979–988.

Cao, S., Lerche, I., and O'Brien, J. J. (1989). Impact on hydrocarbon retention of fracturing and faulting around salt sheets in the Gulf Coast. *Trans.—Gulf Coast Assoc. Geol. Soc.* **39,** 15–26.

Cao, S., Tang, J., Liu, J., and Lerche, I. (1993). The Chukchi Sea Planning Area, Alaska: Structural development, basin analysis and hydrocarbon potential. *Energy Explor. Exploit.* **11,** 235–283.

Carpentier, B., Huc, A. Y., and Bessereau, G. (1989). Wireline logging and source rocks, estimation of organic carbon contents by the CARBOLOG® method. *Rev. Inst. Fr. Pet.* **44,** 699–719.

Cloetingh, S. A. P. L., and Wortel, M. J. R. (1988). On the mechanics of plate boundary formation. In *Mathematical Geophysics* (N. J. Vlaar, G. Nolet, M. J. R. Wortel, and S. A. P. L. Cloetingh, eds.), pp. 363–388. Reidel Pub., Dordrecht, The Netherlands.

Cloos, E. (1968). Experimental analysis of Gulf Coast fracture patterns. *Am. Assoc. Pet. Geol. Bull.* **52,** 420–444.

Coakley, B., and Pratson, L. (1996). A "map-view" model of turbidite transport and deposition. *Numer. Exp. Stratigr.,* Lawrence, KS, *1996,* pp. 73–76.

Coakley, B. J., and Watts, A. B. (1991). Tectonic controls on the development of unconformities: The North Slope, Alaska. *Tectonics* **10,** 101–130.

Cogan, J., Rigo, L., Grasso, M., and Lerche, I. (1989). Flexural tectonics of southeastern Sicily. *J. Geodyn.* **11,** 189–241.

Cogan, J., Dorman, J., Lerche, I., and Kanes, W. (1992). Flexural plate inversion: Application to the Carpathian Foredeep, Czechoslovakia. *Mod. Geol.* **17,** 217–252.

Connan, J. (1974). Time-temperature relation in oil genesis. *AAPG Bull.* **58,** 2516–2521.

Conybeare, C. E. B. (1965). Hydrocarbon generation potential and hydrocarbon yield capacity of sedimentary basin. *Bull. Can. Pet. Geol.* **13,** 509–528.

Cousminer, H. L., and Steinkraus, W. E. (1988). Biostratigraphy of the COST G-2 well (Georges Bank): A Record of Late Triassic synrift evaporite deposition; Liassic doming; and Mid-Jurassic to Miocene postrift marine sedimentation. In *Triassic-Jurassic Rifting-Continental Breakup and the origin of the Atlantic Ocean and Passive Margin* (W. Manspeizer, ed.), pp. 167–184. Elsevier, New York.

Cozzolino, J. M. (1977) A simplified utility framework for the analysis of financial risk. *Soc. Pet. Eng. Econ. Eval. Sym.,* Dallas, TX, *1977,* SPE No. 6359.

Cozzolino, J. M. (1978). A new method for measurement and control of exploration risk. *Soc. Pet. Eng. AIME, 1978,* SPE No. 6632.

Crans, W., Mandl, G., and Haremboure, J. (1980). On the theory of growth-faulting, a geometrical delta model based on gravity sliding. *J. Pet. Geol.* **2,** 265–307.

Cressie, N. A. C. (1991). *Statistics for Spatial Data.* Wiley, New York.

Cross, T. A., ed. (1989). *Quantitative Dynamic Stratigraphy.* Prentice-Hall, Englewood Cliffs, NJ.

Dahl, B., and Yukler, A. (1991). The role of petroleum geochemistry in basin modeling of the Oseberg area, North Sea. In *Source and Migration Processes and Evaluation Techniques* (R. K. Merrill, ed.), Treatise Pet. Geol., pp. 65–85. Am. Assoc. Pet. Geol., Tulsa, OK.

Dahlberg, E. C. (1982). *Applied Hydrodynamics in Petroleum Exploration.* Springer-Verlag, New York.

Davies O. L., and Goldsmith, P. L. (1972). *Applied Statistics in Chemical Engineering.* Plenum Press, New York.

Dickey, P. A. (1975). Possible primary migration of oil from source rock in oil phase. *AAPG Bull.* **59,** 337–345.

Doligez, B., ed. (1986). *Migration of Hydrocarbons in Sedimentary Basins.* Editions Technip, Paris.

Doré, A., ed. (1992). *Basin Modeling: Advances and Applications,* Proc. Norw. Pet. Soc. Meet., Stavanger, Norway, 1991.

Dula, W. F., Jr. (1991). Geometric models of listric normal faults and rollover folds. *AAPG Bull.* **75,** 1609–1625.

Dunbar, J. (1994). Private communication.

Durand, B. (1988). Understanding of hydrocarbon migration in sedimentary basins (present state of knowledge). *Org. Geochem.* **13,** 445–459.

Dutta, N. (1989). Fluid flow in low permeable porous media. In *Migration of Hydrocarbons in Sedimentary Basins* (B. Doligez, ed.), pp. 567–596. Editions Technip, Paris.

Eaton, B. A. (1975). The equations for geopressure prediction from Well Logs. *Soc. Pet. Eng. Pap.* **5544.**

Espitalié, J., Deroo, G., and Marquis, F. (1985). La pyrolyse Rock Eval et ses applications. *Rev. Inst. Fr. Pet.* **41,** 563–579.

Espitalié, J., Ungerer, P., Marquis, F., and Irwin, H. (1987). Primary cracking of kerogens: Experimenting and kinetic modeling. In *Advances in Org. Geochem.*

Ethetton, L. K., and Brumbaugh, W. D. (1995). The economic impact of strategic risk in petroleum ventures: Examples from West Africa. *Soc. Pet. Eng. Hydrocarbon Econ. Eval. Symp.,* Dallas, TX, *1995,* SPE No. 30047, pp. 143–146.

Evamy, B. D., Haremboure, J., Kamerling, P., Knaap, W. A., Molly, F. A., and Rowlands, P. H. (1978). Hydrocarbon habitat of Tertiary Niger Delta. *AAPG Bull.* **62,** 1–39.

Falvey, D. A. (1974). The development of continental margins in plate tectonic theory. *APEA J.* **14,** 95–106.

Falvey, D. A., and Middleton, M. F. (1981). Passive continental margins: Evidence for a prebreakup deep crustal metamorphic subsidence mechanism. *Int. Geol. Cong., 26th, 1980,* Colloq. C3.3, Vol. 4, Suppl., pp. 103–114.

Feller, W. (1957). *An Introduction to Probability Theory and Its Applications,* Vols. 1 and 2. Wiley, New York.

Ferm, J. C. (1970). Allegheny deltaic deposits. *Spec. Publ.—Soc. Econ. Paleontol. Mineral.* **15,** 246–255.

Flemings, P. B., and Carlson, J. (1996). Transport and deposition of heterogeneous-sized sediment in a non-marine sedimentary basin. *Num. Exp. Stratigr.,* Lawrence, KS, *1996,* pp. 83–86.

Freeze, R. A. (1975). A stochastic conceptual analysis of one-dimensional groundwater in nonuniform homogeneous media. *Water Resour. Res.* **11,** 725–741.

Full, W. E., Ehrlich, R., and Klovan, J. E. (1981). EXTENDED QMODEL—objective definition of external end-members in the analysis of mixtures. *Math. Geol.* **13,** 331–344.

Full, W. E., Ehrlich, R., and Bezdek, J. C. (1982). FUZZY QMODEL: A new approach for linear unmixing. *Math. Geol.* **14,** 259–270.

Gibbs, A. D. (1983). Balanced cross-section construction from seismic sections in areas of extensional tectonics. *J. Struct. Geol.* **5,** 153–160.

Gleadow, A. J. W., and Duddy, I. R. (1984). Fission track dating and thermal history analysis of apatites from wells in the North-West Canning Basin. In *Pet. Explor. Soc. Aust., Perth., Symposium on the Canning Basin, Perth, Western Australia,* pp. 377–387.

Gleadow, A. J. W., Duddy, I. R., and Lovering, J. F. (1983). Fission track analysis: A new tool for the evaluation of thermal histories and hydrocarbon potential. *APEA J.* **23,** 93–102.

Glezen, W. H., and Lerche, I. (1985). A model of regional fluid flow: Sand concentration factors and effective lateral and vertical permeabilities. *Math. Geol.* **17,** 297–315.

Gradstein, F. M., Agterberg, F. P., Ogg, J. G., Hardenbol, J., van Veen, P., Thiery, J., and Huang, Z. (1994). A Mesozoic time scale. *J. Geophys. Res.* **99,** 24,051–24,074.

Grans, W., Mandl, G., and Haremboure, J. (1980). On the theory of growth faulting: A geomechanical delta model based on gravity sliding. *J. Pet. Geol.* **2,** 265–307.

Grantz, A., and May, S. D. (1983). Rifting history and structural development of the Continental Margin north of Alaska. *AAPG Mem.* **34,** 77–100.

Grantz, A., Holmes, M. L., and Kososki, B. A. (1975). Geological framework of the Alaskan continental terrace in the Chukchi and Beaufort Seas. *Geol. Surv. Open-file Rep. (U.S.)* **75-124,** 1–31.

Grantz, A., Eittreim, S., and Whitney, O. T. (1979). Geology and physiography of the continental margin north of Alaska and implications for the origin of the Canada Basin. *Geol. Surv. Open-File Rep. (U.S.)* **79-288,** 1–71.

Grauls, D., and Cassignol, C. (1993). Identification of a zone of fluid pressure-induced fractures from log and seismic data: A case history. *First Break Decoarrier* (Elf Aquitaine), pp. 23–27.

Grauls, D., and Kessler, N. (1992). *Role of Fluid Pressure and In-Situ Minimum Stress in Fault Transient Dynamics-Major Implications,* Rep. EP/S/EXP/GDP N°. 92/01RS. Elf Aquitaine.

Grayson, J. F. (1975). Relationship of palynomorph translucency to carbon and hydrocarbons in clastic sediments. In *Petrographie de la matière organique des sediments relations avec la paleotemperature et le potential petrolier,* pp. 261–273. CNRS, Paris.

Green, P. A. (1995). Private communication.

Green, P. F., and Duddy, I. R. (1988). Some comments on paleotemperature estimation from apatite fission track analysis. *J. Pet. Geol.* **12,** 110–115.

Groshong, R. H., Jr. (1989). Half-graben structures: Balanced models of extensional fault-bend folds. *Geol. Soc. Am. Bull.* **101,** 96–105.

Guidish, T. M., Lerche, I., Kendall, C. G. St. C., and O'Brien, J. J. (1984). Relationship between eustatic sea level changes and basement subsidence. *AAPG Bull.* **68,** 164–177.

Hallam, A. (1984). Pre-quaternary sea level changes. *Annu. Rev. Earth Planet. Sci.* **12,** 205–243.

Hamblin, W. K. (1965). Origin of "reverse drag" on the downthrown side of normal faults. *Geol. Soc. Am. Bull.* **76,** 1145–1164.

Hambrey, M. J., and Harland, W. B., eds. (1981). *Earth's Pre-Pleistocene Glacial Record.* Cambridge University Press, Cambridge, UK.

Haq, B. U., Hardenbol, J., and Vail, P. R. (1987). Chronology of fluctuating sea-level since the Triassic (250 million years to present). *Science* **235,** 1156–1167.

Harbaugh, J. W., Doveton, J. H., and Davis, J. C. (1977). *Probability Methods in Oil Exploration.* Wiley, New York.

Harland, W. B., ed. (1971). *The Phanerozoic Time-Scale—A Supplement,* Spec. Publ. No. 5. Geol. Soc. London, London.

Harland, W. B., Smith, A. G., & Wilcock, B., eds. (1964). *The Phanerozoic Time-Scale* (A Symposium Dedicated to Professor Arthur Holmes). *Q. J. Geol. Soc. London* **120,** 458.

Harland, W. B., Cox, A. V., Llewellyn, P. G., Pickton, C. A. G., Smith, A. G., and Walters, R. (1982). *A Geologic Time Scale.* Cambridge University Press, Cambridge, UK.

Harland, W. B., Armstrong, R. L., Cox, A. G., Craigh, L. E., Smith, A. G., and Walters, R. (1989). *A Geologic Time Scale.* Cambridge University Press, Cambridge, UK.

Harrison, M. T., and Be, K. (1983). $^{40}Ar/^{39}Ar$ age spectrum analysis of detrital microclines from the southern San Joaquin basin, California: An approach to determining the thermal evolution of sedimentary basins. *Earth Planet. Sci. Lett.* **64,** 242–256.

Hawley, B. W., Zandt, G., and Smith, R. B. (1981). Simultaneous inversion for hypocenters and lateral velocity variations: An iterative solution with a layered model. *J. Geophys. Res.* **86,** 7073–7086.

Haxby, W. F., Turcotte, D. L., and Bird, J. M. (1976). Thermal and mechanical evolution of the Michigan Basin. *Tectonophysics* **36,** 57–75.

He, Z., and Lerche, I. (1989). Inversion of multiple thermal indicators: Quantitative methods of determining paleoheat flux and geological parameters. IV. Case histories using thermal indicator tomography. *Math. Geol.* **21,** 523–542.

He, Z., and Lerche, I. (1992). Determination of paleoheat flux using multiple thermal indicator tomography. *Math. Geol.* **24,** 825–846.

Hearst, J. R., and Nelson, P. H. (1985). *Well-Logging for Physical Properties.* McGraw-Hill, New York.

Helland-Hausen, W., Kendall, C. G. St. C., Nakayama, K., and Lerche, I. (1988). Simulation of continental basin margin sedimentation in response to crustal movements, eustatic sea level change, and sediment accumulation rates. *Math. Geol.* **20,** 777–802.

Hellinger, S. J., and Sclater, J. G. (1983). Some comments on two-layer extensional models for the evolution of sedimentary basins. *J. Geophy. Res.* **88,** 8251–8269.

Hermanrud, C., Eggen, S., Jacobsen, T., Carlsen, E. M., and Pallesen, S. (1989). On the accuracy of modeling hydrocarbon generation and migration: The Egersund Basin oil find, Norway. *Org. Geochem.* **16,** 389–399.

Hermanrud, C., Cao, S., and Lerche, I. (1990). Estimates of virgin rock temperatures derived from BHT measurements. *Geophysics* **55,** 229–237.

Holmes, A. (1937). *The Age of the Earth.* Nelson, London.

Holmes, A. (1947). The construction of a geological time-scale. *Trans. Geol. Soc. Glasgow* **21,** 117–152.

Holmes, A. (1959). A revised geological time-scale. *Trans. Edinburgh Geol. Soc.* **17,** 183–216.

Hottman, C. E., and Johnson, R. K. (1965). Estimation of formation pressures from log-derived shale properties. *J. Pet. Technol.,* June, pp. 25–32.

Howard, J. (1981). In *Chemistry of Coal Utilization,* 2nd suppl. vol., pp. 733ff.

Hubbert, M. K. (1953). Entrapment of petroleum under hydrodynamic conditions. *Am. Assoc. Pet. Geol. Bull.* **37,** 1954–2026.

Hunt, J. M. (1979). *Petroleum Geochemistry and Geology.* Freeman, San Francisco.

Huntsberger, T. L., and Lerche, I. (1987). Determination of paleo heat-flux from fission scar tracks in apatite. *J. Pet. Geol.* **10,** 365–394.

Irwin, H., Hermanrud, C., Carlsen, E. M., Vollset, J., and Nordvall, I. (1993). Basin modeling of hydrocarbon charge in the Egersund Basin, Norwegian North Sea: Pre- and post-drilling assessments. *NPF Spec. Publ.* **3,** 539–548.

Jackson, J. D. (1975). *Classical Electrodynamics.* Wiley, New York.

Jackson, M. P. A., and Galloway, W. E. (1984). Structural and depositional styles of Gulf Coast Tertiary continental margin: Application to hydrocarbon exploration. *AAPG Contin. Educ. Course Notes Ser.* No. 25, pp. 1–226.

Jackson, M. P. A., Talbot, C. J., and Cornelius, R. R. (1988). Centrifuge modeling of the effects of aggradation and progradation on synsedimentary salt structures. *Rep. Invest.—Univ. Tex. Austin, Bur. Econ. Geol.* **173.**

Jacobi, R. D. (1981). Peripheral bulge—a causal mechanism for the Lower/Middle Ordovician unconformity along the western margin of the Northern Appalachians. *Earth Planet. Sci. Lett.* **56,** 245–251.

James, D. M. D., ed. (1984). *The Geology and Hydrocarbon Resources of Negara Brunei Darussalam: Muzium Brunei.* Bandar Seri Begawan, Brunei.

Jenyon, M. K. (1987). The development by salt diapirs of superficial overhang features, and effects on associated sediments. In *Dynamical Geology of Salt and Related Structures* (I. Lerche and J. J. O'Brien, eds.), pp. 679–709. Academic Press, Orlando, FL.

Johnson, D. D. (1996). Modelling tectonic, climatic and eustatic effects on Orogen/Foreland Basin systems. *Numer. Exp. Stratigr.,* Lawrence, KS, *1996,* pp. 105–108.

Johnson, E. F., and Richard, K. (1990). *Advanced X Window Applications Programming.* Management Information Source Press, Portland, OR.

Johnson, R. C., and Nuccio, V. F. (1986). Structural and thermal history of the Piceance Creek basin, Western Colorado, in relation to hydrocarbon occurrence in the Mesaverde Group. *Stud. Geol.* (*Tulsa, OK*) **24,** 165–205.

Jones, R. W. (1975). A quantitative geologic approach to prediction of petroleum resources. *Stud. Geol. Tulsa, OK* **1,** 186–195.

Jordan, T. E. (1981). Thrust loads and foreland basin evolution, Cretaceous western United States. *AAPG Bull.* **65,** 2506–2520.

Kai, Z., and Lerche, I. (1993). Inversion of dynamical indicators in quantitative basin analysis models II. Synthetic tests and a case history using dynamical indicator tomography. *Math. Geol.* **25,** 107–123.

Kantsler, A. J., Prudence, T. J. C., Cook, A. C., and Zwigulis, M. (1983). Hydrocarbon habitat of the Cooper/Eromanga Basin, Australia. *APEA J.* **23,** 75–90.

Karner, G. D., and Watts, A. B. (1983). Gravity anomalies and flexure of the lithosphere at mountain ranges. *J. Geophys. Res.* **88,** 10449–10477.

Keen, C. E., Beaumont, C., and Boutilier, R. (1981). Preliminary results from a thermomechanical model for the evolution of Atlantic-type continental margins. *Int. Geol. Congr., 26th, 1980* Colloq. C3.3, Vol. 4, Suppl., pp. 123–128.

Kerr, H. G., and White, N. (1996). Application of an inverse method for calculating three-dimensional fault geometries and slip vectors, Nun River Field, Nigeria. *AAPG Bull.* **80,** 432–444.

Kerr, H. G., White, N., and Brun, J.-P. (1993). An automatic method for determining three-dimensional normal fault geometries. *J. Geophys. Res.* **98**(B10), 17,837–17,857.

King, J. D. (1989). Private communication.

Kleinspehn, K. L., and Paola, C., eds. (1988). *New Perspectives in Basin Analysis.* Springer-Verlag, New York.

Klitgord, K. D., Hutchinson, D. R., and Schouten, H. (1988). U. S. Atlantic continental margin; structural and tectonic framework. *GSA, Geol. North Am.* **1–2,** 19–55.

Klovan, J. E., and Imbrie, J. (1971). An algorithm and FORTRAN-IV program for large scale Q-mode factor analysis and calculation of factor scores. *Math. Geol.* **3,** 61–76.

Klovan, J. E., and Miesch, A. T. (1976). EXTENDED CABFAC and QMODEL computer programs for Q-mode factor analysis of compositional data. *Comput. Geosci.* **1,** 161–178.

Kuckelkorn, K., Wang, X., and Lerche, I. (1990). Overthrusting history in the Austrian Alps: Case study of Oberhofen 1. *J. Geodyn.* **12,** 101–120.

Kulp, J. L. (1961). Geologic time-scale. *Science* **133,** 1105–1114.

Lambert, R. St. J. (1971). The pre-Pleistocene Phanerozoic time-scale—a review. *Spec. Publ.—Geol. Soc. London* **5,** 9–34.

Law, B. E., Nuccio, V. F., and Barker, C. E. (1989). Kinky vitrinite reflectance well profiles: Evidence of paleopore pressure in low-permeability, gas-bearing sequences in Rocky Mountain Foreland Basins. *Am. Assoc. Pet. Geol. Bull.* **23,** 999–1010.

Lerche, I. (1987). Direct detention of hydrocarbons: Some preliminary thoughts and questions. *Energy Explor. Exploit.* **5,** 265–285.

Lerche, I. (1988a). Inversion of multiple thermal indicators: Quantitative methods of determining paleoheat flux and geological parameters. I. The theoretical development for paleoheat flux. *Math. Geol.* **20,** 1–36.

Lerche, I. (1988b). Inversion of multiple thermal indicators: Quantitative methods of determining paleoheat flux and geological parameters. II. The theoretical development for chemical, physical and geological parameters. *Math. Geol.* **20,** 73–96.

Lerche, I. (1990a). *Basin Analysis: Quantitative Methods,* Vol. 1. Academic Press, San Diego, CA.

Lerche, I. (1990b). *Basin Analysis: Quantitative Methods,* Vol. 2. Academic Press, San Diego, CA.

Lerche, I. (1991). Inversion of dynamical indicators in quantitative basin analysis models. I. Theoretical considerations. *Math. Geol.* **23,** 817–832.

Lerche, I. (1992a). Probabilistic methods in well logging. *Nucl. Geophys.* **6,** 325–339.

Lerche, I. (1992b). Estimating organic carbon content of source rocks from wireline logs. *Sci. Drill.* **3,** 170–179.

Lerche, I. (1992c). *Oil Exploration: Basin Analysis and Economics.* Academic Press, San Diego, CA.

Lerche, I. (1993). A probabilistic procedure to assess the uncertainty of fractal dimensions from measurements. *Pure Appl. Geophys.* **140,** 503–517.

Lerche, I. (1994). Anatomy of a lease sale: Sub-salt bids and strategies at Gulf of Mexico lease sale 147. *Oil & Gas J.,* October, 3, pp. 68–71.

Lerche, I., and Lowrie, A. (1992). Quantitative models for the influence of salt-associated thermal anomalies on hydrocarbon generation, northern Gulf of Mexico continental margin. *Trans. Gulf Coast Assoc. Geological Soc.* **42,** 213–226.

Lerche, I., and MacKay, J. A. (1995). Model dependences of risk aversion and working interest estimates. *Energy Explor. Exploit.* **14,** 183–196.

Lerche, I., and O'Brien, J. J., eds. (1987a). *Dynamical Geology of Salt and Related Structures.* Academic Press, Orlando, FL.

Lerche, I., and O'Brien, J. (1987b). Modeling of buoyant salt diapirism. In *Dynamical Geology of Salt and Related Structures* (I. Lerche and J. J. O'Brien, eds.), pp. 129–162. Academic Press, Orlando, FL.

Lerche, I., and O'Brien, J. J. (1987c). Modeling of the deformation and faulting of formations overlying an uprising salt dome. In *Dynamical Geology of Salt and Related Structures* (I. Lerche and J. J. O'Brien, eds.), pp. 419–456. Academic Press, Orlando, FL.

Lerche, I., and Petersen, K. (1995). *Salt and Sediment Dynamics.* CRC Press, Boca Raton, FL.

Lerche, I., and Thomsen, R. O. (1994). *Hydrodynamics of Oil and Gas.* Plenum, New York.

Lerche, I., Yarzab, R. F., and Kendall, C. G. St. C. (1984). Determination of paleoheat flux from vitrinite reflectance data. *AAPG Bull.* **68,** 1704–1717.

Lerche, I., Yu, Z., and Liu, J. (1993). Effects of open and shut slant faults in basinal development processes. *GCS-SEPM Found. Res. Conf. Proc.* **14,** 81–97.

Lerche, I., Yu, Z., and Liu, J. (1995). Representation of normal slant faults in a two-dimensional quantitative burial history model. *Terra Nova* **7,** 459–466.

Lessenger, M. A., and Cross, T. A. (1996a). Construction and application of stratigraphic inverse models. *Numer. Exp. Stratigr.*, Lawrence, KS, *1996*, pp. 109–112.

Lessenger, M. A., and Cross, T. A. (1996b). An inverse stratigraphic simulation model—Is stratigraphic inversion possible? *Energy Explor. Exploit.* **14,** 627–638.

Lewis, C. R., and Rose, S. C. (1970). A theory relating high temperatures and overpressures. *J. Pet. Technol.* **22,** 11–16.

Leythaeuser, D., Schaefer, R. G., and Yukler, A. (1982). Role of diffusion in primary migration of hydrocarbons. *AAPG Bull.* **66,** 408–429.

Liu, J., and Lerche, I. (1990). Inverse methods and kinetic models of hydrocarbon generation. II. Case histories for residual kerogen analysis. *Math. Geol.* **22,** 989–1009.

Liu, J., and Lerche, I. (1991). Inverse methods and kinetic models of hydrocarbon generation. III. Case histories for autoclave product analysis. *Math. Geol.* **23,** 325–347.

Liu, J., Yu, Z., and Lerche, I. (1993). Effects of open and shut slant faults in basinal development processes. *GCS/SEPM Found. Res. Conf. Proc.* **14,** 81–95.

Lliboutry, L. A. (1987). *Very Slow Flows of Solids.* Martinus Nijhoff Publishers, The Netherlands.

Lopatin, N. V. (1971). Temperature and geologic time as factors in coalification. *Izv. Akad. Nauk SSSR, Ser. Geol.* **3,** 95–106.

Løseth, H., Lippard, S. J., Saettem, J., Farnavoll, S., Fjerdingstad, V., Leith, T. H., Ritter, U., Smelror, M., and Sylta, Ø. (1993). Cenozoic uplift and erosion of the Barents Sea—evidence from the Svalis Dome area. In *Arctic Geology and Petroleum Potential* (T. Vorren, E. Bergsager, Ø. A. Dahl-Stamnes, E. Holter, B. Johansen, E. Lie, and T. B. Lund, eds.), NPF Spec. Publ. No. 2, pp. 643–664. Elsevier, Amsterdam.

Lowrie, A. (1994). In *Seismic Stratigraphy and Hydrocarbon Traps: Louisiana Onshore and Offshore* (S. N. Domenico, ed.), Course Note Ser., Vol. 5. 213p, Soc. Explor. Geophys., Tulsa, OK.

Lowrie, A., Sullivan, N. M., and Hoffman, K. S. (1990). Regional tectonic and hydrocarbon-trap modeling along the Louisiana Offshore. *Proc.—Annu. Offshore Technol. Conf.*, Pap. No. 6292, pp. 211–223.

Lumley, J. L. (1970). *Stochastic Tools in Turbulence.* Academic Press, New York.

MacKenzie, A. S., and McKenzie, D. (1983). Isomerization and aromatization of hydrocarbons in sedimentary basins formed by extension. *Geol. Mag.* **120,** 417–470.

Magara, K. (1976). Thickness of removed sedimentary rocks, paleopore-pressure and paleotemperature, Southwestern part of Western Canada Basin. *AAPG Bull.* **60,** 554–565.

Magara, K. J. (1978). *Compaction and Fluid Migration—Practical Petroleum Geology.* Elsevier, New York.

Magoon, L. B., and Bird, K. J. (1986). Organic carbon content, hydrocarbon content, visual kerogen and vitrinite reflectance data within NPRA (National Petroleum Reserve in Alaska)—contour maps to evaluate petroleum source rock richness, type, and thermal maturity. *Geol. Surv. Prof. Pap.* (*U.S.*) **139.**

Magoon, L. B., and Sanchez, R. M. O. (1995). Beyond the petroleum system. *AAPG Bull.* **79,** 1731–1736.

Magoon, L. B., Woodward, P. V., Banet, A. C., Jr., Griscom, S. B., and Daws, R. A. (1987). Thermal maturity, richness and type of organic matter of source-rock units. *Geol. Surv. Bull.* (*U.S.*) **1778,** 277–298.

Mandl, G., and Crans, W. (1981). Gravitational gliding in deltas. *Spec. Publ.—Geol. Soc. London.*

Manspeizer, W., and Cousminer, H. L. (1988). Late Triassic-Early Jurassic synrift basins of the U.S. Atlantic margin. *GSA, Geol. North Am.* **1–2,** 197–216.

Maubeuge, F., and Lerche, I. (1993). Representation of a growth fault in two-dimensional quantitative modeling. *Mar. Pet. Geol.* **10,** 467–477.

Maubeuge, F., and Lerche, I. (1994). Geopressure evolution and hydrocarbon generation in a North Indonesian Basin: Two-dimensional modeling. *Mar. Pet. Geol.* **11,** 104–115.

McAuliffe, C. D. (1980). Oil and gas migration: Chemical and physical constraints. *Stud. Geol. (Tulsa, OK)* **10,** 89–108.

McDowell, A. N. (1975). What are the problems in estimating the oil potential of a basin? *Oil & Gas J.,* June 9, pp. 85–90.

McKenna, T., and Lerche, I. (1991). Pollen translucency as a thermal maturation indicator. *J. Pet. Geol.* **14,** 19–36.

McKenzie, D. (1978). Some remarks on the development of sedimentary basins. *Earth Planet. Sci. Lett.* **40,** 25–32.

McMullen, R. J., and Mohraz, B. (1987). Axisymmetric finite element analysis of an elastic lithosphere subjected to thermal loading. *Geophys. J.R. Astron. Soc.* **88,** 325–343.

McMullen, R. J., Hodge, D. S., and Cozzarelli, F. A. (1981). A technique for incorporating geothermal gradients and nonlinear creep into lithospheric flexure models. *J. Geophys. Res.* **86,** 1745–1753.

Megill, R. E. (1971). *Exploration Economics.* Petroleum Publ. Co., Tulsa, OK.

Menke, W. (1984). *Geophysical Data Analysis: Discrete Inverse Methods.* Academic Press, Orlando, FL.

Middleton, G. V. (1966). Experiments on density and turbidity currents. II. Uniform flow of density currents. *Can. J. Earth Sci.* **3,** 627–637.

Middleton, M. F. (1980). A model of intracratonic basin formation, entailing deep crustal metamorphism. *Geophys. J. R. Astron. Soc.* **62,** 1–14.

Miller, B. M. *et al.* (1975). Geological estimation of undiscovered recoverable oil gas resources in the United States. *Geol. Surv. Circ. (U.S.)* **725,** 78.

Moore, C. R. K., and Tucker, J. W. (1995). Ranking exploration opportunities—Some new approaches. *Soc. Pet. Eng. Hydrocarbon Econ. Eval. Symp.,* Dallas, TX, *1995,* SPE No. 30054, pp. 209–218.

Moore, C. V., and Kaufman, R. L. (1981). Your unsuspected problems: Fluid resistivity and water analysis. *Trans. Annu. Logging Symp., SPWLA, 22nd,* pp. C1–C16.

Morton, R. A., and Suter, J. R. (1996). Sequence stratigraphy and composition of Late Quaternary shelf-margin deltas, Northern Gulf of Mexico. *AAPG Bull.* **80,** 505–530.

Mudford, B. (1996). Can biomarker data be used to quantitatively constrain thermal histories? *Energy Explor. Exploit.* **14,** 607–626.

Murtha, J. A. (1995). Estimating reserves and success probability for a prospect with geologically dependent layers. *Soc. Pet. Eng. Hydrocarbon Econ. Eval. Symp.,* Dallas, TX, *1995,* SPE No. 30040, pp. 73–84.

Naeser, N. D., and McCulloh, T. H., eds. (1989). *Thermal History of Sedimentary Basins.* Springer-Verlag, New York.

Nakayama, K. (1987). Two-dimensional basin analysis for petroleum exploration. Ph.D. Dissertation, University of South Carolina, Columbia.

Nakayama, K., and Lerche, I. (1987). Basin analysis by model simulation: Effects of geologic parameters on 1D and 2D fluid flow systems with application to an oil field. *Trans. Gulf Coast Assoc. Geol. Soc.* **37,** 175–184.

Nakayama, K., and Van Siclen, D. C. (1981). Simulation model for petroleum exploration. *AAPG Bull.* **65,** 1230–1255.

Nettleton, L. L. (1955). History of concepts of Gulf coast salt dome formation. *Am. Assoc. Pet. Geol. Bull.* **39,** 2373–2383.

Newendorp, P. D. (1975). *Decision Analysis for Petroleum Exploration.* Petroleum Publ. Co., Tulsa, OK.

Nind, T. E. W. (1981). *Principles of Oil Well Production.* McGraw-Hill, New York.

Norwegian Petroleum Directorate (1993). *Petroleum Resources—Norwegian Continental Shelf.* Norw. Pet. Directorate.

Nunns, A. G. (1991). Structural restoration of seismic and geologic sections in extensional regimes. *AAPG Bull.* **75,** 278–297.

Nyland, L., Jensen, N., Skagen, J., Skarpness, O., and Vorren, T. (1992). Tertiary uplift and erosion in the Barents Sea: Magnitude, timing and consequences. In *Structural and Tectonic Modeling and its Application to Petroleum Geology* (R. M. Larsen, H. Brekke, B. T. Larsen, and E. Talleraas, eds.), Spec. Publ. No. 1, pp. 153–162. Norwegian Petroleum Society, Elsevier, Amsterdam.

O'Brien, J. J., and Lerche, I. (1994). Understanding subsalt overpressure may reduce drilling risks. *Oil & Gas J.,* January 24, pp. 28–34.

Ocamb, R. D. (1961). Growth faults of south Louisiana. *Trans. Gulf Coast Assoc. Geol.* **11,** 137–175.

Oerlemans, J. (1984). Numerical experiments on large scale glacial erosion. *Z. Gletscherkd. Glazialgeol.* **20,** 107–126.

Palciauskas, V. V., and Domenico, P. A. (1980). Microfracture development in compacting sediments: A relation to hydrocarbon maturation kinetics. *AAPG Bull.* **64,** 927–937.

Palmer, A. R., compiler (1983). The decade of North American Geology 1983 Geologic Time Scale. *Geology* **11,** 503–504.

Pantano, J., and Lerche, I. (1990). Stratigraphic age determination from inversion of vitrinite reflectance data and sterane isomerization data. *Math. Geol.* **22,** 281–307.

Passey, Q. R., Creaney, S., Kulla, J. B., Moretti, F. J., and Stroud, J. D. (1990). A practical model for organic richness from porosity and resistivity logs. *AAPG Bull.* **74,** 1777–1794.

Patton, T. L., and O'Connor, S. J. (1988). Cretaceous flexural history of Northern Oman Mountain Foredeep, United Arab Emirates. *AAPG Bull.* **72,** 797–809.

Peters, K. E. (1986). Guidelines for evaluating petroleum source rock using programmed pyrolysis. *AAPG Bull.* **70,** 318–329.

Petersen, H. I., Rosenberg, P., and Andsbjerg, J. (1996). Organic geochemistry in relation to the depositional environments of Middle Jurassic coal seams, Danish Central Graben, and implications for hydrocarbon generative potential. *AAPG Bull.* **80,** 47–62.

Philippi, G. T. (1965). On the depth, time and mechanisms of petroleum generation. *Geochim. Cosmochim. Acta* **25,** 1021–1049.

Pitcher, M. G. (1976). U.S. discovery rate tied to technology. *Oil & Gas J.,* March 22, pp. 34–35.

Pittman, W. C., III (1978). Relationship between sea level change and stratigraphic sequences of passive margins. *Geol. Soc. Am. Bull.* **89,** 1389–1403.

Porter, E. D. (1992). *U.S. Petroleum Supply: History, Prospects and Policy Implications,* Res. Study No. 064. Am. Pet. Inst., New York.

Potential Gas Agency (1993). *A Comparison of Estimates of Ultimately Recoverable Quantities of Natural Gas in the United States,* Gas Resour. Stud. No. 2. Colorado School of Mines, Golden.

Price, L. (1976). Aqueous solubility of petroleum as applied to its origin and primary migration. *AAPG Bull.* **60,** 213–244.

Price, L., Wenger, L. M., Ging, T., and Blount, C. W. (1983). Solubility of crude oil in methane as a function of pressure and temperature. *Org. Geochem.* **4,** 201–221.

Price, L. C. (1982). Time as a factor in organic metamorphism and use of vitrinite reflectances as an absolute paleogeothermometer. *AAPG Bull.* **66,** 619–620.

Quigley, T. M., MacKenzie, A. S., and Gray, J. R. (1987). Kinetic theory of petroleum generation. In *Migration of Hydrocarbons in Sedimentary Basins* (B. Doligez, ed.), pp. 649–665. Editions Technip, Paris.

Quinlan, G. M., and Beaumont, C. (1984). Appalachian thrusting, lithospheric flexure, and the Paleozoic stratigraphy of the Eastern Interior of North America. *Can. J. Earth Sci.* **21,** 973–996.

Rankey, E., and Watney, L. (1996). Of the making of earth science models there is no end. *Numer. Exp. Stratigraphy,* Lawrence, KS, *1996,* pp. 17–33.

Raymer, L. L., Hunt, E. R., and Gardner, J. S. (1980). An improved sonic transit time-to-porosity transform. *Soc. Prof. Well Log Anal., 21st Annu. Logging Symp., 1980.*

Reeckman, S. A., and Mebberson, A. J. (1984). Igneous intrusions in the North-West Canning Basin and their impact on oil exploration. In *Proceedings of the Canning Basin Symposium, Perth, 1984,* pp. 45–52. Petroleum Exploration Society of Australia, Perth.

Roadifer, R. (1975). A probability approach to estimate volumes of undiscovered oil and gas. In *Probability Methods in Oil Exploration* (J. C. Davis, ed.), Am. Assoc. Pet. Geol. Res. Symp., p. 18. Stanford University, Stanford, CA.

Rodgers, D. A. (1980). Analysis of pull-apart basin development produced by en-echelon strike-slip faults. *Spec. Publ. Int. Assoc. Sedimentol.* **4,** 27–41.

Rowan, M. G., and Kligfield, R. (1989). Cross section restoration and balancing as aid to seismic interpretation in extensional terranes. *AAPG Bull.* **73,** 955–966.

Roy, J. J., Procter, R. M., and McCrossan, R. C. (1975). Hydrocarbon assessment using subjective probability. In *Probability Methods in Oil Exploration* (J. C. Davis, ed.), Am. Assoc. Pet. Geol. Res. Symp., pp. 56–60. Stanford University, Stanford, CA.

Royden, L., and Karner, G. D. (1984). Flexure of lithosphere beneath Apennine and Carpathian foredeep basins: Evidence for an insufficient topographic load. *AAPG Bull.* **68,** 704–712.

Royden, L., and Keen, C. E. (1980a). Rifting process and thermal evolution of the continental margin of eastern Canada determined from subsidence curves. *Earth Planet. Sci. Lett.* **51,** 343–361.

Royden, L., Sclater, J. G., and von Herzen, R. P. (1980). Continental Margin subsidence and heat flow: Important parameters in formation of petroleum hydrocarbons. *AAPG Bull.* **64,** 173–187.

Schlee, N. A., and Klitgord, K. D. (1982). Geologic setting of the Georges Bank Basin. *Geol. Surv. Circ. (U.S.)* **861,** 4–10.

Schlumberger Educational Services (1987). *Log Interpretation Principles/Applications.* Houston, TX.

Sclater, J. G., and Christie, P. A. F. (1980). Continental stretching: An explanation of the post-mid-Cretaceous subsidence of the central North Sea basin. *J. Geophys. Res.* **85,** 3711–3739.

Shi, Y., and Wang, C. (1986). Pore pressure generation in sedimentary basins: Overloading versus aquathermal. *J. Geophys. Res.* **91,** 2153–2162.

Skagen, J. I. (1992). Methodology applied to uplift and erosion. *Nor. Geol. Tidsskr.* **72,** 307–311.

Sleep, N. H. (1971). Thermal effects of the formation of Atlantic continental margins by continental breakup. *Geophys. J. R. Astron. Soc.* **24,** 325–350.

Smith, J. D. (1991). *Object-oriented Programming with the X Window System Toolkits.* Wiley, New York.

Smith, M. A., and Van Gijzel, P. (1989). New perspectives on depositional and thermal history of Georges Bank. *AAPG Bull.* **73,** 413–414 (abstr.).

Smith, M. B. (1968). Estimating resources by using computer simulation methods. *Oil & Gas J.,* March 11, pp. 81–84.

Smith, P. J., and Buckee, J. W. (1985). Calculating in-place and recoverable hydrocarbons: A comparison of alternative methods. SPE No. 13776.

Spencer, C. W. (1987). Hydrocarbon generation as a mechanism for overpressuring in Rocky Mountain Region. *AAPG Bull.* **71,** 368–388.

Spencer, C. W., and Keighin, C. W. (1984). Geologic studies in support of the U.S. Department of Energy Multiwell Experiment, Garfield County, Colorado. *Geol. Surv. Open File Rep. (U.S.)* **84–754**.

Statoil (1994). *Suggestions for Future Research in Prediction of Hydrocarbon Composition and Phase.* Statoil, Trondheim, Norway.

Steckler, M. S. (1990). The role of the thermal-mechanical structure of the lithosphere in the formation of sedimentary basins. In *Quantitative Dynamic Stratigraphy* (T. A. Cross, ed.). Prentice-Hall, Englewood Cliffs, NJ.

Steckler, M. S., and Watts, A. B. (1978). Subsidence of the Atlantic-type continental margin off New York. *Earth Planet. Sci. Lett.* **41,** 1–13.

Steffi, D. A. (1991). *Abnormal Formation Pressure in the Navarin Basin, Bering Sea, Alaska,* OCS Rep. MMS91-0034. U.S. Minerals Management Service, Anchorage, AK.

Stocks, A. E., and Lawrence, S. R. (1990). Identification of source rocks from wireline logs. *Spec. Publ. Geol. Soc. London* **48,** 241–252.

Stoian, E. (1965). Fundamentals and applications of the Monte Carlo method. *J. Can. Pet. Technol.* **4,** 120–129.

Sundararaman, P., Merz, P. H., and Mann, R. G. (1992). Determination of kerogen activation energy distribution. *Energy Fuels* **6,** 793–803.

Suppe, J. (1983). Geometry and kinematics of fault-bend folding. *Am. J. Sci.* **283,** 684–721.

Suppe, J. (1985). *Principles of Structural Geology.* Prentice-Hall, Englewood Cliffs, NJ.

Sweeney, J. J. (1990). FORTRAN program to calculate oil and gas generation using a distribution of discrete activation energies. *Geobyte,* April, pp. 37–43.

Sweeney, J. J., and Burnham, A. K. (1990). Evaluation of a simple model of vitrinite reflectance based on chemical kinetics. *AAPG Bull.* **74,** 1559–1570.

Sweeney, J. J., Burnham, A. K., and Braun, R. L. (1986). A model of hydrocarbon maturation in the Uinta Basin, Utah, USA. In *Thermal Modeling in Sedimentary Basins* (J. Burrus, ed.), pp. 547–561. Editions Technip, Paris.

Sweeney, J. J., Burnham, A. K., and Braun, R. L. (1987). A model of hydrocarbon generation from Type I kerogen: Application to Uinta basin, Utah. *AAPG Bull.* **71,** 967–985.

Tang, J., and Lerche, I. (1992). Tertiary flexural evolution of the Beaufort-Mackenzie Basin, Canada. *Mar. Pet. Geol.* **9,** 245–255.

Tang, J., Lerche, I., and Cogan, J. (1990). Elastic flexure with compressive thrusting of the Green River Basin, Wyoming, U.S.A. *Geol. Mag.* **127,** 349–359.

Tang, J., Lerche, I., and Cogan, J. (1992). An inverse method for calculating basement geometries in Foreland basins. *J. Geodyn.* **15,** 85–106.

Tetzlaff, D. M. (1990). SEDO: A simple clastic sedimentation program for use in training and education. In *Quantitative Dynamic Stratigraphy* (T. A. Cross, ed.), pp. 401–415. Prentice-Hall, Englewood Cliffs, NJ.

Tetzlaff, D. (1996). Probabilistic estimates from reservoir-scale sedimentation models. *Numer. Exp. Stratigr.,* Lawrence, KS, *1996,* pp. 145–146.

Thompson, D. T., Lerche, I., and O'Brien, J. J. (1989). Salt basins of Western Europe and Gabon, West Africa: Dynamical aspects and hydrocarbon production. *GCS/SEPM Spec. Publ.,* pp. 122–130.

Thomsen, R. O. (1993). Dynamical models in geology: Sensitivity analysis and scientific risk. *Energy Explor. Exploit.* **11,** 329–356.

Thomsen, R. O., and Lerche, I. (1997). Relative contributions to uncertainties in reserve estimates. *Mar. Pet. Geol.* **14,** 65–74.

Thomsen, R. O., Lerche, I., and Korstgaard, J. (1990). Dynamical hydrocarbon predictions for the Danish Central Graben: An integrated basin analysis assessment. *Mar. Pet. Geol.* **7,** 123–137.

Thomsen, R. O., Maubeuge, F., and Lerche, I. (1992). Simple visualizations of dynamical models. *Geobyte,* October, pp. 53–62.

Thurston, D. K., and Lothamer, R. T. (1991). Seismic evidence of evaporite diapirs in the Chukchi Sea, Alaska. *Geology* **19,** 477–480.

Thurston, D. K., and Theiss, L. A. (1987). *Geologic Report for the Chukchi Sea Planning Area, Alaska: Regional Geology, Petroleum Geology, and Environmental Geology,* OCS Rep. MMS 87-0046. U.S. Department of the Interior, Minerals Management Service, Alaska OCS Region, Anchorage, AK.

Tinker, S. W. (1996). Building the 3-D jigsaw puzzle: Applications of sequence stratigraphy to 3-D reservoir characterization Permian Basin. *AAPG Bull.* **80,** 460–485.

Tissot, B. (1969). Premières données sur les mécanismes et la cinétique de la formation du petrole dans les sediments: Simulation d'un schema réactionnel sur ordinateur. *Rev. Inst. Fr. Pet.* **24,** 470–501.

Tissot, B., and Espitalié, J. (1975). L'évolution thermique de la matière organique des sediments: Applications d'une simulation mathématique. *Rev. Inst. Fr. Pet.* **30,** 743–777.

Tissot, B. P., and Welte, D. H. (1984). *Petroleum Formation and Occurrence,* 2nd ed. Springer-Verlag, New York.

Toth, D. J., Lerche, I., Petroy, D. E., Meyer, R. J., and Kendall, C. G. St. C. (1983). Vitrinite reflectance and the derivation of heat flow changes with time. *Adv. Org. Geochem., Proc. Int. Meet., 10th, 1981,* pp. 589–596.

Turcotte, D. L. (1979). Flexure. *Adv. Geophys.* **21,** 51–87.

Turcotte, D. L., and Ahern, J. L. (1977). On the thermal and subsidence history of sedimentary basins. *J. Geophys. Res.* **82,** 3762–3766.

Turcotte, D. L., and Schubert, G. (1982). *Geodynamics: Applications of Continuum Physics to Geological Problems.* Wiley, New York.

Turner, R. F. (1984). *Geological and Operational Summary, Navarin Basin COST No. 1 Well, Bering Sea, Alaska,* OCS Rep. MMS84-0031. U.S. Department of the Interior, Minerals Management Service, Anchorage, AK.

Turner, R. F., McCarthy, C. M., Seffy, D. H., Lynch, M. B., Martin, G. C., Sherwood, K. W., Flett, T. O., and Adams, A. J. (1984). *Geological and Operational Summary, Navarin Basin COST No. 1 Well, Bering Sea, Alaska,* OCS Rep. MMS 84-0031. U.S. Department of the Interior, Minerals Management Service, Anchorage, AK.

Ungerer, P., and Pelet, R. (1987). Extrapolation of the kinetics of oil and gas formation from laboratory experiments to sedimentary basins. *Nature* (*London*) **307,** 52–54.

Ungerer, P., Doligez, B., Bessis, F., and Chénet, P. Y. (1986). Integrated numerical modeling of heat transfer, hydrocarbon formation, and migration in two dimensions: The Themis model. In *Thermal Modeling in Sedimentary Basins* (J. Burrus, ed.), pp. 174–195. Editions Technip, Paris.

Ungerer, P., Behar, F., Villalba, M., Heum, O. R., and Audibert, A. (1988a). Kinetic modeling of oil cracking. *Org. Geochem.* **13,** 857–868.

Ungerer, P., Espitalie, J., Behar, F., and Eggens, S. (1988b). Modélisation mathématique des interactions entre craquage thermique et migration lors de la formation du petrole et du gaz. *C. R. Sciences Acad. Sci.* (reprint).

Ungerer, P., Burrus, J., Doligez, B., Chénet, P. Y., and Bessis, F. (1990). Basin evaluation by integrated two-dimensional modeling of heat transfer, fluid flow, hydrocarbon generation and migration. *AAPG Bull.* **74,** 309–335.

Vail, P. R., and Hardenbol, J. (1979). Sea level changes during Tertiary. *Oceans* **22,** 71–79.

Vail, P. R., Mitchum, R. M., Jr., Todd, R. G., Widmier, J. M., Thompson, S., III, Sangree, J. B., Bubb, J. N., and Hatleilid, W. G. (1977). Seismic stratigraphy and global changes of sea level. *AAPG Mem.* **26,** 49–212.

Vail, P. R., Hardenbol, J., and Todd, R. G. (1984). Jurassic unconformities, chronostratigraphy and sea level changes from seismic stratigraphy and biostratigraphy. *AAPG Mem.* **36,** 129–144.

Van Eysinga, F. W. B., compiler (1975). *Geological Time Table,* 3rd ed. Elsevier, Amsterdam.

Van Gijzel, P. (1967). Autofluorescence of fossil pollen and spores with special reference to age determination and coalification. *Leidse Geol. Meded.* **40,** 263–317.

Van Gijzel, P. (1989). Transmittance Color Index (TCI) of amorphous organic matter: New thermal maturity indicator for hydrocarbon source rocks, correlation with mean vitrinite reflectance and thermal alteration index (TAI). *AAPG Bull.* **73,** 1177 (abstr.).

Van Gijzel, P., Robison, C. R., Smith, M. A., Bissada, K. K., Liu, J., and Lerche, I. (1992). Thermal history modeling of the Georges Bank, USA: Thermal inversion of Transmittance Color Index (TCI) and vitrinile reflectance (VR) data. *Appl. Geochem.* **7,** 135–143.

van Heek, K. H., and Juntgen, H. (1968). Determination of reaction-kinetic parameters from nonisothermal measurements. *Ber. Bunsenges. Phys. Chem.* **72,** 1223; available as English Translation in Lawrence Livermore National Laboratory Report UCRL-TRANS-10974 (1975).

van Hinte, J. E. (1976a). A Jurassic time scale. *AAPG Bull.* **60,** 489–497.

van Hinte, J. E. (1976b). A Cretaceous time scale. *AAPG Bull.* **60,** 498–516.

van Hinte, J. E. (1978). Geohistory analysis—application of micropaleontology in exploration geology. *AAPG Bull.* **62,** 201–222.

van Krevelen, D. W. (1981). *Coal: Typology, Chemistry, Physics, Constitution.* Elsevier, New York (see, in particular, Chapter 17).

Vorren, T. O., Bergsager, E., Dahl-Stamnes, Ø. A., Holter, E., Johannsen, B., Lie, E., and Lund, T. B., eds. (1993). *Arctic Geology and Petroleum Potential,* NPB Spec. Publ. No. 2. Elsevier, Amsterdam.

Wahlstrom, J. E., Mueller, T. D., and McFarlane, R. C. (1967). Evaluating uncertainty in engineering calculation. *J. Pet. Technol.* **19,** 1595–1603.

Walsh, J. J., and Watterson, J. (1988). Dips of normal faults in British coal measures and other sedimentary sequences. *J. Geol. Soc., London* **145,** 859–873.

Walters, R. (1980). BP International Report.

Waples, D. W. (1980). Time and temperature in petroleum formation: Application of Lopatin's method to petroleum exploration. *AAPG Bull.* **64,** 916–926.

Waples, D. W. (1984). Thermal models for oil generation. *Adv. Pet. Geochem.* **1,** 7–68.

Waples, D. W. (1989). Maturity modeling of sedimentary basins: Approaches, limitations and prediction of future developments. In *Quantitative Dynamic Stratigraphy* (T. A. Cross, ed.), pp. 181–193. Prentice-Hall, Englewood Cliffs, NJ.

Warren, J. E. (1978a). *Exploration Economics Seminar Notes.* Gulf Research and Development Corporation, Pittsburgh, PA.

Warren, J. E. (1978b). The evaluation of oil and gas plays. SPE No. 17554.

Warren, J. E. (1979). Basin evaluation. *Soc. Pet. Eng. Econ. Eval. Symp.,* Dallas, TX, *1979.*

Warren, J. E. (1981). The development decision: Frontier areas. SPE No. 9558.

Warren, J. E. (1983a). A strategic exploration model. SPE No. 11442.

Warren, J. E. (1983b). The development decision: Value of information. SPE No. 11312.

Warren, J. E. (1988). The evaluation of oil and gas plays. SPE No. 17554.

Warren, J. E., and Price, H. S. (1961). Flow in heterogenous porous media. *Soc. Pet. Eng. J.* **1,** 153–169.

Weeks, L. G. (1952). Factors of sedimentary basin development that control oil occurrence. *Am. Assoc. Pet. Geol. Bull.* **36,** 2071–2124.

Wei, Z. P., and Lerche, I. (1988). Quantitative dynamical geology of the Pinedale Anticline, Wyoming, U.S.A.: An application of a two-dimensional simulation model. *Appl. Geochem.* **3,** 423–440.

Welte, D. H., and Yukler, M. A. (1981). Petroleum origin and accumulation in basin evolution—a quantitative model. *AAPG Bull.* **65,** 1387–1396.

Welte, D. H. (1994). Private communication.

West, D. B. (1989). Model for salt deformation on deep margin of Central Gulf of Mexico Basin. *AAPG Bull.* **73,** 1472–1482.

White, N. J., Jackson, J. A., and McKenzie, D. P. (1986). The relationship between the geometry of normal faults and that of the sedimentary layers in their hanging walls. *J. Struc. Geol.* **8,** 897–909.

White, R. E. (1984). Signal and noise estimation from seismic reflection data using spectral coherence methods. *Proc. IEEE* **72,** 1340–1356.

Wilgus, C. K., ed. (1988). *Sea-Level Changes: An Integrated Approach,* SEPM Spec. Publ. No. 42. Soc. Econ. Paleontol. Mineral., Tulsa, OK.

Williams, D. F., Lerche, I., and Full, W. E. (1988). *Isotope Chronostratigraphy: Theory and Methods.* Academic Press, Orlando, FL.

Withers, R. J. (1992). The value of reservoir geophysics. *Geophys. Lead. Edge Explor.,* pp. 35–39.

Worrall, D. M. (1991). Tectonic history of the Bering Sea and the evolution of Tertiary strike-slip basins of the Bering Shelf. *Spec. Pap.—Geol. Soc. Am.* **257,** 1–120.

Worrall, D. M., and Snelson, S. (1989). Evolution of the northern Gulf of Mexico, with emphasis on Cenozoic growth faulting and the role of salt. In *The Geology of North America—An Overview* (A. W. Bally and A. R. Palmer, eds.), DNAG Ser. A, pp. 97–138. Geol. Soc. Am., Boulder, CO.

Wyllie, M. J. R. (1963). *Fundamentals of Well Log Interpretation.* Academic Press, New York.

Wyllie, M. J. R., Gregory, A. R., and Gardner, L. W. (1956). Elastic wave velocities in heterogeneous and porous media. *Geophysics* **21,** 41–70.

Xiao, H., and Suppe, J. (1988). Origin of rollover. *GSA Abstr. Programs* **20,** A109.

Xiao, H., and Suppe, J. (1989). Role of compaction in the listric shape of growth normal faults. *AAPG Bull.* **73,** 777–786.

Xiao, H., and Suppe, J. (1992). Origin of rollover. *AAPG Bull.* **76,** 509–529.

Yu, Z. (1992). Quantitative basin modeling: An application in the Arctic National Wildlife Refuge of Alaska, and salt influences on thermal anomalies and fracturing patterns of sediments in Gulf of Mexico. Ph.D. Dissertation, p. 161. University of South Carolina, Columbia.

Yu, Z., and Lerche, I. (1992). An integrated quantitative basin analysis study of the northeastern part of the Arctic National Wildlife Refuge, northeastern Alaska. *J. Geodyn.* **15,** 107–146.

Yu, Z., and Lerche, I. (1995a). Arcuate and growth faults in basin models with an application to a South Louisiana roll-over structure. *Math. Geol.* **27,** 513–547.

Yu, Z., and Lerche, I. (1995b). Three phase fluid migration with solubilities in a two-dimensional basin simulation model. *Mar. Pet. Geol.* **12,** 3–16.

Yu, Z., Li, Y., and Lerche, I. (1992). Thermal history and hydrocarbon maturity study in the Norton Basin, Alaska. *Energy Explor. Exploit.* **10,** 402–421.

Yu, Z., O'Brien, J. J., and Lerche, I. (1993). Observations and influences for overpressure development under salt sheets in the Gulf of Mexico. *Trans.—Gulf Coast Assoc. Geol. Soc.* **43,** 281–290.

Yu, Z., Liu, J., and Lerche, I. (1995). Representation of normal slant faults in a two-dimensional quantitative burial history model. *Terra Nova* **7,** 459–466.

Yukler, M. A. (1979). Sensitivity analysis of groundwater flow system and an application to a real case. In *Geomathematical and Petrophysical Studies in Sedimentology* (*Computer Geology*) (D. F. Merriam, ed.). Pergamon, New York.

Yukler, M. A., and Kokesh, F. (1984). An overview of models used in petroleum resource estimation and organic geochemistry. *Adv. Pet. Geochem.* **1,** 69–73.

Zhao, K., and Lerche, I. (1993). Inversion of dynamical indicators in quantitative basin analysis models. II. Synthetic tests and a case history using dynamical indicator tomography. *Math. Geol.* **25,** 104–123.

# INDEX